ADVANCES IN DIVERSE INDUSTRIAL APPLICATIONS OF NANOCOMPOSITES

Edited by **Boreddy S. R. Reddy**

Advances in Diverse Industrial Applications of Nanocomposites
Edited by Boreddy S. R. Reddy

Published by InTech

Janeza Trdine 9, 51000 Rijeka, Croatia

Publishing Process Manager Iva Lipovic

Technical Editor Teodora Smiljanic

Cover Designer InTech Design Team

Additional hard copies can be obtained from orders@intechopen.com

Advances in Diverse Industrial Applications of Nanocomposites,
Edited by Boreddy S. R. Reddy
p. cm.
ISBN 978-953-307-202-9

For EU product safety concerns:
IN TECH d.o.o., Prolaz Marije Krucifikse Kozulić 3, 51000 Rijeka, Croatia,
info@intechopen.com or visit our website at intechopen.com.

Meet the editor

Dr. BSR Reddy has been working as a Director Grade Scientist 'G' and Head of the Industrial Chemistry Laboratory in World famous Institute, Central Leather Research Institute, Chennai 600020 (India) for the last 28 years. Professor BSR Reddy's interests include applied synthetic organic chemistry and polymer science / technology, ring-opening metathesis, synthesis and polymerization of a dictionary of several functionally activated acrylates, functional microspheres, polymer drug conjugates, polymer supports, electroactive polymers, phenol-formaldehyde reactions, adhesives, (Nano) composites, organic-inorganic composites and functional polymers. He is the author / co-author of 180 research publications, and 12 patents including a number of book chapters. He is one of the most frequently cited authors in chemical, pharmaceutical and biological sciences.

Contents

Preface

The book "Advances in Diverse Industrial Applications of Nanocomposites" was well thought about knitting several broad disciplines of nanocomposites in mind. A glance through the pages of science and engineering literature shows that the use of nanocomposites for emerging technologies represents one of the most active areas of research and development throughout the fields of chemistry, physics, life sciences, and related technologies. In addition to being of technological importance, the subject of nanocomposites is a fascinating area of interdisciplinary research and a major source of inspiration and motivation in its own right for exploitation to help humanity. The concept of this book is based as much on the serendipity of these new ideas as on the need to offer blueprints for the design of different nanocomposite devices for emerging technologies.

The choice of materials for multi-author complications is always subject to a number of periodical and judicial limitations. These limitations notwithstanding, I have been privileged to benefit from the cooperation of many leading experts in nanocomposites, and also a number of senior researchers of well-known companies at the forefront and nanocomposites not frequently disclosed elsewhere. I must, in particular, acknowledge the courtesy of all the contributors in allowing me to organize their contributions in producing a cohesive and correlated compilation, and to minimize overlaps between closely related chapters.

Nanocomposites are attractive to researchers both from practical and theoretical point of view because of combination of special properties. Many efforts have been made in the last two decades using novel nanotechnology and nanoscience knowledge in order to get nanomaterials with determined functionality. This book focuses on polymer nanocomposites and their possible divergent applications. There has been enormous interest in the commercialization of nanocomposites for a variety of applications, and a number of these applications can already be found in industry.

This book comprehensively deals with the divergent applications of nanocomposites comprising of 23 chapters.

I would like to place on record that one of my Senior Research Fellows, Mr. D. Gnanasekaran, helped me in my editorial work to bring out this book successfully.

BSR Reddy,
Director Grade Scientist 'G' and Head,
Industrial Chemistry Laboratory,
Central Leather Research Institute,
Adyar, Chennai-600 020,
India

1

Microstructure, Mechanical Properties and Corrosion Resistance of Nanocomposite Coatings Deposited by PVD Technology

Krzysztof Lukaszkowicz[1], Leszek A. Dobrzański[1] and Jozef Sondor[2]
[1]*Silesian University of Technology,*
[2]*LISS a.s.,*
[1]*Poland*
[2]*Czech Republic*

1. Introduction

The research issues concerning the production of coatings are the most important directions of surface engineering development, ensuring the obtainment of coatings of high usable properties in the scope of mechanical characteristics and corrosion resistance (Yu et al., 2009; Cheng et al., 2009; Kao, 2009; Mao et al., 2009; Sundararajan et al., 2009). Giving new operating characteristics to commonly known materials is frequently obtained by laying simple monolayer, multilayer or gradient coatings using PVD methods (Dobrzanski et al.,2005; Lukaszkowicz & Dobrzański, 2008).

While selecting the coating material, we encounter a barrier caused by the fact that numerous properties expected from an ideal coating are impossible to be obtained simultaneously. For example, an increase in hardness and strength causes the reduction of the coating's ductility and adherence to the substrate. The application of the nanostructure coatings is seen as the solution of this issue (Voevodin et al., 2005; Yang et al., 2007). According to the Hall-Petch equation, the strength properties of the material rise along with the reduction of the grain size. In case of the coatings deposited in the PVD processes, the structures obtained with grain size ~10 nm cause the obtainment of the maximum mechanical properties. Coatings of such structure present very high hardness >40 GPa, ductility, stability in high temperatures, etc. (Tjong & Chen, 2004; Zhang et al., 2007; Veprek et al., 2005).

The known dependency between the hardness and abrasion resistance became the foundation for the development of harder and harder coating materials. The progress in the field of producing coatings in the physical vapour deposition process enables the obtainment of coatings of nanocrystal structure presenting high mechanical and usable properties. The coatings of such structure are able to maintain a low friction coefficient (self-lubricating coatings) in numerous working environments, maintaining high hardness and increased resistance (Donnet & Erdemir, 2004; Voevodin & Zabinski, 2005).

The main concept in the achievement of high hardness of nanostructure coatings and good mechanical properties and high strength related to it, particularly in case of nanocomposite coatings (Holubar et al., 2000; Rafaja et al., 2007; Carvalho et al., 2004), is the restriction of the rise and the movement of dislocations. High hardness and strength of the nanocomposite coatings are due to the fact that the movement of dislocations is suppressed at small grains and in the spaces between them, which causes the appearance of incoherent deformations. When the grain size is reduced to that of nanometres, the activity of dislocations as the source of the material ductility is restricted. This type of coatings is also characterized with a large number of grain boundaries with a crystalline/amorphous transition across grain–matrix interfaces, restricting the rise and development of cracks. Such mechanism explains the resistance to fragile cracking of nanocomposite coatings (Veprek, 1997; Veprek, 1998; Rafaja et al., 2006). Simultaneously, the equiaxial grain shapes, high angle grain boundaries, low surface energy and the presence of the amorphous boundary phase facilitating the slide along the grain boundaries cause a high plasticity of the nanocomposite coatings (Voevodin et al., 2005).

Nanocomposite coatings comprise at least two phases, a nanocrystalline phase and a matrix phase, where the matrix can be either nanocrystalline or amorphous phase. Various analyses revealed that the synthesized TiAlSiN coatings exhibited nanostructured composite microstructures consisting of solid-solution (Ti,Al,Si)N crystallites and amorphous Si_3N_4. The Si addition caused the grain refinement of (Ti,Al,Si)N crystallites and its uniform distribution with percolation phenomenon of amorphous silicon nitride (Zhang & Ali, 2007; Holubar et al., 2000; Rafaja et al., 2007; Carvalho et al., 2004).

One of the general reasons for depositing by PVD techniques is that protective coatings deposited by PVD tend to have higher corrosion resistance than the substrate material. Ceramic hard coatings increase the life of the coated components, not only due to the protection against aggressive environments, but also during operation involving mechanical contact with abrasive surfaces. This effect results of high hardness resulting from the smaller grain size of the coatings' structure.

2. Experimental procedures

The tests were made on samples of the X40CrMoV5-1 hot work tool steel and the X6CrNiMoTi17-12-2 austenitic stainless steel, deposited by PVD process with TiAlSiN, CrAlSiN, AlTiCrN hard coatings.

The coating deposition process was made in a device based on the cathodic arc evaporation method in an Ar and N_2 atmosphere. Cathodes containing pure metals (Cr, Ti) and the AlSi (88:12 wt%) alloy were used for deposition of the coatings. The base pressure was 5×10^{-6} mbar, the deposition temperature was 500 °C. The deposition conditions are summarized in Table 1.

Coating	Substrate bias voltage [V]	Arc current source [A]	Pressure [Pa]
TiAlSiN	-90	Ti – 80 AlSi – 120	2,0
CrAlSiN	-60	Cr – 70 AlSi – 120	3,0
AlTiCrN	-60	Cr – 70 AlTi – 120	2,0

Table 1. Deposition parameters of the coatings.

Diffraction and thin film structure were tested with the use of the JEOL JEM 3010 UHR transmission electron microscope, at 300 kV bias voltage.
Observations of surface and structures of the deposited coatings were carried out on cross sections in the SUPRA 25 scanning electron microscope. Detection of secondary electron was used for generation of fracture images with 15 kV bias voltage.
Phase identification of the investigated coatings was performed by glancing angle X-ray diffraction (GAXRD).
The cross-sectional atomic composition of the samples (coating and substrate) was obtained by using a glow discharge optical spectrometer, GDOS-750 QDP from Leco Instruments. The following operation conditions of the spectrometer Grimm lamp were fixed during the tests:

- lamp inner diameter - 4 mm;
- lamp supply voltage - 700 V;
- lamp current - 20 mA;
- working pressure - 100 Pa.

Tests of the coatings' adhesion to the substrate material were made using the scratch test on the CSEM REVETEST device. The tests were made using the following parameters:

- load range 0–100 N,
- load increase rate (dL/dt): 100 N/min,
- indenter's sliding speed (dx/dt): 10 mm/min,
- acoustic emission detector's sensitivity AE: 1.

The critical load L_C, causing the loss of the coating adhesion to the material, was determined on the basis of the values of the acoustic emission, AE, and friction force, Ft and observation of the damage (Burnett & Rickerby, 1987; Bellido-Gonzalez et al., 1995) developed in the track using a LEICA MEF4A optical microscope.
The microhardness tests of coatings were made with the SHIMADZU DUH 202 ultra-microhardness tester. The test conditions were selected in order as to be comparable for all coatings. Measurements were made with 50 mN load, to eliminate the substrate influence on the coating hardness.
The thickness of coatings was determined using the "kalotest" method, i.e. measuring the characteristics of the spherical cap crater developed on the surface of the coated specimen tested (Holmberg & Matthews, 1994).
The X-ray line broadening technique was used to determine crystallite size of the coatings using Scherrer formula (Behera et al., 2004) with silicon as internal standard.
Investigation of the electrochemical corrosion behaviour of the samples was done in a PGP 201 Potentiostat/Galvanostat, using a conventional three-electrode cell consisting of a saturated calomel reference electrode (SCE), a platinum counter electrode and the studied specimens as the working electrode. To simulate the aggressive media, 1-M HCl solution was used under aerated conditions and room temperature. The aqueous corrosion behaviour of the coatings was studied first by measuring the open circuit potential (OCP) for 1 h. Subsequently, a potentiodynamic polarization curve has been recorded. The curve started at a potential of ~100 mV below the corrosion potential and ended at +1200 mV or a threshold intensity level set at 100 mA/cm^2. Once this level was reached, the reverse cycle was started. The scan rate was 15 mV/min. The corrosion current densities and the polarization resistance were obtained on the basis of the Tafel analysis after potentiodynamic polarization measurements.

3. Results and discussion

The coatings present a compact structure, without any visible delaminations or defects. The morphology of the fracture of coatings is characterized by a dense structure, in some cases there is a columnar structure (Figs. 1, 2, 3). The fracture surface of the steel samples was examined and the deposited coatings show a sharp transition zone between the substrate and the coating.

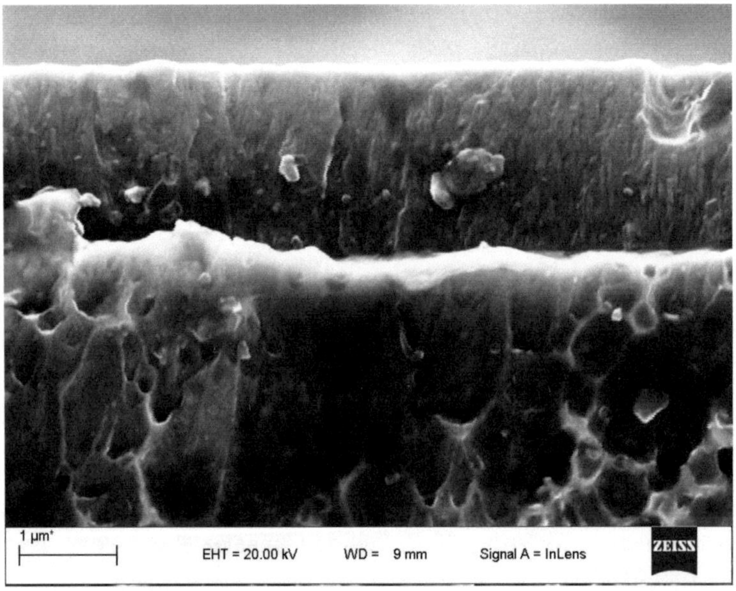

Fig. 1. Fracture of the TiAlSiN coating deposited onto the X40CrMoV5-1 steel substrate.

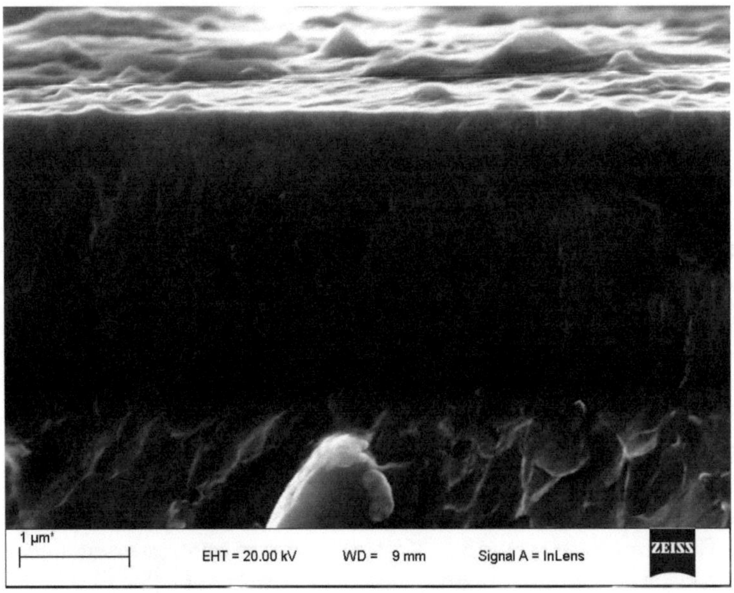

Fig. 2. Fracture of the CrAlSiN coating deposited onto the X40CrMoV5-1 steel substrate.

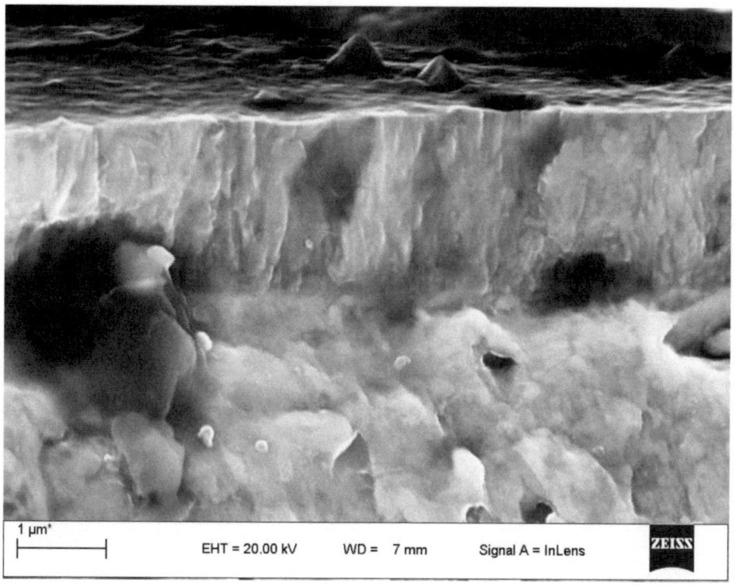

Fig. 3. Fracture of the AlTiCrN coating deposited onto the X40CrMoV5-1 steel substrate.

Transmission electron microscopy (TEM) examination of the coatings showed that they consisted of fine crystallites (Fig. 4), and there was no suggestion of epitaxial growth. Single large grains were only observed in case of the TiAlSiN coating (Fig. 5), which may suggest the occurrence of the epitaxion phenomenon as the consequence of large crystallite occurrence in the coating.

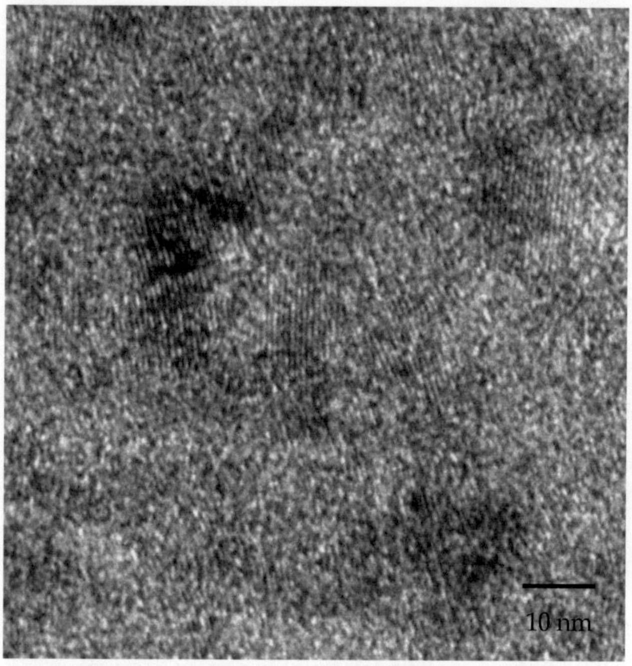

Fig. 4. Structure of the thin foil from the CrAlSiN coating.

Based on the glancing angle X-ray diffraction (GAXRD) of the samples examined (Figs. 6, 7, 8), the occurrence of fcc phases was only observed in the coatings. The hexagonal AlN of wurtzite type was not discovered in the coatings examined, which could have been caused by a low amount of aluminium in the coatings. Based on the results obtained, using Scherer method, the size of crystallites in the coatings examined was determined. The results were presented in Table 2.

The hardness of the coatings tested fits within the range from 40 to 42 GPa. The highest hardness was recorded in the case of the AlTiCrN coating (Table 2).

Coating	Thickness [μm]	Microhardness [GPa]	Crystallite size [nm]	Critical load L_{C1} [N]		Critical load L_{C2} [N]	
				AS	HWTS	AS	HWTS
AlTiCrN	2.4	42	17	15	24	27	54
CrAlSiN	2.9	40	25	8	18	28	49
TiAlSiN	2.1	40	11	9	16	27	46

Table 2. The characteristics of the tested coatings (AS – austenitic steel, HWTS – hot work tool steel).

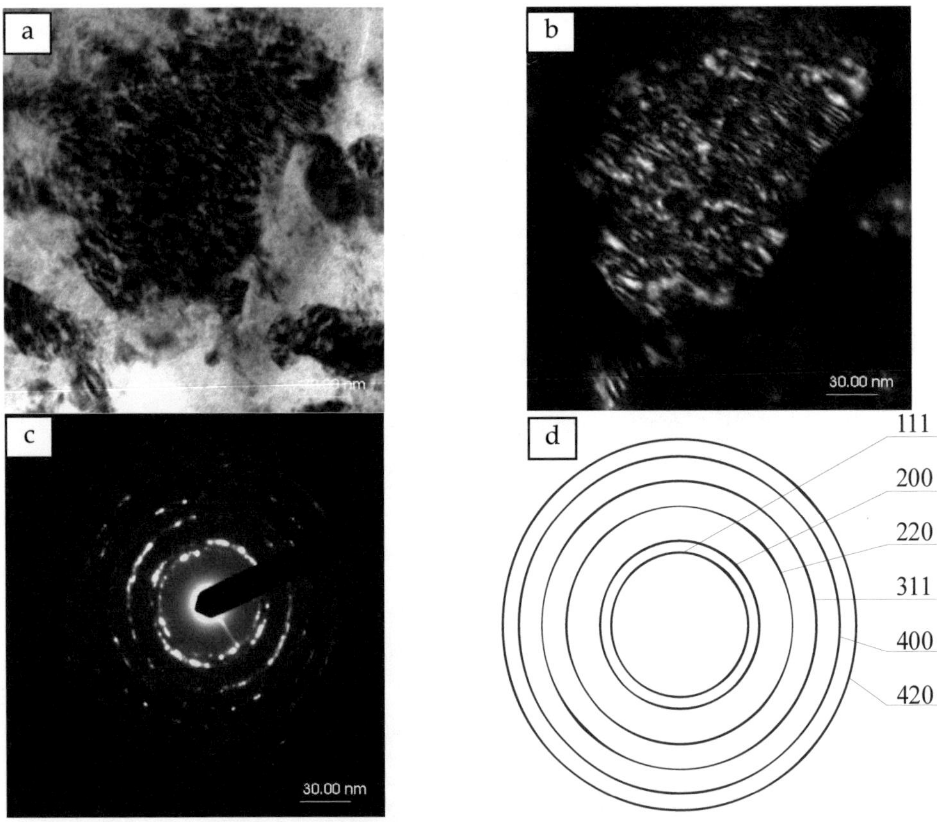

Fig. 5. Microstructure of the thin foil from the TiAlSiN coating, (a) light field, (b) dark field from the (111) reflex, (c) diffraction pattern from the area as in figure a, (d) solution of the diffraction pattern.

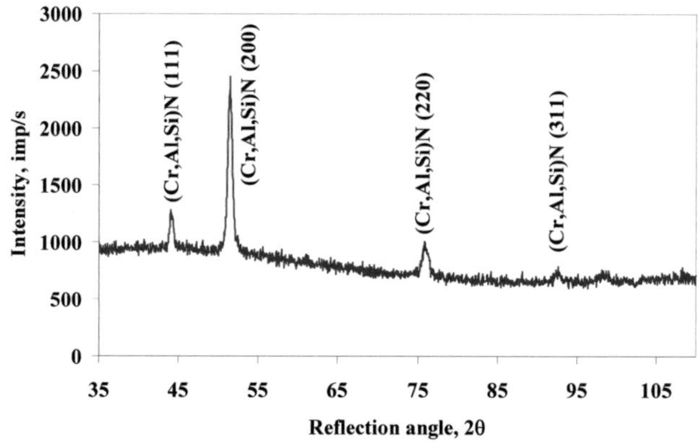

Fig. 6. GAXRD spectra of the CrAlSiN coating at glancing incidence angle α=2°.

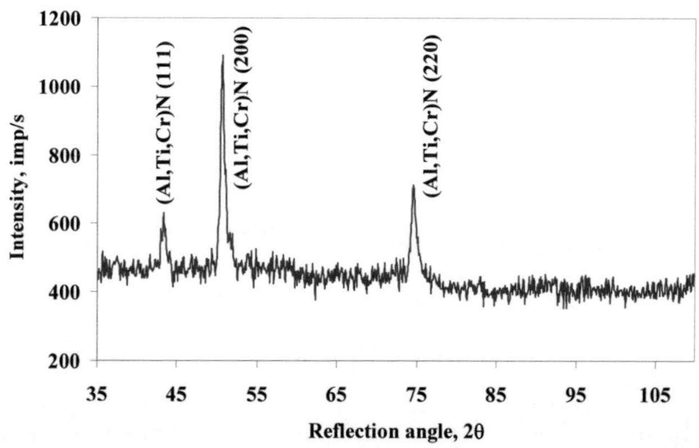

Fig. 7. GAXRD spectra of the AlTiCrN coating at glancing incidence angle α=2°.

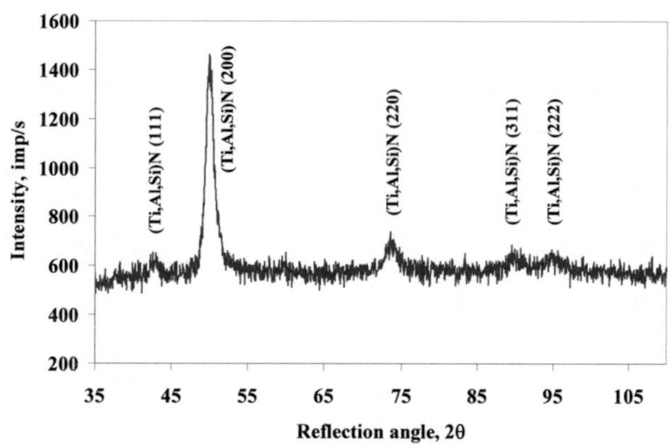

Fig. 8. GAXRD spectra of the TiAlSiN coating at glancing incidence angle α=2°.

The critical load values L_{C1} and L_{C2} were determined by the scratch test method (Figs. 9, 10, 11). The load at which the first coating defects appear is known in the references (Sergici & Randall, 2006; He et al., 2006)) as the first critical load L_{C1}. The first critical load L_{C1} corresponds to the point at which first damage is observed; the first appearance of microcraking, surface flaking outside or inside the track without any exposure of the substrate material-the first cohesion-related failure event (Figs. 12a, 13a, 14a). L_{C1} corresponds to the first small jump on the acoustic emission signal, as well as on the friction force curve (Figs. 9, 10, 11). The second critical load L_{C2} is the point at which complete delamination of the coatings starts; the first appearance of cracking, chipping, spallation and delamination outside or inside the track with the exposure of the substrate material—the first adhesion related failure event (Figs. 12b, 13b, 14b). After this point the acoustic emission graph and friction forces have a disturbed run (become noisier). The cumulative specification of the test results are presented in Table 2.

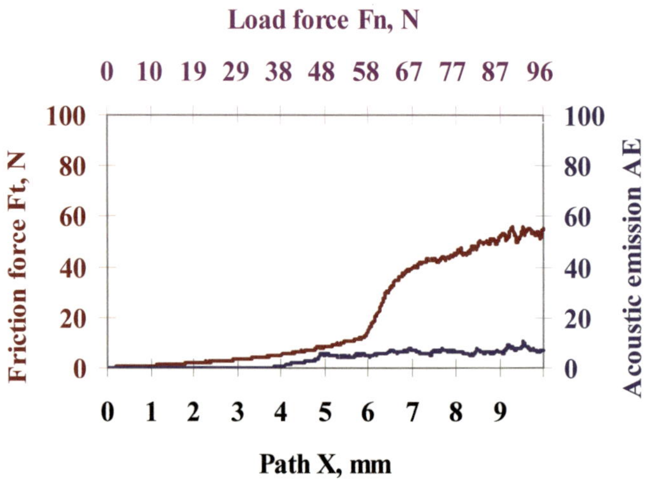

Fig. 9. Diagram of the dependence of the acoustic emission (AE) and friction force F_t on the load for the X40CrMoV5-1 steel with the AlTiCrN coating.

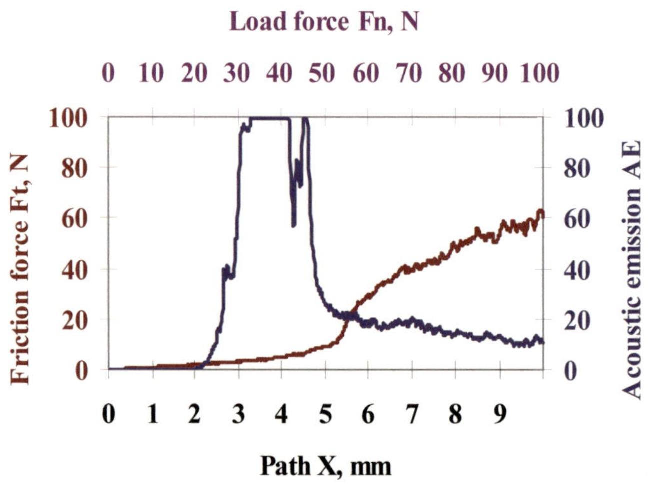

Fig. 10. Diagram of the dependence of the acoustic emission (AE) and friction force F_t on the load for the X40CrMoV5-1 steel with the CrAlSiN coating.

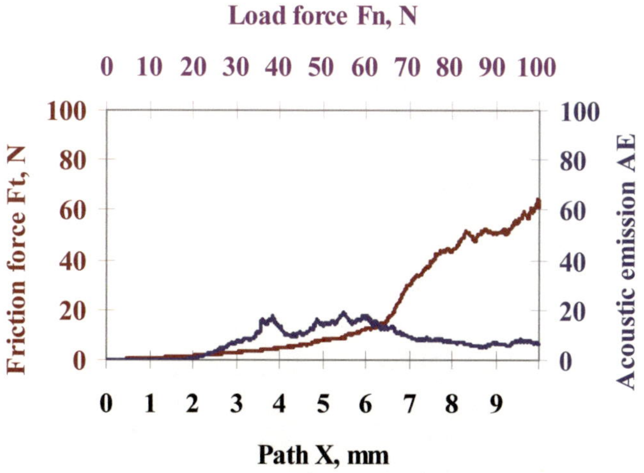

Fig. 11. Diagram of the dependence of the acoustic emission (AE) and friction force F_t on the load for the X40CrMoV5-1 steel with the TiAlSiN coating.

Fig. 12. Scratch failure pictures of the AlTiCrN coating on X40CrMoV5-1 steel substrate at: (a) L_{C1}, (b) L_{C2}.

(a)

(b)

Fig. 13. Scratch failure pictures of the CrAlSiN coating on X40CrMoV5-1 steel substrate at: (a) L_{C1}, (b) L_{C2}.

To establish the nature of damage causing the increase in acoustic emission intensity, the examinations of the scratches that arose during the test were made with the use of the light microscope coupled with a measuring device, thus determining the value of the L_{C1} and L_{C2} critical load on the basis of metallographical observations. In case of the coatings examined, it was found that coating AlTiCrN had the highest critical load value L_{C1} = 24 N and L_{C2} = 54 N deposited on substrate made of the X40CrMoV5-1 steel, whereas CrAlSiN and TiAlSiN coatings deposited on the substrate made of the X6CrNiMoTi17-12-2 steel had the lowest value. In general, the coatings deposited on the substrate made of the X40CrMoV5-1 hot work tool steel show better adherence to the substrate than coatings deposited on the substrate made of the X6CrNiMoTi17-12-2 austenitic steel. This is caused by a significantly higher hardness of the X40CrMoV5-1 steel.

The first symptoms of damage in most of the coatings examined are in the form of arch cracks caused by tension or scaling occurring on the bottom of the scratch that appears during the scratch test (Figs. 12, 13, 14). Occasionally, there are some small chippings on the scratch edges. Along with the load increase, semicircles are formed caused by conformal cracking, leading to delaminations and chippings, resulting in a local delamination of the coating. As a result of the steel fracture test against the coatings deposited, made after prior cooling in liquid nitrogen, no case delaminations were revealed along the substrate-coating separation surface, which indicates a good adhesion of coatings to substrate.

(a)

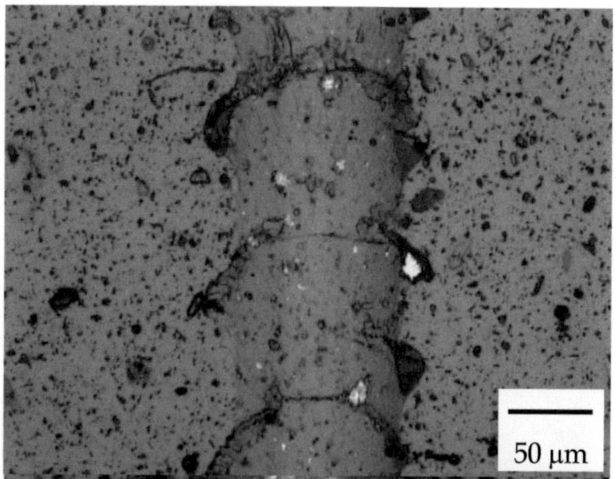

(b)

Fig. 14. Scratch failure pictures of the TiAlSiN coating on X40CrMoV5-1 steel substrate at: (a) L_{C1}, (b) L_{C2}.

Changes of coating component concentration and substrate material made in GDOS were presented in Figs. 15, 16, and 17. The tests carried out with the use of GDOS indicate the occurrence of a transition zone between the substrate material and the coating, which results in the improved adhesion between the coatings and the substrate. In the transition zone between the coatings and the substrate, the concentration of the elements of the substrate increases with simultaneous rapid decrease in concentration of elements contained in the coatings. The existence of the transition zone should be connected with the increase in desorption of the substrate surface and the occurrence of defects in the substrate and the relocation of the elements within the connection zone as a result of a high energy ion reaction. Such results, however, cannot be interpreted explicitly, due to the non-homogeneous evaporation of the material from the sample surface.

The corrosion resistance test results of coatings deposited on substrate made of the X6CrNiMoTi17-12-2 austenitic steel with the method of potentiodynamic polarization curves in 1-M HCl solution were presented in Fig. 18. It was found out, as a result of the electrochemical corrosion investigations, that the coatings deposited by PVD process onto the substrate made of the X6CrNiMoTi17- 12-2 steel may be an effective substrate material protection against corrosive agents. The potentiodynamic polarization curve analysis (Fig. 18) and that of the corrosion rate confirm the better corrosion resistance of the samples with

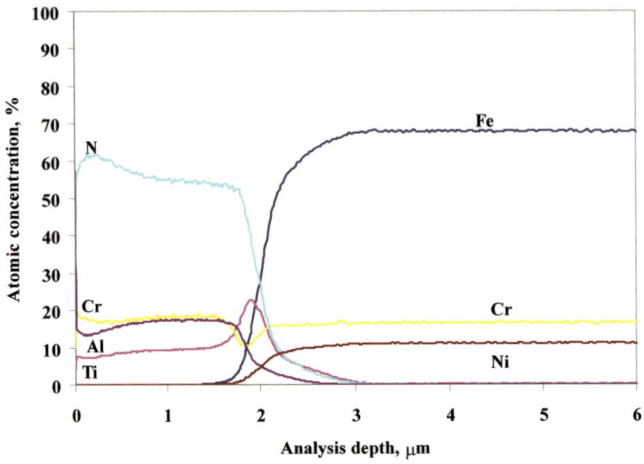

Fig. 15. Changes of constituent concentration of the AlTiCrN and the substrate materials.

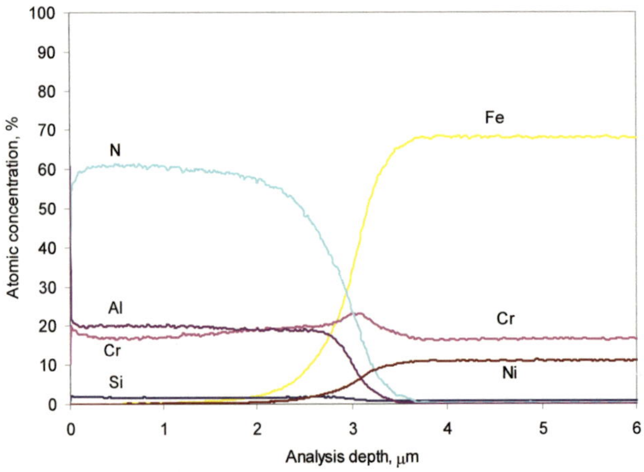

Fig. 16. Changes of constituent concentration of the CrAlSiN and the substrate materials.

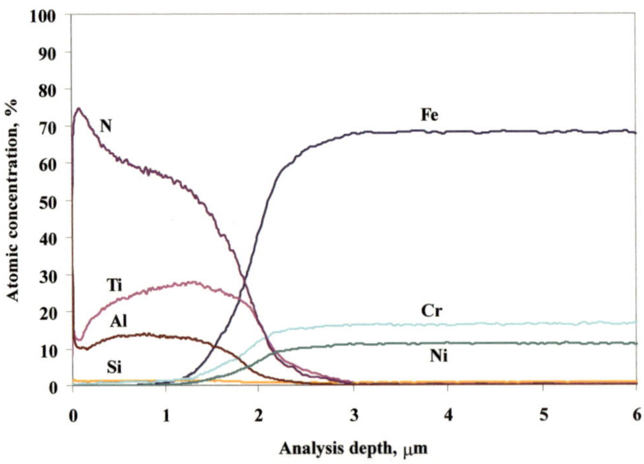

Fig. 17. Changes of constituent concentration of the TiAlSiN and the substrate materials.

coatings layers than the uncovered sample (Table 3). During the anode scanning, the current density is always lower for the sample with a coating deposited on its surface in comparison to the uncovered sample (11.56 μA/cm²), which indicates a good protective effect. The potentiodynamic polarization curve course is the evidence of the active process of the uncoated X6CrNiMoTi17-12-2 steel surface. The lowest corrosion current density of the investigated coatings is obtained (from Tafel plot) for the CrAlSiN coating. This can be explained by the relatively low porosity of this coating. The current density for the other coatings is significantly higher than the one obtained for the CrAlSiN coating. The shape of the curves in the cathode range indicates the strong slowing down of reactions occurring on the coated samples. The behaviour of the systems tested within the anodic range may evidence the porosity or defect of the coatings. Some of the coatings tested within the anodic range were subjected to self-passivation; however, the passive state occurs within a narrow range of the potentials. The growth of the anodic current related to the transpassivation was observed within the 0–0.4 mV potential range. The corrosion current density and corrosion rate were estimated according to the potentiodynamic curve courses (Table 3). The corrosion potential E_{cor} test results confirm the better corrosive resistance of the coatings (Fig. 19) than the uncoated steel samples. The fact that the corrosive potential of the uncoated substrate significantly grows after a 60-min experiment is also worth noting.

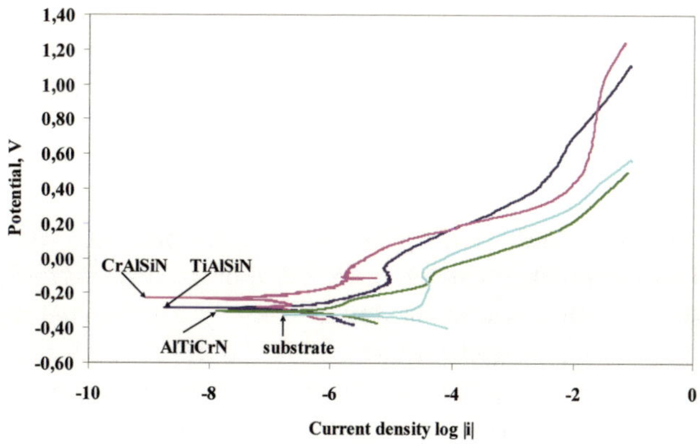

Fig. 18. Potentiodynamic polarization curves of the coatings in 1 M HCl solution.

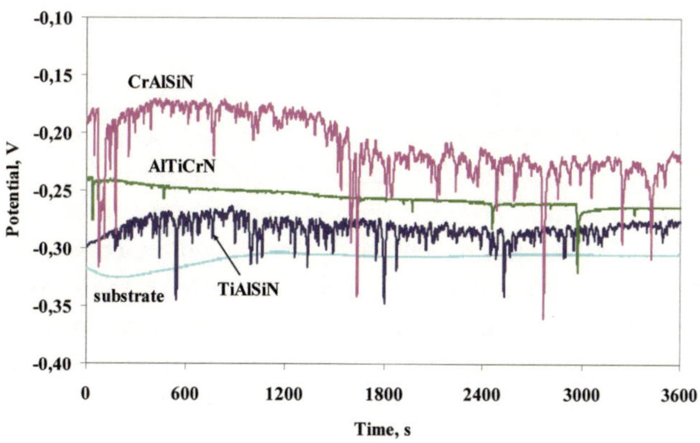

Fig. 19. Open circuit potential curves of the coatings in 1 M HCl solution.

Coating type	Current density i_{cor}, [μA/cm²]	Corrosion potential E_{cor}, [mV]	Corrosion rate, [mm/year]
AlTiCrN	0,77	-0,26	9,0
CrAlSiN	0,15	-0,22	1,7
TiAlSiN	0,48	-0,28	5,6
Substrate	11,65	-0,30	136,2

Table 3. Summary results of the electrochemical corrosion investigation.

Changes of the coating colour and increase in their roughness caused by the intensive dissolving of their surface were observed during the aggressive agent action. Microscope observations make it possible to state that the coating damage process due to electrochemical corrosion proceeds in double way. In the first case, the coating damage develops in many places, whereas the area of these damages is small. In the second case, however, the coating damage caused by the aggressive agent action comprises a big area, leading to changes in its appearance or delamination of the coating parts from the substrate material. The tests show pitting corrosive attacks.

4. Summary

The compact structure of the coatings without any visible delamination was observed in the scanning electron microscope. The fracture morphology of the coatings tested is characterized with a dense structure. Based on the thin film test in the transmission electron microscope, it was observed that the coatings are built of fine crystallites. Their size is 11-25 nm.

The scratch tests on coating adhesion reveal the cohesive and adhesive properties of the coatings deposited on the substrate material. In virtue of the tests carried out, it was found that the critical load L_{C2} fitted within the range 46-54 N for the coatings deposited on a substrate made of hot work tool steel X40CrMoV5-1 and 27-28 N for coatings deposited on the substrate made of the X6CrNiMoTi17-12-2 austenitic steel. The coatings deposited on the substrate made of the X40CrMoV5-1 steel present a better adhesion than the coatings deposited on the substrate made of the X6CrNiMoTi17-12-2 steel. This is caused by a better hardness of the X40CrMoV5-1 steel. The tests made with the use of GDOS indicate the occurrence of a transition zone between the substrate material and the coating, which affects the improved adhesion between the coatings and the substrate.

As a result of the potentiodynamic polarization curve analysis, the corrosion current density-corrosion rate was determined. It confirms the better corrosion resistance of samples coated with the use of the PVD technique to the uncoated samples made of the austenitic steel (11.65 μA/cm²). The corrosion current density for the coatings tested fits within the range 0.15-0.77 μA/cm², which proves their good anticorrosion properties.

In order to evaluate with more detail the possibility of applying these surface layers in tools, further investigations should be concentrated on the determination of the thermal fatigue resistance of the coatings. The very good mechanical properties of the nanocomposite coatings make them suitable in industrial applications. The investigation results will provide useful information to applying the nanocomposite coatings for the improvement of mechanical properties of the hot work tool steels.

5. Acknowledgement

Research was financed partially within the framework of the Polish State Committee for Scientific Research Project No N N507 550738 headed by Dr Krzysztof Lukaszkowicz.

6. References

Behera, S.K.; Sahu, P.K.; Pratihar, S.K. & Bhattacharyya, S. (2004). *Materials Letters*, 58, 29, (November 2004) 3710-3715.

Bellido-Gonzalez, V.; Stefanopoulos, N. & Deguilhen, F. (1995). Friction monitored scratch testing. *Surface and Coatings Technology*, 74–75, 2, (October 1995) 884889.

Burnett, P.J. & Rickerby, D.S. (1987). The relationship between hardness and scratch adhesion. *Thin Solid Films*, 154, 1-2, (November 1987) 403-416.

Carvalho, S.; Ribeiro, E.; Rebouta, L.; Tavares, C.; Mendonca, J.P.; Caetano Monteiro, A.; Carvalho, N.J.M.; De Hosson, J.Th.M. & Cavaleiro, A. (2004). Microstructure, mechanical properties and cutting performance of superhard (Ti,Si,Al)N nanocomposite films grown by d.c. reactive magnetron sputtering. *Surface and Coatings Technology*, 177–178, (January 2004) 459-468.

Cheng, J.B.; Liang, X.B.; Xu, B.S. & Wu, Y.X. (2009). Characterization of mechanical properties of FeCrBSiMnNbY metallic glass coatings. *Journal of Materials Science*, 44, 13, (July 2009) 3356-3363.

Dobrzanski, L.A.; Lukaszkowicz, K.; Zarychta, A. & Cunha, L. (2005). Corrosion resistance of multilayer coatings deposited by PVD techniques onto the brass substrate. *Journal of Materials Processing Technology*, 164-165, (May 2005) 816-821.

Donnet, C. & Erdemir, A. (2004). Historical development and new trends in tribological and solid lubricant coatings. *Surface and Coatings Technology*, 180–181, (March 2004) 76-84.

He, Y.; Apachitei, I.; Zhou, J.; Walstock, T. & Duszczyk, J. (2006). Effect of prior plasma nitriding applied to a hot-work tool steel on the scratch-resistant properties of PACVD TiBN and TiCN coatings. *Surface and Coatings Technology*, 201, 6, (December 2006) 2534-2539.

Holmberg, K. & Matthews, A. (1994). Coatings Tribology - Properties, Techniques and applications in Surface Engineering, In: *Tribology Series*, Dowson, D. (Ed.), 264-268, Elsevier, ISBN 0 444 88870 5, Amsterdam.

Holubar, P.; Jilek, M. & Sima, M. (2000). Present and possible future applications of superhard nanocomposite coatings. *Surface and Coatings Technology*, 133–134, (November 2000) 145-151.

Kao, W.H. (2009). Tribological properties and high-speed drilling performance of Zr-C:H:$N_{x\%}$ coatings with different amounts of nitrogen addition. *Journal of Materials Science*, 44, 13, (July 2009) 3488-3497.

Lukaszkowicz, K. & Dobrzanski, L.A. (2008). Structure and mechanical properties of gradient coatings deposited by PVD technology onto the X40CrMoV5-1 steel substrate. *Journal of Materials Science*, 43, (May 2008) 3400-3407.

Mao, Z.; Ma, J.; Wang, J. & Sun, B. (2009). Comparison of the coatings deposited using Ti and B_4C powder by reactive plasma spraying in air and low pressure. *Journal of Materials Science*, 44, 12, (June 2009) 3265-3272.

Rafaja, D.; Poklad, A.; Klemem, V.; Schreiber, G.; Heger, D. & Sima, M. (2007). Microstructure and hardness of nanocrystalline $Ti_{1-x-y}Al_xSi_yN$ thin films. *Materials Science and Engineering A*, 462, 1-2, (July 2007) 279-282.

Rafaja, D.; Poklad, A.; Klemm, V.; Schreiber, G.; Heger, D.; Sima, M. & Dopita, M. (2006). Some consequences of the partial crystallographic coherence between nanocrystalline domains in Ti-Al-N and Ti-Al-Si-N coatings. *Thin Solid Films*, 514, 1-2, (August 2006) 240-249.

Sergici, A.O. & Randall, N.X. (2006). Scratch Testing of Coatings. *Advanced Materials & Processes*, (April 2006) 1-3.

Sundararajan, G.; Sudharshan Phani P.; Jyothirmayi, A. & Gundakaram, R.C. (2009). The influence of heat treatment on the microstructural, mechanical and corrosion behaviour of cold sprayed SS 316L coatings. *Journal of Materials Science*, 44, 9, (May 2009) 2320-2326.

Tjong, S.C. & Chen, H. (2004). Nanocrystalline materials and coating. *Materials Science and Engineering*, R 45, 1-2, (September 2004) 1-88.

Veprek, S. (1997). Conventional and new approaches towards the design of novel superhard materials. *Surface and Coatings Technology*, 97, 1-3, (December 1997) 15-22.

Veprek, S. (1998). New development in superhard coatings: the superhard nanocrystalline-amorphous composites. *Thin Solid Films*, 317, 1-2, (April 1998) 449-454.

Veprek, S.; Veprek-Heijman, M.G.J.; Karvankova, P. & Prochazka, J. (2005). Different approaches to superhard coatings and nanocomposite. *Thin Solid Films*, 476, 1, (April 2005) 1-29.

Voevodin, A.A. & Zabinski, J.S. (2005). Nanocomposite and nanostructured tribological materials for space applications. *Composites Science and Technology*, 65, 5, (April 2005) 741-748.

Voevodin, A.A.; Zabinski, J.S. & Muratore, C. (2005). Recent Advances in Hard, Tough, and Low Friction Nanocomposite Coatings. *Tsinghua Science and Technology*, 10, 6, (December 2005) 665-679.

Yang, S.M.; Chang, Y.Y.; Wang, D.Y.; Lin, D.Y. & Wu, W.T. (2007). Mechanical properties of nano-structured Ti-Si-N films synthesized by cathodic arc evaporation. *Journal of Alloys and Compounds*, 440, (August 2007) 375-379.

Yu, C.; Wang, S.; Tian, L.; Li, T. & Xu, B. (2009). Microstructure and mechanical properties of CrAlN coatings deposited by modified ion beam enhanced magnetron sputtering on AISI H13 steel, *Journal of Materials Science*, 44, 1, (January 2009) 300-305.

Zhang, S.; Sun, D. & Bui, X.L. (2007). Magnetron Sputtered Hard and Yet Tough Nanocomposite Coatings with Case Studies: Nanocrystalline TiN Embedded in Amorphous SiN_x, In: *Nanocomposite thin films and coatings*, Zhang, S. & Ali, N. (Ed.), 1-110, Imperial College Press, ISBN-13 978-1-86094-784-1, London.

2

Cellulose Nano Whiskers as a Reinforcing Filler in Polyurethanes

Yang Li and Arthur J. Ragauskas
Institute of Paper Science and Technology, School of Chemistry and Biochemistry, Georgia Institute of Technology, Atlanta, USA

1. Introduction

Bio-based polymers and biocomposites are a relatively new and growing market in light of recent societal concerns including dwindling petroluem reserves, environmental and end-of-live disposal issues (Mohanty et al., 2005; Vijay, 2009). Polymers derived from plants, especially those from non-food resources, are gaining the attention of governments, industries and institutes, primarily due to their environmental compatibility, superior physical properties and low stable market prices which are becoming competitive with petroleum-derived polymers. The three major chemical components of biomass, cellulose, hemicellulose and lignin are utilized in diverse fields, such as biofuels, particularly bioethanol and green diesel, biomaterials, including conventional composites and novel nanocomposites, and other value-added chemicals. Among them, cellulose is the most abundant biopolymer in the world with a total annual biomass production of about 1.5×10^{12} tons (Klemm et al., 2005). It has led to a large body of research due to its renewable nature, wide availability, non-food agricultural based economy, low density, high specific strength and modulus, high aspect ratio and reactive surface (Samir et al., 2005). Cellulose is a polydispersed linear polymer of β-(1,4)-D-glucose. A cellulose fiber is composed of bundles of microfibrils where the cellulose chains are stabilized laterally by inter and intra-molecular hydrogen bonding. Microfibrils are comprised of elementary fibrils where monocrystalline domains are linked by amorphous domains. Generally, monocrystallite cellulose has been reported with length ranges from 100 to 300 nm and diameter between 5 and 20 nm. In other words, cellulose monocrystallite has a high aspect ratio of 20-60 (Helbert et al., 1996; Eichhorn et al., 2001; Mathew & Dufresne, 2002; Morin & Dufresne, 2002; Samir et al., 2004). Table 1 summarizes the degree of crystallinity and the lateral dimension of elementary fibrils from several cellulose samples measured by X-ray diffraction (XRD). Tensile strength and modulus of native cellulose crystallites are approximately 10000 MPa and 150 MPa, respectively (Kamel, 2007). Under certain process conditions, transverse cleavage of the cellulose happens primarily in the amorphous zone of the fiber and releases needle-like monocrystals referred to as cellulose nano whiskers. Whisker dimensions depend on both the origin of the cellulose and reaction conditions employed. In general, wood and cotton cellulose nano whiskers have a smaller length and cross section compared to those derived from tunicate, bacterial and algae (Hanley et al., 1992; Terech et al., 1999;

Grunert & Winter, 2002; Beck-Candanedo et al., 2005), which is in agreement with the degree of crystallinity and the lateral dimension of elementary fibrils. Cellulose nano whiskers exhibit not only a high elastic modulus of 143 GPa (Sturcova et al., 2005), but also show significant changes in electrical, optical, and magnetic properties in comparison to native cellulosic fibers (Samir et al., 2005). There has been a growing interest in cellulose nano whisker reinforced composites in the last decade, and improvements in mechanical and thermal properties are readily achieved. (Dufresne et al., 1999; Mathew & Dufresne, 2002; Bondeson & Oksman, 2007).

Sample	C, %	D, nm
Natural softwood/hardwood cellulose	60-62	3-4
Isolated sulfite cellulose	62-63	5-6
Isolated Kraft cellulose	64-65	6-7
Natural cotton cellulose	68-69	5-6
Isolated cotton cellulose	70-72	7-8
Natural flax or ramie cellulose	65-66	4-5
Isolated flax or ramie cellulose	67-68	6-7
Bacterial cellulose	75-80	7-8
Algae cellulose	75-80	10-15

Table 1. Degree of crystallinity (C) and lateral dimension (D) of elementary fibrils from several cellulose samples (Ioelovich, 1993; Ioelovich & Larina, 1999; Grunert & Winter, 2002; Ioelovich, 2009; Ioelovich & Leykin, 2009).

Polyurethane (PU) is any polymer consisting of a chain of organic units joined by urethane linkages (-NHCOO-). It is formed through a step-wise polymerization by reacting a monomer containing at least two isocyanate groups with another monomer containing at least two hydroxyl groups in presence of a catalyst (Pascault et al., 2002). PU has rapidly grown to be one of the most diverse and widely-used materials with a continuously increasing global market since its first lab synthesis in 1937 by Otto Bayer and co-workers (Vermette et al., 2001). Compared to conventional materials, e.g., wood and metals, polyurethane has its own unique merits, such as low density, thermal conductivity and moisture permeability, a high strength to weight ratio, and dimensional stability (Lim et al., 2008). In addition, the formulation and reaction conditions can be readily adjusted to synthesize PUs with desired properties for specific applications. Nowadays, PU is primarily used for construction, packaging, insulation, bedding, upholstery, footwear, and vehicle parts, in forms of rigid, semi-rigid and flexible foams with a wide range of densities, as well as elastomers. Despite the significant benefits of PU, it still exehibits some drawbacks including poor degradability and toxicity due to the use of isocyanates which have evoked researchers to find more environmental friendly starting materials. Moreover, the mechanical and thermal properties of PU are not optimial in comparison to other synthetic polymers like polystyrene. These drawbacks have continued to spur research into PU composites, especially nanocomposites, considering the superior properties that can be acquired by the introduction of nano particles into a PU product. In recent years, cellulose nano whiskers have been used as a reinforcing filler in PU synthesis, and improvements of both thermal and mechanical properties have been reported (Marcovich et al., 2006; Cao et al., 2007; Auad et al., 2008; Cao et al., 2009; Auad et al., 2010; Li et al., 2010a; Wang et al., 2010). Since different types of PU have been investigated through various preparation methods and characterization techniques, a summary and comparation with regard to the PU nanocomposite synthesis and a detailed discussion of the properties, mechanisms and other associated issues will facilitate future applications of cellulose nano whiskers in PU and other related polymers.

2. Cellulose nano whisker

2.1 Preparation of cellulose nano whiskers

During the past twenty years, research on cellulose nano whiskers has been extensively developed. Softwood (SW) kraft pulp (Revol et al., 1994; Araki et al., 1998; Araki et al., 1999; Pu et al., 2007), SW sulfite pulp (Beck-Candanedo et al., 2005), hardwood (HW) ECF (elemental chlorine free) pulp (Beck-Candanedo et al., 2005), recycle pulp (Filson et al., 2009), cotton fiber (Revol et al., 1994; Dong et al., 1996; Dong et al., 1998; Araki et al., 2000; Hasani et al., 2008; Cao et al., 2009; Pei et al., 2010; Tang & Weder, 2010; Wang et al., 2010), sisal fiber (de Rodriguez et al., 2006; Tang & Weder, 2010), flax fiber (Cao et al., 2007), ramie fiber (Habibi et al., 2007; Habibi & Dufresne, 2008; Zoppe et al., 2009), wheat straw (Helbert et al., 1996), bamboo residue (Liu et al., 2010), bacterial microfibrils (Grunert & Winter, 2002), grass fiber (Pandey et al., 2009), tunicate cellulose (Favier et al., 1995; Angles & Dufresne, 2000; Sturcova et al., 2005; Ljungberg et al., 2006; Habibi et al., 2007; Siqueira et al., 2010; Tang & Weder, 2010), microcrystalline cellulose (MCC) (Samir et al., 2004; Samir et al., 2004; Bondeson et al., 2006; Oksman et al., 2006; Bondeson & Oksman, 2007; Bai et al., 2009; Auad et al., 2010; Liu et al., 2010) have all been utilized as cellulose sources for whiskers.

The most common preparation method employed is acid hydrolysis, including acid sulfuric and hydrochloric acid. Other methods, such as enzymatic hydrolysis and mechanical disintegration have also been used. Cellulose fibers are usually disintegrated by a Wiley mill to pass through a 20 mesh screen before acid hydrolysis. Sulfuric acid concentrations of 60-70% (w/w), more often 64%, is preferred (Revol et al., 1994). Acid treatment can range from 10 min at 70 °C to 3 hours at 45 °C at select acid to cellulose ratios, and the reaction is typically quenched by diluting with a 10 fold addition of deionized (DI) water. The sediment, cellulose nano whiskers, is then collected and neutralized by repeated centrifugation and prolonged dialysis against deionized water until the pH of the whiskers suspension does not change. For specific investigations and/or application purposes, all ions except H^+ associated with sulfate groups on the surface of H_2SO_4 generated whiskers need to be removed. This can be achieved by treating the whisker suspension with a mixed-bed ion exchange resin and filtering through a 0.45 μm membrane (Dong et al., 1996). Afterward, ultrasonic treatment is necessary to seperate nano whiskers. A plastic reaction flask is preferred to avoid the release of ions from the glass container and the solution needs to be chilled to avoid overheating which could cause desulfation (Dong et al., 1998). Recently, cellulose nano whiskers were also prepared by sulfuric acid hydrolysis of cotton fiber (Hasani et al., 2008), MCC (Bondeson et al., 2006; Bai et al., 2009), and sisal fiber (Siqueira et al., 2010) which followed the same general procedure described above. Whiskers with a narrow size distribution were obtained through differential centrifugation techniques (Bai et al., 2009). A comprehensive compilation of preparation conditions employing sulfuric acid and the average dimensions of cellulose nano whiskers derived from different sources is shown in Table 2.

During sulfuric acid hydrolysis, esterification of cellulose hydroxyl groups to sulfate groups occurs (Figure 1) which can introduce negative charges to the nano whiskers and this provides improved suspension stability. The sulfate content of cellulose nano whiskers can be determined by a conductimetric method described by Araki et al. (1998). A whiskers suspension (~0.01 g/mL, 45 mL) is mixed with a NaCl solution (0.01 M, 5 mL) before measurement. For samples with poor or no sulfonation, 3 mL of water is replaced by 0.01

Cellulose source	H_2SO_4 conc., % (w/w)	Time, min	T, °C	Acid/cellulose, mL/g	Dimension, nm^2
SW pulp	64	10	70	8.75	~ 200 × 5
	60	50	< 70°C	8.75	~ 200 × 5
	65	10	70	10	185 ± 75 × ~ 3.5
	65	60	45	8.75	185 ± 75 × ~ 3.5
	64	45	45	17.5	100-250 × 5-15
HW pulp	64	25	45	8.75	147 ± 7 × 3-5
	64	25	45	8.75	141 ± 6 × 5.0 ± 0.3
	64	45	45	8.75	120 ± 5 × 4.9 ± 0.3
	64	45	45	17.5	105 ± 4 × 4.5 ± 0.3
Cotton	64	120	45	8.75	~ 200 × 5
	64	60	45	8.75	115 ± 10 × ~ 7
	64	45	45	17.5	176 ± 21 × 13 ± 3
	64	120	60	8.33	70-150 × 10-20
	65	60	45	8.75	100-150 × 5-10
Sisal	65	15	60	16.2	~ 250 × 4
Flax	64	240	45	8.33	327 ± 108 × 21 ± 7
Wheat straw	65	60	25	34.3	150-300 × ~ 5
MCC	63.5	130.3	44	10	200-400 × <10
	64	300	45	8.75	41-320 × < 100
	64	180	45	17.5	60-120 × 8-10
	64	-	45	8.75	100-225 × 10-15

Table 2. Dimensions of cellulose nano whsikers prepared under different sulfuric acid hydrolysis conditions (Revol et al., 1994; Favier et al., 1995; Dong et al., 1996; Helbert et al., 1996; Araki et al., 1998; Dong et al., 1998; Araki et al., 1999; Beck-Candanedo et al., 2005; Bondeson et al., 2006; de Rodriguez et al., 2006; Cao et al., 2007; Hasani et al., 2008; Bai et al., 2009; Cao et al., 2009; Auad et al., 2010; Liu et al., 2010).

M HCl in order to obtain a preferred value (Bondeson et al., 2006). An alkaline solution (0.1 N NaOH) is then added to the suspension at a rate of 0.5 mL per 5 min with continuous stirring. The change in conductivity is recorded by a electric conductometer. Theoretically, the two branches of the titration curve should intercept the volume axis at the same value which is the equivalence-point volume. However, in some cases, the abscissa intercepts due to the dissociation of cellulosic carboxylic acid groups after the stronger sulfate groups have been neutralized (Roman & Winter, 2004). The actual equivalence-point volume is then determined as the average of the two volumes. The degree of sulfonation relies highly on the acid to cellulose ratio and the reaction time as shown in Table 3.

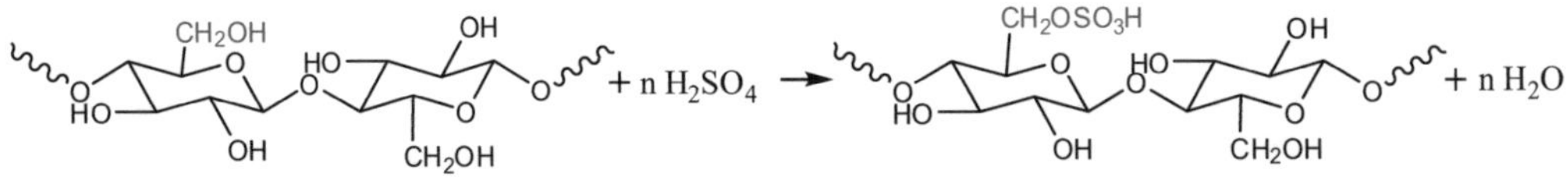

Fig. 1. Esterification of cellulose hydroxyl groups during sufuric acid hydrolysis.

Pulp:acid	Reaction time, min	Sulfur content[a], %	DS[b]
1:4	16	49.540	1.670
	17	59.290	2.000
	45	56.670	1.910
1:6	15	59.110	1.995
	19	59.840	2.020
	18	46.260	1.560
1:8	15	62.260	2.102
	20	59.440	2.013
	45	53.290	1.793
1:10	15	60.700	2.049
	23	63.100	2.129
	22	61.500	2.076

[a] Sulfur content is the percentage of sulfur groups based on the total number of hydroxyl and sulfur groups; [b] degree of substitution (DS) is calculated as the average number of hydroxyl groups in the anhydroglucose that are substituted in the particular product.

Table 3. Effect of sulfuric acid (72%, w/w) hydrolysis condition on the degree of sulfonation of linen pulp (Selim et al., 2004).

The effects of hydrolysis temperature, time, and ultrasonic treatment duration on the properties of cellulose nano whiskers were investigated (Dong et al., 1998) according to an experimental setup illustrated in Table 4. It was shown that with 64% (w/w) sulfuric acid and an acid to cellulose ratio of 8.75 mL/g, 18 hours was required to produce nano whiskers at 26 °C. At 65 °C, hydrolysis could not be controlled easily, a yellow color appeared at the first 10 min, and the sample became black after 1 hour. Side reactions, e.g., dehydration, were presumed to occur under this condition. A temperature of 45 °C was proposed to be optimal and could lead to an efficient hydrolysis yielding an ivory-white colored suspension with a reported yield of 44% after 1 hour. It was also found that the total sulfur content and surface charge of nano whiskers gradually increased when increasing the hydrolysis reaction time from 10 to 240 min at 45°C. The whiskers size decreased in the early stage of the hydrolysis and a relatively stable dimension was achieved after 1 hour as summarized in Table 5. The appearance of the whiskers suspension could be white with some starting pulp particles (low yield), ivory white viscous suspension (optimal), yellowish or even black viscous suspension (over hydrolyzed) (Dong et al., 1998). The particle size decreased within the first 5 min of ultrasonic treatment but no further change was observed upon extended treatment, while the surface charge of cellulose nano whiskers remained constant (Dong et al., 1998). The effect of ultrasonic treatment on whiskers size and surface charge is shown in Table 6. Nevertheless, a 15 min of ultrasonic treatment was reported to produce well separated Korean grass cellulose nano whiskers of 10 nm thickness and several nano-sized lengths, while 5 min of ultrasonic treatment time only resulted in nano-sized (40-50 nm thickness) agglomerates (Pandey et al., 2009). Subsequent studies by Beck-Candanedo et al. (2005) also investigated the effect of hydrolysis time on the properties and behaviors of cellulose nano whiskers. A black spruce sulfite pulp and bleached eucalyptus kraft pulp were treated with 64% (w/w) sulfuric acid at 45 °C. It was shown that longer hydrolysis time produced shorter, less polydispersed whiskers while increased acid to cellulose ratios reduced whiskers dimensions.

To optimize the isolation of cellulose nano whiskers, Bondeson et al. (2006) performed a series of acid hydrolysis experiments with MCC prepared from Norway spruce sulfite pulp using a response surface methodology. Factors included hydrolysis time, temperature, acid concentration, acid to cellulose ratio, and ultrasonic treatment duration. The dimension and yield of whiskers served as the experimental response factors. A summary of this study in terms of experimental conditions and results is shown in Table 7. It was found that MCC required a longer time to be hydrolyzed with to nano whiskers. Briefly, the optimal condition was determined to be sulfuric acid concentration of 63.5% (w/w), acid to cellulose ratio of 10 mL/g, and 130 min hydrolysis at 44 °C followed by approximately 30 min ultrasonic treatment, which produced whiskers of 200 to 400 nm in length and less than 10 nm in width and a yield of 30% of the initial weight. In the above mentioned studies, different ultrasonic treatment times were reported to obtain well separated whiskers. This was probably due to the non-uniform concentration and volume of whiskers suspension as well as various ultrasonic equipment and output.

Sample	T, °C.	Time, h	Appearance of the suspension	Yield, % (w/w)
1	25	1	White, with pulp particles	89.8
2	25	18	Ivory white, viscous	34.4
3	45	1	Ivory white, viscous	43.5
4	65	0.25	Yellow, very viscous	48.1
5	65	1	Black	N/A

Table 4. Effect of hydrolysis condition on the appearance of cellulose nano whiskers suspension (Dong et al., 1998).

Sample	Hydrolysis time, min	Sulfur content[a]	Whisker length, nm
1	10	0.53	390
2	20	0.50	332
3	30	0.58	276
4	45	0.62	226
5	60	0.69	197
6	120	0.74	179
7	240	0.75	177

[a] Sulfur content is the percentage of sulfur groups based on the total number of hydroxyl and sulfur groups.

Table 5. Effect of sulfuric acid hydrolysis time at 45 °C on the sulfur content and length of cellulose nano whiskers (Dong et al., 1998).

Treatment time, min	Whisker length, nm	Sulfur content[a]
1	214	0.484
2	205	0.487
5	182	0.482
10	183	0.489
20	176	0.507
40	182	0.503

[a] Sulfur content is the percentage of sulfur groups based on the total number of hydroxyl and sulfur groups.

Table 6. Effect of ultrasonic treatment on the length and surface charge of cellulose nano whiskers (Dong et al, 1998).

Sample	MCC conc., g/100 mL	H_2SO_4 conc., % (w/w)	Hydrolysis time, min	T, °C	Utrasonic teatment time, min	Whiskers length, µm	Yield, %
1	5	44.1	10	40	30	19.34	93
2	5	44.1	10	80	10	9.12	88
3	5	64.8	10	40	10	38.68	91
4	5	64.8	10	80	30	10.82	47
5	5	44.1	120	40	10	14.30	94
6	5	44.1	120	80	30	7.01	92
7	5	64.8	120	40	30	0.11	34
8	5	64.8	120	80	10	-	0
9	15	44.1	10	40	10	22.36	95
10	15	44.1	10	80	30	8.53	90
11	15	64.8	10	40	30	14.91	73
12	15	64.8	10	80	10	12.69	23
13	15	44.1	120	40	30	13.80	95
14	15	44.1	120	80	10	6.88	86
15	15	64.8	120	40	10	0.18	18
16	15	64.8	120	80	30	-	0
17	10	55.1	65	60	20	7.96	82
18	10	55.1	65	60	20	8.15	78
19	10	55.1	65	60	20	7.84	78

Table 7. Optimization of cellulose nano whiskers isolation

With regards to the hydrochloric acid hydrolysis, cellulose fibers are usually treated with 4 N HCl at 80 °C for approximately 4 hours at an acid to cellulose ratio of 30-35 mL/g. Centrifugation, dialysis, and ultrasonic treatment are followed afterwards with the same conditions used for sulfuric acid hydrolysis. The pH value of HCl-whiskers suspension is ~6, while that of H_2SO_4-whiskers level off at 2-3 which is attributed to the sulfate groups (Araki et al., 1998). The overall mass yield of HCl-whiskers (10-20%) is lower than H_2SO_4-whiskers (70-75%) (Araki et al., 2000). HCl-whiskers of SW kraft pulp were reported to be approximately 180 nm in length and 3.5 nm in width, while cotton yielded HCl-whiskers were around 100 nm in length and 5-10 nm in width (Araki et al., 1998; Araki et al., 1999; Araki et al., 2000).

Compared to the hydrochloric acid procedure, sulfuric acid hydrolysis needs lower temperature, acid to cellulose ratio, and less time to produce whiskers of similar dimensions as illustrated in Table 8. Moreover, it was proposed that the static electronic repulsion between negatively charged sulfates resulted in a more stable suspension instead of easily aggregated whiskers and it was the same reason given for the time independence of suspension viscosity (Kamel, 2007). A HCl-whiskers suspension was found to be thixotropic at concentrations >0.5% (w/v) and anti-thixotropic <0.3% (w/v) (Araki et al., 1998). In order to investigate the influence of sulfate groups on viscosity properties of whiskers suspension, Araki et al. (1999, 2000) introduced sulfate groups to HCl-whiskers by postsulfonation. Briefly, HCl-whiskers precipitated after centrifugation were mixed with 65% (w/w) sulfuric acid to yield a final acid concentration of 55% (w/w) and reacted in a waterbath at 40 °C or 60 °C for 2 hours, which was quenched by adding large amounts of cold water. The number of sulfate groups was controlled by changing experimental factors, such as time and temperature. For example, treatment of an HCl-whiskers suspension with equal weight amount of concentrated sulfuric acid at 60 °C overnight resulted in a sulfate content of 38 mmol/kg, which was much lower than that of H_2SO_4-whiskers (140 mmol/kg) (Araki et al., 2000). An increase of sulfate groups was observed when the postsulfonation reaction was conducted at 40 °C for 2 hours (Araki et al., 1999). It was shown that whiskers with a lower content of sulfate groups showed a slight viscosity increase with time at high solid contents, though it was not as significant as HCl-whiskers. However, HCl-whiskers treated at 40 °C showed nearly the same level of viscosity properties as a H_2SO_4-whiskers suspension. It was also found that viscosity was strongly affected within a surface charge content of 50-60 mmol/kg. In summary, although the microscopic size and shape of whiskers could be the same irrespective of preparation method, the introduction of the surface charge drastically reduced the viscosity of whiskers suspension and removed its time dependencey.

Acid	Acid/cellulose, mL/g	Time, min	T, °C	Whiskers dimension, nm^2
H_2SO_4, 65%(w/w)	8.75	60	45	185 ± 75 × ~3.5
HCl, 4 N	30	225	80	180 ± 80 × 3.5 ± 0.5

Table 8. Comparion between H_2SO_4- and HCl-whiskers (Araki et al., 1999).

Tunicate, a sea animal, has a mantle (tunic) composed of well organized and highly crystalline cellulosic fibrils, and is also widely used to prepared cellulose nano whiskers as an alternative to wood-based materials. The whiskers preparation procedure is more complicated and is summarized in Favier et al. study (1995). In brief, small fragments of tunic were first deproteinized by three succesive bleaching treatments. Each process involved a 1 hour reaction of 5 g tunic at 70-80 °C with 160 mL of water containing 1.5 g of sodium chlorite and 10 drops of glacial acetic acid as described elsewhere (Wise et al., 1946). The bleached tunicin was disintegrated in water, first by using a blender at a concentration of 5 wt% and then via 15 passes through a laboratory homogenizer operated at 400 bar at a lower concentration of 1 wt%. The aqueous tunicin suspension was then mixed with sulfuric acid to achieve an acid concentration of 55% (w/w) which was then vigorously stirred for 20 min at 60 °C. Afterward, dialysis and ultrasonic treatment were followed as described for the preparation of wood-based whiskers. Reported tunicin whiskers have a width on the order of 10-20 nm, and a length ranges from 100 nm to several microns as summarized in Table 9.

Cellulose[a], wt%	Acid[a], wt%	Time, min	T, °C	Whiskers dimension
5	55	20	60	100 nm -several μm × 10-20 nm
5	55	20	60	500 nm - 1-2 μm × 10 nm
-	60	30	65	500 nm - 1-2 μm × 15 nm
5	55	20	60	1 - several μm × 10 -20 nm
-	65	Overnight	Room T	1-2 μm × 8-15 nm

[a] Based on water content.

Table 9. Sulfuric acid hydrolysis conditions and dimensions of tunicate cellulose nano whiskers (Favier et al., 1995; Angles & Dufresne, 2000; Samir et al., 2004; Ljungberg et al., 2006; Habibi et al., 2007).

Enzymatic-mediated production of cellulose nano whiskers was studied by Filson et al. (2009). The highest yield of whiskers was obtained by treatment with 84 EGU of endoglucannase per 200 mg recycled pulp at 50 °C for 60 min assisted by microwave or conventional heating. Microwave treatment was reported to give a higher yield than conventional heating. Enzymatically hydrolyzed whiskers were found to have a relatively larger size than acid hydrolyzed whiskers, typically 30 to 80 nm in width and 100 nm to 1.8 μm in length. In addition, a chemical swelling/ultrasonic separation method was also used to prepare wood-based cellulose nano whiskers (Oksman et al., 2006). MCC (10 wt%) dissolved in a swelling solution of dimethylacetamide (DMAc) with 0.5 wt% of LiCl was treated at 70 °C for 12 hours with stirring. The slightly swollen particles were further separated in an ultrasonic bath for 3 hours over a period of 5 days with long intervals. The whiskers were estimated to be less than 10 nm in width and 200 to 400 nm in length.

2.2 Characterization and properties of cellulose nano whiskers

Transmission electron microscopy (TEM) is a powerful tool for the size measurement of cellulose nano whiskers. A sample is usually prepared by drying a drop of dilute whiskers suspension (~0.1 % w/v) on a carbon coated microscope grid (Dong et al., 1998). The particle dimension can be determined by manual counting from TEM micrographs at a high magnification, and the distributions of particle length and diameter are calculated by counting several individual particles in each sample. On some occasions, the sample can be stained with a uranyl acetate solution to enhance the resolution (de Rodriguez et al., 2006; Bai et al., 2009; Siqueira et al., 2010). Atomic force microscope (AFM) is another useful technique to characterize nano whiskers. In Bech-Candanedo et al.'s study (2005), a cellulose whiskers suspension was diluted to a concentration between 1×10^{-3} and 1×10^{-4} wt% and filtered through a 0.45 μm membrane. A 20 μL drop of 0.1% w/v poly-L-lysine solution was placed on a 1 cm^2 piece of freshly cleaved mica for 3 min, washed off with water and dried. A 10 μL drop of whiskers was allowed to stand on the mica for 1 min, rinsed off with water and dried. The mica was then attached to an AFM specimen disk and analyzed. In addition, particle size distribution, average molecular weight and zeta potential of cellulose nano whisker can also be determined by dynamic light scattering (Filson et al., 2009). As shown in the above tables, whiskers generally have a diameter of 3-20 nm and an aspect ratio of 20-60. Untreated cellulose fibers exhibited many steps of degradation according to the thermal gravimetric analysis (TGA) results, which included an initial moisture loss at 110 °C, and intermediate loss at 280 °C and 352 °C, while H_2SO_4-whiskers showed two well separated degradation processes, one started from 220 to 280 °C and the other was between 330 to 500 °C, with initial moisture loss at 120 °C (Pandey et al., 2008). It was reported that the increase of the moisture loss temperature could be attributed to the strong adhesion of water molecules to the large surface of whiskers. It was also believed that the main degradation between 220 °C and 280 °C was due to depolymerization, dehydration, and decomposition of glycosyl units followed by the formation of a char, while the degradation above 325 °C could be ascribed to the oxidation and breakdown of the char to lower molecular weight gaseous products (Cao et al., 2009). It was argued that the lower degradation temperature of whiskers with sulfate groups was due to the increasing numbers of free end chains and sulfate groups at the surface, which were liable to earlier decomposition (Wang et al., 2007).
A chiral nematic phase is a characteristic of a H_2SO_4-whiskers suspension (Dong et al., 1998; Revol et al., 1992), and phase separation is concentration-driven (Revol et al., 1994). The

critical concentration for anisotropic phase formation can be measured by observing the amounts of isotropic and anisotropic phases in a series of samples with different concentrations (Dong et al., 1996; Dong et al., 1998). The anisotropic phase separates at relatively low concentration, e.g., 4 wt% for a salt-free whiskers suspension. The critical concentration for anisotropic phase formation was found to be slightly increased and the biphasic range became narrower, with longer hydrolysis time and/or higher acid to cellulose ratio which produced shorter, less polydispersed whiskers (Beck-Candanedo et al., 2005). It is not possible to observe any chiral nematic phase of whiskers hydrolyzed by other acids. Postsulfonated whiskers suspension showed a birefringent glassy phase different from the chiral nematic phase (Araki et al., 2000).

3. Cellulose nano whisker/polyurethane nanocomposite

3.1 Starting materials

Starting materials are highly important in view of the desired properties for specific applications of PUs. In general, at least two polyols and an isocyanate are required for a proper performance of PU. Difunctional polymers are preferred for a linear or low cross-linked PU as illustrated in Figure 2. In Marovich et al.'s work (2006), a mixture of di- and multi-functional polyols and polymeric 4,4'-diphenylmethane diisocyanate (MDI) were used. In Auad et al.'s study (2010), polyether glycol (PEG), MDI and 1,4-butanediol (BD) were chosen as raw materials. In case of waterborne PU (WPU), a diol such as polycaprolactone (PCL) and polypropylene glycol (PPG), an isocyanate such as isophorone diisocyanate (IPDI) and 2,4-toluene diisocyanate (TDI), an organic acid, typically dimethylol propionic acid (DMPA), and a catalyst like triethylamine (TEA) were used (Cao et al., 2007; Cao et al., 2009; Wang et al., 2010). In contrast, a highly cross-linked structure exists in rigid PU foam which calls for multifunctional polyols and isocyanates. Moreover, a foaming catalyst should be included in addition to the cross-linking catalyst, and a surfactant is often added to reduce surface tension and obtain a high content of closed cells. For example, sucrose-based and glycerol-based polyols with functionality of 4.4 and 3.0, respectively, polymeric MDI with functionality of 2.7, dimethylcyclohexylamine, 1-methyl-4-(2-dimethylaminoethyl) piperazine, silicone surfactant and pentane were used by Li et al. (2010a).

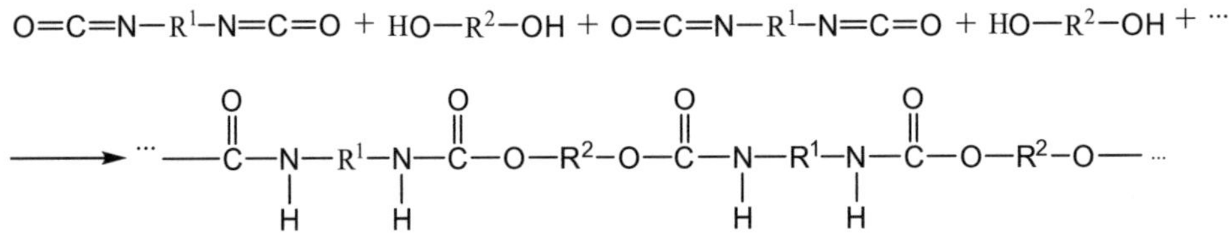

Fig. 2. Chemical reaction of linear polyurethanes synthesis.

3.2 Preparation methods

Due to strong hydrogen bonding interactions between cellulose hydroxyl groups, it is difficult to obtain well separated whiskers in organic solvents, especially for non-polar solvents. Since the dispersion of cellulose nano whiskers in organic solvents is crucial to surface chemical modification of whiskers and preparation of nanocomposites, several methods have been used to solve this issue. Dispersion of freeze-dried whiskers in a polar organic solvents such as dimethylformamide (DMF) or dimethylsulfoxide (DMSO) without surfactant or chemical modification of whiskers has been employed (Samir et al., 2004; Marcovich et al., 2006; Berg et al., 2007). A dilute whiskers suspension was prepared by vigorous mixing and intensive ultrasonic treatment of dried whiskers in organic solvents. In fact, microfibrillated cellulose dispersed in polar solvents such as glycerine, PEG and DMSO after freeze drying were also found to be feasible approaches (Viet et al., 2007). In addition, stable cellulose microfibrils suspensions in toluene and cyclohexane were obtained with the assistance of a phosphoric ester surfactant (Heux et al., 2000). Surface modification, such as partial silylation and grafting of low molecular weight PEG onto the surface of whiskers, is another way to suspend whiskers in non-polar organic solvents (Gousse et al., 2002).

In the past five years, cellulose nano whiskers have been widely used as reinforcing fillers in polyurethanes, including those prepared from MCC (Marcovich et al., 2006; Auad et al., 2008; Auad et al., 2010), flax fiber (Cao et al., 2007), cotton liner pulp (Cao et al., 2009; Wang et al., 2010) and bleached SW pulp (Li et al., 2010a) by acid hydrolysis proposed by Dong et al. (Dong et al., 1996). The dispersion of whiskers in DMF was accomplished either by ultrasonic treatment or a solvent exchange method (Cao et al., 2009).

To prepare a thermoplastic PU film, a whiskers suspension in DMF was first added to a polyol or polyols mixture with stirring which was then followed by ultrasonic treatment and solvent evaporation at 70 °C. Afterward, isocyanate was added to reach a select isocyanate/hydroxyl ratio and mixed throughly. As isocyanate is a very reactive chemical and can easily react with a trace of water present in the starting materials and/or react with it self forming trimers, a molar ratio of isocyanate groups to hydroxyl groups higher than 1.0 is often used (Yao et al., 1995; Xu et al., 2007; Thirumal et al., 2008). The mixture was then cast into an open mold and cured under varying conditions (Marcovich et al., 2006; Auad et al., 2008; Auad et al., 2010). Whiskers content up to 5 wt% was examined in Marcovich's study and up to 1 wt% in Auad's work.

A detailed preparation of WPU can be found in Cao et al.'s study (2007 & 2009). PCL and DMPA were introduced into a reaction flask equipped with a mechanical stirrer and a dropping funnel and heated to 80-85 °C until the PCL melted; IPDI was then added dropwise under a dry nitrogen atmosphere for several hours until the isocyanate content reached a desired value. Subsequently, a whiskers suspension in DMF was added, and the reaction lasted for several hours. Afterward, DMF was removed under reduced pressure at 60 - 80 °C, and acetone was added to reduce the viscosity of the pre-polymer. Carboxylic groups of DMPA were then neutralized with TEA for 30 min, and the product was dispersed in DI water with vigorous stirring at room temperature overnight. The solid content could be further increased to above 25 wt% by rotary vacuum evaporation at 30 °C. The suspension was finally casted in teflon petri dishes and dried in an oven at 40-50 °C for 10-20 hours depending on water content. In Wang et al.'s study (2000), PPG instead of PCL, TDI instead of IPDI, and a whisker suspension in water instead of in DMF were used.

Rigid PU nanocomposite foam reinforced with cellulose nano whiskers has also been prepared (Li et al., 2010a). Whiskers in DMF were mixed with polyols under vigrous stirring followed by removal of DMF under reduced pressure. Catalysts and blowing agent were added and mixed when the mixture cooled down to room temperature. Polymeric MDI was then added and vigorously stirred for ~20s until the self-rising started. Foams were left at room temperature for at least 48 hours before physical testing (Rivera-Armenta et al., 2004).

3.3 Characterizations and discussions

The morphology and fracture surface of PU nanocomposites have been examined by scanning electron microscopy (SEM) (Marcovich et al., 2006; Cao et al., 2007; Auad et al., 2008; Cao et al., 2009; Auad et al., 2010; Li et al., 2010a) and TEM (Wang et al., 2010). The underlying chemical structure can be investigated by FT-IR (Cao et al., 2007; Li et al., 2010a), X-Ray photoelectron spectroscopy (Cao et al., 2009), and wide angle X-ray scattering (Cao et al., 2007; Auad et al., 2008; Cao et al., 2009). Thermomechanical response of specific samples can be determined by using a rheometer in the temperature scan mode while a deformation is applied (Auad et al., 2008), or by a dynamic mechanical analyzer (Marcovich et al., 2006; Cao et al., 2007; Auad et al., 2008; Cao et al., 2009; Auad et al., 2010; Li et al., 2010a; Wang et al., 2010). Differential scanning calorimetry (Marcovich et al., 2006; Cao et al., 2007; Auad et al., 2008; Cao et al., 2009; Auad et al., 2010) and thermal gravimetric analysis (Cao et al., 2009) are both useful tools to study the thermal of PUs stability. Tensile testing can be performed according to ASTM D 1708 (Marcovich et al., 2006; Cao et al., 2007; Auad et al., 2008; Cao et al., 2009; Auad et al., 2010) for PU films and ASTM D 638-08 for PU foams (Li et al., 2010a). ASTM C 365M-05 is a compression test standard for PU foams (Li et al., 2010a).

FT-IR spectra and SEM images of PU nanocomposite films indicated a well dispersion of cellulose nano whiskers in the matrix (Marcovich et al., 2006; Cao et al., 2007; Auad et al., 2008) and a good adhesion in the interfacial area attributed to strong hydrogen bonding (Cao et al., 2007). With respect to the rigid PU foams, it was shown that the closed cells had a homogeneous dispersion in the nanocomposites, and the cell sizes were around 350 μm and decreased slightly with increasing whiskers content (Table 10) (Li et al., 2010b). This was presumably because whiskers served as nucleation sites to facilitate the bubble nucleation process, and the increased number of nucleation sites led to a finer cell structure (Alexandre & Dubois, 2000; Lee et al., 2005). It was found that a higher energy was consumed in the cryo-fracture of nanocomposite films which could possibly improve the tearing characteristics (Marcovich et al., 2006). Tensile modulus of PU films was significantly improved at a small loading of whiskers, i.e., 0.5–5 wt%, and this was more apparent at higher loadings such as 2.5 and 5.0 wt% of whiskers (Table 11) (Marcovich et al., 2006). Contrarily, the creep deformation decreased with increasing whiskers content. For example, incorporation of 1 wt% of whiskers resulted in a tensile modulus improvement of ~53% and creep reduction of ~36%(Auad et al., 2008). The changes are notable considering the relatively small amount of whiskers. With 10 wt% of whiskers, the Young's modulus and strength were significantly enhanced from 1.7 to 107.4 MPa and 4.4 to 9.7 MPa, respectively (Cao et al., 2009). A further increase of whiskers content to 30% resulted in an increase of Young's modulus and tensile strength from 0.5 to 344 MPa and 4.3 to 14.9 MPa, respectively (Cao et al., 2007). This can be ascribed, in part, to the three-dimension network of whiskers reinforced PU linked by intermolecular hydrogen bonding interactions. Additionally, cellulose nano whiskers become chemically bonded to the matrix during curing, through reaction of hydroxyl groups with isocyanate groups which leads to an increase of the glass transition temperature (T_g) and has a positive effect on tensile property of PU (Li et al., 2010b). It was also found that a phase separation of soft and hard domains was favored by whiskers addition which led to an upward shift in melting temperature (T_m) of crystalline phase, an increase in Young's modulus, and a decrease in deformation at break (Auad et al., 2008 & 2010). A remarkable synergistic effect of starch nano crystals (SN) and cellulose nano whiskers (CW) on reinforcing WPU has been observed. Upon incorporation of 1 wt% SN and 0.4 wt% CW, tensile strength, Young's modulus and tensile energy at break of the nanocomposite were significantly improved by 135%, 252% and 136%, respectively, and the elongation at break remained comparable to the pure WPU (Wang et al., 2010). In Cao et al.'s study (2009), it was demonstrated that WPU chains formed crystalline domains on the surface of whiskers which expedited the crystallization of PCL in nanocomposites. This co-crystallization phenomenon is believed to induce the formation of a co-continuous phase between the filler and matrix which significantly enhanced the interfacial adhesion and consequently contributed to an improvement in thermal stability and mechanical strength of the nanocomposites. PU foam reinforced with 0.75 wt% and 1.00 wt% of cellulose nano whiskers showed improved tensile and compressive properties as well as enhanced thermal stability indicated by T_g and T_d (decomposition temperature) (Li et al., 2010b). A detailed comparison of mechanical and thermal properties between the neat PU and PU nanocomposites mentioned above is summarized in Table 11.

Whiskers content, wt%	0	0.25	0.5	0.75	1.00
Cell size, μm	376 ± 41	358 ± 25	345 ± 54	339 ± 28	323 ± 23

Table 10. Cell size of rigid PU nanocomposite foams (Li et al., 2010b).

4. Cellulose fiber and microfibril reinforced polyurethane

Mechanical properties of PU elastomers reinforced with bacterial cellulose microfibers of different aspect ratios were investigated by Bicerano & Brewbaker (1995). It was found that microfibers with a relatively high aspect ratio nearly doubled the Young's modulus and tripled the strength which far surpassed those observed for microfibers of modest aspect ratios. Although cellulose nano whiskers were not discussed, it proved that a high aspect ratio, one of the characteristics of whiskers, had a considerable positive effect on the improvement of mechanical properties. Green algae cellulose fibers of 500 to 600 μm in length and 20 to 200 μm in width were used to prepare a PU composite foam by Johnson & Shivkumar (2004). It was reported that the peak mechanical of properties the composite foams were obtained at a dry fiber content of 5-10 wt%, while a maximum of 40 wt% incorporation, based on the total weight of isocyanate and polyol, were investigated. There was no data on the mechanical improvements of PU composites compared to the neat foam. However, the tensile modulus of the dried algae fibers was measured to be on the order 20-45 MPa, while cellulose nano whiskers have a much higher value of 143 GPa, which makes whiskers a possible stonger reinforcing filler compared to algae fibers. Vegetal cellulose fibers were also used in preparation of PU composites (Silva & Silva, 2005). Hydrogen bonding interactions and covalent linkages between cellulose fibers and polymeric matrix were found by FT-IR but the mechanical properties of nanocomposites were not discussed. Later on, Wu et al. (2007) did a comparison study on cellulose fiber versus nanofibril reinforced high strength elastomeric PU nanocomposites. Cellulose fibers used in the reported study were 1-2 mm in length, while the nanofibrils, prepared by dissolution of MCC (50 μm average particle size) in DMF containing trace amount of LiCl (<0.3 wt%), were 20 to 40 nm in diameter and 450-900 nm in length. Date shown in Table 13 indicated that cellulose nanofibrils had a significant reinforcing effect on the tensile properties of elastomeric PU, while cellulose fibers gave a mild increase of tensile modulus but a decrease of tensile strength and the strain-to-failure. It was stated that the size scale of MCC was typically in the order of tens of microns and above, and it was likely that the less successful reinforcement effect was due to micro- rather than nanoscale dispersion of the fillers (Wu et al, 2007). In Mosievicki et al.'s study (2009), MCC was found to be poorly dispersed into the PU matrix because of the agglomeration of crystalline particles which resulted in lower mechanical and dynamic mechanical properties; however, the thermal stability was enhanced at high temperatures. Generally, the intrinsic high strength, high aspect ratio of cellulose nano whiskers, as well as the nanoscale effect makes it a superior reinforcing filler in comparison with cellulose fibers or microfibrils.

Thermoplastic PUs	Whiskers dimesion, nm^2	Whiskers content, wt%	Young's modulus change,%	Tensile strength change,%	Elongation at break change, %	T_g change, %	T_d change, %	T_m change, %	
								S[c]	H[d]
Polyol	100-225 ×	0.5	+55.1	-18.48[a]	-43.6	-	-	-	-
mixture +	10-15	1.0	+50.6	-48.5[a]	-54.5	-	-	-	-
polymeric		2.5	+102.8	-29.0[a]	-50.9	+53.5[b]	-	-	-
MDI		5.0	+143.6	-37.8[a]	-52.7	-	-	-	-
Commercial	Hundreds	0.1	-	-	-	+0.55	-	+0.73	+2.27
polyester PU	× 10-20	0.5	-	-	-	+1.37	-	+2.93	+4.25
		1.0	-	-	-	+2.19	-	+5.12	+2.05
PEG 650 + MDI	2500-5000	0.1	-	-	-	-	-	-	-
(48 wt %)	× 50-100	0.5	-2.29	-	+11.3	-	-	-	-
		1.0	-1.16	-	-14.2	-	-	-	-
PEG 2000 +		0.1	-	-	-	-	-	+18.3	0
MDI (45 wt%)		0.5	-29.4	-	-92.1	-	-	+60.0	
		1.0	-21.6	-	-81.8	-	-	+35.0	
PEG 2000 +		0.1	-	-	-	-	-	+9.6	0
MDI (40 wt%)		0.5	+37.9	-	-83.3		-	+57.7	
		1.0	+34.0	-	-75.9	-	-	+55.8	
PEG 2000 +		0.1	-	-	-	-	-	+38.2	0
MDI (32 wt%)		0.5	+49.1	-	-82.9	-	-	+70.6	
		1.0	+53.6	-	-89.9	-	-	+32.4	
PEG 2000 +		0.1	-	-	-	-	-	+73.3	-
MDI (23 wt%)		0.5	+42.5	-	-87.9	-	-	+50.7	
		1.0	+44.5	-	-93.4	-	-	+56.0	
Waterborne PU	70-150 ×	2	+218	+43.2	+21.3	+6.65	~15.8	42.51 °C	
	10-20	4	+1224	+70.5	+29.1	+5.67		41.76 °C	
		6	+2335	+102	-2.1	+7.07		42.56 °C	
		8	+3171	+114	-21.2	+5.63		41.66 °C	
		10	+6218	+120	-40.3	+9.38		42.63 °C	
	327 ± 108 ×	5	+40.0	+116	-9.19	-1.75	-	-	-
	21 ± 7	10	+1460	+137	-32.3	-2.92	-	-	-
		15	+940	+181	-43.4	-3.31	-	-	-
		20	+23220	+186	-61.3	-4.67	-	-	-
		25	+47280	+230	-68.6	-4.47	-	-	-
		30	+66780	+247	-82.9	-5.25	-	-	-
	700-800 × 80-100	0.4	+140	+66.7	-4.11	-7.00[b]	+11.1[e]	-	-

[a] yield strength change,%; [b] based on T_a (glass-rubber relaxation temperature); [c] softsegments; [d] hardsegments; [e] based on 30% weight loss temperature .

Table 11. Improvement of mechanical and thermal properties of PU/cellulose nano whisker composite elastomers (Marcovich et al., 2006; Cao et al., 2007; Auad et al., 2008; Cao et al., 2009; Auad et al., 2010; Wang et al., 2010).

Whiskers content, wt%	Tensile modulus change,%	Yield strength change,%	Tensile strength change,%	Compressive strength change,%	Compressive modulus change,%	T_g change, %	T_d change, %
0.25	-30.4	-34.2	-27.0	+29.0	+66.6	+6.8	-1.2
0.50	-20.6	-22.8	-21.4	+131.0	+118.5	+13.6	-0.6
0.75	+36.8	+15.2	+13.8	+143.4	+179.9	+10.2	+0.9
1.00	+227.2	+112.0	+99.2	+269.7	+210.0	+10.2	+3.0

Table 12. Improvement of mechanical and thermal properties of PU/cellulose nano whisker composite foams (Li et al., 2010b).

PU reinforced with filler	Tensile modulus change,%	Tensile strength change, %	Strain-to-failuor change, %
Fiber, 5 wt%	+28.6	-20.5	-62.7
Nanofibril, 5 wt%	+163.3	+110.3	+169.8

Table 13. A comparison between cellulose fibers and nanofibrils reinforced PUs in terms of tensile properties (Wu et al., 2007).

5. Summary and conclusions

The wide availability, renewable and biodegradable features, simple hydrolysis process, and high intrinsic strength and modulus, high aspect ratio and reactivity make cellulose nano whiskers superior to other traditional nano fillers. The reinforcing effect of whiskers in PU is accomplished through both chemical and hydrogen bonding between whiskers and/or the and matrix. Improvements of thermal and mechanical properties of cellulose nano whiskers reinforced nanocomposites are remarkable compared to some other inorganic fillers as well as cellulose fibers and microfibrils. The dispersion difficulty due to hydrophilicity of whiskers has been overcome by either physical or chemical methods. Techniques to produce low polydispersity cellulose nano whiskers are being developed, and it will facilitate improved property control and broaden the applications of cellulose nano whiskers in the future. Incorporation of cellulose nano whiskers in polyurethane and many other polymers has shown huge potential in industrial and consumer uses. Tremendous ongoing work makes it a promising next generation of nano fillers.

6. Acknowledgement

The authors would like to acknowledge the financial support from the PSE Fellowship program at IPST@GT. This work is part of the first author's requirements for the degree of Ph.D. at Georgia Institute of Technology.

7. References

Alexandre M. & Dubois P. (2000) Polymer-layered silicate nanocomposites: preparation, properties and uses of a new class of materials. *Materials science and engineering. R*, 28, 1, 1-63

Angles, M. N. & Dufresne, A. (2000). Plasticized starch/tunicin whiskers nanocomposites. 1. Structural analysis. *Macromolecules*, 33, 22, 8344-8353

Araki, J.; Wada, M.; Kuga, S. & Okano, T. (1998). Flow properties of microcrystalline cellulose suspension prepared by acid treatment of native cellulose. *Colloids and Surfaces a-Physicochemical and Engineering Aspects*, 142, 1, 75-82

Araki, J.; Wada, M.; Kuga, S. & Okana, T. (1999). Influence of surface charge on viscosity behavior of cellulose microcrystal suspension. *Journal of Wood Science*, 45, 3, 258-261

Araki, J.; Wada, M.; Kuga, S. & Okano, T. (2000). Birefringent glassy phase of a cellulose microcrystal suspension. *Langmuir*, 16, 6, 2413-2415

Auad, M. L.; Contos, V. S.; Nutt, S.; Aranguren, M. I. & Marcovich, N. E. (2008). Characterization of nanocellulose-reinforced shape memory polyurethanes. *Polymer International*, 57, 4, 651-659

Auad, M. L.; Mosiewicki, M. A.; Richardson, T.; Aranguren, M. I. & Marcovich, N. E. (2010). Nanocomposites Made from Cellulose Nanocrystals and Tailored Segmented Polyurethanes. *Journal of Applied Polymer Science*, 115, 2, 1215-1225

Bai, W.; Holbery, J. & Li, K. C. (2009). A technique for production of nanocrystalline cellulose with a narrow size distribution. *Cellulose*, 16, 3, 455-465

Beck-Candanedo, S.; Roman, M. & Gray, D. G. (2005). Effect of reaction conditions on the properties and behavior of wood cellulose nanocrystal suspensions. *Biomacromolecules*, 6, 2, 1048-1054

Berg, O.; Capadona, J. R. & Weder, C. (2007). Preparation of Homogeneous dispersions of tunicate cellulose whiskers in organic solvents. *Biomacromolecules*, 8, 4, 1353-1357

Bicerano, J. & Brewbaker, J. L. (1995). Reinforcement of polyurethane elastomers with microfibers having varing aspect ratios. *Journal of the Chemical Society-Faraday Transactions*, 91, 16, 2507-2513

Bondeson, D.; Mathew, A. & Oksman, K. (2006). Optimization of the isolation of nanocrystals from microcrystalline cellulose by acid hydrolysis. *Cellulose*, 13, 2, 171-180

Bondeson, D. & Oksman, K. (2007). Polylactic acid/cellulose whisker nanocomposites modified by polyvinyl alcohol. *Composites Part a-Applied Science and Manufacturing*, 38, 12, 2486-2492

Cao, X. D.; Dong, H. & Li, C. M. (2007). New nanocomposite materials reinforced with flax cellulose nanocrystals in waterborne polyurethane. *Biomacromolecules*, 8, 3, 899-904

Cao, X. D.; Habibi, Y. & Lucia, L. A. (2009). One-pot polymerization, surface grafting, and processing of waterborne polyurethane-cellulose nanocrystal nanocomposites. *Journal of Materials Chemistry*, 19, 38, 7137-7145

de Rodriguez, N. L. G.; Thielemans, W. & Dufresne, A. (2006). Sisal cellulose whiskers reinforced polyvinyl acetate nanocomposites. *Cellulose*, 13, 3, 261-270

Dong, X. M.; Kimura, T.; Revol, J. F. & Gray, D. G. (1996). Effects of ionic strength on the isotropic-chiral nematic phase transition of suspensions of cellulose crystallites. *Langmuir*, 12, 8, 2076-2082

Dong, X. M.; Revol, J. F. & Gray, D. G. (1998). Effect of microcrystallite preparation conditions on the formation of colloid crystals of cellulose. *Cellulose*, 5, 1, 19-32

Dufresne, A.; Kellerhals, M. B. & Witholt, B. (1999). Transcrystallization in Mcl-PHAs/cellulose whiskers composites. *Macromolecules*, 32, 22, 7396-7401

Eichhorn, S. J.; Baillie, C. A.; Zafeiropoulos, N.; Mwaikambo, L. Y.; Ansell, M. P.; Dufresne, A.; et al. (2001). Review: Current international research into cellulosic fibres and composites. *Journal of Materials Science*, 36, 9, 2107-2131

Favier, V.; Chanzy, H. & Cavaille, J. Y. (1995). Polymer nanocomposites reinforced by cellulose whiskers. *Macromolecules,* 28, 18, 6365-6367

Filson, P. B.; Dawson-Andoh, B. E. & Schwegler-Berry, D. (2009). Enzymatic-mediated production of cellulose nanocrystals from recycled pulp. *Green Chemistry,* 11, 11, 1808-1814

Gousse, C.; Chanzy, H.; Excoffier, G.; Soubeyrand, L. & Fleury, E. (2002). Stable suspensions of partially silylated cellulose whiskers dispersed in organic solvents. *Polymer,* 43, 9, 2645-2651

Grunert, M. & Winter, W. T. (2002). Nanocomposites of cellulose acetate butyrate reinforced with cellulose nanocrystals. *Journal of Polymers and the Environment,* 10, 1-2, 27-30

Habibi, Y. & Dufresne, A. (2008). Highly filled bionanocomposites from functionalized polysaccharide nanocrystals. *Biomacromolecules,* 9, 7, 1974-1980

Habibi, Y.; Foulon, L.; Aguie-Beghin, V.; Molinari, M. & Douillard, R. (2007). Langmuir-Blodgett films of cellulose nanocrystals: Preparation and characterization. *Journal of Colloid and Interface Science,* 316, 2, 388-397

Hanley, S. J.; Giasson, J.; Revol, J. F. & Gray, D. G. (1992). Atomic force microscopy of cellulose microfibrils - comparison with transmission electron-microscopy. *Polymer,* 33, 21, 4639-4642

Hasani, M.; Cranston, E. D.; Westman, G. & Gray, D. G. (2008). Cationic surface functionalization of cellulose nanocrystals. *Soft Matter,* 4, 11, 2238-2244

Helbert, W.; Cavaille, J. Y. & Dufresne, A. (1996). Thermoplastic nanocomposites filled with wheat straw cellulose whiskers .1. Processing and mechanical behavior. *Polymer Composites,* 17, 4, 604-611

Heux, L.; Chauve, G. & Bonini, C. (2000). Nonflocculating and chiral-nematic self-ordering of cellulose microcrystals suspensions in nonpolar solvents. *Langmuir,* 16, 21, 8210-8212

Ioelovich, M. Y. (1993). A study on formation of supermolecular structure of cotton cellulose. *Vysokomolekulyarnye Soedineniya Seriya a & Seriya B,* 35, 5, B268-B271

Ioelovich, M. & Larina, E. (1999). Parameters of crystalline structure and their influence on the reactivity of cellulose I. *Cellulose Chemistry and Technology,* 33, 1-2, 3-12

Ioelovich, M. (2009). Accessibility and crystallinity of cellulose. *Bioresources,* 4, 3, 1168-1177

Ioelovich, M. & Leykin, A. (2009). Accessibility and supermolecular structure of cellulose. *Cellulose Chemistry and Technology,* 43, 9-10, 379-385

Johnson, M. & Shivkumar, S. (2004). Filamentous green algae additions to isocyanate based foams. *Journal of Applied Polymer Science,* 93, 5, 2469-2477

Kamel, S. (2007). Nanotechnology and its applications in lignocellulosic composites, a mini review. *Express Polymer Letters,* 1, 9, 546-575

Klemm, D.; Heublein, B.; Fink, H. P. & Bohn, A. (2005). Cellulose: Fascinating biopolymer and sustainable raw material. *Angewandte Chemie-International Edition,* 44, 22, 3358-3393

Lee L. J.; Zeng C.; Cao X.; Han X.; Shen J. & Xu G. (2005) Polymer nanocomposite foams. *Composites science and technology,* 65, 15-16, 2344-2363

Li, Y.; Ren, H. F. & Ragauskas, A. J. (2010a) . Rigid polyurethane foam reinforced with cellulose whiskers: Synthesis and characterization. [journal article]. *Nano-Micro Letters,* 2, 2, 6

Li, Y.; Ren, H. F. & Ragauskas, A. J. (2010b). Rigid polyurethane foam/cellulose whisker nanocomposites: preparation, characterization, and properties. Accepted by *Journal of nanoscience and nanotechnology*

Lim, H.; Kim, S. H. & Kim, B. K. (2008). Effects of silicon surfactant in rigid polyurethane foams. *Express Polymer Letters,* 2, 3, 194-200

Liu, D. G.; Zhong, T. H.; Chang, P. R.; Li, K. F. & Wu, Q. L. (2010). Starch composites reinforced by bamboo cellulosic crystals. *Bioresource Technology,* 101, 7, 2529-2536

Liu, H. Y.; Liu, D. G.; Yao, F. & Wu, Q. L. (2010). Fabrication and properties of transparent polymethylmethacrylate/cellulose nanocrystals composites. *Bioresource Technology,* 101, 14, 5685-5692

Ljungberg, N.; Cavaille, J. Y. & Heux, L. (2006). Nanocomposites of isotactic polypropylene reinforced with rod-like cellulose whiskers. *Polymer,* 47, 18, 6285-6292

Marcovich, N. E.; Auad, M. L.; Bellesi, N. E.; Nutt, S. R. & Aranguren, M. I. (2006). Cellulose micro/nanocrystals reinforced polyurethane. *Journal of Materials Research,* 21, 4, 870-881

Mathew, A. P. & Dufresne, A. (2002). Morphological investigation of nanocomposites from sorbitol plasticized starch and tunicin whiskers. *Biomacromolecules,* 3, 3, 609-617

Mohanty, A. K., Misra M. & Drzal, L. T. (2005). *Natural fibers, biopolymers, and biocomposites,* CRC Press, 978-0-8493-1741-5, Florida

Morin, A. & Dufresne, A. (2002). Nanocomposites of chitin whiskers from Riftia tubes and poly(caprolactone). *Macromolecules,* 35, 6, 2190-2199

Mosiewicki, M. A.; Casado, U.; Marcovich, N. E. & Aranguren, M. I. (2009). Polyurethanes From Tung Oil: Polymer Characterization and Composites. *Polymer Engineering and Science,* 49, 4, 685-692

Oksman, K.; Mathew, A. P.; Bondeson, D. & Kvien, I. (2006). Manufacturing process of cellulose whiskers/polylactic acid nanocomposites. *Composites Science and Technology,* 66, 15, 2776-2784

Pandey, J. K.; Lee, J. W.; Chu, W. S.; Kim, C. S. & Ahn, S. H. (2008). Cellulose nano whiskers from grass of Korea. *Macromolecular Research,* 16, 5, 396-398

Pandey, J. K.; Kim, C. S.; Chu, W. S.; Lee, C. S.; Jang, D. Y. & Ahn, S. H. (2009). Evaluation of morphological architecture of cellulose chains in grass during conversion from macro to nano dimensions. *E-Polymers,* no. 2

Pascault J., Sautereau H., Verdu J. & Williams R. J. J. (2002). *Thermosetting polymers,* Marcel Dekker, 0-8247-0670-6, USA

Pei, A. H.; Zhou, Q. & Berglund, L. A. (2010). Functionalized cellulose nanocrystals as biobased nucleation agents in poly(L-lactide) (PLLA) - Crystallization and mechanical property effects. *Composites Science and Technology,* 70, 5, 815-821

Pu, Y. Q.; Zhang, J. G.; Elder, T.; Deng, Y. L.; Gatenholm, P. & Ragauskas, A. J. (2007). Investigation into nanocellulosics versus acacia reinforced acrylic films. *Composites Part B-Engineering,* 38, 3, 360-366.

Revol, J. F.; Bradford, H.; Giasson, J.; Marchessault, R. H. & Gray, D. G. (1992). Helicoidal self-ordering of cellulose microfibrils in aqueous suspension. *International Journal of Biological Macromolecules,* 14, 3, 170-172

Revol, J. F.; Godbout, L.; Dong, X. M.; Gray, D. G.; Chanzy, H. & Maret, G. (1994). Chiral nematic suspensions of cellulose crystallites - Phase separation and magnetic field orientation. *Liquid Crystals,* 16, 1, 127-134

Rivera-Armenta, J. L.; Heinze, T. & Mendoza-Martinez, A. M. (2004). New polyurethane foams modified with cellulose derivatives. *European Polymer Journal*, 40, 12, 2803-2812

Roman, M. & Winter, W. T. (2004). Effect of sulfate groups from sulfuric acid hydrolysis on the thermal degradation behavior of bacterial cellulose. *Biomacromolecules*, 5, 5, 1671-1677

Samir, M.; Alloin, F.; Sanchez, J. Y.; El Kissi, N. & Dufresne, A. (2004). Preparation of cellulose whiskers reinforced nanocomposites from an organic medium suspension. *Macromolecules*, 37, 4, 1386-1393

Samir, M.; Alloin, F.; Sanchez, J. Y. & Dufresne, A. (2004). Cross-linked nanocomposite polymer electrolytes reinforced with cellulose whiskers. *Macromolecules*, 37, 13, 4839-4844

Samir, M.; Alloin, F. & Dufresne, A. (2005). Review of recent research into cellulosic whiskers, their properties and their application in nanocomposite field. *Biomacromolecules*, 6, 2, 612-626

Selim, I. Z.; Zikry, A. A. F. & Gaber S. H. (2004) Physicochemical properties of prepared cellulose sulfates: II. From linen pulp bleached by the H_2O_2 method. *Polymer-plastics technology and engineering*, 43, 5, 1387-1402

Silva, M. C.,& Silva, G. G. (2005). A new composite from cellulose industrial waste and elastomeric polyurethane. *Journal of Applied Polymer Science*, 98, 1, 336-340

Siqueira, G.; Bras, J. & Dufresne, A. (2010). New Process of Chemical Grafting of Cellulose Nanoparticles with a Long Chain Isocyanate. *Langmuir*, 26, 1, 402-411

Sturcova, A.; Davies, G. R. & Eichhorn, S. J. (2005). Elastic modulus and stress-transfer properties of tunicate cellulose whiskers. *Biomacromolecules*, 6, 2, 1055-1061

Tang, L. M. & Weder, C. (2010). Cellulose Whisker/Epoxy Resin Nanocomposites. *Acs Applied Materials & Interfaces*, 2, 4, 1073-1080

Terech, P.; Chazeau, L. & Cavaille, J. Y. (1999). A small-angle scattering study of cellulose whiskers in aqueous suspensions. *Macromolecules*, 32, 6, 1872-1875

Thirumal, M.; Khastgir, D.; K.Singha, N.; Manjunath, B. S. & Naik, Y. P. (2008). Effect of foam density on the properties of water blown rigid polyurethane foam. *Journal of applied polymer science*, 108, 3, 1810-1817

Vajay, V. K. (2009). Biogas enrichment and bottling technology for vehicular use developed at ITT Delhi. *FITT Forum*, 15, 1, 2-3

Vermette, P.; Griesser, H. J.; Laroche, G. & Guidoin R. (2001). *Biomedical applications of polyurethanes*, Eurekah.com, 1-58706-023-X, Texas

Viet, D.; Beck-Candanedo, S. & Gray, D. G. (2007). Dispersion of cellulose nanocrystals in polar organic solvents. *Cellulose*, 14, 2, 109-113

Wang, N.; Ding, E.; & Cheng R. (2007). Thermal degradation behaviors of spherical cellulose nanocrystals with sulfate groups. *Polymer*, 48, 12, 3486-3493

Wang, Y. X.; Tian, H. F. & Zhang, L. N. (2010). Role of starch nanocrystals and cellulose whiskers in synergistic reinforcement of waterborne polyurethane. *Carbohydrate Polymers*, 80, 3, 665-671

Wise, L.E.; Murphy, M.; D'Adieco, A.A. (1946). Chlorite holocellulose, its fractionation and bearing on summative wood analysis and on studies on the hemicelluloses, *Paper Trade J*, 122, 2, 35–43

Wu, Q. J.; Henriksson, M.; Liu, X. & Berglund, L. A. (2007). A high strength nanocomposite based on microcrystalline cellulose and polyurethane. *Biomacromolecules,* 8, 12, 3687-3692

Xu, Z. B.; Tang, X. L.; Gu, A. J. & Fang, Z. P. (2007). Novel preparation and mechanical properties of rigid polyurethane foam/organoclay nanocomposites. *Journal of applied polymer science,*106, 1, 439-447

Yao Y.; Yoshioka, M. & Shiraishi, N. (1995). Rigid polyurethane foams from combined liquefaction mixtures of wood and starch. *Mokuzai Gakkaishi,* 41, 7, 659-668

Zoppe, J. O.; Peresin, M. S.; Habibi, Y.; Venditti, R. A. & Rojas, O. J. (2009). Reinforcing Poly(epsilon-caprolactone) Nanofibers with Cellulose Nanocrystals. *Acs Applied Materials & Interfaces,* 1, 9, 1996-2004

3

Nanocomposite Based on Natural Materials

Milorad Davidovic[1], Marina Kutin[1], Suzana Linic[1],
Ubavka Mioc[2], Zoran Nedic[2], Svjetlana Sredic[3],
Aleksandra Nikolic[4], Dusan Jovanovic[4], and Polycarpos Pissis[5]
[1]*GOSA Institute*
[2]*University of Belgrade, Faculty of Physical Chemistry*
[3]*Institute of Mining*
[4]*IHTM, Center of Catalysis and Chemical Engineering*
[5]*National Technical University of Athens, Department of Phisics, Athens*
[1,2,4]*Serbia*
[3]*Republic of Srpska, Bosnia and Herzegovina*
[5]*Greece*

1. Introduction

Nanotechnology today is one of the most important ways of industrial manufacture of materials. A great variety of »nanomaterials« with controlled size and composition are developed with the aim of improving optical, mechanical, magnetic, electrical and other properties. With nanocomposite in principal, we will be able to develop structures not previously observed in nature. Also, in this way we expect to produce nanocomposite with better properties using natural raw materials as starting compound.

The most abundant natural materials at the surface of the earth are clay minerals. Clay is material less than 2 μm with specific properties (plasticity, shapeability, rheology, slip formation, ion-exchangeable, etc.) related to the fact that natural clay is nanosized layer. The clay minerals are hydrous aluminum silicates and are classified as phyllosilicates, or layer silicates. Bentonite clay contains a high percentage of montmorillonite.

Montmorillonite clays are known as inorganic ion exchange materials which are composed of alluminosilicate layers stacked one above the other, so called 2:1 layer type silicates.

The layered structure, shown in Fig. 1., presents two silica tetrahedral (corner shared) sheets fused to one (edge shared) octahedral sheet of alumina (alluminosilicate) or magnesia (magnesium silicate)(Manocha, 2006; Wang,1998)

Each layer has a small net negative charge due to an isomorphous substitution of ions in the framework, shown in Fig. 2. This charge is compensated by interlayer hydrated cations, known as exchangeable cations. It is well known that montmorillonite can accommodate various types of compounds in its interlayer spaces to give an intercalation type of inclusion compounds (Moore, 1989; Colomban, 2003; Slade, 1987).

Modification of chemical and structural properties of raw bentonite clay makes possible new nanostructured materials preparation suitable for different applications.

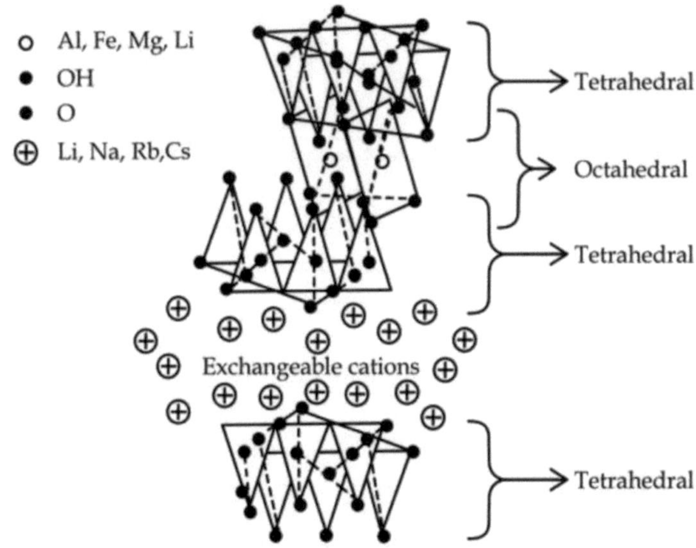

Fig. 1. Structure of montmorillonite clay

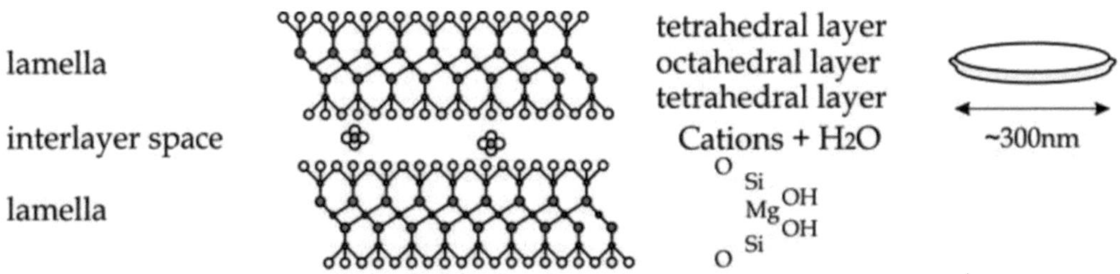

Fig. 2. The structure of montmorillonite, detailed particle view after Karnland (Kröhn, 2003).

With the aim to obtain stable new microporous material based on natural bentonite, we applied pillaring process. This is a multi-step process which involves replacement of the exchangeable interlayer cations by large inorganic polynuclear cations and calcination. Due to the calcination process, intercalated polycations transform into metal oxide clusters (named pillars) linking permanently silicate layers. New material is suitable for different applications (catalyst, catalyst support, etc.) due to their high specific area, chemical and thermal resistivity and micro- and mesoporosity (Ruiz-Hitzky, 1996).

Properties of pillared clays such as surface area, pore size distribution, acidity, depend on the synthesis treatment as well as nature minerals.

To obtain montmorillonite with increased catalytic properties and proton conductivity we have applied treatment which consisted of pillaring procedure and incorporation of different materials (polyoxomethalates -- 12-tungstphosphoric acid and ammonium decavanadate and transition metals Co and Ni) after that. The physicochemical characteristics (porosity, specific surface area, chemical composition, conductivity and catalytic properties) of obtained composite materials have been studded by porosimetry methods, atomic force microscopy, thermal analysis, X-ray diffraction (XRPD), and conductivity measurements. Several experimental techniques were employed to investigate the structure, textural and dynamic, as well as the conductivity properties of the composites.

2. Material preparation - Modification of the raw material

In our investigations we used natural bentonite clay "Šipovo" (Bosnia and Herzegovina) as starting material. The main component part of bentonite is the mineral montmorillonite (75%). Modification of the raw material consists of several steps (Sredić, 2004):

Step 1. Raw material is purified by sieving and sedimentation

Step 2. Ion-exchange step

Aqueous dispersion of purified clay is used and for Na+ exchange, a triple washing with NaCl solution followed by repeated washing until complete elimination of chloride ions is achieved. We expect largely hominid cations exchanged clays.

Step 3. Pillaring

Exchangeable cations can be substituted by large polynuclear cations which, after the thermal treatment, form nanoparticle oxides linking permanently the silicate layers. We used polyhydroxy cation of aluminium as pillaring agent to become thermally stable microporous solids. Al-pillaring reagent was prepared by slowly mixing 0.4 M solutions of $Al(NO_3)_3$ and NaOH until OHAl ratio was fixed at 2.4-2.5. Clay's dispersion was treated with pillaring reagent under continuous magnetic stirring at 80^{0}C over 3 hours and than allowed to cool to room temperature. Resulting sample was washed several times until complete elimination of nitrate ions, air-dried at 60^{0}C and calcinated at 400^{0}C. So we obtained pillared clays (PILCs). Heating causes dehydroxylation of the Al-polyhydroxy cations with formation of alumina clusters, water and protons. It has been proposed that protons generated during these processes migrate to the octahedral layer to charge compensate for the presence of divalent ions in octahedral sites. The exact nature of the alumina clasters that form pillar are not known. Some authors (Ocelli at al.) belive that AlOOH units connect to form the Al-pillar. Informations about dynamic of proton species can provide new knowledges about structure of pillars. There are two different stages of pillars: after drying in air at 100 0C and after calcination at 400^{0}C with different concentracion of hydrogen.

Chemical composition after three steps of material preparation is given in Table 1.

Chemical Characteristics	Samples			
	Starting raw material (Bentonite Sipovo)	Purified (I step)	Na-exchanged Clay (II step)	Al-pillared clay (III step)
SiO2 content (%)	53,50	50,96	49,10	46,32
Al2O3 content (%)	24,39	24,12	25,21	30,09
Na2O content (%)	0,09	0,09	4,37	0,20
K2O content (%)	0,41	0,44	0,24	0,27
CaO content (%)	2,86	2,62	0,60	0,45
MgO content (%)	1,03	1,62	1,62	0,68
Fe2O3 content (%)	4,82	4,69	3,88	3,89
TiO2 content (%)	0,41	0,39	0,67	0,68

Table 1. Chemical composition of the raw material

Experimentally obtained results on Sorptomat (Table 2) show that the chemical changing of the raw montmorillonite (ionic exchange and pillaring) indicate transparent texture changes manifesting as increase in total pore volume (16.3%), total microspore volume (64.0%) and increasing in the specific surface area by 46.8%. At the same by increasing in micro pore volume, total volume of the mezo pore is decreasing by 23.2%, which is consistent with the increase of specific surface area of the modified material. Chemical modification of the raw montmorillonite gave the slight increase in pore width from 3.740 to 3.947nm while the concentration of pores size from 3 to 6nm increased for 27.9% in ratio to started raw material.

	MM (raw)	Ion Exchange MM	Pillared Ion Exchanged MM
Pore volume (cm^3/g)	0.098	0.092	0.107
Micro pore volume (cm^3/g)	0.0364	0.0386	0.0597
Specific surface area (m^2/g)	79.9	85.2	117.3
Maximum pore width (nm)	3.740	3.791	3.947
Pore (3-6 nm) concentration (%)	73.5	78.4	94.0

Table 2. Textural characteristic of the samples after first, second and third treatment

The textural change evident from experimentally obtained results on Sorptomat (Table 2) is also visible on the AFM measurements, Fig. 3.

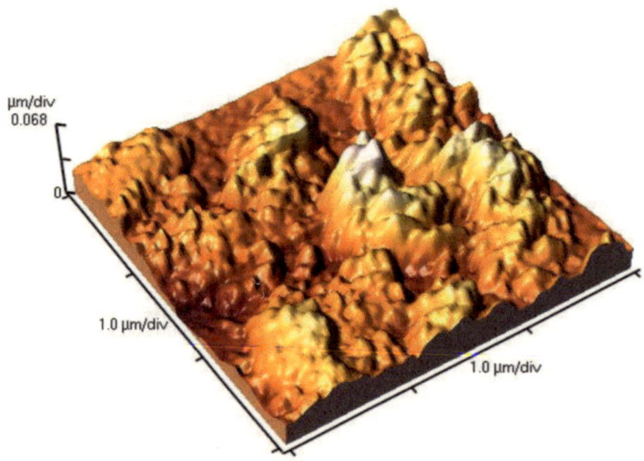

Fig. 3. AFM image of the surface area Al-pillared sample

Step 4. Composite preparation

- A new WPA-PILC material has been prepared by the sol-gel method. The heteropoly acid used in this work was 12-tungstophosphoric acid hydrate $H_3PW_{12}O_{40} \cdot nH_2O$ (WPA-n). The WPA-PILC composite materials were prepared by using water solutions of WPA-6 and pillared clay. We prepared samples with different WPA contents (16, 67 mass %, 28, 57 mass % and 37.50 mass %). The solutions were stirred at room temperature (25^{0}C) for 2 hours, and kept at this temperature for 24 hours to transfer to a gelatine. The gelatine was dried at the temperature of 70^{0}C (Čajkovski, 2005).
- Preparation of new ADV-PILC MM materials by the sol-gel method. The ADV-PILC MM composite materials were prepared by using aqueous ADV

(ammonium decavanadate, solution and MM pillared clay suspension. The ADV content was 2 and 5 mass %. The solutions were stirred at room temperature for 24 hours, and then dried at a temperature of 60 ^{0}C (Sredić, 2008).

Textural characteristics of samples after the first, second and third treatment and of the composite sample with incorporated WPA and ADV are given in Table 3.

	Pillared Ion Exchanged MM	Pillared Ion Exchanged MM+ WPA (28,57 mass%)	Pillared Ion Exchanged MM+ADV (5 mass%)
Pore volume (cm^3/g)	0.107	0.038	0.114
Microspore volume (cm^3/g)	0.0597	0.0291	0.0474
Specific surface area (m^2/g)	117.3	53.46	102.0
Maximum pore width (nm)	3.947	3.857	3.701
Pore (3-6 nm) concentration (%)	94.0	81.6	59.9

Table 3. Textural characteristics of samples after the first, second and third treatment and of the composite sample with incorporated WPA and ADV

3. Dielectric measurements

Polyoxometalates are widely used as industrial catalysts and there is a possibility of using them as materials of specific electric properties. New composite materials have better properties in comparison to starting material.

This resulted especially in a substantial increase of the electrical conductivity.

Investigation of dielectrical and conductivity properties of pillared montmorillonite with incorporated 12-tungstophosphoric acid (PILC-WPA) and ammonium-decavanadate (PILC-ADV) has shown that the new materials exhibits improved electrical conductivity which is higher than the conductivity of both starting compounds.

3.1 PILC-WPA

Using broad band dielectric relaxation spectroscopy (DRS), electrical properties of the obtained material were investigated. Thermally stimulated depolarization current technique was also used due to its high sensitivity and high resolving power. It has been established that pillared montmorillonite doped with WPA-6 ($H_3PW_{12}O_{40}6H_2O$) shows conductivity better than that of the pure components, as shown in Fig. 4.

TSDC plots for PILC montmorillonite are shown in Fig. 5.

The high values of ε'' indicate that these relaxations are caused not only by water molecules but we think also by ion movements affected by the presence of water molecules (Fig. 6.)

In Fig. 7., a comparison of TSDC plots for MM+28.57 mass % of WPA and WPA is shown. We again observe that the conductivity of the MM+28.57 mass % of WPA sample is higher than that of WPA. In the inset of the figure a peak is seen probably indicating phase transition. This phenomenon is observed only for MM+28.57 mass % of WPA, but not for WPA, probably due to its higher conductivity.

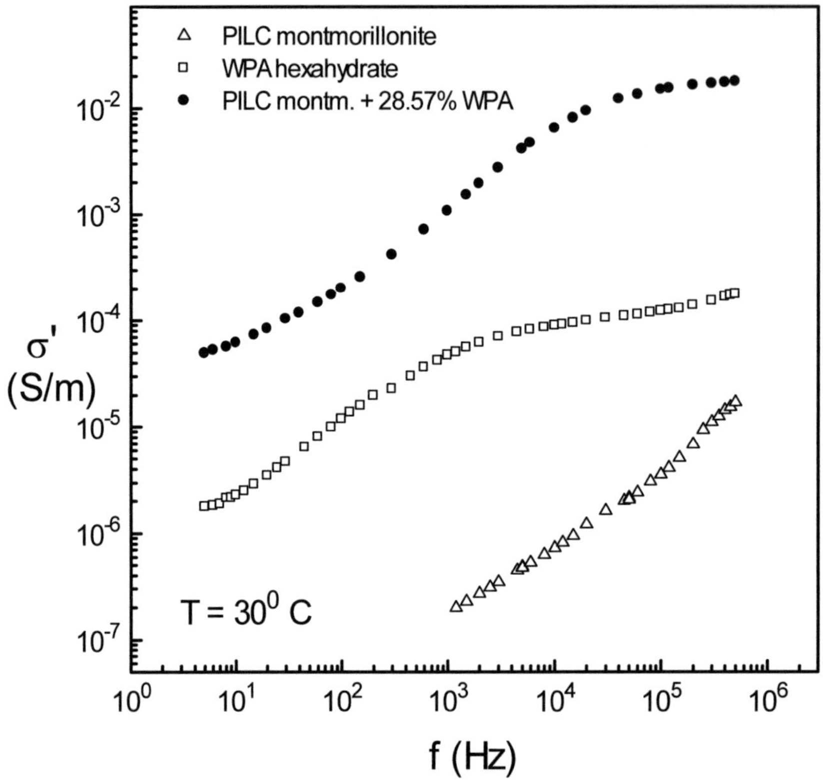

Fig. 4. Conductivity versus frequency for the investigated materials.

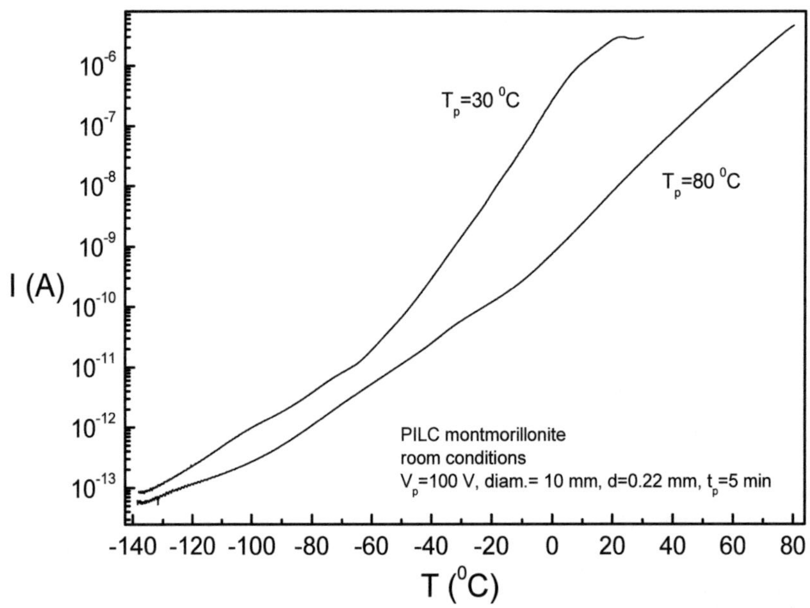

Fig. 5. TSDC for montmorillonite for two polarization temperatures: Tp=30^0 C and Tp=80^0 C.

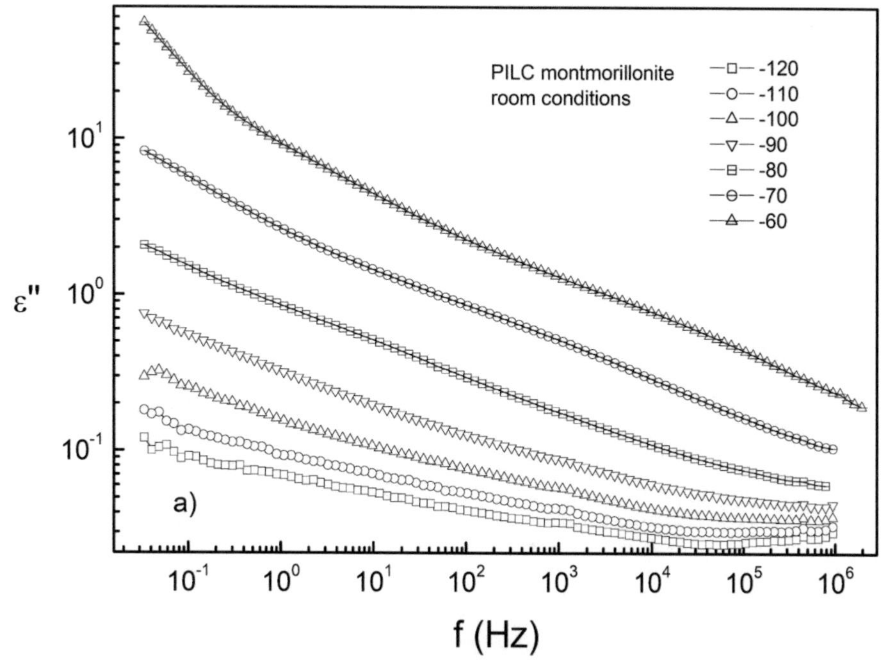

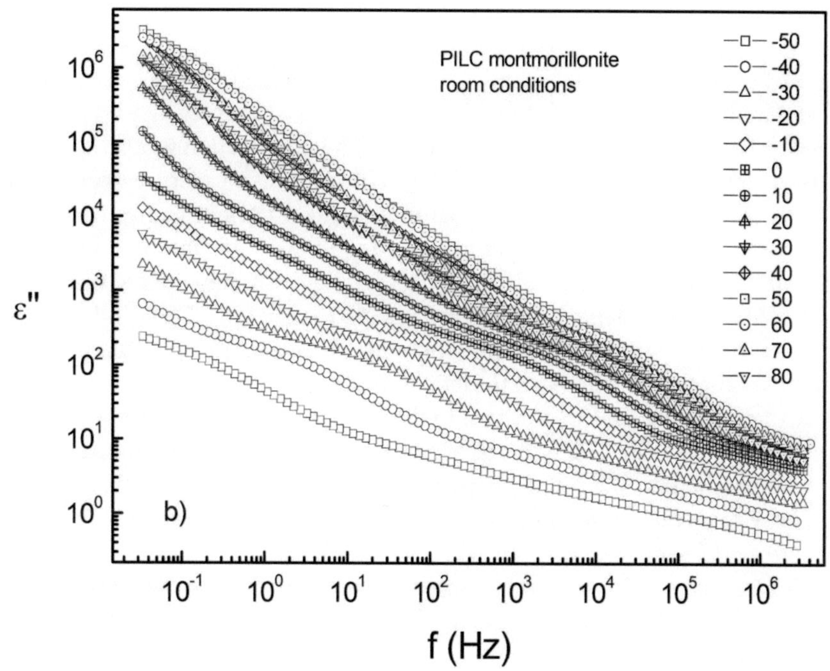

Fig. 6. a) and 6. b) Frequency dependence of imaginary part ε'' of the complex permittivity ε* of PILC montmorillonite at temperatures indicated on the plots.

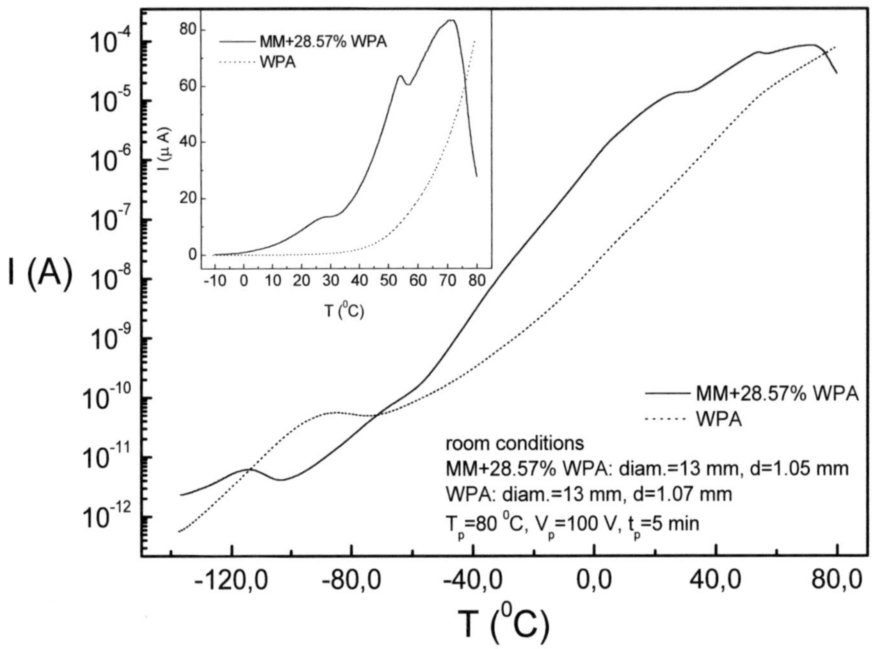

Fig. 7. Comparison of TSDC plots of WPA and MM + 28.57 mass% of WPA.

3.2 PILC-ADV

Comparative TSDC and DRS measurements of all samples were carried out in ambient relative humidity conditions. At this stage, no efforts were made to measure and control the water content in the samples.

Fig. 8. presents the results of TSDC measurements. The polarization temperature was 80 0C. In all samples, the depolarization current rising with rising temperature over several orders of magnitude, indicating rising dielectric activity. Relatively high values of depolarization current, in particular at higher temperatures, suggest that the main contribution to the dielectric response arises from conductivity rather than dipolar effects (Čajkovski, 2005; Sredić, 2008). Above 30⁰C depolarization current is higher for ADV-Al-PILC than for the pure ADV and in all investigated temperature region doped samples show significantly higher responses than pure MM This finding could be explained by greater contribution of ionic conductivity to total one. In addition to the overall increase of dielectric response with increasing temperature, the TSDC thermo grams in Fig.8. have the feature reflecting the presence of relaxations. The high depolarization current values suggest that these relaxations arise from the motion of charge carriers. In the case of doped MM samples three relaxations are evident, located at about -125, -95 and -55⁰C.

For pure MM i.e. pure ADV and the sample doped with 2 mass % ADV the frequency dependence of ac conductivity, σ_{ac}, at 30⁰C, calculated from the ε''(f) plots of Fig. 9. by means of equation (1)

$$\sigma_{ac}(f) = 2\pi f \, \varepsilon_0 \, \varepsilon''(f) \tag{1}$$

where ε_0 is the permittivity of free space.

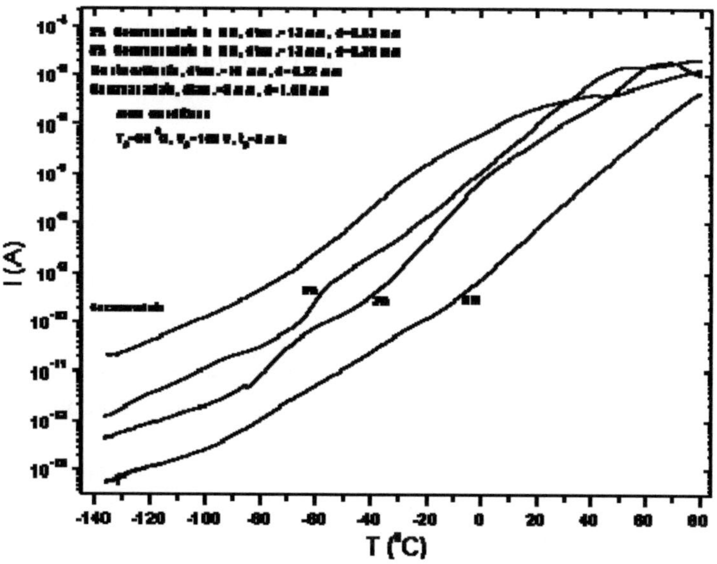

Fig. 8. TSDC thermograms obtained with the samples indicated on the plot

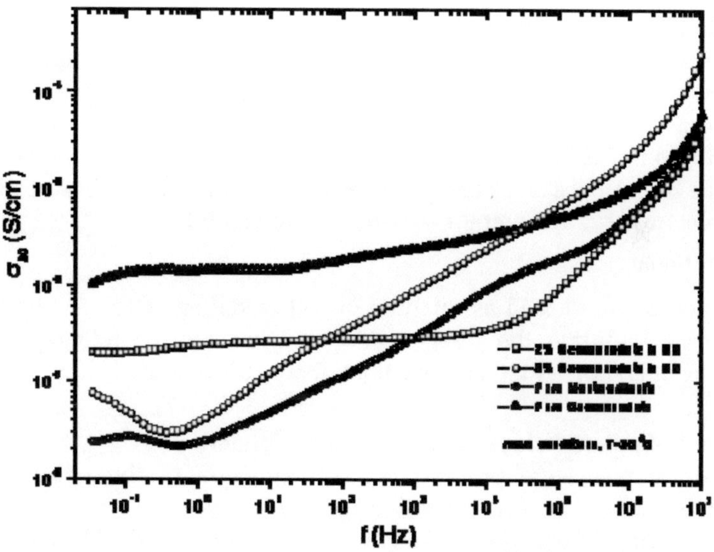

Fig. 9. Shows ac conductivity σ_{ac} versus frequency for the samples indicated on the plot at 30^{0}C and ambient relative humidity

The relaxation processes observed as shoulders in Fig. 9., mask the typical behaviour of an ionic conductor, i.e. the change from a frequency independent value at low frequencies (σ_{dc}, dc conductivity plateau) to a steady increase of σ_{ac} with increasing frequency at higher frequencies (Mioč, 2010). The results allow, however, direct reading of the values of σ_{dc} at the temperature and experimental conditions of the measurements. So, we get the values of about 2x10^{-8} S/cm in pure MM, 1x10^{-6} S/cm in pure ADV and 2x10^{-7} S/cm in the doped sample. It is interesting to note that that σ_{dc} increases by one order of magnitude after doping of MM with only 2 mass % of ADV. This result is interesting also with respect to possible applications. Please note that the absolute values of dc conductivity refer to samples equilibrated in air (ambient relative humidity conditions).

4. Catalytic properties

4.1 Heteropoly compounds supported on montmorillonite as catalysts for wet peroxide oxidation of toluene

Acid and redox properties of heteropoly acids (HPAs) and their salts render them suitable catalysts. One of the disadvantages of HPA catalysts, their low surface area and high solubility, can be overcome by impregnation of porous supports by HPAs or by neutralization of the HPA with different cations. In present study catalytic activity of montmorillonite based catalysts with different heteropoly compounds was further investigated.

Catalytic wet peroxide oxidation (CWPO) of toluene by Al-pillared montmorillonite (PILM) doped with 12-tungstophosphoric acid, $H_3PW_{12}O_40 \cdot nH_2O$ (WPA), before and after calcination and by neutral and acidic Cs salts of WPA was investigated and compared to catalytic activity of Al-Cu pillared bentonite clay (PILM-Cu) as the reference sample.

Raw material obtained from "Šipovo" bentonite deposits, Republic of Srpska, was modified and doped with WPA according to the procedure described elsewere. Cs salts were synthesized by drop wise addition of solution of Cs_2CO_3 (0.1mol/L) to solution of WPA (0.08mol/L). The precipitates were separated and dried at 40^0C overnight. The calcination was performed at 300^0C for three hours. The amount of WPA incorporated into PILM was determined from ICP spectrometric measurements and was 25,3%.

The characterization of catalysts was performed previously by IR spectrophotometry, DTA and TGA, while BET surface areas and pore size distributions were calculated from nitrogen adsorption isotherms at 77 K. Samples were outgassed at 353 K.

Oxidation of toluene

The obtained solids were tested as catalysts in wet oxidation of toluene by H_2O_2 at 40^0C. Filtrates were taken during the reaction and analyzed for the toluene content by gas chromatography. It is shown that after one hour about 80% of toluene was oxidised to CO_2.

4.2 Modification of pillared clays with cobalt functionality

The supported samples were prepared by impregnation of the pillared solids using $Co(NO_3)_2 \times 6H_2O$ solutions as a impregnation agenses.

The amount of the metal salts was equivalent to a content of 1,5,10 and 20 mass.% Co in the final samples. After the impregnation, the samples were dried (24 h, room temperature, 18h, 60^0C) and calcinated (4h, 295^0C).

4.3 XRD characterization

All the samples showed a type IV isotherm with a hysteresis loop, which is usually observed for mesoporous materials. The dimensions of pores in PILM samples are large enough for incorporation of Keggin ion (r~1.2 nm).

Adsorption-desorption isotherms for Cs2.5 H0.5 WPA are shown in Fig. 10.

Reaction kinetics of wet toluene oxidation over various montmorillonite based catalysts is shown in Fig. 11.

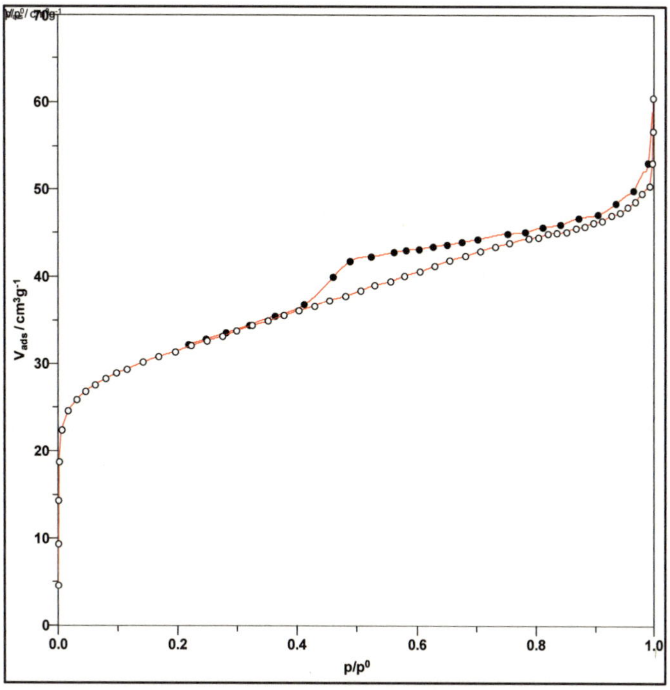

Fig. 10. Adsorption-desorption isotherms for Cs2.5 H0.5 WPA

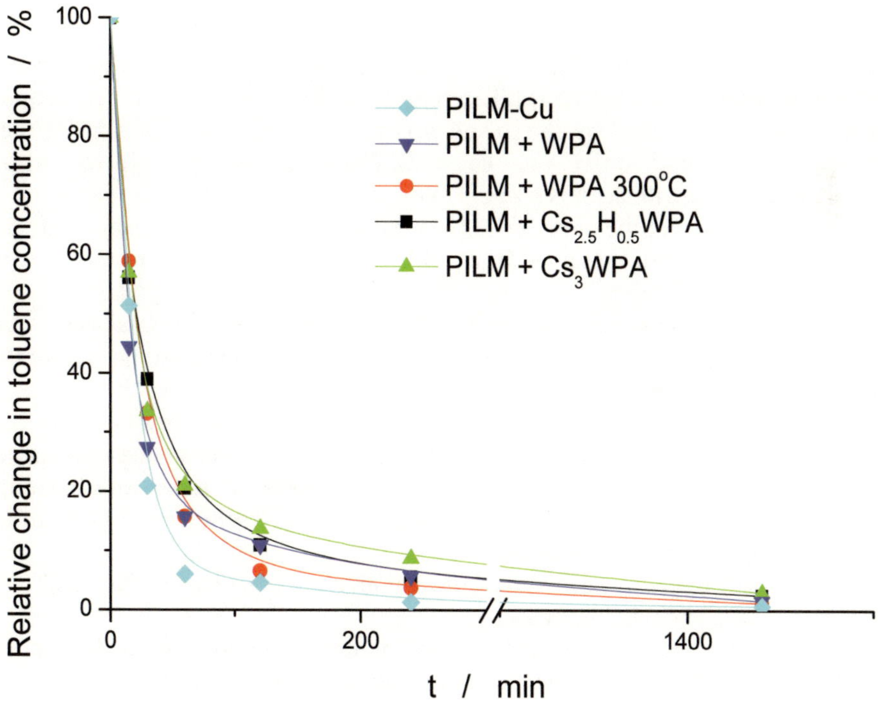

Fig. 11. Reaction kinetics of wet toluene oxidation over pillared montmorillonite catalysts

Statistical parameters of various models are compared in Table 4.
Table 5 presents rate constants for model with parallel first and second order reactions.

Sample	$Tol \xrightarrow{k_1} P1$	$Tol + Tol \xrightarrow{k_2} P2$	$Tol \xrightarrow{k_1} P1$ $Tol + Tol \xrightarrow{k_2} P2$
PILM-Cu	2.7320	9.7614	4.9724
PILM + WPA	7.3759	3.3605	0.9151
PILM + WPA 300oC	4.9623	3.4395	2.3565
PILM + Cs2.5H0.5WPA	7.8425	2.2392	1.1575
PILM + Cs3WPA	8.2171	2.9924	1.7793

Table 4. Standard error of the fit with tested reaction models, for different samples. The first and second order reactions are tested.

Sample	PILM-Cu	PILM+WPA	PILM+WPA (300)	PILM+C2.5H0.5 WPA	PILM+C3WPA
k1	5.6375E-02	9.5279E-02	4.4655E-02	2.3820E-02	5.4794E-02
k2	5.1602E+00	4.6719E+01	2.8530E+01	6.7017E+01	2.9349E+01

Table 5. Rate constants for toluene wet peroxide oxidation over various catalyst samples. Best fitting model with parallel first and second order reactions is used.

According to the results given in Table 1, sample of reference PILM-Cu shows kinetics of the first order reaction, contrary to the Al-pillared montmorillonite doped with WPA and its Cs salts, which can be described by the mixed first and second order reaction kinetics. The highest value of second order rate constant (Table 2) is obtained for Al-pillared montmorillonite doped with acidic Cs salt of WPA (C2.5 H0.5WPA).
The structure properties of the new-doped material have been studied using X-ray powder Diffraction (XRPD) method, presented in Fig. 12. The analyses were performed on Siemens D500 diffractometer.
Data were processed using Siemens Diffrac Plus software and JCP database. Data were collected in the 2θ range from 2 to 80 ° having a continuous scan with the step of 0.02 °2θ and the time of 1 s per step.
Reflections of the Co_2O_3 samples of the pillared clays impregnated with Co are distinguished from individual phase of Co_2O_3 but some numbers of these oxides can be incorporated in montmorillonite. In the sample of Co where the reflex ions in samples of Co_2O_3 are weaker in the case of samples with 20% Co, and reflexion of (100) in case of montmorillonite is widen in direction of d values.
The results of kinetic tests during cyclohexane oxidation are presented in Figure 13.
The oxygen uptake is given in Figure 14.
The increase of cyclohexanone and cyclohexanol is liner after short induction period and both products have similar yield. The further oxidation leads to unwanted byproducts.
All samples gain activity comparable with previously tested catalysts. Unexpectedly, the samples with lower Co content were more active. This observation is another evidence of

the inhibition effects in the cyclohexane oxidation system, and implies the need for the optimization of the catalysts.

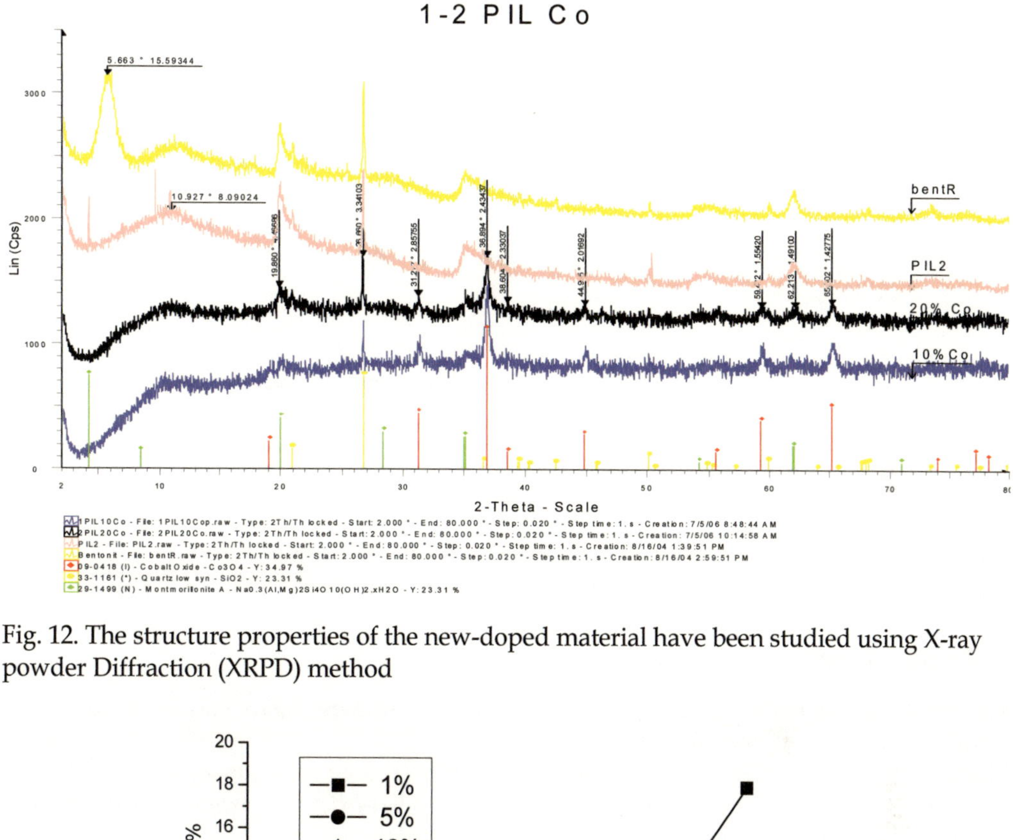

Fig. 12. The structure properties of the new-doped material have been studied using X-ray powder Diffraction (XRPD) method

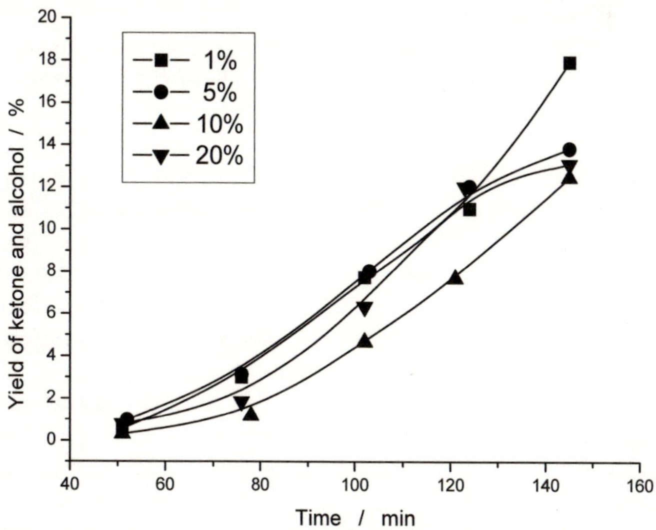

Fig. 13. The yield of useful products of the cyclohexane oxidation (cyclohexanone, cyclohexanol and cyclohexylhydroperoxide), at montmorilonite supported Co catalysts with indicated Co content.

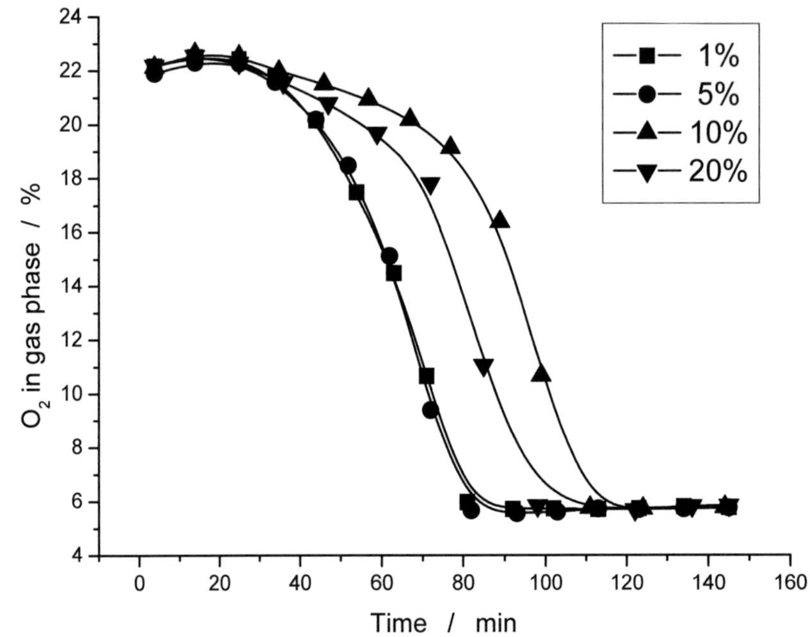

Fig. 14. The oxygen concentration in gas phase during the cyclohexane oxidation, at montmorilonite supported Co catalysts with indicated Co content.

5. Applications

The field of polyoxometalates (POMs), although a mature field, continues to attract significant attention (Katsoulis, 1998).

Montmorillonite structure differs from the others kind of natural minerals by having an interlayer space that gives it a swelling behaviour and restricts the water movement. The other clays are non-swelling.

Nanostructured materials have received much attention due to their potential to achieve properties superior to conventional engineering materials. Nanoscale reinforcements such as nanoclays can be used for fabrication of polymer-clay nanocomposites or polymer matrix composite with nanotubes. Although resulting in enhanced yield stress, tensile strength and Young's modulus in the former case, while in latter case electrical conductivity can be incorporated in the polymer in addition to increase in tensile strength. Thus, incorporation of nanoparticles, nanoclays and nanotubes in polymer matrix or ceramic matrix leads to the formation of composites with better properties for applications.

In recent years significant progress has been achieved in the synthesis of various types of polymer-nanocomposites and in the understanding of the basic principles, that further defines their optical, electronic and magnetic properties. Nanocomposite-based devices, such as light emitting diodes, photodiodes, photovoltaic solar cells and gas sensors, have been developed, often using chemically oriented synthetic methods such as soft lithography, lamination, spin-coating or solution casting. The key moment of an application development, of nanocomposite-based devices, is discovery of the possibility of filling conductive polymer matrices of types: polyaniline, substituted poly (paraphenylenevinylenes) or poly (thiophenes), with semiconducting nanoparticles. The possibility to fill the polymer matrix, both n- and p- conductivity types with nanoparticles, thus providing access to peculiar morphologies, like interpenetrating grid, p-n semiconductors or fractal p-n interfaces, not achievable by traditional microelectronics technology.

The peculiarities in the conduction mechanism through a grid of semiconductor nanoparticle chains provide the basis for the manufacture of highly sensitive gas sensors. These sensors combine the properties of the polymer matrix with those of the nanoparticles. It allows the fabrication of sensor devices selective to some definite components in mixtures of gases.

The basic properties and functions of material structures and systems can be changed as a function of the organization of matter via molecular interactions (such as hydrogen bonds, electrostatic dipole, van der Waals forces, various surface forces, electro-fluidic forces, etc.)

Tables 6 and 7 present potential applications of ceramic-, metal- and polymer-based nanocomposites, respectively. Metal and ceramic nanocomposites are expected to generate a great impact over a wide variety of industries. These tables summarize the possible developments associated with these materials in catalysts, sensors, structural materials, electronic, optical, magnetic, mechanical and energy conversion devices suggested by researchers in the field.

Nanocomposites	**Applications**
SiO2/Fe	High performance catalysis, data storage technologies
Metal oxides/Metal	Catalysis, sensors, optoelectronic devices
SiO2/Ni	Chemical sensors

Table 6. Potential application of ceramic Nanocomposite systems

Nanocomposite	**Application**
Fe/MgO	Catalysis, magnetic devices
Fe/TiN	Catalysis

Table 7. Potential application of metal Nanocomposites systems

Materials with defined pore-size in the nanometer range are of special interest for a wide range of industrial applications because of their outstanding properties with regard to thermal insulations, controllable material separation and release and their applicability as templates or fillers for chemistry and catalysis. One example of nanoporous material is aerogel, which is produced by sol-gel chemistry. A wide range of potential applications of these materials include catalysis, thermal insulation, electrode materials, environmental filters and membranes as well as controlled release drug carriers.

The applications of POMs are based on combinations of so-called "value-adding properties" which are summarized in Table 1. such as:

1. metal oxide like
2. stable (H2O/air, T): processing advantage
3. large size (diameter, 6-25 Å)

4. discrete size/discrete structure (confined geometric factors)
5. anions (charge from -3 to -14)
6. high ionic weight (103 - 104)
7. fully oxidized compounds/reducible
8. variable oxidation numbers for the addenda atoms (E1/2) 0.5 to -1.0 V vs SCE)
9. colour of oxidized forms different from colour of reduced forms
10. photoreducible
11. arhenius acids (pKa < 0)
12. incorporate over 70 elements and form large number of structures: processing advantage
13. acid forms very soluble in H2O & other oxygen carrying solvents (ethers, alcohols, ketones)
14. also soluble or transferable into nonpolar solvents: processing advantage
15. hydrolysable to form deficient structures: processing advantage

From the above listed properties, the applications of POMs are centred primarily on their redox properties, photochemical response, ionic charge, conductivity, and ionic weights. Primarily $H_3PMo_{12}O_{40}$, $H_3PW_{12}O_{40}$, $H_4SiMo_{12}O_{40}$, and $H_4SiW_{12}O_{40}$ are used as the main examples for many applications for different purposes: coatings, analytical chemistry, processing radioactive waste, membranes, sensors, dyes/pigments, electrooptics, electrochemistry/electrodes, capacitors, dopants in sol-gel matrixes, flammability control, etc. The majority of the applications of POMs are found in the area of catalysis.

Coatings

A nanocoating can produce high quality with super properties in the coating layers. Sustainable products can be produced with super ability to resist the stress and loads for different applications.

Of the most widely used methods to combat corrosion of metal surfaces is the utilization of corrosion inhibitors such as chromates, phosphates, or silicates. The main concern with the use of chromate inhibitors is their toxic nature.

Polyoxometalates have relatively low toxicity as compared to chromates; they accept electrons without major changes of their structures and form insoluble salts with large cations. These properties make them attractive as oxidizing and film-forming corrosion inhibitors (Katsoulis, 1998).

The high solubility of heteropolyacids in water is a drawback since it renders them unstable under humid conditions, and thus various hydrophobic hosts are required to preserve their value-adding properties. As one might expect the primary application of composite systems is in coatings where attributes such as optical properties, adhesion, hardness and abrasion resistance are desirable.

Analytical chemistry

The recent explosive growth of sol-gel science and technology offers several examples of utilization of POMs as additives in inorganic or organic matrixes. The number of examples is not as large as one might have anticipated based on the advantages that POMs offer in the ease of incorporation into sol-gel matrixes and on their rich combinations of properties analytical chemistry such as ion conductivity, photochromicity, and electrochromicity. This trend is changing, and more material scientists are investigating the interactions of POMs with organic and inorganic matrixes.

Substances of high electrical conductivity have the potential to function as effective antistatic agents. POM acid (Keggin type) are among the many compounds that have been claimed to function as such. Their function is based on the synergy between their hygroscopic and ionic character.

Processing radioactive waste

The ideal nanocomposite would not only resist radiation damage. It would also not itself become radioactive by absorbing neutrons.
A good number of applied analytical publications deal with the usage of POMs as process aids for the processing of radioactive nuclei. (Katsoulis, 1998)
Their presence renders the processing and solidification (vitrification) of these wastes more difficult because they evaporate to a noticeable degree and leach out of the vitrified products that are intended for permanent storage such as glass blocks, cement blocks, and bitumen blocks.
Limitation of these approaches is the fact that $H_3PMo_{12}O_{40}$ is also soluble in the aqueous phase and causes interference with the subsequent process of MAW such as vitrification.
It could still take years before such materials are approved for use, but the modelling methods will greatly speed up the process.

Membranes and Sensors

Composite membrane has good permeability and selectivity for gases, which depends upon the operating conditions. The solubility and diffusivity of the penetrants in the polymer matrix control the transport through non-porous dense membrane (Vijay, 2006).
A large number of application patents and publications deal with the building of membrane-based devices and sensors that exploit the high ionic conductivity of POMs, their capability to form a plethora of salts with virtually any cation, and their ability to undergo redox processes under many mild conditions. The most common uses of these membranes are in selective electrodes, in gas detection apparatuses, in solid-state electrochromic devices, and in liquid and solid electrolytic cells.
Many of the selective electrodes described in the literature were constructed for use in clinical chemistry.
Campiglio (1994) described an ion selective electrode for the potentiometric microtitration of vitamin B_1.
The sensor biosensor for the determination of the neurotransmitter acetylcholine, based on a two-layer selective poly(vinyl chloride)-poly(vinyl acetate) membrane consisting of a choline phosphotungstate as the electroactive compound (Katsoulis, 1998; Eppelsheim, 1994)
The membranes were used in gas sensors for the detection of hydrogen and gaseous compounds capable of dissociating into or combining with hydrogen ions.
The gas sensors are prepared by organic-inorganic polymer membranes made of water soluble organic polymers (i.e. poly(vinyl alcohol), poly(ethylene oxide), poly(ethylene glycol), cellulose acetate, etc.) and heteropolyacids or their salts.
Improvements of the functional life of such sensors have been achieved with the use of various porous solid-state encapsulants that encapsulate at least one component of the chemical sensor (Katsoulis, 1998; Goldstein, 1996).

Dopants in sol-gel matrixes

Common inorganic matrixes are usually derivatives of Al, Si, Ge, Ti, or Zr alkoxides and to lesser degree derivatives of Sn, Pb, or Nb alkoxides. The acid catalysis properties of POMs

are used in conjunction with their photochromic, electrochromic, and ion conductive properties to produce composite siloxane materials through the sol-gel approach. (Katsoulis, 1998; Katsoulis, 1995; Tasumisago, 1993)
Substances of high electrical conductivity have the potential to function as effective antistatic agents. POMs are among the many compounds that have been claimed to function as such. Their function is based on the synergy between their hygroscopic and ionic character.
The electrodes had excellent optical uniformity due to properties of $H_4SiW_{12}O_{40}$ compacted to a disk in a form of electrochromic display (Katsoulis, 1998).

Dyes/pigments

The ability of POMs to form stable precipitates with cationic dyes has resulted in considerable patent activity from industries related to pigments, dyes, and inks.[83-95] There continues to be a need for better control of particle size, particle size distribution, surface activity, surface area, porosity, and color quality.
Silicomolybdic acid was the most preferred complex heteropolyacid for preparing pigments for printing inks, particularly for quick set offset printing.
The compounds had excellent utility as pigments, primarily in textiles (i.e. denims) and inks (replacement for indigo dyes) (Katsoulis, 1998)

Electrochemistry and electrodes

The storage of energy today is a great technology challenge and ongoing activity, especially the process of charging and recharging of light weight batteries and/or capacitors in compact forms.
For modification and improvement of activating electrodes and their storage charge capacity, the polyoxometalates are used. Subsequently, the polyoxometalates are applied for fabrication of electric double-layer capacitors with optimized energy storage properties.
Due to their capability, reversible redox reactions provide a secondary charge storage mechanism without decreasing the charge storage ability of the activated electrode's materials.
During the past few years, significant developments on electrode modification by POMs have taken place aiming to improve the reaction rates of fuel cells and produce economically sufficient currents for practical uses (Katsoulis, 1998; Ndzebet, 1995; Savadogo, 1990)

Capacitors

Heteropolyacids and their salts have been claimed as electrolytes for capacitors (solids and solutions) (Katsoulis, 1998).
The capacitors also maintained good reliability at high temperature operation up to 150 °C. Aging studies at 125 °C for 310 h showed constant resistivity (Katsoulis, 1998).

Fuel cell

Fuel cell applications include nanocomposite in the proton exchange membrane, binder for the electrodes and matrix for bipolar plates.
Improvement of the mechanical properties for proton exchange membrane and enhanced proton conductivity of it yield the nanoparticles to be incorporated into proton exchange membrane structure.

When the direct methanol fuel cells are considered, nanoparticles are applied for reduction of methanol crossover (Paul, 2008; DeLuca, 2006)
Heteropolyacids like $H_3PW_{12}O_{40}$ are applied to proton exchange membranes to yield improved proton conductivity at higher temperatures while retaining good mechanical properties (Paul, 2008; Kim, 2003; Wang, 2006)

6. Brief conclusion

In the first line of interest, natural nanostructures have potential to overtake the primate of conventional engineering materials by means of properties and application. Nanomaterials as nanoclays can be modified to polymer-clay nanocomposites or matrix composites and for result they form the pre-products with improved characteristics.
In accordance with the properties of natural nanocomposite materials and their possibilities for application, the need of interdisciplinary science approach is arising, involving encompassing physics, chemistry, biology, materials science and engineering. Synergy of knowledge will result in new science discipline and very new technologies and therefore especially materials. Furthermore, development and implementation of nanocomposites will impact the society in various areas: electronic, chemical, transportation, medicine and health care, energy and environmental protection. It is expected that natural nanocomposites will make significant positive influence on life quality in future (Camargo, 2009).

7. Acknowledgments

This work has been supported by the Ministry of Science and Environmental Protection of the Republic of Serbia, Grants No. 142047, 141030G, 45001 and TR-34028.

8. References

Camargo, P.H.C.; Satyanarayana, K.G; Wypych, F. (2009) Nanocomposites: synthesis, structure, properties and new application opportunities, Mat. Res., Vol.12, No.1, December 2010., http://www.scielo.br/scielo.php?script=sci_arttext&pid=S1516-14392009000100002

Colomban, Ph. (2003). Raman Analyses and "Smart" Imaging of Nanophases and Nanosized Materials, *Spectroscopy Europe*, 15/6

Čajkovski, T.; Davidović, M.; Pissis, P.; Polizos, G.; Čajkovski, D.; Likar-Smiljanić, V.;Biljić, R.; Nedić, Z.; Mioč, U.B. (2003) *Solid State Ionics*,162-163, 203.

Čajkovski, T.; Davidović, M.; Pissis, P.; Polizos, G.; Čajkovski, D.; Likar-Smiljanić, V.; Sredić, S.; Mioč, U.B. (2005) *Journal of non-crystalline solids*, 351 2842.

DeLuca, NW; Elabd, YA. (2006) *J Polym Sci Part B Polym Phys*; 44:2201–25.

Holclajtner-Antunović; Čupić, Ž.; Davidović, M.; Mioč, U.B.; Todorović, M.; Jovanović, D.; Banković, P. (2007) "Physicochemical characteristics of Rb+ and Cs+ salts of 12-tungstophosphoric acid" Ninth Yugoslav Materials Research Society Conference, *Book of Abstracts of "YUCOMAT 2007"*, 32, Herceg-Novi, September, 2007, "YUCOMAT 2007", Herceg-Novi,

Katsoulis, Dimitris E.(1998) A Survey of Applications of Polyoxometalates, *Chemical Reviews*,Vol. 98, No. 1, pp.359-387

Katsoulis, D. E.; Keryk, J. R. (1995) *Abstract ACS National Meeting*, Inorganic Chemistry Section, Aug 20-24, Chicago

Kim, Y.S.; Wang, F.; Hickner, M.; Zadwodzinski, T.A.; McGrath, J.E. (2003) J Membr Sci; 212:263–82.

Kröhn, K.P. (2003) *Applied Clay Science* 2325-33

Lončarević, D.; Čupić, Ž. (2005) "Characterization and Catalytic Activity of Poly(4-Vinylpyridine-co-Divinylbenzene)-Co2+ Complex", *Material Sciences Forum*, Vol. 494, Current research in Advanced Materials and Processes, pp.363-368

Lončarević, D.; Čupić, Ž.; Odović, M. (2005) "Inhibition Effects in the Partial Oxidation of Cyclohexane on Polymer Supported Co(II) Catalysts", *J.Serb.Chem.Soc.*, 70, pp.209-221.

Manocha, I.M.; Valand, Jignesh; Patel, Nikesh; Warrier, Asís & Manocha (2006) Nanocomposites for structural application, *Indian Journal of Pure & Applied Phisycs*, Vol 44, pp.135-142

Mioc, U.B.; Davidovic, M.; Tjapkin, N., Colomban, Ph.; Novak, A. (1993) *Solid State Ionics*, 46, 103.

Mioč, Ubavka B., and Davidović, Milorad (Eds.) (2010)*Fast Proton-Ion Transport Compounds*, Transworld Research Network, Kerala, India

Moore, D.M.; Reynolds, R.C. (1989)*X-Ray Diffraction and the Identification and Analysis of Clay Minerals*, Oxford University Press,.

Ndzebet, E.; Savadogo, O. (1995) *Int. J. Hydrogen Energy, 20,* 635.

Occelli, M.L.; Bertrand, J.A.; Gould, S.A.C.; Dominguez, J.M. (2000)Microporous and Mesoporous Materials, 34, pp.195-206

Paul, D.R.; Robeson L.M. (2008) Polymer nanotechnology: Nanocomposites, Polymer, 49,pp.3187-3204

Ruiz-Hitzky, Eduardo; Galvan, Juan, Carlos; Merino, Jesus; Casal, Blanca; Pilar, Aranda, and Jimenez-Morales, Antonia (1996) *Solid State Ionics* 85

Savadogo, O.; Amuzgar, K.; Piron, D. L. (1995) *Int. J. Hydrogen Energy, 15,* 783.

Slade, Robert C. T.; Barker, Jeremy; Hirst Paul R.; Halstead, Thomas. K., and Reid, Paul I. (1987) *Solid State Ionics*, 24239

Sredić, S.; Čajkovski, T.; Davidović, M.; Čajkovski, D.; Likar-Smiljanić, V.; Marinović-Cincović, M.; Mioč, U.B.; Nedić, Z.; and Biljić, R. (2004) *Materials Science Forum*, Vols. 453-454, pp.157-162

Sredić, S.; Davidović, M.; Spasojević-de-Bire; A.; Mioc, U.B.; Todorovic, M.; Segan, D.; Jovanovic, D.; Polizos, G.; Pissis, P. (2008) Inorganic-inorganic composite: Surface and conductive properties, *Journal of Physics and Chemistry of Solids*, 69 1883-1890

Tasumisago, M.; Honjo, H.; Sakai, Y.; Minami, T. (1993) *Solid State Ionics*, 59, pp.171.

Vijay, Y. K.; Kulshrestha, Vaibhav; Awasthi, Kamlendra; Acharya, N. K.; Jain, A.; Singh, M.; Dolia, S. N.;Khan, S. A.; Avasthi, D.K(2006) Characterization of Nanocomposite Polymeric Membrane, *Journal of Polymer Research*, Vol. 13, No 5, pp.357-360

Wang, L.; Long, C.; Wang X.; Li Q & Qi Z (1998) *Journal of Applied Polimere* Science, Vol 69,1557.

Wang, Z.; Ni, H.; Zhao, C.; Li, X.; Fu, T.; Na, H. (2006) *J Polym Sci Part B Polym Phys;* 44:1967–78.

4

Nanocomposites in Food Packaging – A Review

Henriette Monteiro Cordeiro de Azeredo[1],
Luiz Henrique Capparelli Mattoso[2] and Tara Habig McHugh[3]
[1]*Embrapa Tropical Agroindustry - CNPAT,*
[2]*Embrapa Agricultural Instrumentation - LNNA/CNPDIA,*
[3]*Agricultural Research Service - ARS/WRRC/USDA,*
[1,2]*Brazil*
[3]*USA*

1. Introduction

A nanocomposite is a multiphase material derived from the combination of two or more components, including a matrix (continuous phase) and a discontinuous nano-dimensional phase with at least one nano-sized dimension (i. e., with less than 100 nm). The nano-dimensional phase can be divided into three categories according to the number of nano-sized dimensions. Nanospheres or nanoparticles have the three dimensions in the nanoscale. Both nanowhiskers (nanorods) and nanotubes have two nanometric dimensions, with the difference that nanotubes are hollow, while nanowhiskers are solid. Finally, nanosheets or nanoplatelets have only one nano-sized dimension (Alexandre & Dubois, 2000).

Most nano-sized phases have a structural role, acting as reinforcements to improve mechanical properties of the matrix (usually a polymer), since the matrix transfers the tension to the nanoreinforcement through the interface. Nanoreinforcements are especially useful for biopolymers, because of their usually poor performance when compared to conventional petroleum-based polymers. The incorporation of nano-sized reinforcements to biopolymers may open new possibilities for improving not only their properties but also their cost-price-efficiency (Sorrentino et al., 2007).

Besides nanoreinforcements, whose main role is to improve mechanical and barrier properties of polymers, there are nanostructures responsible for other applications related to food packaging. For instance, when incorporated to polymer matrices, they may interact with the food and/or its surrounding environment, thus providing active or "smart" properties to packaging systems. Such properties, when present in food packaging systems, are usually related either to improvements in food safety/stability or information about the safety/stability status of a product.

The main types of nanostructures will be presented according to their primary functions/applications in food packaging systems. Some structures can have multiple applications, and sometimes applications can overlap, such as some immobilized enzymes which can act as antimicrobial components, oxygen scavengers and/or nanosensors.

2. Nanoreinforcements in food packaging materials

Polymer nanocomposites usually have much better polymer/filler interactions than conventional composites (Ludueña et al., 2007). A uniform dispersion of nanofillers into a polymer matrix results in a very large matrix/filler interfacial area, which restricts the mechanical mobility of the matrix, and improves its mechanical, thermal (especially glass transition temperature - T_g), and barrier properties.
The ratio of the largest to the smallest dimension of a filler is an important property known as aspect ratio. Fillers with higher aspect ratios have higher specific surface area, providing better reinforcing effects (Azizi Samir et al, 2005; Dalmas et al, 2007). In addition to the effects of the nanoreinforcements themselves, an interphase region of decreased mobility surrounding each nanofiller results in a percolating interphase network in the composite which plays an important role in improving the nanocomposite properties (Qiao & Brinson, 2009). For a constant filler content, a reduction in particle size increases the number of filler particles, bringing them closer to one another; thus, the interface layers from adjacent particles overlap, altering the bulk properties more significantly (Jordan et al., 2005).

2.1 Nanoclays (layered silicates)

Nanoclays have been the most studied nanofillers, due to their high availability, low cost, good performance and good processability. The first publications about applications of polymer-nanoclays composites to food packaging date from the 1990's (Ray et al., 2006). The clays for nanocomposites usually are bidimensional platelets with very tiny thicknesses (frequently around 1 nm) and several micrometers in length.
In contrast with the typical tactoid structure of microcomposites (conventional composites), in which the polymer and the clay tactoids remain immiscible (Ludueña et al., 2007; Alexandre et al., 2009), the interaction between layered silicates and polymers may produce two types of nanoscale composites (Figure 1), namely: intercalated nanocomposites, which result from penetration of polymer chains into the interlayer region of the clay, producing an ordered multilayer structure with alternating polymer/inorganic layers (Weiss et al., 2006), and exfoliated nanocomposites, which involve extensive polymer penetration, with the clay layers delaminated and randomly dispersed in the polymer matrix (Ludueña et al., 2007). Exfoliated nanocomposites have been reported to exhibit the best properties due to the optimal clay-polymer interactions (Adame & Beall, 2009; Alexandre et al., 2009).
The most studied clay is montmorillonite (MMT), whose chemical general formula is $M_x(Al_{4-x}Mg_x)\ Si_8O_{20}(OH)_4$. MMT is a representative of *2:1* layered *phyllosilicates,* whose platelets have two layers of tetrahedral silica sheets filled with a central octahedral alumina sheet (Weiss et al., 2006). This kind of clay has a moderate negative surface charge that is important to define the interlayer spacing (Alexandre & Dubois, 2000). The imbalance of the surface negative charges is compensated by exchangeable cations (typically Na^+ and Ca^{2+}). The parallel layers are linked together by weak electrostatic forces (Tan et al., 2008). MMT is an excellent reinforcing filler, thanks to its high surface area and large aspect ratio, which ranges from 50 to 1000 (Uyama et al., 2003).
The hydrophilicity of the surface of most clays make their dispersion in organic matrices difficult (Kim et al., 2003). Organoclays, produced by interactions of clays and organic compounds, have found an important application in polymer nanocomposites. An adequate organophilization is essencial for successful exfoliation of clays in most polymeric matrices, since organophilization reduces the energy of clays and improves their compatibility with

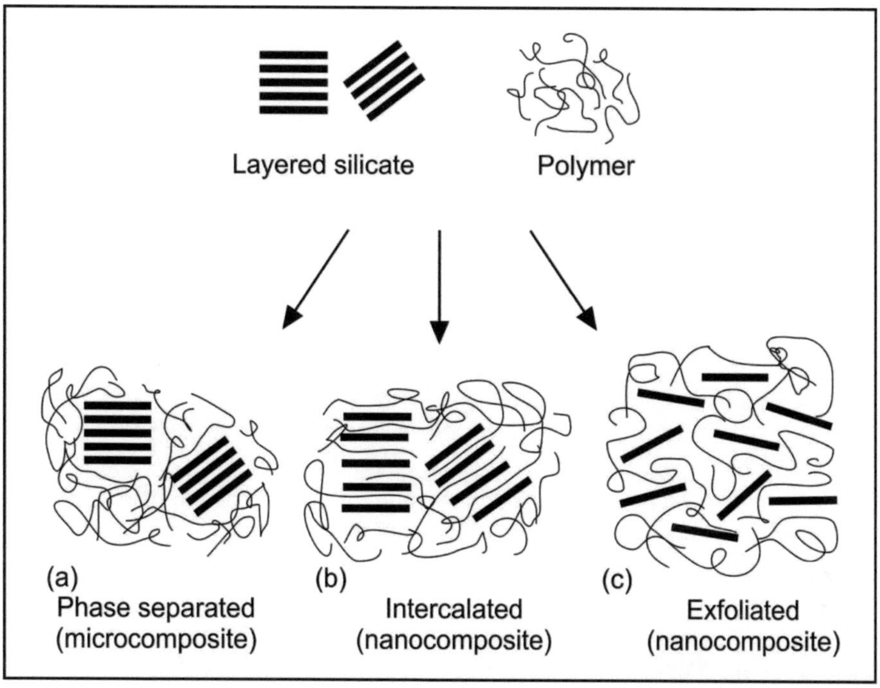

Fig. 1. Types of composites from polymer-clay interactions (Alexandre & Dubois, 2000).

organic polymers (Paiva et al., 2008). Organomontmorillonite (oMMT) have been produced, for example, by exchanging inorganic cations of MMT with organic ammonium ions, improving compatibility of MMT with organic polymers (Osman et al., 2003; Paul et al., 2003), leading to a more regular organization of the layers, and decreasing the water uptake by the resulting nanocomposite (Picard et al., 2007).

The improved barrier properties of polymer-clay nanocomposites seem to be due to a increased tortuosity of the diffusive path for permeants (Figure 2), forcing them to travel a longer path to diffuse through the film. This theory was developed by Nielsen (1967) and was further corroborated by other authors (Mirzadeh & Kokabi, 2007; Adame & Beall, 2009). The increase in path length is a function of the aspect ratio of the clay and the volume fraction of the filler in the composite. Nielsen's model has been used effectively to predict permeability of systems at clay loadings of less than 1%, but some experimental data have reported much lower permeabilities than predicted at higher loadings (Adame & Beall, 2009). Beall (2000) proposed a new model to predict permeability of nanocomposites focused on the polymer-clay interface as an additional governing factor to the tortuous path, thus providing a correction factor to Nielsen's model.

Clays have been also reported to improve the mechanical strength of biopolymers (Chen & Evans, 2005; Russo et al., 2007; Cyras et al., 2008), although they may decrease polymer elongation (Petersson & Oksman, 2006).

2.2 Cellulose nanoreinforcements

Cellulose nanoreinforcements (CNRs) are interesting materials for the preoparation of low cost, lightweight, and high-strength nanocomposites (Helbert et al., 1996; Podsiadlo et al., 2005).

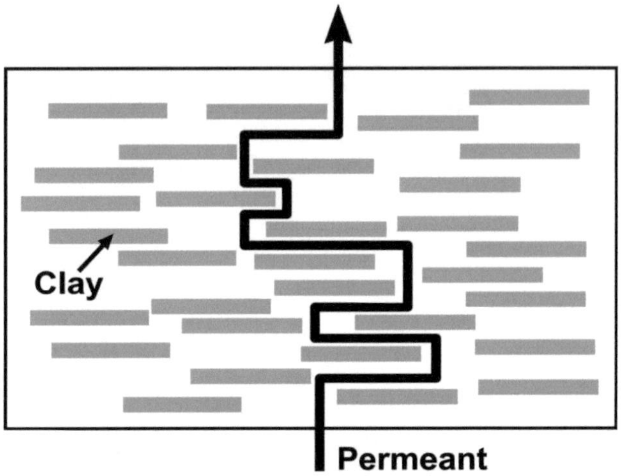

Fig. 2. Tortuous path for a permeant through a polymer-clay nanocomposite, according to Nielsen's model. (Adapted from Adame & Beall, 2009).

Cellulose chains are synthesized in living organisms (mainly plants) as microfibrils (or nanofibers), which are bundles of elongated molecules (with 2-20 nm in diameter and micrometric in length) stabilized by hydrogen bonds (Azizi Samir et al., 2005; Oksman et al., 2006; Mattoso et al., 2009). Each microfibril, formed by elementary fibrils, have crystalline and amorphous regions. The crystalline parts, which may be isolated by procedures such as acid hydrolysis, are the nanocrystals or nanowhiskers (Dujardin et al., 2003; Azizi Samir et al., 2004), whose aspect ratios are related to the origin of the cellulose and processing conditions (Azizi Samir et al., 2005). Thus, a microfibril can be considered as a string of whiskers linked by amorphous domains, which are taken as structural defects.

Our group has studied the influence of cellulose nanofibers on the physical properties of mango puree edible films (Azeredo et al., 2009) and chitosan films (Azeredo et al., 2010).

In our first study (Azeredo et al., 2009), different concentrations of cellulose nanofibers (Novacel® PH-101, provided by FMC BioPolymer, Philadelphia, PA, USA) were added to mango puree edible films. The nanofiller was homogenized with the mango puree at 6500 rpm for 30 minutes, by using a Polytron PT 3000 (Brinkmann, Westbury, NY, USA). A control film was prepared with non-reinforced mango puree. The film-forming dispersions were vacuum degassed, and films were cast on leveled glass plates and allowed to dry for 16 h at 22°C and 42% RH. Samples of the dried films were cut and peeled from the casting surface for analyses. Tensile properties were measured according to standard method D882-97 (ASTM, 1997), by using an Instron Model 55R4502 (Instron, Canton, MA) with a 100 N load cell. The gravimetric Modified Cup Method (McHugh et al., 1993) based on standard method E96-80 (ASTM, 1989) was used to determine water vapor permeability (WVP).

Table 1 presents physical properties of mango puree films containing different CNR concentrations. The addition of at least 10% CNRs was effective to decrease water vapor permeability (WVP) of the films (Table 1), similarly to results reported by Paralikar et al. (2008) and Sanchez-Garcia et al. (2008). The interactions of CNRs with mango polysaccharides may have favored water vapor barrier. The nanofillers were also effective to increase tensile strength and (especially) Young's modulus. The elongation was slightly impaired, but only at nanofiller concentrations above 10%. Several other studies have reported positive effects of CNRs on tensile properties - especially on modulus - of polymers (Helbert et al., 1996; Bhatnagar & Sain, 2005; Wu et al., 2007), although they tend to decrease elongation (Freire et al., 2008; Tang & Liu, 2008; Kim et al., 2009). According to Helbert et al. (1996), the great effect of CNRs on modulus is ascribed not only to the geometry and stiffness of the fillers, but also to the formation of a fibrillar network within the polymer matrix, the CNRs being probably linked through hydrogen bonds.

CNR (%)*	TS (MPa)	EB (%)	YM (MPa)	WVP (g.mm/ kPa.h.m²)
0	$(4.09 \pm 0.12)^{d}$	$(44.07 \pm 0.98)^{a}$	$(19.85 \pm 0.51)^{e}$	$(2.66 \pm 0.06)^{a}$
5	$(4.58 \pm 0.21)^{c}$	$(41.79 \pm 0.44)^{b}$	$(30.93 \pm 1.27)^{d}$	$(2.16 \pm 0.05)^{b}$
10	$(4.91 \pm 0.13)^{c}$	$(43.19 \pm 1.73)^{ab}$	$(40.88 \pm 1.41)^{c}$	$(2.03 \pm 0.11)^{b}$
18	$(5.54 \pm 0.07)^{b}$	$(39.8 \pm 0.53)^{b}$	$(78.82 \pm 5.00)^{b}$	$(1.90 \pm 0.06)^{bc}$
36	$(8.76 \pm 0.11)^{a}$	$(31.54 \pm 2.29)^{c}$	$(322.05 \pm 19.43)^{a}$	$(1.67 \pm 0.11)^{c}$

*On a dry basis. TS: tensile strength (MPa); EB: elongation at break (%); YM: Young's Modulus (MPa); WVP: water vapor permeability ($g.mm/kPa.h.m^2$). Means in same column with different letters are significantly different at $p<0.05$.

Table 1. Physical properties of mango puree films with different concentrations of CNRs.

In our second study (Azeredo et al., 2010), nanocomposite films were also obtained from a chitosan matrix with CNRs (Avicel® PH, also provided by FMC BioPolymer). The experiment was conducted according to a central composite design, with two variables: concentrations (on a dry basis) of CNRs (0-20%) and glycerol (0-30%). A dilute chitosan solution was made by preparing a 3% chitosan (71.3 kDa, 94% deacetylation, from *Polymar Ciência e Nutrição S/A*, Fortaleza, Brazil) in 1.5% acetic acid solution. CNRs and glycerol were added to the chitosan solution, and the dispersions were homogenized at 4500 rpm for 30 minutes, with a Polytron PT 3000. The procedure for degassing, casting, drying and analyses were the same as for the mango puree films. The glass transition temperatures (T_g) of the films were measured by differential scanning calorimetry (DSC) with an mDSC 2910 (TA Instruments, New Castle, DE), from 30°C to 200°C, at a heating rate of 10°C/min.

Figure 3 presents the contour plots for the physical properties of the nanocomposite chitosan films as functions of the CNR and glycerol concentrations. The nanofillers increased the overall tensile properties of the films, but decreased their elongation. The water vapor barrier was improved, as indicated by the decreased water vapor permeability (WVP) with increasing CNR concentrations. Moreover, T_g of the films was increased by CNRs. Thus, most responses (except by elongation) were favored by high CNR concentrations and low glycerol contents.

Several studies have been focused on CNR effects on starch systems, indicating that CNRs reduce the water sensitivity of starch (Dufresne & Vignon, 1998; Dufresne et al., 2000; Lima & Borsali, 2004; Lu et al., 2005), and reduce starch brittleness (Dufresne & Vignon, 1998). The effect of CNRs on starch brittleness is consistent with the transcrystallization phenomenon, i.e., orientation of crystals of a semicrystalline matrix perpendicularly to the cellulose microfibrils, as described by Helbert & Chanzy (1994) and Hulleman et al. (1996). Moreover, some studies have reported that the incorporation of CNRs increases T_g of starch (Anglès & Dufresne, 2000; Alemdar & Sain, 2008). However, Mathew & Dufresne (2002) found inconsistent effects of cellulose whiskers on T_g of a starch matrix. They observed that the T_g of the nanocomposite first increased up to a whisker content of around 10-15 wt % and then

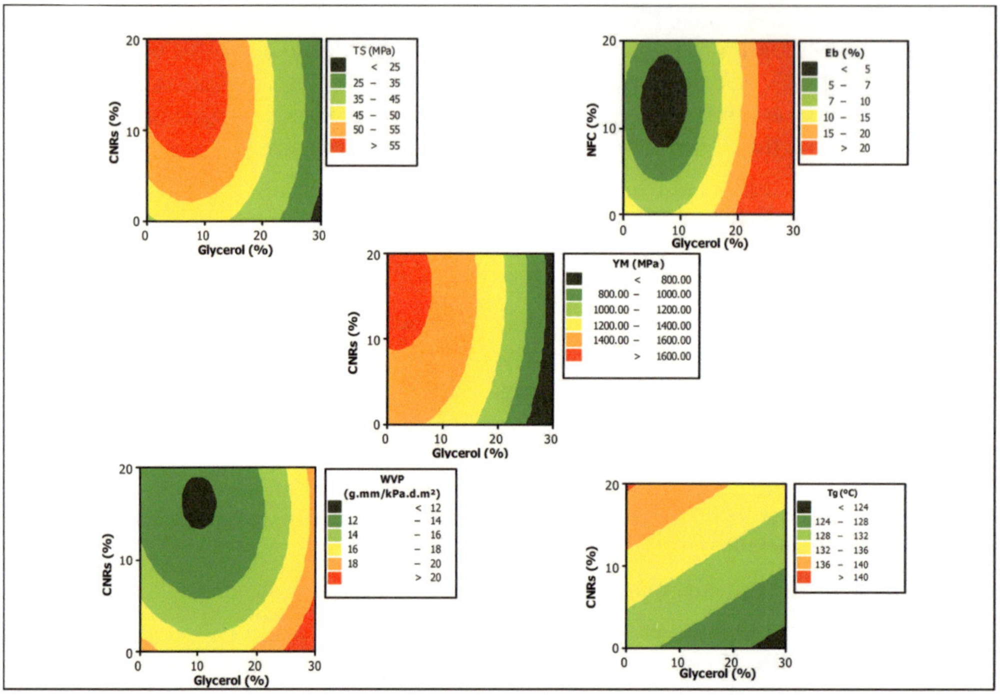

Fig. 3. Physical properties of chitosan films added with cellulose nanofibers and glycerol. TS: tensile strength (MPa); Eb: elongation at break (%); YM: Young's Modulus (MPa); WVP: water vapor permeability (g.mm/kPa.h.m^2); T_g: glass transition temperature (^{o}C).

decreased. The increase of T_g up to 15 wt % of whiskers was ascribed to the increase in the crystallinity of the matrix, the restricted mobility of amorphous amylopectin chains resulting from the physical cross-links induced by the crystallization. For the decrease of T_g above 15 wt % whiskers, a possible explanation is that sorbitol may have been partially ejected from the crystalline domains of the matrix during crystallization, its concentration increasing in the amorphous domains. This phenomenon should compete with the T_g increasing effect of the whiskers themselves, being probably predominant at high loading level inducing a decrease of T_g.

Similarly to nanoclays, the presence of cellulose nanoreinforcements is believed to increase the tortuosity of the diffusivity path for the permeants, lowering the polymer permeability (Sanchez-Garcia et al., 2008). In fact, several studies have reported improvements in barrier properties of polymers by addition of cellulose nanoreinforcements (Paralikar et al., 2008; Sanchez-Garcia et al., 2008; Svagan et al., 2009). The barrier properties are further enhanced if the filler is less permeable, well dispersed in the matrix, and with a high aspect ratio (Lagaron et al., 2004).

The resulting overall properties of polymer composites with cellulose nanoreinforcements have been reported to be strongly related to the dimensions and consequent aspect ratio of the fillers (Chen et al., 2009) as well as to orientation of the nanostructures (Kvien & Oksman, 2007).

Because of the hydrophilic cellulose surface, interactions between CNRs and hydrophilic matrices are usually satisfactory (Bondeson & Oksman, 2007). On the other hand, incorporation of cellulose nanoreinforcements to hydrophobic matrices results frequently in weak filler-matrix interactions (Hubbe et al., 2008) and filler aggregation by hydrogen bonding (Freire et al., 2008). Another limitation to the hydrophilic character of cellulose nanoreinforcements is the high water absorption capacity, which is undesirable in many potential applications (Hubbe et al., 2008). Such problems can be reduced by a variety of modifications (hydrophobization) on cellulose surfaces by several reactions involving hydroxyl groups, such as esterifications (Mohanty et al., 2001) and acylation with fatty acids (Freire et al., 2008).

2.3 Other nanoreinforcements

Carbon nanotubes may consist of a one-atom thick single-wall nanotube, or a number of concentric tubes called multiwalled nanotubes, having extraordinarily high aspect ratios and elastic modulus (Zhou et al., 2004). Several polymers have been found to have their tensile strength/modulus improved by addition of carbon nanotubes, such as polyethylene naphtalate (Kim et al., 2008), polyvinyl alcohol (Chen et al., 2005), polypropylene (López Manchado et al., 2005; Prashantha et al., 2009), and a polyamide (Zeng et al., 2006). According to Brody (2006), researches from Natick indicated that polylactic acid not only had its tensile properties improved by carbon nanotubes, but also had its water vapor transmission rate decreased in 200%.

Silica nanoparticles ($nSiO_2$) have been reported to improve tensile properties of polypropylene (Wu et al., 2002; Vladimiriov et al., 2006), starch (Xiong et al., 2008), starch/polyvinyl alcohol (Tang et al., 2008), besides decreasing water absorption by starch (Tang et al., 2008; Xiong et al., 2008) and improving oxygen barrier of polypropylene (Valdimiriov et al., 2006). Jia et al. (2007) prepared nanocomposites of polyvinyl alcohol with $nSiO_2$ by radical copolymerization of vinyl silica nanoparticles and vinyl acetate. The nanocomposites had improved thermal and mechanical properties when compared to the pure polyvinyl alcohol, due to strong interactions between $nSiO_2$ and the polymer matrix via covalent bonding.

Some studies have been conducted on effects of chitin or chitosan nanostructures on polymer properties. Incorporation of chitin whiskers have greatly improved the tensile properties as well as the water resistance of soy protein isolate thermoplastics (Lu et al., 2004). De Moura et al. (2009) incorporated chitosan-tripolyphosphate nanoparticles into hydroxypropyl methylcellulose films, and observed that they significantly improved tensile and barrier properties of the films.

Other nanostructures have also been reported as good reinforcing agents, such as starch nanocrystals (SNCs), which have increased tensile strength, modulus, and T_g of pullulan films, but decreased their elongation (Kristo & Biliaderis, 2007).

3. Nanocomposite active food packaging

Conventional food packaging systems are supposed to passively protect the food, that is to say, to act as a barrier between the food and the surrounding environment. On the other hand, an active food packaging may be defined as a system that not only acts as a passive barrier but also interacts with the food in some desirable way, e. g. by releasing desirable compounds (antimicrobial or antioxidant agents, for instance), or by removing some detrimental factor (such as oxygen or water vapor). The consequences of such interactions are usually related to improvements in food stability. Some examples of nanocomposite active food packaging systems are presented in this section.

3.1 Antimicrobial systems

Antimicrobial food packaging systems have received considerable attention since they help control the growth of pathogenic and spoilage microorganisms on food surfaces, where microbial growth predominates. Antimicrobial nanocomposite systems are particularly interesting, since materials in the nanoscale range have a higher surface-to-volume ratio when compared with their microscale counterparts. Nanomaterials are thus more efficient, since they are able to attach more copies of microbial molecules and cells (Luo & Stutzenberger, 2008). Nanoscale materials have been investigated for antimicrobial activity as growth inhibitors (Cioffi et al., 2005), killing agents (Stoimenov et al., 2002; Qi et al., 2004; Huang et al., 2005; Kumar & Münstedt, 2005; Lin et al., 2005), or antibiotic carriers (Gu et al., 2003).

Silver is well known for its strong toxicity to a wide range of microorganisms (Liau et al., 1997), besides some processing advantages such as high temperature stability and low volatility (Kumar & Münstedt, 2005). Silver nanoparticles have been shown to be effective antimicrobials (Aymonier et al., 2002; Sondi & Salopek-Sondi, 2004; Son et al., 2006; Yu et al., 2007; Tankhiwale & Bajpai, 2009), even more effective than larger silver particles, thanks to their larger surface area available for interaction with microbial cells (An et al., 2008; Kvítek et al., 2008). In fact, the most common nanocomposites used as antimicrobial films for food packaging are based on silver nanoparticles, whose antimicrobial activity has been ascribed to different mechanisms, namely: (a) adhesion to the cell surface, degradation of lipopolysaccharides and formation of "pits" in the membranes, largely increasing permeability (Sondi & Salopek-Sondi, 2004); (b) penetration inside bacterial cell, damaging DNA (Li et al., 2008); and (c) releasing antimicrobial Ag^+ ions by dissolution of silver nanoparticles (Morones et al., 2005). The latter mechanism is consistent with findings by Kumar & Münstedt (2005), who have concluded that the antimicrobial activity of silver-based systems depends on releasing of Ag^+, which binds to electron donor groups in biological molecules containing sulphur, oxygen or nitrogen. Besides the antimicrobial activity, silver nanoparticles have been reported to absorb and decompose ethylene, which may contribute to their effects on extending shelf life of fruits and vegetables (Li et al., 2009). Nanostructured calcium silicate (NCS) was used by Johnston et al. (2008) to adsorb Ag^+ ions from a solution. The resulting NCS-Ag complex exhibited effective antimicrobial activity at desirably low levels of silver down to 10 $mg.kg^{-1}$, and could be incorporated into food packaging as an antimicrobial agent.

Titanium dioxide (TiO_2) is widely used as a photocatalytic disinfecting material for surface coatings (Fujishima et al., 2000). TiO_2 photocatalysis, which promotes peroxidation of the phospholipids present in microbial cell membranes (Maness et al., 1999), has been used to inactivate food-related pathogenic bacteria (Kim et al., 2005; Robertson et al., 2005). Chawengkijwanich & Hayata (2008) developed a TiO_2 powder-coated packaging film able to reduce *E. coli* contamination on food surfaces. Gelover et al. (2006) demonstrated the efficacy of TiO_2-coated films exposed to sunlight to inactivate fecal coliforms in water. Metal doping improves visible light absorbance of TiO_2 (Anpo et al., 2001), and increases its photocatalytic activity under UV irradiation (Choi et al., 1994). It has been demonstrated that doping TiO_2 with silver greatly improved photocatalytic bacterial inactivation (Page et al., 2007; Reddy et al., 2007). This combination was explored by Cheng et al. (2006), who have obtained effective antibacterial activity from a polyvinyl chloride nanocomposite with TiO_2/Ag^+ nanoparticles.

Qi et al. (2004) have reported antibacterial activity from chitosan nanoparticles, which may be attributed to interactions between the positively charged chitosan and the negatively charged cell membranes, increasing membrane permeability and eventually causing rupture and leakage of the intracellular material. This is consistent with the observation by the same authors (Qi et al., 2004) that both chitosan and its engineered nanoparticles are ineffective at pH lower than 6, which is probably due to the absence of protonated amino groups. Another two antimicrobial mechanisms were proposed by Rabea et al. (2003), namely: chelation of trace metals by chitosan, inhibiting microbial enzyme activities; and (in fungal cells) penetration through the cell wall and membranes to bind DNA and inhibit RNA synthesis.

Carbon nanotubes have also been reported to have antibacterial properties. Direct contact with aggregates of carbon nanotubes have been demonstrated to kill *E. coli*, possibly because the long and thin nanotubes puncture microbial cells, causing irreversible damages and leakage of intracellular material (Kang et al., 2007). On the other hand, there are studies suggesting that carbon nanotubes may also be cytotoxic to human cells, at least when in contact to skin (Shvedova et al., 2003; Monteiro-Riviere et al., 2005) or lungs (Warheit et al., 2004), which would affect people manipulating the nanotubes in processing stages rather than consumers. Anyway, once present in the food packaging material, the nanotubes might eventually migrate into food. Then, it is mandatory to know any eventual health effects of ingested carbon nanotubes.

3.2 Oxygen scavengers

Oxygen (O_2) participates in several forms of food deterioration. Direct oxidation reactions result in browning reactions and rancid flavors, to name only a few examples. Food deterioration by indirect action of O_2 includes food spoilage by aerobic microorganisms. The incorporation of O_2 scavengers into food packaging systems can maintain very low O_2 levels, which is useful for several applications.

Oxygen scavenger films were successfully developed by Xiao-e et al. (2004) by adding TiO_2 nanoparticles to different polymers. The nanocomposite materials could be used as packaging films for a variety of oxygen-sensitive food products. Since TiO_2 acts by a photocatalytic mechanism, its major drawback would be the requirement of UVA light (Mills et al., 2006).

3.3 Enzyme immobilization systems

Enzymes have a variety of applications in food industry. However, their sensitivity to processing conditions and/or to enzyme inhibitors can sometimes restrict the applicability of the direct enzyme addition to foods. Immobilization is usually an effective way to improve enzyme stability to pH and temperature, resistance to proteases and other denaturing compounds, as well as to provide an adequate environment for their repeated use or controlled release (Kandimalla et al., 2006; Lopez-Rubio et al., 2006).

Enzyme immobilization has been considered for packaging applications (Appendini & Hotchkiss, 1997; Soares & Hotchkiss, 1998). The incorporation of enzymes like lactase or cholesterol reductase to packaging materials could increase the value of food products and answer the needs of consumers with enzyme deficiencies (Fernández et al., 2008). Nanoscale enzyme immobilization systems would have enhanced performance when compared to conventional ones, because of their much higher surface contact area and mass transfer rate, which are probably the most important factors affecting the effectiveness of such systems (Fernández et al., 2008). Approaches might be expected dealing with enzyme adsorption into nanoclays incorporated to polymers (Rhim & Ng, 2007), since nanoclays have a high affinity for protein adsorption, and have been reported to be efficient enzyme carriers (Sinegani et al., 2005; Gopinath & Sugunan, 2007). Conductive polymers may also be used as immobilizing matrices for biomolecules (Ahuja et al., 2007), as reported by Sharma et al. (2004), who immobilized glucose oxidase onto films of poly(aniline-co-fluoroaniline). SiO_2 nanoparticles have been modified to immobilize glutamate dehydrogenase and lactate dehydrogenase (Qhobosheane et al., 2001), which have shown excellent enzyme activity upon immobilized.

4. Nanocomposite smart food packaging

A smart food packaging system may be defined as a system that "perceives" some property of the packaged food and uses some mechanism to register and transmit information about the current quality or safety status of the food. In this context, nanostructures can be applied as reactive particles in packaging materials. The so-called nanosensors may be able to respond to environmental changes during storage (e.g., temperature, relative humidity, oxygen exposure), degradation products or microbial contamination (Bouwmeester et al., 2009).

Food expiration dates are estimated by considering distribution and storage conditions which the food product is supposed to be exposed to. However, such conditions are frequently not the real ones. For instance, foods which require a cold chain are often exposed to temperature abuse; micropores or sealing defects in packaging systems can lead food products to an unexpectedly high exposure to oxygen. Nanosensors integrated into food packaging systems may detect spoilage-related changes, pathogens and chemical contaminants, being then useful to eliminate the need for inaccurate expiration dates, providing real-time status of food freshness (Liao et al., 2005). Below are some examples of applications of nanocomposites as smart packaging systems.

4.1 Time-temperature integrators

Time- temperature indicators or integrators (TTIs) are designed to monitor, record and translate whether a certain food product is safe to be consumed, in terms of its temperature history. This is particularly important when food is stored in conditions other than the optimal ones. For instance, if a product is supposed to be frozen, a TTI can indicate whether it had been inadequately exposed to higher temperatures and the time of exposure.

The TTIs are categorized into three basic types, namely, abuse indicators, partial temperature history indicators, and full temperature history indicators. Abuse indicators, or critical temperature indicators, merely indicate whether a reference temperature has been achieved. Partial temperature history indicators integrate the time-temperature history only when the temperature exceeds a critical predetermined value. Finally, full temperature history indicators provide a continuous register of temperature changes with time (Shing, 2000).

The communication is usually manifested by a color development (related to a temperature dependent migration of a dye through a porous material) or a color change (using a temperature dependent chemical reaction or physical change). Timestrip® has developed a system (iStrip) for chilled foods, based on gold nanoparticles, which is red at temperatures above freezing. Accidental freezing leads to irreversible agglomeration of the gold nanoparticles resulting in loss of the red color (Robinson & Morrison, 2010).

4.2 Detection of gases produced by food spoilage

Food spoilage is caused by microorganisms whose metabolism produces gases which may be detected by several types of gas sensors which have been developed to translate chemical interactions between particles on a surface into response signals.

Nanosensors to detect gases are usually based on metal oxides or, more recently, conducting polymer nanocomposites, which are able to quantify and/or identify microorganisms based on their gas emissions.

Sensors based on conducting polymers (or electro active conjugated polymers) consist on conducting particles embedded into an insulating polymer matrix. The resistance changes of the sensors produce a pattern corresponding to the gas under investigation (Arshak et al., 2007). Conducting polymers are very important because of their electrical, electronic, magnetic and optical properties, which are related to their conjugated π electron backbones (Retama, 2005; Ahuja et al., 2007; Wiziack et al., 2007). Polyene and polyaromatic conducting polymers such as polyaniline, polyacetylene, and polypyrrole have been widely studied (Ahuja et al., 2007). Electrochemically polymerized conducting polymers have a remarkable ability to switch between conducting oxidized (doped) and insulating reduced (undoped) states, which is the basis for several applications (Rajesh et al., 2004).

Nanosensors containing carbon black and polyaniline developed by Arshak et al. (2007) have been demonstrated to be able to detect and identify three foodborne pathogens by producing a specific response pattern for each microorganism.

4.3 O_2 sensors

There has been an increasing interest to develop non-toxic and irreversible O_2 sensors to assure O_2 absence in oxygen-free food packaging systems, such as packaging under vacuum or nitrogen.

Lee et al. (2005) developed an UV-activated colorimetric O_2 indicator which uses TiO_2 nanoparticles to photosensitize the reduction of methylene blue (MB) by triethanolamine in a polymer encapsulation medium using UVA light. Upon UV irradiation, the sensor bleaches and remains colorless until it is exposed to oxygen, when its original blue color is restored. The rate of color recovery is proportional to the level of oxygen exposure.

Gutiérrez-Tauste et al. (2007) deposited MB/TiO_2 nanocomposite thin films on glass by liquid phase deposition (LPD), a soft chemical technique which has been applied to deposition of oxides to several substrates. This technique could be used to develop an O_2 indicator packaging system for a variety of oxygen-sensitive foods.

Mills & Hazafy (2009) used nanocrystalline SnO_2 as a photosensitizer in a colorimetric O_2 indicator comprising a sacrificial electron donor (glycerol), a redox dye (MB), and an encapsulating polymer (hydroxyethyl cellulose). Exposure to UVB light led to activation (photobleaching) of the indicator and photoreduction of MB by the SnO_2 nanoparticles. The color of the films varied according to O_2 exposure - bleached when not exposed, and blue upon exposed.

5. Final considerations

Nanotechnology has demonstrated a great potential to provide important changes in the food packaging sector. Nanocomposites are promising to expand the use of biodegradable polymers, since the addition of nanoreinforcements has been related to improvements in overall performance of biopolymers, making them more competitive in a market dominated by nonbiodegradable materials. Moreover, several nanostructures can be useful to provide active and/or "smart" properties to food packaging systems, as exemplified by antimicrobial properties, oxygen scavenging ability, enzyme immobilization, or indication of the degree of exposure to some detrimental factor such as inadequate temperatures or oxygen levels. So, nanocomposites may not only be used to passively protect foods against environmental factors, but also to incorporate desirable properties to the packaging system so it may actually enhance stability of foods, or at least to indicate their eventual inadequation to be consumed.

However, there are important safety concerns about nanotechnology applications to food contact materials. On the one hand, the properties and safety of most starting materials in their bulk form are usually well known. Nano-sized counterparts frequently exhibit different properties from those found at the macro-scale, because the very small sizes of the former, in principle, would allow them to move through the body more freely than larger particles, while their high surface area increases their reactivity, although they form large aggregates in most conditions. There is limited scientific data about migration of nanostructures from packaging materials into food, but because of their tiny dimensions it is reasonable to assume that migration may occur. Few studies have been conducted to assess the risks associated to the presence of such extremely small particles, some of them biologically active, in the human body or dispersed in the environment. Hence, significant research is still required to evaluate the potential toxicity of nanotechnology products, as well as the environmental safety of their use.

6. Acknowledgements

MCT/FINEP, CNPq, CAPES, FAPESP, and Embrapa/LABEX, for the financial support; FMC BioPolymer, for providing cellulose nanofibers for the experiments.

7. References

Adame, D. & Beall, G.W. (2009). Direct measurement of the constrained polymer region in polyamide/clay nanocomposites and the implications for gas diffusion. *Applied Clay Science*, Vol. 42, 545-552, ISSN 0169-1317.

Ahuja, T.; Mir, I. A.; Kumar, D. & Rajesh. (2007). Biomolecular immobilization on conducting polymers for biosensing applications. *Biomaterials*, Vol. 28, 791-805, ISSN 0142-9612.

Alemdar, A. & Sain, M. (2008). Biocomposites from wheat straw nanofibers: Morphology, thermal and mechanical properties. *Composites Science and Technology*, Vol. 68, 557-565, ISSN 0266-3538.

Alexandre, B.; Langevin, D.; Médéric, P.; Aubry, T.; Couderc, H.; Nguyen, Q. T.; Saiter, A. & Marais, S. (2009). Water barrier properties of polyamide 12/montmorillonite nanocomposite membranes: structure and volume fraction effects. *Journal of Membrane Science*, Vol. 328, No. 1-2, 186-204, ISSN 0376-7388.

Alexandre, M. & Dubois, P. (2000). Polymer-layered silicate nanocomposites: preparation, properties and uses of a new class of materials. *Materials Science and Engineering, R: Reports*, Vol. 28, 1-63, ISSN 0927-796X.

An, J.; Zhang, M; Wang, S. & Tang, J. (2008). Physical, chemical and microbiological changes in stored green asparagus spears as affected by coating of silver nanoparticles-PVP. *LWT - Food Science and Technology*, Vol. 41, No. 6, 1100-1107, ISSN 0023-6438.

Anglès, M. N. &; Dufresne, A. (2000). Plasticized starch/tunicin whiskers nanocomposites. 1. Structural analysis. *Macromolecules*, Vol. 33, 8344-8353, ISSN 0024-9297.

Anpo, M.; Kishiguchi, S.; Ichihashi, Y.; Takeuchi, M.; Yamashita, H.; Ikeue, K.; Morin, B.; Davidson, A. & Che, M. (2001). The design and development of second-generation titanium oxide photocatalysts able to operate under visible light irradiation by applying a metal ion-implantation method. *Research on Chemical Intermediates*, Vol. 27, No. 4-5, 459-467, ISSN 1568-5675.

Appendini, P. & Hotchkiss, J. H. (1997). Immobilisation of lysozyme on food contact polymers as potential antimicrobial films. *Packaging Technology and Science*, Vol. 10, p. 271-279, ISSN 1099-1522.

Arshak, K.; Adley, C.; Moore, E.; Cunniffe, C.; Campion, M. & Harris, J. (2007). Characterisation of polymer nanocomposite sensors for quantification of bacterial cultures. *Sensors and Actuators B*, Vol. 126, 226-231, ISSN 0925-4005.

ASTM. 1989. Standard test methods for water vapor transmission of materials, E96-80, In: *Annual book of American Standard Testing Methods*. ASTM, Philadelphia.

ASTM. 1997. Standard test method for tensile properties of thin plastic sheeting, D882-97, In: *Annual book of American Standard Testing Methods*. ASTM, Philadelphia.

Aymonier, C.; Schlotterbeck, U.; Antonietti, L.; Zacharias, P.; Thomann, R.; Tiller, J. C. & Mecking, S. (2002). Hybrids of silver nanoparticles with amphiphilic hyperbranched macromolecules exhibiting antimicrobial properties. *Chemical Communications*, Vol. 2002, No. 24, 3018-3019, ISSN 1364-548X.

Azeredo, H. M. C.; Mattoso, L. H. C.; Avena-Bustillos, R. J.; Ceotto Filho, G.; Munford, M. L.; Wood, D. & McHugh, T. H. (2010). Nanocellulose reinforced chitosan composite films as affected by nanofiller loading and plasticizer content. *Journal of Food Science*, Vol. 75, No. 1, N1-N7, ISSN 0022-1147.

Azeredo, H. M. C.; Mattoso, L. H. C.; Wood, D.; Williams, T. G.; Avena-Bustillos, R. J. & McHugh, T. H. (2009). Nanocomposite edible films from mango puree reinforced with cellulose nanofibers. *Journal of Food Science*, Vol. 74, No. 5, N31-N35, ISSN 0022-1147.

Azizi Samir, M. A. S.; Alloin, F. & Dufresne, A. (2005). Review of recent research into cellulosic whiskers, their properties and their application in nanocomposite field. *Biomacromolecules*, Vol. 6, 612-626, ISSN 1525-7797.

Azizi Samir, M. A. S.; Alloin, F.; Sanchez, J. Y. & Dufresne, A. (2004). Cellulose nanocrystals reinforced poly(oxyethylene). *Polymer*, Vol. 45, 4149-4157, ISSN 0032-3861.

Beall, G. W. (2000). New conceptual model for interpreting nanocomposite behavior, In: *Polymer-clay nanocomposites*, Pinnavaia, T. J. & Beall, G. W. (Ed.), 267-279, Wiley, ISBN 0471637009, New York.

Bhatnagar, A. & Sain, M. (2005). Processing of cellulose nanofiber-reinforced composites. *Journal of Reinforced Plastics and Composites*, Vol. 24, No. 12, 1259-1268, ISSN 1530-7964.

Bondeson, D. & Oksman, K. (2007). Polylactic acid/cellulose whisker nanocomposites modified by polyvinyl alcohol. *Composites: Part A*, Vol. 38, 2486-2492, ISSN 1359-835X.

Bouwmeester, H.; Dekkers, S.; Noordam, M. Y.; Hagens, W. I.; Bulder, A. S.; De Heer, C.; Ten Voorde, S. E. C. G.; Wijnhoven, S. W. P.; Marvin, H. J. P. & Sips, A. J. A. M. (2009). Review of health safety aspects of nanotechnologies in food production. *Regulatory Toxicology and Pharmacology*, Vol. 53, 52-62, ISSN 0273-2300.

Brody, A. L. (2006). *Nano and food packaging technologies converge. Food Technology*, Vol. 60, No. 3, 92-94, ISSN 0737-3937.

Carrero-Sánchez, J.C.; Elías, A.L.: Mancilla, R.; Arrellín, G.; Terrones, H.; Laclette, J.P.; Terrones, M. (2006). Biocompatibility and toxicological studies of carbon nanotubes doped with nitrogen. *Nano Letters*, Vol. 6, No. 8, 1609-1616, ISSN 1530-6992.

Chawengkijwanich, C. & Hayata, Y. (2008). Development of TiO_2 powder-coated food packaging film and its ability to inactivate *Escherichia coli in vitro* and in actual tests. *International Journal of Food Microbiology*, Vol. 123, No. 3, 288-292, ISSN 0168-1605.

Chen, B. & Evans, J. R. G. (2005). Thermoplastic starch-clay nanocomposites and their characteristics. *Carbohydrate Polymers*, Vol. 61, 455-463, ISSN 0144-8617.

Chen, W.; Tao, X.; Xue, P. & Cheng, X. (2005). Enhanced mechanical properties and morphological characterizations of poly(vinyl alcohol)-carbon nanotube composite films. *Applied Surface Science*, Vol. 252, 1404-1409, ISSN 0169-4332.

Chen, Y.; Liu, C.; Chang, P. R.; Cao, X. & Anderson, D. P. (2009). Bionanocomposites based on pea starch and cellulose nanowhiskers hydrolyzed from pea hull fibre: effect of hydrolysis time. *Carbohydrate Polymers*, Vol. 76, No. 4, 607-615, ISSN 0144-8617.

Cheng, Q.; Li, C.; Pavlinek, V.; Saha, P. & Wang, H. (2006). Surface-modified antibacterial TiO_2/Ag^+ nanoparticles: Preparation and properties. *Applied Surface Science*, Vol. 252, 4154-4160, ISSN 0169-4332.

Choi, W.; Termin, A. & Hoffmann, M. (1994). The role of metal ion dopants in quantum size TiO_2: correlation between photoreactivity and charge carrier recombination dynamics. *The Journal of Physical Chemistry*, Vol. 98, No. 51, 13669-13679, ISSN 1520-5215.

Cioffi, N.; Torsi, L.; Ditaranto, N.; Tantillo, G.; Ghibelli, L.; Sabbatini, L.; Bleve-zacheo, T.; D'alessio, M.; Zambonin, P. G. & Traversa, E. (2005). Copper nanoparticle/polymer composites with antifungal and bacteriostatic properties. *Chemistry of Materials*, Vol. 17, 5255-5262, ISSN 1520-5002.

Cyras, V. P.; Manfredi, L. B.; Ton-that, M. T. & Vázquez, A. (2008). Physical and mechanical properties of thermoplastic starch/montmorillonite nanocomposite films. *Carbohydrate Polymers*, Vol. 73, 55-63, ISSN 0144-8617.

Dalmas, F.; Cavaillé, J. Y.; Gauthier, C.; Chazeau, L. & Dendievel, R. (2007). Viscoelastic behavior and electrical properties of flexible nanofiber filled polymer nanocomposites. Influence of processing conditions. *Composites Science and Technology*, Vol. 67, 829-839, ISSN 0266-3538.

De Moura, M. R.; Aouada, F. A.; Avena-bustillos, R. J.; Mchugh, T. H.; Krochta, J. M. & Mattoso, L. H. C. (2009). Improved barrier and mechanical properties of novel hydroxypropyl methylcellulose edible films with chitosan/tripolyphosphate nanoparticles. *Journal of Food Engineering*, Vol. 92, 448-453, ISSN 0260-8774.

Dufresne, A.; Dupeyre, D. & Vignon, M. R. (2000). Cellulose microfibrils from potato tuber cells: processing and characterization of starch-cellulose microfibril composites. *Journal of Applied Polymer Science*, Vol. 76, No. 14, 2080-2092, ISSN 0021-8995.

Dufresne, A. & Vignon, M. R. (1998). Improvement of starch film performances using cellulose microfibrils. *Macromolecules*, Vol. 31, 2693-2696, ISSN 0024-9297.

Dujardin, E.; Blaseby, M. & Mann, S. (2003). Synthesis of mesoporous silica by sol-gel mineralisation of cellulose nanorod nematic suspensions. *Journal of Materials Chemistry*, Vol. 13, No. 4, 696-699, ISSN 1364-5501.

Fernández, A.; Cava, D.; Ocio, M. J.; Lagaron, J. M. (2008). Perspectives for biocatalysts in food packaging. *Trends in Food Science & Technology*, Vol. 19, No. 4, 198-206, ISSN 0924-2244.

Freire, C. S. R.; Silvestre, A. J. D.; Pascoal Neto, C.; Gandini, A.; Martin, L. & Mondragon, I. (2008). Composites based on acylated cellulose fibers and low-density polyethylene: Effect of the fiber content, degree of substitution and fatty acid chain length on final properties. *Composites Science and Technology*, Vol. 68, No. 15-16, 3358-3364, ISSN 0266-3538.

Fujishima, A.; Rao, T. N. & Tryk, D. A. (2000). Titanium dioxide photocatalysis. *Journal of Photochemistry and Photobiology C: Photochemistry Reviews*, Vol. 1, No. 1, 1-21, ISSN 1389-5567.

Gelover, S.; Gómez, L. A.; Reyes, K. & Leal, M. T. (2006). A practical demonstration of water disinfection using TiO_2 films and sunlight. *Water Research*, Vol. 40, 3274-3280, ISSN 0043-1354.

Gopinath, S.; Sugunan, S. (2007). Enzymes immobilized on montmorillonite K 10: Effect of adsorption and grafting on the surface properties and the enzyme activity. *Applied Clay Science*, Vol. 35, No. 1-2, 67-75, ISSN 0169-1317.

Gu, H. W.; Ho, P. L.; Tong, E.; Wang, L. & Xu, B. (2003). Presenting vancomycin on nanoparticles to enhance antimicrobial activities. *Nano Letters*, Vol. 3, 1261-1263, ISSN 1530-6992.

Gutiérrez-Tauste, D.; Domènech, X.; Casañ-Pastor, N. & Ayllón, J. A. (2007). Characterization of methylene blue/TiO_2 hybrid thin films prepared by the liquid phase deposition (LPD) method: Application for fabrication of light-activated colorimetric oxygen indicators. *Journal of Photochemistry and Photobiology A: Chemistry*, Vol. 187, 45-52, ISSN 1010-6030.

Helbert, W.; Cavaillé, C. Y. & Dufresne, A. (1996). Thermoplastic nanocomposites filled with wheat straw cellulose whiskers. Part I: Processing and mechanical behaviour. *Polymer Composites*, Vol. 17, No. 4, 604-611, ISSN 1548-0569.

Helbert, W., & Chanzy, H. (1994). Oriented growth of V amylase n-butanol crystals on cellulose. *Carbohydrate Polymers*, Vol. 24, 119-122, ISSN 0144-8617.

Huang, L.; Li, D. Q.; Lin, Y. J.; Wei, M.; Evans, D. G. & Duan, X. (2005). Controllable preparation of nano-MgO and investigation of its bactericidal properties. *Journal of Inorganic Biochemistry*, Vol. 99, 986-993, ISSN 0162-0134.

Hubbe, M. A.; Rojas, O. J.; Lucia, L. A. & Sain, M. (2008). Cellulosic nanocomposites: A review. *Bioresources*, Vol. 3, No. 3, 929-980, ISSN 1930-2126.

Hulleman, S. H. D.; Helbert, W. & Chanzy, H. (1996). Single crystals of V amylose complexed with glycerol. *International Journal of Biological Macromolecules*, Vol. 18, 115-122, ISSN 0141-8130.

Jia, X.; Li, Y.; Cheng, Q.; Zhang, S. & Zhang, B. (2007). Preparation and properties of poly(vinyl alcohol)/silica nanocomposites derived from copolymerization of vinyl silica nanoparticles and vinyl acetate. *European Polymer Journal*, Vol. 43, 1123-1131, ISSN 0014-3057.

Jordan, J.; Jacob, K. I.; Tannenbaum, R.; Sharaf, M. A. & Jasiuk, I. (2005). Experimental trends in polymer nanocomposites: a review. *Materials Science & Engineering A*, Vol. 393, No. 1-2, 1-11, ISSN 0921-5093.

Kandimalla, V. B.; Tripathi, V. S. & Ju, H. (2006). Immobilization of biomolecules in sol-gels: biological and analytical applications. *Critical Reviews in Analytical Chemistry*, Vol. 36, No. 2, 73-106, ISSN 1547-6510.

Kang, S.; Pinault, M.; Pfefferle, L. D. & Elimelech, M. (2007). Single-walled carbon nanotubes exhibit strong antimicrobial activity. *Langmuir*, Vol. 23, 8670-8673, ISSN 1520-5827.

Kim, K. Y.; Lim, H. J.; Park, S. M. & Lee, S. J. (2003). Synthesis and characterization of high impact polystyrene/organically modified layered silicate nanocomposites. *Polymer (Korea)*, Vol. 27, 377-384, ISSN 0379-153X.

Kim, J. Y., Il Han, S., & Hong, S. (2008). Effect of modified carbon nanotube on the properties of aromatic polyester nanocomposites. *Polymer*, Vol. 49, 3335-3345, ISSN 0032-3861.

Kim, T. Y.; Lee, Y. H.; Park, K. H.; Kim, S. J. & Cho, S. Y. (2005). A study of photocatalysis of TiO_2 coated onto chitosan beads and activated carbon. *Research on Chemical Intermediates*, Vol. 31, No. 4-6, 343-358, ISSN 1568-5675.

Kim, Y.; Jung, R.; Kim, H. S. & Jin, H. J. (2009). Transparent nanocomposites prepared by incorporating microbial nanofibrils into poly(L-lactic acid). *Current Applied Physics*, Vol. 9, No. 1, S69-S71, ISSN 1567-1739.

Kristo, E. & Biliaderis, C. G. (2007). Physical properites of starch nanocrystal-reinforced pullulan films. *Carbohydrate Polymers*, Vol. 68, 146-158, ISSN 0144-8617.

Kumar, R. & Münstedt, H. (2005). Silver ion release from antimicrobial polyamide/silver composites. *Biomaterials*, Vol. 26, 2081-2088, ISSN 0142-9612.

Kvien, I. & Oksman, K. (2007). Orientation of cellulose nanowhiskers in polyvinyl alcohol. *Applied Physics A: Materials Science and Processing*, Vol. 87, 641-643, ISSN 0947-8396.

Kvítek, L.; Panáček, A.; Soukupová, J.; Kolář, M.; Večeřová, R.; Prucek, R.; Holecová, M. & Zbořil, R. (2008). Effect of surfactants and polymers on stability and antibacterial activity of silver nanoparticles (NPs). *The Journal of Physical Chemistry C*, Vol. 112, No. 15, 5825-5834, ISSN 1932-7455.

Lagaron, J. M.; Catalá, R. & Gavara, R. (2004). Structural characteristics defining high barrier polymeric materials. *Materials Science and Technology*, Vol. 20, 1-7, ISSN 0267-0836.

Lee, S. K.; Sheridan, M. & Mills, A. (2005). Novel UV-activated colorimetric oxygen indicator. *Chemistry of Materials*, Vol. 17, No. 10, 2744-2751, ISSN 1520-5002.

Li, H.; Li, F.; Wang, L.; Sheng, J.; Xin, Z.; Zhao, L.; Xiao, H.; Zheng, Y. & Hu, Q. (2009). Effect of nano-packing on preservation quality of Chinese jujube (*Ziziphus jujuba* Mill. var. inermis (Bunge) Rehd). *Food Chemistry*, Vol. 114, No. 2, 547-552, ISSN 0308-8146.

Li, Q.; Mahendra, S.; Lyon, D. Y.; Brunet, L.; Liga, M. V.; Li, D. & Alvarez, P. J. J. (2008). Antimicrobial nanomaterials for water disinfection and microbial control: Potential applications and implications. *Water Research*, Vol. 42, No. 18, 4591-4602, ISSN 0043-1354.

Liao, F.; Chen, C. & Subramanian, V. (2005). Organic TFTs as gas sensors for electronic nose applications. *Sensors and Actuators B*, Vol. 107, No. 2, 849-855, ISSN 0925-4005.

Liau, S. Y.; Read, D. C.; Pugh, W. J.; Furr, J. R. & Russell, A. D. (1997). Interaction of silver nitrate with readily identifiable groups: Relationship to the antibacterial action of silver ions. *Letters in Applied Microbiology*, Vol. 25, 279-283, ISSN 1472-765X.

Lima, M. M. D. & Borsali, R. (2004). Rodlike cellulose microcrystals: Structure, properties, and applications. *Macromolecular Rapid Communications*, Vol. 25, No. 7, 771-787, ISSN 1521-3927.

Lin, Y. J.; Li, D. Q.; Wang, G.; Huang, L. & Duan, X. (2005). Preparation and bactericidal property of MgO nanoparticles on γ-Al2O3. *Journal of Materials Science: Materials in Medicine*, Vol. 16, 53-56, ISSN 1573-4838.

López Manchado, M. A.; Valentini, L.; Biagotti, J. & Kenny, J. M. (2005). Thermal and mechanical properties of single-walled carbon nanotubes-polypropylene composites prepared by melt processing. *Carbon*, Vol. 43, p. 1499-1505, ISSN 0008-6223.

Lopez-Rubio, A.; Gavara, R. & Lagaron, J. M. (2006). Bioactive packaging: turning foods into healthier foods through biomaterials. *Trends in Food Science & Technology*, Vol. 17, 567-575, ISSN 0924-2244.

Lu, Y.; Weng, L. & Cao, X. (2005). Biocomposites of plasticizes starch reinforced with cellulose crystallites from cottonseed linter. *Macromolecular Bioscience*, Vol. 5, 1101-1107, ISSN 1616-5195.

Lu, Y.; Weng, L. & Zhang, L. (2004). Morphology and properties of soy protein isolate thermoplastics reinforced with chitin whiskers. *Biomacromolecules*, Vol. 5, 1046-1051, ISSN 1525-7797.

Ludueña, L. N.; Alvarez, V. A. & Vasquez, A. (2007). Processing and microstructure of PCL/clay nanocomposites. *Materials Science and Engineering: A*, Vol. 460-461, 121-129, ISSN 0921-5093.

Luo, P. G. & Stutzenberger, F. J. (2008). Nanotechnology in the detection and control of microorganisms. *Advances in Applied Microbiology*, Vol. 63, 145-181, ISSN 0065-2164.

Maness, P. C.; Smolinski, S.; Blake, D. M.; Huang, Z.; Wolfrum, E. J. & Jacoby, W. A. (1999). Bactericidal activity of photocatalytic TiO_2 reaction: Toward an understanding of its killing mechanism. *Applied and Environmental Microbiology*, Vol. 65, No. 9, 4094-4098, ISSN 1098-5336.

Mathew, A. P. & Dufresne, A. (2002). Morphological investigation of nanocomposites from sorbitol plasticized starch and tunicin whiskers. *Biomacromolecules*, Vol. 3, No. 3, 609-617, ISSN 1525-7797.

Mattoso, L. H. C.; Medeiros, E. S.; Baker, A. D.; Avloni, J.; Wood, D. F. & Orts, W. J. (2009). Electrically conductive nanocomposites made from cellulose nanofibrils and polyaniline. *Journal of Nanoscience and Nanotechnology*, Vol. 9, 2917-2922, ISSN 1550-7033.

McHugh, T. H.; Avena-Bustillos, R. J.; Krochta, J.M. (1993). Hydrophilic edible films: modified procedure for water vapor permeability and explanation of thickness effects. *Journal of Food Science*, Vol. 58, No. 4, 899–903, ISSN 0022-1147.

Mills, A., Doyle, G., Peiro, A. M., & Durrant, J. (2006). Demonstration of a novel, flexible, photocatalytic oxygen-scavenging polymer film. *Journal of Photochemistry and Photobiology A: Chemistry*, Vol. 177, 328-331, ISSN 1010-6030.

Mills, A. & Hazafy, D. (2009). Nanocrystalline SnO_2-based, UVB-activated, colourimetric oxygen indicator. *Sensor and Actuators B: Chemical*, Vol. 136, No. 2, 344-349, ISSN 0925-400.

Mirzadeh, A. & Kokabi, M. (2007). The effect of composition and draw-down ratio on morphology and oxygen permeability of polypropylene nanocomposite blown films. *European Polymer Journal*, Vol. 43, No. 9, 3757-3765, ISSN 0014-3057.

Mohanty, A. K.; Misra, M. & Drzal, L. T. (2001). Surface modifications of natural fibers and performance of the resulting biocomposites: An overview. *Composite Interfaces*, Vol. 8, No. 5, 313-343, ISSN 0927-6440.

Monteiro-Riviere, N. A.; Nemanich, R. J.; Inman, A. O.; Wang, Y. Y. & Riviere, J. E. (2005). Multi-walled carbon nanotube interactions with human epidermal keratinocytes. *Toxicology Letters*, Vol. 155, No. 13, 377-384, ISSN 0378-4274.

Morones, J. R.; Elechiguerra, J. L.; Camacho, A.; Holt, K.; Kouri, J. B.; Ramirez, J. T. & Yacaman, M. J. (2005). The bactericidal effect of silver nanoparticles. *Nanotechnology*, Vol. 16, 2346-2353, ISSN 1361-6528.

Nielsen, L. E. (1967). Models for the permeability of filled polymer systems. *Journal of Macromolecular Science: Part A*, Vol. 1, No. 5, 929-942, ISSN 1520-5738.

Oksman, K.; Mathew, A. P.; Bondeson, D. & Kvien, I. (2006). Manufacturing process of cellulose whiskers/polylactic acid nanocomposites. *Composites Science and Technology*, Vol. 66, No. 15, 2776-2784, ISSN 0266-3538.

Osman, M. A.; Ploetze, M. & Suter, U. W. (2003). Surface treatment of clay minerals - thermal stability, basal-plane spacing and surface coverage. *Journal of Materials Science*, Vol. 13, 2359-2366, ISSN 1573-4803.

Page, K.; Palgrave, R. G.; Parkin, I. P.; Wilson, M.; Savin, S. L. P. & Chadwick, A. V. (2007). Titania and silver-titania composite films on glass-potent antimicrobial coatings. *Journal of Materials Chemistry*, Vol.17, No.1, 95-104, ISSN 1364-5501.

Paiva, L. B.; Morales, A. R. & Díaz, F. R. V. (2008). Argilas organofílicas: características, metodologias de preparação, compostos de intercalação e técnicas de caracterização. *Cerâmica*, Vol. 54, p. 213-226, ISSN 0366-6913.

Paralikar, S. A.; Simonsen, J. & Lombardi, J. (2008). Poly(vinyl alcohol)/cellulose nanocrystal barrier membranes. *Journal of Membrane Science*, Vol. 320, No. 1-2, 248-258, ISSN 0376-7388.

Paul, M. A.; Alexandre, M.; Degée, P.; Henrist, C.; Rulmont, A. & Dubois, P. (2003). New nanocomposite materials based on plasticized poly(L-lactide) and organo-modified montmorillonites: thermal and morphological study. *Polymer*, Vol. 44, No. 2, 443-450, ISSN 0032-3861.

Petersson, L. & Oksman, K. (2006). Biopolymer based nanocomposites: Comparing layered silicates and microcrystalline cellulose as nanoreinforcement. *Composites Science and Technology*, Vol. 66, 2187-2196, ISSN 0266-3538.

Picard, E.; Gauthier, H.; Gérard, J. F. & Espuche, E. (2007). Influence of the intercalated cations on the surface energy of montmorillonites: Consequences for the morphology and gas barrier properties of polyethylene/montmorillonites nanocomposites. *Journal of Colloid and Interface Science*, Vol. 307, 364-376, ISSN 0021-9797.

Podsiadlo, P.; Choi, S. Y.; Shim, B.; Lee, J.; Cuddihy, M. & Kotov, N. A. (2005). Molecularly engineered nanocomposites: Layer-by-layer assembly of cellulose nanocrystals. *Biomacromolecules*, Vol. 6, 2914-2918, ISSN 1525-7797.

Prashantha, K.; Soulestin, J.; Lacrampe, M. F.; Krawczak, P.; Dupin, G. & Claes, M. (2009). Masterbatch-based multi-walled carbon nanotube filled polypropylene nanocomposites: Assessment of rheological and mechanical properties. *Composites Science and Technology*, Vol. 69, No. 11-12, 1756-1763, ISSN 0266-3538.

Qhobosheane, M.; Santra, S.; Zhang, P. & Tan, W. H. (2001). Biochemically functionalized silica nanoparticles. *Analyst*, Vol. 126, No. 8, 1274-1278, ISSN 0003-2654.

Qi, L. F.; Xu, Z. R.; Jiang, X.; Hu, C. & Zou, X. (2004). Preparation and antibacterial activity of chitosan nanoparticles. *Carbohydrate Research*, Vol. 339, 2693-2700, ISSN 0008-6215.

Qiao, R. & Brinson, L. C. (2009). Simulation of interphase percolation and gradients in polymer nanocomposites. *Composites Science and Technology*, Vol. 69, No. 3-4, 491-499, ISSN 0266-3538.

Rabea, E. I.; Badawy, M. E.; Stevens, C. V.; Smagghe, G. & Steurbaut, W. (2003). Chitosan as antimicrobial agent: applications and mode of action. *Biomacromolecules*, Vol. 4, No. 6, 1457-1465, ISSN 1525-7797.

Rajesh; Takashima, W. & Kaneto, K. (2004). Amperometric phenol biosensor based on covalent immobilization of tyrosinase onto an electrochemically prepared novel copolymer poly(N-3-aminopropyl pyrrole-copolymer) film. *Sensors and Actuators B*, Vol. 102, 271-277, ISSN 0925-4005.

Ray, S.; Easteal, A.; Quek, S. Y. & Chen, X. D. (2006). The potential use of polymer-clay nanocomposites in food packaging. *International Journal of Food Engineering*, Vol. 2, No. 4, artigo 5, 2006.

Reddy, M. P.; Venugopal, A. & Subrahmanyam, M. (2007). Hydroxyapatite-supported Ag-TiO_2 as *Escherichia coli* disinfection photocatalyst. *Water Research*, Vol. 41, p. 379-386, ISSN 0043-1354.

Retama, R. J. (2005). Synthesis and characterization of semiconducting polypyrrole/polyacrylamide microparticles with GOx for biosensor applications. *Colloids and Surfaces A: Physicochemical and Engineering Aspects*, Vol. 270-271, 239-244, ISSN 0927-7757.

Rhim, J. W. & Ng, P. K. W. (2007). Natural biopolymer-based nanocomposite films for packaging applications. *Critical Reviews in Food Science and Nutrition*, Vol. 47, No. 4, p. 411-433, ISSN 1040-8398.

Robertson, J. M. C.; Robertson, P. K. J. & Lawton, L. A. (2005). A comparison of the effectiveness of TiO_2 photocatalysis and UVA photolysis for the destruction of three pathogenic micro-organisms. *Journal of Photochemistry and Photobiology A: Chemistry*, Vol. 175, No. 1, 51-56, ISSN 1010-6030.

Robinson, D. K. R. & Morrison, M. J. (2010). *Nanotechnologies for food packaging: Reporting the science and technology research trends.* ObservatoryNANO, Aug 2010. http://www.observatorynano.eu/project/filesystem/files/Food%20Packaging%20Report%202010%20DKR%20Robinson.pdf

Sanchez-Garcia, M. D.; Gimenez, E. & Lagaron, J. M. (2008). Morphology and barrier properties of solvent cast composites of thermoplastic biopolymers and purified cellulose fibers. *Carbohydrate Polymers*, Vol. 71, 235-244, ISSN 0144-8617.

Sharma, A. L.; Singhal, R.; Kumar, A.; Rajesh; Pande, K. K. & Malhotra, B. D. (2004). Immobilization of glucose oxidase onto electrochemically prepared poly(aniline-*co*-fluoroaniline) films. *Journal of Applied Polymer Science*, Vol. 91, No. 6, 3999-4006, ISSN 0021-8995.

Shing, R.P. (2000). Scientific principles of shelf-life evaluation, In: *Shelf-life evaluation of foods*, Man, D. & Jones, A. (Ed.), 2-22, Aspen, ISBN 978-0834217829, New York..

Shvedova, A.; Castranova, V.; Kisin, E.; Schwegler-berry, D.; Murray, A.; Gandelsman, V.; Maynard, A. & Baron, P. (2003). Exposure to carbon nanotube material: assessment of nanotube cytotoxicity using human keratinocyte cells. *Journal of Toxicology and Environmental Health, Part A*, Vol. 66, No. 20, 1909-1926, ISSN 1528-7394.

Sinegani, A. A. S.; Emzitai, G. & Shariamadari, H. (2005). Sorption and immobilization of cellulase on silicate clay minerals. *Journal of Colloid and Interface Science*, Vol. 290, 39-44, ISSN 0021-9797.

Soares, N. F. F. & Hotchkiss, J. H. (1998). Naringinase immobilization in packaging films for reducing naringin concentration in grapefruit juice. *Journal of Food Science*, Vol. 63, 61-65, ISSN 0022-1147.

Son, W. K.; Youk, J. H. & Park, W. H. (2006). Antimicrobial cellulose acetate nanofibers containing silver nanoparticles. *Carbohydrate Polymers*, Vol. 65, 430-434, ISSN 0144-8617.

Sondi, I. & Salopek-Sondi, B. (2004). Silver nanoparticles as antimicrobial agent: a case study on E. coli as a model for Gram-negative bacteria. *Journal of Colloid and Interface Science*, Vol. 275, 177-182, ISSN 0021-9797.

Sorrentino, A.; Gorrasi, G. & Vittoria, V. (2007). Potential perspectives of bio-nanocomposites for food packaging applications. *Trends in Food Science & Technology*, Vol. 18, No. 2, 84-95, ISSN 0924-2244.

Stoimenov, P.; Klinger, R. L.; Marchin, G. L. & Klabunde, K. J. (2002). Metal oxide nanoparticles as bactericidal agents. *Langmuir*, Vol. 18, 6679-6686, ISSN 1520-5827.

Svagan, A. J.; Hedenqvist, M. S. & Berglund, L. (2009). Reduced water vapour sorption in cellulose nanocomposites with starch matrix. *Composites Science and Technology*, Vol. 69, No. 3-4, 500-506, ISSN 0266-3538.

Tan, W.; Zhang, Y.; Szeto, Y. S. & Liao, L. (2008). A novel method to prepare chitosan/montmorillonite nanocomposites in the presence of hydroxyl-aluminum olygomeric cations. *Composites Science and Technology*, Vol. 68, No. 14, 2917-2921, ISSN 0266-3538.

Tang, C. & Liu, H. (2008). Cellulose nanofiber reinforced poly (vinyl alcohol) composite film with high visible light transmittance. *Composites Part A: Applied Science and Manufacturing*, Vol. 39, No. 10, 1638-1643, ISSN 1359-835X.

Tang, S.; Zou, P.; Xiong, H. & Tang, H. (2008). Effect of nano-SiO_2 on the performance of starch/polyvinyl alcohol blend films. *Carbohydrate Polymers*, Vol. 72, 521-526, ISSN 0144-8617.

Tankhiwale, R. & Bajpai, S. K. (2009). Graft co-polymerization onto cellulose-based filter paper and its further development as silver nanoparticles-loaded antibacterial food packaging material. *Colloids and Surfaces B: Biointerfaces*, Vol. 69, No. 2, p. 164-168, ISSN 0927-7765.

Uyama, H.; Kuwabara, M.; Tsujimoto, T.; Nakano, M.; Usuki, A. & Kobayashi, S. (2003). Green nanocomposite from renewable resources: plant oil-clay hybrid materials. *Chemistry of Materials*, Vol. 15, 2492-2494, ISSN 1520-5002.

Vladimiriov, V.; Betchev, C.; Vassiliou, A.; Papageorgiou, G. & Bikiaris, D. (2006). Dynamic mechanical and morphological studies of isotactic polypropylene/fumed silica nanocomposites with enhanced gas barrier properties. *Composites Science and Technology*, Vol. 66, 2935-2944, ISSN 0266-3538.

Warheit, D. B.; Laurence, B. R.; Reed, K. L.; Roach, D. H.; Reynolds, G. A. M. & Webb, T. R. (2004). Comparative pulmonary toxicity assessment of single-wall carbon nanotubes in rats. *Toxicological Sciences*, Vol. 77, 117-125, ISSN 1096-0929.

Weiss, J.; Takhistov, P. & McClements, D. J. (2006). Functional materials in food nanotechnology. *Journal of Food Science*, Vol. 71, No. 9, R107-116, ISSN 0022-1147.

Wiziack, N. K. L.; Paterno, L. G.; Fonseca, F. J. & Mattoso, L. H. C. (2007). Effect of film thickness and different electrode geometries on the performance of chemical sensors made of nanostructured conducting polymer films. *Sensors and Actuators B*, Vol. 122, 484-492, ISSN 0925-4005.

Wu, C. L.; Zhang, M. Q.; Rong, M. Z. & Friedrick, K. (2002). Tensile performance improvement of low nanoparticles filled-polypropylene composites. *Composites Science and Technology*, Vol.62, 1327-1340, ISSN 0266-3538.

Wu, Q.; Henriksson, M.; Liu, X. & Berglund, L. A. (2007). A high strength nanocomposite based on microcrystalline cellulose and polyurethane. *Biomacromolecules*, Vol. 8, p. 3687-3692, ISSN 1525-7797.

Xiao-e, L.; Green, A. N. M.; Haque, S. A.; Mills, A. & Durrant, J. R. (2004). Light-driven oxygen scavenging by titania/polymer nanocomposite films. *Journal of Photochemistry and Photobiology A: Chemistry*, Vol. 162, p. 253-259, ISSN 1010-6030.

Xiong, H. G.; Tang, S. W.; Tang, H. L. & Zou, P. (2008). The structure and properties of a starch-based biodegradable film. *Carbohydrate Polymers*, Vol. 71, 263-268, ISSN 0144-8617.

Yu, H.; Xu, X.; Chen, X.; Lu, T.; Zhang, P. & Jing, X. (2007). Preparation and antibacterial effects of PVA-PVP hydrogels containing silver nanoparticles. *Journal of Applied Polymer Science*, Vol. 103, 125-133, ISSN 0021-8995.

Zeng, H.; Gao, C.; Wang, Y.; Watts, P. C. P.; Kong, H.; Cui, X. & Yan, D. (2006). In situ polymerization approach to multiwalled carbon nanotubes-reinforced nylon 1010 composites: Mechanical properties and crystallization behavior. *Polymer*, Vol. 47, 113-122, ISSN 0032-3861.

Zhou, X.; Shin, E.; Wang, K. W. & Bakis, C. E. (2004). Interfacial damping characteristics of carbon nanotube-based composites. *Composites Science and Technology*, Vol. 64, No. 15, 2425-2437, ISSN 0266-3538.

5

Hydrogen Storage Properties of Hydrogenated Graphite and Lithium Hydride Nanocomposite

Takayuki Ichikawa, Hiroki Miyaoka and Yoshitsugu Kojima
Hiroshima University,
Japan

1. Introduction

Recently, hydrogen storage and transportation are being studied all over the world as the key technology to establish clean and renewable energy systems for a sustainable society. In the case of an on-board application for a vehicle, hydrogen should be stored in a compact, light, safe, and reasonable vessel. Hydrogen storage materials can safely store higher density of hydrogen compared to the gaseous and liquid hydrogen storage systems. Therefore, the systems using hydrogen storage materials are considered as the most suitable technique (Akiba, 1999, Grochala & Edwards, 2004, Sandrock, 1999, Schlapbach & Züttel, 2001, Züttel, 2007). Particularly, much attention has been paid to materials based on the light elements because these materials are expected to realize high gravimetric and volumetric densities of hydrogen (Orimo & Nakamori et al., 2007, Schüth & Bogdanović et al., 2004, Züttel, 2004).
Carbon is one of the attractive light elements because of abundant resource and low cost. Therefore, a lot of carbon based materials have been investigated as a hydrogen storage material since Dillon *et al.* reported on single-walled carbon nano-tubes in 1997 (Dillon & Jones et al., 1997). Among the carbon based materials, the hydrogenated nano-structural graphite ($C^{nano}H_x$) can stably store large amount of hydrogen. The hydrogen ab/desorption properties of $C^{nano}H_x$ have been investigated so far (Chen et al., 2003, Ichikawa et al., 2004, Majer et al., 2003, Miyaoka et al., 2010, Orimo et al., 1999, Orimo et al., 2001, Stanik et al., 2005). $C^{nano}H_x$ is synthesized from graphite by ball-milling method under hydrogen atmosphere. With respect to the hydrogen absorption site in this product, the hydrogen atoms are chemisorbed as the stable C-H bonds at the graphene edges and defects induced by ball-milling (Fukunaga & Itoh et al., 2004, Itoh & Miyahara et al., 2003, Ogita & Yamamoto et al., 2004, Smith & Miyaoka et al., 2009). On the other hand, this product needs a high temperature of more than 700 °C to release the hydrogen, and desorbs a considerable amount of hydrocarbons, such as methane (CH_4) and ethane (C_2H_6), together with hydrogen. Furthermore, it is quite difficult to recharge this product with hydrogen under moderate conditions of pressure and temperature for the on-board application. In order to improve the hydrogen absorption and desorption properties of $C^{nano}H_x$, Ichikawa *et al.* have paid attention to the chemical reaction between NH_3 and LiH, which is one of the elementary reactions in the Li-N-H system (Ichikawa & Hanada et al., 2004). This reaction proceeds even at room temperature and the hydrogen is released, indicating that the stable ionic crystal of LiH is

destabilized by reacting with polar molecule ammonia (NH_3) to form lithium amide ($LiNH_2$). On the analogy of this reaction, a hydrogen storage system using a reaction between $C^{nano}H_x$ and LiH has been designed, where the C−H groups in $C^{nano}H_x$ are polarized (similar to NH_3) as is clear from the existence of the IR active modes corresponding to the C−H bonding (Ogita et al., 2004). In fact, it was demonstrated that the hydrogen absorption and desorption properties of $C^{nano}H_x$ is improved by synthesizing the nano-composites with LiH (Ichikawa & Fujii et al., 2005, Ichikawa & Isobe et al., 2005). The Li−C−H system would be categorized as a new hydrogen storage system because the hydrogen absorption and desorption reactions would be induced by a characteristic interaction between the polarized C−H groups in nano−structural graphite and LiH, which is not reported so far. Therefore, the hydrogen absorption and desorption reactions are quite different from the conventional hydrogen storage systems, indicating that an understanding of the hydrogen ab/desorption properties of the systems could lead to design a new hydrogen storage system with further high performance.

In this chapter, the details of hydrogen absorption and desorption mechanism on the $C^{nano}H_x$−LiH composite is discussed on the basis of the experimental results obtained by various kinds of analyses. In addition, the similar systems composed of $C^{nano}H_x$ and other alkali (−earth) metal hydrides are also introduced.

2. Experimental details

2.1 Sample preparation

The hydrogenated nano−structural graphite ($C^{nano}H_x$ or $C^{nano}D_x$) was synthesized from high purity graphite powder by the vibrating ball−mill apparatuses. Graphite of 300 mg and 20 ZrO_2 balls with 8 mm in diameter were put into a milling vessel made of Cr steel. Then, the ball−milling was performed under 1.0 MPa H_2 or D_2 pressure at room temperature for 80 h. In order to synthesize the composites, $C^{nano}H_x$ (or $C^{nano}D_x$) and *M*H (*M*=Li, Na, Mg, and Ca) were mechanically milled for 2 h under a 1.0 MPa H_2 atmosphere at room temperature by using a rotating ball−mill apparatus. The molar ratio of $C^{nano}H_x$ and LiH was chosen to be 2:1 as a suitable composition from previous studies (Ichikawa et al., 2005). The dehydrogenating and rehydrogenating treatments for the composites were performed at 350 °C for 8 h under high vacuum condition and 3.0 MPa H_2 pressure in a reactor made of steel, respectively. The details of sample preparation are shown in the previous articles (Ichikawa et al., 2005, Miyaoka et al., 2007, Miyaoka et al., 2009, Miyaoka et al., 2008).

2.2 Experimental techniques

Thermal gas desorption properties of the composites were examined by a thermal desorption mass spectroscopy (TDMS) connected to thermogravimetry (TG) and differential scanning calorimetry (DSC). Phase identification for the composites at each stage was carried out by powder X−ray diffraction (XRD) measurement (CuKα radiation). The neutron diffraction measurements for the deutrated composites ($C^{nano}D_x$−LiD) were carried out using high intensity total scattering spectrometer (HIT−II) at the high energy accelerator research organization (KEK, Tsukuba, Japan). The remaining gases in the reactor used to rehydrogenate the composites by the heat treatment were identified by gas chromatography (GC). The details of above analyses are described in the articles reported before (Fukunaga et al., 2004, Itoh et al., 2003, Miyaoka et al., 2007, Miyaoka et al., 2009, Miyaoka et al., 2008).

3. Results and discussion

3.1 Hydrogen absorption and desorption mechanism of the $C^{nano}H_x$LiH composite

Fig. 1 shows the TDMS and TG profiles of the composites after synthesizing and after each cycle of hydrogen absorption/desorption up to 5 times, where mass number 2, 16, and 28 are assigned to H_2, CH_4, and C_2H_6, respectively (Ichikawa et al., 2005, Miyaoka et al., 2009). In the TDMS profile of the as–synthesized composite shown in Fig. 1, the intensities of CH_4 and C_2H_6 are enlarged to ten times. From the result, it is found that this composite desorbs the hydrogen around 350°C with small amount of hydrocarbon desorption. The hydrogen desorption peak temperature is lower than the decomposition temperatures of $C^{nano}H_x$ and LiH themselves. The TG profile of the as–synthesized composite (dash and dot line in Fig. 1) reveals a weight loss of about 10.0 mass % during heating up to 500°C.

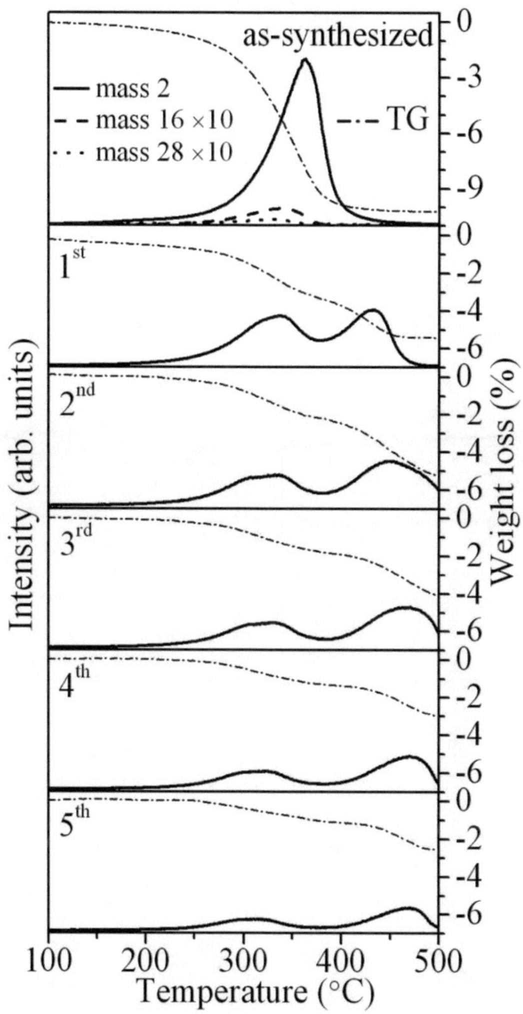

Fig. 1. TDMS and TG profiles (dash and dot line) of the $C^{nano}H_x$-LiH composites after synthesizing and rehydrogenation of 5 times.

This weight loss is mostly due to hydrogen gas, but includes a small amount of hydrocarbon desorption. After the dehydrogenation and rehydrogenation cycles, the hydrogen desorption profile of the composites reveal a two–peak structure, where these peaks correspond to hydrogen desorption are located around 350 and 450°C. These hydrogen desorption temperatures are still lower than the decomposition temperature of each component, suggesting that the hydrogen desorption of cycled composites could take place by the interaction between the polarized C–H groups and LiH. It should be noticed that only hydrogen is desorbed from the rehydrogenated composites without any hydrocarbons emission, suggesting that hydrogen absorption state of the hydrogen recharged composite is slightly different from that of the as–synthesized composite. The hydrogen desorption peak around 450°C in the TDMS profile may be caused by the slow reaction kinetics because this peak almost disappeared in the TDMS after the 2nd dehydrogenation at 350°C. The 1st and 2nd rehydrogenated composites can desorb 5.0 mass % of hydrogen as shown in their TG profiles. However, the hydrogen desorption amount gradually decreases with the hydrogen ab/desorption cycles after the 3rd rehydrogenation although the shape of the hydrogen desorption profiles are almost unchanged.

The structural change with the hydrogenation and dehydrogenation cycles was examined by XRD measurements. The XRD profiles of the composite after synthesis and after each dehydrogenation and rehydrogenation treatment are shown in Fig. 2 (Miyaoka et al., 2009). As a reference, the XRD profiles of LiH (PDF #65–2897) and Li_2C_2 (PDF #21–0484) in the database are shown.

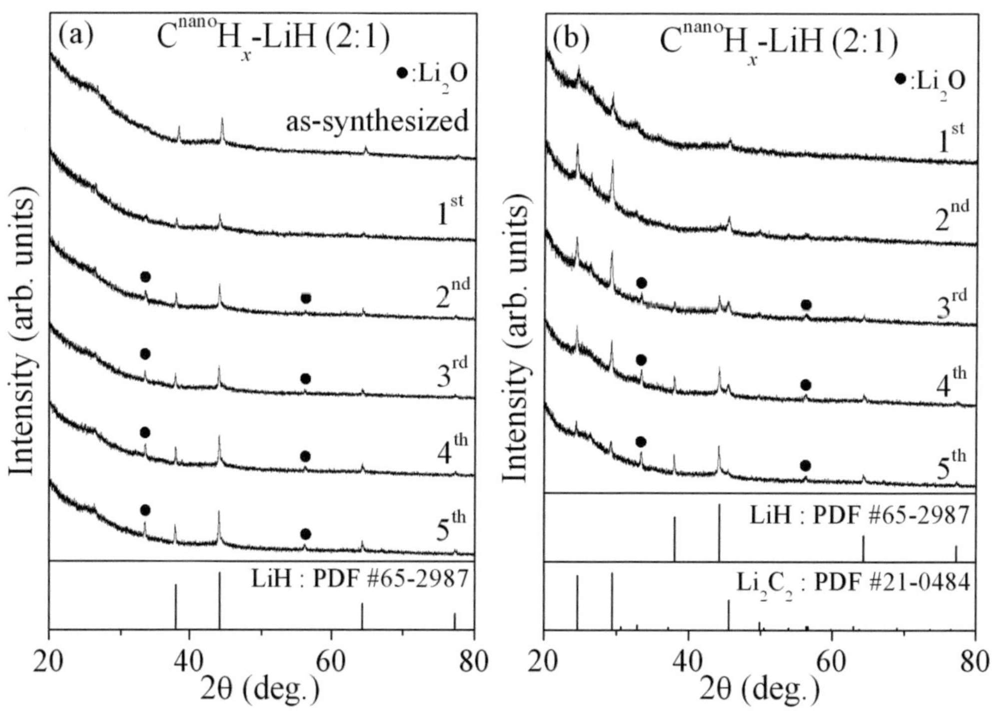

Fig. 2. XRD patterns of (a) the hydrogenated and (b) the dehydrogenated $C^{nano}H_x$-LiH composites at room temperature after each dehydrogenating and rehydrogenating treatments.

In the XRD pattern of the as–synthesized composite in Fig. 2 (a), diffraction peaks corresponding to a LiH phase are observed, and no other peaks related to carbon appear due to the nano–structural feature. After the 1st dehydrogenation treated at 350°C for 8 h, LiH phase disappears. On the other hand, new diffraction peaks assigned to Li_2C_2 appear as shown in Fig. 2 (b) although the peaks are quite small. By the 1st rehydrogenation at 350°C for 8 h under H_2 pressure of 3 MPa, the Li_2C_2 phase disappears, and then, the LiH phase is recovered as shown in Fig. 2 (a). With the re/dehydrogenation treatments after the 3rd cycle, it is clearly found that the peaks assigned to LiH remain in the dehydrogenating sample. The diffraction peaks corresponding to Li_2C_2 are grown by the 2nd dehydrogenation treatment, however they gradually decrease from the 3rd cycle. These results indicate that the formation of Li_2C_2 due to the interaction between the polarized C–H groups in $C^{nano}H_x$ and LiH decreases from the 3rd cycle. The phenomenon is quite consistent with the decrease in the hydrogen desorption amount after the 3rd cycle obtained by TG.

The local chemical bonding states of the as-synthesized, 1st dehydrogenated, and 1st rehydrogenated composites were investigated by the neutron diffraction to understand the variation of the nano-structural graphite. The structure factors (S(*Q*)) of the $C^{nano}D_x$–LiD composite at each state are shown in Fig. 3, where the S(*Q*) of host graphite and LiD measured as reference are also included in the figure (Miyaoka et al., 2008).

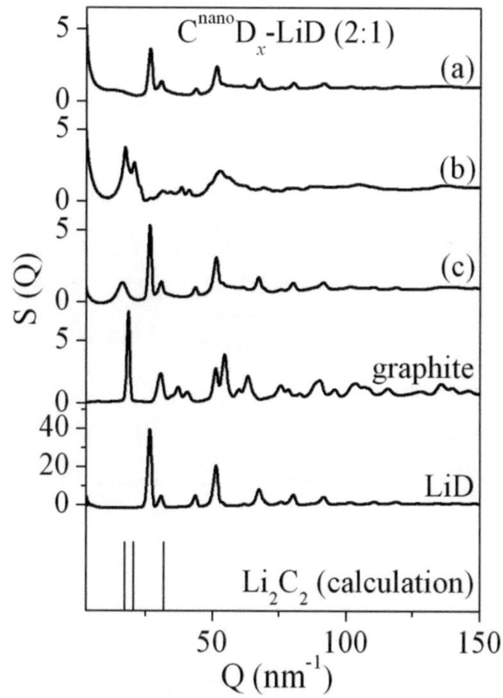

Fig. 3. Structure factors (S(*Q*)) of (a) the as-synthesized, (b) the 1st dehydrogenated, (c) the 1st rehydrogenated $C^{nano}D_x$-LiD composite. The lower spectra are S(*Q*) of host graphite, LiD and main peaks of Li_2C_2 calculated form database (PDF #21-0484) as reference.

As shown in Fig. 3 (a), no peaks corresponding to graphite are observed in the spectrum of the as–synthesized composite although some observed peaks are assigned to LiD by comparing with the reference. Additionally, a large scattering intensity is observed in the small angle scattering region below 5.0 nm^{-1} in S(*Q*). This result indicates that the graphite in the composite possesses the disordered states such as nano-structure or amorphous structure. Fig. 3 (b) shows S(*Q*) of the dehydrogenated composite. Moreover, main peaks of Li_2C_2 calculated from powder X-ray diffraction file of L_2C_2 (PDF #21-0484) in database are also shown in Fig. 3, where intensity of each peak is assumed to be unity. The peaks corresponding to LiD completely disappear after the dehydrogenation. On the other hand, new peaks are observed, in which these are not consistent with the peaks corresponding to graphite shown as a reference. Among them, main peaks observed in S(*Q*) are consistent with the peaks calculated from database of Li_2C_2. Therefore, it is considered that the peaks observed after the dehydrogenation would be assigned to Li_2C_2. As shown in Fig. 3 (c), it is observed that the peaks corresponding to LiD are recovered after rehydrogenation. Moreover, broad peak appears around 1.6 nm^{-1} in S(*Q*), which would be assigned to the graphite (002), where it is noticed that the position of this peak is slightly shifted to lower *Q* compared with that of graphite shown as a reference. It is expected that this peak shift is induced by expansion of the graphite layers due to grain size of nano-meter order. In addition to this result, the intensity in the small angle scattering region decreases compared with those of the as-synthesized and dehydrogenated composite. From these results, it is suggested that a small part of nano-structural graphite would be ordered by rehydrogenating treatment.

The Radial distribution function RDF(*r*) obtained from S(*Q*) of the composites at each stage are shown in Fig. 4 (Miyaoka et al., 2008).

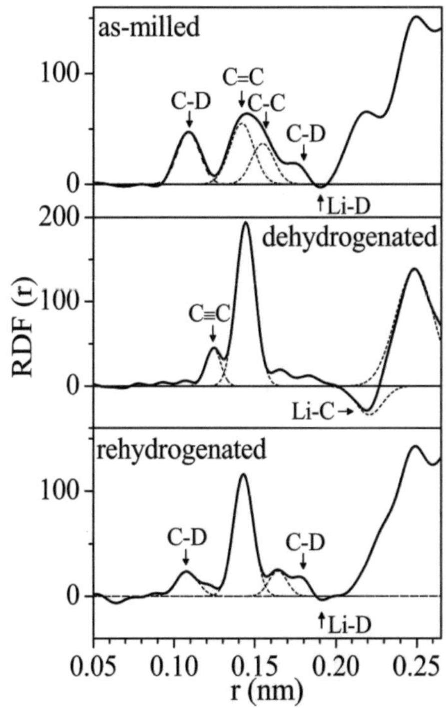

Fig. 4. Radial distribution function RDF(*r*) of the as-synthesized, dehydrogenated, and rehydrogenated $C^{nano}D_x$-LiD composites. Dashed lines represent the fits using Gaussian distribution functions.

For the peaks observed within 0.25 nm, the correlations between atoms are estimated by fitting used Gaussian distribution functions. Furthermore, some C−H groups, which are −CH, $-CH_2$ and $-CH_3$, are shown as models of hydrogen chemisorbed states in Fig. 5, where these models are built by geometric expectations and the previous studies (Miyaoka et al., 2008). In this RDF(r) of the as-milled composite, two kinds of peaks related to C−C correlation are observed at about 0.14 and 0.15 nm. The peak around 0.14 nm would originate in the bonding of sp^2−hybridized carbon atoms (C=C) in hexatomic ring of graphite structure. With respect to another peak corresponding to the C−C correlation, the bond length of 0.15 nm is close to that of sp^3−hybridized carbon atoms (C−C) in crystalline diamond. Thus, it is expected that the 4−fold bonding is generated due to the destruction of graphite structure by using ball−milling as shown in Fig. 5 (d) and (e). It is also possible that some kinds of chained C−H groups with the C−C sp^3−bonding may exist at the edges in the nano−structural graphite. Furthermore, the peaks corresponding to two kinds of the C−D correlations are observed at about 0.11 and 0.18 nm, respectively. The C−D correlation observed around 0.11 nm is ascribed to the covalent bonds, (Fukunaga et al., 2004) suggesting that hydrogen atoms are chemisorbed at dangling bonds such as the edges or as the defects in graphene sheets produced by ball−milling as models shown in Fig. 5. It is considered that the C−D peak around 0.18 nm is assigned to the correlation between hydrogen atoms of the C−D groups at the graphene edges and second neighbor carbon atoms (C^{2nd}−D). For these models of the C−H group, the C^{2nd}−H bond length is estimated by a geometrical calculation as follows. The C^{2nd}−H bond length in case of the covalent bond at zigzag edges in ab−plane as model (a) is estimated to be about 0.22 nm, which is longer than 0.18 nm. On the other hand, it is possible that the C^{2nd}−H bond length of other models is estimated to be around 0.18 nm if the atoms form a suitable C−C−H bond angle. The bond length of 0.18 nm might be realized in a situation when the hydrogen atom in the model (a), (b), and (c) exists at out of ab−plane, where the C−C−H bond angle should be around 91 °. In addition to above C−C and C−D correlations, it is noticed that the negative correlation is revealed at about 0.19 nm. In fact, this peak is not clear because the negative peak overlaps with the positive correlation around 0.18 nm, so that it is quite difficult to perform the fitting by Gaussian distribution functions for both peaks. Here, it is clarified from RDF(r) obtained by neutron diffraction for LiD itself that the Li−D correlation is revealed as negative peak because a coherent scattering length of Li atom is only negative value among those of the atoms included in this composite. Furthermore, the result of the XRD measurements and S(Q) of the $C^{nano}D_x$−LiD composites demonstrated the existence of LiD. From these experimental facts, the negative correlation is assigned to the Li−D bond.

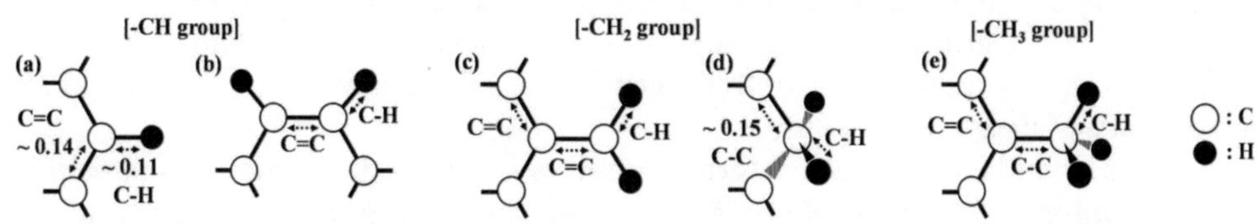

Fig. 5. Models of hydrogen chemisorbed state as -CH, $-CH_2$ and $-CH_3$ groups in the $C^{nano}H_x$-LiH composite; (a) -CH group at zigzag edges and (b) arm-chair edges, (c) $-CH_2$ group with sp^2-hybridyzed carbon (C=C) and (d) sp^3-hybridyzed carbon (C-C), (e) $-CH_3$ group, where shown unit of length is nm.

In the RDF(r) of the dehydrogenated $C^{nano}D_x$–LiD composite, three kinds of characteristic peaks are revealed at about 0.12, 0.14 and 0.22 nm. The peak around 0.14 nm is assigned to the C=C correlation of sp^2–bonding in nano–structural graphite, which is sharper than that of as–synthesized composite. The positive peak around 0.12 nm and the negative peak around 0.22 nm are ascribed to the C–C and C–Li correlation in Li_2C_2, respectively. The crystal structure of Li_2C_2 have been investigated and reported before. (Juza et al., 1967, Ruschewitz & Pottgen, 1999) Among those studies, Ruschewitz *et al.* reported on the crystal structure Li_2C_2 obtained by X–ray diffraction measurements. (Ruschewitz & Pottgen, 1999) In that paper, it is determined that the bond length of the C–C correlation in Li_2C_2 is about 0.123 nm. This C–C bond would be the typical C–C triple bond (C≡C) of sp–hybridization, where this bond length is close to that of 0.120 nm in acetylene. Thus, the C–C correlation around 0.12 nm is assigned to the C–C triple bond in Li_2C_2. The obvious negative peak around 0.22 nm should be ascribed in a correlation related to Li atom. By comparing with the RDF(r) of the as–synthesized composite, this bond length is longer than that of 0.19 nm in case of the Li–D correlation. It is confirmed from XRD pattern and S(Q) that the LiD phase disappeared after dehydrogenation. From these experimental facts, it can be explained that the origin of the peak around 0.22 nm is caused by the C–Li correlation in Li_2C_2. In the case of the rehydrogenated composite, the peaks assigned to Li_2C_2 disappear and the Li–D correlation assigned to LiD phase is recovered at about 0.19 nm, where this phenomenon has been clarified by the XRD measurement already. The peak ascribed to the C=C sp^2–bonding is also observed around 0.14 nm. On the other hand, the peak at 0.15 nm of as–synthesized composite is not revealed in the RDF(r) profile of the rehydrogenated composite. Considering these result, it is expected that the amount of sp^3–hybridyzed carbon atoms decrease due to the structural order by the hydrogen desorption and absorption treatments, where this result is consistent with the behavior in the small angle scattering region in S(Q). It is noticed that the peaks corresponding to the C–D correlation are obviously recovered around 0.11 and 0.18 nm. Therefore, it is considered that the only hydrogen desorption from the rehydrogenated composite is realized by the interaction between the polarized C–H bonding at the dangling bonds of nano–structural graphite and LiH. The intensity of the peak around 0.11 nm is slightly weak, in comparison to that of the as–synthesized composite, indicating that the amount of chemisorbed hydrogen at the dangling bonds is smaller than that in the as–synthesized composite. In addition, new peak is observed at about 0.16 nm, which is longer than that of the C–C bonds of ~0.15 nm or shorter than that of C^{2nd}–H bond of ~0.18 nm. Thus, it seems likely that the origin of this peak might be a new C–D correlation, which is different from the C^{2nd}–D correlation revealed around 0.18 nm, as a one of the possibilities. From the above results of the neutron diffraction measurements, it is clarified that not only Li but also nano–structural carbon (C^{nano}) in this composite can be recharged with hydrogen by the rehydrogenating treatment at 350 °C under 3 MPa of hydrogen pressure.

In order to understand the hydrogen absorption process in further detail, the gases remaining in the reactor, are examined by GC after each rehydrogenation. The gas chromatograms for the composites are shown in Fig. 6 (a), and the results of GC for H_2 and CH_4 gases are appended in Fig. 6 (b) as references (Miyaoka et al., 2009). All the chromatograms of the composites show two peaks at different retention times. The respective peaks are assigned to H_2 and CH_4 because the positions of peaks are quite consistent with those of the reference. The results

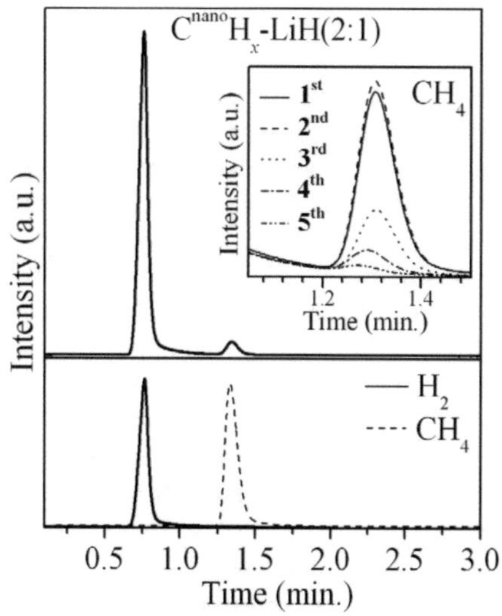

Fig. 6. (a) The chromatograms of the gases remaining in the reactor after each rehydrogenation of the $C^{nano}H_x$-LiH composite by heat treatment at 350°C for 8 h under 3 MPa of hydrogen, and the second peaks observed after each rehydrogenation are shown in inset, in which the peak intensity is normalized by the intensity of first peak. (b) The chromatogram of H_2 and CH_4 itself as reference.

indicate that the hydrocarbons, mainly CH_4, are desorbed with decomposition of Li_2C_2 by recharging the composite with the hydrogen, where a quite small amount of ethane C_2H_6 is also observed in the GC. As shown in the inset of Fig. 6 (a), the normalized intensity of the peak ascribed to CH_4 is gradually weakened by the 3rd rehydrogenation, in which the normalization of the peaks is carried out by using the intensity of H_2.

Fig. 7 shows the amount of carbon lost as hydrocarbon during each rehydrogenating treatment as a function of the number of hydrogen ab/desorption cycles, assuming that the desorbed gas is only CH_4 (Miyaoka et al., 2009).

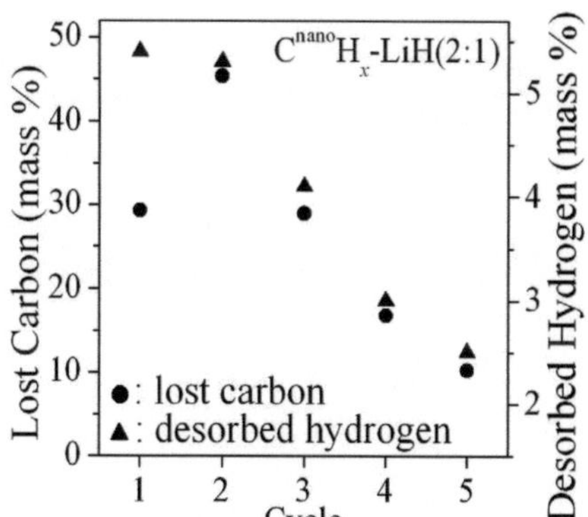

Fig. 7. Amounts of the carbon lost during each rehydrogenation of the $C^{nano}H_x$-LiH composite and the hydrogen desorbed from the each rehydrogenated composite as a function of the hydrogen ab/desorption cycle, which are estimated by the results from GC and TG, respectively.

The amount of carbon in CH_4 and H_2 are estimated respectively by the GC and TG. At the 1st cycle, about 29.0 mass % of the carbon atoms is desorbed as CH_4 and 5.0 mass % of H_2 is desorbed from the composite. The amount of lost carbon is reached up to 45.4 mass % at the 2nd hydrogen recharge, and gradually decreases with the hydrogenating treatment from the 3rd cycle. The decrease in the hydrogen desorption amount is also started from the 3rd cycle. The variation of the amount of lost carbon and desorbed hydrogen by the hydrogenation and dehydrogenation cycles is quite consistent with each other. Therefore, the results of GC indicate that the hydrogen absorption and desorption properties of the composite are strongly related to the hydrocarbon desorption during the hydrogen recharging process.
From the above experimental facts and the following four assumptions,

1. the quite small amount of hydrocarbons desorption from the as-synthesized composite can be ignored (Fig. 1),
2. the hydrogenated state of the ball-milled graphite in the composite after rehydrogenation is defined as the $C^{nano}H_y$ because the rehydrogenated state of the composite is different from that of $C^{nano}H_x$ in the as-synthesized composite , and the $H/C = y$ is not changed with the cycled treatments,
3. the half of the carbons in the Li_2C_2 is desorbed as CH_4 gas with recharging hydrogen, and
4. H_2 or CH_4 desorbed by each treatment is removed from the system,

it is expected that the hydrogen absorption and desorption reaction in the Li-C-H system with the cycles of hydrogenation and dehydrogenation treatments is expressed as follows (Miyaoka et al., 2009):

Initial desorption: $$2C^{nano}H_x + LiH \rightarrow C^{nano} + \frac{1}{2}Li_2C_2 + \left(\frac{4x+2}{4}\right)H_2 \uparrow \quad (1)$$

1st cycle: (abs.) $$C^{nano} + \frac{1}{2}Li_2C_2 + \left(\frac{3y+6}{4}\right)H_2 \rightarrow \frac{3}{2}C^{nano}H_y + \frac{1}{2}CH_4 \uparrow + LiH \quad (2)$$

(des.) $$\frac{3}{2}C^{nano}H_y + LiH \rightarrow \frac{1}{2}C^{nano} + \frac{1}{2}Li_2C_2 + \left(\frac{3y+2}{4}\right)H_2 \uparrow \quad (3)$$

2nd cycle: (abs.) $$\frac{1}{2}C^{nano} + \frac{1}{2}Li_2C_2 + \left(\frac{2y+6}{4}\right)H_2 \rightarrow C^{nano}H_y + \frac{1}{2}CH_4 \uparrow + LiH \quad (4)$$

(des.) $$C^{nano}H_y + LiH \rightarrow \frac{1}{2}Li_2C_2 + \left(\frac{2y+2}{4}\right)H_2 \uparrow \quad (5)$$

3rd cycle: (abs.) $$\frac{1}{2}Li_2C_2 + \left(\frac{y+6}{4}\right)H_2 \rightarrow \frac{1}{2}\left(C^{nano}H_y + CH_4 \uparrow\right) + LiH \quad (6)$$

(des.) $$\frac{1}{2}C^{nano}H_y + LiH \rightarrow \frac{1}{4}Li_2C_2 + \left(\frac{y+1}{4}\right)H_2 \uparrow + \frac{1}{2}LiH \quad (7)$$

4th cycle: (abs.) $$\frac{1}{4}Li_2C_2 + \frac{1}{2}LiH + \left(\frac{y+6}{8}\right)H_2 \rightarrow \frac{1}{4}\left(C^{nano}H_y + CH_4 \uparrow\right) + LiH \quad (8)$$

(des.) $$\frac{1}{4}C^{nano}H_y + LiH \rightarrow \frac{1}{8}Li_2C_2 + \left(\frac{y+1}{8}\right)H_2 \uparrow + \frac{3}{4}LiH \quad (9)$$

5th cycle: (abs.) $$\frac{1}{8}Li_2C_2 + \frac{3}{4}LiH + \left(\frac{y+6}{16}\right)H_2 \rightarrow \frac{1}{8}\left(C^{nano}H_y + CH_4 \uparrow\right) + LiH \quad (10)$$

(des.) $$\frac{1}{8}C^{nano}H_y + LiH \rightarrow \frac{1}{16}Li_2C_2 + \left(\frac{y+1}{16}\right)H_2 \uparrow + \frac{7}{8}LiH. \quad (11)$$

In this reaction model, the non−reacted LiH remains from the 3rd cycle. The hydrogen ab/desorption reactions of the Li−C−H system from the 3rd cycles can be described as following equations ($n = 1, 2, 3\ldots$),

n^{th} cycle: (abs.) $$\frac{1}{2^n}Li_2C_2 + \left(\frac{2^{n-1}-1}{2^{n-1}}\right)LiH + \left(\frac{y+6}{2^{n+1}}\right)H_2 \rightarrow \frac{1}{2^n}\left(C^{nano}H_y + CH_4 \uparrow\right) + LiH \quad (12)$$

(des.) $$\frac{1}{2^n}C^{nano}H_y + LiH \rightarrow \frac{1}{2^{n+1}}Li_2C_2 + \left(\frac{y+1}{2^{n+1}}\right)H_2 \uparrow + \left(\frac{2^n-1}{2^n}\right)LiH. \quad (13)$$

Therefore, it is considered that the model of hydrogen absorption/desorption reactions of the Li−C−H system expressed by the above equations would describe the experimental results obtained in this work.

3.2 Hydrogen desorption properties of the $C^{nano}H_xMH$(M=Na, Mg, and Ca) composite

Fig. 8 shows the results obtained by (a) TDMS, (b) TG-DSC, and (c) XRD measurements of the $C^{nano}H_x$−NaH composite, where the intensity of hydrocarbons is enlarged ten times (Miyaoka et al., 2007). As shown in inset, the TDMS profile indicates that the milled NaH needs 320 °C to release hydrogen gas. On the other hand, the composite exhibits a broad sub−peak at slightly lower temperature than the temperature corresponding to a main peak for hydrogen desorption, where these peaks are located around 260 and 300 °C, respectively. It should be noted that their hydrogen desorption temperatures are lower than those of constituent components. Accompanied by the lower hydrogen desorption corresponding to 260 °C, it is noticed that quite small amount of hydrocarbons are desorbed from the composite as well. However, the hydrocarbons desorption is immediately suppressed with further increasing temperature after showing a broad peak around 260 °C. Fig. 8 (b) shows the TG and DSC profiles for the $C^{nano}H_x$−NaH composite. The DSC spectrum indicates that both the reactions at the sub and the main peaks in the TDMS spectrum at 260 and 300 °C are endothermic, suggesting that the reactions are rechargeable. Additionally, the weight loss of the composite was estimated to be ~5 mass% as shown in the TG profile, though the weight loss due to the decomposition of lone NaH in the composite is theoretically estimated to ~2 mass%. This suggests that hydrogen in $C^{nano}H_x$ is also synchronously desorbed at the same temperatures as the hydrogen desorption. Fig. 8 (c) shows the XRD pattern for the $C^{nano}H_x$ and NaH composite before and after the thermal gas−desorption examination. Mainly, the peaks corresponding to the NaH phase are observed because the $C^{nano}H_x$ phase itself does not show any diffraction peaks (Isobe et al., 2004). The XRD pattern of the composite after increasing temperature up to 350 °C indicates only the appearance of metallic Na phase. This fact shows that the hydrogen desorption is at least caused by the decomposition of NaH. Moreover, since hydrogen in the $C^{nano}H_x$ phase is also destabilized by making composite with NaH, a composite of Na and C^{nano} is supposed to be formed by dehydrogenation. However, the metallic Na phase is clearly observed, indicating that not only the nano-composite phase between Na and C^{nano} but also the disproportionating phases are formed after dehydrogenation. This behavior seems to be originated in a quite low melting point of Na (below 100 °C).

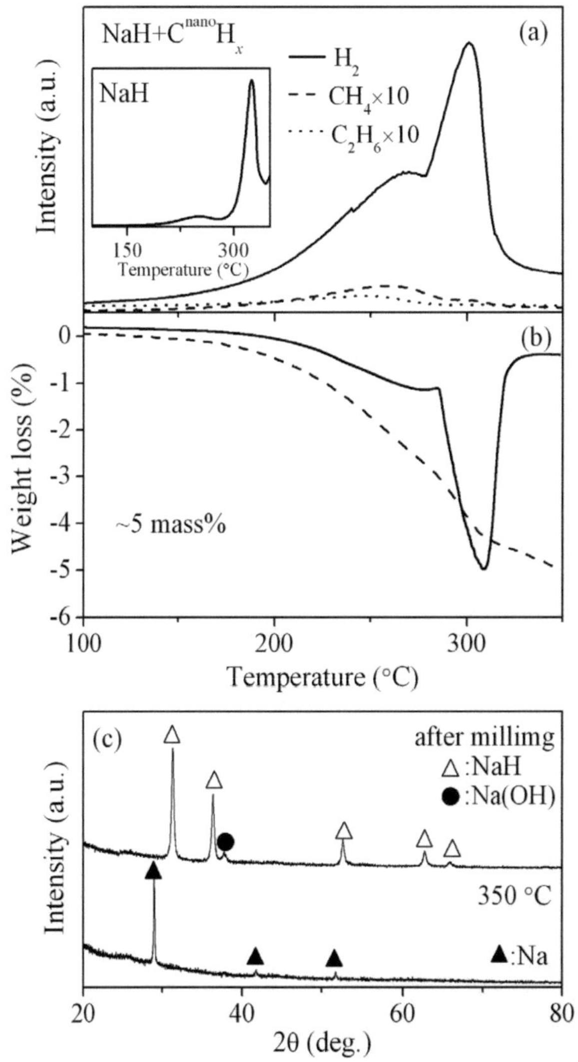

Fig. 8. (a) TDMS profiles of hydrogen and hydrocarbons from the $C^{nano}H_x$–NaH composite, and the inset shows TDMS profile of the milled NaH. (b) TG (dashed line) and DSC (solid line) spectra of the composite. (c) The XRD patterns after milling and after heating up to 350 °C of the composite.

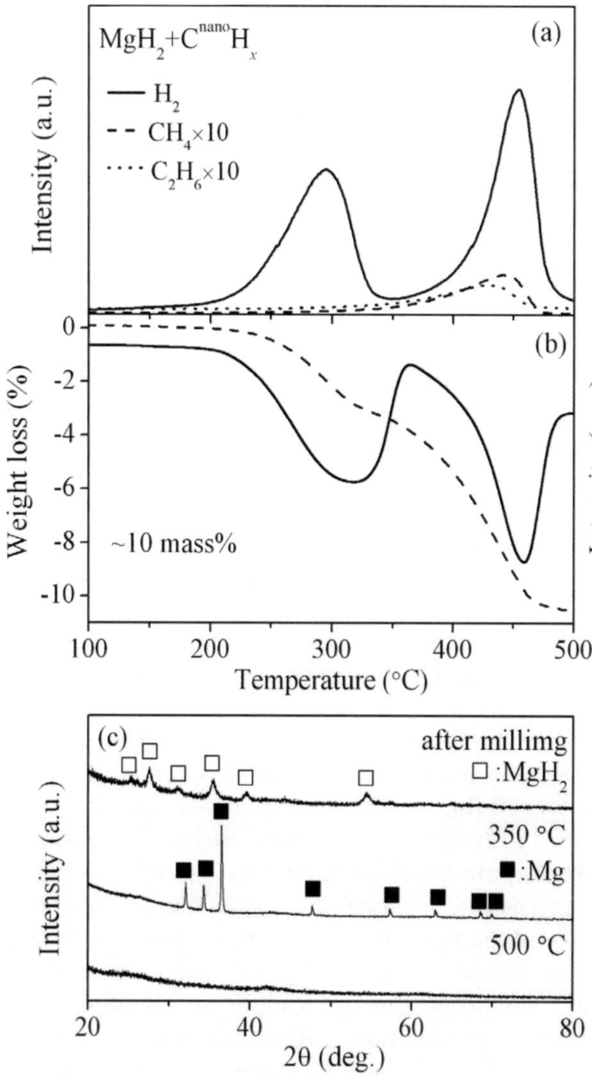

Fig. 9. (a) TDMS profiles of hydrogen and hydrocarbons from the $C^{nano}H_x$–MgH_2 composite, where the intensity of hydrocarbons are enlarged ten times. (b) TG (dashed line) and DSC (solid line) spectra of the composite. (c) The XRD patterns after milling, after heating up to 350 °C and 500 °C of the composite.

The results of thermal and structural analyses for the $C^{nano}H_x$–MgH_2 composite are shown Fig. 9 (Miyaoka et al., 2007). This composite desorbs hydrogen with two peaks. The first and second hydrogen desorption peaks were located around 300 and 450 °C, respectively. The first hydrogen desorption temperature was lower than ~350 °C, which seems to correspond to the typical decomposition temperature of pure MgH_2 milled for 2 h (Hanada et al., 2005, Hanada et al., 2005). Actually, the metallic Mg phase appears in the XRD pattern after desorbing hydrogen at 350 °C as shown in Fig. 9 (c). Moreover, the TG profile of the composite as shown in Fig. 9 (b) reveals ~3 mass% weight loss until 350°C, where the weight loss is consistent with the calculated value of hydrogen desorption due to decomposition of pure MgH_2 in the composite, indicating that only the decomposition of MgH_2 occurs until 350 °C. Therefore, the above results suggest that the MgH_2 solid phase does not such strongly interact with the $C^{nano}H_x$ product. On the other hand, when the temperature was increased up to 500 °C corresponding to a higher temperature than a second hydrogen desorption peak one, no evident peak appears in the XRD pattern, indicating that a nano−structural cluster composed of a metallic Mg and nano−structural carbon is generated at 500 °C after desorbing hydrogen by an interaction between Mg and $C^{nano}H_x$. Actually, the amount of the corresponding hydrocarbons is quite small compared with that from $C^{nano}H_x$ itself. This result indicates that hydrogen is desorbed instead of hydrocarbon due to destabilization of hydrocarbons group in the $C^{nano}H_x$ product by an interaction with the Mg solid phase. The total amounts of weight loss reach up to ~10 mass% with increasing temperature up to 500 °C in the TG profile. The composite is also expected to be rehydrogenated because two endothermic peaks clearly appear at 300 and 450 °C in the DSC spectrum, respectively.

Fig. 10 shows (a) TDMS, (b) TG-DSC, and (c) XRD results of the $C^{nano}H_x$−CaH_2 composite, where the intensity of hydrocarbons are enlarged ten times (Miyaoka et al., 2007). As is shown in Fig. 10 (a), the hydrogen desorption spectrum for the $C^{nano}H_x$ and CaH_2 composite shows a two−peak structure in the temperature range from 300 to 500 °C, where the CaH_2 hydride itself could not be decomposed by heating up to 500 °C. Furthermore, the suppression of hydrocarbons is found in the TDMS profile by making composite of $C^{nano}H_x$ with CaH_2, indicating that the C−H bonding in the hydrocarbon groups in the $C^{nano}H_x$ product is weakened by an interaction between CaH_2 and $C^{nano}H_x$. Then, the composite exhibits a weight loss of ~8 mass% in the TG profile by heating up to 500 °C as shown in Fig. 10 (b), where the weight loss caused by decomposition of lone CaH_2 is theoretically estimated to ~2 mass% in the composite. This suggests that hydrogen in the $C^{nano}H_x$ product is synchronously desorbed at the same temperature as CaH_2 desorbs hydrogen as well. Thermal analysis due to the DSC measurement indicates that the hydrogen desorption reaction of the composite is weakly endothermic as shown in Fig. 10 (b). This result suggests that the rehydrogenation reaction is expected to occur in the $C^{nano}H_x$ and CaH_2 composite similar to the other metal−C−H materials. Fig. 10 (c) shows the XRD patterns on the composite. After increasing temperature up to 500 °C, the intensity corresponding to CaH_2 in the XRD pattern is significantly suppressed compared with that before dehydrogenation. Furthermore, the XRD profile of the composite exhibits no existence of Ca or CaC_2 after dehydrogenation at 500 °C. Therefore, it is suggested that a nano−structural cluster composed of C^{nano} and Ca is formed in the composite after the dehydrogenation.

4. Conclusion

The hydrogen desorption of $C^{nano}H_x$ is lowed to 350 °C by synthesizing the composite with LiH, indicating that the C-H bonds in $C^{nano}H_x$ is destabilized by the interaction with LiH. Such destabilization of the C-H bonds is also clarified in the case of the composite of $C^{nano}H_x$ and other alkali(-earth) metal hydride. The $C^{nano}H_x$−LiH composite rechargeably store 5.0 mass % of hydrogen until the 2nd cycle. The hydrogenated state of this composite is the C-H bonds at the graphene edges and defects in the nano-structural graphite and LiH, and its

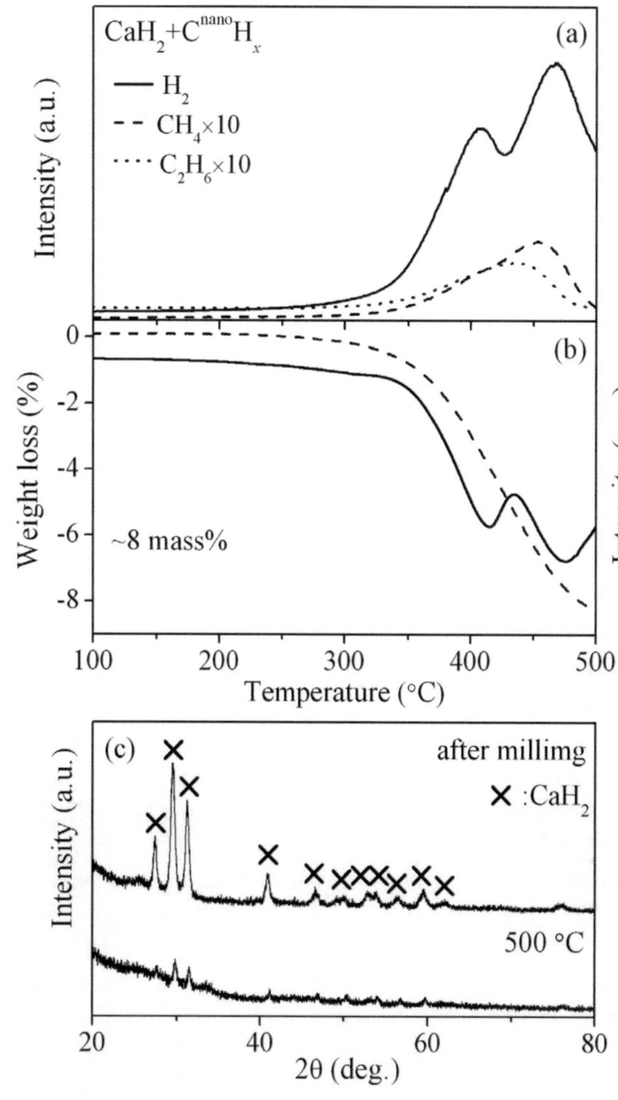

Fig. 10. (a) TDMS profiles of hydrogen and hydrocarbons from the $C^{nano}H_x$–CaH_2 composite. (b) TG (dashed line) and DSC (solid line) spectra of the composite. (c) The XRD patterns after milling, after heating up to 500 °C of the composite.

dehydrogenated state is Li_2C_2 and nano-structural graphite without the C-H bonds. The hydrogen desorption amount gradually decreases from the 3rd cycle. In the hydrogen absorption process, considerable amount of hydrocarbon was generated, indicating that the carbon atoms in the composite are lost with each hydrogen recharging process. Therefore, it is considered that the decrease in the rechargeable hydrogen amount is caused by desorbing hydrocarbons with the decomposition of Li_2C_2 under rehydrogenation process, in other words, the amount of the hydrogen desorption reaction between the polarized groups composed of the C–H bonds and LiH are gradually reduced due to decrease in the C–H groups by the loss of the carbon atoms with the rehydrogenation.

The Li−C−H system would be recognized as one of attractive systems for hydrogen storage. However, the hydrocarbon desorption during hydrogen recharging process is essentially a problem on this system, because it reduces the hydrogen absorption amount. For practical use as hydrogen storage system, an improvement, e.g. usage of a catalyst or scaffold, is needed to prevent the hydrocarbon desorption during the hydrogenation.

5. References

Akiba, E. (1999). Hydrogen-absorbing alloys. *Curr. Opin. Solid State Mater. Sci.*, 4, 3, 267-272, 1359-0286.

Chen, D. M.; Ichikawa, T.; Fujii, H.; Ogita, N.; Udagawa, M.; Kitano, Y. & Tanabe, E. (2003). Unusual hydrogen absorption properties in graphite mechanically milled under various hydrogen pressures up to 6 MPa. *J. Alloys Compd.*, 354, 1-2, L5-L9, 0925-8388.

Dillon, A. C.; Jones, K. M.; Bekkedahl, T. A.; Kiang, C. H.; Bethune, D. S. & Heben, M. J. (1997). Storage of hydrogen in single-walled carbon nanotubes. *Nature*, 386, 6623, 377-379, 0028-0836.

Fukunaga, T.; Itoh, K.; Orimo, S. & Aoki, K. (2004). Structural observation of nano-structured and amorphous hydrogen storage materials by neutron diffraction. *Mat. Sci. Eng. B-Solid State Mat. Adv. Technol.*, 108, 1-2, 105-113, 0921-5107.

Grochala, W. & Edwards, P. P. (2004). Thermal decomposition of the non-interstitial hydrides for the storage and production of hydrogen. *Chem. Rev.*, 104, 3, 1283-1315, 0009-2665.

Hanada, N.; Ichikawa, I. & Fujii, H. (2005). Catalytic effect of Ni nano-particle and Nb oxide on H-desorption properties in MgH_2 prepared by ball milling. *J. Alloys Compd.*, 404-406, 716-719, 0925-8388.

Hanada, N.; Ichikawa, T. & Fujii, H. (2005). Catalytic effect of nanoparticle 3d-transition metals on hydrogen storage properties in magnesium hydride MgH_2 prepared by mechanical milling. *J. Phys. Chem. B*, 109, 15, 7188-7194, 1520-6106.

Ichikawa, T.; Chen, D. M.; Isobe, S.; Gomibuchi, E. & Fujii, H. (2004). Hydrogen storage properties on mechanically milled graphite. *Mat. Sci. Eng. B-Solid State Mat. Adv. Technol.*, 108, 1-2, 138-142, 0921-5107.

Ichikawa, T.; Fujii, H.; Isobe, S. & Nabeta, K. (2005). Rechargeable hydrogen storage in nanostructured mixtures of hydrogenated carbon and lithium hydride. *Appl. Phys. Lett.*, 86, 24, 241914,

Ichikawa, T.; Hanada, N.; Isobe, S.; Leng, H. Y. & Fujii, H. (2004). Mechanism of novel reaction from $LiNH_2$ and LiH to Li_2NH and H_2 as a promising hydrogen storage system. *J. Phys. Chem. B*, 108, 23, 7887-7892, 1520-6106.

Ichikawa, T.; Isobe, S. & Fujii, H. (2005). Hydrogen desorption properties of lithium-carbon-hydrogen system. *Mater. Trans.*, 46, 8, 1757-1759, 1345-9678.

Isobe, S.; Ichikawa, T.; Gottwald, J. I.; Gomibuchi, E. & Fujii, H. (2004). Catalytic effect of 3d transition metals on hydrogen storage properties in mechanically milled graphite. *J. Phys. Chem. Solids*, 65, 2-3, 535-539, 0022-3697.

Itoh, K.; Miyahara, Y.; Orimo, S.; Fujii, H.; Kamiyama, T. & Fukunaga, T. (2003). The local structure of hydrogen storage nanocrystalline graphite by neutron scattering. *J. Alloys Compd.*, 356, 608-611, 0925-8388.

Juza, R.; Wehle, V. & Schuster, H. U. (1967). Zur Kenntnis Des Lithiumacetylids. *Z. Anorg. Allg. Chem.*, 352, 5-6, 252, 0044-2313.

Majer, G.; Stanik, E. & Orimo, S. (2003). NMR studies of hydrogen motion in nanostructured hydrogen-graphite systems. *J. Alloys Compd.*, 356, 617-621, 0925-8388.

Miyaoka, H.; Ichikawa, T. & Fujii, H. (2007). Thermodynamic and structural properties of ball-milled mixtures composed of nano-structural graphite and alkali(-earth) metal hydride. *J. Alloys Compd.*, 432, 1-2, 303-307, 0925-8388.

Miyaoka, H.; Ichikawa, T.; Fujii, T.; Ishida, W.; Isobe, S.; Fujii, H. & Kojima, Y. (2010). Anomalous hydrogen absorption on non-stoichiometric iron-carbon compound. *J. Alloys Compd.*, 507, 2, 547-550, 0925-8388.

Miyaoka, H.; Ichikawa, T. & Kojima, Y. (2009). The reaction process of hydrogen absorption and desorption on the nanocomposite of hydrogenated graphite and lithium hydride. *Nanotechnology*, 20, 20, 204021, 0957-4484.

Miyaoka, H.; Itoh, K.; Fukunaga, T.; Ichikawa, T.; Kojima, Y. & Fuji, H. (2008). Characterization of hydrogen absorption/desorption states on lithium-carbon-hydrogen system by neutron diffraction. *J. Appl. Phys.*, 104, 5, 053511-053517,

Ogita, N.; Yamamoto, K.; Hayashi, C.; Matsushima, T.; Orimo, S.; Ichikawa, T.; Fujii, H. & Udagawa, M. (2004). Raman scattering and infrared absorption investigation of hydrogen configuration state in mechanically milled graphite under H_2 gas atmosphere. *J. Phys. Soc. Jpn.*, 73, 3, 553-555, 0031-9015.

Orimo, S.; Majer, G.; Fukunaga, T.; Züttel, A.; Schlapbach, L. & Fujii, H. (1999). Hydrogen in the mechanically prepared nanostructured graphite. *Appl. Phys. Lett.*, 75, 20, 3093-3095, 0003-6951.

Orimo, S.; Matsushima, T.; Fujii, H.; Fukunaga, T. & Majer, G. (2001). Hydrogen desorption property of mechanically prepared nanostructured graphite. *J. Appl. Phys.*, 90, 3, 1545-1549, 0021-8979.

Orimo, S. I.; Nakamori, Y.; Eliseo, J. R.; Zuttel, A. & Jensen, C. M. (2007). Complex hydrides for hydrogen storage. *Chem. Rev.*, 107, 10, 4111-4132, 0009-2665.

Ruschewitz, U. & Pottgen, R. (1999). Structural phase transition in Li_2C_2. *Z. Anorg. Allg. Chem.*, 625, 10, 1599-1603, 0044-2313.

Sandrock, G. (1999). A panoramic overview of hydrogen storage alloys from a gas reaction point of view. *J. Alloys Compd.*, 295, 877-888, 0925-8388.

Schüth, F.; Bogdanović, B. & Felderhoff, M. (2004). Light metal hydrides and complex hydrides for hydrogen storage. *Chem. Commun.*, , 20, 2243-2248,

Schlapbach, L. & Züttel, A. (2001). Hydrogen-storage materials for mobile applications. *Nature*, 414, 6861, 353-358, 0028-0836.

Smith, C. I.; Miyaoka, H.; Ichikawa, T.; Jones, M. O.; Harmer, J.; Ishida, W.; Edwards, P. P.; Kojima, Y. & Fuji, H. (2009). Electron Spin Resonance Investigation of Hydrogen Absorption in Ball-Milled Graphite. *J. Phys. Chem. C*, 113, 14, 5409-5416, 1932-7447.

Stanik, E.; Majer, G.; Orimo, S.; Ichikawa, T. & Fujii, H. (2005). Nuclear-magnetic-resonance measurements of the hydrogen dynamics in nanocrystalline graphite. *J. Appl. Phys.*, 98, 4, 044302, 0021-8979.

Züttel, A. (2004). Hydrogen storage methods. *Naturwissenschaften*, 91, 4, 157-172, 0028-1042.

Züttel, A. (2007). Hydrogen storage and distribution systems. *Mitig. Adapt. Strat. Glob. Change*, 12, 3, 343-365

6

Nanocomposite Films Deposition by means of Various Filtered Vacuum Arc Systems

Seunghun Lee[1], Do-Geun Kim[1], Igor Svadkovski[2] and Jong-Kuk Kim[1]
[1]Korea Institute of Materials Science,
[2]Belarussian State University of Informatics and Radioelectronics,
[1]Korea
[2]Belarus

1. Introduction

A hard coating is of considerable significance in industrial applications such as high speed cutting tools. The hard coating technologies have been investigated to improve a life time as well as performance of the tools. Up to now, TiN coatings were widely used. However, the substitution of TiN has been required because of its low oxidation temperature of 500 °C. From 1970s, it was revealed that nanocomposite films have their infinite probability enough to replace the previous materials such as TiN. Nanocomposite materials such as nanocrystalline TiN/Si_3N_4, $TiN/TiC/Si_3N_4$, and $TiN/AlN/Si_3N_4$ exhibited superior oxidation temperature (~1000 °C) as well as ultra hardness (45-55 GPa) [1-6].
Many deposition tools like a magnetron sputtering, plasma enhanced chemical vapor deposition (PECVD), arc ion plating (AIP), and filtered vacuum arc (FVA) were introduced for synthesizing nanocomposite films. Table 1 summarized previous works of nanocomposite coatings.
Nanocomposite films based on TiN were dominantly investigated with the incorporation of silicon or carbon. The incorporation methods such as an alloy arc cathode, addition of reactive gas, and additional magnetron sputtering were used to deposit ternary or quaternary composed nanocomposite films. The magnetron sputtering and PECVD have been firstly used to the growth of nanocomposite films due to the simplicity of controlling a composition ratio. Precise composition control of additional components is important. For example, Ti-Si-N nanocomposite films represent the maximum hardness at Si content of 9±1 at.%. After that, a vacuum arc discharge has been applied to the nanocomposite coatings because it can generate dense plasma with energetic ions near a cathode spot (<60 eV), which help a delicate crystallization and rapid growth. Nevertheless, the vacuum arc method cannot avoid the problem of macro particles emitted at an arc spot. The macro particles make some defects in the coatings and result in drastically decreasing corrosion resistance when the coatings are exposed in the corrosive environment. Therefore magnetic filters have been introduced to transport plasma except the macro particles. The filters only transport charged particles using electromagnetic fields and the neutral macro particles collide with a filter wall by an inertia drift. As a result, various FVA methods have been widely applied to the nanocomposite deposition.

Method	Material	Hardness	Ref.
Arc	Ti-Si-N	45 GPa	[7][9]
Arc	Ti-Al-N/Cr-N	37 GPa	[8]
Arc	Ti-Al-Si-N	34 GPa, 42.4 GPa	[10][11]
Arc	Ti-Al-N	35.5 GPa	[2]
Arc, magnetron sputter	TiN-Cu, CrN-Cu, MoN-Cu	27-42 GPa	[12]
Arc, magnetron sputter	Ti-Si-N	45-55 GPa	[13][15]
FVA, magnetron sputter	Ti-Si-N	45 GPa	[1]
FVA, magnetron sputter, E-beam evaporation	Ti-Cr-N,Ti-B-C	43.2 GPa	[14][17]
FVA	Ti-Si-N	40.1 GPa,	[16]
PECVD	Ti-Si-N	3500 HK(kg/mm^2), 40 GPa	[18][19]
PECVD	Ti-Si-C-N	48 GPa	[4]
PECVD	Ti-Si-C-N	52 GPa	[20]
Magnetron sputter	Ti-Si-N	38GPa, 45 GPa	[21][22]

*Arc=Cathodic arc evaporation, Arc ion plating

**FVA=Filtered vacuum arc

Table 1. Nanocomposite coatings by mean of various methods

2. Various FVA applications for nanocomposite coatings

2.1 Theory of FVA

Main issues of FVA are the efficient removal of the macro particles from plasma, and the minimum loss of ions through a filter wall. The effective approach to reduce the macro particles is based on the spatial separation of the trajectories of macro particles and ions [23]. If the magnetic field is curved such as the field inside a curved solenoid, electrons follow the curvature. Such electrons are said to be magnetized. In contrast, ions are usually not magnetized because the gyration radius of ion is much larger than that of electron and the characteristic radius of the filter. Nevertheless, ions are forced to follow the magnetic field lines due to the electric fields between electrons and ions. Therefore, plasma containing ions and electrons is transported along magnetic field lines [24].

It was shown that the plasma transport efficiency is limited by drifts caused by the centrifugal force and by the electric field in plasma [25]. Several studies have been conducted to determine the maximum efficiency and optimum condition for curved magnetic filters. An analysis of plasma motion along the toroidal magnetic field shows that plasma transported by the magnetic field in the plasma guiding duct should satisfy the following relation [26],

$$B > M_i V_o / Zea \quad (1)$$

where M_i is the ion mass, V_o is the translational velocity, Z is the charge multiplicity of the ion, e is the electron charge, and a is the minor radius of the plasma guide. Transport of heavy metal ions having energy of even a few tens of electronvolts requires magnetic fields of above 1 Tesla to fulfill the inequality in Eq. (1). At these magnetic fields, however, it is practically impossible to provide a stable burning of the direct current arc discharge. Besides the plasma injection into this field would also present certain difficulties. Therefore, it is reasonable to investigate heavy-element plasma flow transport in a curvilinear system with crossed electric and magnetic fields using the principles of plasma optics as a guide [27,28]. In this case, the required magnetic field is determined by the following condition, $\rho_e < a < \rho_i$, where ρ_e and ρ_i are the electron and ion Larmor radius, respectively. The required field is significantly lower than the fields defined by the expression in Eq. (1). Electron Larmor radius is

$$\rho_e = (m_e k T_e)^{1/2} / eB \tag{2}$$

where m_e is the electron mass, k is Boltzmann constant, and T_e is the electron temperature.
Note that electrons are only magnetized, while the ions are not. The electrons move along the magnetic field lines. In view of the high longitudinal conductivity of plasma, the magnetic field lines are equi-potentials. Considering a plasma diffusion in vacuum, electrons have higher mobility than ions due to their smaller mass except for a sheath boundary. However, electrons expand with the same velocity as ions because electrostatic forces keep the electrons and ions together. And a cross field diffusion is given by the Bohm formula, $D_B = kT_e/16B$, though the cross field diffusion coefficient, D, is proportional to B^{-2} in the classical theory [33]
Anders mentioned about the criterion of system efficient, K_s, which is generally considered as the ratio of the total ion flow at the exit of the system, I_i, to the arc discharge current, I_a, as follows,

$$K_s = I_i / I_a \ . \tag{3}$$

The system coefficient is typically 1% [24]. There is general agreement that the transport efficiency is maximized by focusing the plasma into the duct and biasing the duct to a potential of 20 V. Predictions of the available maximum transmission vary between 11 and 25% depending on the ion energy [29]. In practice, the transport of plasma produced by pulsed high current arcs (HCA) was showed that the system coefficient was 7% [24]. For linear FVA [30], the maximum value of the system efficiency reached 8% when arc current I_a was adjusted in the range of 100–110 A and the magnetic filter field was ~ 20 mT.
To supplement the accuracy of system coefficient, a particle system coefficient, K_p, is proposed to eliminate the influence of the various ion charge states by considering the mean ion charge state, Z_{av}, of the used metal.

$$K_p = I_i / Z_{av} I_a \tag{4}$$

Because the average ion charge state is taken into account, the particle system coefficient is more closely related to the deposition rate [31]. However this can be particularly insufficient for filter optimization when the system used a graphite cathode that generates solid rebounding macro particles.
The problem was solved by computing the particles trajectories using a two-dimensional approximation [32]. It was assumed that the macro particles were solid and spheres, the inner surfaces of the plasma guide and the intercepting fins were smooth, the repulsion of particles from the walls was partially elastic, the particles were emitted by the cathode spot with equiprobability in any direction, i.e. of angular macro particle flow density distribution

has the form $N(\alpha)=const$, where a is the angle between the normal to the cathode active surface and the direction of macro particle emission. The computation results make it possible to estimate the ratio of the exit macro particle flow N_{ex} to the flow N_{ent} generated by the cathode spot. The ratio N_{ex}/N_{ent} characterizes the likelihood of an macro particle passing through the system. The results of simulations indicate that the absence of a direct line-of-sight between the cathode and the substrate is not always sufficient to provide the required degree of macro particle removal from the plasma. The results of computations performed for various magnetic filters are presented in Table 2 [33].

Filter type	Knee (45°)	Torus (45°)	Rectang.	Dome	Torus (90°)	Retil.	Radial	Wide apert.
N_{ex}/N_{ent} [%] (predicted)	1.7	25.0	17.0	1.7	0	4.4	0	0
Transport [%] (measured)	3.0	2.5	2.5	2.5	1.5	1.8	8.4	~6.0

Table 2. Fitlering (N_{ex}/N_{ent}) and transporting properties of magnetic plasma filters [33]

2.2 Various types of FVA systems [34]

There are various types of filters as shown in Fig. 1. Most types are used magnetic fields to transport plasma without macro particles. Several types are only used the collisional reduction of macro particles. In most cases, the plasma is transported from the cathode to the substrate, and the droplets are eliminated by the plasma transportation wall. Many review papers of the filtered arc system and technology have been reported [35]-[42]. A typical filtered arc system with its different electromagnetic plasma transportation duct or droplet filter configurations is shown in Fig. 1(a)–(h). Electromagnetic coils transporting plasma in the out of line of sight direction can be positioned in the chamber, instead of placing them outside of the filter duct. The off-plane double bend filter [43], [44] is nicknamed FCVA and is now commercially available. Most FAD units have electromagnetic coils outside of the plasma duct and have baffles inside the duct wall. However, some types have freestanding coils inside the plasma duct or the chamber. Other interesting filters have been developed. Examples are shown in Fig. 1(i)–(l). In the Venetian-blind filter, the plasma passes between the vane lamellae, and the droplets are caught or reflected by the lamellae [45], [46]. A coaxial filter is operated with a large current pulse, and the plasma is driven by a self magnetic field [47]. An electrostatic filter can be used with a pulsed arc having a laser trigger [48]. However, recently only the laser triggered arc is used without the electrostatic filter. Mechanical filters can be used in a pulse arc [49], which may also be used in pulsed laser deposition [50].

2.3 FVA systems for nanocomposite films deposition

Various FVA methods were introduced to deposit ternary or quaternary composed nanocomposite films. A hybrid system consisting of FVA and magnetron sputter was widely used. In the hybrid system, FVA evaporates materials to be required a large amount such as Ti and the incorporated material like Si was used to the magnetron sputtering cathode. By controlling the power of magnetron sputter, the additional content of Si can be

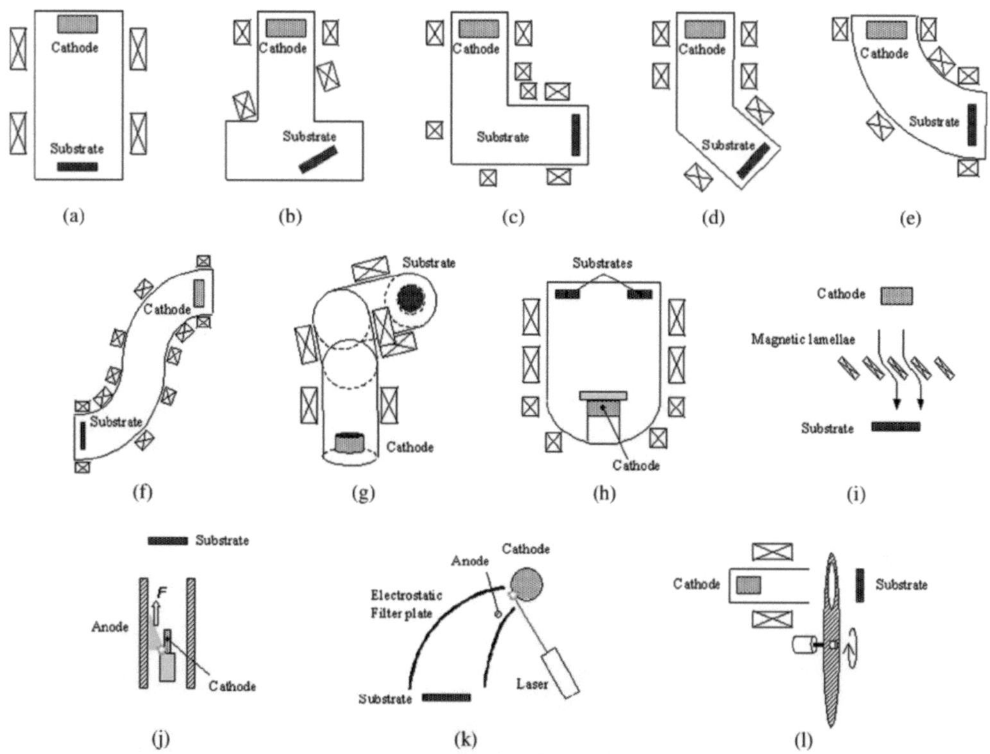

Fig. 1. Various types of filter systems [34]. (a) Rectilinear. (b) Bent. (c) Rectangular. (d) Knee. (e) Torus. (f) S-shape. (g) Off-plane double bend. (h) Dome. (i) Venetian blind. (j) Co-axial (pulse). (k) Electrostatic filter with laser trigger (pulse). (l) Mechanical pulse.

modulated. For example, Fig. 2 shows the schematic of hybrid system for nanocomposite coatings. The magnetron sputtering cathode is placed near the outlet of filter vacuum arc source. The use of two separated targets allowed an independent regulation and enabled the adjustment of Ti/Si ratio by varying the current of magnetron. However, the interference of magnetic field between FVA and magnetron sputter should be considered to obtain sufficient deposition rate. The closed geometry of magnetic field between the two sources will cause the loss of ions and the decrease on deposition rate.

Sintering or alloy arc cathode methods are useful way to deposit nanocomposite films if one knows the optimum composition of additional components and the optimized composition is in the available range for cathode sintering. For instance, Ti-Si sintering cathode can be used for nanocomposite TiN/a-Si_3N_4 coatings because the optimum content of Si for high hardness is in the possible range (~10%) for Ti-Si sintering. And Ti-Al (5:5) alloy cathode, which manufacturing is simple, can be used to nc-TiN/AlN/a-Si_3N_4 coatings since the composition of alloy target is good for realization of superior properties.

Additionally, the injection of additive gas is also another way to deposit nanocomposite films. Especially, metal-organic source such as hexamethyldisiloxane (HMDSO), hexamethyldisilazane (HMDS), and tetramethylsilane (TMS) could be used to Si source. The plasma density and ion energy are enough to dissociate metal-organic materials and continuously supply deposition precursors.

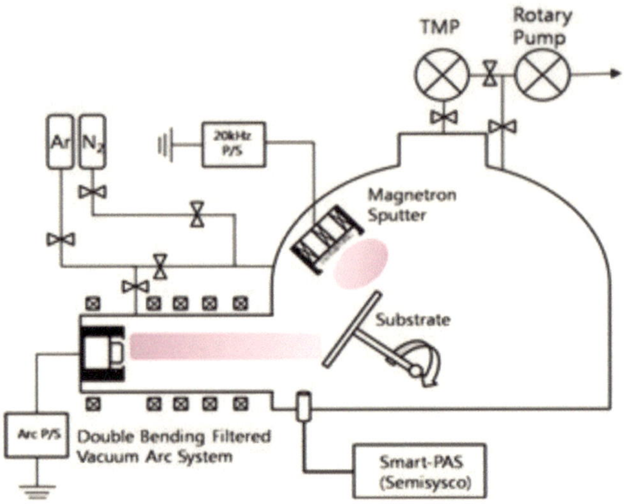

Fig. 2. A schematic of hybrid system.

Fig. 3. S-shape filtered vacuum arc system.

3. Ti-Si-N/Ti-Al-Si-N nanocomposite film prepared by a FVA and unbalanced magnetron sputtering technique

3.1 Ti-Si-N/Ti-Si-Al-N nanocomposite films

Many investigations were carried out Ti-Si-N and Ti-Al-Si-N coatings due to its attractive physical and chemical properties, such as high hardness and good oxidation resistance. The coating layers with Si content of approximately 10 at.% shows very high hardness (more than 40 GPa) and those microstructures were characterized by nanocomposite consisting of nano-sized TiN or AlN crystallites embedded in amorphous silicon nitride matrix [15,51,52]. The nanocomposite coating layer also exhibited significantly improved oxidation resistance compared to that of TiN [53,54]. It was reported that the amorphous SiO_2 oxidized from Si_3N_4 played a major role as the oxidation barrier [55]. And the microstructure characterized by solid-solution of Al into TiN improved oxidation property [56,57]. With a decrease in grain size, the multiplication and mobility of the dislocations are hindered, and the hardness of materials increases. Veprek [58] imbedded 4-11 nm TiN crystals in amorphous Si_3N_4 matrix and obtained a coating hardness of 50-70 GPa. Combination of two phases provides complex boundaries to accommodate coherent strain, which result in the increase of coating hardness [59].

3.2 System configuration for nanocomposite films deposition

Nanocomposite coatings were carried out by a filtered vacuum arc and an unbalanced magnetron (UBM) sputtering technique as shown in Fig.2. The filtered vacuum arc used the S-shape magnetic filter consisting of 5 magnetic coils to eliminate macro particles. The magnetic field inside the plasma duct was optimized using a numerical simulation to prevent the striking of the plasma stream onto the plasma duct. The average strength of the magnetic field was 15 mT and the diameter of the plasma duct was 15 cm. A Ti (99.8%) arc cathode (diameter: 8 cm) was used in Ti-Si-N coatings. A $Ti_{50}Al_{50}$ (99.8%) alloy cathode (diameter: 8 cm) was used in Ti-Al-Si-N coatings. The area of the Si UBM source was 400 cm^2 and placed next to the outlet of the plasma duct. The UBM power was released as a 20 kHz asymmetric pulse. The substrates were high speed steel. The substrate was heated by plasma and the substrate temperature was maintained ~ 200°C for a deposition.
The coating process consisted of 3 steps, an ion bombardment cleaning, an interlayer deposition, and nanocomposite coating. In the ion bombardment cleaning, Ar ions were impinging into the substrate with the substrate bias of -600 V. Ar gas was introduced into the arc cathode while maintaining a pressure of 1 mTorr. In the interlayer deposition, the pressure decreases to 0.3 mTorr to enhance the ion transport into the substrate. After that, multi-components nanocomposite films were coated using a gas mixture of Ar (5 sccm) and N_2 (15 sccm). The dissipated powers of arc and magnetron sputter were 1.8 kW and 0-0.5 kW, respectively. Typical deposition conditions are shown in Table 3.

Base pressure	1.0×10^{-5} Torr	Arc current	60 A
Working pressure	6.0×10^{-4} Torr	Sputter current	0-1.5 A
Gas ratio	N_2:Ar=3:1	Substrate voltage	-20 V
Arc target	Ti or$Ti_{50}Al_{50}$	Rotational speed	25 RPM
Sputter target	Si (99.99%)	Working temperature	< 200 °C

Table 3. Typical deposition conditions for Ti-Si-N nanocomposite coatings

3.3 Characteristics of Ti-Si-N nanocomposite films

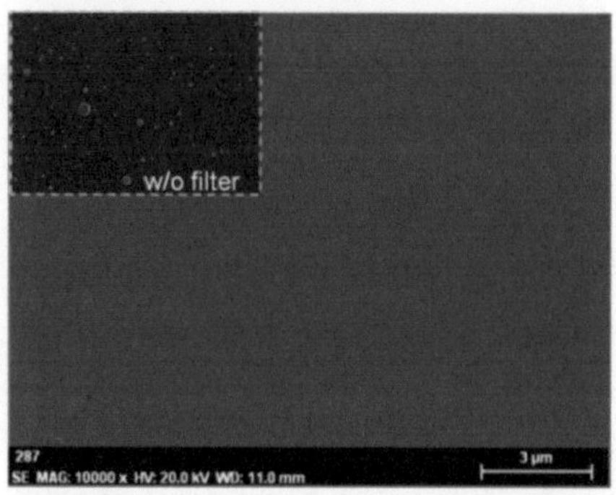

Fig. 4. Image of coated sample by scanning electron microscope without macro particles

Figure 4 shows the surface morphology of the sample prepared using the S-shape magnetic filter which removes macro particles during the deposition and the insert image indicates the surface morphology of the sample deposited by arc ion plating without the magnetic filter.

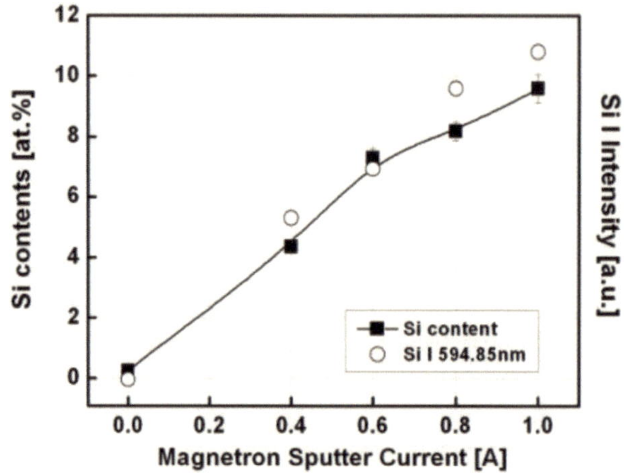

Fig. 5. Si contents as a variation of Si sputtering current and emission intensity of Si neutral (594.85 nm)

Figure 5 represents that Si content of the Ti-Si-N films as a function of magnetron sputtering system current. The composition data have been obtained by electron probe microanalysis. As magnetron sputter current increases from 0 to 1 A, Si contents increase up to level of 9.5%. This represents that Si contents can be easily controlled by the UBM power. Optical emission spectra of the Si neutrals (wavelength: 594.85 nm, energy: 7.17 eV→5.08 eV) from discharge also are proportional to the sputter current.

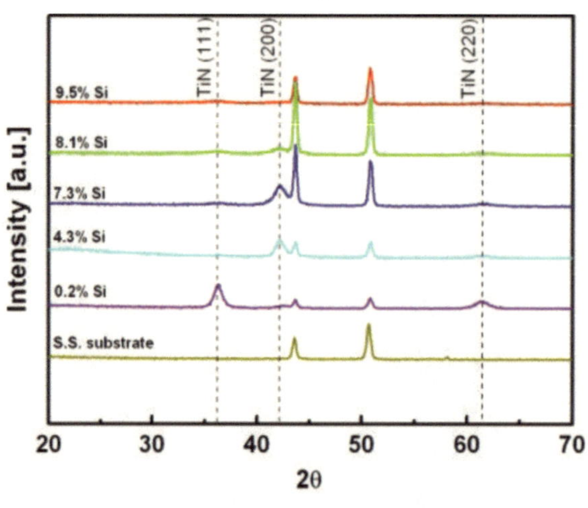

Fig. 6. X-ray diffraction patterns as a function of Si contents

Figure 6 shows the XRD patterns of Ti-Si-N coatings with different Si contents. The XRD patterns of Ti-Si-N coatings showed orientations of (111), (200), and (220) of TiN. These multiple orientations were in agreement with other Ti-Si-N coating layers [13,14]. However, any peaks corresponded to crystalline Si_3N_4 or titanium silicide phase were not observed. It was estimated that Si was present in an amorphous phase of silicon or silicon nitride. The TiN peak intensity gradually reduced and a broadening phenomenon of peaks was observed with increasing Si content in Ti-Si-N films. XRD analysis represents that crystalline TiN (111), (220) is changed to crystalline (200) as Si content increases. Such XRD peaks behavior was believed due to generally decreasing of TiN grain size with increase of Si content. The result indicates that Ti-Si-N coatings have been characterized as nanocomposites, consists of TiN crystallites and amorphous Si_3N_4. If Si contents increases further, crystalline TiN becomes weaker and Si rich amorphous structure becomes dominant structure.

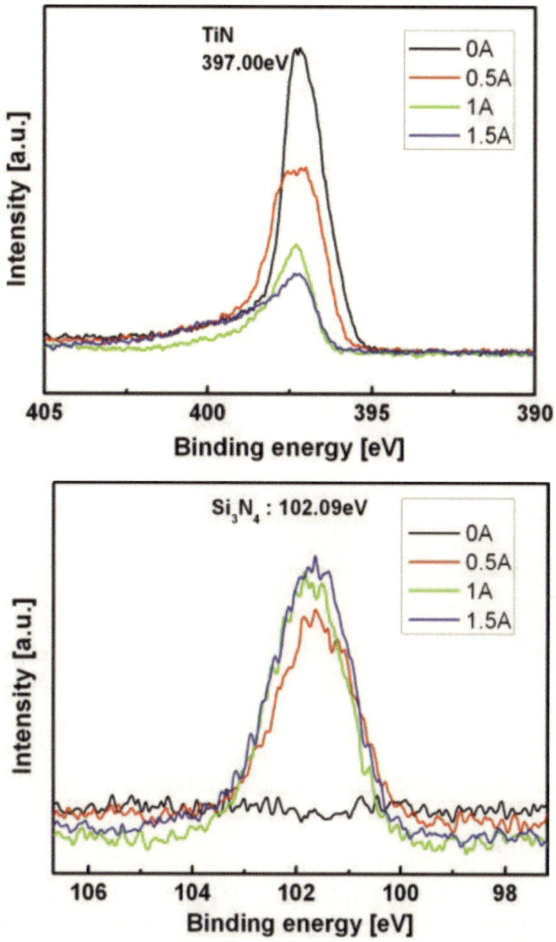

Fig. 7. X-ray photoelectron spectroscopy of Ti-Si-N coatings (1) Si 2p (2) N 1s

Figure 7 shows high-resolution XPS spectra of Si 2p (a) and N 1s (b) for Ti-Si-N coatings with different sputter currents. The binding energy of Si 2p at 101.8 eV corresponded of the data for stoichiometric Si_3N_4 [60, 61]. It should be note that, in spite of peak of 101.8 eV increased with increasing Si content in Ti-Si-N coating, there is not another peak corresponding to free Si at the binding energy of 99.3 eV. XPS analysis also shows that increase of Si content causes the weaker crystalline TiN bond status. This result is similar to XRD peak variation due to Si content variation. Precision control and determination of grain size is very important for nanocomposite coating.

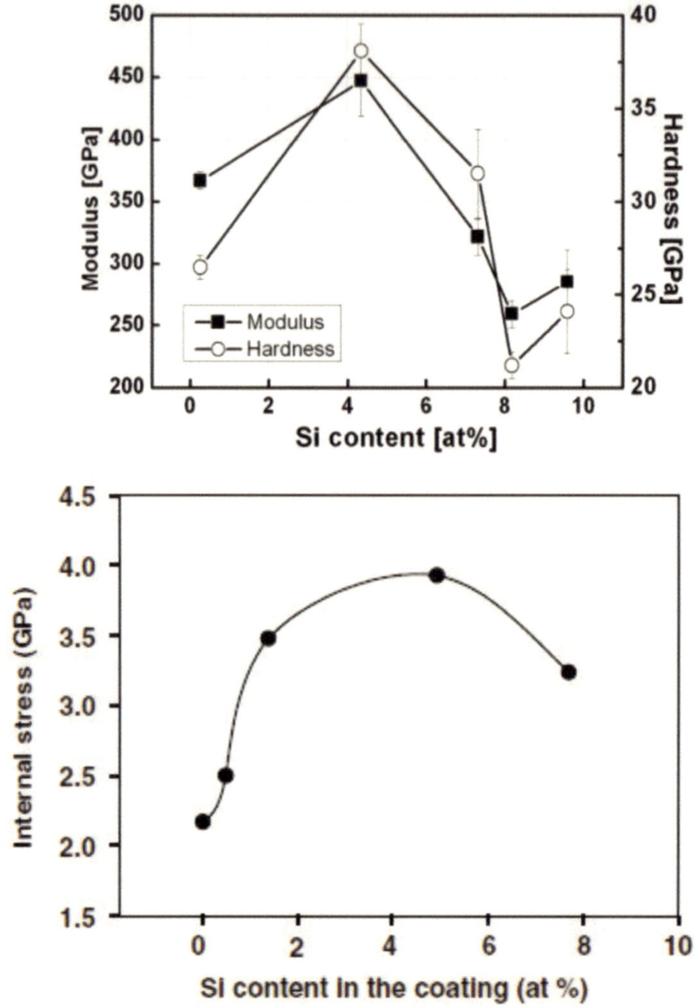

Fig. 8. (a) Micro hardness and modulus (substrate temperature: 200 °C, bias voltage: -75 V) [1] (b) internal stress (substrate temperature: 350 °C, bias voltage: -40 V) [62] as a function of the Si content in the coatings

Figure 8 shows micro hardness and modulus of Ti-Si-N coatings from Ref. [1] and internal stress from Ref [62] as a function of the Si contents. The increase in the Si content in the Ti-Si-N coatings from 0 at% to 4.3 at% results in a drastic increase in the micro hardness from 25 to 38 GPa. Then the micro hardness decreases to 23 GPa with a further increase in the Si content to 9.3 at%. The internal stress also has the maximum value of 4 GPa at Si content of ~ 5 at%. The Si incorporation causes the embedded amorphous Si_3N_4 phase surrounding nano-crystalline TiN phase. This induces the increase in micro hardness as well as the internal stress by the nanocomposite structure. Further Si addition, however, causes the decreases in the hardness. It has been proposed that the internal stress in the coating is strongly correlated with the defects, which cause the deformation and distortion of the lattice [63]

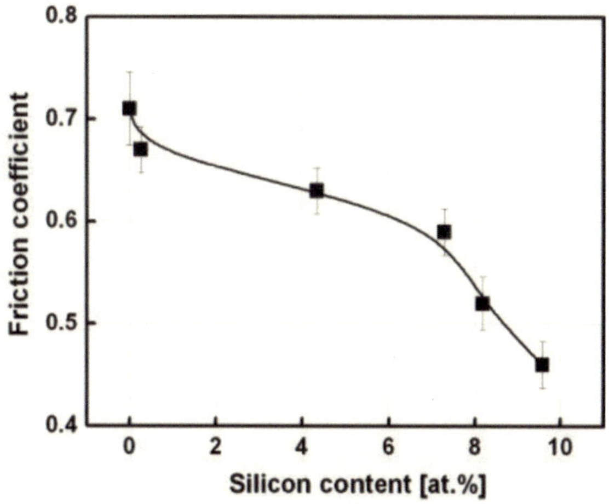

Fig. 9. Friction coefficient of Ti-Si-N films with various Si contents

In Fig. 9, the friction coefficient of Ti-Si-N films has lower value when Si incorporation is increased. The friction coefficient was decreased by the self-lubricating layer effect induced by SiO_2 or $Si(OH)_2$. Note that the self-lubricating effect is dominantly affected by ambient humidity the addition of Si into the coatings led to a continuous decrease in the grain size, which increased the defect density in the coatings.

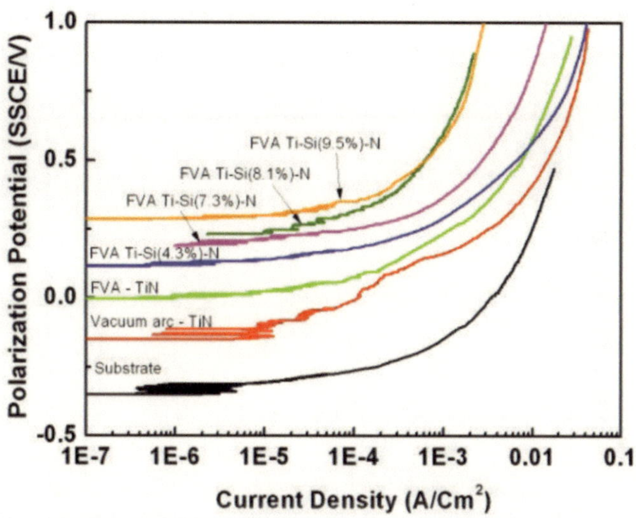

Fig. 10. Polarization voltage (Corrosion resistance) of Ti-Si-N films with various Si contens

Figure 10 shows the corrosion behavior of Ti-Si-N nanocomposite films. The coating samples of vacuum arc shows the worst polarization voltage because macro particles exist on the surface. The crack between the particle and the coating induces worse corrosion resistance. In case of FVA, the polarization voltage is increased due to the lack of macro particles. As Si incorporation was increased, the decreased grain size hinders the corrosion progress and the polarization voltage was increased.

3.4 Characteristics of Ti-Al-Si-N nanocomposite films

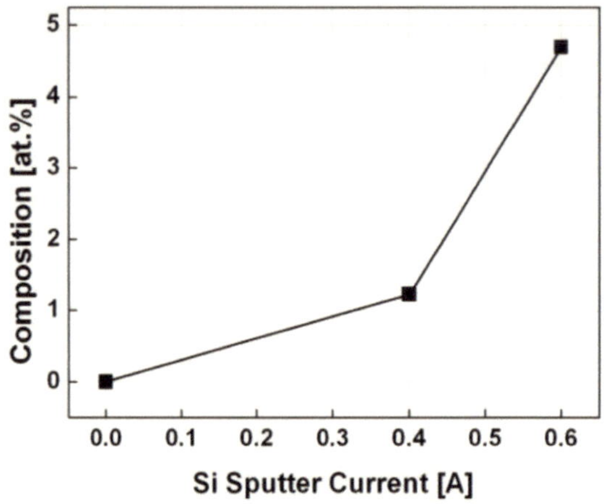

Fig. 11. Si contents as a function of Si UBM current.

Figure 11 shows the EPMA results when the Si contents were 0.04 at%, 1.22 at%, and 4.69 at% as the Si sputtering currents were 0 A, 0.4 A, and 0.6 A, respectively. The compositional analysis of the coating is carried out from electron probe microanalysis (EPMA, EPMA1600, Shimadzu). This means that the controlling of the Si content is available by the modulation of the Si sputter current. In the hybrid coating system consisting of the vacuum arc and the magnetron sputter, the ease of controlling the Si content in the multi components coating has been previously reported in other papers [1,64]. The non-linear increase on the Si contents could be induced by the discharge enhancement near the substrate due to the UBM discharge.

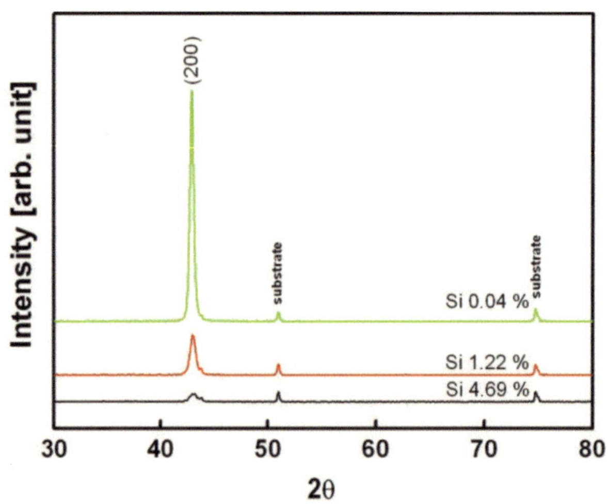

Fig. 12. X-ray diffraction patterns of Ti-Al-Si-N films

Figure 12 represents XRD patterns with various atomic percentages of Si. X-ray diffraction (XRD) with CuK_{α} radiation (X'Pert PRO macro particleD, Philips) was used to confirm the crystal phase. A crystal phase corresponding to a cubic TiN with a NaCl-type B1 structure was observed. There were no signals from silicon nitride or titanium silicide. The crystal plane of

(200) was observed continuously and the peak of the (200) plane becomes broader and smaller as the Si contents were increased from 0.04 to 4.69 at %. The broadening of the diffraction patterns implies the decrease of the grain size. The grain size evaluated by Scherrer's equation decreased from 9 nm to 7.5 nm as the Si content increased. The reduced grain size means the formation of amorphous Si_3N_4 between the crystalline TiAlN structures and this phenomenon is similar to the effect of the Si incorporation on the Ti-Si-N film [1]. In comparison with the Ti-Al-Si-N films prepared by a normal vacuum arc system with a heated substrate, there are no diffraction patterns of (111), (220), (311) and (222) planes in our results [65]. The low energy of incident ions and radicals by the self-bias and non-heated substrate induced the low mobility of the precursors and prevented the crystallization of TiN.

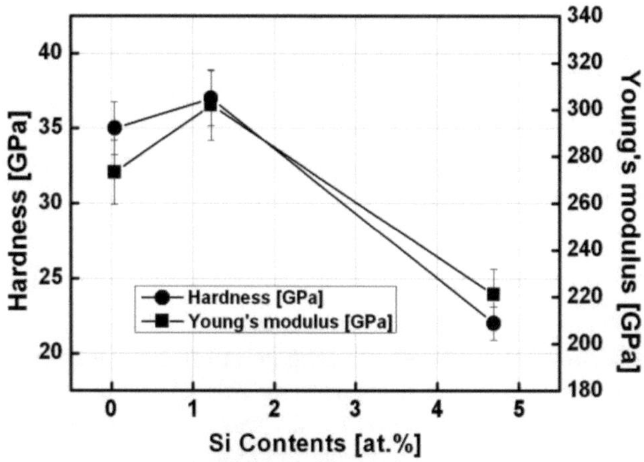

Fig. 13. Hardness and Young's modulus of Ti-Al-Si-N films

The micro hardness and Young's modulus of Ti-Al-Si-N coatings are shown in Fig. 13. A nanoindentation tester (MTS, Nanoindentation XP) equipped with a Berkovich diamond indentor was used to measure the micro hardness, averaged from 25 points. The displacement rate of the indentor was 0.2 nm/s until a depth of 100 nm was reached. As the Si contents increased until 4.69 at.%, the micro hardness and Young's modulus increased from 35 GPa and 270 GPa to 37 GPa and 300 GPa, respectively. However they became lowered to 22 GPa and 220 GPa, respectively, when the Si contents increased to 4.69 at.%. In comparison with the micro hardness of Ti-Al-N films, at ~30 GPa, the hardness of the films enhanced slightly with the incorporation of Si [66]. The nanocomposite structure consisting of the inter phase boundary between the crystalline Ti-Al-N and the amorphous Si_3N_4 phases should help account for the enhancement of hardness. The micro hardness of Ti-Al-Si-N [65] deposited by the normal vacuum arc system is higher than that of Ti-Al-Si-N prepared by our system. As mentioned previously when discussing the XRD patterns, the low mobility at the surface due to the low temperature of the substrate could not be effective to synthesize the crystallized nc-TiAlN.

Figure14 shows the H^3/E^{*2} values of Ti-Al-Si-N films as a function of the Si contents. The effective Young's modulus, $E^*=E/(1-\nu^2)$, and H^3/E^{*2} value were calculated from the results of Fig.13 with the assumption of Poisson ratio, v=0.25. The H^3/E^{*2} value is known to be proportional to the resistance to plastic deformation of hard materials [67]. The highest H^3/E^{*2} value was obtained at the Si content of 4 at.%. The highest resistance to plastic deformation of that film could be related to preventing dislocation formation or movement

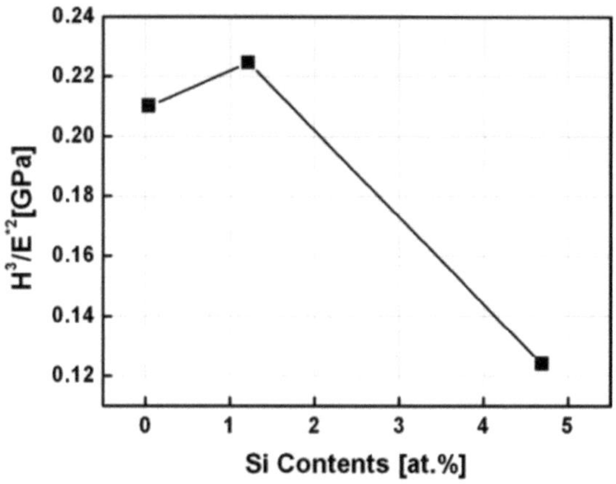

Fig. 14. H^3/E^{*2} ratios of Ti-Al-Si-N films

due to the nano-sized (Ti,Al,Si)N crystallites and the strong inhibition of crack propagation in amorphous Si_3N_4 under a high applied load [65,68].

4. Concluding remark

Nanocomposite materials could be synthesized using filtered vacuum arc applications. TiN based nanocomposite films were actively investigated until now and will be studied further. Another MeN/metal nanocomposite films such as ZrN-Cu were also reported that it has superior hardness and thermal stability [69]. FVA can be used to the other nanocomposite films and has promising probability as a deposition tool.

To use the FVA for industrial applications, a uniform coating technology is required to fulfill the standard of productivity. Methods of sweeping magnetic field and multi FVA systems have been investigated for uniform coatings in large area and more investigations are required continuously.

5. Reference

[1] Do-Geun Kim, Igor Svadkovski, Seunghun Lee, Jong-Won Choi and Jong-Kuk Kim, Current Applied Physics 9 (2009) S179–S181

[2] J. Bujak, J. Walkowicz and J. Kusinski, Surf. Coat. Technol.180 –181 (2004) 150–157

[3] P.J. Martina, A. Bendavida, J.M. Cairneyb and M. Hoffmanb, Surf. Coat. Technol. 200 (2005) 2228– 2235

[4] S.L. Ma, D.Y. Ma, Y. Guo, B. Xu, G.Z. Wu, K.W. Xu and Paul K. Chu, Acta Materialia 55 (2007) 6350–6355

[5] PalDey S, Deevi SC., Mater. Sci. Eng. A 342 (2003) 58–79.

[6] L. Chen, Yong Du, Ai J.Wang, She Q. Wang and Shu Z. Zhou, Int. Journal of Refractory Metals & Hard Materials 27 (2009) 718–721

[7] Sheng-Min Yang, Yin-Yu Chang, Da-Yung Wang, Dong-Yih Lin, WeiTe Wu, Journal of Alloys and Compounds 440 (2007) 375–379

[8] Yin-Yu Chang, Da-Yung Wang, Chi-Yung Hung, Surf. Coat. Technol. 200 (2005) 1702 – 1708

[9] A. Flink, T. Larsson, J. Sjolen, L. Karlsson, L. Hultman, Surf. Coat. Technol. 200 (2005) 1535 - 1542
[10] Y. Tanaka, N. Ichimiya, Y. Onishi, Y. Yamada, Surf. Coat. Technol. 146 -147 (2001) 215-221
[11] Li Chen, Yong Du, Ai J. Wang, She Q. Wang, Shu Z. Zhou, Int. Journal of Refractory Metals & Hard Materials 27 (2009) 718–721
[12] A. Ozturk, K.V. Ezirmik, K. Kazmanl, M. Urgen, O.L. Eryılmaz, A. Erdemir, Tribology International 41 (2008) 49–59
[13] Sung Ryong Choi, In-Wook Park, Sang Ho Kim, Kwang Ho Kim, Thin Solid Films 447 - 448 (2004) 371–376
[14] V.I. Gorokovsky, C. Bowman, P.E. Gannon, D. Van Vorous, A.A. Voevodin, C. Muratore, Y.S. Kang, J.J. Hu, Wear 265 (2008) 741–755
[15] Kwang Ho Kim, Sung-ryong Choi, Soon-young Yoon, Surf. Coat. Technol. 298 (2002) 243–248
[16] Chi-Lung Chang, Jun-Han Chen, Pi-Chuen Tsai, Wei-Yu Ho, Da-Yung Wang, Surf. Coat. Technol. 203 (2008) 619-623
[17] V. Gorokhovsky, C. Bowman, P. Gannon, D. VanVorous, A.A. Voevodin, A. Rutkowski, C. Muratore, R.J. Smith, A. Kayani, D. Gelles, V. Shutthanandan, B.G. Trusov, Surf. Coat. Technol. 201 (2006) 3732–3747
[18] In-Wook Park, Kwang Ho Kim, Journal of Materials Processing Technology 130–131 (2002) 254–259
[19] Eung-Ahn Lee, Kwang Ho Kim, Thin Solid Films 420 -421 (2002) 371–376
[20] Yan Guo, Shengli Ma, Kewei Xu and Tom Bell, Nanotechnology 19 (2008) 215603
[21] Soo Hyun Kim, Jong Kuk Kim, Kwang Ho Kim, Thin Solid Films 420 -421 (2002) 360-365
[22] L. Rebouta, C.J. Tavares, R. Aimo, Z. Wang, K. Pischow, E. Alves, T.C. Rojas, J.A. Odriozola, Surf. Coat. Technol. 133-134 (2000) 234-239
[23] I.I. Aksenov, V.E. Strel'nitskij, V.V. Vasilyev, D.Yu. Zaleskij. Surf. and Coat. Technol. 163 -164 (2003) 118–127
[24] D.M. Sanders, A. Anders. Surf. Coat. Technol.133-134 (2000) 78-90.
[25] R.L. Boxman, V. Zhitomirsky, B. Alterkop, E. Gidalevich, I. Beilis, M. Keidar, S. Goldsmith.. Surf. Coat. Technol. 86-87 (1996) 243-253.
[26] N.A. Khizhnyak, Sov. Phys. Technol. Phys. 35 (1965) 847.
[27] I.I. Aksenov, V.A. Belous, V.G. Padalka, USSR Authors Certificate No. 605425, 1978, (Rus.).
[28] I.I. Aksenov, et al., Prib. Tekhn. Ehksp. 5 (1978) 236 (Rus.).
[29] P.J. Martin, A. Bendavid. Thin Solid Films 394 (2001) 1-15.
[30] I.I. Aksenov, V.V. Vasilyev, B. Druz, A.A. Luchaninov, A.O. Omarov, V.E. Strel'nitskij. Surf. Coat. Technol. 201 (2007) 6084–6089.
[31] E. Byon, J.-K. Kim, S.-C. Kwon, A. Anders. IEEE Trans. Plasma Sci., 23 (2004) 433-439.
[32] I.I. Aksenov, D.Yu. Zaleskij, V.E. Strel'nitskij, 1st International Congress on Radiation Physics, High Current Electronics and Modification of Materials, September, Tomsk, Russia. Proceedings3, 2000, p.130.
[33] A. Anders, S. Anders and I.G. Brown. Plasma Sources Sci. Technol. 4 (1995) 1-12.
[34] Hirofumi Takikawa, and Hideto Tanoue, *IEEE Trans. Plasma Sci.*, vol. 35, no. 4, pp. 992-999, Aug. 2007.
[35] D. A. Karpov, *Surf.Coat. Technol.*, vol. 96, no. 1, pp. 22–33, Nov. 1997.

[36] P. J. Martin, A. Bendavid, and H. Takikawa, *J. Vac. Sci. Technol. A, Vac. Surf. Films*, vol. 17, no. 4, pp. 2351-2359,Jul. 1999.
[37] A. Anders, *Surf. Coat. Technol.*, vol. 120/121, pp. 319–330, 1999.
[38] D. M. Sanders and A. Anders, *Surf. Coat. Technol.*, vol. 133/134, pp. 78–90, 2000.
[39] P. J. Martin and A. Bendavid, *Surf. Coat. Technol.*, vol. 142–144, pp. 7–10, 2001.
[40] A. Anders, *Vacuum*, vol. 67, no. 3/4, pp. 673–686, Sep. 2002.
[41] I. I. Aksenov, V. E. Strel'nitskij, V. V. Vasilyev, and D. Y. Zaleslij, *Surf. Coat. Technol.*, vol. 163/164, pp. 118–127, 2003.
[42] P. J. Martin and A. Bendavid, *Thin Solid Films*, vol. 394, no. 1/2, pp. 1–5, Aug. 2001.
[43] X. Shi, B. K. Tay, H. S. Tan, E. Liu, J. Shi, L. K. Cheah, and X. Jin, *Thin Solid Films*, vol. 345, no. 1, pp. 1–6, May 1999.
[44] B. K. Tay, Z. W. Zhao, and D. H. C. Chua, *Mater. Sci. Eng. R*, vol. 52, no. 1–3, pp. 1–48, May 2006.
[45] A. I. Ryabchikov and I. B. Stepanov, *Rev. Sci. Instrum.*, vol. 69, no. 2, pp. 810–812, Feb. 1998.
[46] O. Zimmer, *Surf. Coat. Technol.*, vol. 200, no. 1–4, pp. 440–443, Oct. 2005.
[47] S. Y. Chun, A. Chayaraha, A. Kinomura, N. Tsubouchi, C. Heck, Y. Horino, and H. Fukui, *Jpn. J. Appl. Phys.*, vol. 38, no. 4B,pp. L467–L469, Apr. 1999.
[48] C. F. Meyer and H.-J. Scheibe, presented at the Int. Conf. Metallurgical Coatings Thin Films (ICMCTF), San Diego, CA, 1999, Paper B4-9.
[49] Y. Taki, T. Kitagawa, and O. Takaki, *J. Mater. Sci. Lett.*, vol. 16, no. 7, pp. 553–556, Apr. 1997.
[50] T. Yoshitake, G. Shiraishi, and K. Nagayama, *Appl. Surf.Sci.*, vol. 197/198, pp. 379–383, 2002.
[51] J. Patscheider, T. Zehnder, M. Disserens, Surf. Coat. Technol. 146–147 (2001) 201–208.
[52] S. Veprek, Surf. Coat. Technol. 97 (1997) 15.
[53] M. Diserens, J. Patscheider, F. Levy, Surf. Coat. Technol. 120–121 (1999) 158.
[54] K.H. Kim, B.H. Park, Chem. Vapor Depos. 5 (6) (1999) 275.
[55] F. Vaz, L. Rebouta, P. Goudeau, J. Pacaud, H. Garem, J.P. Riviere, A. Cavaleiro, E. Alves, Surf. Coat. Technol. 133–134 (2000) 307.
[56]W.-D. Munz, J. Vac. Sci. Technol. A 4(6) (1986) 2717
[57] O. Knotek, M. Bohmer, T. Leyendecker, J. Vac. Sci. Technol. A 4(6) (1986) 2695.
[58] S. Veprek, J. Vac. Sci. Technol. A 17 (5) (1999) 2401.
[59] S. Zhang, D. Sun, Y. Fu, H. Du, Surf. Coat. Technol. 167 (2003) 113–119.
[60] S.H. Kim, J.K. Kim, K.H. Kim, Thin Solid Films 420–421 (2002) 360.
[61] J.B. Choi, K. Cho, M.-H. Lee, K.H. Kim, Thin Solid Films 447–448 (2004) 365–370.
[62] Y H Cheng, T Browne, B Heckerman, P Gannon, J C Jiang, E I Meletis, C Bowman and V Gorokhovsky, J. Phys. D: Appl. Phys. 42 (2009) 125415
[63] Lughi V and Clarke D R 2006 *Appl. Phys. Lett.* 24 241911
[64] M.C. Polo, J.L. Andujar, A. Hart, J. Robertson, W.I. Milne, Diamond Relat. Mater. 9 (2000) 663-667.
[65] Sung Ryong Choi, In-Wook Park, Sang Ho Kim and Kwang Ho Kim, Thin Solid Films 447 -448 (2004) 371–376
[66] H. Han, F. Ryan, M. McClure, Surf. Coat. Technol. 120 (1999) 579-584.
[67] P.C. Johnson, in: M.H. Francombe, J.L. Vossen (Eds.), Contemporary Preparative Techniques, Academic Press, London, 1989, p.129.
[68] S.Veprek, S.Reiprich, L. Shizhi, Appl.Phys. Lett. 66 (1995) 2640.

7

Polymer/Clay Nanocomposites

Ali Olad
University of Tabriz,
Iran

1. Introduction

A composite is any material made of more than one component. Composite materials are solid multiphase materials formed through the combination of materials with different structural, physical and chemical properties. This makes composites different from the other multi component systems such as blends and alloys. Composites are widely used in such diverse as transportation, construction and consumer products (Giannelis, 1996). Composites offer unusual combinations of component materials properties such as weight, strength, stiffness, permeability, electrical, biodegradability and optical properties that is difficult to attain separately by individual components. A composite material can be custom tailored to have specific properties that will meet special requirements. The optimum design of composite component materials and manufacturing process to meet the target properties for specific engineering application is very important.
In a composite, one phase is continues and is called matrix, while the other is filler material which make the dispread phase. Composite materials based on the nature of matrix phase can be divided into polymeric, ceramic and metallic composites. Usually the filler phase is embedded to the host matrix phase to make a composite which has properties far from either phase alone. Polymers often have advantages over other materials such as metals and ceramics. They are widely used in various technical applications because of their unique advantages such as ease of production, light weight and ductility. However they have lower mechanical, modulus and strength properties compared to that of metals and ceramics. The commercial importance of polymers and their increasing use, results to the continuous demand for improvement in their properties to meet the necessary conditions. By the composite technology, polymer properties are improved while maintaining their light weight and ductile nature (Jordan et al., 2005).

2. Polymer composites

A polymer composite is made by the combination of a polymer and synthetic or natural inorganic filler. Fillers are employed to improve the desired properties of the polymer or simply reduce the cost. Nowadays, the application of polymer composites as engineering materials has become state of the art. Polymer composites with improved mechanical, thermal, barrier and fire retardancy properties are widely used in very large quantities in variety of applications. However by the application of conventional fillers such as talc, calcium carbonate, fibers, etc, it often requires to use a large amount of filler in the polymer

matrix to have significant improvements in the composite properties which may result to some other undesired properties such as brittleness or loss of opacity (Zhang & Fridrich, 2003).
The final properties of reinforced polymers (composites) are influenced by the nature, properties and content of components, dimensions of components and micro structure of composite and interfacial interactions between matrix and dispersed phase. The efficiency of properties improvements depends strongly on the properties (mechanical) of the filler, the adhesion between matrix and filler and especially on the aspect ratio of the filler. The aspect ratio of the filler is very important and crucial for many properties in composite such as electrical (Al-Saleh & Sundararaj, 2010; Grossiord et al., 2008), mechanical (Zhang et al., 2007) and thermal (Meneghetti & Qutubuddin, 2006) properties. Polymer composites with the high aspect ratio of nano-fillers such as platelet clays, carbon nanotubes and nanofibers are receiving considerable attention due to their unique multifunctional and highly enhanced properties. Combination of filler nanoscale dimension and high aspect ratio with its nanoscale dispersion within polymer matrix leads to the significant improvements in the polymer properties at very low filler volume fractions. As a result of lower filler using, the macroscopic homogeneity and low density of primary polymer retains as well as it's opacity in the final nanocomposite system. Polymer nanocomposites are the new class of hybrid materials in this category (Mittal, 2009).

3. Polymer nanocomposites

A nanocomposite is defined as a composite material in which at least one dimensions of at least one component is in the nanometer size scale (< 100 nm). Although the term nanocomposite represent a new and exciting field in material science and technology, the nanocomposites have actually been used for centuries by the nature. Using natural materials and polymers such as carbohydrates, lipids and proteins, nature makes strong nanocomposites such as bones, shells and wood (Hussain et al., 2006). However in recent years the characterization and control of structures at the nanoscale have been studied, investigated and exploited by the learning from the natural surroundings. Consequently the nanocomposite technology has emerged as an efficient and powerful strategy to upgrade the structural and functional properties of synthetic polymers. This is the new nanocomposite science, so referred recently in nanotechnology, and was started by the Toyota report (Usuki et al., 1993) on the superior improvement in the properties of nylon-6 by the preparation of exfoliated nylon-6/clay nanocomposites and has been continued by more recent studies with carbon nanotubes, carbon nanofillers, exfoliated graphite, nanocrystaline metals and fibers modified with inorganic fillers used in polymeric composites (Paul & Robeson, 2008).
Polymer nanocomposites have attracted great attention world wide academically and industrially due to the exhibition of superior properties such as modulus, strength, toughness and barrier far from those of conventional micro composites and comparable with those of metals. However polymer nanocomposites have added advantage of lower density and ease of processability. In polymer nanocomposites, the filler has at least one dimension in the nanometer scale and its nanoscale dispersion within the polymer matrix leads to the tremendous interfacial contacts between the polymer and inorganic filler which causes to the superior properties than those of bulk polymer phase. When the dimensions of filler particles are decreased to the nanoscale, their properties change also significantly

(Kumar et al., 2009). This is well-known as nano-effect. In the other word the effect of nanofillers on the polymer properties is different from that of predicted by using the thermodynamical studies for the reduced particle size fillers (Gacitua et al., 2005). Studies and modelings using continuum mechanics reveal that the enhanced properties of nanocomposites are strongly dependent on the particular features of nanofiller system, in particular, its content, aspect ratio and the ratio of filler mechanical properties to those of the matrix (Sheng et al., 2004). The nanoscale is considered where the dimensions of filler particles (diameter), platelets (thickness) or fibers (diameter) are in the size range of 1-100 nm (figure 1).

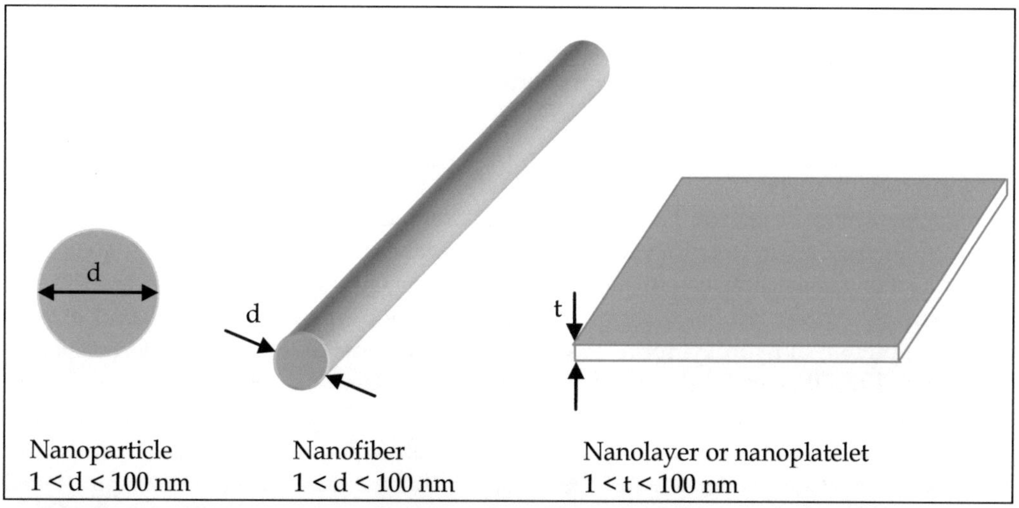

Fig. 1. Scheme of various types of nanofillers or fillers with nanoscale dimensions.

A broad spectrum of polymer properties can be improved by nanocomposite technology such as mechanical, thermal, barrier, durability, chemical stability, flame retardancy, scratch/wear resistance, biodegradability as well as optical, magnetic and electrical properties (Armentano et al., 2010; Cosoli et al., 2008; Ma et al., 2006; Pandey et al., 2005).

The final properties of nanocomposites are determined by the component properties, composition, micro-structure and interfacial interactions. However it has been established that the properties of nanocomposites are strongly influenced by the dimensions and micro structure of filler phase (Sheng et al., 2004). In other word the filler nature has a main effect on the final morphology and properties of the polymer nanocomposite. Clays are one group of nano-fillers which have been widly used for the preparation of polymer nanocomposites. Recently there has been a growing interest for the development of polymer/clay nanocomposites due to their dramatic improved properties compared to the conventional filled polymers in a very low fraction of filler addition (LeBaron et al., 1999; Thostenson et al., 2005). Polymer/clay nanocomposites have received intense attention and research interest driven by the unique properties which can never be obtained by micro size fillers or especially by other nanofillers. The value added properties enhanced without the sacrificing of pure polymer processability, mechanical properties and light weight, make the clays more and more important in modern polymer industry. Clay minerals are belonging to a main group of silicates with layered structure known as layered silicates.

4. Layered silicates

The layered silicates are natural or synthetic minerals consisted of the regular stacks of aluminosilicate layers with a high aspect ratio and high surface area. Layered silicates are easily available and have low cost. The current most popular layered silicates used in polymer nanocomposites preparation are clays.
Clay is referred to a part of soil fraction with the particle size of less than 2 µm. The clay layers have a thickness of about 1 nm which is in the nanoscale. There are many members of clays with some difference in their formula, structure and properties including swelling and exfoliation. Those members who are able to be exfoliated by polymer chains or monomers and distributed as individual clay layers within polymer matrix are suitable for the preparation of polymer nanocomposites. The individual clay layers can cause to the dramatic improvements in polymer properties due to their high aspect ratio and high interfacial interactions with polymer matrix.

4.1 Structure of clays

Clays have layered structure. The layers are built from tetrahedral sheets in which a silicon atom is surrounded by four oxygen atoms and octahedral sheets in which a metal like aluminium or magnesium is surrounded by eight oxygen atoms. The tetrahedral (T) and octahedral (O) sheets are fused together by sharing the oxygen atoms. Unshared oxygen atoms are present in hydroxyl form. Two main arrangements of tetrahedral and octahedral sheets fused together make the one layer structure of clay. One tetrahedral fused to one octahedral (1:1) is known as kaolin group with the general composition of $Al_2Si_2O_5(OH)_5$ and the layer thickness of ~ 0.7 nm. The crystal lattice consisted of one octahedral sheet sandwiched between two tetrahedral sheets (2:1) with the total thickness of 0.94 nm is well known as phyllosilicates. The 2:1 phyllosilicate layers can be electrostatically neutral with no inter layer ion present known as pyrophyllite. Due to not presence of inter layer ions, the layers do not expand in water. When silicon in tetrahedral sheets is substituted by aluminium, the 2:1 structure is called mica. The negative charge induced by this substitution is balanced by the insertion of potassium cations between layers. Due to the equal size of potassium cation and the hole created by Si/Al tetrahedral sheets, the presence of potassium cation makes no inter layer spacing. Consequently the 2:1 layers are held together strongly and the swelling or exfoliation of layers is not possible. When in neutral pyrophyllite, the aluminum cations in the octahedral layers are partially substituted by divalent magnesium or iron cations the smectite clay group is formed. The negative charge created by this substation is balanced by inter layer counter sodium or calcium ions. The charge created on the layers is not locally constant and varies from layer to layer. An average charge value is considered for the layers which can be determined by cation exchange capacity (CEC) measurement. Due to the unequal size of inter layer cations with the holes of tetrahedral sheets, the presence of inter layer cations causes to an inter layer spacing. The layers stay near together with a regular gap between them. The gap is called as inter layer or gallery. The thickness of the repeated units in a regular multilayer structure contained of one layer and one inter layer space is called d-spacing (d_{001}) or basal spacing (figure 2). The basal spacing of clays can be measured or calculated from their X-ray diffraction patterns. The inter layer dimension is also dependent to the clay nature and swelling or hydration degree of inter layer cations. The inter layer distance is about 1 nm for montmorillonite.

The electrostatic and van der waals forces holding the layers together are relatively weak and inter layer distance varies depending on the charge density on the layers, inter layer cation radius and its degree of hydration. Because of the inter layer spacing and weak inter layer forces, the cations present between the layer can be hydrated in aqueous solutions well known as clay swelling. The swelling causes to the more increasing of inter layer space. The charge density on the clay layers is different in various clay groups with different clay structures (Mittal, 2009). Due to the inter layer spacing and weak forces between layers, especially in the hydrated form, other molecules also can be intercalate between layers, leads to the expanding of layered lattice and finally may be caused to the separation of individual layers. The unique intercalation/exfoliation behavior of smectite clay minerals which is responsible to the high aspect ratio of this clay type makes them very important and powerful as reinforcing filler for polymers.

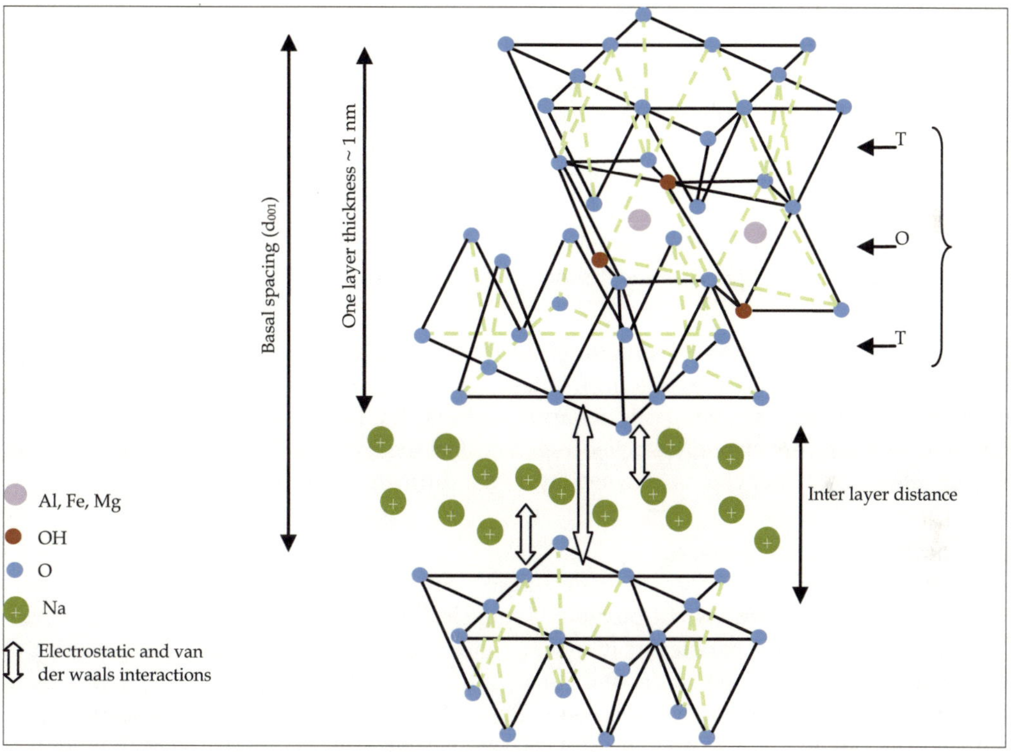

Fig. 2. Scheme of the 2 : 1 smectite clay structure.

The dimension of 2:1 layer structures are about 1 nm in thickness and 30 nm to several microns in lateral depending on the source and preparation method of clay. The clay lateral dimensions and its ability to disperse into separated layers determine the aspect ratio of clay which is strongly important in polymer/clay interfacial interactions. Separation of individual clay layers having about 1 nm thickness display a perfect structure which strongly affects on the improvement of polymer properties. However due to the higher internal surface in high degrees of dispersed structures, the internal forces for agglomeration of individual layers is also increases. Therefore the layers have tendency to hold together or aggregate. Several layers organized parallel together with the total thickness of about 10 nm make "primary clay particles". Primary particles are laterally

joined together and make "micro aggregates". "Aggregates" are formed by randomly joining of micro aggregates and primary particles, and finally the staying of aggregates near together makes "association" in the smectite structures.
Clay minerals are belonging to the nanoplatelet structured filler particles in polymer composite technology. Montmorillonite, hectorite and saponite are the well known smectites which are most commonly used in the preparation of polymer nanocomposites (table 1) (Pavlidou & Papaspyrides, 2008). Montmorillonite (MMT) has the widest acceptability for use in polymer nanocomposites because of its ease availability, well known intercalation/exfoliation chemistry, high surface area and high surface reactivity. The MMT layer aspect ratio can be in the range 1000 in well dispersed state without breaking of layers and surface area of about 750 m^2/g. Usually because of the breaking of clay layers in to small plates during preparation process of nanocomposite, the aspect ratio of MMT layers decreases to about 300-500.

Smectites	Chemical formula
Montmorillonite	$M_x(Al_{4-x}Mg_x)Si_8O_{20}(OH)_4$
Hectorite	$M_x(Mg_{6-x}Li_x)Si_8O_{20}(OH)_4$
Saponite	$M_xMg_6(Si_{8-x}Al_x)O_{20}(OH)_4$

Table 1. Chemical structure of commonly used smectite type clays, M: monovalent cation, x: degree of cations isomorphous substitution in octahedral sheets, (Pavlidou & Papaspyrides, 2008).

Other clay materials such as 1:1 structures, 2:1 pyrophillits and mica are not commonly used in polymer nanocomposites because they have not useful intercalation/exfoliation characteristics or they are not easily separated into layers. Due to the higher charge density on the layers they are strongly stacked together and the hydration of inter layer cations or intercalation of polymer chains between layers are difficult or not possible.

4.2 Modification of clay layers

The preparation of polymer/clay nanocomposites with good dispersion of clay layers within the polymer matrix is not possible by physical mixing of polymer and clay particles. It is not easy to disperse nanolayers in most polymers due to the high face to face stacking of layers in agglomerated tactoids and their intrinsic hydrophilisity which make them incompatible with hydrophobic polymers. Only a few hydrophilic polymers such as poly (ethylene oxide) and poly (vinyl alcohol) can be miscible with clay nanolayers (Pavildou & Papaspyrides, 2008).
The intrinsic incompatibility of hydrophilic clay layers with hydrophobic polymer chains prevents the dispersion of clay nanolayers within polymer matrix and causes to the weak interfacial interactions. Incompatibility and weak interfacial interactions hinders the exfoliation and preparation of dispersed stable nanocomposite with improved properties. Modification of clay layers with hydrophobic agents is necessary in order to render the clay layers more compatible with polymer chains. This is a surface modification which causes to the reduction of surface energy of clay layers and match their surface polarity with polymer polarity. The organoclays with lowered surface energy are more compatible with polymers and polymer molecules are able to intercalate within their interlayer space or galleries under well defined experimental conditions.

The surface modification of clay layers can be achieved through a cation exchange process by the replacement of sodium and calcium cations present in the inter layer space or clay galleries by alkylammonium or alkylphosphonium (onium) cations (Ahmad et al., 2009). In addition to the surface modification and increasing the hydrophobisity of clay layers, the insertion of alkylammonium or alkylphosphonium cations into the galleries causes to some degree of increasing in the inter layer spacing which promotes the following intercalation of polymer chains into the galleries during nanocomposite preparation (Chigwada et al., 2006). Also the alkylammonium or alkylphosphonium cations can provide functional groups which interact with polymer chains or initiate the polymerization and therefore increase the interfacial interactions. Figure 3 schematically shows the organically modification of clay layers using alkylammonium cations via the ion exchange process.

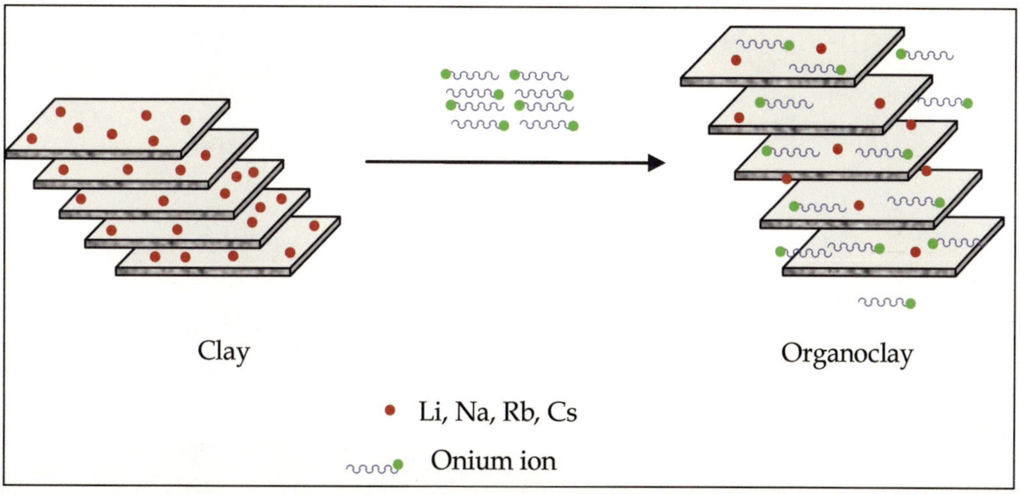

Fig. 3. Scheme of the modification of clay layers by organic onium cations.

The ion exchange process is carried out in aqueous solution with the hydrated inter layer cations. The affinity of monovalent sodium cations to hydration promotes the ion exchange process and increases the efficiency of organic modification of clay layers. Clays containing divalent counter cations such as calcium can not be easily hydrated and therefore their replacement by ion exchange process is not efficient. Therefore, in these clays, the divalent cations should be ion exchanged with easily exchangeable sodium cations prior to modification by onium cations (Ray & Okamoto, 2003). The efficiency of organic modification by ion exchange process in the increasing of basal spacing and consequently the exfoliation of clay and formation of stable nanocomposite systems depends also on the surface charge of clay layers. The surface charge density of clay layers depends on the clay nature and its preparation (growth) conditions. When the charge density of clay layer is high such as about 1 equiv/mol for mica, the electrostatic forces between layers is also high. Therefore the inter layer cations can not to hydrate or swollen and consequently the ion exchange process can not be carried out. The medium charge densities (0.5–0.8 equiv/mol) such as in vermiculite, can make a high potential for partial swelling and hydration of cations. However the divalent cations exist in the inter layer spacing of clays such as vermiculite make the swelling process difficult due to the lower affinity of divalent cations for hydration. The lower charge densities (0.25–0.5 equiv/mol) such as formed in montmorillonite causes to the weak electrostatic forces between layers. Also the presence of sodium cations, with high affinity for hydration and exchanging, between the MMT layers make this kind of clays more suitable for swelling, organic modification and exfoliation. The structure of the organoclays and their basal spacing also depends on the molecular size of organic onium and its configuration in the inter layer spacing after modification. Figure 4 schematically shows the various configurations of the organic modifiers inserted within the clay layers by ion exchange with sodium cations. It is clear that the charge density of clay layers, the efficiency of ion exchange process and the spatial characteristics of onium cation are effective on the configuration of onium cations in the modified clays.

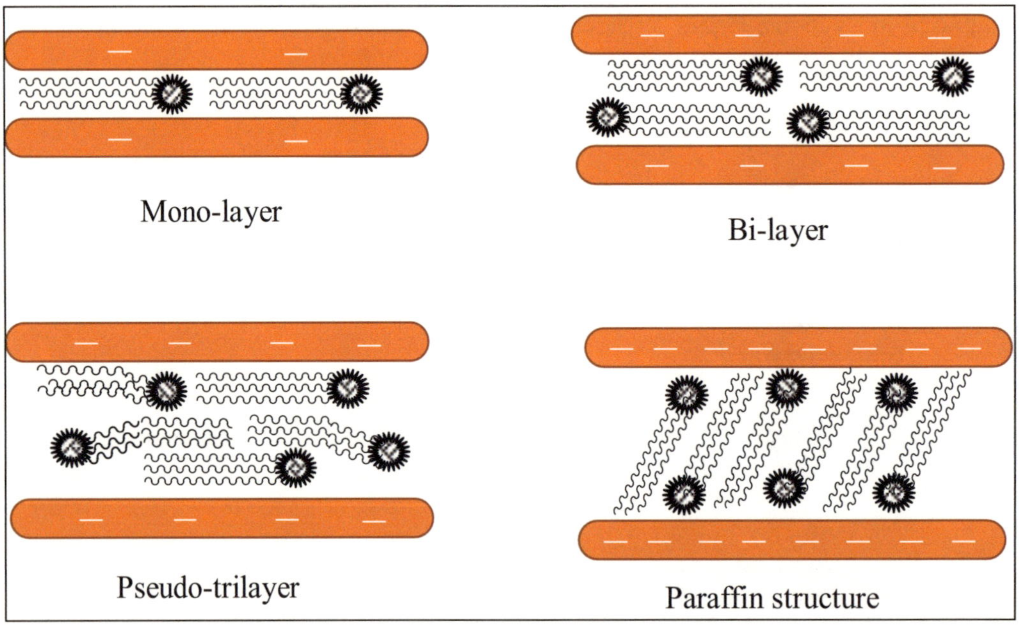

Fig. 4. Various configurations of organic oniums into the clay galleries.

The organic tail of onium laid up parallel to the clay layer can be obtained with lower charge densities. By increasing the charge density of clay layers, bilayer or trilayer structures with more increasing of inter layer spacing can be formed. Finally with the higher charge densities more inter layer spacing is obtained by the formation of paraffin structure. Also increasing of temperature can lead to the structures with more basal spacing by increasing the mobility of onium molecules into the galleries. However in a given temperature two parameters determine the equilibrium layer spacing, the charge density of layers, driving the packing and number of onium chains into the interlayer space, and the size or chain length of organic tail in onium compound.

5. Nanocomposite structure

In general, the structures of polymer/clay nanocomposites are classified according to the level of intercalation and exfoliation of polymer chains into the clay galleries. Various parameters including clay nature, organic modifier, polymer matrix and preparation method are affective on the intercalation and exfoliation level. Therefore depending on the nature and properties of clay and polymer as well as preparation methodology of nanocomposite, different composite micro-structures can be obtained.

5.1 Phase separated structure

When the organic polymer is interacted with inorganic clay (unmodified clay), the polymer is unable to intercalate within the clay layers and the clay is dispersed as aggregates or particles with layers stacked together within the polymer matrix. The obtained composite structure is considered as "phase separated". The properties of phase separated polymer/clay composites are in the range of traditional micro composites.

5.2 Intercalated structure

When one or more polymer chains are inserted into the inter layer space and cause to the increasing of the inter layer spacing, but the periodic array of the clay layer is still exist, the intercalated nanocomposite is formed. The presence of polymer chains in the galleries causes to the decreasing of electrostatic forces between the layers but it is not totally dissolved. A well-ordered multilayer hybrid morphology with a high interference interactions consisted of polymer chains and clay layer is obtained in this configuration.

5.3 Exfoliated structure

Exfoliated or delaminated structure is obtained when the insertion of polymer chains into the clay galleries causes to the separation of the layers one another and individual layers are dispersed within the polymer matrix. At all, when the polymer chains cause to the increasing of interlayer spacing more than 80-100 Å, the exfoliated structure is obtained. Due to the well dispersion of individual clay layers, high aspect ratio is obtained and lower clay content is needed for exfoliated nanocomposites. Also most significant improvement in polymer properties is obtained due to the large surface interactions between polymer and clay. Various polymer/clay structural configurations have been shown in figure 5.

6. Preparation of polymer/clay nanocomposites

Many efforts have been made for the preparation of intercalated and exfoliated polymer/clay nanocomposites with improved properties. A variety of polymer characteristics including polarity, molecular weight, hydrophobicity, reactive groups as well as clay characteristics such as charge density and its modified structure and polarity are affective on the intercalation of polymer chains within the clay galleries. Therefore different synthetic approaches have been used for the preparation of polymer/clay nanocomposites. In general there are four preparation methods including insitu template synthesis, solution intercalation, insitu intercalative polymerization and melt intercalation.

6.1 Insitu template synthesis

In this method the clay layers are synthesized insitu in the presence of polymer chains. The polymer and clay primary materials are dissolved in an aqueous solution. Typically magnesium hydroxide, silica and lithium fluoride as clay building blocks are mixed with polymer in a solvent. The gel or slurry is refluxed usually at high temperatures followed by washing and drying. The nucleation and growth of clay layers are take place on the polymer chains and the polymer chains are trapped in the clay inter layers. Although the clay layers may be well dispersed within the polymer matrix without the modification of clay by onium cations, however this method has serious disadvantages. The high temperature applied for the synthesis of clay layers causes to the decomposition of polymers. Only hectorite clay is

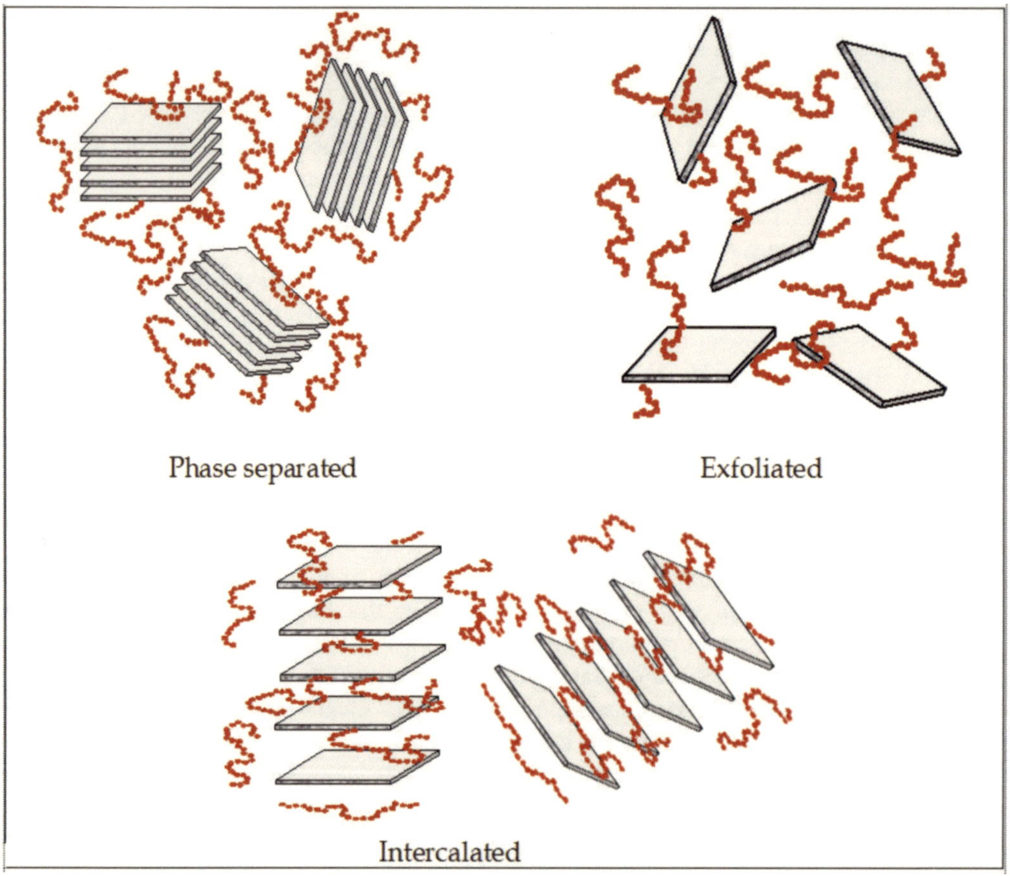

Fig. 5. Various micro-structures of polymer/clay composites.

synthesized at the lower temperatures. Also the synthesized clay crystals, generated by the self-assembly process, have tendency to aggregate. Therefore this method is not widely used and only a few nanocomposites using hectorite clay and poly (vinyl alcohol), polyaniline and polyacrylonitrile have been synthesized by this method (Alexandre & Dubois, 2000; Lagaly, 1999; Zanetti et al., 2000).

6.2 Solution intercalation

In this method the polymer or prepolymer is dissolved in a solvent and the clay is dispersed in the same solution. The clay is swollen in the solvent and the polymer chains intercalate between the layers. The intercalated nanocomposite is obtained by solvent removal through vaporization or precipitation. Clays can be swollen easily in solvents such as water, acetone, chloroform and toluene. The polymer chains are absorbed onto the delaminated sheets. During the solvent evaporation the entropy gained by the exit of solvent molecules from the interlayer spacing, allows the polymer chains to diffuse between the layers and sandwiching. We have used the solution intercalation method for the preparation of epoxy /clay nanocomposites. Diglycidyl ether bisphenol A (DGEBA), (196-208 epoxy equivalent weight, Epiran-05, Khouzestan Petrochemical Company, Iran), was used as the polymer matrix. EPIKURE curing agent 3200, Aminoethylpiperazine (AEP) from Hexion Specialty

Chemicals Inc., was used both as curing agent of epoxy resin and also as intercalating agent for montmorillonite (MMT). The inorganic clay used in this study was K-10 grade MMT obtained from Sigma-Aldrich Co (USA) with a surface area of 130 m^2/g. Organically modified clays, Cloisite 30B (d-spacing = 18.5 Å) and Cloisite 15A (d-spacing = 29.88 Å) were provided by the Southern Clay Products. Tetraethylammonium chloride (TEA) salt was also used as intercalating agent of MMT. Acetone was used as solvent.

Depending on the interactions between polymer and clay sheets, intercalated or exfoliated structures may be obtained in this method. This technique is used for the preparation of epoxy/clay nanocomposites. However due to the need for use solvent, this technique can not be applied in industry. The nanocomposite preparation by emulsion polymerization, with the clays dispersed in the aqueous solution, is also categorized as solvent intercalation method (Rehab & Salahuddin, 2005). Toyota group has been used this technique to produce polyimide/clay nanocomposites (Yano et al., 1993). In our group the emulsion and bulk polymerization methods have been used for the preparation of polystyrene/clay nanocomposites using the Na-MMT, cloisite 30B and cloisite 15A clay materials. The effect of clay swelling method and sonication on the inter layer spacing and intercalation have been investigated. Good dispersion of Na-MMT in the polystyrene (PS) matrix was observed when an emulsion polymerization was employed and good dispersion of cloisite 30B in the PS was observed during bulk polymerization. Cloisite 30B showed better dispersion when this clay was swollen with the monomer during emulsion polymerization. Sonication had good effect on dispersion of the clays in the PS matrix. PS/Cloisite 15A nanocomposites showed minor increase in d-spacing compared to the pure cloisite 15A. TGA analysis showed that the thermal stability of the nanocomposites has been improved compared to the pure polystyrene (table 2).

6.3 Insitu intercalative polymerization

This technique was the first method used for the preparation of polymer/clay nanocomposite by Toyota research group in the preparation of Nylon-6 nanocomposite from caprolactam monomer (Usuki et al., 1993). In this method the organoclay is swollen in monomer liquid or monomer solution. The monomers diffused into the inter layer spacing are polymerized by the heat or radiation, by the diffusion of an initiator or by the organic initiator present on the organic modifier of clay (Hussain et al., 2006). The polymerization is carried out within the clay galleries as well as extra galleries. The growth of polymer chains results to the exfoliation and formation of disordered structure.

This method is suitable for the preparation of thermoset/clay nanocomposites and has been widely used for the epoxies and styrenic polymer nanocomposites (Lan et al., 1995). The polarity of monomer and clay layers determines the diffusion rate and equilibrium concentration of monomer within the clay galleries. Consequently the exfoliation and dispersion of clay layers can be tailored by the clay and monomer chemistry (Pavlidou & Papaspyrides, 2008). Polyaniline/MMT nanocomposites have been prepared by the insitu polymerization of aniline in the presence of MMT (Olad & Rashidzadeh, 2008). Both unmodified and organomodified MMT were used in the preparation of nanocomposites. Due to the using of acidic aqueous solutions for the polymerization of aniline, the polar organic anilinium cations can be interact with both unmodified and modified clays.

6.4 Melt intercalation

Clay is mixed within the polymer matrix in molten temperature. The conventional methods such as extrusion and injection molding are used for dispersion of clay layers within the polymer matrix. This method is effective technique for the preparation of thermoplastic nanocomposites (Kornmann et al., 2001). The polymer chains are intercalated or exfoliated into the galleries. Clays are organically modified and polymer chains are surface modified with more polar functional groups to enhance their compatibility and therefore promote the exfoliation. In melt intercalation method no solvent is required and it has many advantages for the preparation of nanocomposites and is a popular method for industry (Ray et al., 2003).

7. Structural characterization of polymer/clay nanocomposites

In the preparation of a polymer/clay nanocomposite it is important to know the degree of intercalation/exfoliation and its effect on the nanocomposite moderate properties. In other word there is need to analyze the micro structure of the prepared nanocomposite. Two common techniques including X-ray diffraction (XRD) analysis and transmission electron microscopy (TEM) are widely used to characterize the micro structure of nanocomposite as well as pure clay or pure organoclay. Clays and organoclays show a characteristic peak in XRD analysis due to their regular layered structures. The peak is indicative of the platelet separation or d-spacing in clay structure. Using the peak width at half maximum height and peak position (2θ) in the XRD spectra the inter layer space can be calculated utilizing Bragg's law (eq. 1).

$$\text{Sin}\,\theta = n\lambda \;/\; 2d \qquad (1)$$

Where λ is wave length of X-ray radiation used in the diffraction experiments, d is the space between layers in the clay lattice and θ is measured diffraction angle. Any change in the inter layer or d-spacing of a clay lattice by organic modification or polymer intercalation causes to the change in the position, broadness and intensity of the characteristic peak in XRD spectra. According to the Bragg law, increasing of d-spacing results to the broadening and shifting of related XRD peak toward lower diffraction angles (2θ). By monitoring the position (2θ), shape and intensity of the characteristic peak for organoclay in nanocomposite structure it is possible to determine the degree of intercalation/exfoliation. For example figure 6 shows the XRD patterns of pristine Na-MMT as well as polystyrene (PS)/Na-MMT nanocomposites prepared by emulsion polymerization method. The characteristic peak of MMT has been appeared at $2\theta = 7.13°$ corresponding to the inter layer space of 1.24 nm. The XRD patterns of polystyrene/Na-MMT nanocomposites containing different clay percents (2-6 wt %) show diffraction peaks all shifted to lower 2θ degrees against the peak of pure Na-MMT, which indicates the increasing of layers spacing due to the intercalation or exfoliation. The experimental conditions, peak position (2θ) and related calculated clay layer spacing obtained from XRD spectra of pure clays and polystyrene/clay nanocomposites have been shown in table 2.

The results indicate that the Na-MMT was efficiently dispersed in the polystyrene matrix during the emulsion polymerization due to the hydrophilicity of pure montmorillonite. Also the effect of sonication on the dispersion of Na-MMT layers in the polymer matrix was investigated. The results showed that the polystyrene/Na-MMT nanocomposite prepared by the application of sonication has no noticeable diffraction peak in the XRD spectrum at 2θ values of 2° to 10° (figure 6e). This indicates that the clay plates have been dispersed effectively and may be the exfoliated structure has been obtained.

Sample	Experimental polymerization conditions	2θ (°)	d (Å)
pure Na-MMT	-	7.13	12.38
PS/Na-MMT (2wt%)	Emulsion	2.58	34.16
PS/Na-MMT (4wt%)	Emulsion	2.72	32.46
PS/Na-MMT (6wt%)	Emulsion	2.58	34.16
PS/Na-MMT (2wt%)	Bulk	6.00	14.72
Pure clay 30B	-	4.68	18.86
PS/ clay 30B (2wt%)	Emulsion, swelling with water	3.54	24.93
PS/ clay 30B (2wt%)	Emulsion, swelling with water+ sonication	3.02	29.27
PS/ clay 30B (2wt%)	Emulsion, swelling with monomer	2.79	31.67
PS/ clay 30B (4wt%)	Emulsion, swelling with monomer	2.99	29.51
PS/ clay 30B (6wt%)	Emulsion, swelling with monomer	3.03	29.10
PS/ clay 30B (2wt%)	Bulk	2.71	32.55
pure clay 15A	-	2.95	29.88
PS/ clay 15A (2wt%)	Emulsion, swelling with monomer	2.75	32.16
PS/ clay 15A (2wt%)	Bulk	2.69	32.76

Table 2. XRD data and inter layer space for MMT and PS/MMT nanocomposites.

It may be need to use wide-angle X-ray diffraction (WAXD) patterns in the range $1° < 2\theta < 10°$ to identify the fully exfoliated structures which is correspond to the d-spacing of at least 6 nm. The XRD technique offers a conventional method to determine the d-spacing in the pure clay as well as in nanocomposite structure, within 1-4 nm, using the position, broadness and intensity of characteristic peak in WAXD diffractogram. However the disappearance of such a peak is not a conclusive evidence for a highly exfoliated structure in nanocomposite (Hedayati & Arefazar, 2009).

A direct way to visually observe the nanostructure of nanocomposites and clay d-spacing is to the use of transmission electron microscopy (TEM) technique. TEM micrographs allow a qualitative understanding of the internal structure, exfoliation or spatial distribution of layers within the polymer matrix and their structural defects. In the TEM micrographs, the darker lines in the brighter matrix shows the clay layers because of the presence of heavier elements including Al, Si and O in the composition of clay sheets or layers compared the lighter atoms such as C, H, N and Na present in the polymer matrix or inter layer spacing of clay sheets. Therefore the distance between darker liner sections presented in the TEM micrographs can qualitatively show the d-spacing and dispersion status. Therefore the overall structure of the nanocomposite including intercalation, exfoliation, dispersion and defects of clay layer can be conclusively obtained using TEM technique.

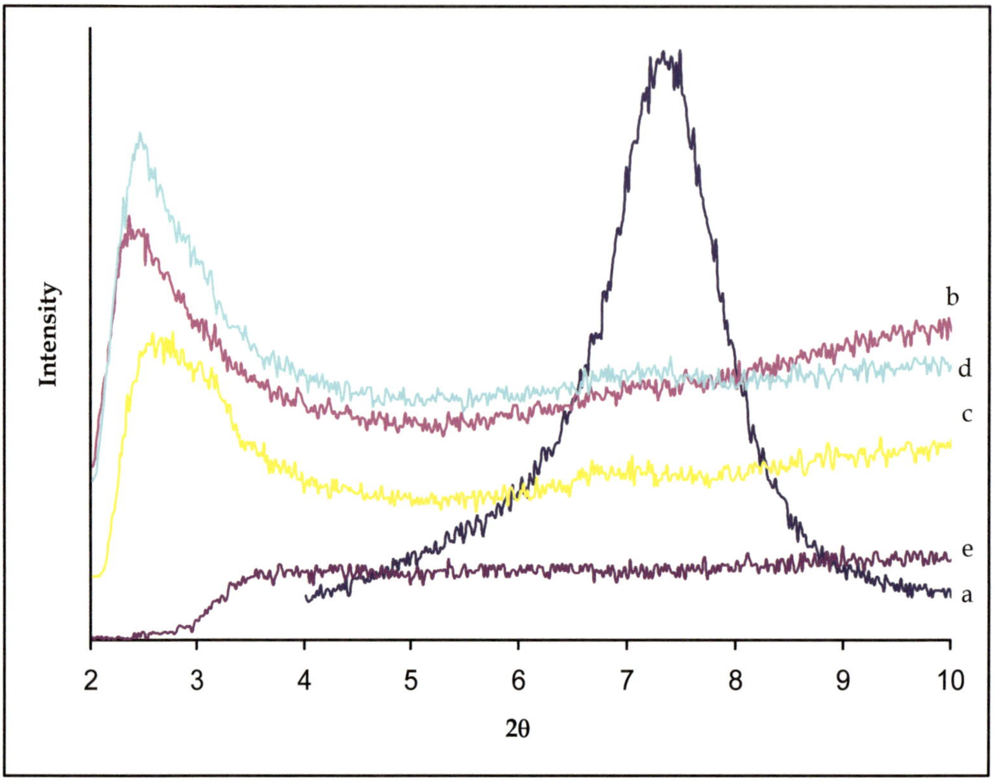

Fig. 6. XRD patterns of pure Na-MMT (a) and PS/Na-MMT nanocomposites prepared by emulsion polymerization method with clay content of 2 wt % (b), 4 wt % (c) and 6 wt % (d) and PS/Na-MMT (2 wt %) with sonication (e).

However both XRD and TEM techniques are essential tools for evaluation of the nanocomposite structure and complementary show the structural features of polymer/clay nanocomposite microstructure. XRD provides almost quantitative and TEM provides qualitative informations about the exfoliation and d-spacing of clay layers in the polymer matrix compared to that of in pure clay material. Also other techniques such as those based on thermal analysis, can be used to evaluate the interfacial interactions between clay layers and polymer chains.

8. Properties of polymer/clay nanocomposites

The aim for the addition of clay minerals to the polymers is to improve the polymer properties and to produce the polymer/clay nanocomposites with demand characteristics for desired applications. Almost it is wanted to obtain a nanocomposite with demand properties and characters or overcome the drawbacks of polymers while remaining the intrinsic advantages of primary polymer matrix.

Because of the low price, availability, high aspect ratio as well as desirable nanostructure and interfacial interactions, clays can provide dramatic and adjustable improved properties at very lower loadings which help to the more remaining of polymer original useful properties. The nature and properties of components as well as preparation methodology and conditions affects on the final properties of polymer/clay nanocomposite (Ray, 2010). In this section the various improved properties of polymer/clay nanocomposites as well as the mechanism and effective parameters are discussed.

8.1 Mechanical properties

The first and important goal in the application of fillers to the polymer is improvement in the mechanical properties and therefore fillers are commonly called as reinforcement agents (Jiang et al., 2005). The mechanism of the reinforcement is based on the higher resistance of rigid filler materials against straining due to their higher module. When a rigid filler is added to the soft polymer matrix, it will carry the major portion of applied load to the polymer matrix under stress conditions, if the interfacial interactions between filler and matrix is adequate (Tortora et al., 2002; Gorrasi et al., 2003).

Therefore the larger the interface between filler and matrix, and also more strong interfacial interactions, the great the reinforcement effect will be. Also the modulus increases by increasing the filler ratio in the composite or nanocomposite composition. Because of the rigid structure of clay layers and their high aspect ratio, they have proven to be very effective in the increasing of modulus and stiffness of the polymer matrix in well dispersed condition. In fact due to the very higher aspect ratio of nano-clay fillers compared to that of regular fillers such as glass fibers, dramatic improvements in the mechanical properties of nanocomposites are achieved even at very lower nanoclay additions (LeBaron et al., 1999). In another approach, polymer chains, adhesived by strong physisorption forces onto the rigid clay mono layers, become equally a portion of rigid material and dramatically exhibit high modulus. Considering the very large interfacial area in well dispersed nanocomposite structures, significant improvement in the modulus can be expected. However any enhancement in the polymer-clay interfacial contact leads to the better stress transfer in the nanocomposite. It has been well demonstrated that the enhancement in the interfacial adhesion properties between clay and polymer by surface modification of polymer chains using a suitable polar compatibilizer causes to the increasing in the mechanical properties of nanocomposite (Osman et al., 2005; Mittal, 2008). It should be noticed that in higher loadings of compatibilzers, they may cause to the some degree of plasticization due to their lower molecular weight, which negatively affect on the modulus of nanocomposite (Mittal, 2008). Also the using of organically modified clay filler increases both the modulus and stiffness of nanocomposite because of the more intercalation/exfoliation achieved. At all, any parameter which helps to the intercalation of polymer molecules within the clay galleries and consequently causes to the more exfoliation and interfacial interactions, results to the greater improvement in the modulus. However it is difficult to achieve the complete exfoliation of clay layers and there are variety of platelet structures with different thicknesses (depending on the number of layers stacked together) in polymer matrix. Fornes and Paul (2003) have analytically formulated the effect of incomplete exfoliation on the nanocomposite properties.

Also it has been reported that the tensile modulus increases by increasing the filler volume fraction in nanocomposite (Mittal, 2007). However clay loading more than threshold limit value, causes to leveling-off in the increasing period of Young's modulus, which is due to the formation of partially exfoliated/intercalated structure after that fully exfoliated structure formation (Ray & Okamoto, 2003; Alexandre & Dubois, 2000).

Also the incorporation of clay minerals usually improves the tensile strength of polymers (Fornes et al., 2001; Shelley et al., 2002; Kojima et al., 1993). However the tensile strength is influenced by the final morphology of nanocomposite. There are a number of reports on the reduction of tensile strength by the addition of clay minerals (Alexandre & Dubois, 2000; Finnigan et al., 2004). Similar to modulus, any factor affecting on the degree of intercalation/exfoliation has an impact on the tensile strength of nanocomposite.

The effect of clay fillers on the stress at break values for polymer nanocomposites depends on the interfacial interactions between polymer and clay layers. The stronger interfacial interaction causes to the increasing of stress at break and the weak interfacial forces may lead to some decreasing of stress at break for nanocomposite. The polymers with more polarity such as nylon and rubber based polymers have more strong interfacial interactions with polar clay layers and their nanocomposites represent improved stress at break values compared to that of pure polymer (Wang et al., 1998). Polyamide/clay nanocomposites have shown highly improved stress at break values, because of their higher degree of exfoliated structures based on the presence of strong ionic bonds between polymer chains and clay layers (Alexandre & Dubois, 2000). The nanocomposites of non-polar polymers such as polyolefines with clays exhibit weak interfacial interactions and low degree of exfoliation and consequently show decreased or slightly improved stress at break values. However the modification of polymer chains or clay layers may be causes to improved stress at break characteristics for non-polar polymers.

Elongation at break for polymer/clay nanocomposites is similarly dependent to the interfacial interactions of polymer/clay system. There are both increasing and decreasing reports of elongation at break for polymer/clay nanocomposites in literature (Finnigan et al., 2004; Yao et al., 2002). The impact strength of polymer/clay nanocomposites was also studied and compared with that of pure polymer systems. The reports are usually emphasis on the increasing of impact strength by the addition of very low fraction of clay loadings such as 0.1 wt% (Zhang et al., 2000; Oya et al., 2000). In these studies organic modified clay fillers were added to the compatibilized polyolefines. There are also reports on the decreasing of impact resistance by the addition of clay materials (Phang et al., 2005). Dynamic mechanical analysis studies on polymer/clay nanocomposites have shown that usually storage modulus and glass transition temperature (T_g) are improved by the addition of clay nanofillers (Nam et al., 2001; Laus et al., 1997; Krikorian & Pochan, 2003; Ray et al., 2002).

8.2 Thermal properties

Thermal stability of polymer/clay nanocomposites have been analyzed and compared with that of pure polymer under various oxidative (air) or non-oxidative (inert gases) conditions. Thermogravimetric analysis (TGA) has been usually used to evaluate the thermal stability of polymer and their nanocomposites. In this technique, the weight loss of the material due to the formation of volatile compounds under degradation because of the heating and temperature rising is monitored.

In general, it has been reported that the polymer/clay nanocomposites are thermally more stable than pure polymers (Ray & Bousima, 2005; Becker et al., 2004; Zanetti et al., 2004; Wang et al., 2002). That is the incorporation of clay fillers into the polymer matrices results to the improvement of their thermal stability. The effect of clay layers has been more explained as superior insulation and mass transport barrier against the volatile compounds generated during the decomposition of polymer under thermal conditions (Zhu et al., 2001). Also it has been reported that clays assist in the formation of layered carbonaceous char during degradation of polymer/clay nanocomposite (Ray et al., 2003). Clay minerals are inorganic materials and are almost stable in the temperature ranges that organic polymers are degraded in to volatile compounds. Therefore in TGA experiments the clay content of nanocomposites is remain as residue after heating program (Gilman, 1999).

Blumstein (1965) first reported the improved thermal stability of PMMA/MMT nanocomposite. TGA showed that PMMA intercalated into the Na-MMT has 40-50 °C higher decomposition temperature. Vyazovkin et al. (2004) reported the thermal stability of polystyrene/clay nano-composite compared to pure PS. They have showed that PS/clay nanocomposites have 30-40 °C higher degradation temperature compared to pure PS under nitrogen and air heating degradation conditions. Other studies have been showed that the nature of the clay modifier can affect on the thermal stability of modified clay and related nanocomposites (Mittal, 2007).

In some other studies the thermal decomposition behavior of nanocomposites and pure polymers under air (oxidative decomposition) and inert gas (non-oxidative decomposition) have been compared (Berta et al., 2006; Zhao et al., 2005; Beyer et al., 2002). Results showed that the formation of insulator layered char in oxidative degradation is achieved better than non-oxidative degradation heating program using some inert gases such as helium (Zhao et al., 2005; Beyer et al., 2002). Therefore under inert gas thermogravimetric experiments the nanocomposites may have the same degradation temperature as pure polymer.

We have analyzed the thermal degradation behavior of PS/MMT nanocomposites as well as pure PS. According to the results, thermal stability of nanocomposite has been improved compared to the pure polystyrene. Thermogravimetric analysis (TGA) of neat PS and PS/clay nanocomposites was obtained by heating program of 10 °C /min. Figure 7 shows the TGA curves for PS and PS/MMT nanocomposites. According to the results, the degradation temperature of the PS in nanocomposites has been shifted to higher temperatures compared to neat PS. The temperature of the 50% degradation of PS/clay nanocomposites has been increased compared to the pure PS. The 50% degradation temperature of pure PS was 411 °C, but that of the PS/Na-MMT nanocomposite was 422 °C and for PS/Cloisite 30B was 416 °C which is indicating respectively 11 °C and 5 °C improvement with 4 wt% of the clay loadings. For PS/Cloisite 30B nanocomposite two stages of weight loss under heating program took place. The first stage of weight loss at temperatures of about 220-312 °C represents the decomposition of modifying agent of Cloisite 30B. The second stage of weight loss at temperatures of about 312-450 °C is related to the decomposition of PS. The TGA data are summarized in table 3.

The effect of polymer compatibilizer on the thermal degradation behavior of compatiblized polymer/clay nanocomposites has also been investigated. Results showed that the incorporation of compatibilizers even with low molecular weight has not a significant unwanted effect on the thermal degradation of nanocomposites (Mittal, 2008).

8.3 Flame retardancy

Because of the large use of polymers especially in domestic applications, there is need to reduce their potential for ignition or burn in order to make them more safer in applications. Conventionally chemical additives as flame retardants are used to retardant the ignition and control burn.

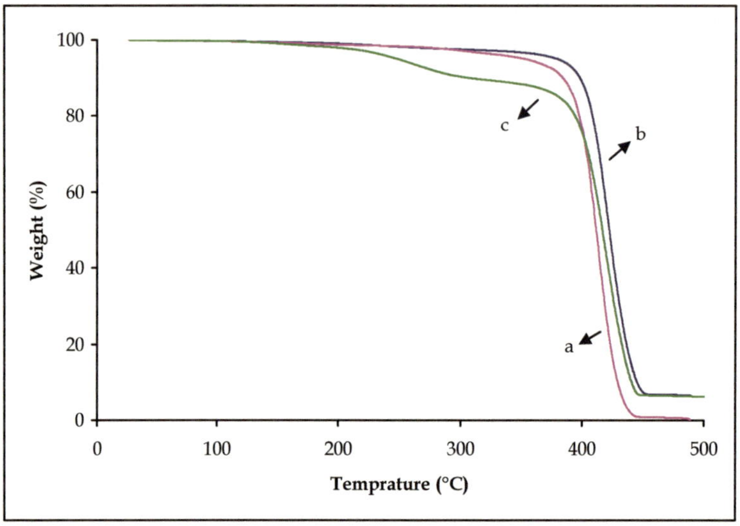

Fig. 7. TGA curves for PS (a), PS/Na-MMT (b) and PS/Cloisite 30B (c).

Traditionally a number of halogen-based as well as phosphorous and some other compounds have the flame retardancy effect in polymers without the reducing of their other properties and quality. However the halogenated compounds suffer from environmental contamination characteristics and have pushed the market trends to halogen-free flame retardants (Kiliaris & Papaspyrides, 2010; Levchik, 2007). Some of inorganic candidates require high level of loading which cause to un-wanted effects on the costs, process ability and quality of the product (Zhao et al., 2005; Laoutid et al., 2009).

Sample	$T_{0.15}{}^{a}$(°C)	$T_{0.5}{}^{b}$(°C)	End Degradation temperature (°C)
Pure PS	392°C	411°C	447°C
PS/Na-MMT (4 wt %)	404°C	422°C	455°C
PS/Cloisite 30B (4 wt %)	380°C	416°C	450°C

Table 3. Thermal degradation data of the PS and PS/clay nanocomposites, based on the TGA results. [a] : 15% Degradation temperature ($T_{0.15}$) and [b] : 50% Degradation temperature ($T_{0.5}$).

Also in other work the thermal properties of epoxy and epoxy/clay nanocomposites were investigated. Figure 8 shows the TGA thermograms of the neat epoxy and epoxy/cloisite 30B nanocomposite with 3% and 5% wt of cloisite 30B. In general, major weight losses are observed in the range of ~300-400 °C for epoxy and nanocomposites, which may be corresponding to the structural decomposition of the polymers. It can be seen in figure 8 that the Td (the temperature of degradation at which weight loss is 20 wt %) for epoxy/cloisite 30B (3 wt %) nanocomposite has shifted toward the higher temperature range against to the pure epoxy in a way that, the Td of the nanocomposite is 319 °C, which is superior to that of the pure epoxy (Td = 313 °C). However the Td of epoxy/cloisite 30B (5 wt

%) nanocomposite is lower than that of neat epoxy due to the catalytic effect of nanoclay on the cross-linking reaction of the epoxy resin with curing agent. After ~500 °C, all of the curves became flat and mainly the inorganic residue remains. The char yields of the nanocomposites are higher than that of the neat epoxy.

Investigations on the polymer/clay nanocomposites have been demonstrated that clay minerals represent some degree of flame retardancy along with the improvement in the physical and mechanical properties for polymers (Kiliaris & Papaspyrides, 2010; Giannelis, 1998). However detailed explanations implies that clay materials are beneficial for retarding the flame spread in developing fires and they not improve the ignition retardancy or control the fully developed fires (Livchik & Weil, 2000; Zanetti et al., 2002; Zanetti, 2006). Therefore the clay minerals are used together with a low fraction of conventional flame retardants (Wang et al., 2004). However the clay minerals and traditional flame retardants have considerable synergistic effect in the reduction of ignitability of polymers and various reports of organically modified clay nanocomposites combined with halogenated flame retardants (Weil et al., 1992; Wang et al., 2005) and phosphorous flame retardants (Zhang & Horrocks, 2003; Hao et al., 2006) were presented in literature. The synergic effect of clays on the flame retardancy of halogenated and phosphorous compounds causes to the reduction of their volume fraction needed in the polymer/clay nanocomposite systems.

In a series of other works attempts have been made to intercalate the various phosphorous flame retardants into the clay galleries. The clays modified with such phosphorous compounds have been used in the preparation of nanocomposites with various polymers. In this approach the phosphorous flame retardants are shielded by the clay sheets which prevent their evaporation during preparation of nanocomposite especially by the melt compounding method and it's remaining for more efficient flame retardancy in the final product (Chigwada & Wilkie, 2003).

Schartel et al. (2006) have used tetra-phenyl phosphonium modified clay for the preparation of nanocomposites based on DGEBA resin. Toldy et al. (2006) have also used phosphorous compounds both as curing agent and flame retardant in the preparation of epoxy/clay nanocomposites.

8.4 Barrier properties

One of the most considerable effects of clays in the polymer matrix properties is the dramatic improvement of barrier properties of polymers. Clay sheets are naturally impermeable. Clays increase the barriers properties of polymers by creating a maze or tortuous path that retards the diffusion of gas molecules through the polymer matrix (figure 9) (Pavlidou & Papaspyrides, 2008; Yano et al., 1993).

The degree of enhancement in the barrier properties depends on the degree of tortuousity created by clay layers in the diffusion way of molecules trough the polymer film. The tortuous factor is determined by the ratio of actual distance which diffusive molecule is walked to the shortest distance to diffuse (polymer film thickness). This factor is affected by the aspect ratio of clay dispersed in the matrix. Increasing the lateral length of clay sheet as well as increasing of exfoliation or dispersion degree cause to the more barrier enhancement in the polymer matrix. Many studies have reported the barrier properties of polymer/clay nanocomposites against the diffusion of gases and vapors (Tortora et al., 2002; Ray et al., 2003; Lange & Wyser, 2003; Giannelis, 1996; Koh et al., 2008).

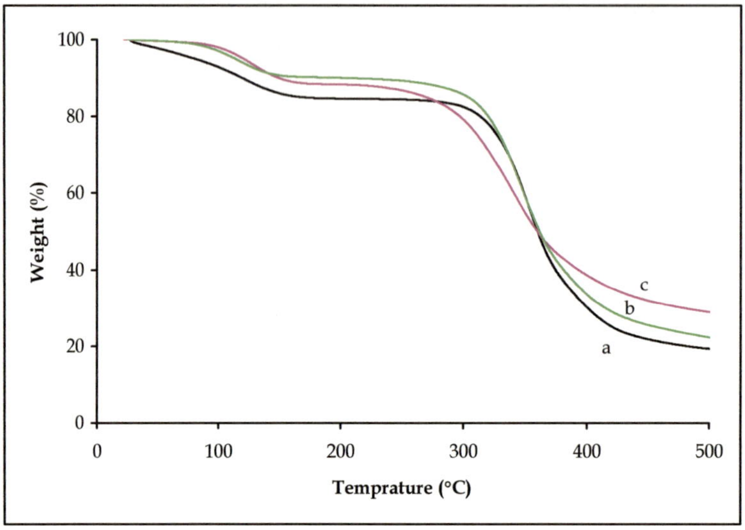

Fig. 8. TGA thermograms of neat epoxy (a) and epoxy/cloisite 30B nanocomposites with 3 wt % (b) and 5 wt % (c) cloisite 30B contents.

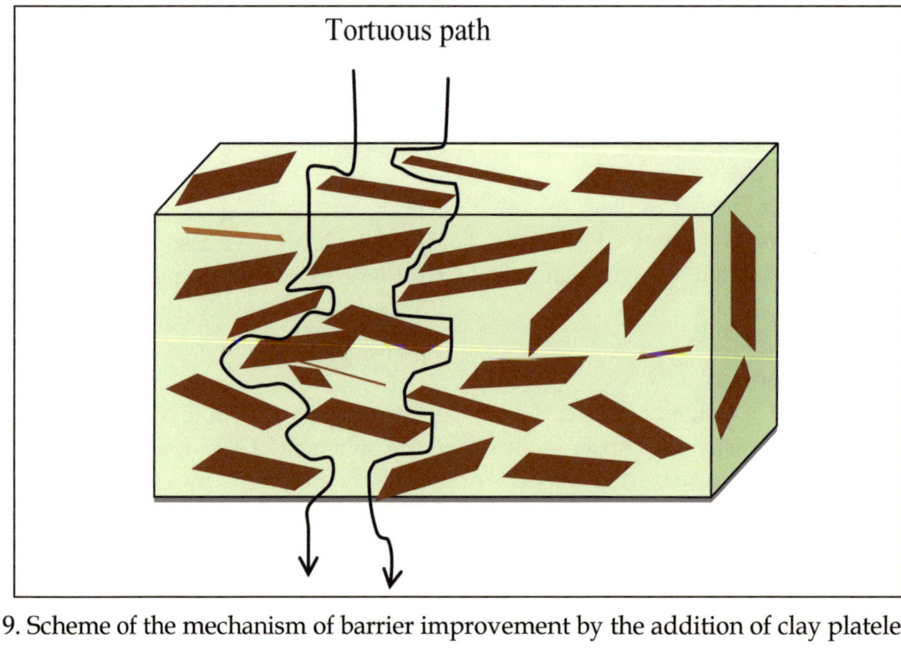

Fig. 9. Scheme of the mechanism of barrier improvement by the addition of clay platelets.

8.5 Anticorrosive properties

Polymers are widely used as anticorrosive coatings on metals to prevent the corrosion. Primarily polymeric coatings act as physical barrier against the diffusion of aggressive species to the metal surface. However most polymer coatings show some degree of permeations especially in long time period of contact with aggressive species. Various techniques have been applied to improve the barrier effect of polymeric coatings. One of the effective methods is based on the preparation of polymer based composite or nanocomposite coatings by the addition of proper fillers to the polymer coating matrix. It has been shown that the addition of platelet fillers like as layered silicates, effectively improves the anticorrosive barrier effect of polymer coatings by increasing the length of the diffusion pathways for aggressive species. Clay materials due to their platelet structure and high aspect ratio, in well dispersed state, decrease the permeability of polymer coating films by increasing the diffusion pathways such as shown in figure 9.

The nanocomposite of various polymeric materials such as polystyrene, poly (styrene-co-acrylonitrile), polyaniline, polypyrrole, polysulfone, polyacryltes, polyimide and epoxy with unmodified and organically modified clays have been investigated as anticorrosive coatings on metals (Yeh et al., 2008; Olad & Rashidzadeh, 2008). According to our results obtained by electrochemical tafel slope analysis, the corrosion current of iron samples coated by the 100 μm thickness of polyaniline/MMT coating is more lower than that of samples coated by pure polyaniline coating with the same thickness in various corrosive environments (Olad & Rashidzadeh, 2008).

9. References

Ahmad, M.B.; Hoidy, W.H.; Ibrahim, N.A.B. & Al-Mulla, E.A.J. (2009). Modification of montmorillonite by new surfactants. *J. Eng. Appl. Sci.*, Vol.4, No. 3, 184-188.

Alexandre, M. & Dubois, P. (2000). Polymer–layered silicate nanocomposites: preparation, properties and uses of a new class of materials. *Mater. Sci. Eng. R Rep.*, Vol. 28, 1–63.

Al-Saleh, M.H. & Sundararaj, U. (2010). Processing-microstructure-property relationship in conductive polymer nanocomposites. *Polymer*, Vol. 51, 2740-2747.

Armentano, I.; Dottori, M.; Fortunati, E.; Mattioli, S. & Kenny, J.M. (2010). Biodegradable polymer matrix nanocomposites for tissue engineering: A review. *Polym. Degrad. Stab.*, In Press.

Becker, O.; Varley, R.J. & Simon, G.P. (2004). Thermal stability and water uptake of high performance epoxy layered silicate nanocomposites. *Euro. Polym. J.*, Vol. 40. 187–195.

Berta, M.; Lindsay, C.; Pans, G. & Camino, G. (2006). Effect of chemical structure on combustion and thermal behavior of polyurethane elastomer layered silicate nanocomposites. *Polym. Degrad. Stab.*, Vol. 91, 1179-1191.

Beyer, G. (2002). Flame retardant properties of EVA-nanocomposites and improvements by combination of nanofillers with aluminium trihydrate. *Fire Mater.*, Vol. 25, 193-197.

Blumstein, A. (1965). Polymerization of adsorbed monolayers: II. Thermal degradation of the inserted polymers. *J. Polym. Sci. A*, Vol. 3, 2665-2670.

Chigwada, C. & Wilkie, C.A. (2003). Synergy between conventional phosphorus fire retardants and organically-modified clays can lead to fire retardancy of styrenics. *Polym. Degrad. Stab.*, Vol. 80, 551-557.

Chigwada, G.; Wang, D.; Jiang, D.D. & Wilkie C.A. (2006). Styrenic nanocomposites prepared using a novel biphenyl-containing modified clay. *Polym. Degrad. Stab.*, Vol. 91, 755-762.

Cosoli, P.; Scocchi, G.; Pricl, S. & Fermaglia, M. (2008). Many-scale molecular simulation for ABS-MMT nanocomposites: Upgrading of industrial scraps. *Microporous and Mesoporous Materials*, Vol. 107, 169-179.

Finnigan, B.; Martin, D.; Halley, P.; Truss, R. & Campell, K. (2004). Morphology and properties of thermoplastic polyurethane nanocomposites incorporating hydrophilic layered silicates. *Polymer*, Vol. 45, 2249-2260.

Fornes, T.D. & Paul, D.R. (2003). Modeling properties of nylon 6/clay nanocomposites using composite theories. *Polymer*, Vol. 44, 4993-5013.

Fornes, T.D.; Yoon, P.J.; Keskkula, H. & Paul, D.R. (2001). Nylon 6 nanocomposites: the effect of matrix molecular weight. *Polymer*, Vol. 42, 9929–9940.

Gacitua, W.E.; Ballerini, A.A. & Zhang, J. (2005). Polymer nanocomposites: synthetic and natural fillers, A review. *Ciencia technologia*, Vol. 7, No. 3, 159-178.

Giannelis, E.P. (1996). Polymer layered silicate nanocomposites. *Adv. Mater.*, Vol. 8, 29–35.

Giannelis, E.P. (1998). Polymer-layered silicate nanocomposites: synthesis, properties and applications. *Appl. Organomet. Chem.*, Vol. 12, 675–680.

Gilman, J.W. (1999). Flammability and thermal stability studies of polymer layered-silicate (clay) nanocomposites. *Appl. Clay Sci.*, Vol. 15, 31-49.

Gorrasi, G.; Tortora, M.; Vittoria, V.; Pollet, E.; Lepoittevin, B. & Alexandre, M. (2003). Vapor barrier properties of polycaprolactone montmorillonite nanocomposites: effect of clay dispersion. *Polymer*, Vol. 44, 2271–2279.

Groosiord. N.; Loos, J.; Laake, L.; Maugey, M.; Zakri, C.; Koning, C.E. & Hart, A.J. (2008). High-conductivity polymer nanocomposites obtained by tailoring the characteristics of carbon nanotube fillers. *Adv. Func. Mater.*, Vol. 18, 3226-3234.

Hao, X.; Gai, G.; Liu, J.; Yang, Y.; Zhang, Y. & Nan, C. (2006). Flame retardancy and antidripping effect of OMT/PA nanocomposites. *Mater. Chem. Phys.*, Vol. 96, 34–41.

Hedayati, A. & Arefazar, A. (2009). Multi-scale analysis of polypropylene based organoclay containing composites, part 1: Morphology. *Polymer Testing*, Vol.28, 128-138.

Hussain, F.; Hojjati, M.; Okamoto, M. & Gorga, R.E. (2006). Review article: Polymer-matrix nanocomposites, processing, manufacturing, and Application: An Overview. *Journal of Composite Materials*, Vol. 40, No. 17, 1511-1565.

Jiang, L.; Lam, Y.C.; Tam, K.C.; Chua, T.H.; Sim, G.W. & Ang, L.S. (2005). Strengthening acrylonitrile-butadiene-styrene (ABS) with nano-sized and micron-sized calcium carbonate. *Polymer*, Vol. 46, 243-252.

Jordan, J.; Jacob, K. I.; Tannenbaum, R.; Sharaf, M. A. & Jasiuk, I. (2005). Experimental trends in polymer nanocomposites-a review. *Mater. Sci. Eng. A*, Vol. 393, 1-11.

Kiliaris, P. & Papaspyrides, C.D. (2010). Polymer/layered silicate (clay) nanocomposites: An overview of flame retardancy. *Prog. Polym. Sci.*, Vol. 35, 902-958.

Koh, H.C.; Park, J.S.; Jeong, M.A.; Hwang, H.Y.; Hong, Y.T.; Ha, S.Y. & Nam, S.Y. (2008). Prearation and gas permeation properties of biodegradable polymer/layered silicate nanocomposite membranes. *Desalination*, Vol. 233, 201-209.

Kojima, Y.; Usuki, A.; Kawasumi, M.; Okada, A.; Kurauchi, T. & Kamigaito, O. (1993). Sorption of water in nylon 6–clay hybrid. *J. Appl. Polym. Sci.*, Vol. 49, 1259–1264.

Kornmann, X.; Linderberg, H. & Bergund, L.A. (2001). Synthesis of Epoxy–Clay Nanocomposites: Influence of the Nature of the Curing Agent on Structure. *Polymer*, Vol. 42, 4493–4499.

Krikorian, V. & Pochan, D. (2003). Poly(l-lactide acid)/layered silicate nanocomposite: fabrication, characterization, and properties. *Chem. Mater.*, Vol. 15, 4317-4324.

Kumar A.P.; Depan, D.; Tomer N.S. & Singh, R.P. (2009). Nanoscale particles for polymer degradation and stabilization-Trends and future perspectives. *Prog. Polym. Sci.*, Vol. 34, 479-515.

Lagaly, G. (1999). Introduction: from caly mineral-polymer interactions to caly mineral-polymer nanocomposites. *Appl. Clay Sci.*, Vol. 15, 1-9.

Lan, T.; Kaviratna, P.D. and Pinnavaia, T.J. (1995). Mechanism of clay tactoid exfoliation in epoxy–clay nanocomposites. *Chem. Mater.*, Vol. 7, No.11, 2144–2150.

Lange, J. & Wyser, Y. (2003). Recent innovations in barrier technologies for plastic packaging-a review. *Packag. Technol. Sci.*, Vol.16, 149–158.

Laoutid, F.; Bonnaud, L.; Alexandre, M.; Lopez-Cuesta, J.M. & Dubois, P. (2009). New prospects in flame retardant polymer materials: from fundamentals to nanocomposites. *Mater. Sci. Eng.*, Vol. 63, 100–125.

Laus, M.; Francesangeli, O. & Sandrolini, F. (1997). New hybrid nanocomposites based on an organophilic clay and poly(styrene-b-butadiene) copolymers. *J. Mater. Res.*, Vol. 12, 3134-3139.

LeBaron, P.C.; Wang, Z. & Pinnavaia, T.J. (1999). Polymer-layered silicate nanocomposites: an overview. *Appl. Clay Sci.*, Vol. 15, 11-29.

Levchik, S.V. (2007). Introduction to flame retardancy and polymer flammability. In: Morgan, A.B. & Wilkie, C.A. editors. *Flame retardant polymer nanocomposites.* Hoboken, NJ: John Wiley & Sons Inc. 1–29.

Levchik, S.V. & Weil, E.D. (2000). Combustion and fire retardancy of aliphatic nylons. *Polym. Int.*, Vol. 49, 1033–73.

Ma, H.; Xu, Z.; Tong, L.; Gu, A. & Fang, Z. (2006). Studies of ABS-graft-maleic anhydride/clay nanocomposites: Morphologies, thermal stability and flammability properties. *Polym. Degrad. Stab.*, Vol. 91, 2951-2959.

Meneghetti, P. & Qutubuddin, S. (2006). Synthesis, thermal properties and applications of polymer-clay nanocomposites. *Thermochimica Acta*, Vol. 442, 74-77.

a) Mittal, V. (2007). Polypropylene-layered silicate nanocomposites: Filler matrix interactions and mechanical properties. *J. Thermoplast. Compos. Mater.*, Vol. 20, 575-599.

b) Mittal, V. (2007). Gas permeation and mechanical properties of polypropylene nanocomposites with thermally-stable imidazolium modified clay. *Euro. Polym. J.*, Vol. 43, 3727-3736.

Mittal, V. (2008). Mechanical and gas permeation properties of compatibilized polypropylene–layered silicate nanocomposites. *J. Appl. Polym. Sci.*, Vol. 107, 1350-1361.

Mittal, V. (2009). Polymer Layered Silicate Nanocomposites: A Review. *Materials*, Vol. 2, 992-1057.

Nam, P.H.; Maiti, P.; Okamoto, M.; Kotaka, T.; Hasegawa, N. & Usuki, A. (2001). A hierarchical structure and properties of intercalated polypropylene/clay nanocomposites. *Polymer*, Vol. 42, 9633-9640.

Olad, A. & Rashidzadeh, A. (2008). Preparation and anticorrosive properties of PANI/Na-MMT and PANI/O-MMT nanocomposites. *Prog. Org. Coat.*, Vol. 62, 293-298.

a) Osman, M.A.; Rupp, J.E.P.& Suter, U.W. (2005). Tensile properties of polyethylene-layered silicate nanocomposites. *Polymer*, Vol. 46, 1653-1660.

b) Osman, M.A.; Rupp, J.E.P. & Suter, U.W. (2005). Effect of non-ionic surfactants on the exfoliation and properties of polyethylene-layered silicate nanocomposites. *Polymer*, Vol. 46, 8202-8209.

Oya, A.; Kurokawa, Y. & Yasuda, H. (2000). Factors controlling mechanical properties of clay mineral/polypropylene nanocomposites. *J. Mater. Sci.*, Vol. 35, 1045-1050.

Pandey, J.K.; Reddy, K.R.; Kumar, A.P. & Singh, R.P. (2005). An overview on the degradability of polymer nanocomposites. *Polym. Degrad. Stab.*, Vol. 88, 234-250.

Paul, D.R. & Robeson L.M. (2008). Polymer nanotechnology: Nanocomposites. *Polymer*, Vol. 49, 3187-3204.

Pavlidou, S. & Papaspyrides, C.D. (2008). A review on polymer-layered silicate nanocompopsites. *Prog. Polym. Sci.*, Vol. 32, 1119-1198.

Phang, I.Y.; Liu, T.; Mohamed, A.; Pramoda, K.P.; Chen, L.; Shen, L. Chow, S.Y.; He, C.; Lu, X. & Hu, X. (2005). Morphology, thermal and mechanical properties of nylon 12/organoclay nanocomposites prepared by melt compounding. *Polym. Int.*, Vol. 54, 456-464.

Ray, S.S. (2010). A new possibility for microstructural investigation of clay-based polymer nanocomposite by focused ion beam tomography. *Polymer*, Vol. 51, 3966-3970.

Ray, S.S. & Bousima, M. (2005). Biodegradable polymers and their layered silicate nanocomposites: in greening the 21st century materials world. *Prog. Mater. Sci.*, Vol. 50, 962–1079.

Ray, S.S.; Maiti, P.; Okamoto, M.; Yamada, K. & Ueda, K. (2002). New polylactide/ layered silicate nanocomposites. 1. Preparation, characterization and properties. *Macromolecules*, Vol. 35, 3104-3110.

Ray, S.S. & Okamoto, M. (2003). Polymer/layered silicate nanocomposites: a review from preparation to processing. *Prog. Polym. Sci.*, Vol. 28, 1539-1641.

Ray, S.S.; Yamada, K.; Okamoto, M.; Ogami, A. & Ueda, K. (2003). New polylactide/ layered silicate nanocomposites. 3. High performance biodegradable materials. *Chem. Mater.*, Vol. 15,1456–1465.

Rehab, A. & Salahuddin, N. (2005). Nanocomposite Materials Based on Polyurethane Intercalated into Montmorillonite Clay. *Mater. Sci. Eng. A*, Vol. 399, 368–376.

Schartel, B.; Knoll, U.; Hartwig, A, & Putz, D. (2006). Phosphonium-modified layered silicate epoxy resins nanocomposites and their combinations with ATH and organo-phosphorus fire retardants. *Polym. Adv. Technol.*, Vol. 17, 281–293.

Shelley, J.S.; Mather, P.T. & DeVries, K.L. (2002). Reinforcement and environmental degradation of nylon 6/clay nanocomposites. *Polymer*, Vol. 42, 5849–58.

Sheng, N.; Boyce, M.C.; Parks, D.M.; Rutledge, G.C.; Abes, J.I. & Cohen R.E. (2004). Multiscale micromechanical modeling of polymer/clay nanocomposites and the effective clay particle. *Polymer*, Vol. 45, 487-506.

Thostenson, E.T.; Li, C. & Chou, T.W. (2005). Nanocomposites in context. *Compos. Sci. Technol.*, Vol. 65, 491-516.

Toldy, A.; Toth, N.; Anna, P.; Keglevich, G.; Kiss, K. & Marosi, G. (2006). Flame retardancy of epoxy resin with phosphorus-containing reactive amine and clay minerals. *Polym. Adv. Technol.*, Vol. 17, 778–781.

Tortora, M.; Vittoria, V.; Galli, G.; Ritrovati, S. & Chiellini, E. (2002). Transport properties of modified montmorillonite/poly(caprolactone) nanocomposites. *Macromol. Mater. Eng.*, Vol. 287, 243-249.

Tortora, M.; Gorrasi, G., Vittoria, V.; Galli, G.; Ritrovati, S. & Chiellini, E. (2002). Structural characterization and transport properties of organically modified montmorillonite/polyurethane nanocomposites. *Polymer*, Vol. 43, 6147–6157.

Usuki, A.; Kawasumi, M.; Kojima, Y.; Okada, A.; Kurauchi, T. & Kamigaito, O. J. (1993). Swelling behavior of montmorillonite cation exchanged for V-amino acids by E-caprolactam. *Mater. Res.*, Vol. 8, No.5, 1174.

Vyazovkin, S.; Dranka, I.; Fan, X. & Advincula, R. (2004). Kinetics of the thermal and thermo-oxidative degradation of a polystyrene–clay nanocomposite. *Macromol. Rapid Commun.*, Vol. 25, 498-503.

Wang, D.; Echols, K. & Wilkie, C.A. (2005). Cone calorimetric and thermogravimetric analysis evaluation of halogen-containing polymer nanocomposites. *Fire Mater.*, Vol. 29, 283–294.

Wang, S.; Hu, Y.; Song, L.; Wang, Z.; Chen, Z. & Fan, W. (2002). Preparation and thermal properties of ABS/montmorillonite nanocomposite. *Polym. Degrad. Stab.*, Vol. 77, 423-426.

Wang, S.; Hu, Y.; Zong, R.; Tang, Y.; Chen, Z. & Fan, W. (2004). Preparation and characterization of flame retardant ABS/montmorillonite nanocomposite. *Appl. Clay Sci.*, Vol. 25, 49–55.

Wang, S.J.; Long, C.F.; Wang, X.Y.; Li, Q. & Qi, Z.N. (1998). Synthesis and properties of silicon rubber organo-montmorillonite hybrid nanocomposites. *J. Appl. Polym. Sci.*, Vol. 69, 1557-1561.

Weil, E.D.; Hirschler, M.M.; Patel, N.G. & Shaki, S. (1992). Oxygen index: correlations to other fire tests. *Fire Mater.*, Vol. 16, 159–167.

Yano, K.; Usuki, A.; Okada, A.; Kurauchi, T. & Kamigaito, O. (1993). Synthesis and properties of polyimide/clay hybrid. *J. Polym. Sci. Polym. Chem.*, Vol. 31, 2493–2498.

Yao, K.J.; Song, M.; Hourston, D.J. & Luo, D.Z. (2002). Polymer/layered clay nanocomposites: 2 polyurethane nanocomposites. *Polymer*, Vol. 43, 1017-1020.

Yeh, J.M. & Chang, K.C. (2008). Polymer/layered silicate nanocomposite anticorrosive coatings. *J. Ind. Eng. Chem.*, Vol. 14, 275-291.

Zanetti, M. (2006). Flammability and thermal stability of polymer/layered silicate nanocomposites. In: Mai YW, Yu ZZ, editors. *Polymer nanocomposites*. Cambridge, UK: Woodhead Publishing Ltd; 256–272.

Zanetti, M.; Bracco, P. & Costa, L. (2004). Thermal degradation behavior of PE/clay nanocomposites. *Polym. Degrad. Stab.*, Vol. 85, 657–665.

Zanetti, M.; Camino, G.; Canavese, D.; Morgan, A.B.; Lamelas, F.J. & Wilkie, C.A. (2002). Fire retardant halogen–antimony–clay synergism in polypropylene layered silicate nanocomposites. *Chem. Mater.*, Vol. 14, 189–193.

Zanetti, M.; Lomakin, S. & Camino, G. (2000). Polymer layered silicate nanocomposites. *Macromol. Mater. Eng.*, Vol. 279, 1-9.

Zhang, Q.; Fu, Q.; Jiang, L. & Lei, Y. (2000). Preparation and properties of polypropylene/montmorillonite layered nanocomposites. *Polym. Int.*, Vol. 49, 1561-1564.

Zhang, R.; Ni, Q.Q.; Natsuki, T. & Iwamoto, M. (2007). Mechanical properties of composites filled with SMA particles and short fibers. *Composite Structures*, Vol. 79, 90-96.

Zhang, S. & Horrocks, A.R. (2003). A review of flame retardant polypropylene fibres. *Progr. Polym. Sci.*, Vol. 28, 1517–1538.

Zhang, Z. & Friedrich, K. (2003). Artificial neural networks applied to polymer composites: a review. *Compos. Sci. Technol.*, Vol. 63, 2029-2044.

Zhao, C.; Qin, H.; Gong, F.; Feng, M.; Zhang, S. & Yang, M. (2005). Mechanical, thermal and flammability properties of polyethylene/clay nanocomposites. *Polym. Degrad. Stab.*, Vol. 87, 183-189.

Zhu, J.; Uhl, F.M.; Morgan, A.B. & Wilkie, C.A. (2001). Studies on the mechanism by which the formation of nanocomposites enhances thermal stability. *Chem. Mater.*, Vol. 13, 4649–4654.

8

A Strategy to Decorate the Surface of NPs and Control their Locations within Block Copolymer Templates

Misang Yoo[1], Joona Bang[1],
Kwanyeul Paek[2] and Bumjoon J. Kim[2]
[1]*Department of Chemical and Biological Engineering, Korea University,*
[2]*Department of Chemical and Biomolecular Engineering,*
Korea Advanced Institute of Science and Technology,
Republic of Korea

1. Introduction

Polymer nanocomposites consisting of polymers and inorganic nanoparticles (NPs) have attracted many attentions due to their promising potentialsof diverse applications such as solar cell, sensors, catalysts, ferroelectric devices, etc. To impart sufficient stability of NPs and integrate NPs into polymer matrix, it is a prerequisite that the surface of the NPs is properly treated with various ligands having desired properties. Up to date, a number of strategies to synthesize the appropriate gold nanoparticles (Au NPs) have been reported. The most popular method that has been for a long time to prepare Au NPsis to use citrate reduction of $HAuCl_4$ in water, which was introduced by Turkevitch (Turkevich et al., 1951). And this method was developed by Frens to obtain controlled size of Au NPs from 10 to 100 nm, via varying the concentration ratio between $HAuCl_4$ and sodium citrate (Frens, 1973). In organic solvent, the so called "two-phase method" developed by Brust et al. has a considerable impact in this field until today due to the facile synthesis of stable Au NPsthat are protected by alkanethiols with controlled size and monodispersity (Brust et al., 1994). Later, they also introduced the "one-phase method" in methanol (Brust et al., 1995) and Yee et al. have expanded this method to various metal NPs such as gold, palladium and iridium in tetrahydrofuran (Yee et al., 1999). Furthermore, Hostetler et al. demonstrated that the functionality of monolayer protected NPs can be further enhanced via ligand exchange method (Hostetler et al., 1999).

Recently, the inorganic NPs which is surface modified with polymeric ligands instead of the small surfactant molecules, such as alkanethiols or citrate, have attracted great interest as they can provide not only the improved stability but also various functionalities, unique structure and characteristics, and compatibility with other matrices. As a versatile approach for the surface modification of NPs, it has been shown that various polymers can be grafted onto the NP surface via "grafting-from", "grafting-to", "ligand exchange" or "templating" methods. In the "grafting-from" method, polymers are usually grown from the NP surfacesvia living-free radical polymerization, which are modified with the initiators. In contrast, when the end-functionalized polymers are synthesized, such as thiol-terminated polymers, they can be directly grafted onto the NP surface. In both methods, it has been demonstrated that the surface of various NPs can be successfully modified with desired polymeric ligands to fabricate the well-defined nanocomposites. For example, the location of the NPs within block copolymers can be precisely controlled. When the NP surface is modified with one block, it was shown that the resulting NPs are located at the corresponding domain. The more interesting issue is to control the location of NPs at the interface of AB block copolymers. In this case, it was shown that the decoration of the NP surface with a mixture of A/B homopolymers can result in the "Janus" type amphiphilic NPs and hence they are located at the interface of AB block copolymers. This "Janus" type NPs provide a practical method for producing nanostructures that are amenable to technological applications. For examples, they can be used as templates to transfer the copolymer structure into the substrate, i.e., nanoreactors where catalysts can be anchored to the base of the pores, and photoelectrochemical applications. In this chapter, we will cover aforementioned issues, NP surface modification, the control of NPs within the block copolymers and their applications.

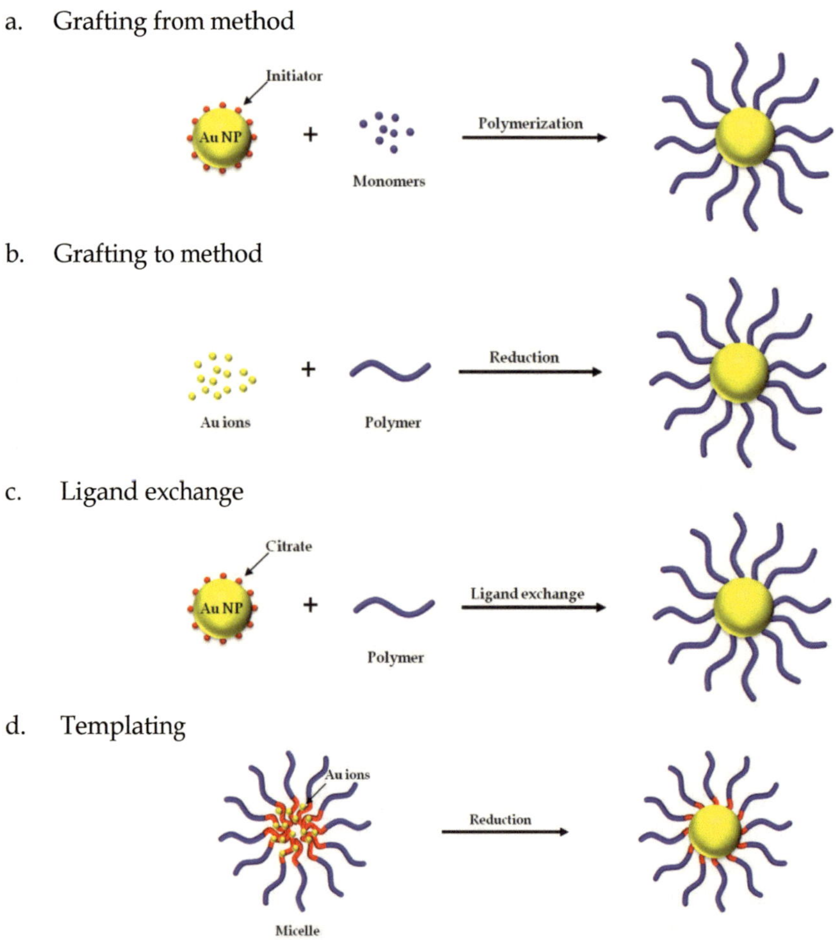

Fig. 1. Schematic illustration to prepare the surface modified Au NPs via (a) grafting from, (b) grafting to, (c) ligand exchange and (d) templating methods.

2. Decoration of nanoparticle surface

2.1 Grafting from approach

In "grafting from"method, the surface initiated NPs are preformed and polymers are grown directly from the surface of the NPs.To synthesize the polymers, various polymerization techniques were reported. The most prevalent technique is the living radical polymerizations,such as atom transfer radical polymerization (ATRP), reversible addition-fragmentation chain transfer (RAFT), due to its simplicity and versatility. Especially, the main advantage of combining the "grafting from" method with the living radical polymerization is not only the control of molecular weight of polymer but also the possibility to achieve higher surface grafting density of polymer brush. In "grafting to" method, it is difficult to obtain high grafting density because of the steric hindrance of polymer ligands. But in case of "grafting from" method, the polymer brushes grow from the initiators on the surface of NPs, and hence the high grafting density of polymer brush can be obtained. The initiator is usually anchored on the surface of NPsusing disulfide initiator or by reacting with active group such as hydroxyl, carboxylic acid, amine, etc. As an example, Raula and coworkerssynthesized Au NPs coated with 11-mercapto-1-undecanol and subsequently modified the Au NPs into RAFT agent which has carboxylic acid at the end (Raula et al., 2003). Then poly(*N*-isopropylacrylamide) (pNIPAM) has been polymerized using RAFT polymerization. They investigated the changes in the surface plasmon of Au NP in different media, and effects of particle concentration, temperature in aqueous solutions. In this case, it was found that the solubility of these Au NPs in water was low due to the low grafting degree of pNIPAM ligands. Kim et al. synthesized the pNIPAM coated Au NPs in aqueous media (Kim et al., 2005). The polymerization initiator, $[BrC(CH_3)_2COO(CH)_{11}S]_2$, was immobilized to the surface of citrate coated Au NPs by ligand exchange method. The cross-linked pNIPAM ligands were synthesized by ATRP as well as brush type pNIPAM with or without ethylene diacylate as a cross-linker. They compared the thermo-responsiveness of the Au NP/pNIPAM hybrid as function of degree of cross-linking. A similar approach was reported using a different cross-linker (Li et al., 2007). The preformed Au NPs were modified by disulfide initiator and the pNIPAM chains were grown in the presence of *N,N*-methylenebisacrylamide (MBAA) as cross-linker via ATRP. As a result, the slightly cross-linked Au NPs can be used to trap and encapsulate other NPs, biomolecules, dyes, or drugs by a temperature-introduced "breathing" process. Later, they also fabricated pH responsive Au NPs which were modified with poly(4-vinylpyridine) (P4VP) using the similar approach(Li et al., 2007). At low pH, the polymer layer of Au NPs is loosely swollen because of the pyridyl groups, which are facially protonated and positively charged. At high pH, the Au NPs were aggregated due to the collapse of polymer layers. Such Au-P4VP nanocomposites as a smart supporter can entrap transition metal ions by their efficient coordinating segments, and the metal ions can be reduced *in situ* to construct the novel pH-stimuli bimetallic catalysts.

Recently, the Au NPs which are conjugated with biomolecules like DNA and protein have attracted great interest due to the potential use of the programmability for diverse application from biosensing, biodiagnostics to nanoelectronics. Especially, DNA was chosen as a polymeric material in many researches due to the specificity of DNA base-pairing, the predictability of inter- or intramolecular interactions, its physicochemical stability, and mechanical rigidity. Zhao et al. polymerized long single-stranded DNA on Au NPs via rolling circle amplification (RCA) reaction catalyzed by a special DNA polymerase known as Φ29 DNA polymerase (Zhao et al., 2006).Thiol-modified DNA primers were functionalized onto the 15 nm diameter Au NPs prepared by the classical citrate reduction. A 63-nucleotide-long circular DNA template was annealed with the Au NPsand then RCA was carried out for long single-stranded DNA polymerization. The resulting DNA-Au NP conjugates were used as 3D scaffolds for further nano-assemblies by incubating long DNA-Au NP with 5 nm Au NPs that were modified with a 25-nt antisense oligonucleotide. Also, Au NPs having peptide chains on the surfaces have been prepared by Higuchi et al. via ring-opening polymerization of γ-methyl L-glutamate *N*-carboxyanhydride with fixed amino groups on the NP surface as an initiator (Higuchi et al., 2007). The peptide-Au NPs, whose peptide chains haveα-helical conformation, formed a fibril assembly owing to the dipole-dipole interaction between the peptide chains on the distinct NPs. On the other hand, when the peptide chains on the surface adopt a random coil conformation, individual peptide-Au NPs existed in dispersed form.

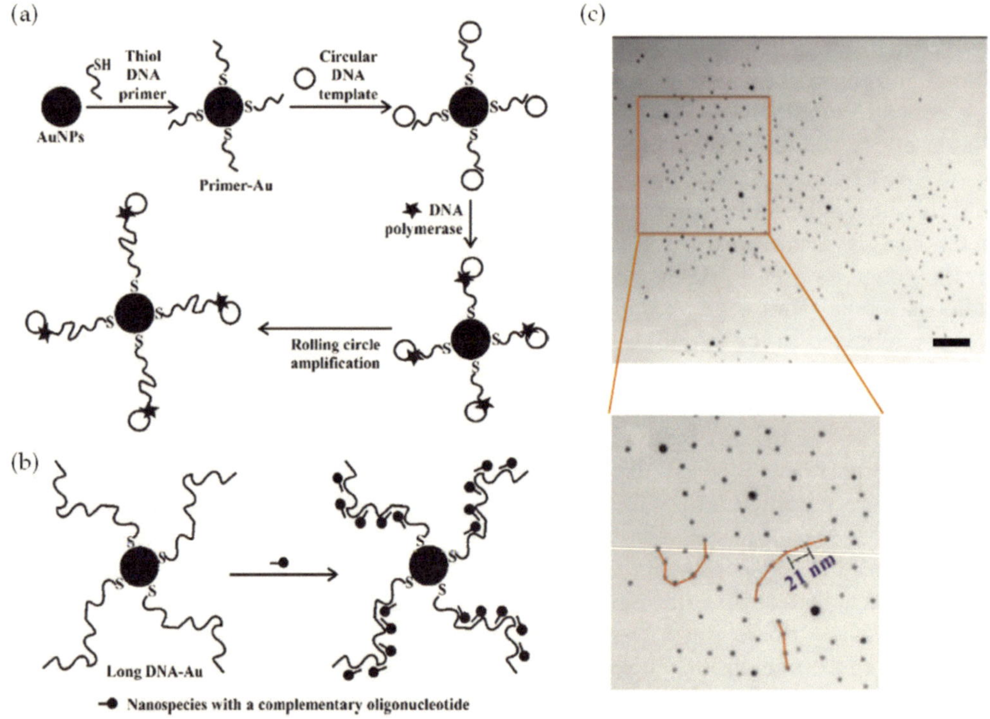

Fig. 2. Schematic illustration of (a) rolling circle amplification (RCA) reaction on AuNPs. (b) DNA–Au NP conjugates produced by RCA as a scaffold forthe formation of 3D nanostructures. (c) TEM images of nano-assembled superstructures prepared from 5nm Au NPs modified with antisense DNA and long DNA–Au scaffolds. Scale bar: 100 nm (Zhao et al., Angew. Chem. Int. Edit., 2006, 45, 2409. Copyright Wiley-VCH Verlag GmbH & Co. KGaA. Reproduced with permission.)

Until now, the special affinity between thiol (-SH) and gold is mainly used to attach polymers to the Au NPs surface. However the reversible nature of the Au-thiol bond, which allows ligand escape above 60 °C, significantly hampers the stability of the Au NPs. Many groups tried to enhance the thermal stability of Au NPs through the introduction of cross-linking of polymeric shell or protective inorganic shell. Kotal et al. synthesized Au-PMMA NPs by ATRP at the elevated temperature (Kotal et al., 2005). The cross-linked siloxane primer layer on the Au NP surface, resulting from the hydrolysis of trimethoxysilane tail group of 3-mercaptopropyltrimethoxysilane, was used to attach the initiators and stabilized the initiator-containing surface to high temperature ATRP. Matyjaszewski group recently demonstrated the one pot synthesis of stable Au NPs protected with a cross-linked shell and linear tethered polymer brushes by "graftingfrom" approach (Dong et al., 2008).The synthesis of cross-linked polymeric shell is accomplished by adding *n*-butyl acrylate monomer and the dimethacrylate as cross-linker at the beginning of ATRP. The reactivity of the monomer is much lower than that of the cross-linker, therefore, the cross-linker was consumed faster than the monomer and was preferentially incorporated into the cross-linked shell. The subsequent polymerization of the remaining monomer formed the linear brush layer. The resulting Au NPs solution was heated up to 110 °C to test the thermal stability and exhibit almost same plasmon band in UV-vis spectra before and after heating.

2.2 Grafting to approach

In "grafting to" method, polymers end-capped with thiol groups or containing disulfide units have been used to directly attach polymer ligands on Au NPs via the two-phase method (Brust method) or one-phase method. Since the polymer ligandsusually prepared via living/controlled polymerization prior to being employed to synthesis of Au NPs, the desirable properties of polymer such as chain length, architecture, composition and functionality can be precisely designed. Recently, the RAFT polymerization has been more widely used than other techniques, such as NMP and ATRP, as it has a wide choice of monomer to polymerize and doesn't require a special metal catalyst. Especially, the main advantage of RAFT polymerization is that the dithioester group existed at the end of polymer can be readily converted to the thiol group by reduction reaction with base. Therefore,the resulting polymers can be used for the surface modification of various metal NPs such as CdSe, Pt and Au. In this case, both the two-phase and one-phase methods were modified by introducing polymer ligands instead of dodecanethiol or octadecanethiol in the original procedure.Corbierre et al. synthesized Au NPs modified with tetradecanethiol (C_{14}-SH) or thiolated polymers (PS-SH and PEO-SH) by one-phase method in THF (Corbierre et al., 2004). They compared the grafting densities of polymers on Au NPs and either on flat gold surface. In this case, increasing the chain length leads to an increase in the steric hindrance of ligands, resulting in decreasing the corresponding surface density both on Au NPs and on flat Au surfaces. Overall, it was found that the grafting density on the curved NPswas higher than that on the flat surface, and, this effect is larger in polymer ligands than small ligands.This is because the presence of the voids at the facet edges results in extra volume for grafting the polymeric ligand and the percentage of void volume is higher for long polymeric ligand than small ligand. Lowe et al. also demonstrated a route for preparing of water soluble Au NPs coated with four different polymers (Lowe et al., 2002). Four (co)polymers have been synthesized by RAFT polymerization in aqueous media: poly(sodium 2-acrylamido-2-methyl propane sulfonate) (PAMPS), poly((*ar*-vinylbenzyl)-trimethylammonium chloride) (PVBTAC), poly(*N,N*-dimethylacrylamide) (PDMA), and poly(3-[2-*N*-methylacrylamido)-ethyldimethyl ammonio propane sulfonate-*b*-*N,N*-dimethylacrylamide) (PMAEDAPS-*b*-PDMA). These (co)polymers are composed of an anionic, cationic, neutral, and zwitterionic (betaine) species. The dithioester group at the end of polymers was reduced to thiols in aqueous media in the presence of a suitable transition metal complex or metal sol and stabilized the Au NPs. When the reduction takes place in the presence of the PVBTAC (anionic polymer), PDMA and PMAEDAPS-*b*-PDMA, stable colloids are formed. But In case of PAMPS, the Au NPs are not as stable as the others and flocculate readily within 48 h. It was also noted that the NPs coated by PMAEDAPS-*b*-PDMA are only stable in the presence of salt (NaBr in this case) due to the betaine nature. The dithioester end-capped polymer was also used to prepare the Au NPs by Shan et al (Shan et al., 2003). They synthesized the Au NPs coated by three types of pNIPAMligands having different end groups or using different synthesis method. The first type was comprised of pNIPAM with narrow molar mass distributions, synthesized by RAFT polymerization and thus bearing a dithioester at the chain end. These polymers were used directly to passivate the Au NPs upon the one-phase method. The second type of ligand was derived from the first one through hydrazinolysis, and they therefore contained a thiol end group. The third type of ligand was pNIPAMs obtained through conventional radical polymerization, postmodified to contain thiol end groups. But the modification efficiency was not high and the resulting Au NPsprotected with this type of pNIPAM tended to cross-link into a network. On the other hand, either by directly using pNIPAM containing dithioester at chain end or by using thiolated pNIPAM obtained via hydrolysis of the first type, the Au NPs were successfully prepared. They also investigated thermal properties of pNIPAM coated Au NPs in water (Shan et al., 2004). They observed the two separate phase transitions of the pNIPAM chains on Au NPs, different from that of the corresponding free pNIPAM dissolved in water. The first transition with a sharp and narrow endothermic peak occurs at lower temperature, while the second one with a broader peak occurs at higher temperature. The observations are tentatively rationalized by assuming that the pNIPAM brush can be subdivided into two zones, the inner shell and the outer shell. In the inner shell, the pNIPAM segments are close to the gold surface, densely packed, less hydrated, and undergo the first transition. In the outer shell, the pNIPAM segments are looser and more hydrated, adopt a restricted random coil conformation, and show a phase transition, which is dependent on both concentration of Au NPs and the chemical nature of the end groups of the pNIPAM chains.

Recently, several groups have demonstrated the preparation of the amphiphilic Au NPs coated by two different polymeric ligands. Shan et al. prepared the amphiphilic Au NPs grafted with a mixture of pNIPAM and polystyrene (PS) by one-phase method in a homogenous THF phase (Shan et al., 2005). The pNIPAM and PS chains have dithioester group because they were synthesized by RAFT polymerization and directly used to passivate the Au NPs without further modification as their previous work. Monolayer of the amphiphilic Au NPs at the air/water interface was built up by employing a Langmuir trough. During the compression of the monolayer, they observed several characteristic regions that can be ascribed to the pNIPAM conformational transition affecting the optical properties. They concluded that the chemically different pNIPAM and PS chains grafted on the gold core tend to be phase separated through the contact angle measurements on the upper and lower surfaces of the Au NPs monolayer. They also investigated the surface plasmon resonance and the thermal responsiveness of the amphiphilic Au NPs monolayer (Shan et al., 2006). The surface plasmon resonance band of the monolayer undergoes a blue-shift as the surface pressure increases which is closely related to the conformational change of pNIPAM chains. In contrast, the UV-vis absorption spectra of the Au NPs multilayers transferred at different temperatures onto slides have been measured and showed a red-shift in the surface plasmon resonance peak with increasing of the deposition cycle due to the presence of a weak interparticle coupling in the multilayers. However, during the synthesis of the amphiphilic Au NPs, the distribution and exact ratio of polymers attached to the surface of Au NPs may be difficult to control by using two separated polymers as the aforementioned method. Zubarev et al. and Genson et al. prepared the Au NPs which were functionalized with amphiphilic polybutadienen-poly(ethylene glycol) (PB-PEG) V-shaped arms (Zubarev et al., 2006; Genson et al., 2006). First, they prepared a V-shaped PB-PEG amphiphile containing a carboxyl group at its junction point and then the diblock copolymers were attached to the hydroxyl-terminated 2 nm Au NPs by the esterification. Consequently, 65 V-shaped PB-PEG arms were grafted on a given Au NP, resulting in a high grafting density of 2.94 chains/nm^2. The amphiphilic nature of the Au NPs is expressed in their unique ability to dissolve in any conventional solvent from hexane, benzene, dichloromethane, THF, methanol to water. And these Au NPs formed stable Langmuir monolayers at the air-water and the air-solid interfaces. At these interfaces, the binary arms vertically segregated into a dense polymer corona, which is surrounded the Au NPs, preventing their large-scale agglomeration and keeping individual NPs well-separated from each other and forming flattened, pancake nanostructures. Wan et al. synthesized the Au NPs using a hyperbranched polyglycerol that hasa hydrophilic moiety as well as a hydrophobic moiety in one polymer(Wan et al., 2006). The amphiphilic polymer was composed of hydrophobic thioether containing shell and hydrophilic polyglycerol core and used to synthesize the Au NP by two-phase method.

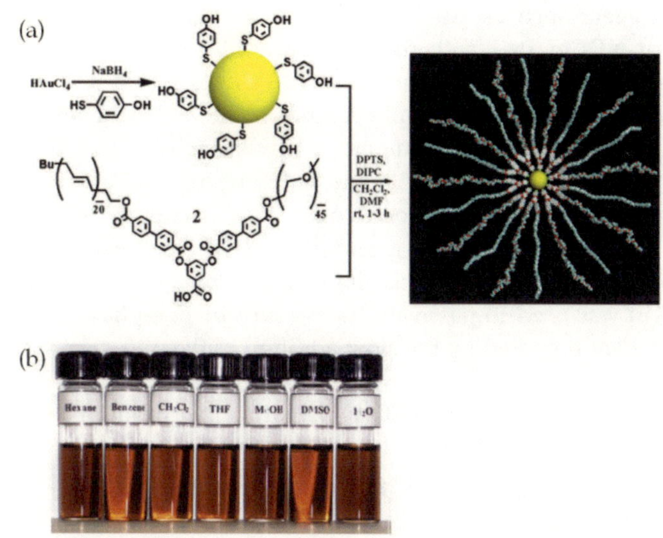

Fig. 3. (a) Synthesis of amphiphilic Au(PB-PEG)n NPs. (b) solutions of amphiphilic Au(PB-PEG)n NPs in various solvents (Adapted in part with permission from J. Am. Chem. Soc., 2006, 128, 4958; Langmuir, 2006, 22, 7011. Copyright 2006 American Chemical Society.)

One of most great challenges in the nanotechnology is to synthesize of highly monodisperse Au NPs as well as control the size of NPs. Digestive ripening, heating a colloidal suspension at or near the solvent boiling point in the presence of a surface-active ligand, was applied to polydisperse colloidal gold in toluene using a series of alkylthiols, viz., octyl-, decyl-, dodecyl-, and hexadecylthiols (Prasad et al., 2002). For all cases, the digestive ripening reduced the average particle size and polydispersity significantly. Adjusting the interparticle separation of the Au NPs by the alkyl chain length controls their solubility behavior, leading to the formation of exclusive 2D or 3D superlattices. In case of the Au NPs coated with octanethiol and decanethiol, the stronger and larger 3D superlattices were observed and also lead to large shift in the optical spectra into the near-infrared region. Hussain et al. have demonstrated the size-controlled synthesis of near-monodisperse Au NPs in the 1-4 nm range (Hussain et al., 2005). The particle size was controlled by the ratio of gold to capping ligand, a water soluble alkyl thioether end-functionalized poly(methacrylic acid) (PMAA). The Au NPs are readily obtainable in both aqueous solution and nonaqueous solutions. Even though they couldn't give a precise rationalization for their new approach, this simple protocol for the gram scale preparation of monodisperse Au NPs below 5 nm has a practical value in various fields where the precise size control and the monodispersity are required.

Recently, the Au NPs conjugated with biomaterials led to new and great development with infinite potential in biotechnology and nanotechnology. Combining the inorganic Au NPs and organic biomaterials brings the synergistic effects that incorporate the highly selective catalytic and recognition properties of biomaterials, such as proteins/enzymes and DNA, with the unique electronic, photonic, and catalytic features of NPs. The excellent biocompatibility and biodegradability of poly(ε-Caprolactone) (PCL) make them well suited for drug and gene delivery and tissue engineering. But their fate *in vivo* cannot be directly observed by TEM in histological section. Therefore the labeling of PCL biocarrier is desirable by a contrasting agent such as Au NPs which have high electron density. Aryal et al. fabricated the Au NPs grafted with the thiolated PCL by two-phase method as well as ligand exchange (Aryal et al., 2006). The thiolated PCL (PCL-SH) was synthesized through the ring opening bulk polymerization of ε-Caprolactone and removal of protecting group under slightly basic condition subsequently.PCL was also employed forming diblock copolymer to stabilize the Au NPs (Azzam & Eisenberg, 2007). Poly(ethylene oxide)-*b*-poly(ε-caprolactone) (PEO-*b*-PCL) copolymer was synthesized by ring opening polymerization, and the hydroxyl end group of the PCL block was modified with thioctic acid using dicyclohexyl carbodiimide as the coupling reagent. The resulting PEO-*b*-PCL-thioctate ester (TE) was used to passivate the Au NPs by one-phase method in THF. Then micelles in water were prepared by the slow addition of the dilute Au NP solution in DMF into a large excess of water. Under these conditions, the PCL block forms the shell surrounding the gold core, while the PEO block forms the corona. They observed the only micelles with a single gold core per micelle based on the TEM.

On the other hand, DNA or polypeptide grafted Au NPs were investigated by "grafting to" method but there are only a few examples because the Au NPs are usually synthesized through the "ligand exchange" in water for the biocompatibility. Higashi et al. described the preparation of disulfide terminated polypeptide (poly(γ-benzyl-L-glutamate)) (PBLG) monolayercovered Au NPs via two-phase method(Higashi et al., 2005). The PBLG segments attached on Au NPadopt α-helix conformation with nearly 100% content which is much higher value compare to the free PBLG chains in chloroform. They explained that the observed helix enhancement is probably due to PBLG side chain interactions such as π–π stacking of benzyl groups, resulting from characteristic alignment of PBLG segments on Au NPs, at which they are forced to assemble densely.Recently, Cheng et al. applied the DNA coated Au NPs to fabricate the free-standing Au NP superlattice sheets (Cheng et al., 2009). First, they prepared monodisperse 13 nm diameter Au NP capped with 5'-thiolated single-stranded DNA (ssDNA) through the two-phase method. They used the DNA as a dry ligand, similar to alkyl molecular ligands, without the requirement of base-pairing. For further application of superlattices in free-standing electronics, the rigorous control over their internal structures and the overall shapes of discrete features are required. However, in alkyl based fabrication of superlattices, the interparticle spacing is constrained to a small range (~1.2 - ~2.3 nm) owing to synthetic limitations on maximum molecular ligand length. Using the ssDNA as a dry ligand is one of methods that can circumvent the above limitations, by combining a microhole-confined, drying-mediated self-assembly process. They obtained discrete, freestanding superlattice sheets in which both structure (interparticle spacings) and functional properties (plasmonic and mechanical) can be rationally controlled by adjusting DNA length. In particular, the edge-to-edge interparticle spacing for monolayered superlattice sheets can be tuned up to 20 nm, which is a much wider range than has been achieved with alkyl molecular ligands.

In common with the grafting from method, many efforts have been devoted to increase the thermal stability of Au NPs through grafting to method.The instability is resulted from the dissociation of polymeric ligand from the surface of Au NPs under high temperature. The stability of Au NPs could be improved through the introduction of multiple thiol-anchoring groups at the polymer chain ends or shell cross-linking of polymeric ligands encapsulating the Au NPs. Huang et al. prepared the Au NPs coated with poly(*p*-methylstyrene) containing multiple thioether groups on the chain side (Huang et al., 2005). The synthesis of poly(*p*-methylstyrene) (PMS) was accomplished via anionic polymerization. The resulting polymer was chlorinated with sodium hypochloriteand the extent of chlorination was determined as 30% from the ^{1}H NMR.Since a PMS molecule (with a molecular weight of 15,000) comprised 127 repeating *p*-methylstyrene units, the 30% chlorination amounted to 38 chloromethyl groups on each polymer molecule. The chlorinated poly(*p*-methylstyrene) was then reacted with methylthiomethyllithium, which had been prepared via a metalation reaction of dimethyl sulfide with n-butyllithium, to form a styrenic polymer containing thioether groups on the sides of the molecule, PMS-$(CH_2SCH_3)_n$. Au NPs were synthesized *in situ* in the presence of PMS-$(CH_2SCH_3)_n$ by borohydride reduction of gold precursor in chloroform phase. Although the PMS-$(CH_2SCH_3)_n$ coated Au NPs were relatively stable below the T_g of PMS (~102 °C), the particles were free to move and start clustering at temperatures above the T_g. In TEM micrograph of PMS–$(CH_2SCH_3)_n$ coated Au NPs after being heated at 150 °Cfor 24 h, the clustering of particles was significant. These methods still suffer from a lack of thermal stability and require multistep synthetic approaches.Very recently, Yoo and coworkers reported a simple, yet powerful strategy for preparing thermally stable Au NPs by introducing photo-cross-linkable azide groups ($-N_3$) into the polymeric ligands attached to the gold core (Yoo et al., 2010). Thiol-terminated block copolymers composed of polystyrene and azido-polystyrene, P(S-*b*-S-N_3)-SH, were prepared via RAFT polymerization. They designed the ligands in whichthe short PS-N_3 block is placed adjacent to the thiol group.Consequently, the gold core can be protected by cross-linked shells after *in situ* photo-cross-linking to give core-shell nanostructures with well-defined linear polymer chains surrounding a cross-linked polymer shell. The P(S-*b*-S-N_3)-SH coated Au NPs were synthesized by the two-phase method, and the resulting Au NPs were efficiently cross-linked by exposure to UV light subsequently. The cross-linked Au

NPsexhibit excellent thermal stability and no aggregation was observed on heating to 200 °C for extended periods of time in both solutions and nanocomposite films. Furthermore, it was demonstrated that these Au NPscan serve as compatibilizers for thermally annealed PS/PMMA blends. The cross-linked Au NPs were located at the PS/PMMA interface resulting in the reduction in the phase separated droplet size of PS or PMMA. Since the design of the photo-cross-linkable polymeric ligands is modular, the surface property of the Au NPs can be readily tuned and applied to other functional metal NPs. This facile synthetic method can provide a versatile route to the fabrication of a variety of nanocomposite systems.

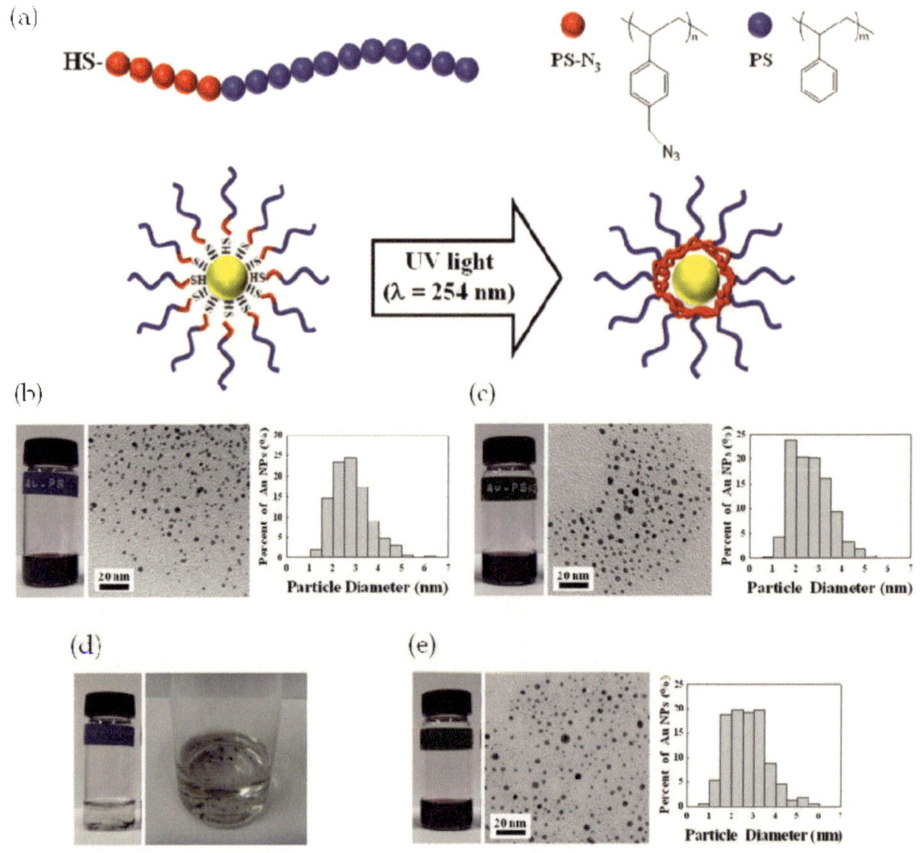

Fig. 4. Schematic illustration for the design of shell-protected Au NPs via UV cross-linking reaction of azido (-N_3) group. TEM images and the size distribution of (b) PS-Au NPs and (c) cross-linked P(S-*b*-S-N_3)-Au NPsbefore heating. (d) Photograph image of PS-Au NPs in DBP solution after heating (e) TEM images and the size distribution of cross-linked P(S-*b*-S-N_3)-Au NPs after heating (Adapted in part with permission from Macromolecules, 2010, 43, 3570. Copyright 2010 American Chemical Society.)

2.3 Ligand exchange approach

In order to improve the stability of given NPs, the ligands on the surface can be exchanged by others containing new properties or functionalities. In most cases, the incoming ligands should bind more strongly to the inorganic NP surface. Au NPs are usually preformed via citrate reduction which is one of the oldest ways to synthesize the Au NPs (Turkevich et al., 1951), prior to the ligand exchange. And the resulting citrate coated Au NPs have high degree of monodispersity with size ranging from 5 to 250 nm. Thus, Au NPs by this method are widely used in not only the areas that require the monodisperse NPs such as electrochemical quantized capacitance charging and single-electron transistor assembly but also in the biotechnology that requires the biocompatibility rather than the two- or one-phase method performed in organic media. Mirkin and coworkers have pioneered the DNA linked Au NPs using ligand exchange(Mirkin et al., 1996). The 13 nm Au NPs were prepared via citrate reduction and they were readily modified with two kinds of thiol-terminated DNA oligonucleotides containing 8-base pair sticky ends which are complementary each other. After an addition of the DNA grafted Au NP aqueous solution to another DNA grafted Au NP solution, an immediate color change is observed from red to purple and a precipitation of the Au NPs ensues resulting from the linking of complementary oligomers anchored on the Au NPs surface. This process can be reversed by thermal denaturation generating the unlinked Au NPs which are soluble in the aqueous solution.

Poly(ethylene glycol) (PEG) is well soluble in a number of organic polar and nonpolar solvents, as well as in water where it is heavily hydrated, forming random coils. Its simple structure and chemical stability make them inert and biocompatible. The inertness and non-toxic properties of PEG combined with NPs give rise to a number of applications in medicine, chemistry, or biotechnology. PEG modified NPs are less toxic as well as more stable at high salt concentrations and in biological environments. Niidome et al. developed a technique to modify the gold nanorods with PEG for medical applications (Niidome et al., 2006). Since gold nanorods stabilized with hexadecyltrimethylammonium bromide (CTAB) show strong cytotoxicity, it is not suitable for biological applications. They achieved the PEG modification by adding thiolated PEG in the CTAB coated gold nanorod solution through the ligand exchange, then, excess CTAB was removed by dialysis. PEG-modified gold nanorods showed a nearly neutral surface and had little cytotoxicity *in vitro*. Liu et al. reported the assembly of PEG and mixed peptide/PEG monolayers on Au NP surfaces(Liu et al., 2007). The PEG coated or mixed peptide/PEG coated Au NPs were prepared through incubation of citrate coated Au NPs with thiolated PEG and cysteine-terminated peptide. The stability of the resulting bioconjugates in high ionic strength media was characterized as function of NP size, PEG length, and monolayer composition. They found that the Au NP stability increases with increasing PEG length, decreasing particle size and increasing PEG mole fraction. Also, mixed monolayers prepared via the sequential addition of PEG followed by peptide were more stable than particles prepared via simultaneous co-adsorption, because the highly cationic, lysine-rich peptide may associate with the anionic Au NPs more rapidly than the thiolated PEG. As a consequence, the charge on the NPs would be neutralized prior to the formation of a PEG steric barrier.

The modification of other molecules with PEG is often referred to as 'PEGylation'. PEGylation can provide water solubility to hydrophobic materials and PEGylated ligands can be introduced by ligand exchange to Au NPs. Becker et al. demonstrated an efficient way to prepare protein-functionalized Au NPs through the*N*-terminal modification of the small protein, GTPase Rab6A, with a thiolated PEG linker by native chemical ligation(Becker et al., 2007). The PEGylated proteins were used for immobilization on DNA-functionalized Au NPs. A ligand exchange reaction with the modified proteins was used to displace thiolated oligonucleotides from Au NPs in order to efficiently immobilize protein on the Au NPs. The protein-functionalized Au NPs have potentials for applications as probes in bioanalytical assays and as building blocks for the bottom-up assembly of functional hybrid nanoarchitectures. Bergen et al. presented a model system based on Au NPs grafting with thiolated PEG and thiolated PEG-galactose for determining optimal physicochemical characteristics for cell-specific delivery *in vivo*(Bergen et al., 2006). Negatively-charged Au NPs by capping with citrates and positively-charged Au NPs by capping with aminoethanethiol were prepared first, and thenNPs surface was modified by reaction with either PEG-thiol or galactose-PEG-thiol. Modification of the NPs did not significantly affect particle size based on the measurement of dynamic light scattering. The Au NPs were then exposed tosalt solutions to observe the salt stability. In the presence of salt, it was found that unmodified NPswere aggregated,while PEG modified NPs does not show the aggregation due to the steric stabilization. This platform was applied to screen for NP formulations that demonstrate hepatocyte targeted delivery *in vivo*. They found that the NP size and the presence of galactose ligands significantly impact the targeting efficiency.

Conjugation of Au NPs with smart polymers such as pNIPAM makes it possible to tailor the electronic and optical properties of hybrid system in response of temperature. Seker et al. have demonstrated a method for fabrication of nanostructured composites featuring simultaneous switching of several optical properties such as color, transparency and metallic luster through remote control of classical coupling in dense pNIPAM coated AuNP arrays (Seker et al., 2005). The pNIPAMs were prepared by RAFT polymerization and subsequently modified chain end group to thiol or disulfide. Au NPs were prepared by the citrate reduction in water then pNIPAM was modified by exchange reaction with weakly bound surface citrate ions. The thin films of pNIPAM/Au NPs can be prepared by simple solution-casting methods. Based on UV-vis spectra of the thin films, they observed that the change in interparticle separation occurs throughout the film. Unaggregated core/shell particles with stimulus-responsive pNIPAM shells of controlled thickness and architecture could generate controlled interparticle separations that govern the optical properties of composite materials. Singh and Lyon demonstrated a synthetic route toward the hollow pNIPAM nanogel via a "seed and feed" method combined with the sacrificial metal NP templating method (Singh & Lyon, 2007). They prepared citrate capped Au NPs as seeds for polymerization of pNIPAM. Seeds compatible with the precipitation polymerization of NIPAM were formed by the adsorption of a layer of amino-terminated pNIPAM onto the Au NPs by ligand exchange. Onto these pNIPAM stabilized Au NPs, further pNIPAM shells were synthesized by free-radical polymerization with an addition of acrylic acid as a comonomer to form pH-responsive nanogel particles. To obtain hollow nanogels from the gold core nanogels, the gold core was dissolved by the KCN solution. The resulting sub-50 nm hollow pNIPAM nanogels are ideal for the drug and gene delivery, cellular imaging and nanoreactors.

Block copolymers also have been used to stabilize a variety of NP materials by forming steric or ionic barriers around the particles. Kang and Tatonprepared Au NPs surrounded with shell of amphiphilic poly(styrene-*b*-acrylic acid) (PS-*b*-PAA) in water and permanently fixed them by cross-linking the outer PAA block(Kang & Taton, 2005). In almost of previous works, the specific chemical interaction of the particle surface and the surface-bound polymer should be explicitly tailored in order to form the shell. But in this case, they used the block copolymer to form shell around the gold core without chemical anchoring. The PS-

b-PAA copolymers were synthesized by ATRP and dissolved in DMF, which is a good solvent for both hydrophobic (PS) and the hydrophilic (PAA) blocks. Citrate-capped Au NPs were redispersed in DMF. The samples of polymer and NPs in DMF were combined in the presence of 1-dodecanethiol, which hydrophobically functionalizes the surfaces of theAu NPs. Then, water was gradually added to the mixture tosimultaneously desolvate both the particles and the hydrophobicpolymer block from solution and induce the formation of micelles around the NPs. In the absence ofPS-*b*-PAA, the dodecanethiol-modified NPsprecipitated quickly from DMF as water was added. Inthe absence of dodecanethiol, the NPs remainedoutside the assembled block copolymer micelles. Finally the assembled copolymer micelles containing Au NPs were permanently fixed by cross-linking the PAA block with addition of cross-linkers in water.They also demonstrated the influence on the shell thickness of block copolymer respect to the size of gold core, concentration and composition of block copolymer in this system (Kang & Taton, 2005). When the feed ratio of block copolymer to the total surface area of gold core is fixed, two different results were observed. For same or lager size of Au NPs than the block copolymer radius of gyration, each core-shell structure consisted of exactly one NP surrounded by a shell of fairly consistent thickness. However, for smaller Au NPs, some structures contained multiple Au NPs that act as solutes and dissolved within the micelle core. When the size of Au NPs is fixed, the shell thickness of block copolymer gradually increases with increasing content of the block copolymer and chain length of block copolymer, as expected. Another type of block copolymer encapsulated Au NP was also reported (Luo et al., 2005). Luo et al. fabricated core/shell Au NPs stabilized with a monolayer of double hydrophilic block copolymer, poly(2-(dimethylamino)ethyl methacrylate)-*b*-poly(ethylene oxide) (PDMA-*b*-PEO), and demonstrated their stimuli responsiveness before and after shell cross-linking. The PDMA-*b*-PEO diblock copolymer was prepared via RAFT polymerization and the dithioester group was reduced to a thiol group at the chain end. The thiolated PDMA-*b*-PEO was exchanged with citrate grafted onto the preformed Au NPs and 1,2-bis(2-iodoethoxy)ethane (BIEE) was utilized to selectively cross-link the PDMA residues in the inner shell. While both cross-linked and uncross-linked Au NPsexhibited the reversible pH responsiveness, the shell cross-linked Au NPsexhibited a robust core-shell nanostructure with high colloidal stability.

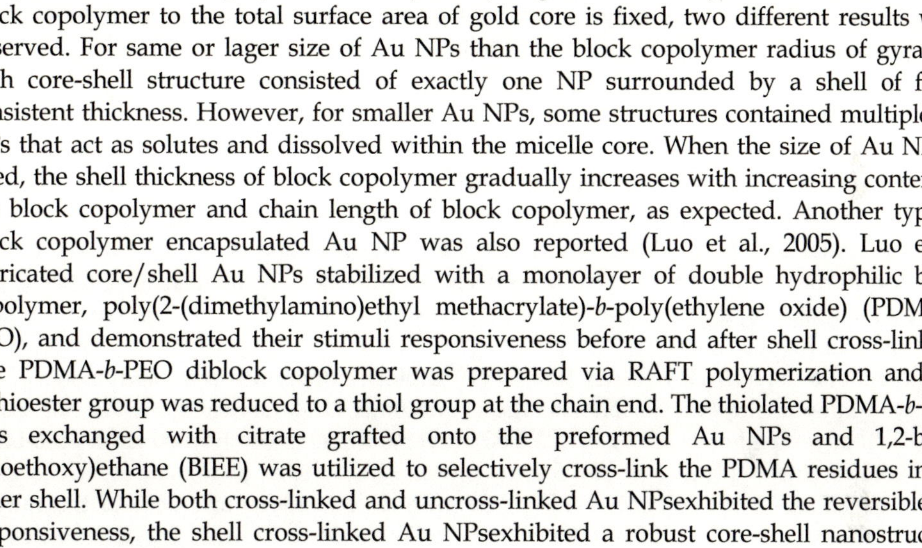

Fig. 5. Preparation of core/shell Au NPs using PS-*b*-PAA (Kang & Taton, Angew. Chem. Int. Edit., 2005, 44, 409. Copyright Wiley-VCH Verlag GmbH & Co. KGaA. Reproduced with permission.)

As an another example of the ligand exchange,the NPs in the original phase can be transferred to the second phase, e.g., by exchanging hydrophilicinto hydrophobic ligands. Commonly used ligand molecules include thiol groups that bind strongly to inorganic surfaces of NPs such asAu, Ag and CdSe, replacing the weaker bound ligands that the NPsusually have from synthesis.Underwood and Mulvaney have reported the phase transferred Au NPsfrom aqueous solution to organic solutionssuch as butyl acetate, carbon disulfide using a copolymer consisting of a backbone of methyl methacrylate and glycidyl methacrylate with pendant side chains of poly(12-hydroxystearic acid) (Underwood & Mulvaney, 1994). Kumar et al. demonstrated the phase transfer of various NPs, e.g. CdS into petroleum ether using octadecanethiol (Kumar et al., 2000), Ag into hexane using octadecaneamine (Kumar et al., 2003) and Pt into hexane using octadecaneamine (Kumar et al., 2004) from aqueous solution. In some applications that require the compatibility with biological systems, it is desirable that NPs synthesized in organic solvents are transferred to aqueous solutions. Kanaras et al. have transferred theAu NPs from toluene to aqueous phase using monohydroxy (1-mercaptoundec-11-yl) tetraethylene glycol (Kanaras et al., 2002) and Algar and Krull succeed in transferring the CdSe/ZnS quantum dots from toluene to aqueous solution with mercaptoacetic acid (MAA), mercaptosuccinic acid (MSA), and dihydrolipoic acid (DHLA) (Algar & Krull, 2006).

2.4 Templating approach

In "templating" method, polymers such as di- or triblock copolymer micelles, star polymers and dendrimers are used as the templates during the synthesis of Au NPs. These polymers are also referred to as "nanoreactor", since this approach uses the affinity between specific part of polymer and the gold precursor, $HAuCl_4$ or $KAuCl_4$, and thus reduced Au NPs are fixed within the nano-templates. This approach can provide controlled sizes, a narrow size distribution and long term stability of Au NPs as well as fabrication of single or multi gold core/polymer hybrid.

Poly(styrene)-*b*-poly(vinylpyridine) (PS-*b*-PVP) diblock copolymer micellesare one of the most well-known templates for NPs synthesis. Mössmer et al. fabricated the Au NPs using PS-*b*-P2VP as template (Mössmer et al., 2000). Inverse mecelles are formed from PS-*b*-P2VP diblock copolymer with a shell of PS and core of P2VP chains in toluene which is good solvent to PS and poor solvent to P2VP. By adding tetrachloroauric acid to the micelle solution, the complexation of P2VP block and $AuCl_4^-$ could be induced and reduced theAu NP at the core of micelle formed with subsequent addition of anhydrous hydrazine. However, without an addition of HCl to neutralize the excess hydrazine, Au NPs are coagulated to groups of two to four particles which can be attributed to the coagulation of micelles from the swelling of core volume owing to the excess hydrazine. Lu and Yi also prepared the Au NPs using PS-*b*-P2VP as template (Lu & Yi, 2006). They prepared $HAuCl_4$ loaded PS-*b*-P2VP solution in similar way and monolayer of the micelles was produced by spin-casting onto a silicon substrate. In contrast to previous examples, the reducing step has not been used during the process to generate Au NPs in thezero-valent state. Without any post or prior reduction treatments, the Au NPsare formed after treatment with oxygen plasmathat alsoremove the block copolymer templates. These Au NPs have been demonstrated as catalyst system for growing small-diameter silicon nanowires. PS-*b*-P4VP micelles are also used as template to prepare Au NPs in similar approach with PS-*b*-P2VP (Hou et al., 2007). The micellization of (PS-*b*-P4VP) in chloroform can be induced by the interaction between P4VP blocks and $HAuCl_4$, forming micelles with PS as the shell and the P4VP/$HAuCl_4$ complex as the core. Subsequent reduction of $HAuCl_4$ by hydrazine hydrate leads to hybrid polymeric micelles containing Au NPs in the core. The resultant hybrid polymeric micelles are quite stable in chloroform when the P4VP chains are fully protonated

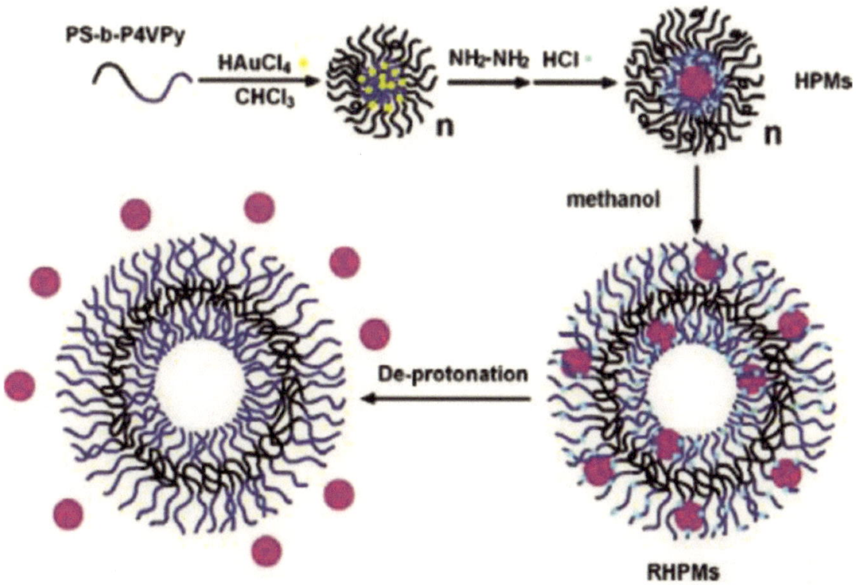

Fig. 6. Schematic description of the preparation of Au NPs in PS-*b*-P2VP micelles, vesicle-like reverse hybrid polymeric micelles, and subsequent release of Au NPs from the vesicle-like reverse hybrid polymeric micelles after deprotonation (Adapted in part with permission from Macromolecules, 2007, 40, 2134. Copyright 2007 American Chemical Society.)

by addition of hydrochloric acid. Continuous addition of methanol, where P4VP is soluble while PS is insoluble, to the solution of the hybrid micelle until the volume ratio of methanol to chloroform being 9/1 leads to a core-shell reversion of micelles, forming vesicle-like aggregates with PS as the wall and P4VP/Au NPs as the shell. Further protonation of P4VP block can induce the release of the Au NPs from the shell.

Various type of poly(ethylene oxide)-*b*-poly(propylene oxide)-*b*-poly(ethylene oxide) (PEO-*b*-PPO-*b*-PEO) amphiphilic triblock copolymer micelles have been examined in water to prepare Au NPs. Sakai and Alexandridis demonstrated a sing-step synthesis of Au NPsin aqueous PEO-*b*-PPO-*b*-PEO solution (Sakai & Alexandridis, 2004). The PEO-*b*-PPO-*b*-PEO block copolymers act as both reductants and stabilizers and they could obtain the stable Au NPs without any other reducing agent after 2 hours later at ambient temperature. They also investigated the effect of the overall molecular weight and the PEO and PPO block length on the reaction activity of Au NPs formation. The reaction activity increases with increasing overall molecular weight and concentration of the block copolymer, PEO block length and PPO block length. They suggested that the $AuCl_4^-$ions bound to PEO blocks are reduced via the oxidation of PEO by the metal center and PPO blocks cannot directly contribute to the reduction of metal ions but can help enhance the reactivity of the PEO blocks with metal ions, which is consistent with their observation. Regarding the effect of PEO-*b*-PPO-*b*-PEO block copolymer on the synthesis of Au NPs, they proposed that while both PEO and PPO blocks contribute to the $AuCl_4^-$ reduction, the PEO contribution appears to be dominant (Sakai & Alexandridis, 2005).Chen et al. presented the effect of PEO-*b*-PPO-*b*-PEO block copolymer micelles and their hydrophobicity on the stabilization of Au NPs (Chen et al., 2006). The stabilization capacities of the micelles were modulated by tuning the block copolymer concentration and composition and an amount of salts added. As the block

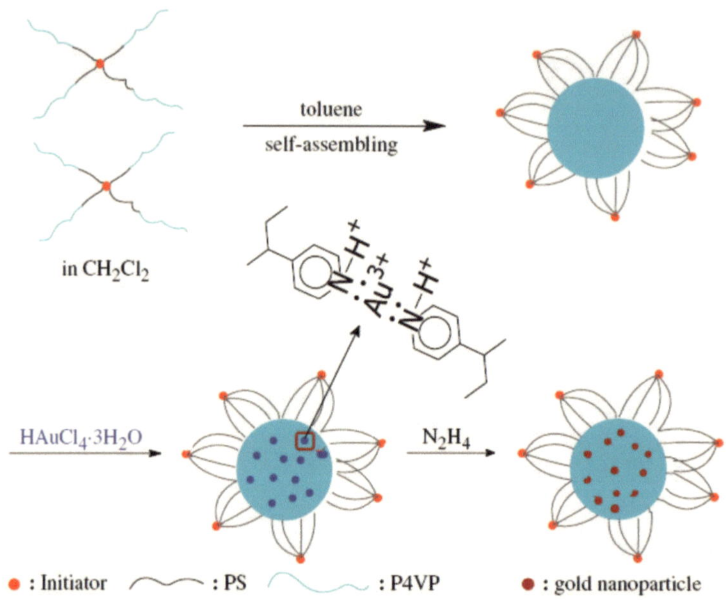

Fig. 7. Schematic representation for the preparation of reversed micelles of (PS-*b*-P4VP)$_4$four-arm star block copolymer and Au NPs (Reprinted from Polymer, 2006, 47, 8480 with permission from Elsevier.)

copolymer concentration and/or molecular weight increase, the micelles became more efficient in stabilizing Au NPs. When the stabilization was optimized, Au NPs were spherical and uniform in size (5-10 nm in diameter). Otherwise, Au NPs were aggregated and irregular in shape such as triangular, hexagonal, and rod-like. They also found that the presence of a small amount of NaF significantly increases the stabilization capacity of the micelles and consequently modified the quality of the Au NPs.

Star block copolymers resembled diblock copolymer micelles physically but the core segments that are covalently bound can be used in the synthesis of Au NPs to overcome the drawbacks of micelle instability which can be affected by concentration, temperature, solvent and pressure as stated above. Youk et al. synthesized the PS-*b*-P2VP star-block copolymer by living anionic polymerization and coupling with ethylene glycol dimethacrylate, forming P2VP block as core and PS block as shell (Youk et al., 2002). The $HAuCl_4$ loaded within the P2VP core segments was reduced to single Au NPs by anhydrous hydrazine. Differing from the synthesis of Au NPs using micelles, this approach does notrequire an extra step for long-term stability such as removal of excess hydrazine and cross-linking. Fustin et al. also prepared the Au NPsin the core of poly(ethylene glycol)-*b*-poly(ε-caprolactone) (PEG-*b*-PCL) five-arm starblock copolymer in DMF (Fustin et al., 2006). The PEG core was swollen with $KAuCl_4$ in DMF and Au NPs were subsequently obtained by reduction with $NaBH_4$. The size distribution and long-term stability of resulting Au NPs were dependent on the PCL block length. However, these Au NPs could not be dispersed in water due to the hydrophobic PCL shell. To circumvent this problem, they grafted the hydrophilic PAA containing a trithiocarbonate moiety as another corona grafted onto the NPsand this leads to the phase transfer of the Au NPs from DMF to water.Li et al. have demonstrated the star block copolymer as nanoreactors for the preparation of Au NPs in different way (Li et al., 2006). They prepared the multi Au NPs in a reverse micelle of star polymers, instead of single star polymer as template for single Au NP. The (PS-*b*-P4VP)$_4$four-arm starblock copolymer was synthesized by ATRP ina form of four PS chain at the ends. The reverse micelles were produced by adding toluene into the dichloromethane solution of (PS-*b*-P4VP)$_4$ and the structure of the micelle is kinetically frozen as dichloromethane is removed from the solution with P4VP chains (external block) as the core and PS chains (inner block) as the petal-like shell. Compared with the brush-like shell of the micelle formed by diblock copolymer, this shell has a restricted and closed structure because the chains of the same copolymer spread from the core and join at a single point. $HAuCl_4$ was incorporated into the micelle core by forming complex with the P4VP block and reduced with anhydrous hydrazine. They observed the hybrid micelle containing dozens of small Au NPs in the core by TEM. On the contrary, Au NPs decorated in brush layer were prepared by Zhang et al. (Zhang et al., 2007). They synthesized poly(2-(dimethylamino) ethyl methacrylate) (PDMAEMA) brushes with high density on the surface of polystyrene latex by ATRP. The PDMAEMA domains were used as the nanoreactors to generate Au NPs on the surface of colloid PS particles. As indicated in aforementioned examples of "templating" method, the location and morphology of Au NPs in the templating polymer can be controlled preciously by adjusting the type, chain length, structure of polymer and the environments during the synthesis. Zheng et al. have demonstrated the template synthesis of various morphological Au NPs using a thermoresponsive and pH responsive coordination triblock copolymer of PEG-*b*-P4VP-*b*-PNIPAM (Zheng et al., 2006). They synthesized the PEG-*b*-P4VP-*b*-PNIPAM triblock copolymer which exists as three morphologies at different pH values and temperatures by ATRP. At pH 2.0 and at 25 °C,below the lower critical solution temperature (LCST) of the PNIPAM block, all three blocks are hydrophilic, and the triblock copolymer exists as unimers in aqueous solution. At pH 6.5 and at 25 °C, the pH-responsive P4VP block is insoluble, and the triblock copolymer self-assembles into core-shell micelles with the hydrophobic P4VP block as the core and the hydrophilic PNIPAM and PEG blocks as the shell. At pH 2.0 and at 50 °C, above the LCST of thePNIPAMblock, the triblock copolymer self-assembles into micellar clusters. The $HAuCl_4$made the complex with the P4VP block of the triblock copolymer, which exhibits unimer, core-shell micelles or micellar cluster at given pH and temperature and then reduced by $NaBH_4$ to form discrete Au NPs, Au NPs in the core of core-shell micelle and Au NPs in single micellar cluster respectively.

A dendrimer was also employed to preparing the small sized Au NPs having narrow size distribution (Kim et al., 2004).They used the forth- and sixth-generation poly(aminoamine) (PAMAM) dendrimers, having about 50% of their peripheral amine groups modified with ammonium chloride, as template to prepare the Au NPs. Since they are quaternized, the dendrimers have permanent positive charges at their periphery for the extraction of aqueous $AuCl_4^-$ into the dendrimer interior. By following reduction with $NaBH_4$, monodisperse 1-2 nm diameter Au NPs were formed.

3. Controlling the nanoparticle location within block copolymer matrix

3.1 Interactions between nanoparticle and block copolymer matrix

Block copolymer can be promising candidates for templates asthis material provides supramolecular control over the size, particle density, and spatial location of inorganic NPs, which possesses interesting photonic, magnetic, electrical, and catalytic properties.

Therefore, hybrid materials of inorganic NPs and block copolymers have been explored as an efficient route for the fabrication of novel functional materials such as nanostructured solar cells, photonic band gap materials, highly efficient catalysts, and high density magnetic storage media. Since most applications of polymer nanocomposites described above ultimately require the integration of nanoscale components into micro- and macroscopic objects, the precise control of the spatial arrangement of inorganic NPs into two or three dimensions within polymer matrices is very crucial. The particle location within polymer matrices is determined by the interaction between the particles and the block copolymer domains, which can be divided into two different interactions of entropic and enthalpic contributions.

Balazs group is among the pioneer groups to show the importance of entropic interaction for the control of particle assembly (Thompson et al., 2001). They demonstrated thatthe control in enthalpic interaction between mesophase-forming of inorganic-organic hybrid materials can lead to well-aligned nanostructures. Thesimulation results utilizeda self-consistent field theory (SCFT) for polymers and a density functional theory (DFT) for NPs to investigate the effect of NP size on the NP location within a block copolymer template.They claimed that conformational entropic penalty of polymer chain stretching and the translational entropy of NPs played a major role in the placementof particles within block copolymer matrix by particle size. For larger particles, they are localizedat the chain ends in the center of domain to minimize the loss of conformational entropy. In contrast, for small particles, translational entropy dominates. Therefore, they are dispersed in the preferred domain.It is shown in a 2D system representing a mixture of NPs within a lamellar block copolymer that the location of particles at the interface is expected to occur for particle size of $d/L < 0.2$, whereas location of the particles at the center of the domain is expected of$d/L > 0.3$, where d is the diameter of particle, and L is the respective domain dimension.

Bockstaller et al.demonstrated hierarchical pattern formation using block copolymers and binary mixtures of different size of hydrophobic NPs (Bockstaller et al., 2003). The system consisted of a symmetric poly(styrene-*b*-ethylene propylene) (PS-PEP) copolymer with the molecular weight of the respective blocks of 4×10^5 g/mol as well as aliphatic coated Au ($d_{core} = 3.5 \pm 1$ nm) and silica ($d_{core} = 21.5 \pm 2.5$ nm) NPs. The localization of AuNPs was along the interface between the PS and PEP domains, whereas larger silica NPs are mostly located at the center of the PEP domains. Another work by Spontak group supports the entropic effect on controlling the NP location in the system of different sized polystyrene coated AuNPs within microphase-ordered PS-PEP-PS triblock copolymers (Spontak et al., 2006). Larger AuNPs (d > 5 nm) are localized in the center of the PS domains, whereas smaller particles (d < 3 nm) distributed throughout the PS domains. The results are correctly predicted by the SCFT model assuming weak block selectivity of the particles. They predicted that small particles within the PS domains tend to being segregated along the interfaces, whilst larger particles are preferentially localized to the center of the PS domains. Both theory and experiment explained the control of particle location based on the competition of two different entropies, the entropy loss from the chain stretching penalty and the translational entropy of particles in polymer domains.

A much more straightforward and practical way to control the particle location is to tune the enthalpic interactions between the block copolymer and the functionalized NP surface. The most important factor lies in the compatibility of the NPs with the block copolymers, which is dominated by the enthalpic interaction between the NP surface and the block copolymer domain.

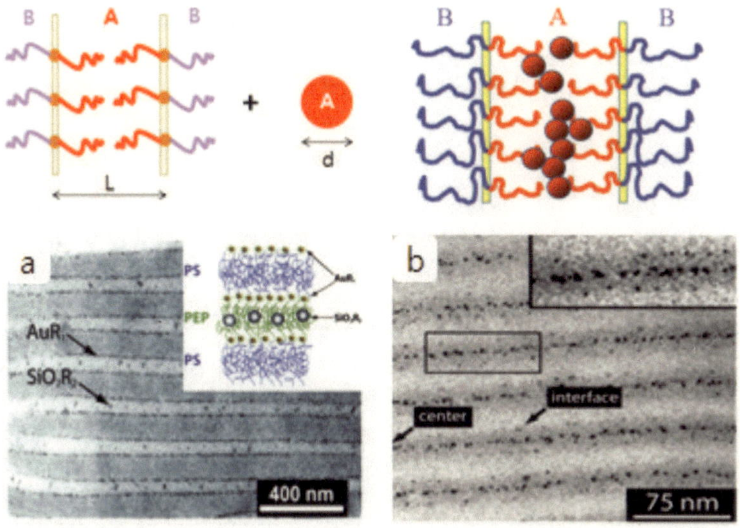

Fig. 8. Schematic representation of the entropy theory, TEM images of (a)Au and silica NPs within PS-*b*-PEP templates (Adapted in part with permission from J. Am. Chem. Soc., 2003, 125, 5276. Copyright 2003 American Chemical Society.),(b) PS coated AuNPs within PS-*b*-PEP-*b*-PS templates (Adapted in part with permission from Nano Lett., 2006, 6, 2115. Copyright 2006 American Chemical Society.).

The enthalpy interaction can be represented by Flory-Huggins interaction parameter (χ).For the system of NPs and A-*b*-B diblock copolymers, there will be two additional interaction parameters, $\chi_{AP}$and χ_{BP}, which represent chemical compatibility between A block and NPs and then between B block and NPs, respectively. If the NP surface is coated with A polymer and this particle is compatible with A block of the block copolymers, χ_{AP} will be zero whereas $\chi_{BP}>$ 0.Therefore, the NPs are segregated in A domains. On the other hand, once particles are "neutral" and the interfacial tension between two different domains of diblock copolymers is sufficiently large, particles will segregate to the interface between two phases. Since the NPs are generally coated by surfactant to shield the particles from agglomeration, the interfacial properties between the NPs and the block copolymer domains can be easily tailored by using various surfactants that provide compatibility between the NPs and polymers. As exemplified by the recent work by Kim and Kramer, a recent strategy for controlling the location of NPs in a block copolymer or polymer blend involves tuning the surface properties of NPs by end-attaching ligands such as organic small molecules, homopolymers, a mixture of homopolymers, random copolymers, and diblock copolymers to the NP surface. Kramer group showed that 100 % polystyrene (M_n ~ 1300 g/mol) coated AuNPs in poly(styrene-*b*-2 vinyl pyridine) (PS-*b*-P2VP) segregated to the center of the PS domain(Chiu et al., 2005). Particles coated with a given short homopolymer lower their enthalpy by segregating into the corresponding domain of the block copolymer. Furthermore, by concentrating particles near the center of the compatible domain where the polymer ends are located, the chains can accommodate particles by moving apart rather than by stretching. Localizing particles near the center of the compatible domain thus sacrifices translational entropy of the particles but avoids an even larger chain stretching penalty incurred by distributing particles throughout the domain. In contrast, NPs that are coated with a mixture of PS and P2VP (1:1) are localized at the interfaces between the PS-*b*-

P2VP domains. This result can be explained by interfacial tension between PS-*b*-P2VP, Au-PS and Au-P2VP ($\chi_{PS\text{-}P2VP}$, $\chi_{Au\text{-}PS}$, $\chi_{Au\text{-}P2VP}$). The adsorption energy of AuNPs to the interface between PS and P2VP blocks (E_a) can be expressed as the following equation.

$$E_a = \pi R^2 \gamma_{PS-P2VP}\left(1-|\cos\theta|\right)^2 \quad (1)$$

where*R* is a radius of NP, and cos θ is the ratio of the interfacial tension between NPs and polymer matrix to that between PS and P2VP. Because $|\cos\theta| << 1$ and Ea ≈ $\pi R2\gamma$PS-P2VP. The Ea of 4 nm radius NPs with γPS-P2VP = 2.8 mN m-1 is 10 *k*BT. Therefore, PS and P2VP coated AuNPs are unlikely to be segregated along the interface.

Kim et al. also demonstrated a simple procedure to control the location of PS coated AuNPs within PS-*b*-P2VP block copolymer domains simply by changing the areal chain density (Kim et al., 2006). The areal chain density of Au particles is varied by changing the mole ratio of PS chains to Au atoms. The NPs completely shielded by PS chainswith high areal chain densityare segregated toward the center of the PS domain. In contrast, NPs partially shielded by PS chains withlow areal chain density are adsorbed at the interfaces between the PS and P2VP blocks. Interestingly, the particle location is sharply changed as the areal chain density is from 1.6 to 1.2 chains/nm^2.This is due to the preferential interaction between the P2VP block and the imperfectly shielded Au surface that is responsible for the segregation of AuNPs imperfectly shielded by PS chains to the PS and P2VP interface.

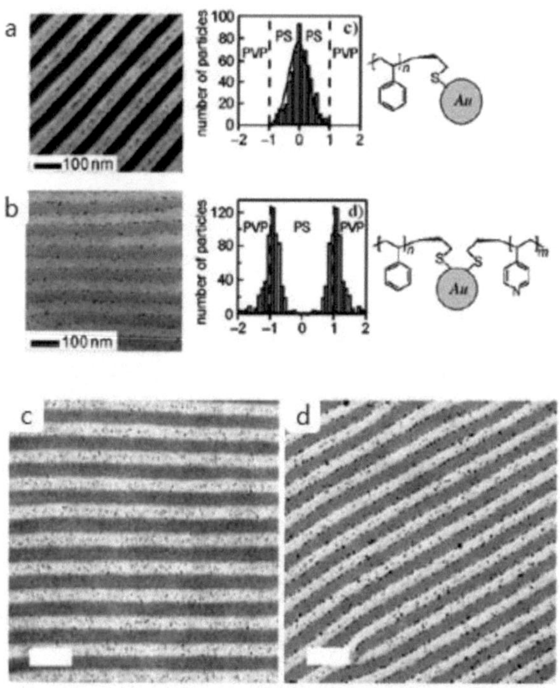

Fig. 9. (a), (b) PS or PS and P2VP mixed AuNPs within PS-*b*-P2VP (Adapted in part with permission from J. Am. Chem. Soc., 2005, 127, 5036. Copyright 2005 American Chemical Society.), Effects of areal chain densities on their position within PS-*b*-P2VP (c) 1.64, (d) 1.22 chains/nm^2 (Adapted in part with permission from Macromolecules, 2006, 39, 4108. Copyright 2006 American Chemical Society.).

3.2 Approaches for controllingthe nanoparticle distribution within block copolymer matrix

Over the years, two approaches have been developed for incorporating inorganic NPs into polymeric nanostructures. The *in-situ*approach is that NPs are directly synthesized within a block copolymer domain from metal precursors. In that case, a specific and strong interaction between the precursors and the block copolymers is required to guarantee the incorporation of metal precursors into the domain. The other approach is *ex-situ* method where pre-synthesized and tailored NPs are incorporated into a preferred domain. In that case, tailoring materials on the surface of NPs play a critical role in particle location.

In the *in-situ* approach, preformed micelles of block copolymers containing metal precursors are used as nanoreactors to synthesize NPs selectively in block copolymers. Due to its chemical affinity, the salt selectively infiltrates the hydrophilic copolymer domain. The NPs then form selectively, upon reduction within the precursor-loaded domains.Schrock et al. first demonstrated the fabrication a bulk film consisting of block copolymers and inorganic NPs (Chan et al., 1992). First, the NPs are selectively incorporated in various morphologies of microdomains by ring-opening metathesis polymerization of organometallic complexes composed of norbornene and η^3-1-phenyallyl. Thenthe complex is annealed to form specific morphology.Finally, after reduction, NPis formed within block copolymer.They prepared Pt and Pd NPs in block copolymer micelles by reducing the metal salts with hydrogen while Pb salts incorporated in block copolymer micelles could be converted into PbS quantum dots through the reaction with H_2S.Saito et al. showed the synthesis of silver NPs in the lamellar and spherical domains of PS-*b*-P2VP block copolymer using iodine as reducing agent by ion exchange in P2VP domains (Saito et al., 1992). They found that iodine ion did not exist in the PS phase but in the crosslinked P2VP phase of the crosslinked PS-*b*-P2VP film swollen in water. Then the silver NPs reduced from AgI by photo and chemical treatments arelocalized in the P2VP domains in the microphase separated PS-*b*-P2VP films. They also controlled the amount of silver NPs into the P2VP domains of PS-*b*-P2VP block copolymers by changing the amount of silver nitrate in the feed (Saito & Ishizu, 1995).

Following this approach, Cohen et al., have developed twodimensional hexagonal patterns of metal NPs in PS-*b*-poly(acrylic acid) (PS-*b*-PAA) matrix (Boontongkong & Cohen, 2002). The carboxyl-functionalized cores in the PAA micelle could serve as the nanoreactors for synthesis of metal NPs such as Pd, Cu, Au and Ag, and semiconductor NPs such as PbS. And Stadler et al. demonstrated the interaction between metal salts with poly(2-vinylpyridine)-*b*-poly(ethylene oxide) (P2VP-*b*-PEO) where the size and morphology of the NPs can be controlled by pH change (Bronstein et al., 1997). Only a small part of the vinyl double bonds on P2VP domain has been complexed with Pd, while most of the double bonds are reacted with the Fe complex. Using same *in-situ* method, Sohn et al. fabricated a multilayered nanostructure of alternating pure polymeric lamellae and AuNP-containing lamellae by utilization of thin films of symmetric polystyrene-*b*-poly(4-vinyl pyridine) (PS-*b*-P4VP) (Sohn & Seo, 2001). Strong interaction between the P4VP block and the silicon wafer substrates and lower surface energy of the PS block produced multilayers of lamellae parallel to the substrate. After annealing, AuNPs are synthesized selectively in the parallel lamellae of the P4VP block using $NaBH_4$ as reducing agents. Recently, Fahmi et al. developed a strategy to generate AuNPs within the P4VP domains of a PS-*b*-P4VP block copolymer in solid state (Mendozaet al., 2009). The Au precursor ($HAuCl_4$) is coordinated in solution to the pyridine groups by hydrogen bonding, which induced its selective location within the P4VP block. Then, thermal reduction process generates metallic AuNPs, where the P4VP domain acts as a nanoreactor.

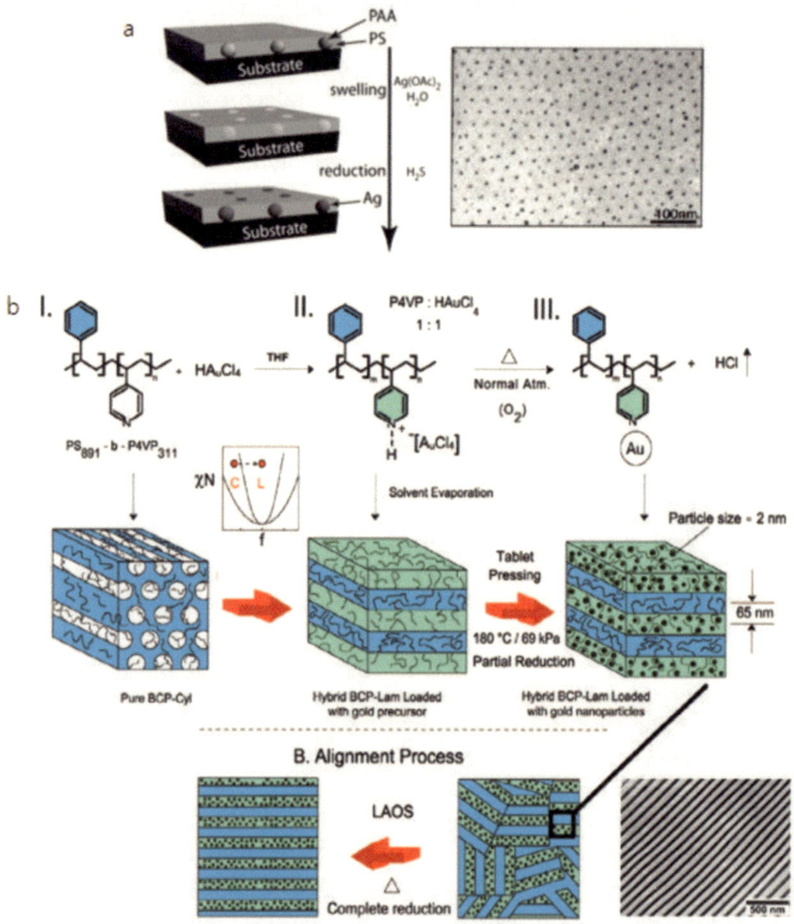

Fig. 10. (a) Synthesis of silver NPs in PS-*b*-PAA micelle (Adapted in part with permission from Macromolecules, 2002, 35, 3647. Copyright 2002 American Chemical Society.), (b) Process of incorporating AuNPs and its subsequent macroscopic alignment (Adapted in part with permission from Macromolecules, 2009, 42, 1203. Copyright 2009 American Chemical Society.).

The *in-situ* approach is simple and versatile, which produces high density NP assembly within one of the block copolymer domains. However, the approach suffers from several limitations. First, controlling the arrangement of the NPs within the periodic structure of the block copolymer is difficult. Second the approach usually requiresstrong and specific interaction to bind metal precursors into one block of the block copolymer chains, which requires limited number of combinations of polymer and metal precursors or NPs. Third,the control of particle size and its uniformity could not be achieved. This aspect is critical in the sense that many applications require inorganic and/or metallic NP with well controlled size and uniformity. For example, in the case of semiconductor NPs such as quantum dot, since the photoluminescence (band gap) of a quantum dot is strongly dependent on the particles size, the particle size itself is a critical factor to determine their optical property. Therefore, it is difficult to obtain the application with desired optical propertiesby the *in-situ*method.

The *ex-situ* method of controlling the structural characteristics of the sequestered component uses cooperative self-organization of premade NPs and block copolymers. Through this method, NPs with desired size and shape coated with oligomers or polymers can be incorporated in the block copolymer matrix.Once phase separation of the copolymer has occurred, the NPs can be confined in one of the segregated domains. Clinard et al. pioneered the field of the *ex-situ* approach for precise placement of NPs within the block copolymer templates (Hamdoun et al., 1996). They synthesized3.5 nm diameter iron oxide NPs grafted with PS chains, which aresubsequently solubilized in the PS domain of a lamellar poly(styrene-*b*-butylmethacrylate) (PS-*b*-PBMA). However, the periodicity of the copolymer's lamellar microstructure was too small to clearly resolve the metal particles within domains. Lauter-Pasyuk introduced the small-angle neutron scattering technique to resolve the γ-Fe_2O_3NP location within the PS-*b*-PBMA matrix (Lauter-Pasyuk et al., 1997). They found that smaller particle sizes (d = 4 nm) tend to localize close to the interface, whereas larger particles (d = 6 nm) tend to locate at the center of the block copolymer domains.Recently, Russell groupshowed the control of block copolymer orientation in thin film by adding the surface active cadmium selenide (CdSe) NPs. Surface modified CdSe NPs with tri-n-octylphosphine oxide (TOPO) aremixed withPS-*b*-P2VP copolymer in toluene solution (Lin et al., 2005). After being casted to produce the thin films, these NPs are selectively segregated into the P2VP domain. TheCdSe NPs located at the surface of the thin films compensate the difference of surface energies between PS and P2VP domains, which resulted in the perpendicular orientation of PS-*b*-P2VP block copolymers in thin films.

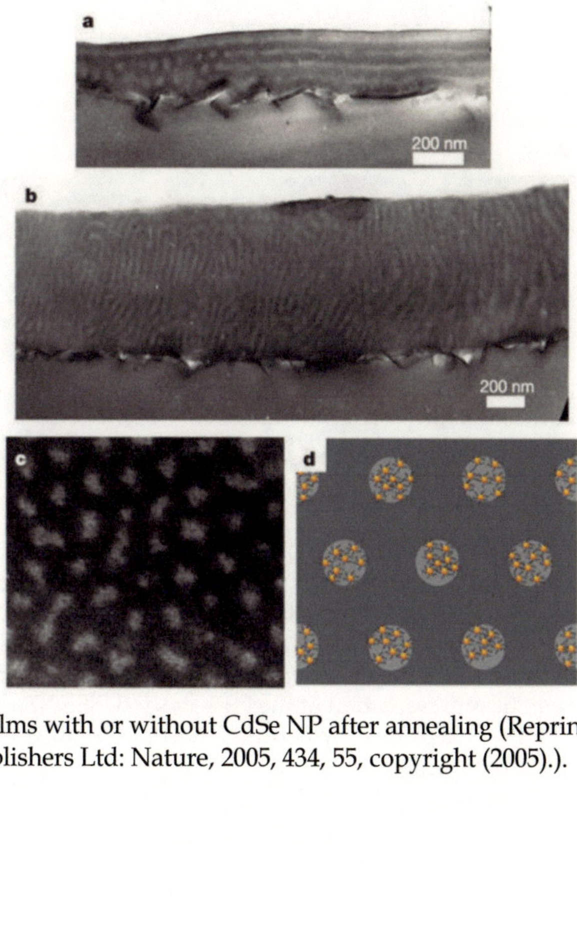

Fig. 11. PS-*b*-P2VP films with or without CdSe NP after annealing (Reprinted by permission from Macmillan Publishers Ltd: Nature, 2005, 434, 55, copyright (2005).).

Since the *ex-situ* approach utilizes pre-made NPs that have desired size and shape, and uniformity, thisapproach is suitable for many applications that require inorganic and/or metallic NP with well controlled size and uniformity. Furthermore, the location of NPs within block copolymer matrix can be easily and precisely controlled by tailoring the surface properties of the NPs. However, due to the steric hindrance between the core-shell type NPs and block copolymer chains, the density of NPs within the templates is expected to be much lower compared to that by the *in-situ* approach.In addition, if the surface modification of NPs is necessary for such precise control, the intrinsic properties of NP can be deteriorated by increase in defect sites on the surface of NPs.

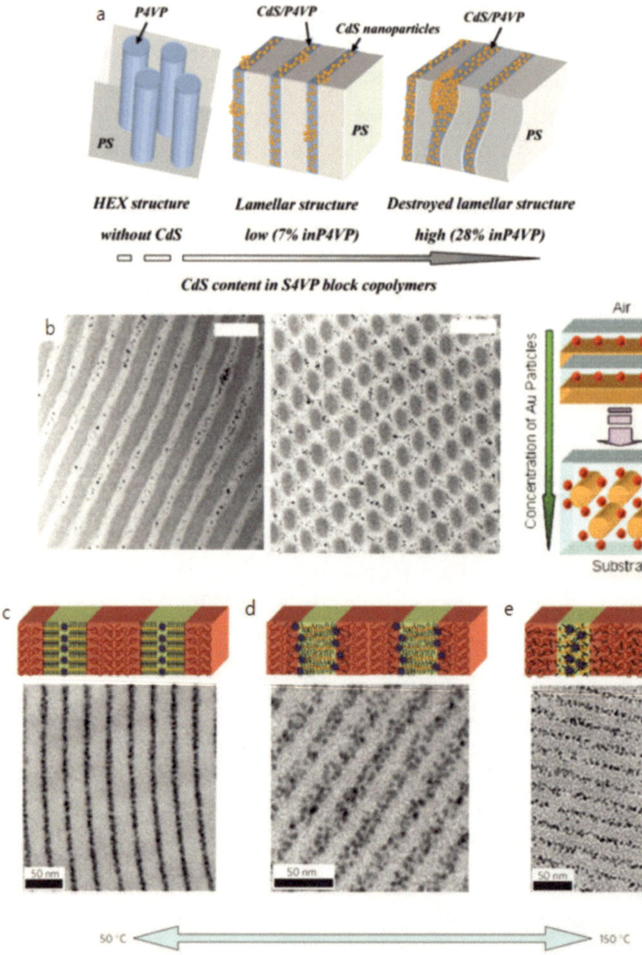

Fig. 12. (a) Schematic representation of the morphological transformation in PS-*b*-P4VP block copolymer with various ratios of CdS NPs (Adapted in part with permission from Macromolecules, 2005, 38, 6559. Copyright 2005 American Chemical Society.), (b) Phase transition in PS-*b*-P2VP with various concentrations of AuNPs (Kim et al., Adv. Mater., 2005, 17, 2618. Copyright Wiley-VCH Verlag GmbH & Co. KGaA. Reproduced with permission.) TEM images and schematic representation of the PS-*b*-P4VP/CdSe NPs annealed at (c) 50 °C, (d) 110 °C, and (e) 150 °C (Reprinted by permission from Macmillan Publishers Ltd: Nat. Mater., 2009, 8, 979, copyright (2009).).

The NPs with precisely controlled location can influence on the morphology of block copolymers. For example, the NPs selectively localized in one specific domain of the block copolymer templates can behave as homopolymer additives, thus causing the transition of the phase morphology of block copolymers. For example, Yeh et al. have reported morphology transformation from cylinders to spheres in the case of PS-*b*-PEO thin films by added cadmium sulfide (CdS) NPs localized in PEO domain (Yeh et al., 2004). Here, the selective placement of the CdS particles is achieved by strong interaction of thehydrogen bonding between the surface hydroxylated cadmium sulfide (CdS) NPs and the PEO phase. They also reported a morphology change from cylinder to lamellae when more CdS NPs areadded to PS-*b*-P4VP block copolymer film, in which there exists the strong hydrogen bond formation between CdS and the minor block copolymer phase (P4VP) (Yeh et al., 2005). In other work, hexagonal morphology is changed to spherical morphology by adding C_{60} fullerenes to PS-*b*-P4VP block copolymer system. In this case, the morphology change is attributed to C_{60} forming charge-transfer complexes with multiple pyridine groups from different P4VP chains (Laiho et al., 2006). Similarlythe morphology of PS-*b*-P4VP with PS coated AuNPs depends on the particle concentration and composition of block copolymers is reported by Lo et al. (Lo et al., 2007). The PS domains are filled with NPs, while at the same time,the addition of NPsexpand PS domain and change the interfacial curvature between PS and P4VP, which induced order-order transitions. However, at higher concentration, the excess particles form macrophase separation, which leads to a disordered phase.Interestingly, Kim et al. showeda phase transition of PS-*b*-P2VP block copolymers as a function offilm depthin single 100 μm thick film (Kim et al., 2005). Here, the NPs are coated with PS homopolymers and therefore are selectively localized within PS domain. As the depth increases, the concentration of PS NPs within the block copolymer domains increases. Therefore, at low concentrations near the film surface, the block copolymer morphology is lamellar, just as it was in the absence of NPs. In contrast, hexagonal or spherical micellar structures with higher particle concentration are observed near the substrate layers.
In addition to the Flory-Huggins type enthalpic interaction, strong interaction such as hydrogen bonding can be applied to strengthen the interaction between NPsurface and block copolymer matrix, which resulted in better control of the NP assembly. One good example has been recently shown by Zhao et al. (Zhao et al., 2009). They utilized the strong hydrogen bonding interaction to fabricate hierarchical self-assembly of CdSe NPs within PS-*b*-P4VP block copolymer. They usedtwo different molecules, 3-n-pentadecylphenol (PDP) and 4-(4'-octylphenyl)azophenol (OPAP), for coating theNP surface, wherethe NPs hydrogen bond to pyridine rings of P4VP domains. Therefore, without any further chemical modification, the CdSe NPs can be selectively sequesterated into the P4VP domain. And, interestingly, the strength of hydrogen bonding is affected by the temperature.The particle array can be tuned upon the change of the temperature, which makes the nanocomposite thermo responsive materials.

3.3 Applications in technology

Self-assembly of NPs into well-ordered block copolymer structures provides many interesting properties, which are different from either individual NP or polymer. Thesehybrid nanocomposites promise to be candidates for next-generation materials and provide wide applications on nanocomposite-based devices.
Since NPs such as Au, Ag have different colors with varied sizes, polymer-inorganic NPs can produce certain colors.The simulation results by Balazs group showed that the polymer-

particle interactions affected the spatial distribution within the polyme matrix and their optical properties changed according to the particle distribution (Buxton et al., 2003). When the particles are added, band gaps are more defined with 100 % reflectance with wider frequency gap. Refractive index is also crucial factor for optical application of hybrid nanocomposites. Common organic polymers have the value of 1.3~1.7, but it can be extended further if inorganic NPs are combined. For example, lead sulfide (PbS)-gelatin nanocomposites tend to increase their refractive index with increasing PbS, and the highest values areobtained between 82-86 % in 2.4–2.5 for PbS fractions, whereas native gelatin has a refractive index of 1.54 (Caseri, 2000). And, Bockstaller et al. prepared metallodielectric nanocomposite materials using oligo(styrene) coated AuNP and poly(styrene-*b*-ethylene/propylene) (PS-*b*-PEP) (Bockstaller et al., 2003). The photonic properties of mesoscopically ordered nanocomposites are changed with various Au filling fractions. With increasing Au filling fraction, the wavelength of maximum reflectivity is red-shift. Kang et al. also showed a facile way of fabricating organic/inorganic hybrid 1D photonic crystal that utilized self-assembled PS-*b*-P2VP as photonic gels which takes advantage of a swelling and

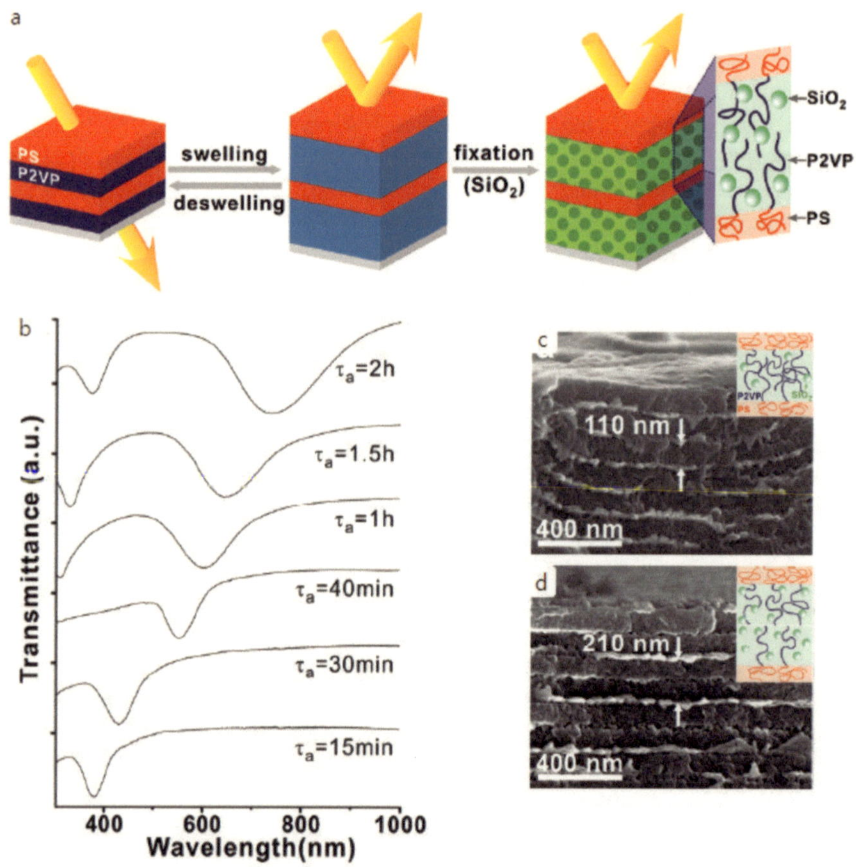

Fig. 13. (a) Schematic representation of the fabrication of photonic gels, (b) Transmittance spectra with various aging time and SEM images of photonic gels at (c) 15 min, (d) 45 min (Adapted in part with permission from J. Am. Chem. Soc., 2009, 131, 7538. Copyright 2009 American Chemical Society.).

freezing process ofsilica NPs within P2VP domains(Kang et al., 2009). These nanocomposites cancontrol the position of stop bands in transmittance spectra with increasing aging time. At longer aging time, the amount of silica NP within the P2VP domains gradually increased, and accordingly the domain size and the overall periodicity also increased.

Some polymers can change their morphology or chain length in different environment such as pH or temperature, which makes them suitable for bio-applications. Especially, NPs within polymeric micelles formed from amphiphilic block copolymers in water are regarded as vehicle systems for drug delivery because the micellar core segregated from the aqueous phase can serve as a reservoir of hydrophobic drugs. Kakizawa et al. showed the first example of the organic-inorganic hybrid NPs intended to be utilized as synthetic DNA carriers using calcium phosphate crystals (CaP) and poly(ethylene glycol)-*b*-poly(aspartic acid) (PEG-*b*-PAA) (Kakizawa & Kataoka, 2002). In this study, CaP surrounded by

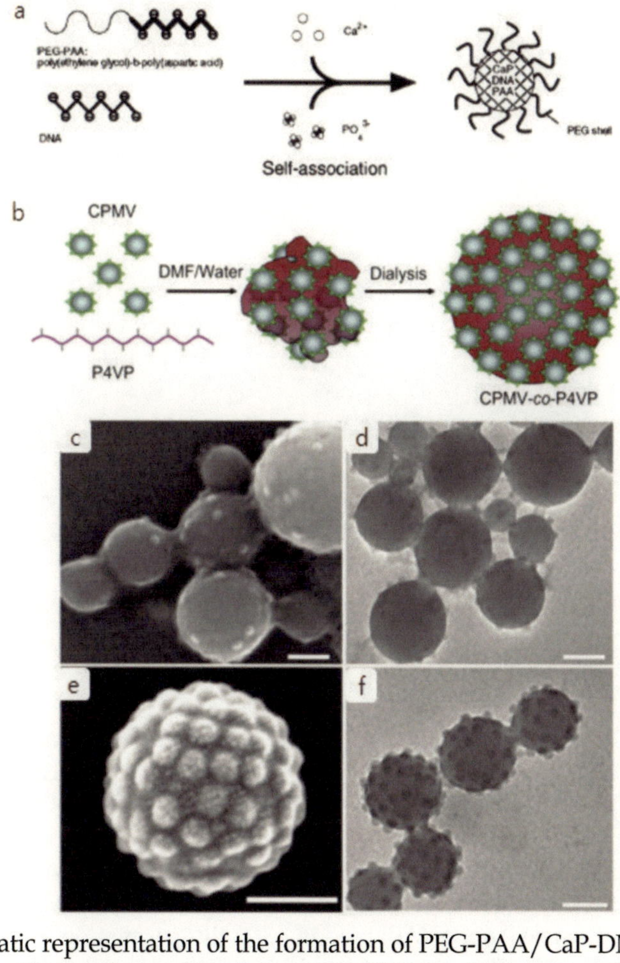

Fig. 14. (a) Schematic representation of the formation of PEG-PAA/CaP-DNA particles (Adapted in part with permission from Langmuir, 2002, 18, 4539. Copyright 2002 American Chemical Society.), (b) Schematic representation of the formation of CPMV-*b*-P4VP and (c)-(f) their TEM images (Li et al., Small, 2008, 4, 1624. Copyright Wiley-VCH Verlag GmbH & Co. KGaA. Reproduced with permission.).

hydrophilic polymer palisades with block copolymers has the ability to incorporate DNA in the core. Li et al. demonstated the raspberry-like assemblies of virus-polymer nanocomposites based on noncovalent interactions between NPs and polymers (Li et al., 2008). They used cowpea mosaic virus (CPMC) with 29 nm in diameter as NP, and P4VP as polymer. This core/shell type biocomposites can be applied in drug delivery by controlling pH-sensitive P4VP polymer resulting in release of drugs from the core.Willner et al.showed for the first time that the semiconductor NPs and enzyme complex can be applied as biosensor. (Pardo-Yissar et al., 2003). He worked on enzyme restrainer using acetylcholine esterase (AChE) / CdS complex, where the AChE/CdS complex determinedthe quantity of photocurrent according to concentration of acetylthio-choline. Since the enzyme restrainer can reduce the photocurrent, AChE/NP is possible for biosensor application.

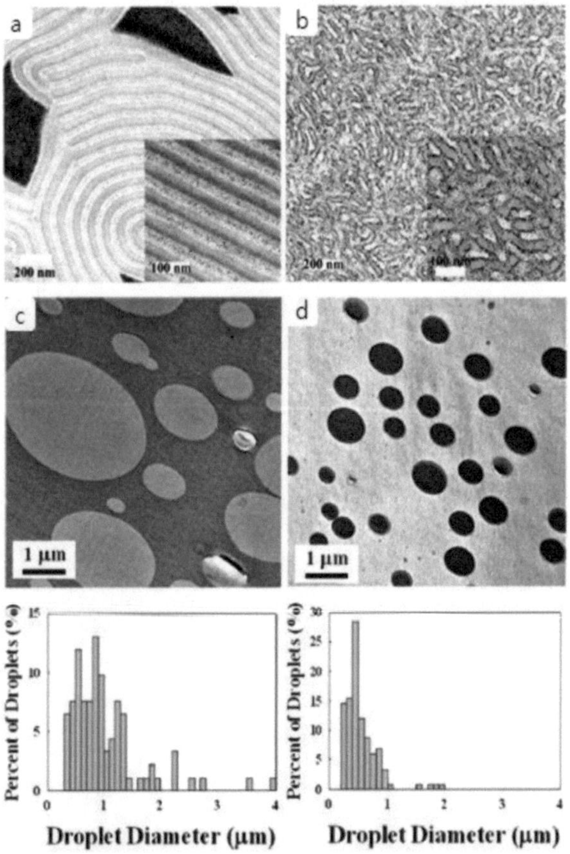

Fig. 15. PS-*b*-P2VP block copolymer with PS coated AuNPs at various volume fractions (a) 0.20, (b) 0.14 Kim (Adapted in part with permission from Langmuir, 2007, 23, 7804. Copyright 2007 American Chemical Society.), TEM images of PS/PMMA blends (c) without or (d) with thermally stable AuNPs (Adapted in part with permission from Macromolecules, 2010, 43, 3570. Copyright 2010 American Chemical Society.)

Colloidal particles are widely used as surfactant, which blends immiscible liquids such as water and oil. Such colloidal particles are located at the interface between water and oil to lower surface energy, restrain collision between droplets composed of water or oil, and stabilize the system. By same principle, if NPhaving similar affinity toboth water and oil is synthesized, it can be located at the interface and used as surfactant. For example, when nanometer-scale CdSe quantum dot is added to water/toluene blend, it stabilizes the dropletsby suppressing the coalescence between the droplets and lowering the interfacial tension between the two phases (Lin et al., 2003).Park et al. showedthe controlled orientation of PS-*b*-PMMA polymers depending on the amount of PEO-coated Au NP in PS-*b*-PMMA polymers (Park et al., 2007). In this case, mutual affinity between PEO-Au particle and PMMA is selective, and thus PEO-Au is located within the PMMA domain. Therefore, PS-*b*-PMMA/PEO-Au complex has a morphology of vertically well-arranged cylinder when NPs are added, just like PS-*b*-PEO in benzene/water solvent annealing. Furthermore, the

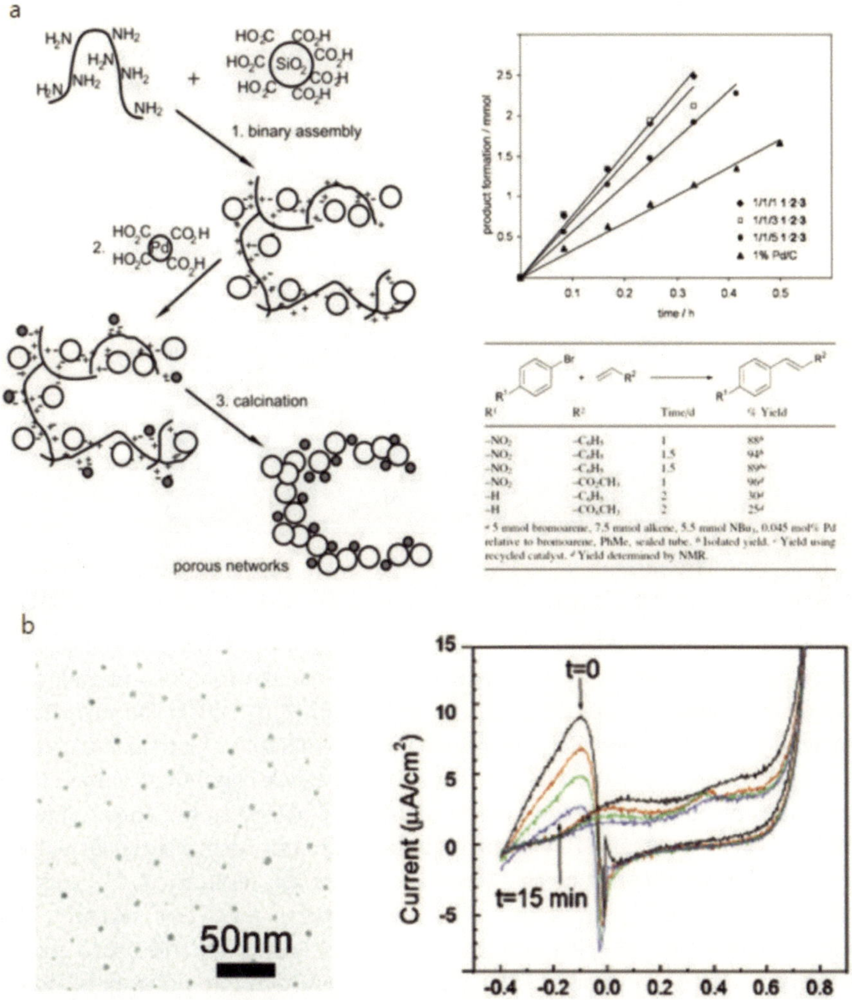

Fig. 16. (a) Schematic representation of the formation of Pd catalyst and its hydrogenation reaction for catalytic activity (Chem. Commun., 2002, 1076 – Reproduced by permission of The Royal Society of Chemistry.), (b) Electro-oxidation of carbon monoxide with AuNPs (Adapted in part with permission from J. Am. Chem. Soc., 2003, 125, 7148. Copyright 2003 American Chemical Society.)

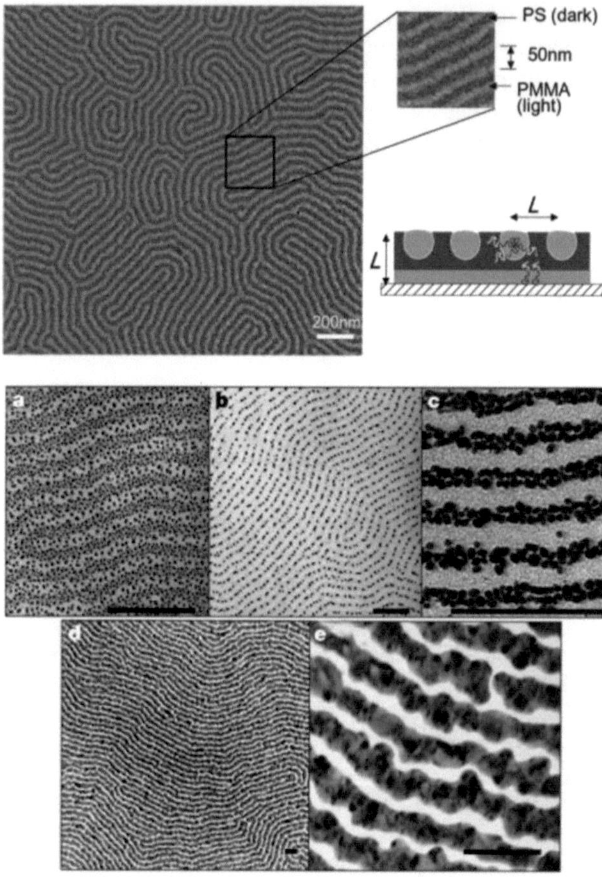

Fig. 17. Metal nanowire formation on PS-*b*-PMMAblock copolymer (Reprinted by permission from Macmillan Publishers Ltd: Nature, 2001, 414, 735, copyright (2001).)

PMMA domain of PS-*b*-PMMA/PEO-Au complex can be selectively removed to produce nanoporous template. Kim et al. showedthat the NPs at the interface of two domains of block-copolymer could be used as nanosurfactants (Kim et al., 2007). The surfactant effect is observed when AuNP is selectively introduced at the interface of lamellae-structured PS-*b*-P2VP having various molecular weights. Lamellae of block-copolymer is maintained with small quantities of NPs, but as the amount of AuNPs acting as the surfactants increases, immiscible two phases (PS and P2VP) turn into bicontinuous form. Particularly, the domain spacing of block-copolymer decreases dramatically as the amount of NP located at the interface increases, which proves that NPs act as surfactant. More recently, Yoo et al. proposed thermally stable AuNPs by introducing photo-cross-linkable azide groups as the shell on the NP surface (Yoo et al., 2010). When NPs are added into polymer blends, they are segragated at the interface between two polymers to reduce the interfacial tension between two polymer phases. As a result, the size of droplet is 0.92 ± 0.33 μm without NP, whereas 0.46 ± 0.14 μm and 0.32 ± 0.09 μm for addition of 5.0 and 10.0 wt % of NPs, respectively. These results suggested that thermally stable AuNPs can act as surfactants for thermally annealed PS/ PMMA blends.

The hybrid NP/ block copolymer materials have been shown to produce the nanowire fabrication if NPs are selectively located in one domain of blockcopolymer. Particularly, if the conductive NPs are used for the fabrication of nanowire, conductivity can be improved dramatically. Jaeger et al. located Ag and Au NPs within cylindrical PS-*b*-PMMA blockcopolymer parallel to the substrate and achieved the selective arrangement (Lopes et al., 2001). First, the metal NPsare located randomly on PS-*b*-PMMA block copolymer and then placed selectively in PS domainparticle when heat is applied. Further annealing procedure is applied to the NP/PS-*b*-PMMA nanocomposites, where Au particles within PS domain hold together to form nanowire and consequently have high conductivity.

4. Conclusions

In this chapter, we described various approaches to decorate the surface of NPs, including "grafting from", "grafting to", ligand exchange, and templating approaches. Then, as an application of these NPs in nanocomposites, several efforts to control the location of NPs within block copolymer templates and their applications in various devices are reviewed. In "grafting from" approach, the desired polymers are grown from the surfaces of preformed NPs, while in "grafting from"approach,the end-functional of pre-synthesized polymers are directly used as ligands to synthesize the NPs. The other choices to modify the NPs surface can be either ligand exchange or templating methods. In ligand exchange, the requirement is that the desired ligands should have stronger interaction with NPs than the existing ligands. This method has been widely used to various NPs, such as Au, Ag, CdSe, especially when "grafting to" or "grafting from" methods are not feasible. The NPs prepared from aforementioned methods can have the desired surface properties, size, shape and uniformity, which determines the enthalpic and entropic interactionbetween the NPs and block copolymer chains and thus control their assembly within the block copolymer matrix. In this case, NPs/block copolymer hybrids are prepared via either *in-situ* or *ex-situ* methods. Interestingly, the NPs with precisely controlled location can be used as nanosurfactants between block copolymers and/or polymer blends, optical and electronic devices, and biological and catalytic applications. From these advances, it is evident thatstrategies for well-definedNPs/block copolymer hybrids promise to be candidates for next-generation materials and provide wide applications on nanocomposite-based devices.

5. Acknowledgements

This work was supported by the KoreaResearch Foundation Grant funded by the Korean Government (MOEHRD) (2010-0016304, 2010-0015294, 2010-0027751, 2009-0088551, 2010-0011033).

6. References

Algar, W. R. & Krull, U. J. (2007). Luminescence and stability of aqueous thioalkyl acid capped CdSe/Zns quantum dots correlated to ligand ionization. *Chem. phys. chem.*, Vol. 8, 561-568, 1439-4235

Aryal, S.; Bahadur, K. C. R.; Bhattarai, N.; Lee, B. M. & Kim, H. Y. (2006). Stabilization of gold nanoparticles by thiol functionalized poly(ε-caprolactone) for the labeling of pcl biocarrier. *Mater. Chem. Phys.*, Vol. 98, 463-469, 0254-0584

Azzam, T. & Eisenberg, A. (2007). Monolayer-protected gold nanoparticles by the self-assembly of micellar poly(ethylene oxide)-*b*-poly(ε-caprolactone) block copolymer. *Langmuir,* Vol. 23, 2126-2132, 0743-7463

Becker, C. F. W.; Marsac, Y.; Hazarika, P.; Moser, J.; Goody, R. S. & Niemeyer, C. M. (2007). Functional immobilization of the small GTPase Rab6A on DNA-gold nanoparticles by using a site-specifically attached poly(ethylene glycol) linker and thiol place-exchange reaction. *Chembiochem,* Vol. 8, 32-36, 1439-4227

Bergen, J. M.; Von Recum, H. A.; Goodman, T. T.; Massey, A. P. & Pun, S. H. (2006). Gold nanoparticles as a versatile platform for optimizing physicochemical parameters for targeted drug delivery. *Macromol. Biosci.,* Vol. 6, 506-516, 1616-5187

Beletskaya, I. P.; Kashin, A. N.; Litvinov, A. E.; Tyurin, V. S.; Valetsky, P. M. & Kotenll, G. (2006). Palladium colloid stabilized by block copolymer micelles as an efficient catalyst for reactions of C-C and C-heteroatom bond formation. *Organometallics,*Vol. 25, 154-158, 0276-7333

Bockstaller, M. R.; Lapetnikov, Y.; Margel, S. & Thomas, E. L. (2003). Size-selective organization of enthalpic compatibilized nanocrystals in ternary block copolymer/particle mixtures. *J. Am. Chem. Soc.,*Vol. 125, 5276-5277, 0002-7863

Bockstaller, M. R. & Thomas, E. L. (2003). Optical properties of polymer-based photonic nanocomposite materials. *J. Phys. Chem. B,*Vol. 107, 10017-10024, 1520-6106

Boontongkong, Y. & Cohen, R. E. (2002). Cavitated block copolymer micellar thin films: lateral arrays of open nanoreactors. *Macromolecules,*Vol. 35, 3647-3652, 0024-9297

Bronstein, L.; Seregina, M.; Valetsky, P.; Breiner, U.; Abetz, V. & Stadler, R. (1997). Transition metal complex induced morphology change in an ABC-triblock copolymer. *Polymer Bull.,*Vol. 39, 361-369 , 0170-0839

Brust, M.; Fink, J.; Bethell, D.; Schiffrin, D. J.& Kiely, C. G. (1995). Synthesis and reactions of functionalised gold nanoparticles. *J. Chem. Soc., Chem. Commun.,* 1655-1656

Brust, M.; Walker, M.; Bethell, D.; Schiffrin, D. J. & Whyman, R. J. (1994). Synthesis of thiol-derivatized gold nanoparticles in a two phase liquid-liquid system. *J. Chem. Soc., Chem. Commun.,* 801-802

Buxton, G. A.; Lee, J. Y.& Balazs, A. C. (2003). Computer simulation of morphologies and optical properties of filled diblock copolymers. *Macromolecules,*Vol. 36, 9631-9637, 0024-9297

Caseri, W. (2000). Nanocomposites of polymers and metals or semiconductos: historical backgraound and optical properties. *Macromol. Rapid Commun.,* Vol. 21, 705-722, 1022-1336

Chan, Y. C.; Craig, G. S. W.;Schrock, R. R. & Cohen, R. E. (1992). Synthesis of palladium and platinum nanoclusters within microphase-separated diblock copolymers. *Chem. Mater.,* Vol.4, 885-894, 0897-4756

Chen, S.; Guo, C.; Hu, G. H.; Wang, J.; Ma, J. H.; Liang, X. F.; Zheng, L. & Liu, H. Z. (2006). Effect of hydrophobicity inside PEO-PPO-PEO block copolymer micelles on the stabilization of gold nanoparticles: experiments. *Langmuir,* Vol. 22, 9704-9711, 0743-7463

Cheng, W. L.; Campolongo, M. J.; Cha, J. J.; Tan, S. J.; Umbach, C. C.; Muller, D. A. & Luo, D. (2009). Free-standing nanoparticle superlattice sheets controlled by DNA. *Nat. Mater.*, Vol. 8, 519-525, 1476-1122

Chiu, J. J.; Kim, B. J.;Kramer, E. J. & Pine, D. J. (2005). Control of nanoparticle location in block copolymers. *J. Am. Chem. Soc.*, Vol. 127, 5036-5037, 0002-7863

Corbierre, M. K.; Cameron, N. S. & Lennox, R. B. (2004). Polymer-stabilized gold nanoparticles with high grafting densities. *Langmuir,* Vol. 20, 2867-2873, 0743-7463

Dong, H. C.; Zhu, M. Z.; Yoon, J. A.; Gao, H. F.; Jin, R. C. & Matyjaszewski, K. (2008). One-pot synthesis of robust core/shell gold nanoparticles. *J. Am. Chem. Soc.,* Vol. 130, 12852-12853, 0002-7863

Fahmi, A. W. &Stamm, M. (2005). Spatially correlated metallic nanostructures on self-assembled diblock copolymer templates. *Langmuir,*Vol. 21, 1062-106, 0743-7463

Frens, G. (1973). Controlled nucleation for regulation of particle-size in monodisperse gold suspensions. *Nature: Phys. Sci.,* Vol. 241, 20-22

Fustin, C. A.; Colard, C.; Filali, M.; Guillet, P.; Duwez, A. S.; Meier, M. A.; Schubert, U. S. & Gohy, J. F. (2006). Tuning the hydrophilicity of gold nanoparticles templated in star block copolymers. *Langmuir,* Vol. 22, 6690-6695, 0743-7463

Galow, T. H.; Drechsler, U.; Hanson, J. A. & Rotello, V. M. (2002). Highly reactive heterogeneous heck and hydrogenation catalysts constructed through 'bottom-up' nanoparticle self-assembly. *Chem. Commun.,*1076-1077, 1395-7345

Genson, K. L.; Holzmueller, J.; Jiang, C. Y.; Xu, J.; Gibson, J. D.; Zubarev, E. R. & Tsukruk, V. V. (2006). Langmuir-blodgett monolayers of gold nanoparticles with amphiphilic shells from V-shaped binary polymer arms. *Langmuir,* Vol. 22, 7011-7015, 0743-7463

Hamdoun, B.; Ausserrė, D.; Joly, S.; Gallot, Y.; Cabuil, V. & Clinard, C. (1996). New nanocomposite materials. *J.Phys. II France,* Vol. 6, 493-502, 1155-4312

Higashi, N.; Kawahara, J. & Niwa, M. (2005). Preparation of helical peptide monolayer-coated gold nanoparticles. *J. Colloid Interf. Sci.,* Vol. 288, 83-87, 0021-9797

Higuchi, M.; Ushiba, K. & Kawaguchi, M. (2007). Structural control of peptide-coated gold nanoparticle assemblies by the conformational transition of surface peptides. *J. Colloid Interf. Sci.,* Vol. 308, 356-363, 0021-9797

Hostetler, M. J.; Templeton, A. C. & Murray, R. W. (1999). Dynamics of place-exchange reactions on monolayer-protected gold cluster molecules. *Langmuir,* Vol. 15, 3782-3789, 0743-7463

Hou, G. L.; Zhu, L.; Chen, D. Y. & Jiang, M. (2007). Core-shell reversion of hybrid polymeric micelles containing gold nanoparticles in the core. *Macromolecules,* Vol. 40, 2134-2140, 0024-9297

Huang, H.-M.; Chang, C.-Y.; Liu, I.-C.; Tsai, H.-C.; Lai, M.-K. & Tsiang, R. C.-C. (2005). Synthesis of gold nanocomposite via chemisorption of gold nanoparticles with poly(*p*-methylstyrene) containing multiple bonding groups on the chain side. *J. Polym. Sci.: Part A: Polym. Chem.,* Vol. 43, 4710-4720, 1099-0518

Hussain, I.; Graham, S.; Wang, Z.; Tan, B.; Sherrington, D. C.; Rannard, S. P.; Cooper, A. I. & Brust, M. (2005). Size-controlled synthesis of near-monodisperse gold nanoparticles in the 1-4 nm range using polymeric stabilizers. *J. Am. Chem. Soc.,* Vol. 127, 16398-16399, 0002-7863

Jaramillo, T. F.; Baeck, S.;Cuenya, B. R. & McFarland, E. W. (2003). Catalytic activity of supported Au nanoparticles deposited from block copolymer micelles. *J. Am Chem. Soc.*, Vol. 125, 7148-8149, 0002-7863

Kakizawa, Y.;&Kataoka, K. (2002). Block copolymer self-assembly into monodispersive nanoparticles with hybrid core of antisense DNA and calcium phosphate. *Langmuir*,Vol. 18, 4539-4543, 0743-7463

Kanaras, A. G.; Kamounah, F. S.; Schaumburg, K.; Kiely, C. J. & Brust, M. (2002). Thioalkylated tetraethylene glycol: A new ligand for water soluble monolayer protected gold clusters. *Chem. Commun.*, 2294-2295, 1359-7345

Kang, C.; Kim, E.;Baek, H.; Hwang, K.; Kwak, D.; Kang, Y. & Thomas, E. L. (2009). Full color stop bands in hybrid organic/inorganic block copolymer photonic gels by swelling-freezing. *J. Am. Chem. Soc.*,Vol. 131, 7538-7539, 0002-7863

Kang, Y. J. & Taton, T. A. (2005). Controlling shell thickness in core-shell gold nanoparticles via surface-templated adsorption of block copolymer surfactants. *Macromolecules*, Vol. 38, 6115-6121, 0024-9297

Kang, Y. J. & Taton, T. A. (2005). Core/shell gold nanoparticles by self-assembly and crosslinking of micellar, block-copolymer shells. *Angew. Chem. Int. Edit.*, Vol. 44, 409-412, 1433-7851

Kim, B. J.; Bang, J.; Hawker, C. J. & Kramer, E. J. (2006). Effect of areal chain density on the location of polymer-modified gold nanoparticles in a block copolymer template. *Macromolecules*,Vol. 39, 4108-4114, 0024-9297

Kim, B. J.; Chiu, J. J.;Yi, G. R.; Pine, D. J. & Kramer. E. J. (2005). Nanoparticle-induced phase transitions in diblock-copolymer films. *Adv. Mater.*,Vol. 17, 2618-2622, 0935-9648

Kim, B. J.; Fredrickson, G. H.;Hawker, C. J. & Kramer, E. J. (2007). Nanoparticle surfactants as a route to bicontinuous block copolymer morphologies. *Langmuir*,Vol. 23, 12693-12703, 0743-7463

Kim, D. J.; Kang, S. M.; Kong, B.; Kim, W. J.; Paik, H. J.; Choi, H. & Choi, I. S. (2005). Formation of thermoresponsive gold nanoparticle/pNIPAAM hybrids by surface-initiated, atom transfer radical polymerization in aqueous media. *Macromol. Chem. Physic.*, Vol. 206, 1941-1946, 1022-1352

Kim, Y. G.; Oh, S. K. & Crooks, R. M. (2004). Preparation and characterization of 1-2 nm dendrimer-encapsulated gold nanoparticles having very narrow size distributions. *Chem. Mater.*, Vol. 16, 167-172, 0897-4756

Kotal, A.; Mandal, T. K. & Walt, D. R. (2005). Synthesis of gold-poly(methyl methacrylate) core-shell nanoparticles by surface-confined atom transfer radical polymerization at elevated temperature. *J. Polym. Sci.: Part A: Polym. Chem.*, Vol. 43, 3631-3642, 0887-624X

Kumar, A.; Joshi, H.; Pasricha, R.; Mandale, A. B. & Sastry, M. (2003). Phase transfer of silver nanoparticles from aqueous to organic solutions using fatty amine molecules. *J. Colloid. Interf. Sci.*, Vol. 264, 396-401, 0021-9797

Kumar, A.; Joshi, H. M.; Mandale, A. B.; Srivastava, R.; Adyanthaya, S. D.; Pasricha, R. & Sastry, M. (2004). Phase transfer of platinum nanoparticles from aqueous to organic solutions using fatty amine molecules. *J. Chem. Sci.*, Vol. 116, 293-300, 0253-4134

Kumar, A.; Mandale, A. B. & Sastry, M. (2000). Phase transfer of aqueous CdS nanoparticles by coordination with octadecanethiol molecules present in nonpolar organic solvents. *Langmuir,* Vol. 16, 9299-9302, 0743-7463

Lauter-Pasyuk, V.; Lauter, H. J.;Ausserre, D.; Gallot, Y.; Cabuil, V.; Kornilove, E. I. & Hamdoun, B. (1997). Effects of nanoparticle size on the internal structure of copolymer-nanoparticles composite thin films studied by neutron relection. *Phys. B: Condensed Matter,*Vol. 241-243, 1092-1094, 0921-4526

Laiho, A.; Ras, R. H. A.;Valkama, S.; Ruokolainen, J.; Östernacka, R. & Ikkala, O. (2006). Control of self-assembly by charge-transfer complexation between C_{60}fullerene and electron donating units of block copolymers. *Macromolecules,* Vol. 39, 7648-8653, 0024-9297

Li, D. X.; He, Q.; Cui, Y. & Li, J. B. (2007). Fabrication of pH-responsive nanocomposites of gold nanoparticles/poly(4-vinylpyridine). *Chem. Mater.,* Vol. 19, 412-417, 0897-4756

Li, D. X.; He, Q.; Cui, Y.; Wang, K. W.; Zhang, X. M. & Li, J. B. (2007). Thermosensitive copolymer networks modify gold nanoparticles for nanocomposite entrapment. *Chem.-Eur. J.,* Vol. 13, 2224-2229, 0947-6539

Li, J. B.; Shi, L. Q.; An, Y. L.; Li, Y.; Chen, X. & Dong, H. J. (2006). Reverse micelles of star-block copolymer as nanoreactors for preparation of gold nanoparticles. *Polymer,* Vol. 47, 8480-8487, 0032-3861

Li, T.; Niu, Z.;Emrick, T.; Russell, T. P. & Wang, Q. (2008). Core/shell biocomposites from the hierarchical assembly of bionanoparticles and polymer. *Small,*Vol. 4, 1624-1629, 0957-1329

Lin, Y.; Bőker, A.;He, J.; Sill, K.; Xiang, H.; Abetz, C.; Li, X.; Wang, J.; Emrick, T.; Long, S.; Wang, Q.; Balazs, A. & Russell, T. P. (2005). Self-directed self-assembly of Nanoparticle/copolymer mixtures. *Nature,*Vol. 434, 55-59, 0028-0836

Lin, Y.; Skaff, H.;Emrick, T.; Dinsmore, A. D. & Russell, T. P. (2003). Nanoparticle assembly and transport at liquid-liquid interfaces. *Science,*Vol. 229, 226-229, 0036-8075

Liu, Y. L.; Shipton, M. K.; Ryan, J.; Kaufman, E. D.; Franzen, S. & Feldheim, D. L. (2007). Synthesis, stability, and cellular internalization of gold nanoparticles containing mixed peptide-poly(ethylene glycol) monolayers. *Anal. Chem.,* Vol. 79, 2221-2229, 0003-2700

Lo, C.; Lee, B.;Pol, V. G.; Rago, N. L. D.; Seifert, S.; Winans, R. E. & Thiyagarajan, P. (2007). Effect of molecular properties of block copolymers and nanoparticles on the morphology of self-assembled bulk nanocomposites. *Macromolucules,* Vol. 40, 8302-8310, 0024-9297

Lopes, W. A. & Jaeger, H. M. (2001). Hierarchical self-assembly of metal nanostructures on diblock copolymer scaffolds. *Nature,*Vol 414, 735-738, 0028-0836

Lowe, A. B.; Sumerlin, B. S.; Donovan, M. S. & McCormick, C. L. (2002). Facile preparation of transition metal nanoparticles stabilized by well-defined (co)polymers synthesized via aqueous reversible addition-fragmentation chain transfer polymerization. *J. Am. Chem. Soc.,* Vol. 124, 11562-11563, 0002-7863

Lu, J. Q. & Yi, S. S. (2006). Uniformly sized gold nanoparticles derived from PS-*b*-P2VP block copolymer templates for the controllable synthesis of Si nanowires. *Langmuir,* Vol. 22, 3951-3954, 0743-7463

Luo, S.; Xu, J.; Zhang, Y.; Liu, S. & Wu, C. (2005). Double hydrophilic block copolymer monolayer protected hybrid gold nanoparticles and their shell cross-linking. *J. Phys. Chem. B,* Vol. 109, 22159-22166, 1520-6106

Mednoza, C.; Pietsch, T.; Cutmann, J. S.; Jehnichen, D.; Gindy, N. & Fahmi, A. (2009). Block copolymers with gold nanoparticles: correlation between structural characteristics and mechanical properties. Macromolecules, Vol. 42, 1203-1211, 0024-9297

Mirkin, C. A.; Letsinger, R. L.; Mucic, R. C. & Storhoff, J. J. (1996). A DNA-based method for rationally assembling nanoparticles into macroscopic materials. *Nature,* Vol. 382, 607-609, 0028-0836

Mossmer, S.; Spatz, J. P.; Moller, M.; Aberle, T.; Schmidt, J. & Burchard, W. (2000). Solution behavior of poly(styrene)-block-poly(2-vinylpyridine) micelles containing gold nanoparticles. *Macromolecules,* Vol. 33, 4791-4798, 0024-9297

Niidome, T.; Yamagata, M.; Okamoto, Y.; Akiyama, Y.; Takahashi, H.; Kawano, T.; Katayama, Y. & Niidome, Y. (2006). PEG-modified gold nanorods with a stealth character for in vivo applications. *J. Control. Release,* Vol. 114, 343-347, 0168-3659

Park, S. C.; Kim, B. J.; Hawker, C. J.; Kramer, E. J.; Bang, J. & Ha, J. S. (2007).Controlled ordering of block copolymer thin films by the addition of hydrophilic nanoparticles. *Macromolecules,*Vol. 40, 8119-8124, 0024-9297

Prasad, B. L. V.; Stoeva, S. I.; Sorensen, C. M. & Klabunde, K. J. (2002). Digestive ripening of thiolated gold nanoparticles: The effect of alkyl chain length. *Langmuir,* Vol. 18, 7515-7520, 0743-7463

Raula, J.; Shan, J.; Nuopponen, M.; Niskanen, A.; Jiang, H.; Kauppinen, E. I. & Tenhu, H. (2003). Synthesis of gold nanoparticles grafted with a thermoresponsive polymer by surface-induced reversible-addition-fragmentation chain-transfer polymerization. *Langmuir,* Vol. 19, 3499-3504, 0743-7463

Saito, R.; Okamura, S.& Ishizu, K. (1992). Introduction of colloidal silver into a poly(2-vinyl pyridine) microdomain of microphase separated poly(styrene-*b*-2-vinyl pyridine) film. *Polymer,*Vol. 33, 1099-1101, 0032-3861

Saito, R. & Ishizu, K. (1995). Introduction of colloidal silver into the poly(2-vinylpyridine) microdomain of microphase-separated poly(styrene-*b*-2-vinylpyridine) film: 4. one-step method. *Polymer,*Vol. 36, 4119-4124, 0032-3861

Sakai, T. & Alexandridis, P. (2004). Single-step synthesis and stabilization of metal nanoparticles in aqueous pluronic block copolymer solutions at ambient temperature. *Langmuir,* Vol. 20, 8426-8430, 0743-7463

Sakai, T. & Alexandridis, P. (2005). Mechanism of gold metal ion reduction, nanoparticle growth and size control in aqueous amphiphilic block copolymer solutions at ambient conditions. *J. Phys. Chem. B.,* Vol. 109, 7766-7777, 1520-6106

Seker, F.; Malenfant, P. R. L.; Larsen, M.; Alizadeh, A.; Conway, K.; Kulkarni, A. M.; Goddard, G. & Garaas, R. (2005). On-demand control of optoelectronic coupling in gold nanoparticle arrays. *Adv. Mater.,* Vol. 17, 1941-1945, 0935-9648

Shan, J.; Chen, H.; Nuopponen, M.; Viitala, T.; Jiang, H.; Peltonen, J.; Kauppinen, E. & Tenhu, H. (2006). Optical properties of thermally responsive amphiphilic gold nanoparticles protected with polymers. *Langmuir,* Vol. 22, 794-801, 0743-7463

Shan, J.; Chen, J.; Nuopponen, M. & Tenhu, H. (2004). Two phase transitions of poly(*n*-isopropylacrylamide) brushes bound to gold nanoparticles. *Langmuir,* Vol. 20, 4671-4676, 0743-7463

Shan, J.; Nuopponen, M.; Jiang, H.; Kauppinen, E. & Tenhu, H. (2003). Preparation of poly(*n*-isopropylacrylamide)-monolayer-protected gold clusters: Synthesis methods, core size, and thickness of monolayer. *Macromolecules,* Vol. 36, 4526-4533, 0024-9297

Shan, J.; Nuopponen, M.; Jiang, H.; Viitala, T.; Kauppinen, E.; Kontturi, K. & Tenhu, H. (2005). Amphiphilic gold nanoparticles grafted with poly(*n*-isopropylacrylamide) and polystyrene. *Macromolecules,* Vol. 38, 2918-2926, 0024-9297

Singh, N. & Lyon, L. A. (2007). Au nanoparticle templated synthesis of pNIPAM nanogels. *Chem. Mater.,* Vol. 19, 719-726, 0897-4756

Sohn, B. H.&Seo, B. H. (2001). Fabrication of the multilayered nanostructure of alternating polymers and gold nanoparticles with thin films of self-assembling diblock copolymers.*Chem. Mater.,* Vol. 13, 1752-1757, 0897-4756

Spontak, R. J.; Shankar, R.; Bowman, M. K.;Krishnan, A. S.; Hamersky, M. W.; Samseth, J.; Bockstaller, M. R. & Rasmussen, K. (2006). Selectivity-and size-induced segregation of molecular and nanoscale species in microphase-ordered triblock copolymers. *Nano Lett.,*Vol. 6, 2115-2120, 1530-6984

Thompson, R. B.; Ginzburg, V. V.;Matsen, M. W. & Balazs, A. C. (2001). Predicting the mesophases of copolymer-nanoparticle composites. *Science,*Vol. 292, 2469-2472, 0036-8075

Turkevich, J.; Stevenson, P. C. & Hillier, J. (1951). A study of the nucleation and growth processes in the synthesis of colloidal gold. *Discuss. Faraday Soc.,* Vol. 11, 55 - 75

Underwood, S. & Mulvaney, P. (1994). Effect of the solution refractive-index on the color of gold colloids. *Langmuir,* Vol. 10, 3427-3430, 0743-7463

Wan, D. C.; Fu, Q. & Huang, J. L. (2006). Synthesis of amphiphilic hyperbranched polyglycerol polymers and their application as template for size control of gold nanoparticles. *J. Appl. Polym. Sci.,* Vol. 101, 509-514, 0021-8995

Yee, C. K.; Jordan, R.; Ulman, A.; White, H.; King, A.; Rafailovich, M. & Sokolov, J. (1999). Novel one-phase synthesis of thiol-functionalized gold, palladium, and iridium nanoparticles using superhydride. *Langmuir,* Vol. 15, 3486-3491, 0743-7463

Yeh, S.; Chang, Y.;Chou, C. & Wei, K. (2004). Effect of surface-hydroxylated CdS nanoparticles on the morphological transformation of polystyrene-block-poly(ethylene oxide) thin films. *Macromol. Rapid Commun.,*Vol. 25, 1679-1686, 1022-1336

Yeh, S.; Wei, K.; Sun, Y.; Jeng, U. & Liang, K. S. (2005). CdS nanoparticles induce a morphological transformation of poly(styrene-*b*-4-vinylpyridine) from hexagonally packed cylinders to a lamellar structure. *Macromolecules,*Vol. 38, 6559-6565, 0024-9297

Yoo, M.; Kim, S.; Lim, J.; Kramer, E. J.; Hawker, C. J.; Kim, B. J. & Bang, J. (2010). Facile synthesis of thermally stable core-shell gold nanoparticles via photo-cross-linkable polymeric ligands. *Macromolecules,* Vol. 43, 3570-3575, 0024-9297

Youk, J. H.; Park, M. K.; Locklin, J.; Advincula, R.; Yang, J. & Mays, J. (2002). Preparation of aggregation stable gold nanoparticles using star-block copolymers. *Langmuir,* Vol. 18, 2455-2458, 0743-7463

Zhang, L. & Lin, J. (2009). Hierachically ordered nanocomposites self-assembled from linear-alternating block copolymer/nanoparticle mixture.*Macromolecules,*Vol. 42, 1410-1414, 0024-9297

Zhang, M. M.; Liu, L.; Wu, C. L.; Fu, G. Q.; Zhao, H. Y. & He, B. L. (2007). Synthesis, characterization and application of well-defined environmentally responsive polymer brushes on the surface of colloid particles. *Polymer,* Vol. 48, 1989-1997, 0032-3861

Zhao, W. A.; Gao, Y.; Kandadai, S. A.; Brook, M. A. & Li, Y. F. (2006). DNA polymerization on gold nanoparticles through rolling circle amplification: Towards novel scaffolds for three-dimensional periodic nanoassemblies. *Angew. Chem. Int. Edit.,* Vol. 45, 2409-2413, 1433-7851

Zhao, Y.; Thorkelsson, K.;Mastroianni, A. J.; Schilling, T.; Luther, J. M.; Rancatore, B. J.; Matsunaga, K.; Jinnai, H.; Wu, Y.; Poulsen, D.; Fréchet, J. M. J., Alivisatos, A. P. & Xu, T. (2009). Small-molecule-directed nanoparticle assembly towards stimuli-responsive nanocomposites. *Nat. Mater.,*Vol. 8, 979-985, 1476-1122

Zheng, P. W.; Jiang, X. W.; Zhang, X.; Zhang, W. Q. & Shi, L. Q. (2006). Formation of gold@polymer core-shell particles and gold particle clusters on a template of thermoresponsive and pH-responsive coordination triblock copolymer. *Langmuir,* Vol. 22, 9393-9396, 0743-7463

Zubarev, E. R.; Xu, J.; Sayyad, A. & Gibson, J. D. (2006). Amphiphilic gold nanoparticles with V-shaped arms. *J. Am. Chem. Soc.,* Vol. 128, 4958-4959, 0002-7863

9

Multifunctional Nanocomposites Based on Mesoporous Silica: Potential Applications in Biomedicine

Andreza de Sousa, Karynne Cristina de Souza, Nelcy D. S. Mohallem, Ricardo Geraldo de Sousa and Edésia Martins Barros de Sousa
Centro de Desenvolvimento da Tecnologia Nuclear - CDTN
Serviço de Nanotecnologia - SENAN
Av. Antonio Carlos 6.627, Campus da UFMG, Pampulha
31270-901 Belo Horizonte, MG,
Brazil

1. Introduction

In the last decade, significant research efforts were devoted to obtaining materials with well-defined nanostructures for a wide range of applications (Hamley, 2003; Soler-Illia et al., 2002; Paul & Sharma, 2006). Mesoporous silica materials like M41S, HMS, SBAn, FSM, and MSU, among others, are a fairly new type of material that has pores in the mesoscopic range of 2–50 nm (Wan et al., 2007). The characteristic features of ordered mesoporous materials are their monodispersed and adjustable pore size in an inert and biocompatible matrix with an easily modified surface.

Procedures to obtain ordered mesoporous silicates rely on the micelle-forming properties of a surfactant, whose chemical composition, size, and concentration control the structural dimensions of the final material (Zhao et al., 1998). In most cases, ionic and neutral surfactants have been employed as templates to direct mesophase formation based on the electrostatic and hydrogen-bonding interaction. Polymerization of the inorganic precursor and further removal of the surfactants result in a rigid silica shell that delimits the structural shape of the mesopores. In such processing routes, the resultant material presents an ordered hexagonal arrangement of unidirectional mesoporous channels and a high surface area, above 800 m^2/g, depending on the synthesis conditions. Figure 1 shows the scheme of structure formation mediated by inorganic mesoporous structure-directing agent.

The intrinsic uniform porous structure of this class of compounds with their large specific surface area and pore volume, associated with surface silanol groups, give these materials a significant potential for applications as matrices of many chemical species, such as organic molecules, metals, and polymeric materials. The combination of different materials to obtain nanocomposites is of great research interest due to their potential medical applications, such as tissue engineering, drug delivery devices, and hyperthermia, among others (Vallet-Regi et al., 2008; Souza et al., 2010).

In this work, the recent developments in nanocomposites based on mesoporous materials (also referred to as hybrid materials) will be overviewed. Two categories of nanocomposites will be described: (1) mesoporous silica with magnetic nanoparticles and (2) mesoporous silica with polymers included. The recent progress and applications of each one of these materials in the biomedical field will be reported.

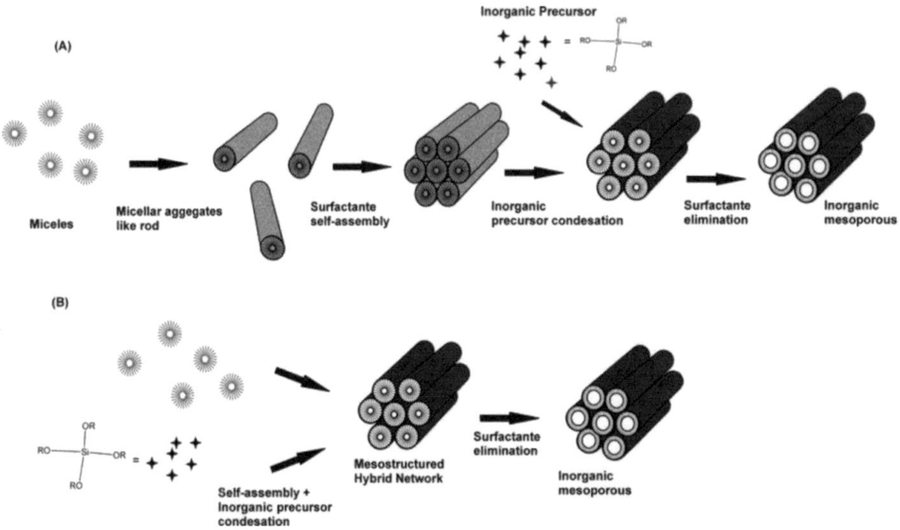

Fig. 1. Scheme of structure formation mediated by inorganic mesoporous structure-directing agent. (A) Formation of liquid crystal from surfactant and (B) self-organizing cooperative process.

2. General aspects

The performance of these materials in many fields of applications depends directly on the silica network porosity. Because of their large pores, high hydrothermal stability, and easy preparation, SBA-15 materials have been considered very promising for hosting and further delivery under appropriate conditions of a variety of molecules of pharmaceutical interest (Sousa et al., 2004). The delivery of these molecules was once considered difficult, because of the difficulty associated with the diffusion of large molecules through the materials of conventional drug delivery systems (DDS). These organic substances are normally very large in size and mesoporous silica is a potential candidate to encapsulate such molecules by utilizing ordered mesopores. Indeed, the properties of the formed solids are extremely dependent on the synthesis temperature, and the properties of surfactants used as templates in the synthesis are reflected in several properties of the final solid. In this way, structural characteristics such as pore size and porosity can be altered by change in some of the synthesis parameters. Consequently, the release kinetics from the matrix, due to its well-arranged structure, can be altered by change in the textural characteristics. Nevertheless, there is one major problem for the mesoporous systems as far as drug release is concerned, and that is the limited control of the drug release whose major mechanism is diffusion. The need to create new materials with optimized, predetermined characteristics has spurred an increasing interest in hybrid materials, especially in organic and inorganic nanocomposites.

The development of new delivery systems has proven to be of great interest for pharmaceutical technology. Ceramic materials generally release drugs by the following mechanisms: (i) diffusion, (ii) chemical reaction, or (iii) solvent activation. Beyond these, some systems can by externally activated to release more drugs when necessary, using external forces such as magnetism or temperature.

In the case of magnetism as an external force, the first type of nanocomposites that will be presented in this chapter, small magnetic beads are uniformly dispersed within a matrix. When these nanoparticles are exposed to a biological system, normal diffusion of the drug occurs due to a concentration gradient. However, upon exposure to an external oscillating magnetic field, larger quantities of drug can be released quickly. This nanocomposite combines the special structural characteristics of mesoporous silica matrix with the advantages of the unique magnetic properties of magnetic nanoparticles, because the unprotected magnetic nanoparticles can easily form aggregates and react with oxygen present in the air. Therefore, a suitable coating with mesoporous silica is essential to prevent such limitation. These materials seem suited for numerous *in vivo* applications, such as tumor hyperthermic treatment and magnetic drug targeting (Souza et al., 2009; Souza et al., 2010). Magnetic drug targeting allows the release of drugs at a defined target site with the aid of a magnetic field. The concept of hyperthermia is to heat a region of the body affected by cancer to temperatures between 43 and 45 °C. At these temperatures, the growth of cancerous cells can be halted.

In the case of temperature as an external force for diffusion, which is the second category that will be presented in this chapter, the systems are hybrid. Several studies have reported the encapsulation of polymers into mesoporous silica hosts. Encapsulation of stimuli-responsive polymers has been of particular interest in the drug delivery area. Among these stimuli-responsive polymers, temperature-responsive hydrogels such as poly(*N*-isopropylacrylamide) [P(*N*-iPAAm)] are a well-studied class of drug delivery systems, as they can respond strongly to temperature changes. In water, P(*N*-iPAAm) exhibits a phase transition at a lower critical solution temperature (LCST) of approximately 33°C (Freitas & Cussler, 1987). Below the LCST, the hydrogel incorporates water and swells, whereas the release of water in response to an increase in temperature causes shrinkage. Thus, the development of hybrid functional nanosystems based on silica-P(*N*-iPAAm) has drawn much attention to the control of molecular transport, including drug release, because self-regulated delivery allows for drug release when it is needed (Sousa et al., 2010).

3. Magnetic nanocomposites and bioapplications

The nanostructured silica has gained importance for biomedical applications and has been studied extensively. Silica has well established chemistries that allow surface modification with various functional groups (e.g., amine, thiol, carboxyl, and methacrylate). These nanoparticles can be further modified with biomolecules such as enzymes, proteins, DNA molecules, etc., for biological applications. The surface modification with biomolecules allows for the specific recognition of immobilized and free species, can provide diagnostic information, and serve as part of a purification or separation process (Jin et al., 2003).

Another type of material that gained prominence in biological applications is magnetic nanoparticles. The synthesis of nanostructured magnetic materials has been intensively pursued because of their broad applications, including in magnetic storage media, catalysis, ferrofluids, contrast materials for magnetic resonance imaging, magnetic bioseparation, and magnetic carriers for drug targeting (Pankhurst et al., 2003), among others.

The application of the different forms of iron oxides for diagnostic procedures has gained wide acceptance in radiological practice, but therapeutic applications are still under investigation. Such applications have exploited two major advantages of iron oxides: their low toxicity to humans and the possibility of controlling their magnetization. For example, many of the particles used are superparamagnetic that can be easily magnetized with an external magnetic field, and when the field is removed, they do not exhibit residual magnetization. If exposed to an alternating magnetic field, the iron oxide particles become powerful heat sources by transforming the energy from the magnetic field into heat (Chastellain et al., 2004; Jordan et al., 1997). The most important characteristic of magnetic nanoparticles for heat generation is particle size (Kalambur et al., 2005). The greatest difficulty in the synthesis of ultrafine particles is size control on the nanometric scale. Thus, the research of easy and flexible magnetic synthetic routes to produce nanoparticles with desired sizes and acceptable size distribution without particle aggregation is of extreme importance to fulfill the potential of these materials in biomedical applications.

Iron oxide (either Fe_3O_4 or γ-Fe_2O_3) can be synthesized through the co-precipitation of Fe^{2+} and Fe^{3+} aqueous salt solutions by addition of a base. The control of size, shape, and composition of nanoparticles depends on the type of salts used (e. g. chlorides, sulphates, nitrates, perchlorates, etc.), Fe^{2+} and Fe^{3+} ratio, pH, and ionic strength of the media (Gupta & Gupta, 2005).

Other methods to prepare magnetite nanoparticles are microemulsion (Liu & Wang, 2004a), electrochemical synthesis (Franger et al., 2004), pyrolysis (Gun'ko et al., 2001), and hydrothermal synthesis (Wu et al., 2005). However, most methods present some synthesis problems, such as difficulties in the prevention of flocculation, the control of Fe^{2+} and Fe^{3+} ratio in the co-precipitation process, and the removal of surfactants in the micron-emulsion process. Recently, an oxidation-precipitation method for the preparation of magnetite nanoparticles has been used by some researchers (Thapa et al, 2004). The formation of Fe_3O_4 follows the nucleation-growth mechanism when the $Fe(OH)_2/Fe(OH)_3$ molar ratio is 1:2, which is the characteristic value of the magnetite structure. However, it does require the careful adjustment of the solution pH value for particle formation and stabilization, and it is difficult to control particle size and size distribution.

As discussed, for an efficient performance, the particles must remain non-aggregated, be stable against oxidation, and display high magnetization during application. Thus, appropriate nanoparticle coating is essential to overcome this limitation. Silica/iron oxide nanocomposites are of particularly interest, since the protective layer afforded by silica can prevent dipolar magnetic attraction between magnetite particles, and consequently produce quite uniform particle dispersion.

The approaches for the synthesis of SiO_2–Fe_2O_3 nanocomposites are varied (Ma et al., 2006; Chen et al., 2008). Such nanocomposites are composed of magnetic nanoparticles embedded or encapsulated in an inorganic oxide matrix. In addition, to coat the surface of magnetic nanoparticles, silica is obtained by classical sol–gel methods, but in this case, the final product may present a wide pore size distribution (Singh, 2008). The new class of nanocomposites, which combine special structural characteristics of mesoporous silica matrix with the advantages of the unique magnetic properties of iron nanoparticles, seems suited for numerous *in vivo* applications, such as tumor hyperthermic treatment and drug delivery. Most importantly, particle growth can be effectively confined to the channels of mesoporous silica, limiting particle size to the nanoscale.

Studies concerning the growth of Fe_3O_4 and γ-Fe_2O_3 nanoparticles inside mesoporous silica MCM-41 and MCM-48, using an ionic surfactant as a template, have also been reported (Köhn et al., 2003; Fellenz et al., 2006a). More recently, Fe_3O_4 nanoparticles modified covalently with triethoxysilane having a quaternarydicetylammonium ion were used as building blocks and incorporated in the structure (Alvaro et al., 2006). The surface of these nanoparticles was modified to become compatible with the structure-directing agent (cetyltrimethylammonium). However, these approaches result in materials presenting two main limitations: (i) due to the use of ionic surfactants, they present typical wall thicknesses around 1 nm (Doadrio et al., 2006) and consequently, low stability; (ii) the functionalization of the mesoporous surface and nanoparticle surface leads to a decrease in the pore size and surface area (Alvaro et al., 2006), facts that could affect the applicability of such materials in adsorption systems, like drug delivery systems.

In this context, we reported the synthesis of mesoporous SiO_2-coated Fe_3O_4 nanoparticles by using a neutral surfactant and without the application of any functionalization method (Souza et al., 2008). To accomplish this, the Fe_3O_4 nanoparticles were synthesized by oxidation–precipitation and coated with mesoporous silica (SBA-15) by using nonionic block copolymer surfactants as the structure-directing agent. We observed that the magnetite nanoparticles were completely and homogeneously coated by well-ordered mesoporous silica with stable (~8 nm thick) pore walls, Figure 2. The applied synthesis route did not affect the structural and magnetic properties of the Fe_3O_4 nanoparticles.

In the other report, we presented a wet impregnation method which proved efficient to obtain well-dispersed Fe_3O_4 nanoparticles inside the pores of silica (Souza et al. 2010). The proposed method involved the preparation of an iron oxide precursor in ethanol and its subsequent transfer to SBA-15 mesoporous hexagonal silica. Iron oxide was formed inside the porous structure to obtain a magnetic material, Figure 3. The sample was heated to 550 °C for 2 h in air for the formation of iron oxide and then was heated to 550 °C for 8 h in an N_2 atmosphere to obtain Fe_3O_4.

The main methods of synthesis of composites using magnetic oxides dispersed in silica matrix and their various applications will be discussed below, having in mind the growing interest in using these materials in the biological area.

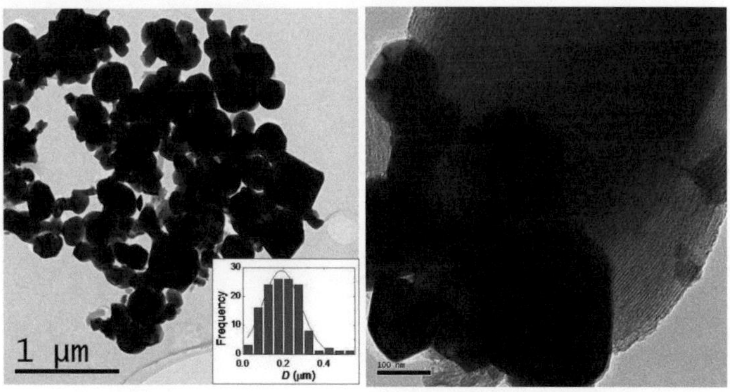

(a) (b)

Fig. 2. Transmission electron micrographs: (a) Low resolution and (b) high resolution images showing the magnetite nanoparticles (dark region) covered by the mesoporous silica; the inset in (a) is the quantification of the size distribution of the Fe3O4 particles.

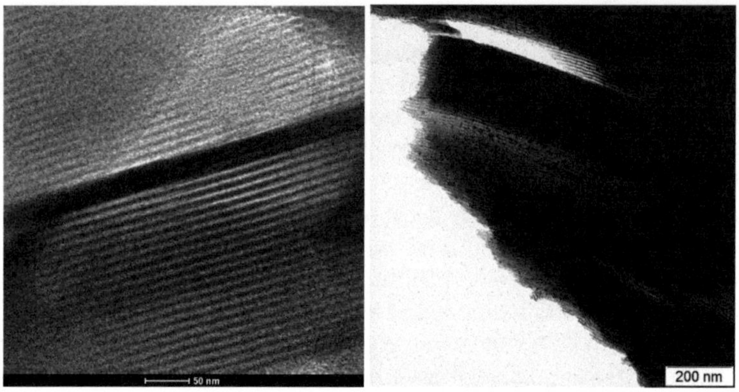

(a) (b)

Fig. 3. TEM of unidirectional canals of the open mesoporous structure, and image showing the magnetite nanoparticles (dark region) covered by mesoporous silica.

3.1 Bioseparation

Magnetic separation provides a convenient method for the removal of magnetizable particles by applying an appropriate magnetic field. In 1973 Robinson et al. used magnetic separation for the first time in a biotechnology context. In their work, silica-coated magnetic iron oxide and cellulose-coated magnetic iron oxide were used to immobilize two enzymes, α-chymotrypsin and β-galactosidase, for applications in bioreactors. Since then, magnetic separation has become an increasingly popular tool for the separation of biological molecules and cells (Bruce et al., 2004).

If the magnetic material surface is selective for a specific protein, it will be able to connect to this surface, and its separation from the biological environment could be done by using a magnet (Figure 4). Once the surface has been modified, a system comprised of a magnetic core coated with silica can be used for bioseparation.

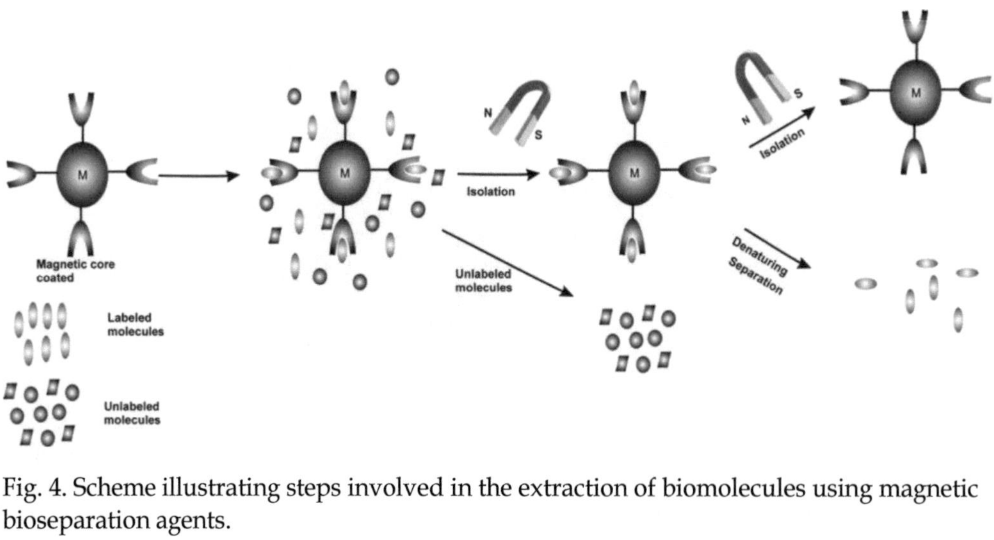

Fig. 4. Scheme illustrating steps involved in the extraction of biomolecules using magnetic bioseparation agents.

The first bioseparation work onto a functionalized silica support was performed by Zhao and coworkers in 2003 (Zhao et al., 2003). The authors developed a novel genomagnetic nanocapturer (GMNC) for the collection, separation, and detection of trace amounts of DNA/RNA molecules with one single-base difference. The GMNC was constructed by bioconjugating molecular beacon DNA probes onto magnetic nanoparticle surfaces. The method showed an excellent ability to differentiate single-base mismatched DNA/mRNA samples by combining the exceptional specificity of molecular beacons and the separation power of magnetic nanoparticles.

Later, Liu and coworkers synthesized magnetite (Fe_3O_4) nanoparticles by the coprecipitation method and coated their surface with silica, obtaining well dispersed magnetic silica nanospheres of about 50–80 nm in diameter. The surface of these nanospheres was functionalized with amino-silane coupling agent, *N*-(2-aminoethyl)-3-aminopropyltrimethoxysilane (AEAPS) (Liu et al., 2004b). The functionalized silica has undergone a series of experiments performed to investigate the protein immobilization in a magnetic matrix. Bovine serum albumin (BSA) was chosen as a model protein and was covalently immobilized onto the amino-silane modified magnetic silica supports. The result shows that such amino-silane modified magnetic silica is a well-dispersed and effective superparamagnetic support for bioseparation with maximum BSA immobilization capacity (up to 86 mg/g), presenting, thus, potential use as immune-magnetic support for protein purification.

Sen et al. (2006) reported the template-assisted fabrication of a magnetic mesoporous silica-magnetite nanocomposite and its potential for application in magnetic bioseparations, evaluating its ability (i) to bind and elute DNA and (ii) to extract RNA from bacterial cells. Magnetite nanoparticles previously synthesized were used as a core material for the fabrication of a mesoporous magnetic nanocomposite. Both core magnetite and silica-magnetite nanocomposite demonstrated a high capacity for binding the DNA. Nearly 100% recovery of DNA was obtained from the surface of the nanocomposite, whereas <10% was recovered from the magnetite core. It was possible to bind and elute approximately the same quantity of DNA using half the amount of the mesoporous silica-magnetite composite, as compared to classical amorphous silica-magnetite materials (Bruce et al., 2004).

Specifically in enzyme immobilization, it is known that because of the low enzyme loading on the conventional magnetic beads, more and more attention was paid to the magnetic mesoporous support. Magnetite mesoporous silica hybrid support was fabricated by depositing magnetite and MCM-41 nanoparticles onto polystyrene beads using the layer-by-layer (LBL) method. The incorporation of magnetite gives an additional magnetic property to the hollow mesoporous silica shells. This perfect combination of mesoporous material properties with a magnetic property improves the enzyme immobilization (Xie et al., 2009).

Another field of interest in applications that has gained prominence in the biological area is the use of these systems in studies of drug delivery.

3.2 Controlled release of drugs

The idea of controlled release of biologically active substances began to be used in the 1950s with the advent of polymeric materials. The potential advantages of controlled release systems include: (i) localized release of drugs in a particular compartment of the body, thereby decreasing the systemic level of the drug, (ii) maintaining constant levels of drug in the body, resulting in greater efficiency in the use of the agent (i.e., it is necessary to lower the concentration of the drug to produce the same effect as in conventional systems), (iii) preservation of the medicines that are quickly destroyed by the body (this is particularly important for biologically sensitive molecules such as proteins), (iv) lower frequency of administration of the active agent, enhancing patient comfort and effectiveness of treatment (Langer, 1990; Ogawa & Plepis, 2002). Besides the polymeric matrix, other controlled release systems based on different materials have been intensively studied in recent years, such as bioceramic and composites. Some examples are antibiotics like gentamicin, released from the matrix composed of calcium phosphates and poly(DL-lactide) (PLA) (Baro et al., 2002) and from the composite of hydroxyapatite/collagen (Martins & Goissis, 2000), and anti-neoplastic agents like cisplatin, released from calcium phosphate (CaP) nanoparticles (Barroug et al., 2004).

The porosity of conventional matrices is highly heterogeneous due to complex chemical composition, which presents as a great disadvantage the difficulty of ensuring homogeneous drug distribution through the matrix, affecting the release rate. Therefore, the need to address this disadvantage has led to improvements in this field through the use of ordered mesoporous materials, chemically homogeneous, which have well defined porosity.

In this context, we have reported the influence of magnetic nanoparticles on drug release kinetics of three model drugs: cisplatin, carboplatin, and atenolol under *in vitro*. In this work (Souza et al., 2009), magnetic nanoparticles embedded into mesoporous silica were prepared in two steps: first, magnetite was synthesized by oxidation-precipitation method, and next, the magnetic nanoparticles were coated with mesoporous silica by using nonionic block copolymer surfactants as structure-directing agents. The SEM images of magnetite in Figure 5 show that the material has uniform morphology and particle size distribution. In addition, some particle clusters may be observed due to the magnetic interaction between the particles. The regular octahedral morphology indicates that the particles are well crystallized. Figure 5.b shows the morphology of SBA-15. It consists of vermicular-shaped particles. In contrast, the nanocomposite presents spherical morphology. This may be attributed to the presence of the magnetite silica-coated nanoparticles. The average size of the spherical particles varied from 0.139 to 0.476 μm. Particles smaller than the human body capillary diameter (8 μm) are suitable for intravenous drug-delivery injection. The desired particle size range for a specific application may be selected by nanoparticle centrifugation at a controlled rate.

Drug release was studied with and without a magnetic field in two experiments: a) with a constant magnetic field and b) with an alternating magnetic field. To simulate drug release in the presence of a magnetic field, a NdFeB magnet (0.250 T) was used in the first experiment, and in the second one, a low-frequency alternating magnetic field was obtained

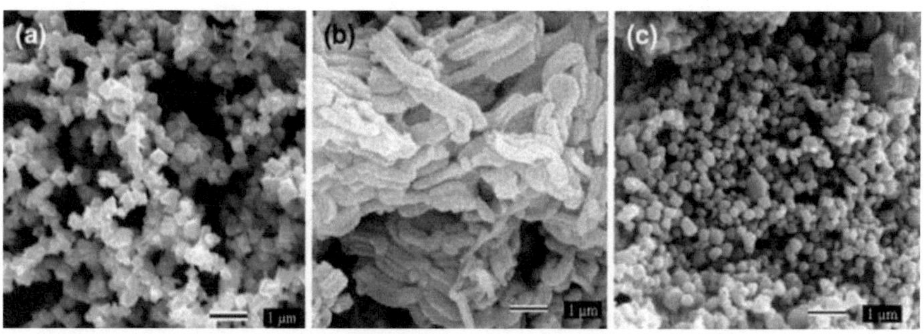

Fig. 5. SEM of (a) magnetite, (b) pure SBA-15, and (c) SBA- 15/Fe_3O4

with a magnetic cavity composed of a Helmholtz coil connected to a direct current source coupled to a frequency generator and a Hall probe for measuring the magnetic field. The following frequencies were used: 15, 30, 45, 60, 100, and 300 Hz with a 17 mT magnetic field. The application of an alternating magnetic field provokes the vibration of the magnetic particles and the quick release of large quantities of drugs. The measures presented here are designed to assess the influence of the application of low frequency (≤300 Hz) alternating magnetic fields (170 Oe) on drug release. Figure 6a shows the release profiles of cisplatin without a magnetic field, with a constant magnetic field, and an alternating magnetic field up to 300 Hz. Drug release was more effective and faster in the alternating magnetic field, Figure 6. The alternating magnetic field showed a large influence on the cisplatin release profile. These results can be explained by the possible interaction of the drugs with the magnetite nanoparticles, possibly due to the incorporation of cisplatin into the mesopores and its interaction with the surface of the magnetic nanoparticles.
Another promising application for these materials is in magnetic hyperthermia.

3.3 Magnetic hyperthermia

Hippocrates (460-370 BC) believed that any disease could be cured by heating the patient's body (Ito et al., 2005). Since 1957, hyperthermia, a modality of cancer treatment with an elevated temperature between 41 and 45 °C with a treatment time of at least 30 min, has received considerable attention due to its clinical efficacy, such as minimizing clinical side effects and the possibility of selectively destroying a localized or a deeply seated malignant cancer tumor (Bae et al., 2009).
Cancer cells generally perish at around 43 °C because their oxygen supply via the blood vessels is not sufficient, whereas normal cells are not damaged at even higher temperatures. In addition, tumors are more easily heated than the surrounding normal tissues, since the blood vessels and nervous systems are poorly developed in the tumor. Therefore, hyperthermia is expected to be a very useful cancer treatment with few side effects. Various techniques for heating the tumors, such as treatments with hot water, infrared rays, ultrasound, and microwaves have been attempted. However, deep-seated tumors cannot be heated effectively and locally using these techniques (Kawashita et al., 2005).

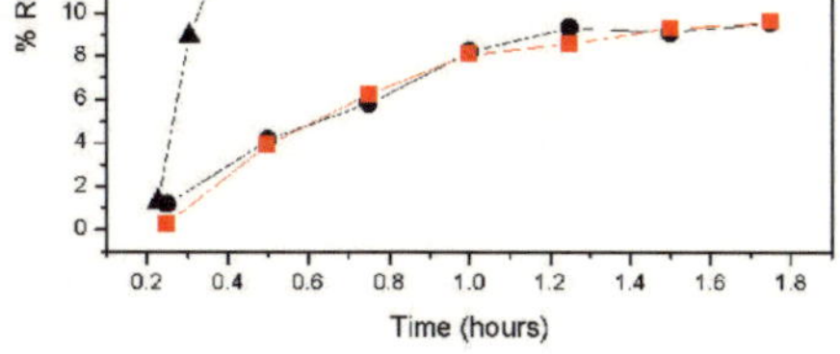

Fig. 6. Release profile of the nanocomposite/cisplatin system without a field, in a constant field, and in an alternating field.

The biggest problem with this treatment is that it is not selective, i.e., when subjected to hyperthermia for tumor treatment, it also affects normal cells adjacent to the tumor. Thus, there is a need to develop a tumor-specific hyperthermia for tumor cells.

Some researchers have proposed the concept of "intracellular" hyperthermia and have developed submicron magnetic particles for inducing hyperthermia. This concept is based on the principle that under an alternating magnetic field (AMF), a magnetic particle can generate heat by hysteresis loss (Ito et al., 2005).

There are at least three different mechanisms by which magnetic materials can generate heat in an alternating field (Kalambur et al., 2005): (i) generation of eddy currents in bulk magnetic materials, (ii) hysteresis losses in bulk and multi-domain magnetic materials, and (iii) relaxation losses in 'superparamagnetic' single-domain magnetic materials. The mechanisms (i) and (ii) contribute very little to the heating system of particles with a single-domain. The mechanism that contributes to significant heating of these particles, therefore, is the relaxation mechanism (iii). Relaxation losses in single-domain magnetic nanoparticles fall into two modes: rotational (Brownian) mode and Néel mode (Figure 7). In the Néel mode, the magnetic moment originally locked along the crystal easy axis rotates away from that axis towards the external field. The Néel mechanism is analogous to the hysteresis loss in multi-domain magnetic particles, whereby there is an 'internal friction' due to the movement of the magnetic moment in an external field that results in heat generation. In the Brownian mode, the whole particle oscillates towards the field with the moment locked along the crystal axis under the effect of a thermal force against a viscous drag in a suspending medium. This mechanism essentially represents the mechanical friction component in a given suspending medium.

Kawashita and coworkers (Kawashita et al., 2005) studied magnetite microspheres with capacity heat generation. Two kinds of microspheres were synthesized: (a) magnetite microspheres formed by melting powders in high-frequency induction thermal plasma, and (b) silica glass microspheres formed by precipitation from an aqueous solution. In case (a), the magnetite microsphere diameter was 20 to 30 μm, with heat generation capacity of 10 W/g, under 300 Oe and 100 kHz; in case (b), the silica glass microspheres coated with magnetite presented average size of 25 μm, while the average size of magnetite crystallites deposited on the silica surface was estimated at 50 nm. In the latter case, the heat generation was estimated to be 41 W/g, under 300 Oe and 100 kHz, showing that the silica/magnetite microsphere system is a promising one for hyperthermic treatment of cancer.

There is a possibility of a second application for this system. The controlled drug delivery could be triggered by thermomagnetic action, adjusting the magnetic field and frequency.

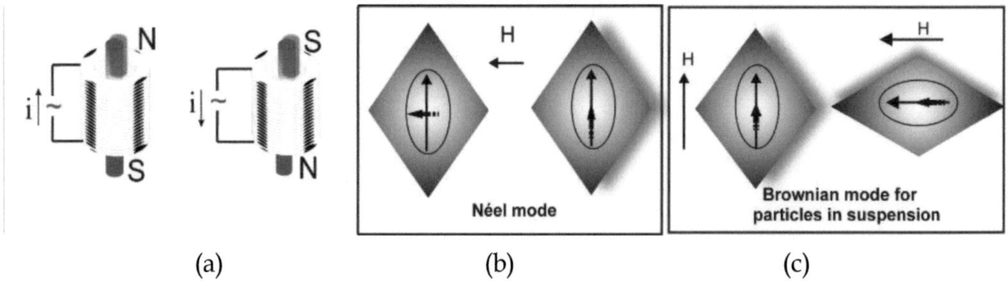

Fig. 7. (a) Induction Field schema for magnetic material in an AC coil. (b) and (c) Relaxational losses leading to heating in an alternating field (H): (b) Néel mode; (c)Brownian mode.

Considering these points, Hu and coworkers studied magnetic-sensitive silica nanospheres for ibuprofen release by controlled bursting to a therapeutically effective concentration by a high-frequency magnetic field (HFMF) (Hu et al., 2008). The acceleration of the rotation of magnetic nanoparticles deposited on the silica matrix generates heat energy; the HFMF enlarges the nanostructure of the silica matrix, producing porous channels that cause the drug to be released easily. Thus, nanospheres showed a marked ibuprofen release profile, with 180-190 min of release, when the system was triggered for 10 min in a high-frequency magnetic field (HFMF). This profile was compared with a control group of silica nanospheres without magnetic iron incorporation. The cumulative release of ibuprofen from silica nanospheres (Si:Fe = 1:1) was measured, both with and without magnetic stimulation of 50 kHz applied at the four specific times, and showed that after the magnetic stimulation, a significant increase in the amount of drug released was observed at each time interval.

A small number of studies have been conducted with mesoporous silica/magnetic particles systems for hyperthermia applications. Recently, we reported the synthesis of nanocomposite-based SBA-15 and magnetite nanoparticles by the wet impregnation method of an iron precursor (Fe3(SO4)3) into a silica framework (Souza et al., 2010). Iron oxide was formed inside the porous structure, thus producing the magnetic device.

The spontaneous magnetization of the nanocomposite is 10.31 $Am^2\ kg^{-1}$, which is lower than that of magnetite (83.98 ± 2.2 $Am^2\ kg^{-1}$), because the nanoscale particles form a single magnetic domain. The magnetic properties obtained by ^{57}Fe Mössbauer spectroscopy (figure not given) showed that the nanocrystals present superparamagnetic characteristics.

Tests of magnetic hyperthermia were performed. For this, AC magnetic-field-induced heating was measured both in the solid material and in an aqueous solution under different applied magnetic fields: 105 Oe (8.36 kA m^{-1}) and 168 Oe (13.37 kA m^{-1}).

Figure 8 shows the time-dependent temperature curves of the dry nanocomposite for 198 kHz and magnetic fields of 105 and 168 Oe AC. When the sample was exposed to an AC magnetic field for 30 min, its temperature ranged from 23 to 32.87 °C in the 168 Oe AC magnetic field and from 23 to 25.16 °C in the 105 Oe, showing its dependence on the magnetic field. This figure shows a relatively sharp increase in temperature in the beginning of heating. Moreover, a nearly steady state regime is established after 10 min under 168 Oe, reaching a temperature range of ΔT_{max} = 9.87 °C.

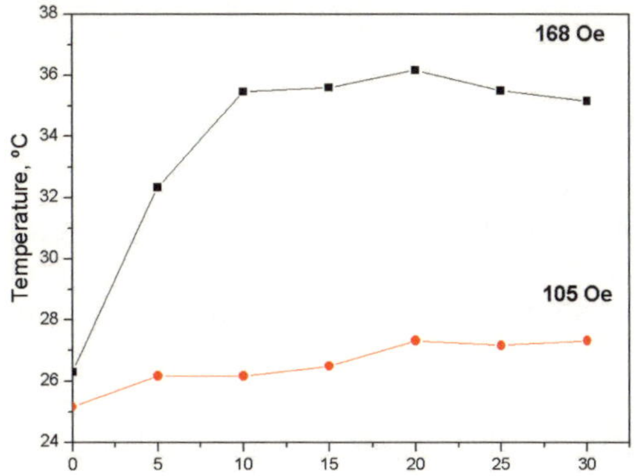

Fig. 8. Time-dependent temperature curves of the nanocomposite under 198 kHz and magnetic fields of 168 Oe (13.37 $kA.m^{-1}$) and 105 Oe (8.36 $kA.m^{-1}$).

When dispersed in water, the difference in the temperature of the nanoparticles between the starting ambient temperature and after 30 min of exposure to the magnetic field was 21 °C under 168 Oe and 3 °C under 105 Oe. Considering the results obtained for the dry nanoparticles and those dispersed in water in the 168 Oe magnetic field, Figures 9a and 9b, we notice a small increase in temperature for the former, resulting in ΔT_{max} = 9.87 °C, and 21 °C for the latter. This behavior can be explained based on the total surface area of the magnetic nanoparticles and their ability to generate heat. If the magnetic nanoparticles are well dispersed in water, their specific surface area increases, as does the convective heat transfer rate to the fluid.

The amount of heat generated by pure water in 168 Oe AC magnetic fields was measured to investigate more precisely the effect of water on the increase in the temperature of the dispersed nanoparticles. According to Figure 9b, although pure water contributes to a slight rise in temperature, from 23 to 29 °C, heat generation is affected by the sample medium. This property of the nanoparticles dispersed in water was in fact measured after sonicating the solution for 30 min. The measured temperatures of the nanoparticle suspension after sonication and under 168 Oe AC magnetic field increased to 47.5 °C after 30 min of assay, presenting a ΔT_{max} of 24.5 °C.

The principle of hyperthermia is to heat a body region affected by cancer to temperatures between 43 and 45 °C. At these temperatures, the growth of cancerous cells can be halted. Thus, a temperature variation of ΔT_{max} = 9.87 °C would be sufficient, based on a body temperature of 37 °C. Considering all the measurement conditions used in this work, heat generation can be controlled in selected magnetic particles by adjusting the magnetic field and choosing an appropriate medium.

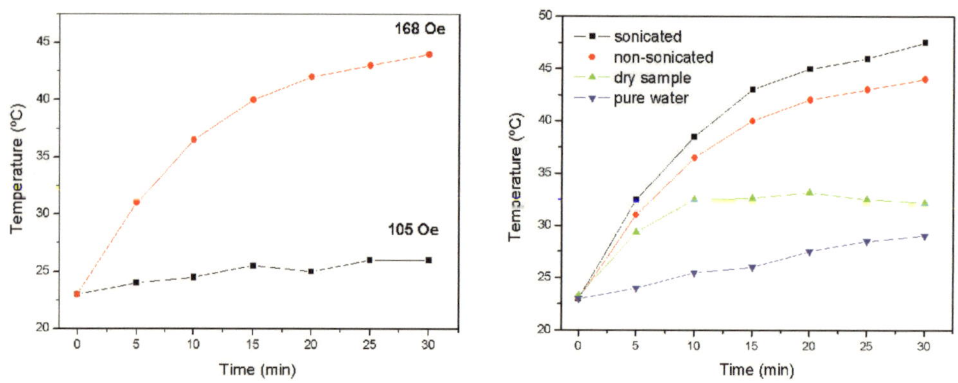

Fig. 9. (a) time-dependent temperature curves of the nanocomposite under 198 kHz and magnetic fields of 168 Oe and 105 Oe; (b) heating-induced by magnetic field of nanocomposite in different situations to 168 Oe compared with pure water, used as a control.

4. Functional hybrid mesoporous silica/poly(n-isopropylacrilamide) useful for drug delivery systems

The second category of multifunctional materials deals with organic-inorganic hybrids systems based on mesoporous materials and stimuli-responsive polymers. Recent studies have made fundamental proposals for development of hybrids materials based on polymer and mesoporous silica. Many different polymers have been encapsulated into the pores structure of the mesoporous materials, and different procedures were used for encapsulation. Recently, strategies for development of hybrids based on ordered mesoporous silica and stimuli-responsive polymers have been reported in the literature (Tian & Yang, 2009; Fu et al., 2007; Liu et al., 2009; Gao et al., 2009; Sousa et al. 2010).These stimuli-responsive polymers show abrupt changes in their swelling behavior in response to small changes in environmental parameters such as temperature, pH, solvent composition, electric fields, etc. They have received much attention due to their potential applications in numerous fields. Among these stimuli-responsive polymers, the most important systems from a biomedical point of view are those sensitive to pH and/or temperature, because variables such as low pH and elevated temperatures are found in the body (Qui & Park, 2001).

One of the most studied temperature-responsive polymers is poly(*N*-isopropylacrylamide) [P(*N*-iPAAm)], which swells to a large extent in water, at a low temperature, and shrinks as the temperature is increased, showing a first-order phase transition around 33 °C, behaving as a polymer solution with a lower critical solution temperature (LCST) (Freitas & Cussler, 1987). Figure 10 shows this phase behavior schematically. This interesting phase behavior of a dramatic dependence of the swelling volume on temperature and a discontinuous transition has attracted the interest of many researchers all over the world in past years. Studies on the thermodynamic behavior of this system (Marchetti et al., 1990; Sousa, 1993; Sousa, 1995; Nakamoto, et al., 1996), different potential applications (Wu et al., 1996; Jin, et al., 1995), and permeability to small solutes (Palasis & Gehrke, 1992; Mukae, et al., 1993) have been reported.

The thermodynamics of polymer gels describes the swelling equilibrium as a result of two opposite contributions: the change in the free energy of mixing as the gel is in contact with a solvent and the change in the elastic free energy due to the elongation of the network as the gel swells, absorbing solvent (Ogata, 1995). The understanding of the gels' elastic structure is, therefore, fundamental. This structure is directly related to the crosslinking density and polymer-polymer interactions.

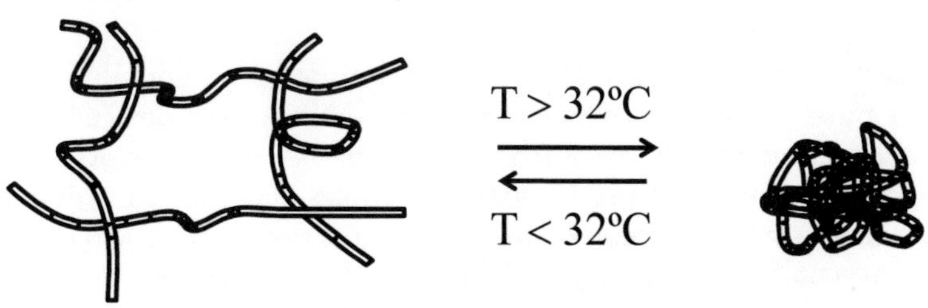

Fig. 10. Schematic representation of the phase behavior of P(*N*-iPAAm).

By utilizing this interesting phase behavior, P(*N*-iPAAm) can be used in biomedical applications such as DDS (Hoffman, 1987; Peppas & Langer, 2003). Furthermore, the interest in the use of P(*N*-iPAAm) in biomedical applications originates from a number of advantages that this presents for use as a biomaterial, such as biocompatibility, specific diffusive properties, elastomeric consistency, and low interfacial tension, among others.

Hybrid functional nanosystems based on silica- P(*N*-iPAAm) have attracted much attention for control of molecular transport, including drug release, because a self-regulated delivery allows a drug to be released when it is needed. Thus the encapsulation of P(*N*-iPAAm) in ordered mesoporous silica has been of particular interest in the drug delivery area, since that combine the advantages of the mesoporous silica (thermal and mechanical stability, biocompatibility and textural properties) with the P(*N*-iPAAm) ability to swell or shrink in temperatures close to physiological temperature. Compared to the sustained-release system, the thermo-responsive controlled release system can realize a controlled release pattern, which can improve the therapeutic efficacy. However, studies on ordered mesoporous silica-thermoresponsive P(*N*-iPAAm) hybrids for use as DDS are few. To the best of our knowledge, only a small number of papers concerning a responsive-carrier-system-based mesoporous silica and P(*N*-iPAAm) for use in the drug delivery area has been reported up to the present (Zhou et al. 2007; Yang et al., 2008; You et al., 2008; Liu et al., 2009; Zhu, et al., 2009; Sousa et al., 2010).

Chang and co-authors (Chang et al., 2004) report the systematic application of the ordered mesoporous materials to smart controlled drug release using the thermoresponsive P(*N*-iPAAm) hybrid mesoporous structures. The P(*N*-iPAAm) hybridized mesoporous materials were prepared by the radical-initiated polymerization with *N*-iPAAm monomers on modified mesoporous silica surface with different pore sizes (10, 17, 30 nm). Indomethacin was used to study the thermosensitive drug release from hybrid. The drug release from hybrid materials was investigated during stepwise temperature changes between 25 and 40 °C and measurement of the concentration in a UV-Vis spectrophotometer at different time intervals. The release profile showed a sustained positive thermoresponsive drug release. When the temperature was increased and maintained at 40 °C, a nearly constant release pattern was observed and a rapidly decreased release was observed on the temperature change to 25 °C. According to the nanodiffusion mechanism proposed by the authors at a low temperature, the drug is trapped in the polymer and in the porous structures, and the polymers swell to prevent the drug from being significantly released into the media. When the temperature is increased above the LCST of P(*N*-iPAAm), the polymers shrink to squeeze the drug into the porous channels, and open the pore structure. Although a large amount of drug may be squeezed out of the polymer above the LCST, the amount released into the media did not show a peak, because diffusion of the drug through the porous channels is the limiting mechanism to control the drug release, rather than the phase behavior of the P(*N*-iPAAm).

In this same sense, Zhou et al. investigated a system based on stimulus-responsive P(*N*-iPAAm) inside a mesostructured cellular foam (MCF-PNIPA) as a device for drug delivery (Zhou et al., 2007). Ibuprofen (IBU) was loaded in MCF-PNIPA samples at 45 °C during 48 h. Drug release studies were performed by soaking samples in a solution with pH maintained at 7.4 and stepwise changing the temperature between 10 and 60 °C. The release process was followed spectroscopically. Loading and release data suggested that the multilayer polymers inside the pores of the MCF apparently formed an internal cavity for drug molecules increasing the IBU storage capacity, in addition to responding to changes in external temperature. A possible mechanism of the drug storage and release profile from the system was developed by the authors, and according to this mechanism, at a low temperature, the drugs are confined in the pores owing to expansion of the P(*N*-iPAAm) molecular chains and the formation of hydrogen bonding between P(*N*-iPAAm) and IBU. With an increasing temperature, the polymer chains are hydrophobic and swell within the pore network, resulting in the collapse of hydrogen bonds and pushing the drug molecules from the pores. Figure 11 shows this dependence of IBU amount released on temperature.

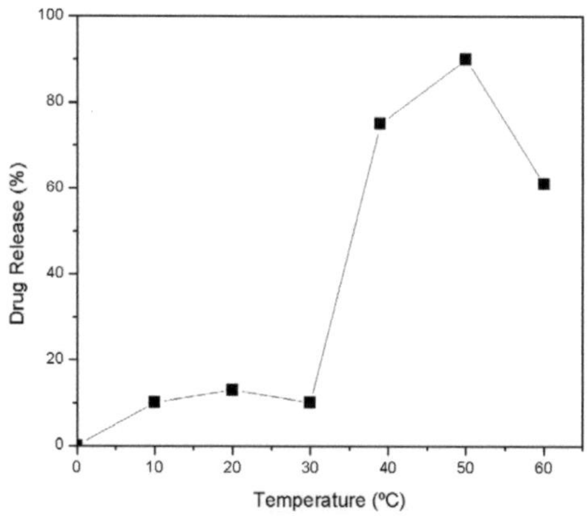

Fig. 11. The dependence of release amounts of IBU on temperature. Data obtained from (Zhou et al., 2007).

Furthermore, in 2007, a novel synthesis of a multi-functional delivery system based on ordered mesoporous silica SBA-15 with magnetic particles formed in situ and thermo-sensitive P(*N*-iPAAm) - m-SBA–PNIPA was reported. They employed hydrophilic $FeCl_2$ as an additional precursor to fabricate magnetic particles inside SBA-15 via surfactant template sol–gel process, and the polymerization of *N*-iPAAm was implemented inside the pores of SBA-15 with a magnetic particle system. Ibuprofen also was encapsulated inside the sample m-SBA–PNIPA and the in vitro delivery assays were performed according to the procedure previously described by Zhou et al., stepwise changing the temperature between 9 and 45 °C. The system exhibited a pronounced transition at approximately 17 °C, and when the temperature was between 7 and 11 °C, a drug release of 40.5% from the m-SBA–PNIPA was observed. As the temperature rose to 17 °C, the polymer shrank, leading to the progressive increase of the release rate. Similar results were found for the systems previously reported for Chang and Zhou et al., where a positive thermosensitive release profile was observed. However, this delivery system did not show characteristic bulk behavior with respect to their LCST, and as mentioned by the authors, this system is not suitable for in vivo application but could probably be used in many others fields.

Another kind of magnetic mesoporous silica/P(*N*-iPAAm) nanocomposite was also prepared recently by Liu et al. (Liu et al., 2009), who reported a strategy to prepare core/shell-structured magnetic mesoporous silica microspheres combining three advantages into one single entity. This strategy involves a Fe_3O_4 magnetic nanoparticle as the core, mesoporous silica as a sandwiched layer, and cross-linked thermosensitive poly(*N*-isopropylacrylamide/*N*-hydroxymethyl acrylamide) - P(NIPAM-co-NHMA) copolymer as an outer shell. Zn(II) phthalocyanine tetrasulfonic acid (ZnPcS4), a photodynamic therapy (PDT) drug, was used as a model drug to assess the drug loading and controlled release behavior of the core/shell composite microspheres. Drug release studies were performed by soaking loaded microsphere hybrids in a solution with pH maintained at 7.4 and measuring the cumulative release amount of ZnPcS4 after 24 h at different temperatures. These composite microspheres with different amounts of the hydrophilic co-monomer exhibited a temperature-dependent drug release behavior (Figure 12) where the cumulative release amount after 24 h at 39 °C was almost three times higher than that at 37 °C for two samples prepared with different monomer ratios. According to the authors, the significant difference

in cumulative release amount at different temperatures indicated a potential application in the therapy of malignant tumors. For instance, the M-MSN/P(NIPAM-co-NHMA) composite microspheres with superparamagnetic cores and temperature-sensitive shells could be guided by an external magnetic field to the target site, undergo a swelling-shrinking transition, and release the loaded drug rapidly because of the temperature difference between malignant tumor tissue and normal tissue, or because of the temperature difference caused by the magnetocaloric effect in the alternating-current magnetic field.
Yang et al. (2008) synthesized a potential material for applications in intercellular imaging, as biomarkers and biosensors. This material is based on mesoporous silica nanoparticles (MSNs) coated with poly(*N*-isopropylacrylamide) - MSN@PNIPAM. The MSN@PNIPAM was synthesized by atom transfer radical polymerization (ATRP) on the surface of MSN through the "grafting-from" method. The thermoresponsive properties of the MSN@PNIPAM were studied by turbidimetry and dynamic light scattering (DLS) measurements and the measured LCST of MSN@PNIPAM material was about 32°C, consistent with pure PNIPAM in water. A fluorescent dye (FITC) was used as a model guest molecule to test the encapsulation and release ability of the MSN@PNIPAM particles. FITC was entrapped by the stretched polymer chains below LCST, and locked inside the MSNs above LCST. Additionally, the authors took MCF-7 cells to investigate the behavior of FITC MSN@PNIPAM in cells. The confocal laser scanning microscopy images showed that some particles had been internalized into the cells and localized in the cell cytoplasm. FITC was locked inside the MSN particles during the cell endocytosis process because the cell culture temperature was higher than LCST. The cytotoxicity of MSN@PNIPAM was also investigated and the results suggested that MSN@PNIPAM particles had negligible influence upon the growth of cells showing very low cytotoxicity.

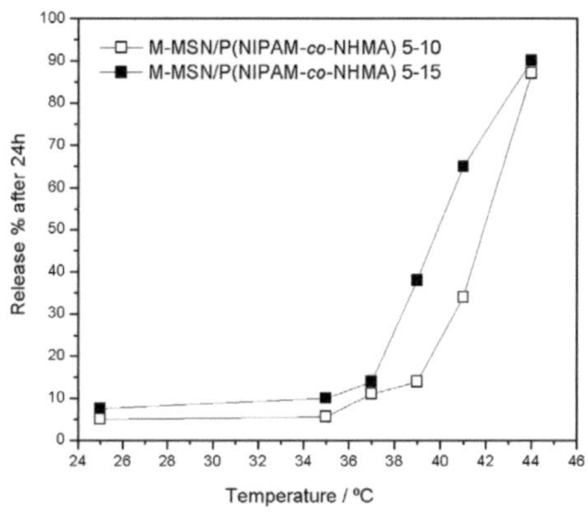

Fig. 12. Cumulative drug release of drug-loaded systems in PBS solution as a function of temperature. Data obtained from (Liu et al., 2009).

We recently reported the control of drug release in response to temperature in a system based on a temperature-responsive hydrogel, poly(*N*-isopropylacrylamide) [P(*N*-iPAAm)], inside a mesoporous silica SBA-15 (Sousa et al., 2010). The incorporation of polymer phase in the mesoporous silica led to a significant change in the structural properties of this system. However, the polymerization and pore filling did not destroy the ordered hexagonal structure of SBA-15, as shown by TEM images (Figure 13), but the surface area, pore size, and pore volume of the hybrid decreased due to the introduction of the polymer. The primary mesopore volume and total pore volume of [SBA-15/P(*N*-iPAAm)] (0.42 and 0.48 cm^3 g^{-1}) were lower than the SBA-15 values (0.83 and 0.96 cm^3 g-1), respectively. The pore diameter decreased from 5.70 to 3.76 nm (~34%).

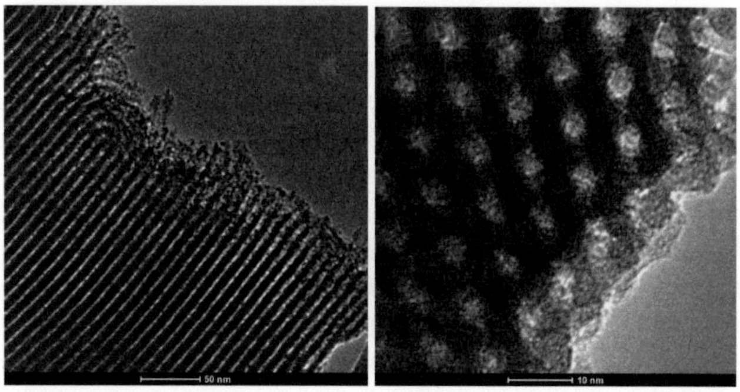

(a) (b)

Fig. 13. TEM images of [SBA-15/P(*N*-iPAAm)] (a) viewed along the pore axis and (b) viewed perpendicular to the pore axis.

Figure 14 shows the nitrogen adsorption isotherms and the pore size distribution for pure SBA-15 and for the [SBA-15/P(*N*-iPAAm)] hybrid. The analysis of the isotherms and the respective adsorption and desorption-derived curves reveal a reduction in symmetry and the occurrence of a stepwise desorption phenomenon in the hybrid sample, suggesting bimodal porosity. This behavior is often referred to as a pore blocking phenomenon and is attributed to the formation of porous plugs in the ordered channels of SBA-15. After polymerization, the shape of the hysteresis loop in the N_2-sorption isotherm displays a broadening suggestive of a reduction in pore size uniformity. Indeed, the desorption branch shifted towards lower relative pressure P/P_0 with a concurrent decrease in the pore diameters. Such a decrease can be verified by analyzing the inflection points of the desorption isotherms and the pore size distribution graphs. The desorption isotherm of mesoporous silica presented an inflection point of 0.64 at P/P_0. On the other hand, the hybrid sample presented two inflection points of 0.60 and 0.47 around P/P_0; this means that the presence of an organic phase in the silica matrix provoked a shift to a smaller relative pressure related to a decrease in pore size.

Atenolol was used to study thermosensitive drug release by hybrid sample. A continued flow was employed to assay the atenolol release profile of this system. Release studies were conducted at four different temperatures: 10, 25, 30, 35, and 40 °C in a simulating body fluid (SBF). A control experiment was conducted in a similar manner for SBA-15. Drug loading was determined through UV–Vis analysis in SBF solution by measuring the absorbance of atenolol at 274 nm. The cumulative drug release profiles (Figure 15) showed clear differences between the studied materials as a function of the temperature. In this figure, the discontinuity in the release profile of the hybrid system as a function of temperature is clearly observed (indicated by arrows above the respective curves). This response in the hybrid system is strongly indicative of the polymer phase behavior. A quasilinear behavior in the release profile of SBA-15 as a function of temperature is also displayed in Fig. 15b. As expected, the release of atenolol from pure SBA-15 is temperature dependent (Fig. 15a). This behavior can be explained considering that the overall drug delivery into the medium is controlled by diffusion through the porous channels. However, as shown by the release profile, delivery seems to be less effective at temperatures as low as 10 °C, since this system presents a comparatively lower delivery rate at this temperature. In this case, the drug was trapped in the porous structures of the silica network. This may be explained by two simultaneous behaviors: the diffusion rate in the mesoporous material, which is severely affected at this temperature, and the drug interaction with the mesoporous silica through hydrogen bonds due to the affinity of the functional groups amine, amide, and hydroxyl of the atenolol molecules and the silanol groups present in the mesoporous silica, which is ascribed to the relatively polar hydrophilic character of the drug. At this temperature (10 °C), the thermodynamic conditions are not enough to destabilize the hydrogen interactions between the hydrophilic groups of the drug and the silanol groups of silica. The atenolol, then, has a tendency to remain in the silica nanostructure. When the temperature was raised to 25 °C, the interactions between the drug molecules and the silica surface were affected, causing its rupture and a major release of atenolol into the medium. At 40 °C, this phenomenon persists and is more evident, as demonstrated by the release profile in Fig. 15a.

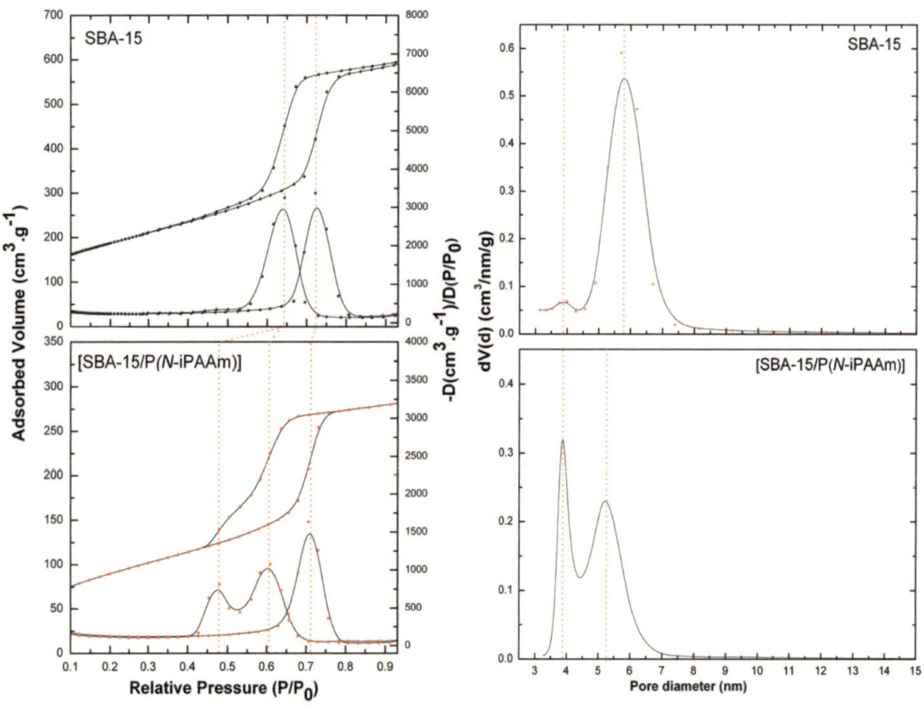

(a) (b)

Fig. 14. (a) N_2 adsorption-desorption isotherms and (b) Pore size distribution of SBA-15 and [SBA-15/P(*N*-iPAAm)].

The analysis of the performance of the hybrid system as a drug delivery device (Fig. 15b) shows that the release profiles of [SBA-15/P(*N*-iPAAm)]vary with temperature, as compared to that of pure SBA-15. At 25 °C, the release rate was high, followed by a sustained release rate; a similar behavior was observed at 10 °C. However, when the temperature was increased to and maintained at 40 °C, a slower and sustained release rate was observed. Nevertheless, it rapidly decreased when the temperature was reduced to 35 °C. This behavior suggests that factors other than temperature affect drug release. These observations lead us to suggest that the drug release response of hybrid systems depends on temperature and the polymer phase behavior. The LCST of P(*N*-iPAAm) is around 35 °C; when the temperature was lower than the LCST, the polymer swelled and drug release was faster. When the temperature was above the LCST, the polymer collapsed, acting as a diffusion barrier, and the drug release from the pores was significantly blocked. With the increase in temperature, an increase in drug diffusion would be expected; however, the increase in temperature provokes a contraction of the polymer network in the hybrid nanostructure. This contraction creates a physical barrier to the drug diffusion, thus decreasing its release to the medium. This temperature-dependent behavior becomes critical upon the complete collapse of the polymer chains at around 35 °C, when the structure barrier to diffusion is the largest. This fact explains the great reduction in the release rate of atenolol from the hybrid system from 25 and 30 to 35 °C. Above the LCST, the polymer does not present phase change anymore, remaining contracted without structural variation. The effect of temperature is illustrated in Fig. 15b, when it varies from 35 to 40 °C, presenting a larger release of atenolol. Even though the release rate is high at this temperature, it is lower than at 25 and 30 °C, as the polymer expanded and consequently presented a lesser barrier to the diffusion of atenolol. Therefore, below the LCST, the contraction phenomenon of the polymeric framework in the silica exerts a more significant influence than the temperature on the release of atenolol to the medium. Above the LCST, the conformational aspect of the polymer does not affect the process and the temperature becomes the dominant variable in the release process.

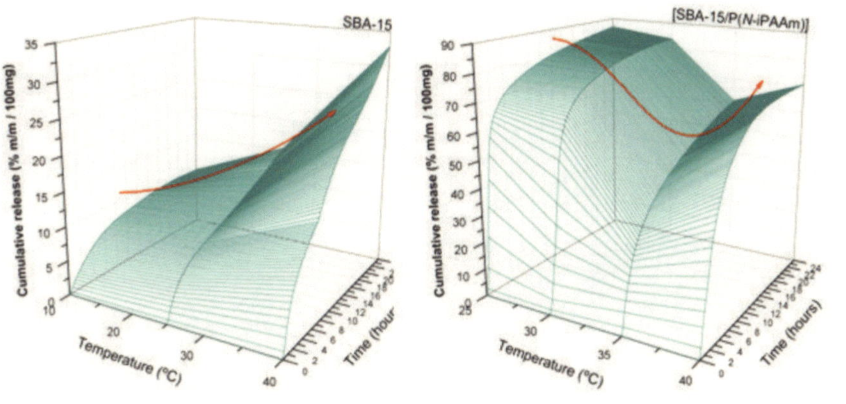

(a) (b)

Fig. 15. Cumulative release profile of atenolol plotted against time and temperature for (a) SBA-15 and (b) [(SBA-15)/P(*N*-iPAAm)].

A mechanism to explain the thermosensitive release profile was developed by the authors (Figure 16). Below the LCST, the polymer swelled and drug release was faster (situations a and b). With the increase in temperature, a decrease in release was observed. It was speculated that the suppression of the release is a result of the contraction of the hydrogel network with heating, which creates a physical barrier to the drug diffusion. This temperature-dependent behavior becomes critical upon the complete collapse of the polymer chains at around 33 °C, when the structure barrier to diffusion is the largest (situation c). Above the LCST (situation d), the polymer does not present phase change anymore, remaining contracted without structural variation, and the temperature becomes the dominant variable in the release process, increasing the release of atenolol.

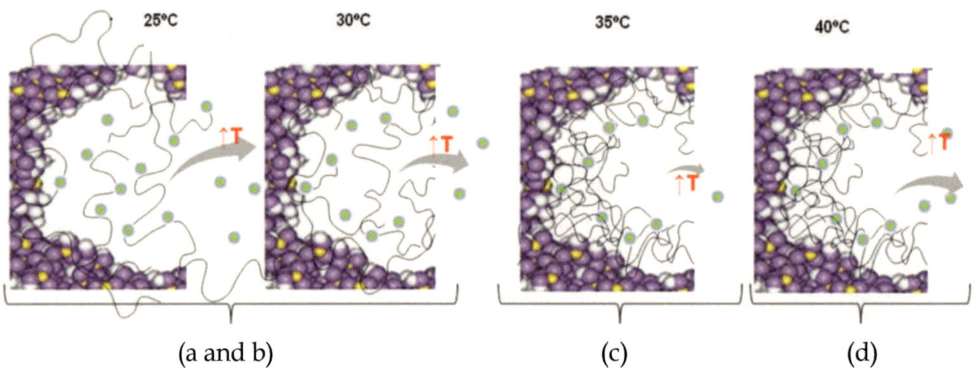

Fig. 16. Schematic representation of the proposed mechanism to explain the thermosensitive release profile.

It is important to note that this release mechanism is clearly different from those previously reported (Zhou et al., 2007; Zhu et al., 2007), since the major release rate was observed below the LCST, rather than above it (a negative thermosensitive release profile was observed). However, similar results were presented by You et al. (You et al., 2008), who studied the release of fluorescein as a function of the temperature from micro to mesoporous silica nanoparticle (MSN)/P(*N*-iPAAm) composites and obtained a greater release rate at 25 °C, below LCST (Figure 17).

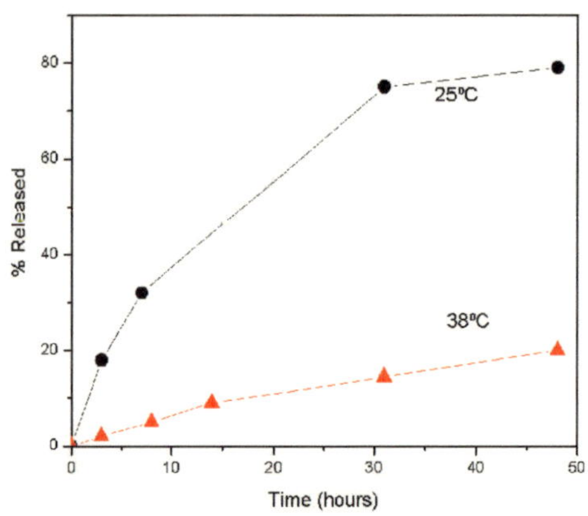

Fig. 17. Release studies conducted at 20 and 38 °C in PBS. Data obtained from (You et al., 2008).

5. Conclusion

This review summarized the recent developments in the methods to synthesize polymer/mesoporous silica and iron nanoparticle/mesoporous silica nanocomposites. We developed an easy and direct synthesis route to obtain nanocomposites and hybrid functional nanosystems based on mesoporous silica. The applicability of this system as a matrix for controlled drug delivery and also in magnetic hyperthermia was studied. The results indicate that different combinations of the ordered mesoporous silica have a potential to encapsulate bioactive molecules to use in biomedicine. The above results show that the composition and morphology of carrier materials and the external agents are important factors in influencing the drug delivery performance. However, a direct comparison of these results with those described by other research groups cannot be performed, since the systems were prepared from mesoporous matrices and by different procedures, including hybrid systems and with magnetic particles. Furthermore, the release assays were conducted differently, using drugs with distinct properties.

In spite of this, we have shown that mesoporous silica-coated magnetite nanoparticles with these structural and magnetic characteristics, unobstructed pores and easy functionalization of the silica surface, seem to be promising material for hosting and further delivery, under appropriate conditions, of a variety of molecules of pharmaceutical interest. Hyperthermia tests indicated that the nanocomposite can be used as a matrix for localized hyperthermia treatment of cancers. Heat generation depended on the intensity of the AC magnetic field and the nanoparticle medium.

On the other hand, even though there have been significant advances, studies involving the potential use of responsive hybrids in the area of controlled release of drugs are still incipient, and many properties of these materials are in the process of analysis, and synthesis procedures are being modified in order to gain greater control over these morphological and structural materials. Even though the pure mesoporous materials are potentially suitable for biomedical applications because they are biocompatible and resistant to biodegradation in cell media, the study of the biological response of responsive hybrids based on mesoporous materials is incipient, as in vivo assays depend heavily on parameters which are not well established, such as structural and morphological properties and their dosage forms.

A substantial effort has been made recently to develop hybrids based on mesoporous silica nanospheres combined with iron nanoparticles as well responsive polymers that can be directly targeted at the cells, releasing a given active principle in response to variations in some external stimulus. A more in depth study considering our systems with regard to morphological parameters and biological behavior is planned in the future.

6. Acknowledgments

This work has been supported by CAPES, CNPq, FAPEMIG and LNLS (Campinas - Brazil).

7. References

Alvaro, M.; Aprile, C.; Garcia, H. & Gómez-García, C. J. (2006). Synthesis of a hydrothermally stable, periodic mesoporous material containing magnetite nanoparticles, and the preparation of oriented films. *Advanced Functional Materials,* Vol. 16, No. 12, (August 4, 2006) 1543-1548, ISSN: 1616-301X

Bae, S.; Lee, S. W.; Hirukawa, A.; Takemura, Y.; Jo, Y. H. & Lee, S. G. (2009). AC magnetic-field-induced heating and physical properties of ferrite nanoparticles for a hyperthermia agent in medicine. *IEEE Transactions on Nanotechnology,* Vol. 8, No. 1, (January, 2009) 86-94, ISSN: 1536-125X

Baro, M.; Sánchez, E.; Delgado, A.; Perera, A. & Évora, C. (2002). *In vitro-in vivo* characterization of gentamicin bone implants. *Journal of Controlled Release,* Vol. 83, No. 3, (October 30, 2002) 353-364, ISSN: 0168-3659

Barroug, A.; Kuhn, L. T.; Gerstenfeld, L. C. & Glimcher, M. J. (2004). Interactions of cisplatin with calcium phosphate nanoparticles: *in vitro* controlled adsorption and release. *Journal of Orthopaedic Research,* Vol. 22, No. 4, (July, 2004) 703-708, ISSN: 0736-0266

Bruce, I. J.; Taylor, J.; Todd, M.; Davies, M. J.; Borioni, E.; Sangregorio, C. & Sem, T. (2004). Synthesis, characterisation and application of silica-magnetite nanocomposites. *Journal of Magnetism and Magnetic Materials,* Vol. 284, (December, 2004) 145-160, ISSN: 0304-8853

Chang, J. H.; Kim, K. J. & Shin, Y-K. (2004). Sustained drug release on temperature-responsive polymer hybrid nanoporous silica composites. *Bulletin of Korean Chemical Society*, Vol. 25, No. 8, (August 20, 2004) 1257-1260, ISSN: 0253-2964

Chastellain, M.; Petri, A.; Gupta, A.; Rao, K. V. & Hofmann, H. (2004). Superparamagnetic silica-iron oxide nanocomposites for application in hyperthermia. *Advanced Engineering Materials,* Vol. 6, No. 4, (April, 2004) 235-241, ISSN: 1438-1656

Chen, F. H.; Gao, Q. & Ni, J. Z. (2008). The grafting and release behavior of doxorubincin from $Fe_3O_4@SiO_2$ core-shell structure nanoparticles via an acid cleaving amide bond: the potential for magnetic targeting drug delivery. *Nanotechnology,* Vol. 19, No. 16, (April 23, 2008) 165103 (9pp), ISSN: 0957-4484

Doadrio, J. C.; Sousa, E. M. B.; Izquierdo-Barba, I.; Doadrio, A. L.; Pérez-Pariente, J. & Vallet-Regí, M. (2006). Functionalization of mesoporous materials with long alkyl chains as a strategy for controlling drug delivery pattern. *Journal of Materials Chemistry,* Vol. 16, No. 5, (February 7, 2006) 462-466, ISSN: 0959-9428

Fellenz, N. A.; Marchetti, S. G.; Bengoa, J. F.; Mercader, R. C. & Stewart, S. J. (2006a). Synthesis and magnetic characterization of magnetite particles embedded in mesoporous MCM-41. *Journal of Magnetism and Magnetic Materials,* Vol. 306, No. 1, (November 1, 2006) 30-34, ISSN: 0304-8853

Franger, S.; Berthet, P. & Berthon, J. (2004). Electrochemical synthesis of Fe_3O_4 nanoparticles in alkaline aqueous solutions containing complexing agents. *Journal of Solid State Electrochem,* Vol. 8, No. 4, (March, 2004) 218-223

Freitas, R. F. S. & Cussler, E. L. (1987). Temperature sensitive gels as extraction solvents. *Chemical Engineering Science,* Vol. 42, No. 1, 97-103. ISSN: 0009-2509

Fu, Q.; Rama Rao, G. V.; Ward, T. L.; Lu, Y. & Lopez, G. P. (2007). Thermoresponsive transport through ordered mesoporous silica/PNIPAAm copolymer membranes and microspheres. *Langmuir,* Vol. 23, No. 1, (January 2 2007) 170-174, ISSN: 0743-7463

Gao, Q.; Xu, Y.; Wu, D.; Sun, Y. H. & Li, X. A. (2009). pH responsive drug release from polymer-coated mesoporous silica spheres. *Journal of Physical Chemisty C,* Vol. 113, No. 29, (July 23, 2009) 12753–12758, ISSN: 1932-7447

Gun'ko, Y. K.; Pillai, S. C. & Mcinerney, D. (2001). Magnetic nanoparticles and nanoparticle assemblies from metallorganic precursors. *Journal of Materials Sciences: Materials in Electronic,* Vol. 12, No. 4-6, (June, 2001) 299-302, ISSN: 0957-4522

Gupta, A. K. & Gupta M. (2005). Synthesis and surface engineering of iron oxide nanoparticles for biomedical applications. *Biomaterials,* Vol. 26, No. 18, (December 10, 2004) 3995-4021, ISSN: 0142-9612

Hamley, I. W. (2003). Nanostructure fabrication using block copolymers. *Nanotechnology,* Vol. 14, No. 10, (October, 2003) R39-R54, ISSN: 0957-4484

Hoffman, A. S. (1987). Applications of thermally reversible polymers and hydrogels in therapeutics and diagnostics. *Journal of Controlled Release,* Vol. 6, No. 1, (December, 1987) 297-305, ISSN: 0168-3659

Hu, S. H.; Liu, T. Y.; Huang, H. Y.; Liu, D. M. & Chen, S. Y. (2008). Magnetic-sensitive silica nanospheres for controlled drug release. *Langmuir,* Vol. 24, No. 1, (January 1, 2008) 239-244, ISSN: 0743-7463

Ito, A.; Shinkai, M.; Honda, H. & Kobayashi, T. (2005). Medical application of functionalized magnetic nanoparticles. *Journal of Bioscience and Bioengineering,* Vol. 100, No. 1, (July, 2005) 1-11, ISSN: 1389-1723

Jin, M. R.; Wu, C. F.; Lin, P. Y. & Hou, W. (1995). Swelling of and solute exclusion by poly(*N*-alkylacrylamide) gels. *Journal of Applied Polymer Science.* Vol. 56, No. 2, (April 11, 1995) 285-288. ISSN: 0021-8995

Jin, L.; Horgan, A. & Levicky, R. (2003). Preparation of end-tethered DNA monolayers on siliceous surfaces using heterobifunctional cross-linkers. *Langmuir,* Vol. 19, No. 17, (May 22, 2003) 6968-6975, ISSN: 0743-7463

Jordan, A.; Scholz, R.; Wust, P.; Fahling, H.; Krause, J.; Wlodarczyk, W.; Sander, B.; Vogl, T. & Felix, R. (1997). Effects of magnetic fluid hyperthermia (MFH) on C3H mammary carcinoma *in vivo. International Journal of Hyperthermia,* Vol. 13, No. 6, (November-December, 1997) 587-605, ISSN: 0265-6736

Kalambur, V. S.; Han, B.; Hammer, B. E.; Shield, T. W. & Bischof, J. C. (2005). *In vitro* characterization of movement, heating and visualization of magnetic nanoparticles for biomedical applications. *Nanotechnology,* Vol. 16, No. 8, (May 20, 2005) 1221-1233, ISSN: 0957-4484

Kawashita, M.; Tanaka, M.; Kokubo, T.; Inoue, Y.; Yao, T.; Hamada, S. & Shinjo, T. (2005). Preparation of ferrimagnetic magnetite microspheres for *in situ* hyperthermic treatment of cancer. *Biomaterials,* Vol. 26, No. 15, (May, 2005) 2231-2238, ISSN: 0142-9612

Köhn, R.; Paneva, D.; Dimitrov, M.; Tsoncheva, T.; Mitov, I.; Minchev, C. & Fröba, M. (2003). Studies on the state of iron oxide nanoparticles in MCM-41 and MCM-48 silica materials. *Microporous Mesoporous Materials,* Vol. 63, No. 1-3, (September 18, 2003) 125-137, ISSN: 1387-1811

Langer, R. (1990). New methods of drug delivery. *Science,* Vol. 249, No. 4976, (September 28, 1990) 1527-1533, ISSN: 0036-8075

Liu, Z. L.; Wang, X.; Yao, K. L.; Du, G. H.; Lu, Q. H.; Ding, Z. H. Tao, J.; Ning, Q.; Luo, X. P.; Tian, D. Y.; Xi, D. (2004a). Synthesis of magnetite nanoparticles in W/O microemulsion. *Journal of Materials Sciences,* Vol. 39, No. 7, 2633-2636, ISSN: 0022-2461

Liu, X. Q.; Xing, J. M.; Guan, Y. P.; Shan, G. B. & Liu, H. Z. (2004b). Synthesis of amino-silane modified superparamagnetic silica supports and their use for protein immobilization. *Colloids and Surfaces A-Physicochemical and Engineering Aspects,* Vol. 238, No. 1-3, (May 4, 2004) 127-131, ISSN: 0927-7757

Liu, C.; Guo, J.; Yang, W.; Hu, J.; Wang, C & Fu, S. (2009). Magnetic mesoporous silica microspheres with thermo-sensitive polymer shell for controlled drug release. *Journal of Materials Chemistry,* Vol. 19, No. 27, 4764-4770, ISSN: 0959-9428

Ma, Z.; Guan, Y. & Liu, H. (2006). Superparamagnetic silica nanoparticles with immobilized metal affinity ligands for protein adsorption. *Journal of Magnetism and Magnetic Materials,* Vol. 301, No. 2, (August 25, 2005) 469-477, ISSN: 0304-8853

Marchetti, M.; Prager, S. & Cussler, E. L (1990). Thermodynamic predictions of volume changes in temperature-sensitive gels 1. Theory. *Macromolecules,* Vol. 23, No. 6, (March 19, 1990) 1760-1765. ISSN: 0024-9297

Martins, V. C. A. & Goissis, G. (2000). Nonstoichiometric hydroxyapatite-anionic collagen composite as support for the double sustained release of gentamicin and norfloxacin/ciprofloxacin. *Artificial Organs,* Vol. 24, No. 3, (March 2000) 224-230, ISSN: 0160-564X

Mukae, K.; Sakurai, M.; Sawamura, S.; Makino, K.; Kim, S. W.; Ueda, I. & Shirahama, K. (1993). Swelling of poly(N-isopropylacrylamide) gels in water-alcohol (C1-C4) mixed solvents. *Journal of Physical Chemistry,* Vol. 97, No. 3, (January 21, 1993), 737-741. ISSN: 0022-3654

Nakamoto, C.; Kitada, T. & Kato, E. (1996). Pressure dependence on the Flory-Huggins interaction parameter of poly(N-isopropylacrylamide) gels. *Polymer Gels and Networks,* Vol. 4, No. 1, 17-31. ISSN: 0966-7822

Ogata, T.; Nonaka, T. & Hurihara, S. (1995). *Journal of Membrane Science,* Vol. 103, No. 1-2, (July 14, 1995), 159-165. ISSN: 0376-7388

Ogawa, C. A. & Plepis, A. M. G. (2002). Liberação *in vivo* de cloridrato de ciprofloxacina em compósitos hidroxiapatita:colágeno. *Polímeros: Ciência e Tecnologia,* Vol. 12, No. 002, 115-122, ISSN: 0104-1428

Palasis, M. & Gehrke, S. H. (1992). Permeability of responsive poly(N-isopropylacrylamide) gel to solutes. *Journal of Controlled Release.* Vol. 18, No. 1, (January, 1992) 1-11. ISSN: 0168-3659

Pankhurst, Q. A.; Connolly, J.; Jones, S. K. & Dobson, J. (2003). Applications of magnetic nanoparticles in biomedicine. *Journal of Physics D: Applied Physics,* Vol. 36, No. 13, (July 7, 2003) R167-R181, ISSN: 0022-3727

Paul, W. & Sharma, C. P. (2006). Nanoceramic matrices: biomedical applications. *American Journal of Biochemistry and Biotechnology,* Vol. 2, No. 2, 41-48, ISSN: 1553-3468

Peppas, N.A. & Langer, R. (2003). Advances in biomaterials, drug delivery, and bionanotechnology. *Bioengineering, Food, and Natural Products,* Vol. 49, No. 12, (December, 2003) 2990-3006, ISSN: 0001-1541

Qiu, Y. & Park, K. (2001). Environment-sensitive hydrogels for drug delivery. *Advanced Drug Delivery Reviews,* Vol. 53, No. 3, (December 31, 2001) 321-339, ISSN: 0169-409X

Sen, T.; Sebastianelli, A.; Bruce, I. J. (2006) Mesoporous Silica−Magnetite Nanocomposite: Fabrication and Applications in Magnetic Bioseparations. Journal of the American Chemical Society, Vol. 128, No. 22, (Jun 7, 2006) 7130-7131. ISSN: 0002-7863

Singh, P. S. (2008). High surface area nanoporous amorphous silica prepared by dodecanol assisted silica formate sol-gel approach. *Journal of Colloid and Interface Science,* Vol. 325, No. 1, (September 1, 2008) 207–214, ISSN: 0021-9797

Soler-Illia, G. J. A. A.; Sanchez, C.; Lebeau, B. & Patarin, J. (2002). Chemical strategies to design textured materials: From microporous and mesoporous oxides to nanonetworks and hierarchical structures. *Chemical Reviews,* Vol. 102, No. 11, (November, 2002) 4093-4138, ISSN: 0009-2665

Sousa, A.; Maria, D. A.; Sousa, R. G. & Sousa, E. M. B (2010). Synthesis and characterization of mesoporous silica/poly(N-isopropylacrylamide) functional hybrid useful for drug delivery. *Journal of Materials Science*, Vol. 45, No. 6, (March, 2010) 1478-1486. ISSN: 0022-2461

Sousa, E. M. B.; Doadrio; A. L.; Doadrio, J. C.; Perez-Pariente, J.; Izquierdo-Barba, I. & Vallet-Regi, M. (2004). Mesoporous SBA-15 HPLC evaluation for controlled gentamicin drug delivery. *Journal of Controled Release*, Vol. 97, No. 1 (May 31, 2004) 125-132, ISSN: 0168-3659

Sousa, R. G. & Freitas, R. F. S. (1995). Determinação do diagrama de fases do gel termossensível poli(N-lsopropilacrilamida). *Polímeros: Ciência e Tecnologia*, Vol. 3, 32- 37, ISSN: 0104-1428

Sousa, R. G., M. Sc. Thesis, Universidade Federal de Minas Gerais, MG, 1993

Souza, K. C.; Ardisson, J. D. & Sousa, E. M. B. (2009). Study of mesoporous silica/magnetite systems in drug controlled release. *Journal of Materials Science: Materials in Medicine*, Vol. 20, No. 2, (February, 2009) 507-512, ISSN: 0957-4530

Souza, K. C.; Mohallem, N. D. S. & Sousa, E. M. B. (2010). Mesoporous silica-magnetite nanocomposite: facile synthesis route for application in hyperthermia. *Journal of Sol-Gel Science and Technology*, Vol. 53, No. 2, (February, 2010) 418-427, ISSN: 0928-0707

Souza, K. C.; Salazar-Alvarez, G.; Ardisson, J. D.; Macedo, W. A. A. & Sousa, E. M. B. (2008). Mesoporous silica-magnetite nanocomposite synthesized by using a neutral s urfactant. *Nanotechnology*, Vol. 19, No. 18, (May 7, 2008) 185603 (7pp), ISSN: 0957-4484

Thapa, D.; Palkar, V. R.; Kurup, M. B. & Malik, S. K. (2004). Properties of magnetite nanoparticles synthesized through a novel chemical route. *Materials Letters*, Vol.58, No.x, (August, 2004) 2692-2694, ISSN: 0167-577X

Tian, B-S. & Yang, C. (2009). Temperature-responsive nanocomposites based on mesoporous SBA-15 silica and PNIPAAm: synthesis and characterization. *Journal of Physical Chemistry C*, Vol. 113, No. 12, (March 26, 2009) 4925-4931, ISSN: 1932-7447

Vallet-Regi, M.; Balas, F.; Colilla, M. & Manzano, M. (2008). Bone-regenerative bioceramic implants with drug and protein controlled delivery capability. *Progress in Solid State Chemistry*, Vol. 36, No. 3, (August, 2008) 163-191, ISSN: 0079-6786

Wan, Y.; Shia, Y. & Zhao, D. (2007). Designed synthesis of mesoporous solids via nonionic-surfactant-templating approach. *Chemical Communications*, No. 9, (March 7, 2007) 897–926, ISSN: 1359-7345

Wu, J.; Sassi, A. P.; Blanch, H. W. & Prausnitz, J. M (1996). Partitioning of proteins between an aqueous solution and a weakly-ionizable polyelectrolyte hydrogel. *Polymer*, Vol. 37, No. 21, (October, 1996) 4803-4808. ISSN: 0032-386.1

Wu, M.; Xiong, Y.; Jia, Y.; Niu, H.; Qi, H.; Ye, J. & Chen, Q. (2005). Magnetic field-assisted hydrothermal growth of chain-like nanostructure of magnetite. *Chemical Physics Letters*, Vol. 401, No. 4-6, (December, 2004) 374-379, ISSN: 0009-2614

Xie, T.; Wang, A.; Huang, L.; Li, H.; Chen, Z. ; Wang, Q. & Yin, X. (2009). Recent advance in the support and technology used in enzyme immobilization. *African Journal of Biotechnology*, Vol. 8, No. 19, (October 5, 2009) 4724-4733, ISSN: 1684-5315

Yang, Y.; Yan, X.; Cui, Y.; He, Q.; Li, D.; Wang, A.; Feia, J. & Li, J. Preparation of polymer-coated mesoporous silica nanoparticles used for cellular imaging by a "graft-from" method. *Journal of Materials Chemistry*, Vol. 18, No. 74, (2008) 5731–5737, ISSN: 0959-9428

You, Y.-Z.; Kalebaila, K.K.; Brock, S.L.; Oupicky, D (2008). Temperature-controlled uptake and release in PNIPAM-modified porous silica nanoparticles. *Chemistry of Materials*, Vol. 20, No. 10, (May 27, 2008) 3354-3359. ISSN: 0897-4756

Zhao, D.; Huo, Q.; Fena, J.; Chemelka, B. F. & Stucky, G. D. (1998). Nonionic triblock and star diblock copolymer and oligomeric surfactant syntheses of highly ordered, hydrothermally stable, mesoporous silica structures. *American Chemical Society*, Vol. 120, No. 24, (June 9, 1998) 6024-6036, ISSN: 0002-7863

Zhao, X. J.; Tapec-Dytioco, R.; Wang, K. M. & Tan, W. H. (2003). Collection of trace amounts of DNA/mRNA molecules using genomagnetic nanocapturers. *Analytical Chemistry*, Vol. 75, No. 14, (July 15, 2003) 3476-3483, ISSN: 0003-2700

Zhou, Z.; Zhu, S.; Zhang, D (2007). Grafting of thermo-responsive polymer inside mesoporous silica with large pore size using ATRP and investigation of its use in drug release. *Journal of Materials Chemistry*, Vol. 17, No. 23, (Jun 21, 2007) 2428-2433. ISSN: 0959-9428

Zhu, Y.; Kaskel, S.; Ikoma, K. & Hanagata, N. (2009). Magnetic SBA 15/poly(N-isopropylacrylamide) composite: preparation, characterization and temperature-responsive drug release property. *Microporous and Mesoporous Materials*, Vol. 123, No. 1-3, (July 1, 2009) 107–112. ISSN: 1387-1811

10

Selective Laser Sintered Poly(L-Lactide)/Carbonated Hydroxyapatite Nanocomposite Scaffolds: A Bottom-up Approach

Wen You Zhou[1], Min Wang[2] and Wai Lam Cheung[2]
[1]*Discipline of Orthodontics, Faculty of Dentistry, The University of Hong Kong, 34 Hospital Road, Hong Kong,*
[2]*Department of Mechanical Engineering, The University of Hong Kong, Pokfulam Road, Hong Kong, China*

1. Introduction

The main objective of this research is to study the feasibility of using the selective laser sintering (SLS) technology to fabricate 3D porous scaffolds from poly(L-lactide) (PLLA) and poly(L-lactide)/carbonated hydroxyapatite (PLLA/CHAp) nanocomposite for bone tissue engineering applications. There are great demands for tissue engineering (TE) and ideal tissue engineering scaffolds should possess physical, mechanical, chemical and biological properties to fulfill the requirements for tissue regeneration. These properties basically depend on two key factors; namely, material composition and scaffold architecture. To address the first issue, biocomposites seem to be a better choice than single matrix. In this study, a biocomposite, which consists of PLLA microspheres filled with CHAp nanoparticles, is developed. PLLA is chosen because it is an FDA-approved, biocompatible and biodegradable polymer which has been widely used in many biomedical applications. Meanwhile carbonated hydroxyapatite is a promising material for bone substitution as it is bioresorbable and also more bioactive *in vivo* than stoichiometric hydroxyapatite. In terms of scaffold architecture, modern rapid prototyping (RP) technologies such as stereolithography apparatus (SLA), fused deposition modeling (FDM), 3D printing and SLS offer excellent flexibility. However, materials used for SLA are typically acrylics and epoxies, which are non-biodegradable. At present, only very limited choices of materials are available for FDM because the materials have to be in the form of filament. 3D printing is also limited by the availability of suitable binders to meet the biological and strength requirements of tissue engineering scaffolds. In contrast, SLS has already been used to produce porous poly(ε-caprolactone) (PCL) bone tissue engineering scaffolds based on actual model of minipig and human condyle (Partee, Hollister et al. 2006), therefore it has a great potential for tissue engineering scaffold fabrication and has been chosen for this project. However, SLS has been developed primarily for industrial applications. At present, it is not financially viable to process most biopolymers or their composites in commercial SLS machines because the amount of material required is quite substantial and the costs of biopolymers are very high.

Furthermore, most available biopolymers are not in the appropriate powder form for the SLS process. In order to reduce powder consumption, Lee (2006) has modified a Sinterstation® 2000 SLS machine to allow small quantities of biopolymer powders to be processed. In this study, three major issues were addressed. Firstly, the nanoemulsion method was applied to synthesize CHAp nanospheres (~20 nm) as the osteoconductive filler for building the nanocomposite scaffolds (Zhou, Wang et al. 2008). Secondly, PLLA microspheres and PLLA/CHAp nanocomposite microspheres (~5-30 µm) were prepared by the oil-in-water and solid-in-oil-in-water emulsion solvent evaporation procedures respectively (Zhou, Wang et al. 2007). Finally, with the modified SLS platform, prototypes of bone tissue engineering scaffolds (~10 mm) with meso-porous structure were successfully built from the PLLA microspheres and PLLA/CHAp nanocomposite microspheres (Zhou, Lee et al. 2007; Zhou, Lee et al. 2008). This study demonstrated a bottom-up nanotechnology approach in tissue engineering which assembling smaller scale simple biomaterials (nanoemulsion synthesized nanoscale particles, emulsion solvent evaporation prepared microscale nanocomposite spheres) into larger scale complex macro-scaffolds by laser technology. Fig. 1 shows the summary of relative scales of substances/scaffolds concerned within this study.

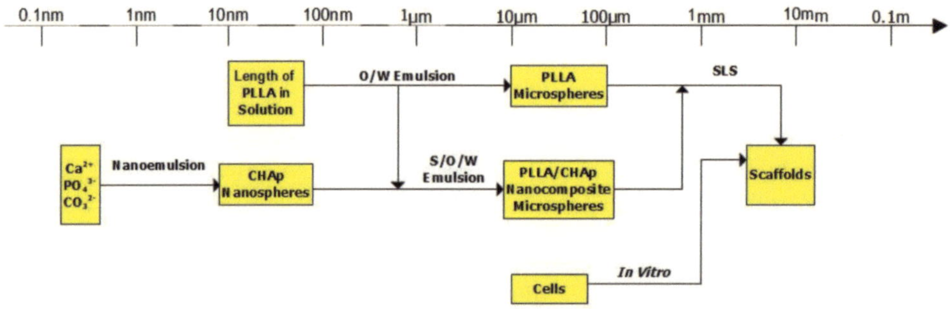

Fig. 1. Summary of relative scales of substances/scaffolds concerned with this study.

2. Synthesis and characterization of carbonated hydroxyapatite nanospheres through nanoemulsion

2.1 Introduction of nanoemulsion synthesis

Nanoemulsions, similar to microemulsions, are a new class of emulsions having very fine and uniform droplet sizes, typically in the range of 20-200 nm (Solans, Izquierdo et al. 2005). Like microemulsions, nanoemulsions can be either transparent or translucent. Some nanoemulsions appear milky with droplet sizes up to 500 nm. Nanoemulsions, in contrast to microemulsions, are not thermodynamically stable but highly kinetically stable because of their small droplet sizes which make them stable against sedimentation and creaming (Uson, Garcia et al. 2004). One major disadvantage of microemulsions is that they require large amounts of surfactants and co-surfactants (typically over 20 wt%) to prepare. This can pose problems for the development of new biomaterials when one considers meeting the paramount requirement of biocompatibility. On the other hand, nanoemulsions offer the possibility of obtaining microemulsion-like dispersions without the need to use high surfactant concentrations or even without the use of any surfactant at all. Nanoemulsions have already found diverse applications in the chemical, cosmetic, pharmaceutical and other industries (Izquierdo, Esquena et al. 2002).

Nanoemulsions can be prepared by spontaneous low-energy emulsification at constant temperature (Forgiarini, Esquena et al. 2001; Bouchemal, Briancon et al. 2004), the phase inversion temperature (PIT) concept (Izquierdo, Esquena et al. 2004), or by the use of a high shear device such as a high pressure homogenizer (aided by appropriate surfactants and co-surfactants). The latter allows a better control of droplet size and also a large choice of compositions. The polymer nanoprecipitation method firstly described by Fessi *et al.* (1989) is one of the spontaneous low-energy emulsification techniques for nanoemulsion formulation which has these distinctive advantages: (a) the use of potentially toxic components (such as chlorinated solvents) can be minimized or avoided; and (b) reproducible nanoparticle size with a narrow size distribution can be achieved without an external energy source. The principle of this method resides in the fact that a preformed material can precipitate as nanospheres when a solution in a water-miscible organic solvent (the organic phase) is mixed with water containing surfactants (the aqueous phase). Nanoemulsion has since been considered to be an ideal nano-reactor for the production of metallic nanoparticles such as GeO_2 (Kawai, Usui et al. 1999) and CdS (Curri, Leo et al. 2001).

Hydroxyapatite has attracted much attention because it has extensive applications as bone substituting material due to its similarity, both chemical and structural, to the mineral portion of bone (LeGeros and LeGeros 1993). Conventionally, HAp powders are synthesized using various methods based on dry processes such as solid-state reaction (Rao, Roopa et al. 1997) or wet processes such as hydrolysis of calcium phosphates (Ishikawa and Eanes 1993) and the precipitation method (Landi, Tampieri et al. 2000). Microemulsion has been shown to be one of the few techniques which could be used to synthesize fine HAp particles with smaller size and higher surface area (Uota, Arakawa et al. 2005). One major disadvantage of this route, as stated previously, is the required use of large amounts of surfactants and co-surfactants. Nanoemulsions can eliminate this problem. Among various HAp-based bioceramics, carbonated hydroxyapatite (CHAp) is a promising material for bone substitution as it is bioresorbable and also more bioactive *in vivo* than stoichiometric HAp. CHAp is used mostly as powders and its usefulness depends on the powder properties such as mean particle size, surface area, and morphology. Nanostructured CHAp particles with a high surface area are desirable for their use in many fields including tissue engineering. The synthesis of nano-CHAp was mostly based on the precipitation reaction developed by Nelson and Featherstone (1982) and studied in details by Barralet *et al.* (1998; 2000; 2002). Doi *et al.* have also successfully produced sintered CHAp which can be resorbed by osteoclasts both under *in vitro* and *in vivo* conditions, whereas sintered stoichiometric HAp cannot be resorbed (Doi, Koda et al. 1993; Doi, Iwanaga et al. 1999; Hasegawa, Doi et al. 2003). In developing bioactive composites (porous or non-porous) for human tissue repair, apart from particle size and size distribution, the shape of the bioactive and reinforcing particles for the composite is also important (Wang 2003). For theoretical analysis of mechanical behavior of a particulate bioceramic-polymer composite, the bioceramic particles in the composite are normally assumed to have a spherical shape. The nanoemulsion method can produce spherical nanoparticles (Chorny, Fishbein et al. 2002). The synthesis of nano-sized CHAp particles of the preferred spherical shape is a critical step in developing composite scaffolds for bone tissue engineering. The aim of the current study is to investigate the use of oil-in-water (O/W) nanoemulsion reactor to synthesize CHAp nanospheres with nearly monodispersed size which will be used subsequently for constructing PLLA/CHAp nanocomposite bone tissue engineering scaffolds.

2.2 Materials

The precursors used for synthesizing CHAp nanospheres were analytical grade calcium nitrate 4-hydrate ($Ca(NO_3)_2{\cdot}4H_2O$, purity 99.0%), di-ammonium hydrogen phosphate ($(NH_4)_2HPO_4$, purity 98.5%) and ammonium hydrogen carbonate (NH_4HCO_3, purity 96.0%) from Uni-Chem® (Orientalab, China). The organic solvent used for making nanoemulsions was analytical grade acetone (AnalaR®, BDH, UK). Water used throughout the investigation was de-ionized (D. I.) through a reverse osmosis water purification system (Barnstead, Diamond™ RO, USA).

2.3 Synthesis of CHAp through nanoemulsion

An acetone solution of $Ca(NO_3)_2{\cdot}4H_2O$ was mixed with an aqueous solution of $(NH_4)_2HPO_4$ and NH_4HCO_3 at a molar ratio of Ca^{2+}: PO_4^{3-}: CO_3^{2-} =1.67:1:0.5 using a magnetic stirrer. The aqueous solution was adjusted to pH 11 with sodium hydroxide (1 M) prior to mixing and the mixed solution was stirred for 0.5 min only. No surfactant was used in all synthesis processes. The resultant CHAp nanoprecipitates in the solution were immediately collected by centrifugation using a Sigma 3-18K refrigerated centrifuge and then washed three times using de-ionized water. Finally, the nanoprecipitates slurry was freeze-dried using a Labconco FreeZone freeze-drying system to obtain dry powder.

When studying the effect of reaction temperature on the microstructure of CHAp, the oil/water volume ratio was fixed at 10/100. The reaction temperatures chosen were 4, 25, 37 and 55 °C (which is below 56.3 °C, the boiling point of acetone.). When studying the effect of oil/water ratio, the reaction temperature was fixed at 25 °C. The oil/water volume ratios chosen were 5/100, 20/100, 40/100 and 80/100. A small quantity of the dry powder synthesized at 25 °C was also calcined at 900 °C for 4h in air using a Carbolite HTF 18/8 high temperature furnace.

2.4 Characterization of CHAp

X-ray diffraction

The phase purity and crystallinity of the synthesized apatite powders were examined using X-ray powder diffraction (Raguku Model D/max III diffractometer, Japan). The X-ray beam was nickel-filtered CuKα_1 radiation (λ=0.1540 nm, operating at 40 kV and 30 mA). XRD data were collected from 20° to 40° (2θ) at a scanning rate of 2 °/min.

Fourier transform infrared spectroscopy

Fourier transform infrared spectroscopy (FTIR, Bruker Vector 33, USA) was performed to determine the presence of functional groups such as OH^- groups and CO_3^{2-} groups in the partially substituted apatite. FTIR samples were prepared from 1:150 CHAp-KBr mixtures, which were ground and pestle for 15 min and pressed into pellets using a cold press.

Electron microscopy

The morphology of the dried nanoprecipitates was examined using a field emission scanning electron microscope (FE-SEM, LEO 1530, Germany) with an energy dispersive X-ray spectrometer (EDX, Inca 300, UK). All samples were sputter coated with gold/palladium to increase conductivity prior to SEM examination. A transmission electron microscope (TEM, Philips Tecnai 20, The Netherlands) with an energy dispersive X-ray spectrometry attachment (EDX, Inca 300, UK) was also used to study the morphology, surface and fine crystalline structure of the nanoprecipitates. TEM samples were prepared by ultrasonically dispersing the powders in ethanol prior to collection on carbon-coated copper grids. The mean particle size of the nanoprecipitates was estimated from TEM dark field images. Selected area electron diffraction (SAD) patterns were also taken in order to confirm the amorphous or crystalline nature.

Specific surface area measurement

The specific surface area of the apatite powders was measured using the BET method (Micromeritics, ASAP 2010 N, USA). BET stands for Brunauer, Emmett, and Teller, the three scientists who optimized the theory for measuring surface area (Brunauer, Emmett et al. 1938). The concept of BET theory is an extension of the Langmuir theory, which is a theory for monolayer molecular adsorption, to multilayer adsorption with some hypotheses. BET specific surface area is an important material property of solids which measures the total surface area per unit of mass (with units of m^2/g). The adsorption gas used was nitrogen and the values of specific surface area of CHAp nanospheres were calculated from the linear relationship of the BET plot.

Thermogravimetric analysis

Thermal stability of the as-synthesized CHAp nanospheres was studied using a thermogravimetric analyzer (TGA, model 2050, TA Instruments, USA). The thermogram was recorded from room temperature to 750 °C at a heating rate of 20 °C/min and in a flow of nitrogen gas (80 mL/min).

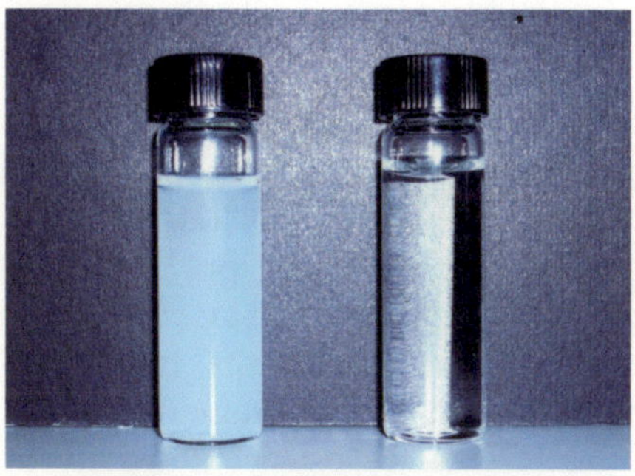

Fig. 2. CHAp nanoemulsion (left) as compared to pure water (right).

Fig. 3. As-synthesized wet slurry nanoparticles before freeze-drying.

2.5 Nanoemulsion and nanoparticles

As shown in Fig. 2, the nanoemulsions prepared in the current study appeared slightly milky. The tinge of white color increased with increasing nanoprecipitate content. Due to the characteristics of nanoemulsion, the translucent solutions prepared in the nanoemulsion process possessed good stability against sedimentation. Fig. 3 shows the as-synthesized nanoparticles before freeze-drying. The freeze-dried powders could flow easily. It appears that the degree of agglomeration caused by this nanoemulsion and freeze-drying technique is comparatively lower than the wet precipitation and conventional oven drying method which tends to give large agglomerates.

Acetone and ethanol are two common solvents which can dissolve calcium nitrate 4-hydrate. Liu *et al.* (2001) used ethanol as one reaction phase to synthesize HAp in their sol-gel production route. Ethanol is a protic solvent, and when dissolved with calcium nitrate 4-hydrate, ethoxide groups could replace some of the nitrate groups to form $Ca(OEt)_y(NO_3)_{2-y}$ which could be incorporated finally into the apatite structure. In the current study, acetone was chosen as the oil phase because it is a nonprotic solvent and thus the solvent would not affect the purity of CHAp synthesized. Acetone is a popular industrial solvent because of its volatility and miscibility with water and many organic solvents. It has been widely utilized for tissue dehydration and deoiling of lecithin during its preparation for the food and drug industry (Kuksis, Ravandi et al. 2005). Meanwhile, acetone is present in very small quantities in normal urine and blood and as a product of the breakdown of body fat (Reisman 1998). The plasma concentration of acetone in human was 15±5 μM detected by gas chromatographs (GC) and 75±21 μM detected by high-performance liquid chromatographic measurement (HPLC) (Kalapos 2003). Using acetone as a reaction medium to synthesize apatite may also mimics the natural process of apatite formation. In the nanoprecipitation process for polymers, acetone acts as a common solvent due to its ability to dissolve many biodegradable polymers such as poly(D, L-lactide) (PLA), poly(D, L-lactide-co-glycolide) (PLGA) and polycaprolactone (PCL). Combining the CHAp nanoemulsion method reported here with the polymer emulsion solvent evaporation method, acetone can be used as a preparation medium for forming spheres, either in nanometer size or in micrometer size, of bioceramic/biodegradable polymer nanocomposite. The incorporation of nano-sized, amorphous CHAp particles into polymer composite is expected to greatly enhance osteoconductivity for bone tissue engineering application.

The high degree of flowability of freeze-dried, as-synthesized CHAp powders is another distinctive feature. This flow property is important when the CHAp powders are used as raw materials to produce non-porous polymer/bioceramic composites or porous composite scaffolds for human tissue repair. The adequate ability of the powders to flow is required so that the nanospheres could be dispersed sufficiently in the matrix polymer, which could be in the molten state or in the state of polymer solutions, thus enabling the eventual achievement of obtaining high performance (both mechanical and biological) biomaterials.

2.6 Identification, morphology and structure of CHAp nanoparticles

Fig. 4 displays the XRD patterns of the nano-sized CHAp particles that were synthesized at different temperatures (4, 25, 37 and 55°C). All diffraction patterns of samples A to D show a very broad hump, indicating a nearly amorphous structure or a very low degree of crystallinity of the nanoparticles. It appears that the reaction temperature had little effect on the crystallinity of the particles synthesized through nanoemulsion. For comparison, the XRD pattern of some calcined particles (sample E) is also included in Fig. 4. It exhibits sharp peaks of crystalline HAp (JCPDS 9-432), suggesting that the nanoparticles synthesized at the different temperatures were hydroxyapatite (HAp). Meanwhile, Fig. 5 exhibits the FTIR spectra of the nanoparticles synthesized at 25°C before and after calcination at 900°C, with the bands labeled accordingly for different functional groups. In both spectra, bands for OH^- and PO_4^{3-} groups are present. In the spectrum of the as-synthesized nanoparticles, there is a band at 1647 cm^{-1}, suggesting the existence of a small amount of occluded water. The spectrum also exhibits bands for the CO_3^{2-} group at 870 cm^{-1} and 1467-1412 cm^{-1}, confirming that the as-synthesized hydroxyapatite is indeed carbonated hydroxyapatite (CHAp). However, these bands disappear after calcination.

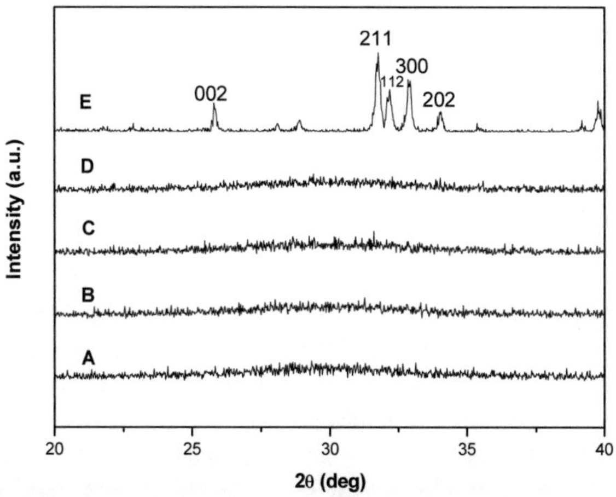

Fig. 4. XRD patterns of CHAp nanoparticles synthesized at: (A) 4°C, (B) 25°C, (C) 37°C, (D) 55°C, (E) synthesized at 25°C and calcined at 900°C for 4 hours in air.

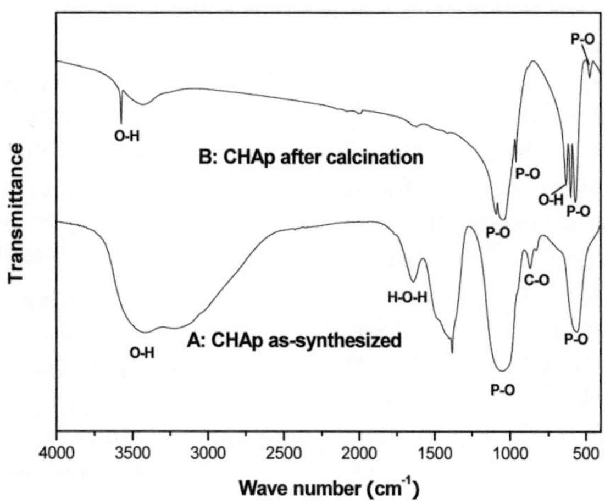

Fig. 5. FTIR spectra of carbonated hydroxyapatite: (A) synthesized at 25 °C, (B) synthesized at 25 °C and calcined at 900 °C for 4 hours in air.

The amorphous structure of the as-synthesized CHAp nanoparticles was probably due to the immediate filtration after nanoprecipitate formation and the subsequent freeze-drying process. Tadic and co-workers (2002) reported similar XRD results when they used high mixing rate and low reaction temperature to form CHAp through a wet precipitation process. There are two processes involved in the formation of apatite in the wet synthesis process: first, precipitation from solution in an amorphous state and second, crystallization within the solid. The proportion of these two processes determines the degree of crystallinity of the apatite obtained. In the current study, the second process was prevented by the fast filtration and freeze-drying processes. Furthermore, carbonate has been known to inhibit HAp crystallization (LeGeros, Trautz et al. 1969). Therefore, the as-synthesized CHAp remained in a generally amorphous or low crystallinity state.

The FTIR spectra shown in Fig. 5 also give some information on the crystallinity of the CHAp particles. Tadic *et al.* (2002) demonstrated in their IR study that the crystallinity of CHAp would affect the phosphate bands at 590-610 cm^{-1} and around 1000 cm^{-1}. Broad and unresolved bands indicated a poor crystallinity, whereas sharp and splitting peaks at 565/605 and 1070/1150 cm^{-1} indicated a high crystallinity. In the current study, the absorption bands of $v_3PO_4^{3-}$ and $v_4PO_4^{3-}$ are broad and the bands of $v_1PO_4^{3-}$ and $v_2PO_4^{3-}$ are absent for the as-synthesized CHAp (spectrum A in Fig. 5), which indicate the amorphous state of the CHAp. On the issue of CO_3^{2-} substitution in CHAp, the carbonate ions can substitute for both the hydroxyl (A-type CHAp) and phosphate (B-type CHAp) groups in the HAp structure (LeGeros, Trautz et al. 1969). The band at 870 cm^{-1} is attributed to $v_2CO_3^{2-}$ substituting for phosphate positions in the HAp lattice (i.e., B-type CHAp). The typical peak of the A-type CHAp at 880 cm^{-1} (Landi, Celotti et al. 2003) did not appear for the CHAp synthesized in the current study. The double peaks of $v_2CO_3^{2-}$ in 1467-1412 cm^{-1} are broad and unresolved, also indicating the amorphous state of the CHAp. On spectrum B, the absorption bands at 602/571 and 474 cm^{-1} are attributed to $v_4PO_4^{3-}$ and $v_2PO_4^{3-}$ of the calcined, crystalline HAp, respectively. Absorptions at 1092/1043 and 962 cm^{-1} are due to $v_3PO_4^{3-}$ and $v_1PO_4^{3-}$. The sharp peaks of O-H stretching and bending at 3574 and 633 cm^{-1}, respectively, indicate that the material had a high degree of crystallinity. As the calcination temperature used in the current study was high, the peaks of the CO_3^{2-} group totally disappeared from spectrum B, suggesting that the carbonate ion had evolved into CO_2 gas during the calcination process.

The carbonate content in bone mineral is around 4 - 8 wt% and it has been shown to vary depending on the age of the individual, with an increase of A-type substituted apatite in the old bone (Rey, Renugopalakrishnan et al. 1991). Hence, B-type CHAp is the most abundant apatite in bones of young people. Astala and Stott (2005) used the first principles simulations to study the different carbonate substitution mechanisms in bulk HAp. Their results showed that B-type or PO_4^{3-} substitution is energetically preferred to A-type or OH^- substitution. This can probably explain why most CHAp synthesized, reported in the literature as well as found in the current study, are the B-type substituted HAp. Carbonate in the B site has been found to reduce apatite's crystallinity and size, making it more difficult to characterize the mechanism of carbonate substitution (Barralet, Best et al. 1998). Nelson and Featherstone (1982) studied the precipitation of tri-ammonium orthophosphate and calcium nitrate 4-hydrate. They pointed out that the mechanism of carbonate substitution could be described by:

$$Ca_{10-y}Na_y(PO_4)_{6-y}(CO_3)_{4y/3}(OH)_{(2-2y/3)} \quad (1)$$

In order to determine the value of y, more precise chemical analysis of CHAp, such as ICP-AES (Inductively Coupled Plasma - Atomic Emission Spectrometer) would be needed for the Ca, Na, P and C wt%.

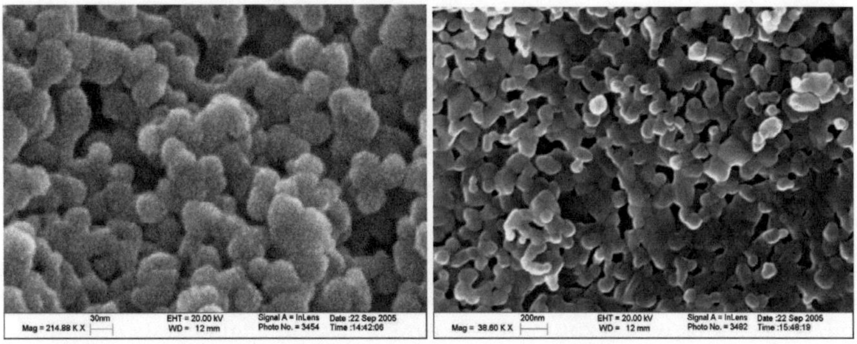

Fig. 6. Morphology of CHAp nanospheres synthesized at 25 °C with an oil/water ratio=10/100 (left) and calcined at 900 °C for 4 hours in air (right).

Fig. 6 shows the FE-SEM micrographs of the as-synthesized and calcined CHAp powders respectively. The as-synthesized CHAp powder consists of tiny agglomerates of nanoparticles (~20 nm in diameter) in the freeze-dried state whereas the calcined CHAp powder has coarsened to particles in the micrometer range. Furthermore, the calcined CHAp particles have undergone necking as a result of the high-temperature calcination process. This observation is in agreement with the BET specific surface area measurements of the CHAp powders. The original specific surface area of the as-synthesized nanoparticles is 43.67±0.03 m^2/g. It drops to 13.24±0.09 m^2/g after calcination at 900 °C. The large reduction in surface area likely affects the properties of the CHAp particles.

Since there was no surfactant used in the nanoemulsion processes in the current study; therefore, the specific surface area of the CHAp nanoparticles is not as high as surfactant-aided systems (Bose and Saha 2003). The calcination temperature selected in the current study was high and hence caused large reductions in the specific surface area of the CHAp particles. The purpose of using such a high calcination temperature was to crystallize the nanoparticles in order to demonstrate that the as-synthesized nanoparticles through the nanoemulsion method were actually apatite. However, in post-synthesis treatment of the precipitated CHAp nanospheres for their actual use as a biomaterial on its own or incorporated into other materials, if calcination is required, the calcination temperature should be kept as low as possible in order not to lose the CO_3^{2-} functional group nor to reduce the characteristic high surface area of the nanoparticles. Calcined apatite crystals on a nanoscale have been difficult to obtain because the precursor particles will be sintered randomly into micro-sized agglomerates consisting of polycrystals (Barralet, Best et al. 2000; Landi, Tampieri et al. 2000; Okada and Furuzono 2007).

Fig. 7 shows the CHAp nanospheres synthesized at different reaction temperatures with an oil/water ratio of 10/100. The primary particles are similar to those synthesized at room temperature, i.e. generally spherical in shape and about 20 nm in diameter. The as-synthesized CHAp nanospheres also tend to form tiny agglomerates in the freeze-dried state. The overall results suggest that the microstructure of the CHAp nanoparticles did not change significantly within the range of reaction temperatures. Therefore, room temperature was chosen to be the reaction temperature in the subsequent studies.

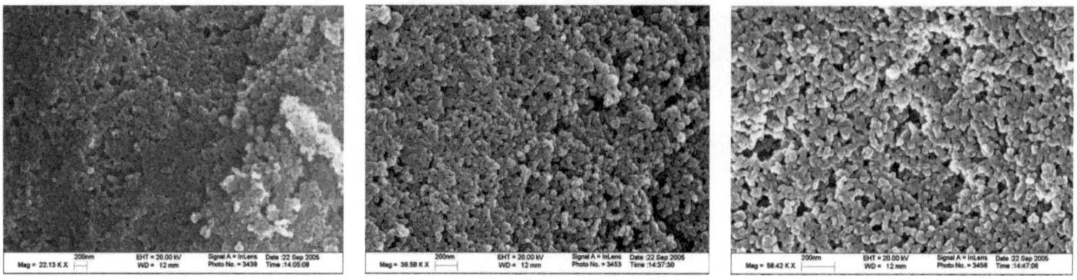

(a) (b) (c)

Fig. 7. SEM micrographs of CHAp nanospheres synthesized with an oil/water ratio=10/100 and synthesis temperatures at: (a) 4 °C, (b) 37 °C, (c) 55 °C.

Fig. 8 shows the CHAp synthesized at room temperature but with different oil/water ratios. Apparently, CHAp nanospheres could only be obtained in low acetone/water ratios and higher oil/water ratios would cause the CHAp particles to agglomerate excessively. Based on the result, the oil/water volume ratio for subsequent studies was chosen to be 10/100.

(a) (b) (c)

Fig. 8. SEM micrographs of CHAp synthesized at 25 °C with oil/water ratios of: (a) 5/100, (b) 20/100, (c) 40/100.

It has been reported that in the wet synthesis processes, HAp mostly forms needle-shaped crystallites that have preferential growth in the (002) direction (Puajindanetr, Best et al. 1994; Bose and Saha 2003). In the current study, the CHAp nanoparticles produced were spherical. This can be attributed to the fact that the precipitates were confined in nano-sized emulsion droplets. LeGeros and co-workers demonstrated that the CO_3^{2-} in substituted HAp caused changes in the size and shape of apatite crystal: from acicular crystals to rods and then to equiaxed crystals with an increasing carbonate content from 2.5 to 17.25 wt% (LeGeros and LeGeros 1993). In the current study, the theoretical carbonate content was 12.8 wt% as calculated from the chemical formula $Ca_{10}(PO_4)_4(CO_3)_2(OH)_2$. The formation of spherical CHAp nanoparticles is due to two reasons: one is the relatively high carbonate content in the CHAp structure, which agrees with LeGeros' observation; the other is the nanoemulsion reactor used in the current synthesis process which is caused by the Ouzo effect (Ganachaud and Katz 2005). The Ouzo effect is a spontaneous emulsification, which causes the formation of acetone/water nanoemulsion. The mechanism of the Ouzo effect is a liquid-liquid nucleation process which arises when mixing with water, the water miscible oil becomes greatly supersaturated, resulting in the nucleation of oil droplets. Following this, the oil immediately begins to diffuse to the nearest droplet so that the level of supersaturation decreases and no further nucleation of droplets occurs. The final metastable emulsion is an ideal nano-reactor to form either ceramic or polymer nanospheres. The formation of CHAp nanospheres with a low oil/water ratio only can be explained by the Ouzo effect too. When using a higher acetone/water ratio to prepare the nanoemulsion, the acetone droplets tend to become larger in order to reduce the level of supersaturation. Sometimes these acetone droplets can become interconnected, forming a continuous CHAp network such as shown in Fig. 8 (b) and (c).

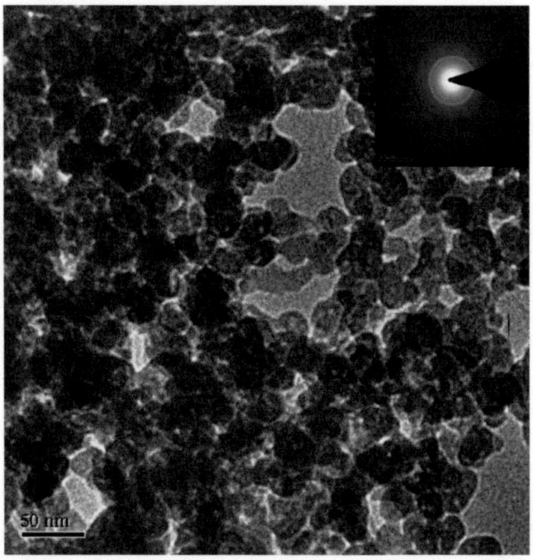

Fig. 9. TEM image and SAD pattern (inset) of the CHAp powder synthesized at 25 °C.

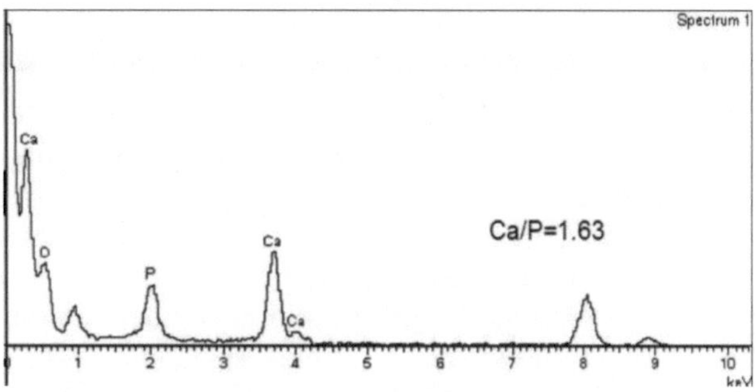

Fig. 10. EDX spectrum of the CHAp powder synthesized at 25 °C.

Fig. 9 is a TEM micrograph, together with the SAD pattern, of the as-synthesized CHAp nanoparticles. Clearly, they are spherical in shape and their sizes are in the range of 10-30 nm. Image analysis of TEM micrographs using the ImageJ® software reveals that these CHAp nanospheres are 16.8 ± 2.6 nm in diameter. The SAD pattern shown on the top right corner is typical of those obtained from the nanosphere agglomerates. It exhibits much-diffused electron diffraction rings, confirming a low degree of crystallinity of the as-synthesized CHAp nanoparticles, and which agrees well with the XRD results in Fig. 4. However, the crystallization process could not be totally stopped, and hence the SAD patterns do indicate some degree of crystallinity of the CHAp nanoparticle. Fig. 10 shows the EDX spectrum of the as-synthesized CHAp. The Ca/P ratio determined from this spectrum is 1.63, which suggests that the CHAp is calcium-deficient.

2.7 Thermal property of CHAp nanospheres

The thermal stability of CHAp powder is important for its biological application. Heat-treatment experiments of CHAp were performed in various gas atmospheres, including nitrogen, carbon dioxide, air, water vapor and wet oxygen (Barinov, Rau et al. 2006). The gas atmosphere was shown to affect the CHAp decomposition significantly. CHAp decomposition occurred at temperature as low as 750°C when heat-treated in an air atmosphere. However, when sintered in a wet carbon dioxide atmosphere, the CHAp did not show any evidence of decomposition at temperature up to 1300 °C (Barralet, Best et al. 2000; Barralet, Knowles et al. 2002). Besides the effect of heat-treatment atmospheres, the thermal stability of CHAp will also depend on the synthesis technique and route.

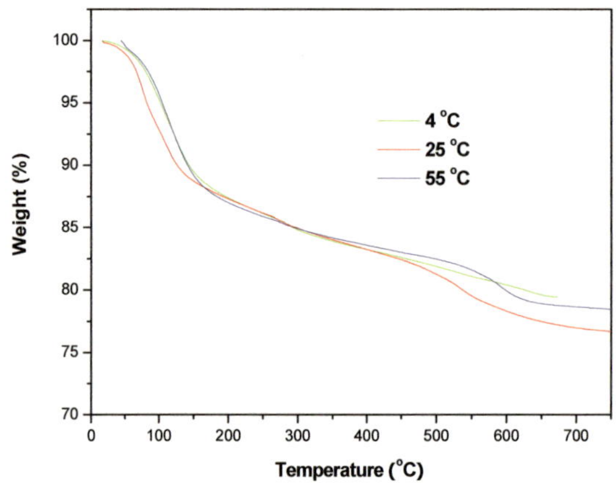

Fig. 11. TGA traces of CHAp nanospheres synthesized at different temperatures.

Fig. 11 shows the TGA traces of CHAp synthesized at different temperatures in terms of weight loss as a function of temperature. In general, there is a major first phase weight loss in the temperature range around 100 °C. The second phase of weight loss occurs more gradually between 150 to 550 °C. After that, another more rapid weight loss is observed between 550 and 650 °C. The weight loss near 100°C is likely due to the evaporation of surface absorbed and lattice (occluded) water molecules. Such weight loss accounts for about 12 wt% of the CHAp synthesized at 4, 25 and 55 °C, indicating a highly absorbent nature of the CHAp nanospheres. The phenomenon can be attributed to the high specific surface area of the nanospheres. In fact, this high water content was also detected in the FTIR spectrum of the as-synthesized CHAp (see O-H band at around 3500 cm^{-1} in Fig. 5). The second phase weight loss up to about 550 °C is probably caused by the condensation reaction of HPO_4^{2-} ion in the following reactions:

$$PO_4^{3-} + H_2O \rightarrow HPO_4^{2-} + OH^- \quad (2)$$

$$2HPO_4^{2-} + 2CO_3^{2-} \rightarrow 2PO_4^{3-} + H_2O\uparrow + 2\,CO_2\uparrow \quad (3)$$

The condensation reaction of HPO_4^{2-} occurs through the proton transfer to labile carbonate ion and the bicarbonate ion (HCO_3^-) decomposition at low temperatures (Landi, Tampieri et al. 2005). Further decrease of TGA traces after 550 °C mainly corresponds to the decomposition of carbonate ion substituted for phosphate groups (B-type substitution) which is confirmed by the CO_2 peak from FTIR on-lined with TGA (Barinov, Rau et al. 2006). There may be a small amount of water released from HPO_4^{2-} reaction, such as (Guo and Xiao 2006)

$$2HPO_4^{2-} \rightarrow P_2O_7^{4-} + H_2O\uparrow \quad (4)$$

$$P_2O_7^{4-} + 2OH^- \rightarrow 2PO_4^{3-} + H_2O\uparrow \quad (5)$$

Furthermore, the TGA traces beyond 500 °C are more wide spread. One possible explanation is that the amounts of carbonate substituted in the samples are different, causing the decomposition rate to change. This indicates that synthesis temperature will affect the amount of substituted carbonate and hence the thermal stability of the CHAp nanospheres at relatively high temperatures. In order to fully understand the thermal characteristics of the CHAp nanospheres, STA (simultaneous thermal analysis) or DTA (differential thermal analysis) on-lined with FTIR would be required and the studying temperature should up to 1400 °C if possible.

3. Fabrication and evaluation of poly(L-lactide) microspheres and poly(L-lactide)/carbonated hydroxyapatite nanocomposite microspheres

3.1 Introduction of nanocomposite microspheres

Bone itself is a naturally occurred inorganic-organic nanocomposite. Therefore, the new generation of biomaterials for bone tissue engineering should include hybrid materials that combine the strength and stiffness of bioactive inorganic fillers with the flexibility and toughness of biocompatible organic matrices. The aim of this part of research is to fabricate and characterize PLLA microspheres and PLLA/CHAp nanocomposite microspheres which will be used in building bone tissue engineering scaffolds by selective laser sintering. CHAp more closely resembles the mineral component of bone than stoichiometric HAp and providing PLLA microspheres with bioactivity or osteoconductivity. PLLA is an FDA-approved, biocompatible and biodegradable polymer which is used in many biomedical applications. But the degradation of pure PLLA implants in the human body is known to produce intermediate acidic products that often lead to the bacteria-free inflammatory responses. CHAp in the PLLA composite could help to buffer such by-products (Agrawal and Athanasiou 1997; Yang, Cui et al. 2006).

Fabrication of desirable micro-sized polymer powders is a critical step for the SLS process in building TE scaffolds with appropriate macro-porosity and surface morphology. Biodegradable microspheres have been prepared by various fabrication methods based on the modifications of three basic techniques: emulsion solvent extraction/evaporation, phase separation (coacervation) and spray-drying (Freitas, Merkle et al. 2005) and widely used in controlled drug delivery systems. Phase separation is frequently impaired by residual organic solvents and coacervation agents found in the microspheres and this method is not suitable for producing small size microspheres (Thomasin, Johansen et al. 1996). Spray-drying is relatively simple but needs special equipment. Moreover, it is difficult to control the size of microsphere and small batches yields are moderate (Mu and Feng 2001). The main disadvantage of spray-drying is its cost, in terms of both equipment and operation (Re 2006). Emulsion solvent extraction/evaporation neither requires high temperatures nor phase separation inducing agents, so it is usually favored at lab scale or small scale. Controlled microsphere sizes in the nano to micrometer range can be achieved, but careful selection of preparation conditions is needed to obtain desired size and size distribution with a low residual solvent content (Freitas, Merkle et al. 2005).

Polymer microsphere preparation by emulsion solvent extraction/evaporation used for controlled delivery of drugs basically consists of four major steps: (i) dissolution or suspension of the bioactive compound either in water phase or in oil phase; (ii) emulsification of oil phase with water phase; (iii) extraction of the solvent from oil phase by water phase which resulting the solidified polymer microspheres; (iv) harvesting and drying the polymer microspheres. As steps (i) – (iii) are all conducted in beakers, this microspheres fabrication technique can be also called as "beaker method" (Siepmann and Siepmann 2006). The major advantage of this "beaker method" is that it does not need high cost equipment. The disadvantages are the problem for the scale-up and the broad size distribution of the microspheres.

3.2 Materials

The PLLA used was 100L 1A with an intrinsic viscosity of 1.9 dL/g (Lakeshore Biomaterials, Birmingham, AL, USA). It was supplied in the form of macro-sized pellets, 1 mm in diameter and 3 mm in length, for conventional extrusion and moulding processes. The emulsifier used was poly(vinyl alcohol) which was cold water soluble and with an average molecular weight of 30,000-70,000 (PVA, m. p.=200 °C, Sigma-Aldrich, USA). Dichloromethane or methylene dichloride (DCM, Uni-Chem®, Orientalab, China) was used as the organic solvent to dissolve PLLA. The analytic reagent (A. R.) grade DCM was used for initial study of the preparation condition of PLLA microspheres, while the chemically pure (C. P.) grade was used for the batch production of large amount of PLLA microspheres for the economical concern.

3.3 Preparation of PLLA microspheres

PLLA microspheres were prepared using a normal oil-in-water (O/W) emulsion solvent evaporation technique. In brief, the PLLA was dissolved in DCM to form a solution first and then rapidly added to aqueous PVA solution and stirred for sometime to allow the solvent to evaporate. The hardened microspheres were centrifuged, washed three times with D. I. water and lyophilized to get dry powder. The powder was stored in the drying cabinet before use. The effects of three preparation conditions, i.e. polymer concentration, emulsifier concentration and mixing speed on the size and size distribution of PLLA microspheres were examined.

3.4 Preparation of PLLA/CHAp nanocomposite microspheres

The PLLA/CHAp nanocomposite microspheres were prepared using a solid-in-oil-in-water (S/O/W) emulsion solvent evaporation method. Briefly, the CHAp nanospheres were dispersed in the PLLA-dichloromethane solution by ultrasonification (Barnstead Lab-Line ultrasonic cleaner, USA) and homogenization (Ultra-Turrax® T25 basic, IKA, Germany) to form an S/O nanosuspension. The nanosuspension was mixed with 1 wt% PVA solution to form S/O/W emulsion. The resultant S/O/W emulsion was magnetic stirred overnight at 800 rpm and then filtered, washed three times with de-ionized water, and freeze-dried to yield a white PLLA/CHAp biocomposite powder. Different contents of CHAp, ranging from 2.5 to 50 wt%, were used to prepare the composite microspheres. Finally, biocomposite microspheres of 10 wt% CHAp in PLLA were used in the SLS process.

3.5 PLLA microspheres

3.5.1 Effect of PLLA concentration

Fig. 12 shows the effect of PLLA concentration on the mean particle size of the PLLA microspheres. It is obvious that the mean particle size increases with increasing PLLA concentration. Similar observations were made by other researchers. Freitas and Marchetti (2005) investigated the influence of this factor, covering a range of PLA concentrations between 0.05% and 0.15% (w/v) in the organic phase. They found that the particle size increased with increasing polymer concentrations, which was in agreement with other contribution (Jeffery, Davis et al. 1991). The phenomenon can be explained by the emulsification efficiency of PVA. The same concentration of PVA would only stabilize larger emulsion droplets with the increasing of PLLA content.

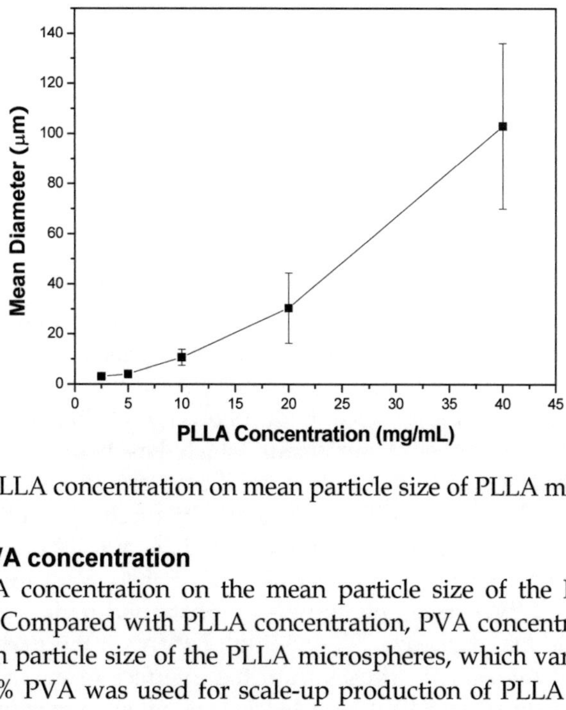

Fig. 12. Effect of PLLA concentration on mean particle size of PLLA microspheres.

3.5.2 Effect of PVA concentration

The effect of PVA concentration on the mean particle size of the PLLA microspheres is shown in Fig. 13. Compared with PLLA concentration, PVA concentration only has limited effect on the mean particle size of the PLLA microspheres, which varies between 17 and 46 μm. Finally, 1 wt% PVA was used for scale-up production of PLLA microspheres, by that most of the particles obtained were between 15 and 45 μm and suitable for the SLS process.
The role of emulsifier is to form a protecting thin layer around the oil drops, with the aim to reduce the coagulation and to stabilize the emulsion. The most commonly employed emulsifiers in emulsion solvent evaporation technique are hydrophilic polymers and anionic/cationic surfactants. PVA is the most widely used emulsifier; however polyvinylpyrrolidone (PVP), alginate, gelatin, methyl cellulose or lecithin can also be employed. Recent studies suggest that a residual of the emulsifier always linked to the particles despite of the washing and purification processes (Shakesheff, Evora et al. 1997) because PVA forms an interconnected network with the polymer microsphere at the surface (Sahoo, Panyam et al. 2002). Freitas and Marchetti (2005) have observed that the amount of surfactant may affect the microspheres properties. They added 1.3% and 8% of PVA to the aqueous solution and determined that the lower and higher concentrations of surfactant

promoted the coagulation of the particles. Thus, the adequate amount of PVA proposed by these authors was 3%. Sahoo *et al.* (2002) reported that the amount of residual PVA that remains associated with the PLGA nanoparticles can be controlled by altering the PVA concentration or the type of organic solvent used in the emulsion formation, and is an important factor that influences the cellular uptake of nanoparticles.

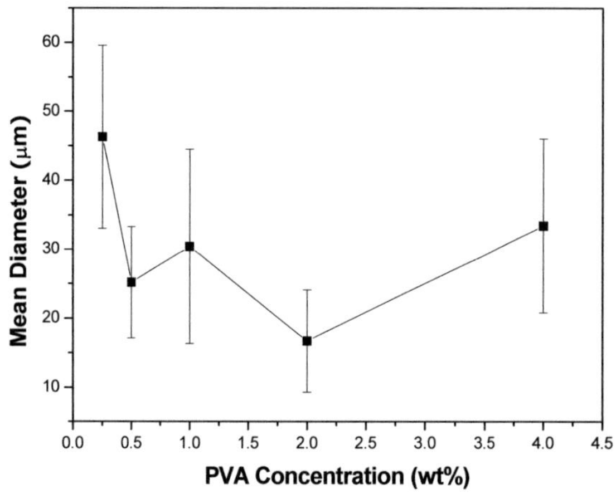

Fig. 13. Effect of PVA concentration on mean particle size of PLLA microspheres.

In this study, the determination of appropriate PVA concentration is based on two considerations: one is the size and size distribution of microspheres which would be suitable for SLS process; the other is the residual amount of PVA on the microsphere surface. The residual PVA can also alter the surface hydrophobicity of PLLA and PLLA/CHAp microspheres since PVA serves as a surfactant which has both the hydrophobic and hydrophilic segments. Imparting more hydrophilicity on PLLA and PLLA/CHAp microsphere surface may benefit the cell attachment. The mechanism of PVA binding has been proposed to be due to the interpenetration of PVA and PLLA molecules during microspheres formulation (Boury, Ivanova et al. 1995; Sahoo, Panyam et al. 2002). The hydrophobic segments of PVA penetrate into the oil phase and remain entrapped into the polymeric matrix of the microspheres. The binding of PVA on the particle surface likely happens when the organic solvent diffuses from the interface of emulsion droplets with water phase. Partially hydrolyzed PVA is a block copolymer of poly(vinyl acetate) and poly(vinyl alcohol). The hydrophobic poly(vinyl acetate) part serves as an anchor polymer at the oil phase for binding to the surface of PLLA during the formulation. Thus, the higher PVA concentration of the continuous phase could lead to an increase in PVA molecule density at the O/W interface of the emulsion droplet, which might increase the amount of PVA on the droplet surface. Previous reports showed that the size of PLGA microspheres decreases with increasing PVA concentration in the external aqueous phase (Zambaux, Bonneaux et al. 1998; Lee, Oh et al. 1999; Sahoo, Panyam et al. 2002). In the current study; however, this trend was not clear and it might be due to the high inherent viscosity (1.6 dL/g) and semicrystalline nature of the PLLA. The inherent viscosity of PLGA usually lies between 0.15 and 0.8 dL/g (lakeshorebio.com). Furthermore, PLGA are amorphous copolymers which make it easier for PVA to break them into emulsion droplets than the semicrystalline PLLA.

3.5.3 Effect of mixing speed

The effect of mixing speed on the mean particle size of PLLA microspheres is shown in Fig. 14. It can be seen that the mixing speed did not affect the mean size of PLLA microspheres significantly. However, the higher speed seemed to give a broader size distribution.

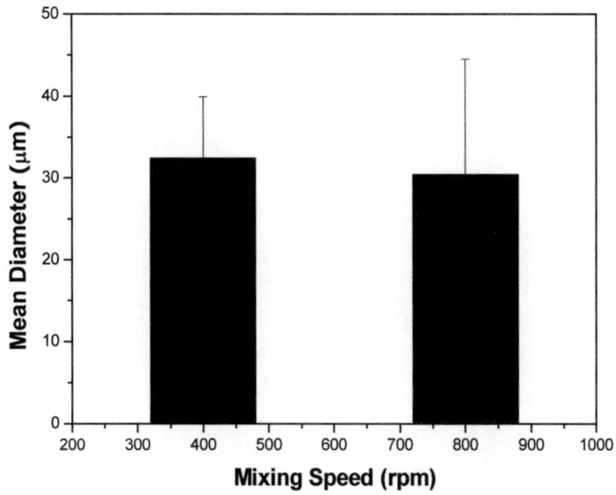

Fig. 14. Effect of mixing speed on mean particle size of PLLA microspheres.

3.5.4 Effect of residual solvent

Dichloromethane (DCM) was used as the oil phase in this study because of the ability to dissolve large amounts of PLLA as well as its low solubility in water (2.0 wt%, w/v) and low boiling point (39.8 °C), which is compatible with the evaporation step (Tewes, Boury et al. 2006). DCM may be retained in the PLLA microspheres as a residual impurity. DCM belongs to the Class 2 solvents according to the International Conference on Harmonization (ICH) because they are suspected of carcinogenicity as well as neurotoxicity and teratogenicity. The maximum limit for DCM imposed by the 2002 edition of the United States Pharmacopeia (USP) and the guidelines of the ICH is 600 ppm while the maximum limit for chloroform is 60 ppm (USP25-NF20 2002). DCM is one of the less harmful of the chlorocarbons based on above data. Benoit *et al.* (1986) measured residual DCM content in progesterone-loaded microparticles prepared by the emulsification solvent evaporation technique. They found that vacuum drying led to a reduction of the solvent content from 18,000 to 360 ppm for PLA microparticles. Bitz and Doelker (1996) showed that the residual amount of DCM in dried PLLA microspheres is 4.2±1.0 ppm by the W/O/W emulsion technique. These results indicate that applying long time of vacuum drying process will dramatically reduce the amount of residual DCM in polymer microspheres and fulfill the guidelines of the Pharmacopeia.

For a typical emulsion solvent evaporation/extraction process, the polymer/ organic solvent dispersed oily phase is emulsified in an aqueous phase, in order to make an O/W emulsion. During mechanical stirring, the organic solvent in emulsion droplets diffuses to the aqueous phase and evaporates to air or extracts to the quenching medium, thereby leading to the formation of microspheres. Dichloromethane, having high volatility and water immiscibility, is the most frequently used dispersed solvent for the emulsion process. In general, less solubility of the organic solvent in water results in a more stable emulsion

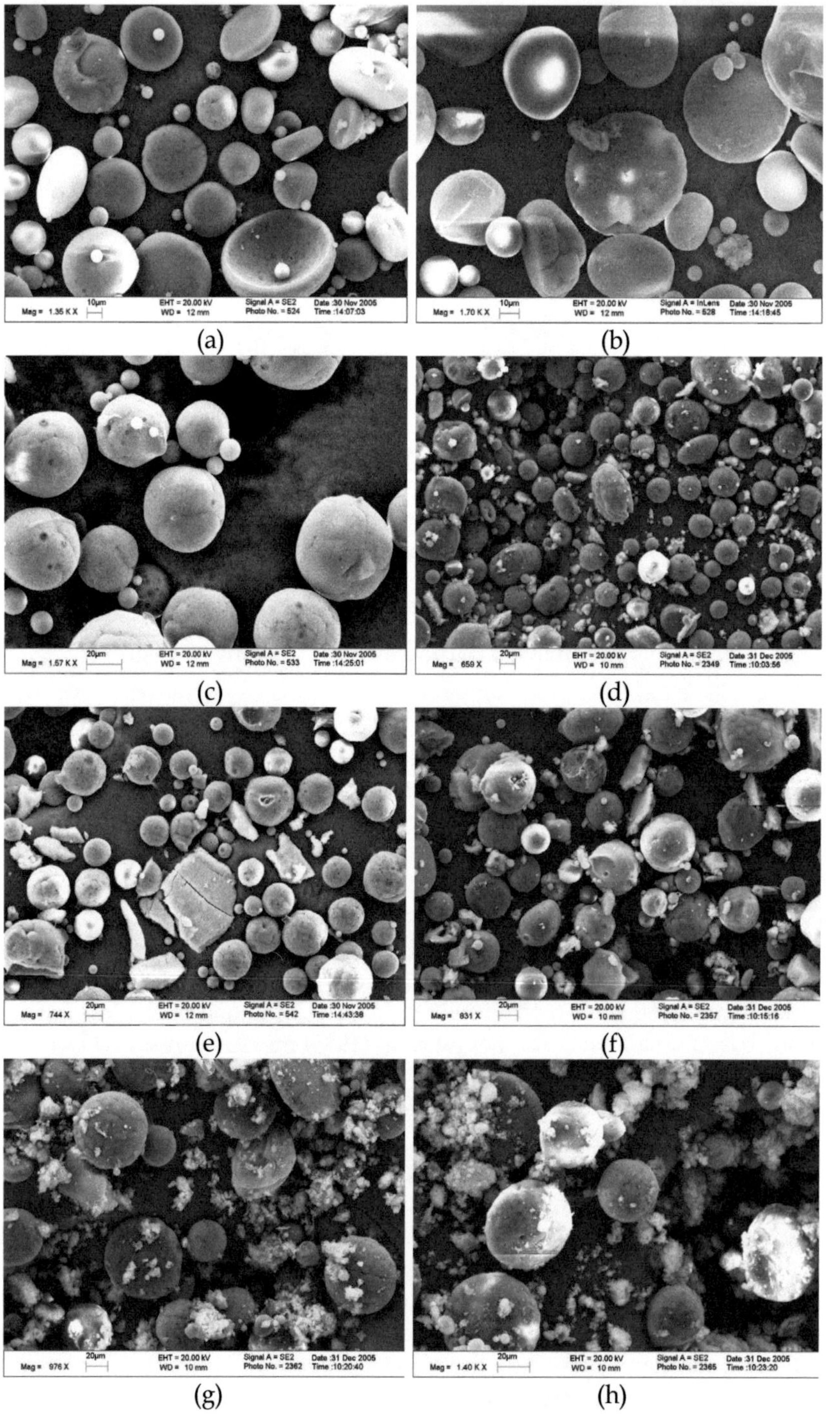

Fig. 15. SEM images of PLLA/CHAp biocomposite microspheres with different CHAp nanosphere content: (a) 2.5 wt%, (b) 5 wt%, (c) 10 wt%, (d) 15 wt%, (e) 20 wt% (f) 30 wt%, (g) 40 wt% and (h) 50 wt%.

that leads to spherical particles with nonporous surface and a better size distribution (Freitas and Marchetti 2005). There are two processes of solvent removal which influence microsphere size and morphology. One is the evaporation process in which the emulsion is maintained at reduced (such as in rotary evaporator) or atmospheric pressure under stirring. Another is the extraction process in which the emulsion is transferred into a large volume of water (with or without surfactant) or another quench medium, where the solvent diffuses out. The similarity for these two processes is in that the solvent must first diffuse out. So the rate of solvent removal influences the properties of the microspheres. Rapid solvent removal leads to the formation of nano-pores on the surface of microspheres and decreases the crystallinity of polymer. The extraction process is fast and generally less than 30 min which forming the porous and more amorphous microspheres (Tewes, Boury et al. 2006). However, the extraction process needs large amount of quench medium which makes it unpractical in the large-scale production.

3.6 PLLA/CHAp nanocomposite microspheres

3.6.1 Effect of CHAp content

Despite changes in the CHAp content, the biocomposite microspheres exhibited a generally constant range of particle sizes between 5 and 30 μm, Fig. 15. Particles of these sizes are suitable for selective laser sintering to produce porous scaffolds for bone tissue engineering. When the CHAp nanosphere content was less than 10 wt%, Fig. 15 (a)-(c), no bare CHAp nanospheres were observed among the composite microspheres and the CHAp nanospheres were believed to be fully encapsulated within the microspheres to form a nanocomposite structure. When the CHAp nanosphere content was above 15 wt%, Fig. 15 (d), a small amount of CHAp agglomerates appeared among the composite microspheres. When the CHAp content exceeded 20 wt%, Fig. 15 (e)-(h), substantial amounts of CHAp agglomerates were seen among the PLLA/CHAp microspheres and major Ca and P peaks could be detected from these agglomerates by EDX.

3.6.2 Internal structure

In order to study the distribution of CHAp nanospheres inside some nanocomposite microspheres, sectional views of the PLLA/CHAp nanocomposite microspheres were observed by different methods. First, some microspheres were embedded in epoxy resin and then sectioned with a microtome. The difficulty of this method was the determination of thickness needed to be removed from the microspheres. After some careful trials, it yielded some satisfactory results. Fig. 16 shows a sectional view of a PLLA/CHAp nanocomposite microsphere. It can be seen that large amount of CHAp nanospheres are encapsulated within the microsphere.

Second, FIB-milling was used to examine the internal structure. FIB-milling is an advanced technology which can remove material at extremely precise location. As stated earlier, polymer and apatite are non-conductive materials. So after milling, the samples need to be coated before the SEM examination. The re-location of the milled sample may be a problem. The SEM image was taken immediately after FIB-milling and it is not so clear due to the low conductivity of PLLA and CHAp. Fig. 17 shows an improved SEM image of the FIB-milled PLLA/CHAp nanocomposite microsphere after sputtering coating with Au. The CHAp nanospheres are generally well distributed both on and inside the microsphere, forming a nanocomposite structure. The rough surface of the nanocomposite microsphere likely facilitates cell attachment.

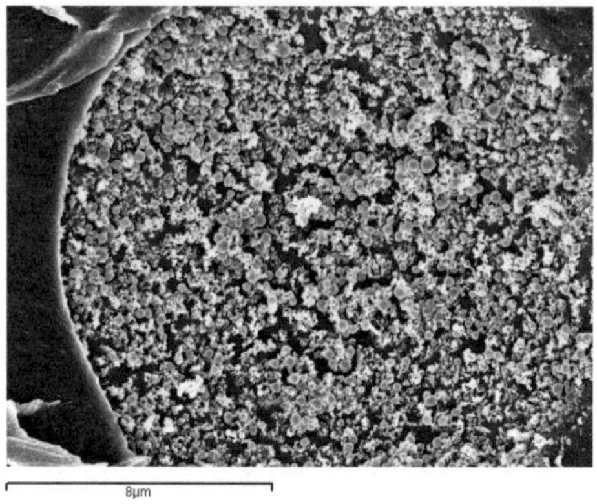

Fig. 16. Cross-sectional view of a PLLA/CHAp nanocomposite microsphere.

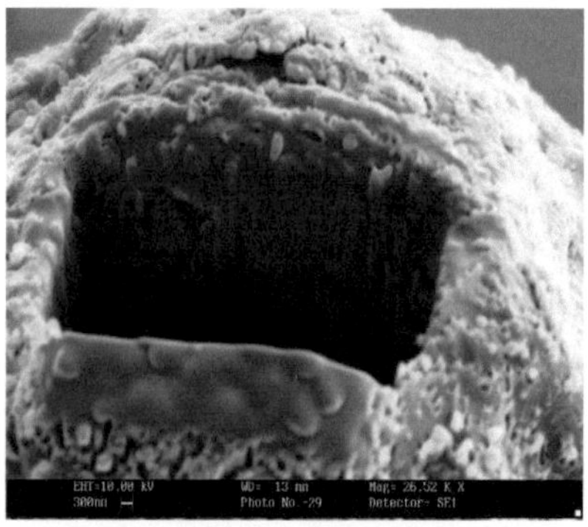

Fig. 17. SEM image of an FIB-milled PLLA/CHAp nanocomposite microsphere.

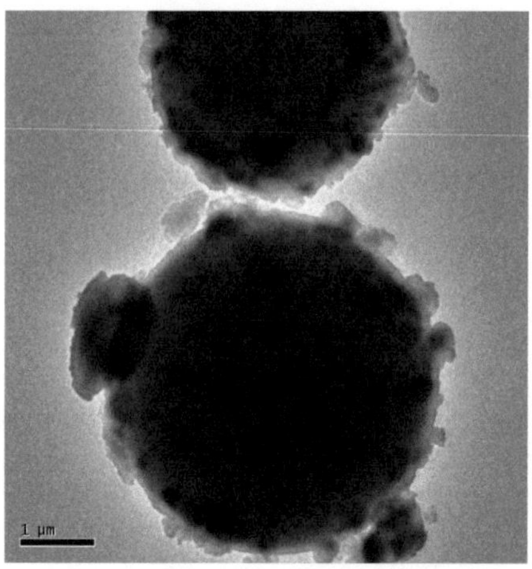

Fig. 18. TEM image of PLLA/CHAp nanocomposite microspheres.

The third method was using TEM to view through the nanocomposite structure. However, TEM uses a very high voltage electron beam so it may damage the polymer matrix and move the CHAp particles. Also, the size of the nanocomposite microspheres could be detected was only several microns. Fig. 18 shows two tiny microspheres (~3-5 μm) under the

TEM. They have a hybrid structure with two phases; namely the PLLA matrix and CHAp agglomerates more or less evenly distributed in the matrix. Recently, Wang *et al.* (2007) used TEM to characterize PHBV/HAp nanocomposite microspheres and they also obtained similar images as that shown in Fig. 18.

3.7 Batch production for SLS

The batch production was conducted on a five position magnetic stirrer and the beakers used had a capacity of 1 liter. The production conditions are listed in Table 1. Some PLLA microspheres produced are shown in Fig. 19 (a). They have a mean particle size of 30 μm which is suitable for the SLS process.

Production Condition	Value
PLLA Concentration	20 mg/mL
PVA Concentration	1 wt%
Mixing Speed	800 rpm

Table 1. Conditions for batch production of PLLA microspheres for SLS.

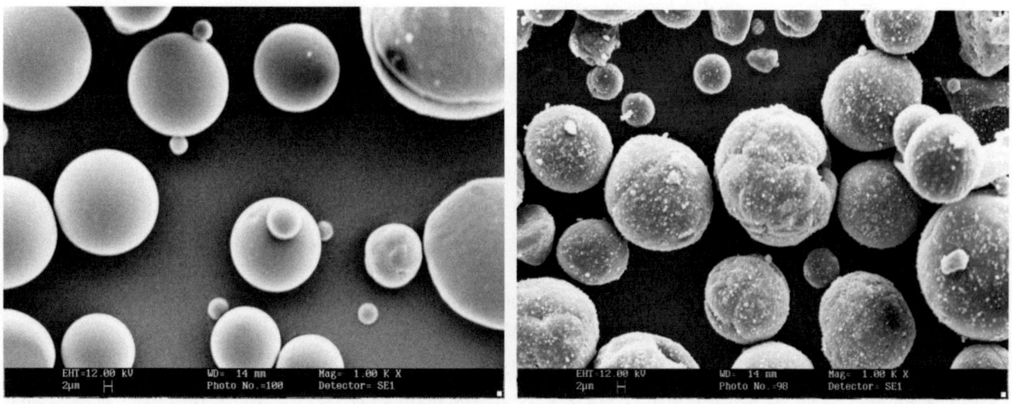

(a) (b)

Fig. 19. SEM images of PLLA microspheres (a) and PLLA/CHAp nanocomposite microspheres (b) prepared in batch production.

Fabrication Condition	Value
PLLA Concentration	20 mg/mL
CHAp:PLLA ratio	1:9
PVA Concentration	1 wt%
Mixing Speed	800 rpm

Table 2. Conditions for batch production of PLLA/CHAp nanocomposite microspheres for SLS.

Table 2 shows the production condition of the PLLA/CHAp nanocomposite microspheres for the SLS process. Basically, it follows the preparation condition of the neat PLLA microspheres, only with the addition of 10 wt% CHAp into the organic phase. Some batch produced PLLA/CHAp microspheres are shown in Fig. 19 (b). Most of them are between 6 and 30 μm in diameter and which is suitable for the SLS process. In the batch production, a homogenizer was used to disperse the CHAp nanospheres in the PLLA/DCM solution before mixing with the PVA aqueous solution to form emulsion. Also, it can be seen that some CHAp nanospheres are partially embedded on the PLLA/CHAp nanocomposite microspheres and this may impart bioactivity (osteoconductivity) for the scaffolds built.

3.8 Discussion

Among the numerous biomaterials developed for bone repair so far, polymer/apatite nanocomposite offers great potential for the successful regeneration of defective bony tissues. The advantage of this nanocomposite system is due to the fact that it mimics the structure and composition of extracellular bone matrix at the lowest structural level (Duan, Wang et al. 2008). The nanocomposites have been processed in different forms for specific applications, such as compressed or extruded molded part, thin film, and porous foam. Among the different formulations, polymer microspheres composed of apatites are of special interest in the treatment of bone defects. Practically, microspheres with a wide range of material compositions have been developed as a delivery vector for drugs and cells which can be released in a controlled manner for specific targeting. Moreover, the microspheres can be effectively injected into defects with minimal surgical invasion (Lemperle, Morhenn et al. 2004). For these reasons, the polymer/apatite nanocomposite microspheres are considered to be particularly useful for the repairing of bone defects by being implanted directly into the defective sites to repopulate osteoprogenitor cells or after *ex vivo* hybridization with osteogenic factors such as growth factors and cells. In such cases, the nanocomposite microspheres should have high potential in terms of their cell population and osteogenic stimulation (Kim, Gu et al. 2007). Moreover, the sintered microspheres matrix can be used as bone TE scaffolds.

The fabrication method for PLLA/CHAp nanocomposite microspheres is S/O/W emulsion solvent evaporation which is similar to the W/O/W emulsion solvent evaporation process used for the delivery of hydrophilic drug contained microspheres. The size of the PLLA/CHAp nanocomposite microspheres can be adjusted by the polymer and surfactant concentration and the mixing speed same as the PLLA microspheres. Meanwhile, the PLLA/CHAp microspheres are a suitable candidate for controlled delivery of recombinant human bone morphogenetic proteins (BMP) in solid form. The protein drugs could be protected by a layer of surfactant in order to avoid the denaturation of proteins by organic solvent. Thus the proteins may be incorporated along with the apatites into polymer microspheres to form a nanocomposite structure.

CHAp nanospheres may act as a colloid stabilizer (co-emulsifier) in the fabrication of PLLA/CHAp nanocomposite microspheres. Pickering (1907) stabilized emulsions by fine solid particles of different materials. This principle was known in ancient times when mustard powder was used to prepare mayonnaise by dispersing oil in water. These kinds of emulsions are termed Pickering emulsions, honoring the eponymous researcher who first described these systems a century ago. A wide variety of solid materials was reported as colloidal stabilizers. Examples are iron oxides and other oxides, hydroxides and oxyhydroxides, basic salts of metals, silica, barium sulfate, carbons, colloidal silver or solid organic materials such as glycerol stearate (Lagaly, Reese et al. 1999). Stabilization by solid materials requires the colloidal particles to form a dense film around the dispersed droplets which impedes coalescence when two droplets approach (the first mechanism). Enrichment of the fine particles at the oil/water interface is related to the contact angle θ at the water–oil–solid line. In general, stabilization is optimal when θ is somewhat below or above 90°. When θ is too small ($\ll 90°$) or too large ($\gg 90°$), the particles leave the interface and move into the continuous medium. A general rule is that O/W emulsions are formed at $\theta<90°$ and W/O emulsions at $\theta>90°$. A second stabilization mechanism is when the particles aggregate and build up a three-dimensional network in the coherent phase, the oil droplets can be trapped in the array of particles. The mechanical stability of the network structure reduces the rate of coalescence. Also, the network reduces the mobility of the particles and, in this way, enhances emulsion stability. The mechanism of CHAp nanosphere as the co-emulsifier is more likely the first mechanism which the stabilization is obtained by envelopes of CHAp nanospheres around the PLLA droplets. The mechanism of main emulsifier, PVA, is more likely as the second mechanism which the stabilization is obtained by forming a 3D network of PVA polymer chains. In fact, a novel surfactant-free process was proposed recently by modifying oil-in-water emulsion method to produce PLA/HAp composite microspheres (Nagata, Miyajima et al. 2003; Nagata, Miyajima et al. 2006; Nagata, Teraoka et al. 2006). Several analyses suggested that the end group of PLA would play as the nucleation site for HAp precipitates at the oil/water interface. The precipitated HAp might play a role as stabilizer for composite microsphere fabrication.

4. Production of 3D porous laser sintered poly(L-Lactide) and poly(L-Lactide)/carbonated hydroxyapatite nanocomposite scaffolds

A custom-made miniature sintering platform was installed in a commercial Sinterstation® 2000 SLS machine as shown in Fig. 20. This platform allowed the use of small quantities of

Fig. 20. Modification of Sinterstation® 2000 system: miniature sintering platform before installation (left) and after installation (right).

biomaterials for TE scaffold production. The effects of laser power; scan spacing and part bed temperature were investigated and optimized, the details were discussed in another book chapter (Zhou, Wang et al. 2010). Finally, porous scaffolds were successfully fabricated from the PLLA microspheres and PLLA/CHAp nanocomposite microspheres as shown in Fig. 21 (a) and (b). In particular, the PLLA/CHAp nanocomposite microspheres appeared to be promising for porous bone TE scaffold production using the SLS technique. The structure and properties, e.g. porosity, compression properties, *in vitro* degradation behavior and cytocompatibility (such as cell adhesion and growth), of these scaffolds were evaluated (Zhou, Lee et al. 2008; Zhou, Wang et al. 2010). The nanocomposite scaffolds provide a biomimetic environment for osteoblastic cell attachment, proliferation and differentiation and have great potential for bone tissue engineering applications (Duan, Wang et al. 2010).

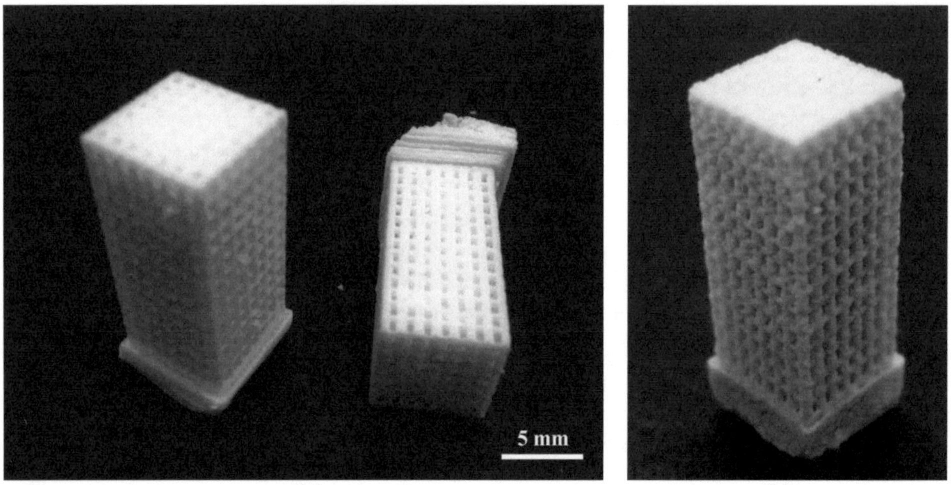

(a) (b)

Fig. 21. (a) Selective laser sintered porous scaffolds from PLLA microspheres and (b) PLLA/CHAp nanocomposite microspheres .

5. Conclusions

The main part of this research focused on the synthesis and characterization of CHAp nanospheres through nanoemulsion; and on the fabrication and evaluation of the morphology of the PLLA microspheres and PLLA/CHAp nanocomposite microspheres as the bottom-up materials to build tissue engineering scaffolds. The custom-made miniature sintering platform allowed small quantities of the biomaterial powders to be processed in a commercial Sinterstation® 2000 SLS machine. With this platform, prototypes of bone tissue engineering scaffolds with meso-scale feature were successfully built from PLLA microspheres and PLLA/CHAp nanocomposite microspheres.

6. Acknowledgements

This work was supported by a GRF research grant (HKU 7118/05E) from the Hong Kong Research Grants Council. Support by Prof. Bakr Rabie in the Faculty of Dentistry, HKU, is acknowledged.

7. References

Agrawal, C. M. and K. A. Athanasiou (1997). Technique to control pH in vicinity of biodegrading PLA-PGA implants. *Journal of Biomedical Materials Research* 38(2): 105-114.

Astala, R. and M. J. Stott (2005). First principles investigation of mineral component of bone: CO_3 substitutions in hydroxyapatite. *Chemistry of Materials* 17(16): 4125-4133.

Barinov, S. M., J. V. Rau, et al. (2006). Carbonate release from carbonated hydroxyapatite in the wide temperature rage. *Journal of Materials Science: Materials in Medicine* 17(7): 597-604.

Barralet, J., S. Best, et al. (1998). Carbonate substitution in precipitated hydroxyapatite: An investigation into the effects of reaction temperature and bicarbonate ion concentration. *Journal of Biomedical Materials Research* 41(1): 79-86.

Barralet, J., J. C. Knowles, et al. (2002). Thermal decomposition of synthesised carbonate hydroxyapatite. *Journal of Materials Science: Materials in Medicine* 13(6): 529-533.

Barralet, J. E., S. M. Best, et al. (2000). Effect of sintering parameters on the density and microstructure of carbonate hydroxyapatite. *Journal of Materials Science: Materials in Medicine* 11(11): 719-724.

Benoit, J. P., F. Courteille, et al. (1986). A physicochemical study of the morphology of progesterone-loaded poly (L-lactide) microspheres. *International Journal of Pharmaceutics* 29(2-3): 95-102.

Bitz, C. and E. Doelker (1996). Influence of the preparation method on residual solvents in biodegradable microspheres. *International Journal of Pharmaceutics* 131(2): 171-181.

Bose, S. and S. K. Saha (2003). Synthesis and characterization of hydroxyapatite nanopowders by emulsion technique. *Chemistry of Materials* 15(23): 4464-4469.

Bouchemal, K., S. Briancon, et al. (2004). Nano-emulsion formulation using spontaneous emulsification: solvent, oil and surfactant optimisation. *International Journal of Pharmaceutics* 280(1-2): 241-251.

Boury, F., T. Ivanova, et al. (1995). Dilatational properties of adsorbed poly (D, L-lactide) and bovine serum albumin monolayers at the dichloromethane/water interface. *Langmuir* 11(5): 1636-1644.

Brunauer, S., P. H. Emmett, et al. (1938). Adsorption of gases in multimolecular layers. *Journal of the American Chemical Society* 60(2): 309-319.

Chorny, M., I. Fishbein, et al. (2002). Lipophilic drug loaded nanospheres prepared by nanoprecipitation: effect of formulation variables on size, drug recovery and release kinetics. *Journal of Controlled Release* 83(3): 389-400.

Curri, M. L., G. Leo, et al. (2001). CdS nanocrystals from a quaternary water-in-oil microemulsion: Preparation and characterization of self-assembled layers. *Journal of Colloid and Interface Science* 243(1): 165-170.

Doi, Y., H. Iwanaga, et al. (1999). Osteoclastic responses to various calcium phosphates in cell cultures. *Journal of Biomedical Materials Research* 47(3): 424-433.

Doi, Y., T. Koda, et al. (1993). Influence of carbonate on sintering of apatites. *Journal of Dental Research* 72(9): 1279-1284.

Duan, B., M. Wang, et al. (2008). Synthesis of Ca-P nanoparticles and fabrication of Ca-P/PHBV nanocomposite microspheres for bone tissue engineering applications. *Applied Surface Science* 255(2): 529-533.

Duan, B., M. Wang, et al. (2010). Three-dimensional nanocomposite scaffolds fabricated via selective laser sintering for bone tissue engineering. *Acta Biomaterialia*: 6(12): 4495-4505.

Fessi, H., F. Puisieux, et al. (1989). Nanocapsule formation by interfacial polymer deposition following solvent displacement. *International Journal of Pharmaceutics* 55(1): R1-R4.

Forgiarini, A., J. Esquena, et al. (2001). Formation of nano-emulsions by low-energy emulsification methods at constant temperature. *Langmuir* 17(7): 2076-2083.

Freitas, M. N. and J. M. Marchetti (2005). Nimesulide PLA microspheres as a potential sustained release system for the treatment of inflammatory diseases. *International Journal of Pharmaceutics* 295(1-2): 201-211.

Freitas, S., H. P. Merkle, et al. (2005). Microencapsulation by solvent extraction/evaporation: reviewing the state of the art of microsphere preparation process technology. *Journal of Controlled Release* 102(2): 313-332.

Ganachaud, F. and J. L. Katz (2005). Nanoparticles and nanocapsules created using the Ouzo effect: Spontaneous emulsification as an alternative to ultrasonic and high-shear devices. *Chemphyschem* 6(2): 209-216.

Guo, X. and P. Xiao (2006). Effects of solvents on properties of nanocrystalline hydroxyapatite produced from hydrothermal process. *Journal of the European Ceramic Society* 26(15): 3383-3391.

Hasegawa, M., Y. Doi, et al. (2003). Cell-mediated bioresorption of sintered carbonate apatite in rabbits. *Journal of Bone and Joint Surgery, British Volume* 85B(1): 142-147.

Ishikawa, K. and E. D. Eanes (1993). The hydrolysis of anhydrous dicalcium phosphate into hydroxyapatite. *Journal of Dental Research* 72(2): 474-480.

Izquierdo, P., J. Esquena, et al. (2002). Formation and stability of nano-emulsions prepared using the phase inversion temperature method. *Langmuir* 18(1): 26-30.

Izquierdo, P., J. Esquena, et al. (2004). Phase behavior and nano-emulsion formation by the phase inversion temperature method. *Langmuir* 20(16): 6594-6598.

Jeffery, H., S. S. Davis, et al. (1991). The preparation and characterisation of poly (lactide-co-glycolide) microparticles. I: Oil-in-water emulsion solvent evaporation. *International Journal of Pharmaceutics* 77(2-3): 169-175.

Kalapos, M. P. (2003). On the mammalian acetone metabolism: from chemistry to clinical implications. *Biochimica et Biophysica Acta* 1621(2): 122-139.

Kawai, T., Y. Usui, et al. (1999). Synthesis and growth mechanism of GeO2 particles in AOT reversed micelles. *Colloids and Surfaces A: Physicochemical and Engineering Aspects* 149(1-3): 39-47.

Kim, H. W., H. J. Gu, et al. (2007). Microspheres of collagen-apatite nanocomposites with osteogenic potential for tissue engineering. *Tissue Engineering* 13(5): 965-973.

Kuksis, A., A. Ravandi, et al. (2005). Covalent binding of acetone to aminophospholipids in vitro and in vivo. *Annals of the New York Academy of Sciences* 1043(1): 417-439.

Lagaly, G., M. Reese, et al. (1999). Smectites as colloidal stabilizers of emulsions: I. Preparation and properties of emulsions with smectites and nonionic surfactants. *Applied Clay Science* 14(1-3): 83-103.

Landi, E., G. Celotti, et al. (2003). Carbonated hydroxyapatite as bone substitute. *Journal of the European Ceramic Society* 23(15): 2931-2937.

Landi, E., A. Tampieri, et al. (2005). Nucleation of biomimetic apatite in synthetic body fluids: dense and porous scaffold development. *Biomaterials* 26(16): 2835-2845.

Landi, E., A. Tampieri, et al. (2000). Densification behaviour and mechanisms of synthetic hydroxyapatites. *Journal of the European Ceramic Society* 20(14-15): 2377-2387.

Lee, S. C., J. T. Oh, et al. (1999). Quantitative analysis of polyvinyl alcohol on the surface of poly(d,l-lactide-co-glycolide) microparticles prepared by solvent evaporation method: effect of particle size and PVA concentration. *Journal of Controlled Release* 59(2): 123-132.

Lee, S. H. (2006). M.Phil. Thesis: Feasibility Study of Selective Laser Sintering of Biopolymer Scaffolds for Tissue Engineering, the University of Hong Kong.

LeGeros, R. Z. and J. P. LeGeros (1993). Dense Hydroxyapatite. *An Introduction to Bioceramics*. L. L. Hench and J. Wilson, World Scientific 139-180.

LeGeros, R. Z., O. R. Trautz, et al. (1969). Two types of carbonate substitution in the apatite structure. *Experientia* 25(1): 5-7.

Lemperle, G., V. B. Morhenn, et al. (2004). Migration studies and histology of injectable microspheres of different sizes in mice. *Plastic and Reconstructive Surgery* 113(5): 1380-1390.

Liu, D.-M., T. Troczynski, et al. (2001). Water-based sol-gel synthesis of hydroxyapatite: process development. *Biomaterials* 22(13): 1721-1730.

Mu, L. and S. S. Feng (2001). Fabrication, characterization and in vitro release of paclitaxel (Taxol) loaded poly (lactic-co-glycolic acid) microspheres prepared by spray drying technique with lipid/cholesterol emulsifiers. *Journal of Controlled Release* 76(3): 239-254.

Nagata, F., T. Miyajima, et al. (2003). Surfactant-free preparation of poly(lactic acid)/hydroxyapatite microspheres. *Chemistry Letters* 32(9): 784-785.

Nagata, F., T. Miyajima, et al. (2006). A method to fabricate hydroxyapatite/poly (lactic acid) microspheres intended for biomedical application. *Journal of the European Ceramic Society* 26(4-5): 533-535.

Nagata, F., K. Teraoka, et al. (2006). Double layered microshells composed of calcium phosphate and poly (lactic acid). *Bioceramics 18, Pts 1 and 2*. 309-311: 915-918.

Nelson, D. G. A. and J. D. B. Featherstone (1982). Preparation, Analysis, and Characterization of Carbonated Apatites. *Calcified Tissue International* 34: S69-S81.

Okada, M. and T. Furuzono (2007). Nano-sized ceramic particles of hydroxyapatite calcined with an anti-sintering agent. *Journal of Nanoscience and Nanotechnology* 7(3): 848-851.

Partee, B., S. J. Hollister, et al. (2006). Selective laser sintering process optimization for layered manufacturing of CAPA (R) 6501 polycaprolactone bone tissue engineering scaffolds. *Journal of Manufacturing Science and Engineering* 128(2): 531-540.

Pickering, S. U. (1907). Emulsions. *J. Chem. Soc. Trans.* 97: 2001-2021.

Puajindanetr, S., S. M. Best, et al. (1994). Characterization and sintering of precipitated hydroxyapatite. *British Ceramic Transactions* 93(3): 96-99.

Rao, R. R., H. N. Roopa, et al. (1997). Solid state synthesis and thermal stability of HAP and HAP - beta-TCP composite ceramic powders. *Journal of Materials Science: Materials in Medicine* 8(8): 511-518.

Re, M.-I. (2006). Formulating drug delivery systems by spray drying. *Drying Technology* 24(4): 433-446.

Reisman, D. J. (1998). Acetone : first draft. Geneva, World Health Organization.

Rey, C., V. Renugopalakrishnan, et al. (1991). Fourier-transform infrared spectroscopic study of the carbonate ions in bone-mineral during aging. *Calcified Tissue International* 49(4): 251-258.

Sahoo, S. K., J. Panyam, et al. (2002). Residual polyvinyl alcohol associated with poly (,-lactide-co-glycolide) nanoparticles affects their physical properties and cellular uptake. *Journal of Controlled Release* 82(1): 105-114.

Shakesheff, K. M., C. Evora, et al. (1997). The adsorption of poly(vinyl alcohol) to biodegradable microparticles studied by X-ray photoelectron spectroscopy (XPS). *Journal of Colloid and Interface Science* 185(2): 538-547.

Siepmann, J. and F. Siepmann (2006). Microparticles Used as Drug Delivery Systems. *Smart Colloidal Materials*: 15-21.

Solans, C., P. Izquierdo, et al. (2005). Nano-emulsions. *Current Opinion in Colloid & Interface Science* 10(3-4): 102-110.

Tadic, D., F. Peters, et al. (2002). Continuous synthesis of amorphous carbonated apatites. *Biomaterials* 23(12): 2553-2559.

Tewes, F., F. Boury, et al. (2006). Biodegradable microspheres: advances in production technology. *Microencapsulation: methods and industrial applications*. S. Benita. New York Taylor & Francis: 2-5.

Thomasin, C., P. Johansen, et al. (1996). A contribution to overcoming the problem of residual solvents in biodegradable microspheres prepared by coacervation. *European Journal of Pharmaceutics and Biopharmaceutics* 42(1): 16-24.

Uota, M., H. Arakawa, et al. (2005). Synthesis of high surface area hydroxyapatite nanoparticles by mixed surfactant-mediated approach. *Langmuir* 21(10): 4724-4728.

Uson, N., M. J. Garcia, et al. (2004). Formation of water-in-oil (W/O) nano-emulsions in a water/mixed non-ionic surfactant/oil systems prepared by a low-energy emulsification method. *Colloids and Surfaces A: Physicochemical and Engineering Aspects* 250(1-3): 415-421.

USP25-NF20 (2002). Organic volatile impurities. *The United States Pharmacopeia (USP 25); The National Formulary (NF 20)* Rockville, MD., United States Pharmacopeial Convention, Inc.

Wang, M. (2003). Developing bioactive composite materials for tissue replacement. *Biomaterials* 24(13): 2133-2151.

Wang, Y., X. Wang, et al. (2007). Fabrication, characterization and long-term in vitro release of hydrophilic drug using PHBV/HA composite microspheres. *Materials Letters* 61(4-5): 1071-1076.

Yang, F., W. Cui, et al. (2006). Poly(l,l-lactide-co-glycolide)/tricalcium phosphate composite scaffold and its various changes during degradation in vitro. *Polymer Degradation and Stability* 91(12): 3065-3073.

Zambaux, M. F., F. Bonneaux, et al. (1998). Influence of experimental parameters on the characteristics of poly(lactic acid) nanoparticles prepared by a double emulsion method. *Journal of Controlled Release* 50(1-3): 31-40.

Zhou, W. Y., S. H. Lee, et al. (2007). Selective laser sintering of tissue engineering scaffolds using poly(L-Lactide) microspheres. *Key Engineering Materials* 334-335: 1225-1228.

Zhou, W. Y., S. H. Lee, et al. (2008). Selective laser sintering of porous tissue engineering scaffolds from poly(L-lactide)/carbonated hydroxyapatite nanocomposite microspheres. *Journal of Materials Science: Materials in Medicine* 19(7): 2535-2540.

Zhou, W. Y., M. Wang, et al. (2008). Synthesis of carbonted hydroxyapatite nanospheres through nanoemulsion. *Journal of Materials Science: Materials in Medicine* 19(1): 103-110.

Zhou, W. Y., M. Wang, et al. (2007). Fabrication and characterization of composite microspheres containing carbonated hydroxyapatite nanoparticles. *Key Engineering Materials* 334-335: 1221-1224.

Zhou, W. Y., M. Wang, et al. (2010). Selective Laser Sintering of Poly(L-Lactide)/Carbonated Hydroxyapatite Nanocomposite Porous Scaffolds for Bone Tissue Engineering. *Tissue Engineering*. D. Eberli. Vienna, Austria, IN-TECH: 179-204.

11

Isothermal and Non-isothermal Crystallization Kinetics of Poly(L-Lactide)/Carbonated Hydroxyapatite Nanocomposite Microspheres

Wen You Zhou[1], Bin Duan[2], Min Wang[2] and Wai Lam Cheung[2]
[1]*Discipline of Orthodontics, Faculty of Dentistry, The University of Hong Kong, 34 Hospital Road, Hong Kong,*
[2]*Department of Mechanical Engineering, The University of Hong Kong, Pokfulam Road, Hong Kong, China*

1. Introduction

Medical-profession-accepted and the US Food and Drug Administration (FDA)-approved biodegradable polymers have been used for tissue engineering applications over the last two decades due to their good biocompatibility and acceptable biodegradation properties. Poly(L-lactide) (PLLA) is a linear aliphatic biodegradable polymer and has been widely studied for use as a scaffolding material for human body tissue regeneration (Wei and Ma 2004; Chen, Mak et al. 2006; Wang 2006). The enzymatic and non-enzymatic hydrolysis rate of PLLA strongly depends on its chemical properties (such as molecular weight and weight distribution) and physical properties (such as crystallinity and morphology). Crystallinity plays an important role in the degradation behavior of biodegradable polymers. It is well known that the crystallinity and morphology of semicrystalline polymers such as PLLA are greatly influenced by their thermal history. Therefore, the crystallization kinetics of PLLA should be carefully studied and correlated to its processing method as it forms a basis for the interpretation of the scaffold properties. The isothermal bulk crystallization kinetics of PLLA has been studied by a number of research groups, covering a temperature range from 70 to 165 °C (Marega, Marigo et al. 1992; Iannace and Nicolais 1997; Miyata and Masuko 1998; Di Lorenzo 2005). But only a few studies were conducted on the non-isothermal crystallization kinetics of neat PLLA. Miyata and Masuko (1998) reported that PLLA could not crystallize and remained amorphous when the cooling rate was higher than 10 °C/min. The knowledge on non-isothermal crystallization kinetics is useful for modelling real industrial processes such as cast film extrusion, which generally takes place at a non-constant cooling rate (Piorkowska, Galeski et al. 2006).

Particulate bioceramic reinforced polymer composites can combine the strength and stiffness of bioactive inorganic fillers with the flexibility and toughness of biodegradable organic matrices. Carbonated hydroxyapatite (CHAp) is a desirable bioactive material for bone substitution as it is bioresorbable and also more bioactive in vivo than stoichiometric hydroxyapatite. PLLA/CHAp nanocomposite has been developed and used for constructing bone tissue engineering scaffolds through selective laser sintering (SLS) (Zhou, Lee et al. 2007; Zhou, Lee et al. 2008). In the SLS process, the laser beam selectively fuses powdered material by scanning cross-sections generated from a 3D digital description of the part (e.g. from a CAD file or scan data) on the surface of a powder bed. After each cross-section is scanned, the powder bed is lowered by one layer thickness, a new layer of material is applied on top, and the process is repeated until the part is completed. Sintering is a thermal fusion process for bonding particles into solid structures (German 1996). The physical process of SLS may involve multiple cycles of melting (full or partial) and crystallization of polymer to produce solid parts. In SLS, individual microspheres are melted by laser beam to different degrees. Most microspheres comprise an un-melted core, surrounded by a melted and crystallized polymer that bonds with other microspheres. The mechanical properties of tissue engineering scaffolds are strongly related to the thermal properties of semi-crystalline biodegradable polymers. In a separate development, Ignjatovic et al. (2004) used hot pressing to produce PLLA/hydroxyapatite (HAp) biocomposite for medical applications. They found that the crystallinity of PLLA decreased after the process time of hot pressing was increased. However, the crystallization kinetics of PLLA/HAp composite was not fully evaluated. Currently, little is known about the effects of nano-sized HAp or CHAp on PLLA crystallization behavior under isothermal or non-isothermal conditions. In order to better understand the in vitro behavior of the PLLA/CHAp nanocomposite scaffolds produced by SLS, it is necessary to study the crystallization kinetics of the nanocomposite and the neat PLLA. In the current investigation, the overall crystallization kinetics and spherulitic morphologies of neat PLLA and PLLA/CHAp nanocomposite were studied by means of differential scanning calorimetry (DSC) and polarized optical microscopy (POM). In the isothermal crystallization study, the sample was rapidly cooled from the melt and allowed to crystallize at a pre-fixed temperature. In the non-isothermal crystallization study, the sample was allowed to crystallize upon cooling at various rates from the melt to room temperature.

2. Materials and methods

2.1 Materials

The PLLA used was Medisorb® 100L 1A (Lakeshore Biomaterials, AL, USA) with an inherent viscosity of 1.9 dL/g. It was supplied in the pellet form: 1 mm in diameter and 3 mm in length. The Mn and Mw of this polymer were determined to be 1.23×10^5 and 2.21×10^5, respectively, by gel permeation chromatography (GPC) using N-methyl pyrrolidone as the solvent. The CHAp nanospheres were synthesized in-house using a nanoemulsion method without surfactants (Zhou, Wang et al. 2008). The mean particle size of the resultant CHAp nanoparticles was about 20 nm. Poly(vinyl alcohol) (PVA, Sigma-Aldrich, cold water soluble) was used as the emulsifier and dichloromethane (DCM, A.R.) used as the organic solvent to dissolve PLLA for microsphere fabrication.

2.2 Fabrication of PLLA microspheres

PLLA microspheres were fabricated using a conventional oil-in-water (O/W) emulsion/solvent evaporation technique (Zhou, Wang et al. 2007). The resultant PLLA microspheres were washed and lyophilized to obtain dry powders.

2.3 Fabrication of PLLA/CHAp nanocomposite microspheres

PLLA/CHAp nanocomposite microspheres were produced using a solid-in-oil-in-water (S/O/W) emulsion/solvent evaporation method as reported previously (Zhou, Wang et al.

2007). Briefly, the CHAp nanoparticles were dispersed in the PLLA-dichloromethane solution by ultrasonification and homogenization to form an S/O nanosuspension. The nanosuspension was mixed with PVA solution to fabricate PLLA/CHAp nanocomposite microspheres. PLLA/CHAp microspheres containing 10 wt% of CHAp nanoparticles was used in this investigation.

2.4 Thermal property measurement and crystallization study

The thermal properties of PLLA and PLLA/CHAp nanocomposites and their crystallization behavior were studied using a Perkin Elmer Pyris 6 differential scanning calorimeter (DSC, MA, USA) and with the heating and cooling rates of 10 °C/min. The apparatus was calibrated with pure indium and zinc standards at various scanning rates. Dry nitrogen gas at a flow rate of 20 mL/min was used to purge through the DSC cell during all measurements. A new sample was used for each measurement in order to eliminate the effect of thermal degradation. Each test was repeated three times to ensure accuracy. The effect of previous heat treatments (thermal history) can significantly affect the shape of the DSC curve for semicrystalline polymers. In order to compare the thermal data of PLLA with values found in the literature, a standardized thermal history is desirable and can be achieved by a heat-cool-reheat DSC method (Lever 2007). In this method, the first heating process destroys any previous thermal history (assuming the maximum temperature is sufficient to remove any remaining nuclei without causing sample degradation), the cooling process imposes a known thermal history on the sample, and the second heating process allows the sample to be measured with a known thermal history. In this investigation, the samples were heated from room temperature to 200 °C with a heating rate of 10 °C/min (step 1) and held there for 5 min to eliminate the thermal history (step 2, annealing). Then samples were quenched to room temperature at cooling rate of 40 °C/min (step 3) and re-heated to 200 °C to probe the melting point (step 4). The glass transition temperature (T_g) was determined before annealing; while the cold crystallization and melting temperature (T_{cc} and T_m, respectively) and the enthalpy of cold crystallization and fusion (ΔH_{cc} and ΔH_m, respectively) were determined after annealing. The crystallinity (X_c) of the PLLA matrix was calculated from the reheating DSC data using the following equation (Sosnowski 2001; Arnoult, Dargent et al. 2007):

$$X_c(\%) = \frac{\Delta H_m - \Delta H_{cc}}{X_{PLLA}\Delta H_m^0} \times 100 \qquad (1)$$

where ΔH_m is the measured enthalpy of fusion, ΔH_{cc} is the cold crystallization enthalpy of PLLA during the heating process, X_{PLLA} is the PLLA weight percentage in the composite. ΔH_m^0 is the enthalpy change of 100% crystalline PLLA, which is 135 J/g, as was estimated by Miyata and Masuko (1998) from the linear relationship between exothermic enthalpy change with density.

2.5 Isothermal crystallization

To investigate the overall kinetics of isothermal melt crystallization, PLLA samples (weighing between 10 to 15 mg) were heated in the DSC from 30 to 200 °C at a rate of 80 °C/min and held at 200 °C for 5 min to allow through melting. They were then cooled at 50 °C/min to the predetermined crystallization temperatures (T_c) and allowed to crystallize. The high cooling rate of 50 °C/min was used to minimize crystallization of PLLA during cooling (Di Lorenzo 2005). The heat evolved during crystallization was recorded as a function of time.

2.6 Polarized optical microscopy

The spherulitic morphologies of neat PLLA and PLLA/CHAp nanocomposites were observed using a polarized optical microscope (POM, Metallux II, Leitz, Germany) equipped with a hot stage (Leitz 350, Germany) and a temperature controller (Partlow, MIC 8200, USA). Samples weighing about 5 mg were sandwiched between two microscope cover slides and melted at 200 °C to form thin films. The samples were held at the same temperature for 5 min to destroy any thermal history and then cooled to isothermal crystallization temperatures of 110, 120 and 130 °C. The temperature of the hot stage was kept constant within 0.1 °C and optical photographs were taken using a digital color camera (Samsung, SCC-101 BP, China).

2.7 Non-isothermal crystallization

For non-isothermal melt crystallization, the samples were quickly heated from 30 to 200 °C at a heating rate of 80 °C/min and maintained at 200 °C for 5 min in the DSC cell to destroy any nuclei that might act as seed crystals. The samples were then cooled to 30 °C at constant rates of 0.5, 1.0, 2.5, 5.0, 7.5 and 10 °C/min, respectively. The exothermic crystallization peaks were recorded as a function of temperature.

3. Results and discussion

3.1 Thermal properties

The thermal properties of neat PLLA and PLLA/CHAp were needed for the crystallization kinetics analysis and therefore determined through DSC. Typical DSC curves of neat PLLA and PLLA/CHAp nanocomposites were shown in Fig. 1 (a), (b) and thermal characteristics were listed in Table 1. It can be seen that the addition of CHAp nanoparticles decreased the glass transition temperature (T_g) and cold crystallization temperature (T_{cc}) of PLLA and slightly increased its melting temperature (T_m). These results suggest that the CHAp nanoparticles promoted initial glass transition and cold crystallization of the PLLA matrix, indicating an enhanced crystallization ability of PLLA in the presence of CHAp, which might behave as nucleating agents. Similar observations were made recently in the PLLA/nano-clay system (Wu, Wu et al. 2007) with a decrease of T_g value of 1-2 °C. The addition of 10 wt% CHAp nanospheres reduces the T_g of PLLA by about 2.89 °C. It has suggested that nanosheet particles are much favorable for the improvement of thermal stability compared to more isotropic geometry fillers, such as nanospheres (Murariu, Da Silva Ferreira et al. 2007). However, the detailed mechanisms for the decreased T_g value of PLLA after addition of CHAp are not clear. ΔH_{cc} and ΔH_m represent the cold crystallization enthalpy and enthalpy of fusion, respectively, from which the absolute degrees of crystallinity, X_c, of melt-quenched PLLA and PLLA/CHAp can be calculated using Equation (1). In theory, the lowest possible value for X_c is zero, which is for totally amorphous polymers. The negative value for the neat PLLA is most likely a result of the accuracy level of the DSC experiment. A more realistic interpretation is that $X_c \approx 0$, i.e. the melt quenched PLLA was more or less amorphous. In fact, our result has confirmed Miyata and Masuko's result (1998) that PLLA could not crystallize and remained amorphous when cooled at rates higher than 10 °C/min. In the current investigation, the samples were quenched at the rate of 50 °C/min. In comparison, the PLLA/CHAp nanocomposite gave a small but non-negligible X_c value of 4%. One possible explanation is that the CHAp nanoparticles acted as a nucleating agent and promoted limited amount of crystallization even under such rapid cooling condition. During the DSC tests, cold crystallization occurred, giving rise to ΔH_{cc}. The value of ΔH_{cc} for PLLA/CHAp is slightly lower than that of the neat PLLA. This is probably because some crystallizable material had been consumed during the cooling process and the amount left for the cold crystallization process had dropped. Finally, the slightly higher ΔH_m value of the nanocomposite, compared to neat PLLA, indicates that the CHAp nanoparticles enhanced the overall crystallinity of the PLLA matrix.

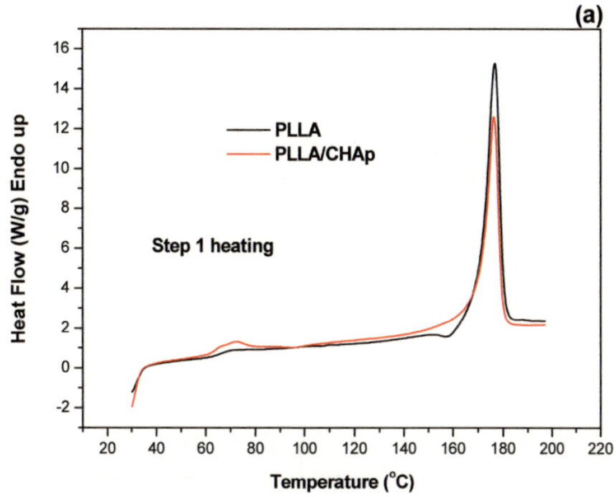

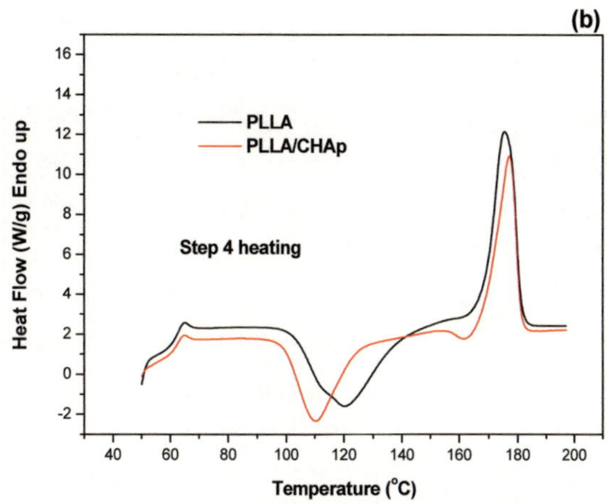

Fig. 1. DSC curves for the neat PLLA and PLLA/CHAp nanocomposites: (a) heating before annealing (step 1) and (b) heating after annealing (step 4).

Samples	T_g (°C)	T_{cc} (°C)	ΔHcc (J/g)	T_m (°C)	ΔHm (J/g)	X_c(%)
PLLA	66.74±0.15	120.13±0.11	37.98±2.75	175.22±0.24	36.79±0.43	-0.9
PLLA/CHAp	63.85±0.30	110.31±0.10	34.27[a]±2.91	177.1±0.33	39.59[a]±1.93	4.0

[a]Values of ΔH_{cc} and ΔH_m for PLLA/CHAp have been normalized to unit mass of PLLA.

Table 1. Thermal properties of melt-quenched neat PLLA and PLLA/CHAp nanocomposite.

3.2 Isothermal crystallization kinetics

The isothermal crystallization kinetics of neat PLLA and PLLA/CHAp nanocomposite was studied by cooling the melt rapidly (50 °C/min) to the crystallization temperature, ranging from 90 to 140 °C. The exothermal curves were then recorded as a function of crystallization time and the results are shown in Fig. 2. The values of crystallization enthalpy (ΔH_c) were obtained from the exothermal curves and are listed in Table 2. Clearly, ΔH_c increased with increasing crystallization temperature for both neat PLLA and PLLA/CHAp. At low isothermal crystallization temperatures, the addition of CHAp seemed to have little influence on the crystallization enthalpy of the PLLA matrix. At higher isothermal crystallization temperatures, however, it decreased the crystallization enthalpy significantly compared with that of neat PLLA. Such change may imply a drop in the amount of crystals formed or a lower degree of perfection of the crystals or both. The presence of a large amount of nano-sized CHAp particles (10 wt%) was likely to increase the melt viscosity, thus making it more difficult for the macromolecular chains to pack into perfect crystals.

Samples	T_c (°C)	ΔH_c (J/g)	ΔH_c^* (J/g)	n	k (min^{-n})	r^2	$t_{0.5}^*$ (min)	$t_{0.5}$ (min)
PLLA	90	22.31		2.60	1.66×10^{-3}	0.99998	10.23	10.18
	100	22.60		2.51	3.16×10^{-2}	0.99994	3.47	3.42
	110	32.95		2.67	2.51×10^{-2}	0.99988	3.51	3.46
	114	33.48		2.60	1.51×10^{-2}	1.00000	4.33	4.35
	116	36.97		2.59	1.26×10^{-2}	0.99998	4.63	4.70
	118	35.78		2.73	7.59×10^{-3}	1.00000	5.20	5.23
	120	42.28		2.80	5.25×10^{-3}	1.00000	5.65	5.72
	130	53.44		2.83	1.82×10^{-4}	0.99982	17.83	18.42
	140	51.34		2.91	7.94×10^{-6}	0.99994	48.67	49.87
PLLA/ CHAp	90	19.38	21.53	2.43	8.32×10^{-3}	0.99996	6.30	6.17
	100	22.27	24.74	2.47	1.15×10^{-1}	0.99998	2.10	2.07
	110	23.30	25.89	2.42	1.10×10^{-1}	0.99996	2.12	2.14
	114	27.88	30.98	2.52	4.90×10^{-2}	0.99996	2.83	2.86
	116	29.12	32.36	2.51	4.27×10^{-2}	0.99998	2.98	3.04
	118	30.94	34.38	2.57	3.10×10^{-2}	0.99994	3.30	3.35
	120	31.91	35.46	2.72	1.95×10^{-2}	1.00000	3.68	3.72
	130	37.23	41.37	2.83	2.29×10^{-3}	1.00000	7.55	7.53
	140	38.06	42.29	2.86	1.66×10^{-4}	0.99996	18.35	18.45

Notes: ΔHc^*: denotes the crystallization enthalpy normalized to unit mass of PLLA
r^2: denotes the coefficient of determination for the Avrami fit
$t_{0.5}^*$: denotes the half-life crystallization time obtained from the experiment (symbols in Fig. 4) $t_{0.5}$: denotes the half-life crystallization time calculated by Equation (7)

Table 2. Properties and parameters obtained from the isothermal crystallization of neat PLLA and PLLA/CHAp nanocomposite.

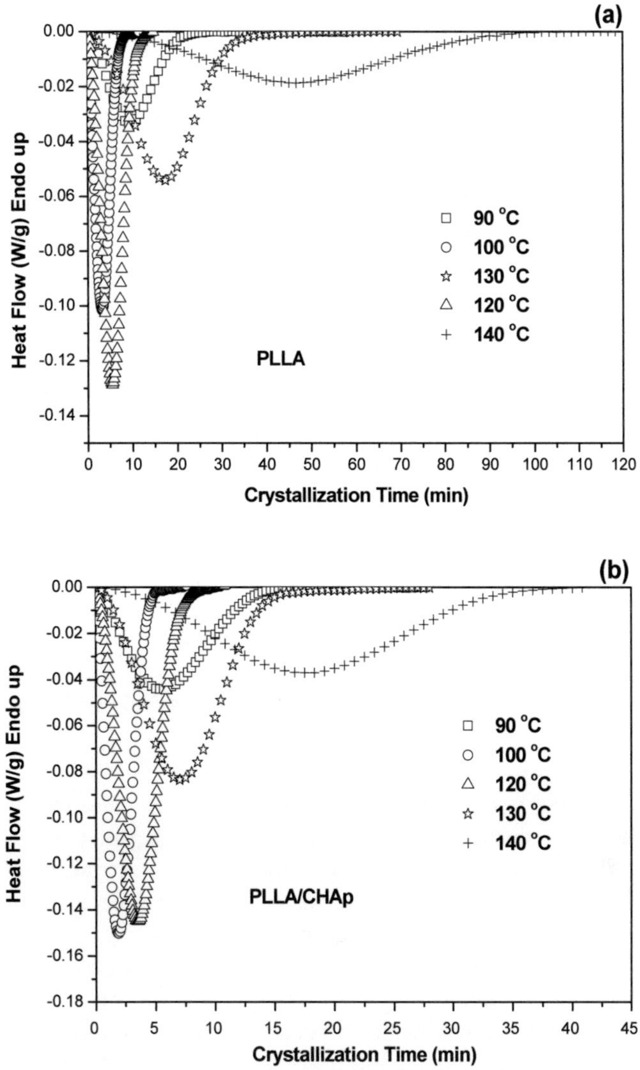

Fig. 2. DSC thermograms obtained from isothermal crystallization of (a) neat PLLA, and (b) PLLA/CHAp nanocomposite.

The kinetics of isothermal crystallization can be described by the well-known Avrami Equation (Piorkowska, Galeski et al. 2006). A time-dependent relative volumetric crystallinity X_t for an isothermal crystallization process can be expressed as:

$$X_t = 1 - \exp(-kt^n) \tag{2}$$

or

$$\log[-\ln(1 - X_t)] = \log k + n \log t \tag{3}$$

where t is the time, n is the Avrami exponent and k is the overall crystallization rate constant which contains contributions from both nucleation and growth. Parameters n and k can be obtained from the slope and intercept, respectively, of the Avrami plot of $\log[-\ln(1-X_t)]$ versus log t. According to Lorenzo et al., (2007), the relative volumetric crystallinity (X_t) can be calculated as:

$$X_t = \frac{W_c}{W_c + (\rho_c/\rho_a)(1-W_c)} \tag{4}$$

where ρ_c and ρ_a are the fully crystalline and fully amorphous polymer densities, respectively. For commonly existed α form of PLLA, ρ_c = 1.283 g/cm^3 and ρ_a = 1.248 g/cm^3, (Oca and Ward 2007). The determination of the absolute crystallinity is not needed for the analysis of the crystallization kinetics, and the degree of relative mass crystallinity, W_c, can be calculated as:

$$W_c = \frac{\Delta H_t}{\Delta H_{total}} = \frac{\int_0^t (dH/dt)dt}{\int_0^\infty (dH/dt)dt} \tag{5}$$

taking ΔH_t as the enthalpy variation as function of the time spent at a given crystallization temperature, while ΔH_{total} is the maximum enthalpy value reached at the end of the isothermal crystallization process. Both quantities can be obtained from the isothermal curve by integration.

Fig. 3 displays the typical Avrami double-logarithmic plots for neat PLLA and PLLA/CHAp. The Avrami parameters n and k were obtained from the plots and are listed in Table 2. Fig. 4 shows the corresponding curves of relative degree of crystallinity with time (symbols) for neat PLLA and PLLA/CHAp obtained from the experimental crystallization isotherms as exhibited in Fig. 2. In a recent investigation (Lorenzo, Arnal et al. 2007), it was suggested that the choice of relative crystallinity range was very important for a good Avrami fit. The initial data points (X_t<3%) may not be accurate due to experimental errors during the primary crystallization stage and hence do not need to be under serious consideration. The secondary crystallization process produces nonlinearity in the Avrami plot so the relative crystallinity range should be chosen during the primary crystallization process. According to their study, a relative crystallinity range of 3 to 20% is sufficient for a good Avrami fit. In order to check the validity of the Avrami method for studying the isothermal crystallization kinetics of neat PLLA and PLLA/CHAp, the variations of the relative crystallinity (X_t) for the respective isothermal crystallization temperatures were calculated using the values of n and k listed in Table 2 and the results are shown as solid lines in Fig. 4. The calculated results fit very well with the experimental data, indicating that the Avrami method can be used to study the isothermal crystallization process in the current investigation in case some of the common problems were avoided (Lorenzo, Arnal et al. 2007). It can be seen from Table 2 that the coefficient of determination of the Avrami plot, r^2, is near unity, which also suggests a good fit between the theoretical and experimental results in the relative crystallinity range of 3 to 20% as recommended by Lorenzo et al. (2007). The valid values of n and k are essential to the Lauritzen-Hoffman analysis by using $t_{0.5}$ data.

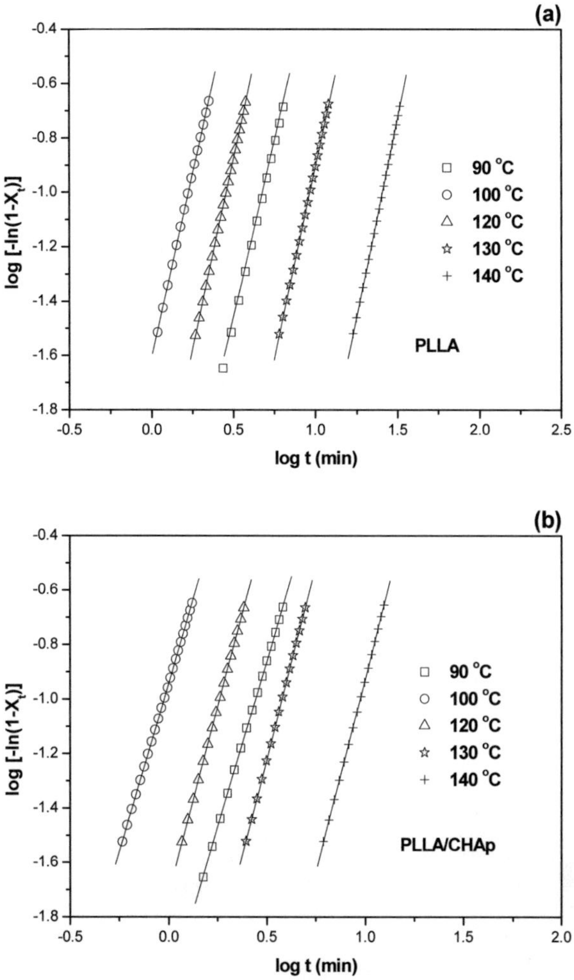

Fig. 3. Avrami plots for isothermal crystallization of (a) neat PLLA and (b) PLLA/CHAp nanocomposite (symbols: experimental data; solid lines: Avrami linear fit).

The Avrami exponent (n) is composed of two terms:

$$n = n_d + n_n \tag{6}$$

where n_d represents the dimensionality of the growing crystals and this quantity can only have, as values, the integer numbers 1, 2 or 3, corresponding to one-, two- or three-dimensional entities that are formed. In the case of polymers, only 2 and 3 are commonly obtained as they represent axialites (two dimensional lamellar aggregates) and spherulites (three dimensional aggregates of radial lamellae), respectively. The time dependence of the nucleation is represented by n_n. In principle, its value should be either 0 or 1, where 0 corresponds to instantaneous or heterogeneous nucleation and 1 to sporadic or homogenous nucleation. However, since in many cases the nucleation may be somewhere between completely instantaneous and completely sporadic nucleation, a noninteger Avrami exponent can be sometimes explained in this way (Lorenzo, Arnal et al. 2007). The Avrami n for neat PLLA is from 2.6 to 2.9 in the isothermal temperature range of 90 to 140 °C; the Avrami n for PLLA/CHAp is from 2.4 to 2.9 in the same temperature range. The Avrami exponent close to 3 at high isothermal crystallization temperatures indicates a changing trend of PLLA crystal growth from two- to three-dimensional with instantaneous nucleation and athermal (Iannace and Nicolais 1997). The obtained n values were similar to those reported for neat PLLA (2.4-3.2 at T_c=90-125 °C by Tsuji et al. (2006), 2.8-3.2 at T_c=90-130 °C by Iannace and Nicolais (1997), 2.5-3.3 at T_c=90-130 °C by Kolstad (1996)). However, Miyata and Masuko (1998) reported an n value ca. 4 at T_c=110-132.5 °C. The differences can be attributed to the determination of the onset of crystallization or induction time, the establishment of the baseline and incomplete isothermal crystallization data, the effect of the cooling rate from the melt to the isothermal crystallization temperature and the relative degree of crystallinity range employed for the Avrami fitting (Lorenzo, Arnal et al. 2007).

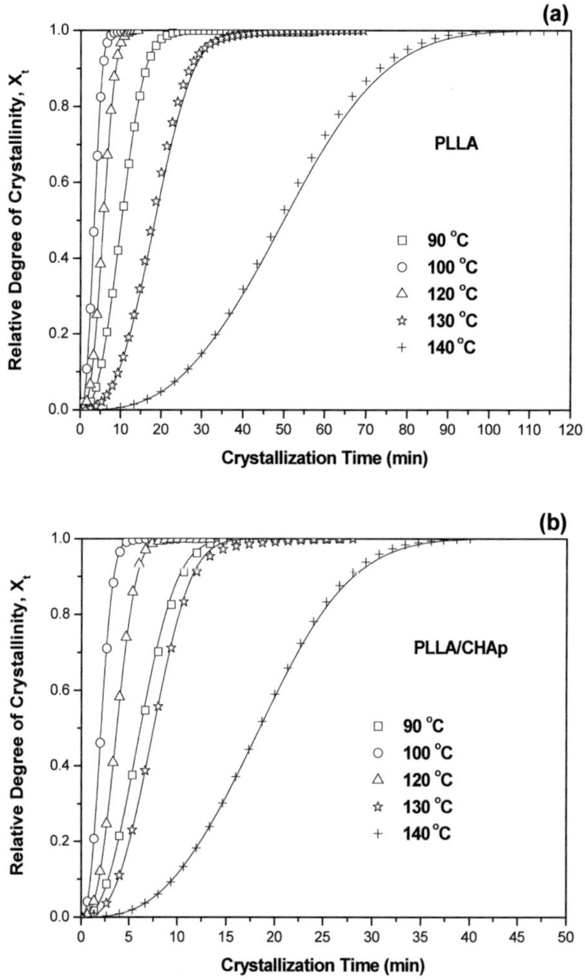

Fig. 4. Development of relative degree of crystallinity as a function of crystallization time for (a) PLLA, and (b) PLLA/CHAp nanocomposite. (Symbols: experimental data; solid lines: calculated crystallinity using Equation (2) and Avrami parameters n and k in Table 2.)

Also, the nucleation is seldom either athermal or simple thermal (Pethrick 2007). It can be seen from Table 2 that the values of Avrami exponent (n) of the PLLA/CHAp nanocomposite are lower than those of the neat PLLA for the same crystallization temperatures. One possible reason is that the large amount of CHAp nanoparticles might have affected the nucleation and growth mechanisms of the PLLA crystals. Better interpretation of the Avrami exponent requires more information about the specific nucleation and growth mechanisms of PLLA crystal, but at this stage we do not have such information.

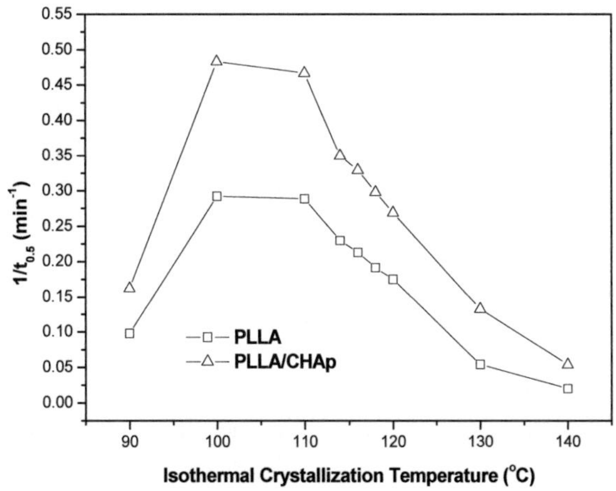

Fig. 5. Dependence of $1/t_{0.5}$ of PLLA and PLLA/CHAp nanocomposite on isothermal crystallization temperature.

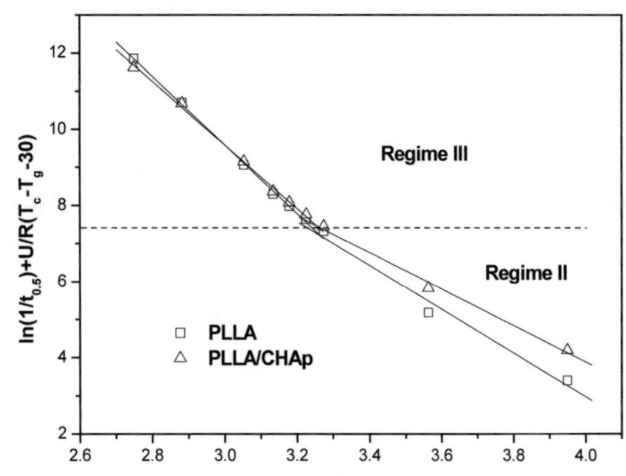

Fig. 6. Hoffman-Lauritzen plots for the estimation of nucleation parameters of neat PLLA and PLLA/CHAp nanocomposite (symbols: calculated data; solid lines: fitting curves).

The half-life crystallization time $t_{0.5}$, which is defined as the time at $X_t = 0.5$, is an important parameter for the discussion of crystallization kinetics. It can also be calculated using the equation:

$$t_{0.5} = (\frac{\ln 2}{k})^{1/n} \tag{7}$$

The half-life crystallization time $t_{0.5}$ can be either obtained directly from the curve of relative degree of crystallinity with time (Fig. 4) or calculated from isothermal parameter n and k by using Equation (7). These two sets of data are compared in Table 2. They are very close to each other for both neat PLLA and PLLA/CHAp within the range of crystallization temperatures. The value of $1/t_{0.5}$ can be used to describe the crystallization rate and is plotted in Fig. 5 as a function of isothermal crystallization temperature. As a typical semicrystalline polymer, the crystallization of PLLA is slow in temperature ranges close to the melting point and glass transition point. Di Lorenzo (2005) reported a discontinuity around 116-118 °C in a similar plot, which was ascribed to a sudden acceleration in spherulite growth. However, such discontinuity is not clear in Fig. 5. It can be seen that the crystallization rate of PLLA/CHAp was faster than neat PLLA. This implies that the nano-sized CHAp particles acted as an efficient nucleating agent. The highest overall isothermal crystallization rate was found between 100 to 110 °C. The increased crystallization rate due to the addition of CHAp into PLLA matrix was also confirmed by the higher values of isothermal crystallization parameter k listed in Table 2. Similar observations of inorganic particle reinforced PLLA composites exhibited increased bulk crystallization rates relative to neat polymer (Krikorian and Pochan 2004; Wu, Wu et al. 2007). This phenomenon is usually attributed to the higher heterogeneous nucleation rates, as demonstrated by increases in the isothermal crystallization parameter k. However, the effects of inorganic fillers on radial spherulite growth rates vary, depending on the properties and content of the fillers.

3.3 Crystal growth analysis by Lauritzen-Hoffman theory

The isothermal crystallization data obtained can be examined in terms of secondary nucleation or Lauritzen-Hoffman (LH) theory of polymer crystal growth (Hoffman, Davis et al. 1976). The growth face undergoes two different but related processes during polymer crystallization. One is the deposition of secondary nuclei on the growth face, and the other is the subsequent growth along the face at the sites where the secondary nuclei are formed. Therefore, there are two competing rates involved in the process that determine the regime in which the polymer crystallizes. On the basis of the secondary nuclei formation rate, *i*, and the lateral growth rate or the surface spreading rate, *g*, the crystallization regimes can be defined as follows:

$i << g$ Regime I

$i \sim g$ Regime II

$i >> g$ Regime III

Regime I, where i is much smaller than g, occurs at very low supercoolings (ΔT); Regime II, where i is in the order of g, occurs at moderate ΔT; and Regime III, where i is much greater than g, occurs at very high ΔT (Patki, Mezghani et al. 2007). Thus Regime I is characterized by mononucleation of the nucleus on a substrate, followed by lateral spreading or growth of a chain-folded crystal lamella. The linear growth rate is controlled by secondary nucleation in Regime I. In contrast, Regime II occurs by multiple nucleations at lower crystallization temperatures (Tjong and Bao 2005). As the temperature is further lowered, prolific and multiple nucleation can occur in Regime III (Pethrick 2007). Most polymers exhibit different regime behavior, depending on the crystallization condition.

LH theory assumes that a free energy barrier associated with nucleation has an energetic origin and it provides the general expression for the growth rate (G) of a linear polymer crystal with folded chains:

$$G = G_0 \exp[\frac{-U^*}{R(T_c - T_\infty)}]\exp(\frac{-K_g}{T_c \Delta T f}) \tag{8}$$

where K_g is the nucleation constant (for details see Equation (11)); ΔT is the undercooling defined by $T_m^0 - T_c$; T_m^0 is the equilibrium melting point which is 479.2K for PLLA (Iannace and Nicolais 1997); f is a factor given as $2T_c / (T_m^0 + T_c)$; U^* is the activation energy for polymer diffusion which means the transportation of segments to the crystallization site, and the universal value of U^* = 1500 cal/mol (6276 J/mol) was used in order to compare the results from literature (Iannace and Nicolais 1997); R is the gas constant [8.314 J/(mol K)]; T_∞ is the hypothetical temperature where all motion associated with viscous flow ceases and normally chosen as T_g-30K; and G_0 is the front factor.

Using a theoretical approach, it can be shown that the linear growth rate G can be considered proportional to $1/t_{0.5}$. Based on the LH theory (Hoffman and Miller 1997), the temperature variation of $1/t_{0.5}$ can be written as:

$$(\frac{1}{t_{0.5}}) = (\frac{1}{t_{0.5}})_0 \exp[\frac{-U^*}{R(T_c - T_\infty)}]\exp(\frac{-K_g}{T_c \Delta T f}) \tag{9}$$

It should be pointed out that this is only an approximation. Ideally, the value of spherulite growth rate (G) should be measured directly by optical microscopy and equation (8) be applied. However, the spherulites of the PLLA/CHAp nanocomposite were too small and accurate measurements of their growth rate were not possible. Second, the G values measured by optical microscopy are obtained under a confined 2D environment, i.e. the spherulites will grow between two microscope cover slides. Also, there is doubt whether limited number of 2D measurements can really represent the true 3D spherulite growth in large number. On the other hand, DSC analysis allows 3D polymer spherulite growth and the data will give a more macro-scale representation of the real situation.

For practical convenient use, Equation (9) is usually rewritten as

$$\ln(\frac{1}{t_{0.5}}) + \frac{U^*}{R(T_c - T_\infty)} = \ln(\frac{1}{t_{0.5}})_0 - \frac{K_g}{T_c \Delta T f} \tag{10}$$

The plot of the left-hand side of Equation (10) vs. $1/T_c\Delta T_f$ gives the slope of $-K_g$ from which the nucleation constant K_g can be evaluated. In the current investigation, U^*= 6276 J/mol, as stated earlier, was used in the plot (Fig. 6). For both neat PLLA and PLLA/CHAp, there were two crystallization regimes, i.e. Regimes II and III, in the isothermal temperature range. From Fig. 6, the K_g (II) values for neat PLLA and PLLA/CHAp were estimated to be 5.74×10^5 and 4.76×10^5, respectively, within the T_c range of 120-140 °C (Table 3). The K_g (II) value for neat PLLA was higher than those reported in the literature, including 2.92×10^5

based on isothermal crystallization parameters (Iannace and Nicolais 1997) and 3.01×10^5 on spherulite growth rate *G* values (Tsuji, Takai et al. 2006). K_g (III) for neat PLLA obtained here was 9.02×10^5 which was similar to 8.91×10^5 as reported by Iannace and Nicolais (1997) and 9.00×10^5 by Tsuji et al. (2006). According to Iannace and Nicolais (1997), the theoretical value of the ratio K_g (III)/K_g (II) is 2. In the current investigation, the respective ratios for neat PLLA and PLLA/CHAp were 1.57 and 1.74. The lower values obtained here could be related to either the different sets of values utilized for U^* and T_∞ or the difference in molecular weight of polymers used.
The transition from Regime II to Regime III was observed at about 120 °C for both neat PLLA and PLLA/CHAp, which was in agreement with the results obtained using *G* values (Iannace and Nicolais 1997; Abe, Kikkawa et al. 2001; Tsuji, Miyase et al. 2005; Tsuji, Takai et al. 2006). This indicates that the LH theory is a valid method to predict nucleation constant if the Avrami fit is good. The current investigation shows that the addition of CHAp nanoparticles did not alter the transition temperature of PLLA crystallization from Regime II to Regime III but caused a reduction in the value of K_g. The lower K_g for PLLA/CHAp suggests that less energy was needed for the formation of critical size of PLLA nuclei in the nanocomposite than in the neat PLLA.
The nucleation constant K_g in Equation (8) is given by:

$$K_g = \frac{Zb_0\sigma\sigma_e T_m^0}{\Delta h_f k_B} \tag{11}$$

where σ is the lateral surface free energy, σ_e is the fold surface energy , b_0 is the layer thickness of the crystal, Δh_f is the volumetric heat of fusion and k_B is the Boltzmann constant. The value of Z is dependent on the crystallization regime and equal to 4 for Regimes I (high temperatures) and III (low temperatures) and values 2 for Regime II (intermediate temperatures) (Di Lorenzo 2005). In the current investigation, the K_g values were used to calculate the fold surface free energy according to Equation (11). The lateral surface energy was determined by the Thomas-Stavely empirical equation (Hoffman, Davis et al. 1976):

$$\sigma = \alpha_0 \Delta h_f \sqrt{a_0 b_0} \tag{12}$$

where α_0 is an empirical constant and usually assumed to be 0.25 which is appropriate to high melting polyesters (Qiu and Yang 2006); represents the cross-sectional area of PLLA chains, with $=5.97\times10^{-10}$ m and $=5.17\times10^{-10}$ m from the literature; and Δh_f=1.11×10^8 J/m^3 (Di Lorenzo 2001). The σ value was then calculated to be 15.4×10^{-3} J/m^2, and which was close to the reported data (Krikorian and Pochan 2004). The values of fold surface energy σ_e were obtained by solving Equation (11) and are listed in Table 3. For neat PLLA, the value of σ_e varies in the range of 91-117 erg/cm^2 (1 erg/cm^2= 1×10^{-3} J/m^2), which agrees well with the result (107 erg/cm^2) by Miyata and Masuko (1998) but higher than 40.5 erg/cm^2 by Di Lorenzo (2001) and 43.5 erg/cm2 by Wu et al. (2007). It has been shown that the value of σ_e is dependent on the molecular weight of PLLA (Miyata and Masuko 1998) and the cooling or heating rate to the isothermal crystallization temperature (Wu, Wu et al. 2007). Miyata and Masuko (1998) implied that the value of σ_e increases with increase in the loose-loop chains on the fold surface of high molecular weight PLLA. The decrease of σ_e with increasing heating rate was considered by Wu et al. (2007) due to the change of nucleation mechanism from Regime III to Regime II by Lauritzen-Hoffman-Miller analysis. However, they did not show the transition temperature from Regime III to Regime II.

Samples	K_g (III) (K^2)	σ_e (III) (J/m^2)	q (III) (kJ/mol)	K_g (II) (K^2)	σ_e (II) (J/m^2)	q (II) (kJ/mol)
PLLA	9.02×10^5	91×10^{-3}	33.8	5.74×10^5	117×10^{-3}	43.5
PLLA/CHAp	8.36×10^5	84×10^{-3}	31.2	4.79×10^5	96×10^{-3}	35.7

Table 3. Lauritzen-Hoffman parameters for isothermal crystallization of neat PLLA and PLLA/CHAp nanocomposite.

It can be seen from Table 3 that addition of CHAp nanoparticles decreased the values of σ_e for PLLA by 18% in Regime II and 7.7% in Regime III. The lower values of the free energy of chain folding of the lamellar crystals in PLLA/CHAp suggest that the CHAp nanoparticles facilitated the crystallization of PLLA. According to Wittmann and Lotz (1990), the incorporation of nucleating agents into polymers changes the polymer crystallization behavior. Heterogeneous nucleating agents, such as filler particles and fibers, tend to promote the nucleation of spherulites on their surfaces, decrease the lamellar thickness and lead to epitaxial growth of crystallites. In this study, the nucleating ability of the CHAp has been clearly shown in Fig. 7 but the latter two mechanisms have not been examined.
Finally, the work for chain folding, q, which means by bending the polymer chain back upon itself in the appropriate configuration and is apparently correlated with molecular structure, can be expressed as follows (Patki, Mezghani et al. 2007):

$$q = 2a_0 b_0 \sigma_e \tag{13}$$

The values of q in Regime II and Regime III for neat PLLA and PLLA/CHAp nanocomposite were determined and are listed in Table 3. It was found that the values of q for neat PLLA were higher than those for PLLA/CHAp in both Regime II and Regime III. q is a measure of the inherent stiffness of polymer chain: more flexible chains have smaller values of q, and vice versa (Runt, Miley et al. 1992). The results show that the CHAp nanoparticles reduced the work needed for PLLA chains to fold into the crystal. Such phenomenon may arise due to a reduced size or a lower degree of perfection of the crystals formed.

3.4 Spherulite morphology

The spherulitic morphologies of neat PLLA and PLLA/CHAp nanocomposite were investigated using polarized optical microscopy at selected crystallization temperatures of 110, 120 and 130 °C and the results are shown in Fig. 7 (a-f). Since sufficient time was given for crystallization, according to the isothermal crystallization kinetics study, both neat PLLA and PLLA/CHAp were completely crystallized. It can be seen from Fig. 7 that the spherulites of neat PLLA were distinctive and of considerable size while those of PLLA/CHAp were much smaller and less distinctive. This clearly shows that the nano-sized CHAp acted as an effective nucleating agent to induce a great number of additional nuclei but, on the other hand, it also limited the growth space for each nucleus, leading to the formation of smaller spherulites. In the theoretical approach for growth rate analysis, the Lauritzen-Hoffman equation can be applied using half-time of crystallization ($t_{0.5}$) under the assumption that the crystallization rates are inversely proportional to $t_{0.5}$. This theoretical approach has been widely used in crystallization study of both neat polymers and their composite systems (Iannace and Nicolais 1997; Tsuji, Takai et al. 2006; Chen, Yao et al. 2007; Liao, Yang et al. 2007; Wu, Wu et al. 2007). It is shown in Fig. 5 that the values of $1/t_{0.5}$ for the PLLA/CHAp nanocomposite were higher than those of the neat PLLA, which means the nanocomposite has a higher crystallization rate than the neat polymer.

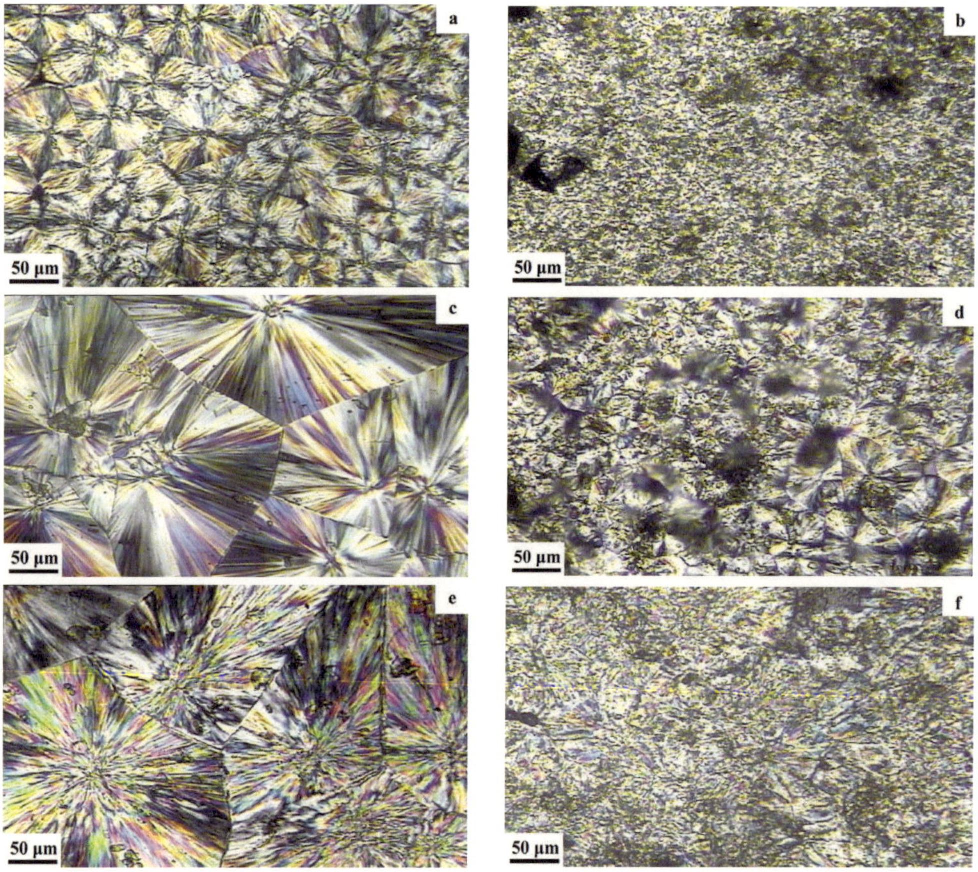

Fig. 7. Polarized optical micrographs showing the spherulitic morphologies of neat PLLA and PLLA/CHAp crystallized at various temperatures: (a) neat PLLA at 110 °C for 50 min, (b) PLLA/CHAp at 100 °C for 30 min, (c) neat PLLA at 120 °C for 90 min, (d) PLLA/CHAp at 120 °C for 40 min, (e) neat PLLA at 130 °C for 180 min, (f) PLLA/CHAp at 130 °C for 70 min. (* Fig c was selected in the book "Physical Metallurgy and Advanced Materials", eds by R.E. Smallman & A.H.W. Ngan, 7th edition, P557.)

3.5 Non-isothermal crystallization kinetics

The non-isothermal crystallization thermograms for neat PLLA and PLLA/CHAp nanocomposites obtained at six cooling rates are shown in Fig. 8. From these curves, useful crystallization parameters such as the peak temperature (T_p), at which the sample has the fastest crystallization, can be determined and used for further calculations. It can be seen that the crystallization enthalpy first increased and then decreased with increasing cooling rates. This phenomenon was also reported by Chen et al. (Chen, Fei et al. 2002) in the non-isothermal crystallization study of maleated poly(3-hydroxybutyrate). The crystallization enthalpy reached a maximum at about 2.5 °C/min for neat PLLA and 5 °C/min for PLLA/CHAp is caused by crystallization in the α′ polymorph, that causes a drastic increase of crystallization rate below 120 °C.

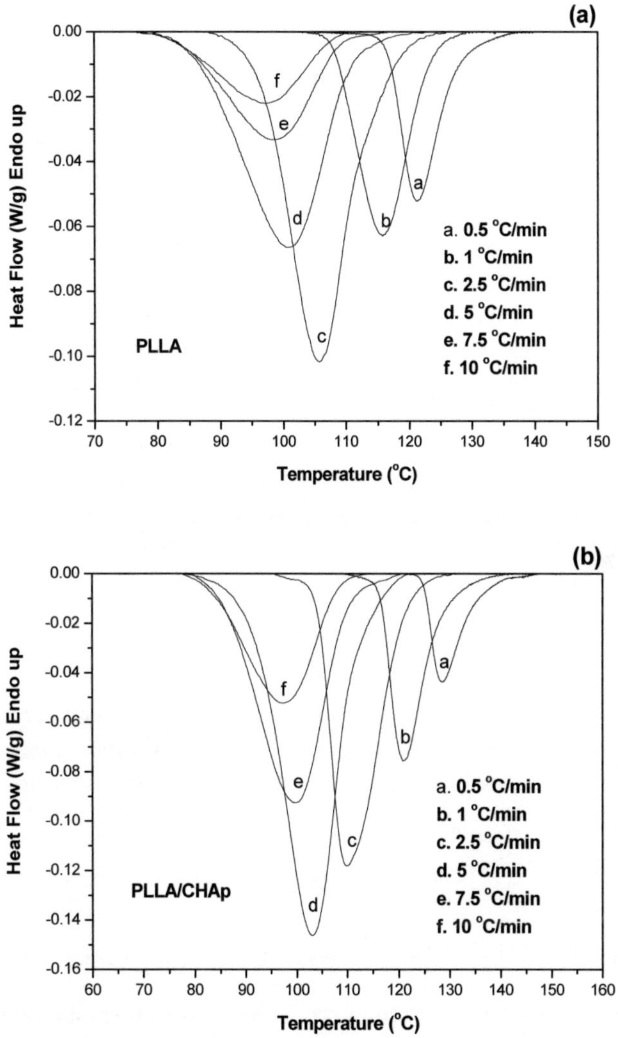

Fig. 8. DSC thermograms obtained from non-isothermal crystallization of (a) neat PLLA, and (b) PLLA/CHAp nanocomposites. (Cooling rates are indicated in the graphs.)

Ozawa (1971) extended the Avrami equation for the non-isothermal crystallization analysis. Assuming that the non-isothermal crystallization process is composed of infinitesimally small isothermal crystallization steps, the relative volumetric crystallinity X_t at temperature T can be calculated as follows:

$$X_t = 1 - \exp[\frac{-K(T)}{\phi^m}] \tag{14}$$

or

$$\log[-\ln(1 - X_t)] = \log K(T) + m \log \phi^{-1} \tag{15}$$

where *m* is the Ozawa exponent, which depends on the dimensions of the crystal growth, and *K(T)* is a function of cooling rate ϕ and indicates how fast crystallization occurs. If the Ozawa method is valid, plots of $\log[-\ln(1-X_t)]$ versus $\log \phi^{-1}$ should be straight lines, and kinetic parameters *K(T)* and *m* should be obtainable from the intercept and slope of the lines, respectively. However, when the cooling rates vary in a large range, the poor linearity of the plots renders the results calculated from Equation (14) questionable. Moreover, the assumption of constant cooling rates may cause problems in modeling the change of crystallinity during polymer processing (Di Lorenzo and Silvestre 1999). The Ozawa theory neglects the slow secondary crystallization and the dependence of lamellar thickness on the temperature so that it cannot describe the full process of non-isothermal crystallization of polymers in general (Lee and Cakmak 1998; Di Lorenzo and Silvestre 1999; Kim, Ahn et al. 2003; Jain, Goossens et al. 2005; Wu, Wu et al. 2007).

Aiming to find a method to describe exactly the non-isothermal crystallization process, a combination of Avrami and Ozawa equations was proposed recently (Liu, Mo et al. 1997). During the non-isothermal crystallization process, the relationship between crystallization time t and temperature T is given by

$$t = \frac{T_0 - T}{|\phi|} \tag{16}$$

where T is the temperature at time t, T_0 is the initial temperature when crystallization begins (t = 0). The Avrami equation relates X_t with time t, and the Ozawa equation relates X_t with cooling rate ϕ, thus the relationship between ϕ and t can be established to connect these two equations as follows:

$$\log \phi = \log F(T) - \alpha \log t \tag{17}$$

where the rate parameter $F(T) = [\frac{K(T)}{k}]^{1/m}$ and the physical meaning is the necessary value of cooling rate to reach a defined degree of crystallinity at unit crystallization time; α is the ratio of the Avrami exponent n to the Ozawa exponent m, i.e., α = n/m. According to Equation (17), at a given degree of crystallinity, the plot of $\log \phi$ as a function of log t gives a straight line with log *F(T)* as the intercept and $-\alpha$ as the slope. The combined Ozawa-Avrami model actually is a modified model in which crystallization functions are related to certain X_t values. This method is more convenient in the analysis of non-isothermal crystallization process, but it needs to be considered with care about its physical meaning (Buzarovska, Bogoeva-Gaceva et al. 2007).

Fig. 9 presents the relative degree of crystallinity as a function of time for neat PLLA and PLLA/CHAp crystallized at various cooling rates. The higher the cooling rate, the shorter time range within which crystallization occurs. The retardation effect of cooling rate on the crystallization is observed only at low cooling rates (<5 °C/min), below which the X_t–t curve shows an obvious S shape. At higher cooling rates, melted PLLA evolves into the glassy state quickly and hence the X_t–t curve tends to straight.

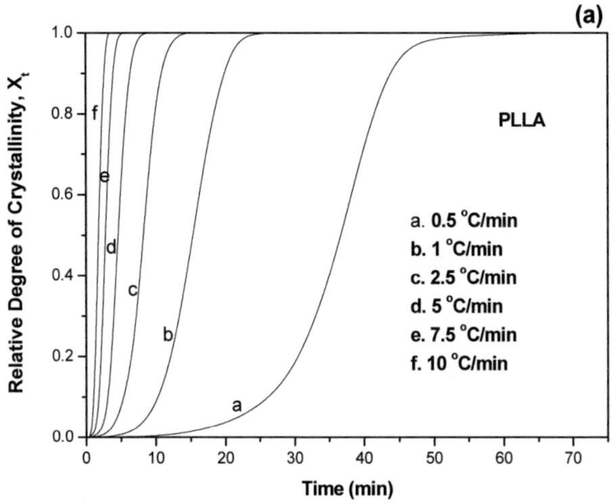

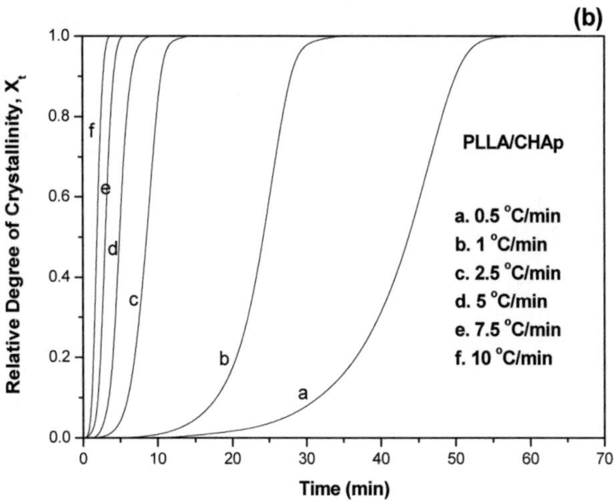

Fig. 9. Plots of relative degree of crystallinity as a function of time for (a) neat PLLA, and (b) PLLA/CHAp nanocomposite.

Fig. 10 presents the plots of log ϕ as a function of log t for neat PLLA and PLLA/CHAp. The good linearity of the plots demonstrates the good applicability of the combined Avrami-Ozawa method in the current investigation. Values of $F(T)$ and α are listed in Table 4, from which it can be seen that the values of *F(T)* increase systematically with an increase in the relative degree of crystallinity. Thus, at a unit crystallization time, a higher cooling rate is needed to achieve a higher degree of crystallinity. *F(T)* is considered as a parameter that indicates the polymer crystallization rate. A lower *F(T)* value means a higher crystallization rate under non-isothermal crystallization condition (Huang, Gu et al. 2006). The PLLA/CHAp nanocomposite had a larger *F(T)* value than neat PLLA in the low X_t range (<0.40), while the trend was opposite in the high X_t range (≥0.40). This indicates that the addition of CHAp only accelerated the PLLA crystallization rate in the initial crystallization stage. The values of parameter α are nearly constant and close to 1 which means that the Avrami exponent and Ozawa exponent are more or less the same for both neat PLLA and PLLA/CHAp.

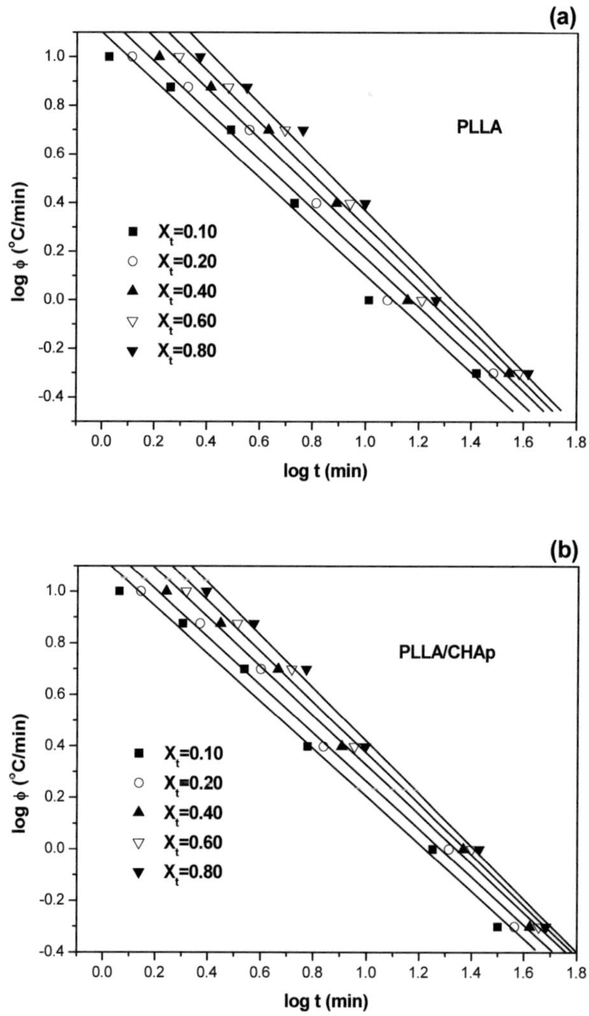

Fig. 10. Plots of log ϕ as a function of log t for (a) neat PLLA and (b) PLLA/CHAp nanocomposite based on combined Ozawa-Avrami equation.

Fig. 11 shows the relation between cooling rate (ϕ) and peak crystallization temperature (T_p) for neat PLLA and PLLA/CHAp. The T_p values for both neat PLLA and PLLA/CHAp shift to a lower temperature as the cooling rate increases. T_p value of PLLA/CHAp is higher than that of neat PLLA at the same cooling rate, indicating that the addition of CHAp accelerates the PLLA crystallization.

X_t	PLLA			PLLA/CHAp		
	α	F(T)	r^2	α	F(T)	r^2
0.10	1.00	12.59	0.9820	0.92	13.49	0.9898
0.20	1.01	15.14	0.9837	0.93	15.49	0.9917
0.40	1.04	19.50	0.9867	0.96	19.06	0.9940
0.60	1.06	23.44	0.9880	0.99	22.91	0.9951
0.80	1.10	29.51	0.9887	1.02	27.54	0.9956

r^2 denotes the coefficient of determination of Fig. 10.

Table 4. Values of *F(T)* and α obtained from the combined Avrami-Ozawa equation for neat PLLA and PLLA/CHAp nanocomposite.

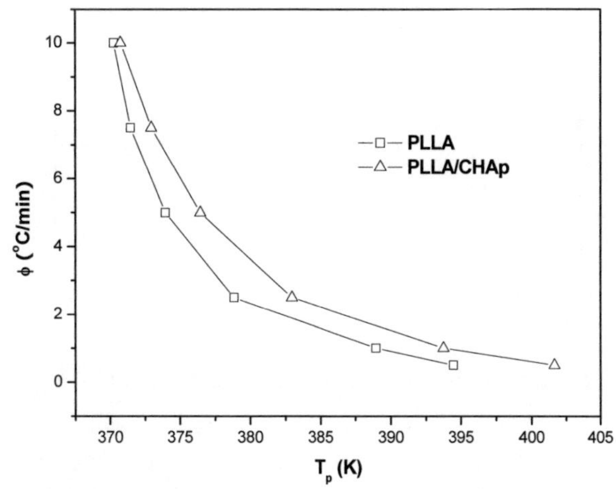

Fig. 11. Plots of cooling rate ϕ as a function of peak crystallization temperature T_p for PLLA and PLLA/CHAp nanocomposites.

3.6 Effective activation energy of non-isothermal crystallization

The crystallization activation energy during non-isothermal processes for neat PLLA and PLLA/CHAp can be evaluated using the Kissinger equation (Kissinger 1956):

$$\frac{d[\ln(\varphi/T_p{}^2)]}{d(1/T_p)} = -\frac{\Delta E_C}{R} \tag{18}$$

where ΔE_C is the crystallization activation energy and R is the gas constant. The activation energy ΔE_C can be obtained from the slope of the plot of $\ln(\phi/T_p^2)$ versus $1/T_p$ which is presented in Fig. 12 for neat PLLA and PLLA/CHAp, respectively. The calculated ΔE_C for neat PLLA and PLLA/CHAp are -149.67 and -124.91 kJ/mol, respectively. ΔE_C values are negative, indicating that the rate of crystallization increased with decreasing temperature and the crystallization process of polymer is a barrierless and spontaneous process (the lower value of ΔE_C, the faster of crystallization rate) (Ma, Hu et al. 2007). The obtained crystallization activation energy of neat PLLA fits well with the reported value of -146.86 kJ/mol (Hao, Li et al. 2005).

There are other methods for obtaining the activation energy from non-isothermal crystallization, such as the Augis-Bennett method (Augis and Bennett 1978) and the Takhor method (Takhor 1971). But the Kissinger method appears to be the most popular method for evaluating ΔE_C. However, all these three methods involve the cooling rate ϕ but its negative sign has been omitted in the logarithm formulation. Recently, Vyazovkin (2002) demonstrated that dropping the negative sign for ϕ is a mathematically invalid procedure that generally makes the Kissinger type equation inapplicable to melt crystallization.

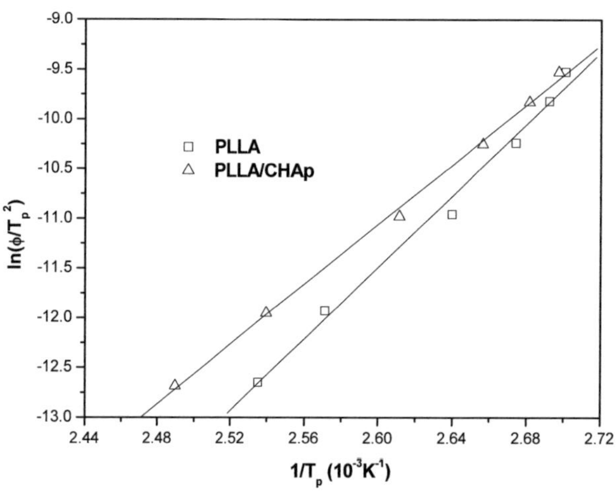

Fig. 12. Plots of $\ln(\phi/T_p^2)$ as a function of $1/T_p$ for neat PLLA and PLLA/CHAp nanocomposite.

Another disadvantage is that only a single value of activation energy is used for multiple non-isothermal crystallization processes by Kissinger-type methods. However, the crystallization rate is determined by the rates of nucleation and growth, whose activation energies are likely to be different. Sánchez-Jiménez et al. (2008) demonstrated that the Kissinger method only valid when the reaction obeys a first order kinetic law. In order to overcome the disadvantages of Kissinger-type methods, an isoconversion method can be applied to non-isothermal crystallization for evaluating the dependence of the activation energy on crystallinity and temperature. The representative methods include the differential isoconversion method (Friedman 1964) and the advanced integral isoconversion method (Vyazovkin and Sbirrazzuoli 2006). In the current investigation, the numerical differential method by Friedman (1964) was used. According to Friedman, different effective activation energies are calculated for every degree of crystallinity using the following equation:

$$\ln\left(\frac{dX}{dt}\right)_{X,i} = Const - \frac{\Delta E_X}{RT_{X,i}} \tag{19}$$

where dX/dt is the instantaneous crystallization rate as a function of time at a given crystallinity X, ΔE_X is the effective activation energy at given crystallinity X, $T_{X,i}$ is the set of temperatures related to a given crystallinity X at different cooling rates and the subscript i refers to every individual cooling rate used. The instantaneous crystallization rate, dX/dt, can be obtained from Fig. 9 by differentiation. Furthermore, by selecting appropriate degrees of crystallinity (i.e., from 10 to 90%) the values of dX/dt at a specific X are correlated to the corresponding crystallization temperature, T_X. Then by plotting dX/dt with respect to $1/T_X$, a straight line should be obtained with a slope equal to $\Delta E_X/R$.

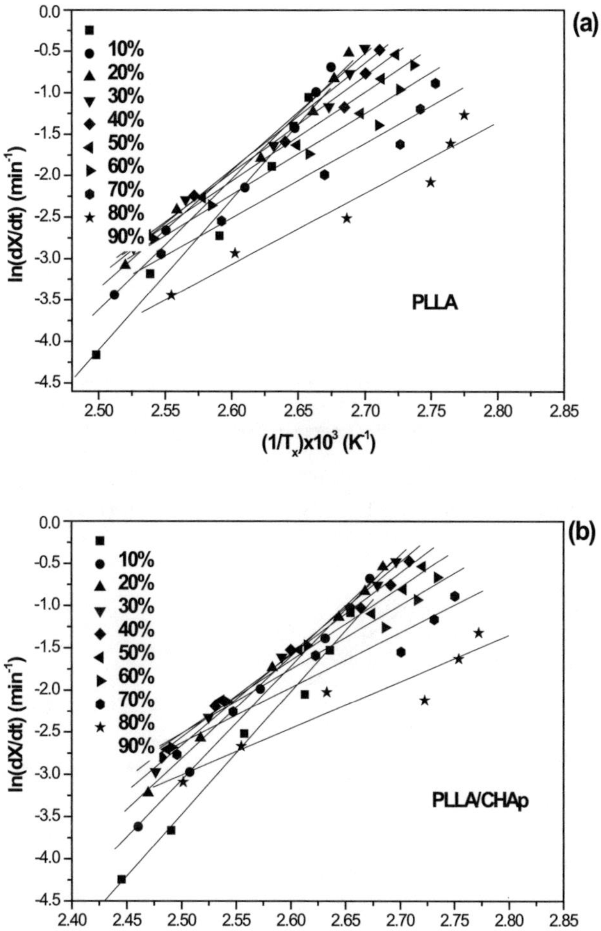

Fig. 13. Friedman plots of ln(dX/dt) vs. $1/T_x$ for (a) neat PLLA and (b) PLLA/CHAp nanocomposite at different relative degrees of crystallinity. The solid lines represent the linear fits.

Fig. 13 shows Friedman plots for neat PLLA and PLLA/CHAp nanocomposites at different relative degrees of crystallinity. The straight lines obtained were used to calculate the effective activation energies of neat PLLA and PLLA/CHAp which are shown in Fig. 13. As can be seen, the effective activation energy increased with the increase in the relative degree of crystallinity for all neat PLLA and PLLA/CHAp. In all cases, the absolute values of ΔE_X for PLLA was higher than that for PLLA/CHAp, indicating that the CHAp nanoparticles (10 wt%) lower the non-isothermal crystallization rate of PLLA. Interestingly, the values of activation energy obtained by the Kissinger method for both PLLA and PLLA/CHAp also fall on the curves in Fig. 14 at the relative degree of crystallinity near 10%. This indicates that the Kissinger method may represent one case in the ΔE_X dependence of crystallinity by the Friedman method.

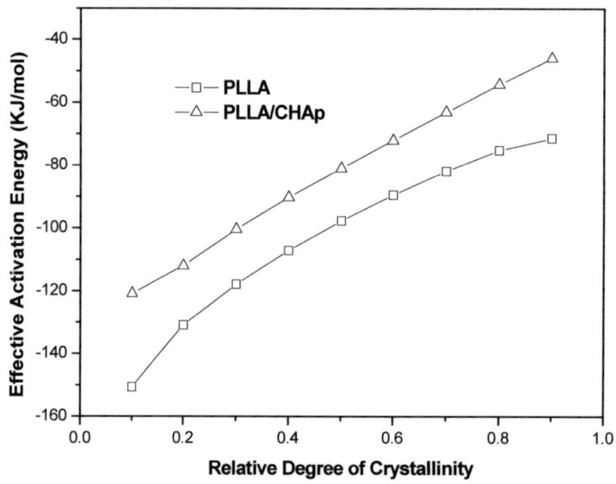

Fig. 14. Dependence of effective activation energy on relative degree of crystallinity in non-isothermal crystallization of PLLA and PLLA/CHAp nanocomposite.

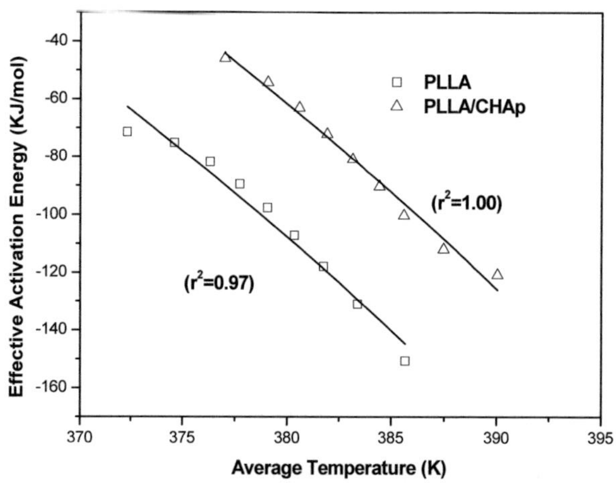

Fig. 15. Dependence of effective activation energy on average temperature for PLLA and PLLA/CHAp nanocomposite. The solid lines represent the nonlinear fits.

Furthermore, the relative degree of crystallinity is dependent on the non-isothermal crystallization temperature. Thus the effective activation energy can be plotted as a function of temperature by taking an average crystallization temperature associated with a certain relative degree of crystallinity (Papageorgiou, Achilias et al. 2007). The symbols in Fig. 15 display the ΔE_X -T relationship for neat PLLA and PLLA/CHAp, respectively. Vyazovkin and Sbirrazzuoli (2004) recently derived an equation to correlate the dependence of effective activation energy on temperature in terms of the Lauritzen-Hoffman parameters (U^* and K_g):

$$E_X(T) = U^* \frac{T^2}{(T - T_\infty)^2} + K_g R \frac{T_m^{0\,2} - T^2 - T_m^0 T}{(T_m^0 - T)^2 T} \tag{20}$$

In the current investigation, the graphics software Origin® (Microcal Software, Inc) was employed to perform the nonlinear fits of Equation (20) and the results are shown as the solid lines in Fig. 15. The values of T_∞ and T_m^0 used were the same as those in Section 3.3. The values of U^* and K_g yielded by the fits are shown in Table 5. The coefficients of determination (r^2) were found to be 0.97 and 1.00 for neat PLLA and PLLA/CHAp, respectively. The K_g values were similar to the values obtained from the isothermal crystallization analysis for Regime III, i.e. for crystallization temperatures below 120 °C (refer to Section 3.3). However, the regime transition was not observed due to the limited temperature intervals in the non-isothermal crystallization experiment. The obtained values of U^* were much lower than the commonly used value of 6276 kJ/mol. This indicates that the commonly used value of U^* may not be applicable to all polymers. The current analysis has demonstrated that the parameters of the Lauritzen-Hoffman equation can be obtained from DSC data on the overall rates of non-isothermal crystallization. In fact, this method have been successfully applied to the non-isothermal crystallization of a poly(ethylene terephthalate) (PET) with medium to slow crystallization rates by Vyazovkin and Sbirrazzuoli (2004), poly(propylene terephthalate) (PPT) and poly(butylene 2,6-naphthalate) (PBN) with fast crystallization rates by Achilias et al (2005) and nanocomposites of polyamide 6/halloysite nanotube recently by Guo et al (2009). In the current investigation, the validity of using the Vyazovkin-Sbirrazzuoli method for PLLA and PLLA/CHAp nanocomposite has been demonstrated.

Samples	U^* (J/mol)	K_g (III) (K^2)	Coefficient of Determination, r^2
PLLA	1846	7.52×10^5	0.97
PLLA/CHAp	2565	6.37×10^5	1.00

Table 5. Lauritzen-Hoffman parameters for neat PLLA and PLLA/CHAp nanocomposite obtained through isoconversion analysis of their non-isothermal crystallization.

4. Conclusions

The following conclusions can be drawn from the investigation into the effects of inclusion of CHAp nanoparticles on thermal properties, isothermal and non-isothermal melt crystallization kinetics of PLLA (Zhou, Duan et al. 2009):

1. The addition of 10 wt% of CHAp decreased the glass transition temperature and cold crystallization temperature of PLLA and slightly increased the melting temperature of PLLA. When cooled rapidly (50 °C/min) from the melt, the neat PLLA remained amorphous while PLLA/CHAp exhibited a crystallinity of 4%, which was probably caused by induced heterogeneous nucleation due to the presence of CHAp nanoparticles.
2. At high isothermal crystallization temperatures, the addition of CHAp decreased the crystallization enthalpy significantly compared with that of neat PLLA. Such changes imply a drop in the amount of crystals formed or a lower degree of perfection of the crystals. The Avrami equation described the isothermal crystallization kinetics well for both neat PLLA and PLLA/CHAp. The Avrami exponent n for neat PLLA and PLLA/CHAp approached 3 when increasing the isothermal crystallization temperature to 140°C, which indicates a three-dimensional crystal growth. The maximum isothermal crystallization rate was found to be between 100 to 110 °C.
3. By using Lauritzen-Hoffman theory, the nucleation constant (K_g), the fold surface energy (σ_e) and the work of chain folding (q) for neat PLLA and PLLA/CHAp could be obtained. The transition temperature from Regime II to Regime III was found to be about 120 °C for both neat PLLA and PLLA/CHAp. The CHAp nanoparticles acted as an efficient nucleating agent, thus increasing the nucleation rate and decreasing the fold surface energy of PLLA. The nucleating agent effect of CHAp was also confirmed by the observations of smaller spherulites in PLLA using polarized optical microscopy.
4. The non-isothermal crystallization kinetics of neat PLLA and PLLA/CHAp nanocomposite were investigated with DSC at cooling rates range from 0.5 to 10 °C/min. The combined Avrami-Ozawa equation was applied to analyze the non-isothermal crystallization process and the Ozawa exponent for neat PLLA and PLLA/CHAp were found to be very close to their Avrami exponent. According to the data obtained, the addition of CHAp only promoted the PLLA crystallization rate in the initial crystallization stage due to crystal germination under the non-isothermal condition. While in the isothermal condition, the addition of CHAp nanoparticles had influences on both crystal germination and growth mainly in Regime II.
5. Using the Kissinger equation, the non-isothermal crystallization activation energies of neat PLLA and PLLA/CHAp were found to be -149.67 and -124.91 kJ/mol, respectively. The differential isoconversional method by Friedman was applied to further estimate the dependence of the effective activation energy on the relative crystallinity and temperature for PLLA and PLLA/CHAp under non-isothermal crystallization. The Lauritzen-Hoffman parameters were obtained from the non-isothermal crystallization data as well by using the Vyazovkin-Sbirrazzuoli equation, and the values of U^* and K_g for neat PLLA and PLLA/CHAp were found to be 1846 J/mol and 7.52×10^5 K^2, 2565 J/mol and 6.37×10^5 K^2, respectively.

5. Acknowledgements

This work was supported by a GRF research grant (HKU 7118/05E) from the Hong Kong Research Grants Council. Support by Prof. Bakr Rabie in the Faculty of Dentistry, HKU, is acknowledged.

6. References

Abe, H., Y. Kikkawa, et al. (2001). Morphological and kinetic analyses of regime transition for poly[(S)-lactide] crystal growth. *Biomacromolecules* 2(3): 1007-1014.

Achilias, D. S., G. Z. Papageorgiou, et al. (2005). Evaluation of the Isoconversional Approach to Estimating the Hoffman-Lauritzen Parameters from the Overall Rates of Non-Isothermal Crystallization of Polymers. *Macromolecular Chemistry and Physics* 206(15): 1511-1519.

Arnoult, M., E. Dargent, et al. (2007). Mobile amorphous phase fragility in semi-crystalline polymers: Comparison of PET and PLLA. *Polymer* 48(4): 1012-1019.

Augis, J. A. and J. E. Bennett (1978). Calculation of the Avrami parameters for heterogeneous solid state reactions using a modification of the Kissinger method. *Journal of Thermal Analysis and Calorimetry* 13(2): 283-292.

Buzarovska, A., G. Bogoeva-Gaceva, et al. (2007). Crystallization behavior of poly(hydroxybytyrate- co -valerate) in model and bulk PHBV/kenaf fiber composites. *Journal of Materials Science* 42(16): 6501-6509.

Chen, C., B. Fei, et al. (2002). Nonisothermal crystallization and melting behavior of poly(3-hydroxybutyrate) and maleated poly(3-hydroxybutyrate). *European Polymer Journal* 38(8): 1663-1670.

Chen, J. H., B. X. Yao, et al. (2007). Isothermal crystallization behavior of isotactic polypropylene blended with small loading of polyhedral oligomeric silsesquioxane. *Polymer* 48(6): 1756-1769.

Chen, Y., A. F. T. Mak, et al. (2006). PLLA scaffolds with biomimetic apatite coating and biomimetic apatite/collagen composite coating to enhance osteoblast-like cells attachment and activity. *Surface and Coatings Technology* 201(3-4): 575-580.

Di Lorenzo, M. L. (2001). Determination of spherulite growth rates of poly (l-lactic acid) using combined isothermal and non-isothermal procedures. *Polymer* 42(23): 9441-9446.

Di Lorenzo, M. L. (2005). Crystallization behavior of poly (l-lactic acid). *European Polymer Journal* 41(3): 569-575.

Di Lorenzo, M. L. and C. Silvestre (1999). Non-isothermal crystallization of polymers. *Progress in Polymer Science* 24(6): 917-950.

Friedman, H. L. (1964). Kinetics of thermal degradation of char-forming plastics from thermogravimetry. *J. Polym. Sci. Part C* 6: 183.

German, R. M. (1996). Introduction to Sintering. *Sintering Theory and Practice*. R. M. German. New York, Wiley: 8.

Guo, B., Q. Zou, et al. (2009). Crystallization behavior of polyamide 6/halloysite nanotubes nanocomposites. *Thermochimica Acta* 484(1-2): 48-56.

Hao, Q., F. Li, et al. (2005). Preparation and Crystallization Kinetics of New Structurally Well-Defined Star-Shaped Biodegradable Poly(L-lactide)s Initiated with Diverse Natural Sugar Alcohols. *Biomacromolecules* 6(4): 2236-2247.

Hoffman, J. D., G. T. Davis, et al. (1976). The rate of crystallization of linear polymers with chain folding. *Treatise on Solid State Chemistry*. N. B. Hannay. 3: 497.

Hoffman, J. D. and R. L. Miller (1997). Kinetics of crystallization from the melt and chain folding in polyethylene fractions revisited: theory and experiment. *Polymer* 38(13): 3151-3212.

Huang, H., L. Gu, et al. (2006). Non-isothermal crystallization and thermal transitions of a biodegradable, partially hydrolyzed poly (vinyl alcohol). *Polymer* 47(11): 3935-3945.

Iannace, S. and L. Nicolais (1997). Isothermal crystallization and chain mobility of poly(L-lactide). *Journal of Applied Polymer Science* 64(5): 911-919.

Ignjatovic, N., E. Suljovrujic, et al. (2004). Evaluation of Hot-Pressed Hydroxyapatite/Poly-L-lactide Composite Biomaterial Characteristics. *Journal of Biomedical Materials Research* 71(2): 284-294.

Jain, S., H. Goossens, et al. (2005). Effect of in situ prepared silica nano-particles on non-isothermal crystallization of polypropylene. *Polymer* 46(20): 8805-8818.

Kim, S. H., S. H. Ahn, et al. (2003). Crystallization kinetics and nucleation activity of silica nanoparticle-filled poly (ethylene 2, 6-naphthalate). *Polymer* 44(19): 5625-5634.

Kissinger, H. E. (1956). Variation of peak temperature with heating rate in differential thermal analysis. *Journal of Research of the National Institute of Standards and Technology* 57(4): 217-221.

Kolstad, J. J. (1996). Crystallization kinetics of poly(L-lactide-co-meso-lactide). *Journal of Applied Polymer Science* 62(7): 1079-1091.

Krikorian, V. and D. J. Pochan (2004). Unusual crystallization behavior of organoclay reinforced poly (L-lactic acid) nanocomposites. *Macromolecules* 37(17): 6480-6491.

Lee, S. W. and M. Cakmak (1998). Growth habits and kinetics of crystallization of poly (ethylene 2, 6-naphthalate) under isothermal and nonisothermal conditions. *Journal of Macromolecular Science, Part B* 37(4): 501-526.

Lever, T. (2007). Optimizing DSC experiments. *Thermal Analysis of Pharmaceuticcals*. D. Q. M. Craig and M. Reading. Boca Raton, CRC Press/Taylor & Francis: 30.

Liao, R. G., B. Yang, et al. (2007). Isothermal cold crystallization kinetics of polylactide/nucleating agents. *Journal of Applied Polymer Science* 104(1): 310-317.

Liu, T., Z. Mo, et al. (1997). Nonisothermal melt and cold crystallization kinetics of poly (aryl ether ether ketone ketone). *Polymer Engineering & Science* 37(3): 568-575.

Lorenzo, A. T., M. L. Arnal, et al. (2007). DSC isothermal polymer crystallization kinetics measurements and the use of the Avrami equation to fit the data: Guidelines to avoid common problems. *Polymer Testing* 26(2): 222-231.

Ma, Y., G. Hu, et al. (2007). Non-isothermal crystallization kinetics and melting behaviors of nylon 11/tetrapod-shaped ZnO whisker (T-ZnOw) composites. *Materials Science and Engineering: A* 460: 611-618.

Marega, C., A. Marigo, et al. (1992). Structure and Crystallization Kinetics of Poly(L-Lactic Acid). *Makromolekulare Chemie-Macromolecular Chemistry and Physics* 193(7): 1599-1606.

Miyata, T. and T. Masuko (1998). Crystallization behaviour of poly(L-lactide). *Polymer* 39(22): 5515-5521.

Murariu, M., A. Da Silva Ferreira, et al. (2007). Polylactide compositions. Part 1: Effect of filler content and size on mechanical properties of PLA/calcium sulfate composites. *Polymer* 48(9): 2613-2618.

Oca, H. M. d. and I. M. Ward (2007). Structure and mechanical properties of poly(L-lactic acid) crystals and fibers. *Journal of Polymer Science Part B: Polymer Physics* 45(8): 892-902.

Ozawa, T. (1971). Kinetics of non-isothermal crystallization. *Polymer* 12(3): 150-156.

Papageorgiou, G. Z., D. S. Achilias, et al. (2007). Crystallization Kinetics of Biodegradable Poly(butylene succinate) under Isothermal and Non-Isothermal Conditions. *Macromolecular Chemistry and Physics* 208(12): 1250-1264.

Patki, R., K. Mezghani, et al. (2007). Crystallization kinetics of polymers. *Physical Properties of Polymers Handbook.* J. E. Mark. New York, Springer: 626.

Pethrick, R. A. (2007). Polymer crystal growth. *Polymer structure characterization: from nano to macro organization.* Cambridge, UK, RSC Publishing: 158.

Piorkowska, E., A. Galeski, et al. (2006). Critical assessment of overall crystallization kinetics theories and predictions. *Progress in Polymer Science* 31(6): 549-575.

Qiu, Z. B. and W. T. Yang (2006). Crystallization kinetics and morphology of poly (butylene succinate)/poly (vinyl phenol) blend. *Polymer* 47(18): 6429-6437.

Runt, J., D. M. Miley, et al. (1992). Crystallization of poly (butylene terephthalate) and its blends with polyarylate. *Macromolecules* 25(7): 1929-1934.

Sánchez-Jiménez, P., J. Criado, et al. (2008). Kissinger kinetic analysis of data obtained under different heating schedules. *Journal of Thermal Analysis and Calorimetry* 94(2): 427-432.

Sosnowski, S. (2001). Poly (l-lactide) microspheres with controlled crystallinity. *Polymer* 42(2): 637-643.

Takhor, R. L. (1971). Advances in Nucleation and Crystallization of Glasses. *American Ceramics Society, Columbus*: 166-172.

Tjong, S. C. and S. P. Bao (2005). Crystallization regime characteristics of exfoliated polyethylene/vermiculite nanocomposites. *Journal of Polymer Science Part B: Polymer Physics* 43(3): 253-263.

Tsuji, H., T. Miyase, et al. (2005). Physical properties, crystallization, and spherulite growth of linear and 3-arm poly(L-lactide)s. *Biomacromolecules* 6(1): 244-254.

Tsuji, H., H. Takai, et al. (2006). Isothermal and non-isothermal crystallization behavior of poly(L-lactic acid): Effects of stereocomplex as nucleating agent. *Polymer* 47(15): 5430-5430.

Vyazovkin, S. (2002). Is the Kissinger Equation Applicable to the Processes that Occur on Cooling? *Macromolecular Rapid Communications* 23(13): 771-775.

Vyazovkin, S. and N. Sbirrazzuoli (2004). Isoconversional approach to evaluating the Hoffman-Lauritzen parameters(U* and Kg) from the overall rates of nonisothermal crystallization. *Macromolecular Rapid Communications* 25(6): 733-738.

Vyazovkin, S. and N. Sbirrazzuoli (2006). Isoconversional kinetic analysis of thermally stimulated processes in polymers. *Macromolecular Rapid Communications* 27(18): 1515-1532.

Wang, M. (2006). Composite Scaffolds for Bone Tissue Engineering. *American Journal of Biochemistry and Biotechnology* 2(2): 80-84.

Wei, G. and P. X. Ma (2004). Structure and properties of nano-hydroxyapatite/polymer composite scaffolds for bone tissue engineering. *Biomaterials* 25(19): 4749-4757.

Wittmann, J. C. and B. Lotz (1990). Epitaxial crystallization of polymers on organic and polymeric substrates. *Progress in Polymer Science* 15(6): 909-948.

Wu, D., L. Wu, et al. (2007). Nonisothermal Cold Crystallization Behavior and Kinetics of Polylactide/Clay Nanocomposites. *Journal of Polymer Science: Part B: Polymer Physics* 45: 1100-1113.

Zhou, W. Y., B. Duan, et al. (2009). Crystallization kinetics of poly(l-lactide)/carbonated hydroxyapatite nanocomposite microspheres. *Journal of Applied Polymer Science* 113(6): 4100-4115.

Zhou, W. Y., S. H. Lee, et al. (2007). Selective Laser Sintering of Tissue Engineering Scaffolds Using Poly(L-Lactide) Microspheres. *Key Engineering Materials* 334-335: 1225-1228.

Zhou, W. Y., S. H. Lee, et al. (2008). Selective laser sintering of porous tissue engineering scaffolds from poly(L-lactide)/carbonated hydroxyapatite nanocomposite microspheres. *Journal of Materials Science: Materials in Medicine* 19(7): 2535-2540.

Zhou, W. Y., M. Wang, et al. (2007). Fabrication and characterization of composite microspheres containing carbonated hydroxyapatite nanoparticles. *Key Engineering Materials* 334-335: 1221-1224.

Zhou, W. Y., M. Wang, et al. (2008). Synthesis of carbonted hydroxyapatite nanospheres through nanoemulsion. *Journal of Materials Science: Materials in Medicine* 19(1): 103-110.

12

Threshold Optical Nonlinearity of Dielectric Nanocomposite

Yu. Kulchin, V. Dzyuba and S. Voznesenskiy
Institute of Automation and Control Processes, FEB RAS, Russia

1. Introduction

Optical properties of small particles are being investigated more than 100 years if dating from the Mie's works. However, the intense investigations of these properties have been started in the last decades with intense studies of optical properties of nanocomposites, including heterogeneous liquid-phase nanocomposites (HLPN) based upon semiconductor, metal or structured nanoparticles with both a core and electron shell of different electrical properties (Mixeeva & Sidorov, 2004, Fedorov & Baranov, 2005, Kul'chin(a) et.al.,2008, Kul'chin(b) et.al.,2009, Kul'chin(c) et.al.,2009, Irimpan et.al.,2008, Krishnana et.al.,2008, Ryasnyanskiy et.al.,2009, Kreibig & Vollmer, 1995). It is concerned with the fact that nanocomposites based upon semiconductor and dielectric nanoparticles are of special interest for those who develop illumination control means for optical information systems, because of such nanoparticles' great nonlinear-optical response in the near infrared and visible ranges of light.

Nonlinear optical properties of nanocomposites are directly connected with physical processes which occur inside nanoobjects under an influence of a laser radiation, and they depend on electronic structure of nanoparticles. In particular, in metal nanoparticles as well as in dielectric ones with a metal shell, on condition of plasmonic resonance, an increase of effective nonlinearity of the nanocomposite's medium and a decrease in energetic threshold of nonlinear response can occur.

Low nonlinear optical threshold in the ranges of near infrared and visible light can be observed in wide-gap semiconductor nanoparticles as well. For example, the investigation of dependence of the density of energy, passed through a nanoparticle array in transparent dielectric liquid matrix, on the density of incident energy have shown that threshold of nonlinear optical response for the sample based upon TiO_2 illuminated with impulses of $\tau = 10$ ns on the wavelength of $\lambda = 1.06$ μm is no greater than 0.15 nJ/sm^2, impulses with the wavelength of 0.53 μm provided the response less than 0.1 nJ/sm^2. Nonlinear optical properties of nanocomposites based upon dielectric nanoparticles with higher energy-gap width E_r were investigated to a less extent. It was concerned, presumably, with the considerations that such energy-gap width could cause the nonlinear optical response of the medium in the ultraviolet range of light. However, recent studies (Kul'chin et.al., 2008, Kul'chin et.al.,2009) have shown that HLPN based upon dielectric nanoparticles can feature low-threshold nonlinear optical response to an external illumination, impulse and continuous alike, in the near infrared or visible ranges of light.

As it shown in the work (Mixeeva & Sidorov, 2004, Irimpan et.al.,2008, Krishnana et.al.,2008), such media based upon liquid-phase matrix with linear optical properties, being under the influence of nanosecond impulses of near infrared and visible light ranges, display low-threshold (less than 0.5 nJ/sm^2) nonlinear optical response. This response is represented in either limitation of passed energy and greater medium transparency. The character of this response in these media was tightly bound with a refraction index and an energy gap width of nanoparticles material. The results of the work (Kul'chin(a) et.al.,2008, Kul'chin(b) et.al.,2009) show that the same effects in such media can be witnessed in case of use of continuous laser radiation as well. In both cases modulation of optical properties of liquid heterogeneous medium can be caused by a combination of the next processes: light-induced absorption change, light scattering by nanoparticles, thermal and photo-induced modulation of refraction index.

The light intensity threshold value must depend both upon properties of nanoparticles material and heterogeneous medium matrix material. It gives us an ability to change their nonlinear optical properties by selection of appropriate materials as well as concentration, size and shape of the nanoparticles. The existance of such nonlinear optical properties of nanocomposites indicates that electronic structure of nanoparticles is considerably different from that of a bulk sample. At least these differences are caused, firstly, by formation of areas of allowed energy of charge carriers inside the forbidden band (these areas' structure is connected with a great density of surface defects of the crystal structure and a complex shape of the nanoparticles). Secondly, by the effect of size quantization which causes a change of an exciton energy spectrum and generation of a discrete energy spectrum in a conduction band of free charge carriers, caused by a space limitation of their wave functions localization. Thirdly, electric dipole moments of transitions within such quasi-zero dimensional objects can be higher than such moments of volume samples (Kul'chin (c) et.al., 2009, Pjkutniy, 2006). As it was shown during the experimental studies (Kul'chin(a) et.al., 2009), a substantial factor which determines nonlinear optical properties of HLPN is a character of dependence of liquid matrix refraction index on the temperature of the matrix medium.

Unfortunately, because of shortage of experimental studies together with even less amount of theoretical studies on nonlinear optical properties of solid and liquid dielectric nanocomposites based upon different dielectric nanoparticles, it is impossible to develop a well-set outlines of physical processes occur within the aforementioned nanocomposites under an influence of an external illumination.

Thus this paper, being based upon the results of our works (Kul'chin(a) et.al.,2008, Kul'chin (b) et.al.,2009, Kul'chin (c) et.al., 2009, Dzyuba (a) et.al., 2010, Kul'chin (d) et.al. 2010, Dzyuba (b) et.al.,2010, Dzyuba (d) et.al.,2010), deals with experimental and theoretical studies of physical characteristics of nanoparticles responsible for specific optical properties of dielectric nanocomposites. During theoretical description and explanation of experimental data obtained, we considered nanoparticle as a multi-particle quantum system of charges which combines elements of quantum structures of a polyatomic molecule and a bulk crystal. Considering nanocomposites with quite a low concentration of nanoparticles (fractions of a percent), we didn't take into account interparticle interaction.

2. Peculiarities of electronic structure of dielectric nanoparticles

It is discovered nowadays that differences between optical properties of dielectric nanoparticles and that of bulk samples are to a substantial degree bounded to differences between energy spectra of nanoparticle's charge carriers and bulk samples. Mainly, such peculiarities of electronic structure of nanoparticles are caused by their small size and smaller areas of localization of wave functions of their charge carriers. It leads to pronounced quantum-sized effects not only within electronic subsystem of a nanoparticle but within other oscillating subsystems, e.g. phononic ones. Secondly, the peculiarities are caused by nanoparticles' shape which directly affects the energy spectrum of charge carriers and other oscillating subsystems of the nanoparticle (Dzyuba (b) et.al.,2010, Dzyuba () et.al.,2010). The third reason is the existence and a character of nanoparticle's surface defects, inner structure and impurity centers, which influence upon energy spectrum as well as dynamics and relaxation mechanisms of states of charge carriers and other elementary excitations.

The effect of size quantization leads to a radical transformation of quasi-continuous energy spectra and wave functions of charge carriers and elementary excitations of a nanoparticle when its size is decreased. Describing the states, one must take into account a shape and a limited size of a surface which has to satisfy boundary conditions imposed upon wave functions of charge carriers and other elementary excitations. This procedure was carried out carefully enough for the most symmetrical surfaces, e.g. for spherical or parallelepipedic ones. It was discovered that discrete levels energy and energy gap width, regardless of a shape, is inversely proportional to a v power of linear dimension of the nanoparticle when $1 \le v \le 2$ (Miyata & Ichikawa 2004). Experimental verification of this dependence has shown that there is only a trend to such dependence which is more pronounced for nanoparticles with size less than 2-3 nm. One of the main reasons of this dependence's disturbance is a deviation of a nanoparticle's shape when compared to the symmetrical one. A modeling of processes of influence of nanoparticle's shape upon an electronic structure of charge carriers and other elementary excitations is a very uncommon issue. However, by using the next approach, we can investigate an influence of nanoparticle's shape upon quantum-size states of its charge carriers (Dzyuba (b) et.al.,2010). Let us consider an energy spectrum as a functional $E[s(\vec{r})]$ determined along the space of functions which describe a shape of a nanoparticle's surface $s(\vec{r})$. Presenting this functional as variational series with variations of the surface $\delta s(\vec{r})$. We obtain

$$E[s(\vec{r})] = E_0[s_0(\vec{r})] + \int_V \left.\frac{\delta E[s(\vec{r})]}{\delta s(\vec{\xi})}\right|_{S=S_0} \delta s(\vec{\xi}) d\vec{\xi} + \frac{1}{2!}\iint_V \left.\frac{\delta^2 E[s(\vec{r})]}{\delta s(\vec{\xi}_1)\delta s(\vec{\xi}_2)}\right|_{S=S_0} \delta s(\vec{\xi}_1)\delta s(\vec{\xi}_2) d\vec{\xi}_1 d\vec{\xi}_2) + \ldots$$

where V – the nanoparticle's volume, $\frac{\delta}{\delta s(\vec{\xi})}$ – variational derivative and $\delta s(\vec{\xi}) = s(\vec{\xi}) - s_0(\vec{\xi})$ – variation of the nanoparticle's surface. Taking into account only the two terms of the variational derivative $\frac{\delta E[s(\vec{r})]}{\delta s(\vec{\xi})} = \frac{dE[s(\vec{r})]}{ds(\vec{r})}\delta(\vec{r} - \vec{\xi})$ we have

$$E[s(\vec{r})] \approx E[s_0(\vec{r})] + \left.\frac{dE[s(\vec{r})]}{ds(\vec{r})}\right|_{s=s_0} \delta s(\vec{r}) \cdot \quad (1)$$

It is convenient to choose a $s_0(\vec{r})$ surface of a simple and the most symmetrical form in order to obtain a well-known energy spectrum of charge carriers. Let us assume that an undisturbed surface is a sphere - $s_0(\vec{r}) = R_0^2$ and proceed into a spherical coordinate system (SCS). In this case the distance between the centre of the sphere and a point upon its surface $R(\theta,\phi)$ will satisfy the condition $R_0^2(\theta,\varphi) = R_0^2$. Within our spherical coordinate system a variation of the surface is determined by the next expression $\delta s(R,\theta,\phi) = s(R,\theta,\phi) - s_0(R,\theta,\phi) = \delta R^2(\theta,\phi) = R^2(\theta,\phi) - R_0^2$ where $R^2(\theta,\phi)$ - the square distance between the centre of the sphere and a point upon disturbed surface $s(R,\theta,\phi)$. It is known that energy of a free charge carrier inside the infinitely deep spherical hole with radius R_0 is determined by the expression $E_{n,l}(R_0) = \frac{\alpha_{n,l}^2 \hbar^2}{2R_0^2 m}$, where $\alpha_{n,l}$ - roots of the Bessel function, m – the mass of a charge carrier. Introducing the designation $E_{n,l}(R_0) = E_{n,l}^0$ from Eq. (1) we obtain -

$$E_{nl}(\theta) = E_{nl}^0 \left(1 - \frac{\delta R^2(\theta,\varphi)}{R_0^2}\right). \tag{2}$$

The angle θ in Eq. (2) should be considered as an angle between a vector of the particle's angular momentum and a chosen spatial direction. There is a preferential spatial direction which corresponds to OZ axis. When the particle's plane of motion makes an angle θ with OZ axis then $\cos^2\theta = \frac{m^2}{l(l+1)}$, where ℓ and m – orbital and magnetic quantum numbers correspondingly. Quantum states density is equal to, $N(E) = J_g \frac{4\pi\rho^2}{(2\pi\hbar)^3}\frac{dP}{dE}$ where J_g - amount of states that are not associated with spatial movement of the particle, P– impulse of the particle. In case of electrons or electron holes $P=\sqrt{2mE}$, $J_g = 2$ $N(E) = 2\frac{4\pi\rho^2}{(2\pi\hbar)^2}\frac{dP}{dE} = \frac{\sqrt{2}m^{\frac{3}{2}}}{\pi^2\hbar^3}V\sqrt{E}$, V - nanoparticle's volume, m- effective mass of electron or electron hole. Density of states is a functional defined in the space of functions which describe a geometrical shape of a nanoparticle. Thus we can consider N(E) as $N(E) = N(E)\big|_{s(\vec{r})=s_0} + \int_V \frac{\delta N(E)}{\delta s(\vec{\xi})}\bigg|_{s_0} \delta s(\vec{\xi})d\vec{\xi} + ...$ and we can observe that for a small disturbance of the surface $\delta s(\vec{r})$ the change of the quantum states density is determined as $\Delta N(E) \approx \int_V \frac{\delta N(E)}{\delta s(\vec{\xi})}\bigg|_{S_0} \delta s(\vec{\xi})d\vec{\xi}$. For quantum-size states we obtain the next expression

$$\Delta N(E) = \frac{m^{3\backslash 2}v}{\sqrt{2}\pi^2\hbar^3\sqrt{E[s_0(\vec{r})]}}\frac{\delta E[s(\vec{r})]}{\delta s(\vec{\xi})}\bigg|_{S_0}\delta s(\vec{\xi}) = \frac{m^{3\backslash 2}v}{\sqrt{2}\pi^2\hbar^3}\frac{\Delta E[\delta s(\vec{r})]}{\sqrt{E[s_0(\vec{r})]}}, \tag{3}$$

where $\Delta E\left[\delta s(\vec{r})\right]=E\left[s(\vec{r})\right]-E\left[s_0(\vec{r})\right]$.

Let us consider several examples of the approach presented. In a general case we represent an axial-symmetrical surface of a nanoparticle with the next equation $R^2(\theta)=R_0^2[1+p^2\cos^{\gamma}(k\theta)]$. Let $p^2=0{,}2$, $\gamma=2$, $k=3$. This nanoparticle has the shape shown in Fig. 1. Relationships between energy spectrum and the density of quantum-size states of this nanoparticle and the nanoparticle of a spherical shape with radius R_0 are equal to

$$\frac{E\left[R^2\right]}{E\left[R_0^2\right]}=\left\{1-0{,}2\left[\frac{4m^2}{\ell(\ell+1)}-3\left(\frac{m^2}{\ell(\ell+1)}\right)^{\frac{1}{2}}\right]\right\}^2,\quad \frac{N\left[R_0^2\right]}{N\left[R^2\right]}=\left(\frac{1}{2}+\frac{E\left[R^2(\theta)\right]}{2E\left[R_0^2\right]}\right).$$

The plots displaying such these relationships are shown in Fig.(1).

When $\gamma=3$, $k=2$ the abovementioned relationships have the next form: $\dfrac{E\left[R^2\right]}{E\left[R_0^2\right]}=1-0{,}2\left[\dfrac{2m^2}{\ell(\ell+1)}-1\right]^3$, $\dfrac{N\left[R_0^2\right]}{N\left[R^2\right]}=\left(1-0{,}1\left[\dfrac{2m^2}{\ell(\ell+1)}-1\right]^3\right)$. The corresponding plots are shown in Fig. (2).

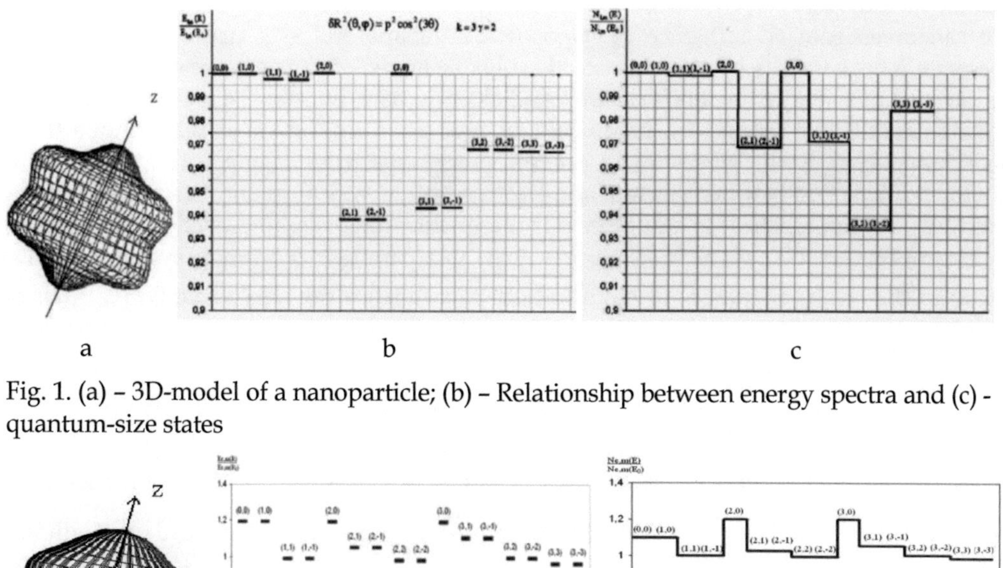

a b c

Fig. 1. (a) – 3D-model of a nanoparticle; (b) – Relationship between energy spectra and (c) - quantum-size states

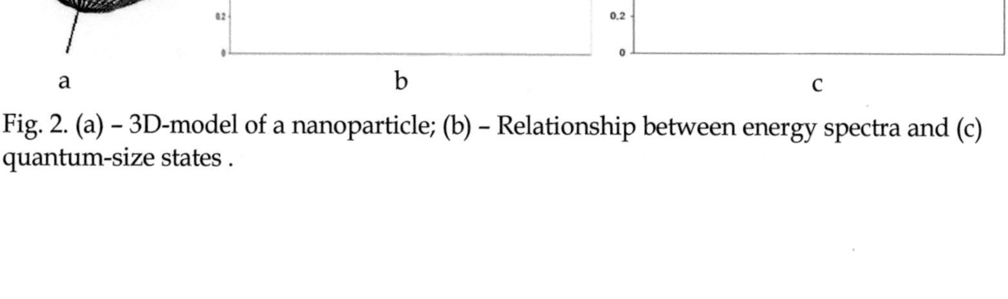

a b c

Fig. 2. (a) – 3D-model of a nanoparticle; (b) – Relationship between energy spectra and (c) quantum-size states .

The examples demonstrate that nanoparticle's shape deviation from the spherical one lessens the degree of magnetic quantum number degeneracy of charge carriers energy. In addition, this deviation leads to a formation of allowed energy minizones which lie above (Fig. 2) as well as below (Fig. 1) non-split levels of a spherical nanoparticle. It causes a corresponding change of quantum-states density which can be greater (Fig.1) or lesser (Fig.2) than a base level.

Previously we considered single-particle states of charge carriers, but formation of optical properties of dielectric and semiconductor nanostructures depends mostly upon double-particle states also known as exciton states. A contribution of exciton states into optical properties is substantial if their Bohr radius is comparable to nanoparticle's size (or less than the size - the weak confinement regime). In contrast to the single-particle states, the accurate description of exciton states is impossible even for spherical nanoparticles. Use of approximate approaches which utilize a model of a nanoparticle represented as a deep potential hole of a spherical shape, and modeling the influence of a boundary upon charge carriers by a quantum hole potential allowed us to make some overall conclusions on energy spectrum of exciton within a nanoparticle (Fedorov & Baranov, 2005).

First of all, every size of a nanoparticle is associated with a finite amount of levels with negative exciton binding energy; this amount is decreasing when an exciton is lessening. There is a critical nanoparticle's size such that these levels disappear. Secondly, the orbital quantum state degeneracy of a relative motion, peculiar to Coulomb motion in infinite medium is removed.

We can investigate an influence of nanoparticles' shape and size upon exciton energy spectrum more carefully by using an exactly solvable model of a nanoparticle represented as a system of two charge carriers - electron and electron hole, which exist inside an infinitely deep potential hole limited by paraboloid of revolution and sizes of real nanoparticles (Kul'chin(c) et.al.,2009). We estimate exciton energy spectrum taking into account only Coulomb interaction between electron and electron hole as well as quantum size effect in the effective mass approach. This assumption is reasonable due to comparatively small sizes of nanoparticles. Exciton's wave function $\Psi(\eta,\xi,\varphi)$ in a parabolic coordinate system with the centre in the centre of gravity of electron-hole pair satisfies the next equation (Landau & Lifshich (a) 2002)

$$\frac{4}{\xi+\eta}\left[\frac{\partial}{\partial\xi}\left(\xi\frac{\partial\Psi}{\partial\xi}\right)+\frac{\partial}{\partial\eta}\left(\eta\frac{\partial\Psi}{\partial\eta}\right)\right]+\frac{1}{\xi\eta}\frac{\partial^2\Psi}{\partial\varphi^2}+2\left(E+\frac{2}{\xi+\eta}\right)=0\cdot \tag{4}$$

Eq. (4) utilizes relative units where the Plank constant and a charge are equal to 1, masses of electron m_e, electron hole m_h and exciton $\mu=\frac{m_e m_h}{m_e+m_h}$ are chosen in accordance with the effective in accordance with the effective mass approach; as a length unit Bohr radius of the exciton. Parabolic coordinates are connected with Cartesian ones as it follows $x=\sqrt{\xi\eta}\cos\phi$, $y=\sqrt{\xi\eta}\sin\phi$, $z=\frac{1}{2}(\xi\text{-}\eta)$.

Paraboloid of revolution around the OZ axis in parabolic coordinates system is represented with the equation $\eta=\eta_0$ (smaller values of η_0 correspond to a narrower paraboloid); a surface $z=z_0$ which is a base of nanoparticle is represented with the equation $\xi=2z_0+\eta$. Coordinates of nanoparticle's surface points satisfy the equations $\eta=\eta_0$ and $\xi=2z_0+\eta$.

The corresponding boundary condition imposed upon the wave function on the surface of the potential hole has the next form $\Psi(\eta_0, \xi = 2z_0 + \eta, \varphi) = 0$. A solution of this equation which is finite in the centre of coordinate system and zero-equal on the infinity is expressed by using degenerated hypergeometric functions $F(\alpha, |m|+1, \rho)$. This solution has the next form

$$\Psi(\eta,\xi,\varphi) = C\frac{1}{n^2|m|!}\sqrt{\frac{(\alpha+|m|)!(\beta+|m|)!}{\alpha!\beta!}} \times$$
$$\times F\left(-\alpha, |m|+1, \frac{\xi}{n}\right) F\left(-\beta, |m|+1, \frac{\eta}{n}\right) \exp\left(-\frac{\xi}{2n} - \frac{\eta}{2n} + im\varphi\right)\left(\frac{\xi\eta}{n^2}\right)^{|m|}$$

The n number defines exciton's binding energy levels $E_n = -\frac{\mu e^4}{2\hbar^2\varepsilon_2^2 n^2} = -\frac{\mu}{m_o\varepsilon_2^2 n^2} 13{,}6\text{eV}$.

There is a ground quantum state with $\beta = 1, \alpha = 0$ and $n(|m|+1) = \eta_0$ which satisfies the boundary condition upon the side surface of the paraboloid of revolution. In this case $n = \frac{|m|+1}{2} \pm \sqrt{\frac{(|m|+1)^2}{4} + \frac{\eta_0}{|m|+1}}$ and every value of quantum number m is associated with two levels. The value of $m = 0$ corresponds to an exciton which has a motion plane parallel to OZ. The effective masses of electron and electron hole in, e.g. Al_2O_3 are equal to $0.4m_0$ and $6.2m_0$. Given these masses within the nanoparticle of 40 nm height and 40 nm width of a bottom the value of η_0 is equal to 13 a.u. and the binding energies of the exciton's ground state are equal to 0.07eV and -0.04eV correspondingly. Assuming the kinetic energy of exciton's motion to be negligibly small, we come to the conclusion that energy levels of the exciton lie inside the forbidden band at a ranges of 0.07eV and 0.04eV from the bottom of conduction band.

The second duplet of levels corresponding to $m = 1$ lies at range 0.05 eV and 0.21 eV from the conduction band to the direction of forbidden band. These values for $m = 2,3,4$ are equal to 0.03 eV and 0.58 eV, 0.04 eV and 1.36 eV, 0.02 and 4.06 eV correspondingly. One can observe the effect of geometrical amplification of electron-hole interaction in the spectrum of the abovementioned lines. The boundary condition upon the surface $\xi = 2z_0 + \eta$ is assosiateld with excited state with $n = \frac{|m|+1}{2} \pm \sqrt{\frac{(|m|+1)^2}{4} + \frac{2z_0+\eta}{|m|+1}}$. One can witness in this state a dependence of energy levels on coordinate η hich varies in the range from 0 to η_0 The dependence demonstrates an influence of a surface shape of a potential hole upon exciton's energy spectrum. Generally, the exciton's energy spectrum remains constant. It consists of two areas: the first one is virtually continuous and it lies next to the bottom of the conduction band, the second one is discrete and it lies in the deep of the forbidden band.

In the effective mass approach the wave function $\Psi(\eta,\xi,\varphi)$ of a free motion of the nanoparticle within the potential hole in parabolic system of coordinates satisfies the Eq. (4) considered without the Coulomb's term. This equation's solution, which is finite in the centre of the coordinate system, is known and is expressed by using Bessel functions $\Psi(\eta,\xi,\varphi) = CJ_{m/2}\left(\frac{k}{2}\eta\right)J_{m/2}\left(\frac{k}{2}\xi\right)\exp(m\varphi)$. The spectrum of free particles is

$$E_{n,m} = \frac{2\hbar^2}{\mu_{e,h} m_0}\alpha_{n,m}^2\left[\frac{1}{\eta_0^2} + \frac{1}{(2z_0+\eta)^2}\right],$$

$\mu_{e,h}$ - relative mass of an electron or an hole.

Energy spectrum of free charge carriers like energy spectrum of an exciton is dependent of η. Thus, taking into account natural and thermal broadening of levels, we can say that energy spectrum of the exciton is formed as a virtually continuous band with a width of 0.1 eV approx., adjacent from the bottom to the conduction band, and broadened discrete levels which lie within the forbidden band at the range about 0.2eV from the bottom of conduction band. Energy spectrum of charge carriers in the conduction band can have discrete levels as well.

The above results may be summed up as follows: we can build up a model of a diagram of energy levels of single- and double-particles charge carriers states in dielectric nanoparticles (Fig. 3) by using Al_2O_3 nanoparticles as an example. This particle is of a complex shape and its sizes are equal to 40-50 nm approx. (Fig. 4).

Unlike volume samples, the nanoparticles charge carriers can have an energy spectrum with broadened quantum size levels, broad band of exciton states and a subzone of allowed energies, lying within forbidden band, caused by surface and internal defects of nanoparticle's material, impurity centers etc., It should be noted that electronic structure of nanoparticle within a matrix depends to a great extent on relationship of dielectric permittivities $\frac{\varepsilon_1}{\varepsilon_2}$ of materials of matrix and nanoparticles.When $\frac{\varepsilon_1}{\varepsilon_2}>1$ polarization interaction leads to positive charges (holes) attraction to the inner surface of a nanoparticle and to exciton destruction. Internal surface states are established and their contribution to the subzone of allowed energies causes its drift towards valence band. When $\frac{\varepsilon_1}{\varepsilon_2}<1$ polarization interaction leads to a repulsion of charges from the nanoparticle's surface

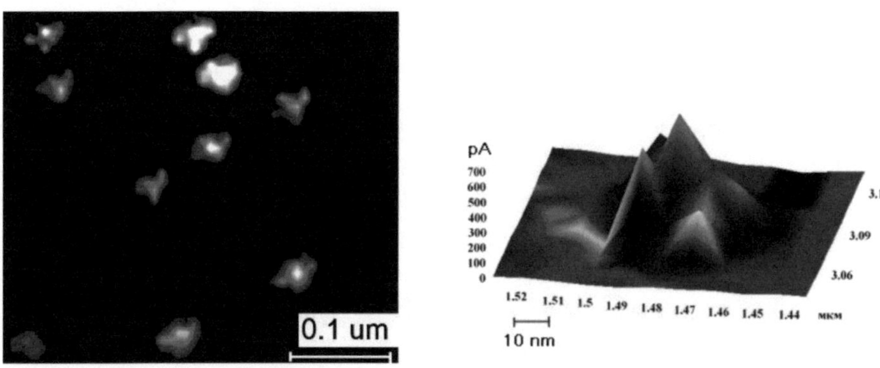

Fig. 3. Images of Al2O3 nanoparticles obtained by using SAFM.

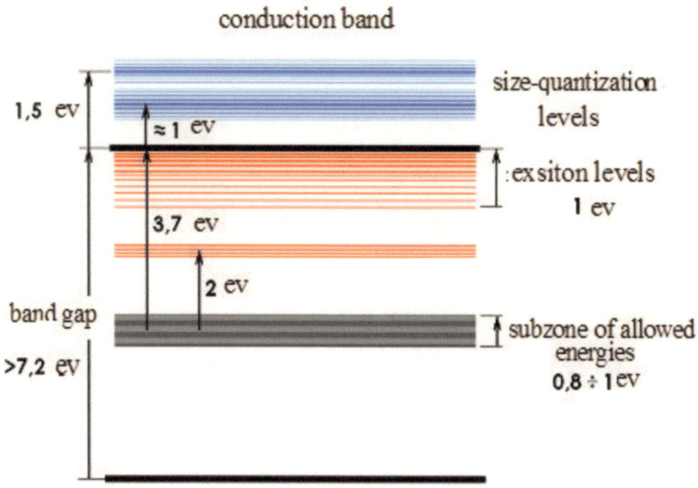

Fig. 4. The diagram of energy levels of single- and double-particles states of charge carriers in dielectric nanoparticles.

towards the inside so the exciton states are formed. However, if nanoparticle's sizes are small, we have $a< a_c=6\left|\frac{\varepsilon_1+\varepsilon_2}{\varepsilon_1-\varepsilon_2}\right|a_b$, where a_b - Bohr radius of charge carrier in the nanoparticle. Otherwise the energy spectrum will be defined by quantum-size states.

2.1 Absorption and scattering cross-sections of dielectric nanoparticles and transmission spectrum of HLPN

One of the most important and informative optical characteristics of electronic structure and physical mechanisms responsible for optical properties of HLPN in near infrared and ultraviolet light are cross-sections of absorption $\sigma_a(\omega,a)$ and scattering $\sigma_s(\omega,a)$ of light by nanoparticles as well as light transmission spectrum of HLPN - $T(\omega,a,N)$. Cross-sections $\sigma_s(\omega,a)$ and $\sigma_a(\omega,a)$ and their dependence upon nanoparticle's shape and light frequency ω can be defined in terms of polarizability of nanoparticle's volume unit $A(\omega,a)$ with the next expressions .

$$\sigma_a(\omega,a)=\frac{4\pi\omega}{c}V\,\mathrm{Im}\,A(\omega,a),\quad d\sigma_s(\omega,a)=\frac{\omega^4}{c^4}V^2\left|A(\omega,a)\right|^2\sin^2\theta d\Omega, \tag{5}$$

where θ - angle between a direction of a scattering vector and electric field vector, c - the speed of light in vacuum. Within the framework of dipole approach a polarizabilty of nanoparticle's volume unit is defined by $D_{ij}(a)$ elements of the matrix of electric dipole moment of transition between $|i\rangle$ and $|j\rangle$ states -

$$A(\omega,a)=\frac{2}{V\hbar}\left[\sum_{i,j}\frac{\left|D_{ij}(a)\right|^2(\omega_j^2-\omega^2)}{(\omega^2-\omega_j^2)^2+\omega^2\Gamma_j^2}\omega_j+i\sum_{i,j}\frac{\left|D_{ij}(a)\right|^2\Gamma_j}{(\omega^2-\omega_j^2)^2+\omega^2\Gamma_j^2}\omega\omega_j\right], \tag{6}$$

where summation is carried out over all dipole moments excited by a laser radiation, ω_j - transition frequency, Γ_j - a width of excited level, $\hbar$ - Plank's constant. It follows from the Eq. (5) and (6) that expressions for absorption cross-section and scattering cross-section of nanoparticle should be written as

$$\sigma_a(\omega,a)=\frac{8\pi}{c\hbar}\sum_{i,j}\frac{\left|D_{ij}(a)\right|^2\omega_j\Gamma_j}{\left(\omega^2-\omega_j^2\right)^2+\omega^2\Gamma_j^2}\omega^2 \tag{7}$$

$$\sigma_s(\omega,a)=\frac{4P\omega^4}{c^4\hbar^2}\sum_{i,j}\sum_{i,k}\left\{\frac{\left|D_{ij}(a)\right|^2\left|D_{ik}(a)\right|^2\omega_j\omega_k[\left(\omega_j^2-\omega^2\right)\left(\omega_k^2-\omega^2\right)+\Gamma_j\Gamma_k\omega^2]}{\left[\left(\omega_k^2-\omega^2\right)^2+\omega^2\Gamma_k^2\right]\left[\left(\omega_j^2-\omega^2\right)^2+\omega^2\Gamma_j^2\right]}\right\}, \tag{8}$$

where $P=\int_{4\pi}\sin^2\theta d\Omega$. Dependence of the cross-sections on nanoparticle's size can be found knowing a relation between values of $D_{ij}(a)$, ω_j , Γ_j and nanoparticle's size. The value of matrix element of a dipole moment of a state $|i\rangle$ to a state $|j\rangle$ transition is $D_{ij}(a)=\Lambda(\varepsilon_1,\varepsilon_2)e\langle j|\vec{r}\vec{\gamma}|i\rangle$, where $\vec{r}$ - radius-vector of a charge, $\vec{\gamma}$ - vector of polarization direction of the electromagnetic wave which induced the abovementioned transition. The factor $\Lambda(\varepsilon_1,\varepsilon_2)$ is responsible for nanoparticle's shape as well as for dielectric permittivities of nanoparticle and medium surrounding this nanoparticle. The theorem on a mean value of definite integral as well as from wave function normalization imply that $D_{ij}(a)=\Lambda(\varepsilon_1,\varepsilon_2)e\Psi^*_{[j]}(\vec{r}_1)\Psi_{[i]}(\vec{r}_1)\int_V\vec{r}\vec{\gamma}d\vec{r}=d_{ij}a$, where $\vec{r}_1$ - radius-vector of a point inside the nanoparticle, value of d_{ij} is directly related to a shape of nanoparticle. The linear dependence of $D_{ij}(a)$ upon a shape of nanoparticle is valid for any a smaller than a_c. Reaching this limit we must take into account polarization interaction between charge carriers and charge induced at the nanoparticle-medium interface [19]. On condition that nanoparticle's size exceeds a_c, the $D_{ij}(a)$ is defined by a size of localization area of dipole transitions states. In case of a transition to an area with continuous spectrum a matrix transition element will be proportional to an area of charge carrier's delocalization - a approx., i.e. as in case of bound states. Thus, considering dipole moments of transition towards quantum-size states $D_{ij}^q(a)$, exciton states $D_{ij}^{ex}(a)$ and continuous spectrum states $D_{ij}^c(a)$, we can write: $D_{ij}^q(a)=d_{ij}^q a$, $D_{ij}^{ex}(a)=d_{ij}^{ex}a$ and $D_{ij}^c(a)=d_{ij}^c a$. Frequencies of transitions to quantum size states $\omega_j=\frac{E_j}{\hbar}=\frac{\hbar\phi_j^2}{2m_{e,h}a^2}\chi=\frac{c_j}{a^2}$, where ϕ_j^2 - roots of Schrodinger equation for free electrons (electron holes) inside the potential hole, defined from the conditions upon the boundaries of the hole, m_e and m_h - effective masses of an electron and a hole correspondingly, χ - factor defined by a shape of a nanoparticle.

The expressions obtained allow us to determine $\sigma_a(\omega,a)$ and $\sigma_s(\omega,a)$ for the next charge carriers transitions: towards quantum size states

$$\sigma_a^q(\omega,a)=\frac{8\pi}{c\hbar}\sum_{i,j}\frac{d_{ij}^{q2}c_j\Gamma_j}{\left((\omega^2-\frac{c_j^2}{a^4})\right)^2+\omega^2\Gamma_j^2}\omega^2, \tag{8}$$

$$\sigma_s^q(\omega,a)=\frac{4P\omega^4}{c^4\hbar^2}\sum_{i,j}\sum_{i,k}\left\{\frac{d_{ij}^2d_{ik}^2c_jc_k\left[\left(c_j^2-a^4\omega^2\right)\left(c_k^2-a^4\omega^2\right)+a^8\Gamma_j\Gamma_k\omega^2\right]}{\left[\left(c_k^2-a^4\omega^2\right)^2+a^8\omega^2\Gamma_k^2\right]\left[\left(c_j^2-a^4\omega^2\right)^2+a^8\omega^2\Gamma_j^2\right]}a^8\right\}; \tag{9}$$

towards exciton states

$$\sigma_a^{ex}(\omega,a)=\frac{8\pi}{c\hbar}\sum_{i,j}\frac{d_{ij}^{ex2}a^2\omega_j\Gamma_j}{\left(\omega^2-\omega_j^2\right)^2+\omega^2\Gamma_j^2}\omega^2 \tag{10}$$

$$\sigma_s^{ex}(\omega,a)=\frac{4P\omega^4}{c^4\hbar^2}\sum_{i,j}\sum_{i,k}\left\{\frac{d_{ij}^2d_{ik}^2\omega_j\omega_k\left[\left(\omega_j^2-\omega^2\right)\left(\omega_k^2-\omega^2\right)+\omega^2\Gamma_j\Gamma_k\right]}{\left[\left(\omega_k^2-\omega^2\right)^2+\omega^2\Gamma_k^2\right]\left[\left(\omega_j^2-\omega^2\right)^2+\omega^2\Gamma_j^2\right]}a^4\right\}.$$

For transitions towards continuous spectrum with $\omega_j=\omega$

$$\sigma_a^c(\omega,a)=\frac{8\pi}{c\hbar}\sum_{i,j}\frac{d_{ij}^{c2}a\omega}{\Gamma_j},\quad \sigma_s^c(\omega,a)=\frac{4P\omega^4}{c^4\hbar^2}\sum_{i,j}\sum_{i,k}\left\{\frac{d_{ij}^2d_{ik}^2}{\Gamma_j\Gamma_k}a^4\right\}. \tag{11}$$

The expression for transmission spectrum T of a flat layer with a thickness of h ($h>>\lambda$) of a nanocomposite during a normal light incidence is

$$T(\omega,a,N)=\frac{\left(1-R^2\right)^2\exp(-\alpha h)}{1-R^2\exp(-2\alpha h)}, \tag{12}$$

where R - coefficient of light reflection from the layer's edge which is, as a rule, much smaller than 1 on experimental conditions. Extinction coefficient α is expressed by abovementioned cross-sections of absorption and scattering, a number N of nanoparticles per volume unit - $\alpha=N\left[\sigma_a(\omega,a)+\sigma_s(\omega,a)\right]+\alpha_m(\omega)$ and coefficient of laser light attenuation by matrix medium $\alpha_m(\omega)$. The ratio of cross-sections of absorption to scattering, on condition that the width of the j-th excited level is only slightly dependent of j, is $\frac{\sigma_s(\omega,a)}{\sigma_a(\omega,a)}=\frac{P\omega^3}{2\pi c^3\hbar\Gamma}\sum_{i,j}\left|D_{ij}(a)\right|^2\sim\omega^3a^2$. For nanoparticles with size within range from 10 to 100 nm in the frequency range from 10^{13} to 10^{16} Hz and when $\Gamma=10^9$ Hz, the ratio of the cross-sections doesn't exceed 10^{-7} , so the main factor which defines $T(\omega,a,N)$ is a light absorption by nanoparticles.

Fig. 5 shows the light transmission spectrum of Al_2O_3 nanoparticles ensemble obtained by using Hitachi U2010 spectrophotometer. Energy gap width of a volume sample of aluminium oxide is within range from 6 eV to 8.8 eV, its refraction index is equal to 1.7 approx. Through the use of this sample and a transparent liquid dielectric matrix a heterogenous liquid-phase nanocomposite (HLPN) was prepared. A liquid matrix was based upon the vacuum oil VM-4 (a colorless transparent and viscous liquid with refraction index equal to 1.4 approx.). The vacuum oil has linear optical properties in infrared and visible light. The nanoparticles interacted very weakly, it was possible due to a very low mass concentration of the nanoparticles in the HLPN (0.03% approx.).

The nanocomposite prepared was placed inside a parallelepiped cuvette made of a 160 μm optical glass, cuvette width was 20 mm along the light propagation direction. The reference sample was made of a similar cuvette with a pure vacuum oil. According to a common procedure we obtained the light transmission spectrum of Al_2O_3 nanoparticles ensemble– it was carried out by division of the nanocomposite transmission spectrum by transmission spectrum of the reference sample. Wavelength resolution was $\Delta\lambda$ = 0.5 nm. It follows from the results obtained, that there are two shallow absorption bands within spectral ranges 220-225 nm and 265-307 nm as well as relatively deep absorption band within spectral range 308-400 nm. The corresponding minimum values of transmission coefficient $T(\lambda = 237\text{nm}) = 0.425$, $T(\lambda = 287\text{nm}) = 0.445$ and $T(\lambda = 337\text{nm}) = 0.178$.

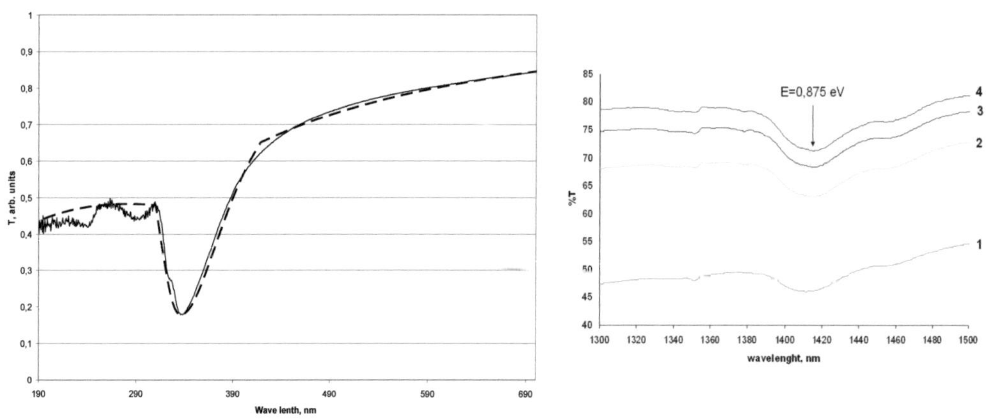

Fig. 5. The light transmission spectrum of Al_2O_3 nanoparticles ensemble. Wavelength resolution $\Delta\lambda$ = 0.5 nm. Mass concentration of nanoparticles in the HLPN is 0.03% approx. a) Infrared light transmission spectrum: experimental (solid line) and theoretical (dashed line) curves. Curves from 1 to 4 correspond to different nanoparticles concentration – from a maximum concentration (curve 1) to a minimum one (curve 2).

The $T(\lambda)$ plot in the vicinity of transmission spectrum minimum (308-400 nm) is of an asymmetrical shape. Long wavelength (in comparison with central wavelength λ = 337 nm) portion of absorption band curve is more sloping than that of a short wavelength, it counts in favours of exciton states existence in the nanoparticles under study. Photon energy which corresponds to the minimum transmission wavelength $\lambda = 337$ nm is equal to 3.7 eV. It is

considerably less than E_g energy gap width of a bulk crystal, but is approximately equal to the value of E_g = 3.6 eV for Al_2O_3 nanoparticles stated in the appropriate work (Mixeeva & Sidorov, 2004). Difference between these values and the values of E_g of a bulk crystal are attributable to a considerable change of electronic structure of Al_2O_3 crystals caused by small size of nanoparticles and a substantial number of near-surface defects of nanoparticle's crystal structure due to the complex structure of nanoparticles. In this case the value of $E_g = 3.7$ ev, and it will correspond with the energy of a transition from defect zone centre to conduction band. Photon energies associated with the other minima equal to 4.1eV and 5.2 eV correspond to transitions to quantum-size states and energy of 0.875 eV corresponds to a transition between quantum-size states with minimal energy (Fig. 4).
By using the expressions (8-11) the model of 5 the Al_2O_3 nanoparticles ensemble transmission spectrum which relates the transmission spectrum to the light frequency and a size of nanoparticles was developed. Depending upon the light frequency the expression for transmission spectrum has the next form

$$T(\omega,a) \approx \exp\left[-Nh\left(A_1^q\omega^{-2} + A_1^{ex}a^2\omega^{-2} + A_1^c a\omega\right)\right], \text{ when } \omega > \omega_j, \quad (13)$$

$$T(\omega,a) \approx \exp\left[-Nh\left(A_3^q a^8\omega^2 + A_3^{ex}a^2\omega^2 + A_3^c a\omega\right)\right], \text{ when } \omega < \omega_j. \quad (14)$$

In the vicinity of resonance frequency ω_j ($\lambda = 337$nm), introducing the quntity $\Delta\omega = \omega - \omega_j$, we obtain

$$T(\omega,a) \approx \exp\left[\left[-Nh\sum_{i,j}\left(\frac{d_{ij}^{q2}c_j\Gamma_j}{4\Delta\omega^2 + \Gamma_j^2} + \frac{d_{ij}^{ex2}\omega_j\Gamma_j a^2}{4\Delta\omega^2 + \Gamma_j^2} + A_2^c a\omega\right)\right)\right]\right], \quad (15)$$

where summation is carried out over the all state $|i\rangle$ to state $|j\rangle$ transitions.
It was supposed that the main mechanisms responsible for the transmission spectrum witnessed were the charge carriers transitions from defect energy zone, localized within forbidden band, to the exciton states ($\lambda >$ 337 nm), with their subsequent light-induced destruction and the transition to quantum-size states. In case of wavelength $\lambda <$ 337 nm the main mechanisms are the charge carriers transition from additional allowed energy bands to a conduction band including quantum-size states. The intersections of curves correspond to an experimentally obtained values of light transmission spectrum which were used to derive the set of equations in unknowns A and Γ.
As it follows from the Fig. 5 there is a good correspondence between experimental curve and the theoretical one which counts in favor of proposed mechanisms of generation of the light transmission spectrum reasoned by the electronic structure of dielectric nanoparticles. It should be noticed that the same transmission spectra can be observed in case of using many other semiconductor and dielectric nanoparticles (Mixeeva & Sidorov, 2004).

2. Basics of theory of optical nonlinearity of dielectric nanocomposites

Over the last years, experimental studies of nonlinear optical properties of nanocomposites [1-8] with small concentrations of wide band semiconductor and dielectric nanoparticles embedded into dielectric matrix have revealed, that these media, influenced by either of nanosecond pulses and continuous illumination of near infrared and visible light, demonstrate a low threshold (pulse regime $P_{thr} < 0.5\ \mathrm{nJ/cm^2}$) response. This response disappears when the light intensity becomes high, and the dependence of intensity of the output light (passed through the nanocomposite) on the input light (incident upon the nanocomposite) intensity $P_{out}(P_{in})$ becomes a linear one. Secondly, despite the broad energy gap of volume samples of nanoparticles, e.g. Al_2O_3 has this gap equal to 7.2 eV, the nonlinear response can be witnessed in near infrared and visible light ranges. Thirdly, this response exists when the light transmission spectrum of nanoparticles has broad light absorption bands which are absent in volume samples, and there exists such frequency ω_P of the light that is, being different from the central frequency of absorption band, responsible for change of a sign of nonlinear addition to the nanocomposite's refraction index when this frequency is overcome.

The theory of optical nonlinearity of effective refraction index of dielectric nanocomposites in the field of a weak laser radiation is based upon the next assumptions: the main reasons of low threshold nonlinearity of refraction index are: 1) a dipole electrical moment of a nanoparticle, induced by an external optical radiation and caused by the difference of charge carrier states populations which is distinct from the thermal one; 2) the direction of polarization vector of nanoparticles along the polarization vector of the external optical field. The most simple, from the point of view of describing media with nonlinear optical properties, is a «gas» of nanoparticles. Liquid and solid nanocomposites have a more complex structure but they can be prepared easier. Theoretical descriptions of refraction indices of these media have very much in common. Because of this, we consider a nanocomposite which consists of dielectric nanoparticles embedded into transparent liquid matrix made of isotropic dielectric material. The nanocomposite obtained is characterized by linear optical properties and a small coefficient of viscosity of nanoparticle. Let the amount N of particles per unit volume be small enough to neglect the interaction between nanoparticles. Let $\alpha = \{\alpha_{ij}\}$ -effective polarization tensor of nanoparticle in such matrix, so the components of electric dipole moment of the nanoparticle, which is induced by a monochromatic and linearly polarized electromagnetic field with frequency ω and vector of electric field intensity, are defined directly by the external field rather than by the local one. Let introduce two coordinate systems with the common centre. The first one is a coordinate system of principal axes of polarization tensor of an arbitrary nanoparticle $\{\alpha_1, \alpha_2, \alpha_3\}$ with $(\vec{n}_1, \vec{n}_2, \vec{n}_3)$ basis vectors, and the second one is a laboratory Cartesian coordinate system with $(\vec{n}_x, \vec{n}_y, \vec{n}_z)$ basis vectors. The $\{\alpha_1, \alpha_2, \alpha_3\}$ coordinate system defines polarization vector of the nanoparticle as $\vec{P} = \sum_j^3 \alpha_{jj}(\vec{n}_j\vec{E})\vec{n}_j$. If $\vec{E}$ vector is aligned with Z axis so the projection of the polarization vector upon this axis is equal to

$$P_z = \sum_j^3 \alpha_{jj} E(\vec{n}_j \vec{n}_z)^2 = \sum_j^3 \alpha_{jj} E \cos^2\theta_j, \qquad (16)$$

Where $\theta_1, \theta_2, \theta_3$ - angles between vector $\vec{E}$ and $\{\alpha_1, \alpha_2, \alpha_3\}$ axis correspondingly. In the lab coordinate system $P_z = \alpha_{zz} E$. Comparing this expression with the Eq.(16) we obtain

$\alpha_{zz} = (\alpha_{11}\cos^2\theta_1 + \alpha_{22}\cos^2\theta_2 + \alpha_{33}\cos^2\theta_3)$ .Polarizabilty tensor $\chi = \{\chi_{ij}\}$ of such medium to a first approximation can be assumed diagonal and polarization vector of voiume unit is $\vec{D} = \chi_{zz}\vec{E}$. In this expression $\chi_{zz} = N < \alpha_{zz} >$ where brackets mean averaging over angles of directions of nanoparticle's polarization vector. Making little manipulations and taking into account the expression $1 = \cos^2\theta_1 + \cos^2\theta_2 + \cos^2\theta_3$ one can obtain the next expression for - $\chi_{zz} = N(\alpha_0 + \Delta\alpha_1 Q_1 + \Delta\alpha_2 Q_2)$ where $Q_1 =< \mathrm{Cos}^2\theta_1 - \frac{1}{3} >$ and $Q_2 =< \mathrm{Cos}^2\theta_2 - \frac{1}{3} >$ actually are parameters of orientation order of nanoparticles in the external field, and $\alpha_0 = \frac{\alpha_{11} + \alpha_{22} + \alpha_{33}}{3}$, $\Delta\alpha_1 = \alpha_{11} - \alpha_{33}$, $\Delta\alpha_2 = \alpha_{22} - \alpha_{33}$.

Nanoparticle's reorientation in the external field is connected with an alteration of its energy $U = -\frac{1}{2}\mathrm{Re}(\vec{p}\cdot\vec{E}^*)$ of interaction with this field. It causes the angle distribution function and hence order parameters Q_1 and Q_2 to be dependent upon the field intensity, and, through the mediation of α_{ij} component, upon the light frequency. The values of these parameters lie within range from -1/3 to 2/3 independently of light frequency and its intensity. In case of uniform angular distribution these values are equal to zero. If a one of the Q parameters is equal to 2/3 it indicates that vector $\vec{p}$ of nanoparticle's dipole moment is directed along the one axis $(Q = \frac{2}{3})$ and is directed perpendicularly the other ones $(Q = -\frac{1}{3})$. In the frequency range where light transmission spectra of nanoparticles have broad absorption bands or sets of relatively narrow bands, the components of tensor α_{ij} should reach maximum values associated with charge carriers dipole state $|n\rangle$ to state $|g\rangle$ transitions. It is known that diagonal elements of the α_{ij} tensor within the abovementioned frequency range can be represented as (Shen 1983)

$$\alpha_{ii}(\omega) = \sum_{n<g}\sum_{g} \frac{|<n|er_i|g>|^2}{\hbar(\omega - \omega_{ng} + i\Gamma_{ng})}\Delta\rho_{ng}\,, \tag{17}$$

where summation is carried out over all allowed optical transitions of nanoparticle's charge carriers with transition frequency ω_{ng}, half-width of the transition line Γ_{ng} and component of dipole electric moment equal to $p_{ng}^i =< n|er_i|g >$. Population difference $\Delta\rho_{ng}$ between states $|n\rangle$ and $|g\rangle$ induced by a light is a function of the incident light intensity. To a double-particles system approximation [21] this function is

$$\Delta\rho_{ng}(I) = (1 - \sum_{n}\sum_{g} \frac{I/I_s}{(\omega - \omega_{ng})^2 + \Gamma_{ng}^2(1 + I/I_s)}\Gamma_{ng}^2)\Delta\rho_{ng}^0\,, \tag{18}$$

where $\Delta\rho_{ng}^0$ - equilibrium thermal population difference, I_s - saturation intensity associated with $\frac{\Delta\rho_0}{2}$ amount of charge carriers transitioned from state $|n\rangle$ to state $|g\rangle$.

Complex nonlinear refraction index $n(\omega, I)$ of a nanocomposite in the field of linearly polarized light, in view of Eq. (17) and (18), can be expressed as

$$n(\omega,I) \approx n_0 + \frac{2\pi\chi_{zz}(\omega,I)}{n_0} = n_0 + \sum_n \sum_g \Delta\rho_{ng}(\omega,I)\left[\frac{\omega-\omega_{ng}}{(\omega-\omega_{ng})^2+\Gamma_{ng}^2} + i\frac{\Gamma_{ng}}{(\omega-\omega_{ng})^2+\Gamma_{ng}^2}\right]A_{ng}(Q_1,Q_2), \quad (19)$$

where $A_{ng}(Q_1,Q_2) = \frac{2\pi N}{\hbar n_0}[\frac{1}{3}|\vec{p}_{ng}|^2 + Q_1(|p_{ng}{}^1|^2 - |p_{ng}{}^3|^2) + Q_2(|p_{ng}{}^2|^2 - |p_{ng}{}^3|^2)]$ is a monotonic rising function of I. Isotropic nanoparticles have a polarization vector always directed along the vector $\vec{E}$ and a $A_{ng}(Q_1,Q_2)$ of a maximum value, which is equal to $A_{ng}(Q_1,Q_2) = \frac{2\pi N}{\hbar n_0}|\vec{p}_{ng}|$. When the light intensity is equal to zero, due to a lack of own dipole moments of nanoparticles $A_{ng}(Q_1,Q_2)=0$ and the refraction index $n(\omega, I) = n_0$. The real part of the refraction index responsible for light refraction is

$$n'(\omega,I) = n'_0 + \sum_n \sum_g \left\{ [1-(\frac{(I/I_s)\Gamma_{ng}{}^2}{(\omega-\omega_{ng})^2+\Gamma_{ng}{}^2(1+I/I_s)})]\frac{A_{ng}(Q_1,Q_2)(\omega-\omega_{ng})}{(\omega-\omega_{ng})^2+\Gamma^2_{ng}}\Delta\rho^0_{ng} \right\}. \quad (20)$$

Electronic structure of dielectric nanoparticles is characterized by broad light absorption bands, absent within volume samples, a wide energy gap, a subzone of allowed energy (exciton, doped, etc.) of electrons within forbidden band and adjacent to the bottom of conduction band and broadened quantum-size levels (minizones). Allowance of the electronic structure of nanoparticles can be made by substituting in the Eq. (5) the integration within limits from $(\omega-\Delta\omega_1)$ to $(\omega+\Delta\omega_2)$ with state densities g_1 and g_2 correspondingly for the summation over $|g\rangle$ states. Here ω_n is the frequency of interzone transition from state $|n\rangle$ to a quantum-size zone, $\Delta\omega_1$ - the width of allowed energies subzone within forbidden band, $\Delta\omega_2$ - the width of subzone of quantum size levels which corresponds to $|g\rangle$ states. For the sake of simplicity of the expressions to obtain, let us assume the state densities and to be independent of frequency and $\Gamma_{ng} = \Gamma_n$. Going from summation to integration over frequencies in Eq. (20) we obtain

$$n'(\omega,I) = n'_0 + \frac{\hbar}{2}\sum_n A_n(Q_1,Q_2)\Delta\rho^0 \left[g_1 \ln\frac{(\omega-(\omega_n-\Delta\omega_1))^2+\Gamma^2{}_n(1+\frac{I}{I_s})}{(\omega-\omega_n)^2+\Gamma^2{}_n(1+\frac{I}{I_s})} + g_2 \ln\frac{(\omega-\omega_n)^2+\Gamma^2{}_n(1+\frac{I}{I_s})}{(\omega-(\omega_n+\Delta\omega_2))^2+\Gamma^2{}_n(1+\frac{I}{I_s})} \right]. \quad (21)$$

In case of nanoparticles a value of Γ_n can be much higher than Γ_{ng} as in case of molecules. This is supported by the fact of considerable broadening of absorption bands in nanoparticles in comparison with that of volume samples. Eq. (21) shows that nonlinear part of refraction index $n_n(I)$, being the function of external illumination intensity, is described by product of $A_n(Q_1,Q_2)$ and logarithms within square brackets. The behavior of these multipliers with increase of light intensity is the opposite one as it shown in Fig. 6.

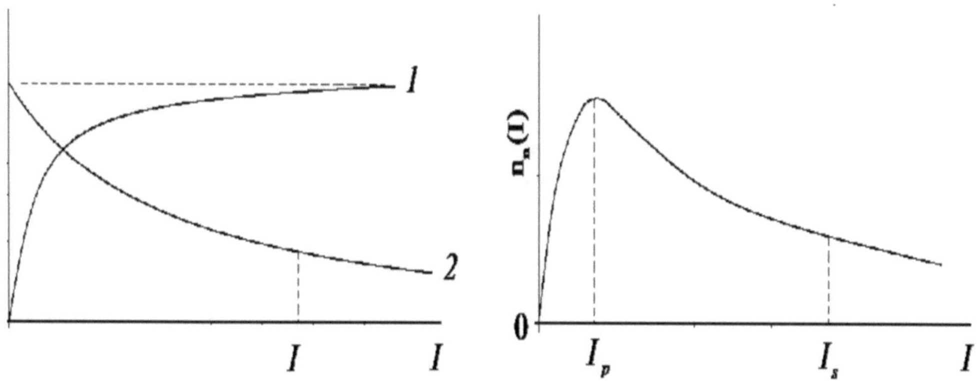

Fig. 6. The behavior of: a) curve 1 - value of $A_n(Q_1,Q_2)$ per unit, curve 2 value of $\Delta\rho_{ng}(I)$ per unit; b) nonlinear part of the refraction index.

In case of solid nanocomposites the value of $A_n(Q_1,Q_2)$ is close to zero when external illumination intensity is less than nonlinearity threshold, and the $A_n(Q_1,Q_2)$ is a constant when the intensity exceeds the threshold value. In particular, isotropic nanoparticles, e.g. of a spherical shape, have $A_n(Q_1,Q_2)$ of a maximum value. When the light intensity increases, the value of $A_n(Q_1,Q_2)$ rises up rapidly (Fig. 1a) reaching its maximum (corresponding with a full orientation of nanoparticles along the field) and the values of logarithms in the nonlinear part of the refraction index $n'(\omega,I)$ go down to zero (Fig. 6a). From the physical point of view it corresponds with population differenc $\Delta\rho_{ng}$ e equal to zero.

The competition between these two opposedly directed processes leads to a rather rapid increase

of nonlinear part of refraction index of a nanocomposite, when the external illumination intensity grows up, reaching up its maximum with a subsequent fall down to zero (Fig. 6b). During experiments it was discovered that a similar dependence of refraction index emerged as a selfrefraction of the laser beam, and a nonlinearity of dependence of the output light intensity on the input light intensity $P_{out}(P_{in})$. The next important result follows from the Eq. (21). If $(\omega-\omega_1)>0$ than the nonlinear part of refraction index is positive, if $(\omega-\omega_1)<0$ so such light frequency ω_p will exist, that will be responsible for the change of sign of nonlinear part when this frequency is exceeded. In particular, if $g_1=g_2$, so $\omega_p=\omega_n-\frac{\Delta\omega_1-\Delta\omega_2}{2}$ and if $\omega>\omega_p$ it has a negative value. This effect of a sign changing was observed during experiments described in works [3,4] when the frequencies corresponded to (2-2.3) eV were used. According to Eq. (21), the change of sign must occur when the difference $(\Delta\omega_1-\Delta\omega_2)$ corresponds to 3-3.4 eV. According to works (Mixeeva & Sidorov, 2004, Kul'chin(a) et.al.,2008, Kul'chin(b) et.al.,2009) , in this case ω_n is equal to 3.7 eV, and absorption band has a very smear longwavelength edge.

The other experimentally discovered fact follows from the Eq.(21). When $(\omega-\omega_n)>(\Delta\omega_1,\Delta\omega_2)$ the refraction index of the nanocomposite is linear. It means that low threshold nonlinear optical response is observed within the frequency range inside the absorption band.

We can refer to the only one work (Vituxnovski et. al. 2008) which contains indirect estimation of nonlinear addition to the refraction index n_n^{ex} based upon experimental results. Thus, for the purpose of comparison of theoretical values of n_n, we use the experimental results described in works (Vituxnovski et. al. 2008, Kul'chin(a) et.al.,2008) which allow us to make an estimation of n_n^{ex}. The abovementioned experiments shown a photoinduced non-thermal selfrefraction of the laser beam passing through a cuvette with a liquid nanocomposite as well as formation of interference rings of number M which is sufficient for the estimation of a nonlinear addition to a refraction index. The experimental results of the work (Kul'chin(a) et.al.,2008) demonstrate from 2 to 4 bright rings in dependence on the 633 nm light intensity. If the nanocomposite layer thickness ℓ is equal to 20 mm so, by using the - $M = \frac{|n_n^{ex}|\ell}{\lambda}$ it is easy to determine that $n_n^{ex} \approx 0.13 \cdot 10^{-3}$ for $M = 4$ and $n_n^{ex} \approx 0.65 \cdot 10^{-4}$ for $M = 2$. The nanoparticles concentration is $N \approx 10^{13}$ cm^{-3} and their size varies from 40 to 50 nm. The matrix refraction index $n_0 \approx 1.4$, the refraction index of Al_2O_3 nanoparticles is 1.7 approx. Photon energy corresponding to the maximum light absorption by nanoparticles is 3.7 eV (Fig. 5), and the energy quantum of laser radiation is equal to 1.96 eV. In this case a dipole transition of electron to an exciton state is most probable due to the quantum energy which is less than energy gap of the nanoparticle. Due to its small volume (10^{-17} cm^{-3}), the nanoparticle will host only a small quantity of such transitions. Thus, we use the Eq.(20) to estimate the value of n_n. For small intensities of laser radiation and a small amount of possible transitions we have $n_n(I) \sim \frac{A_{ng}(Q_1,Q_2)(\omega - \omega_{ng})}{(\omega - \omega_{ng})^2 + \Gamma^2_{ng}} \Delta\rho^0_{ng}$, where $\Delta\rho^0_{ng} \approx 1$. The dipole electric moment, induced by transition, for such nanoparticles is proportional to their size $p_{ng} \approx \Lambda ea$ where form-factor Λ ~ 1. The nanoparticles used in the experimental studies [3] were of wide absorption band; wavelength $\lambda = 633$ nm lies within absorption band, so the ω_{ng} is equal to central frequency corresponding to $\lambda = 335$ nm wavelength. And $(\omega - \omega_{ng})^2 \gg \Gamma^2_{ng}$. Using the expression for $A_{ng}(Q_1,Q_2)$ we obtain $n_n \approx 0.32 \cdot 10^{-4}$ for small light intensities (Q_1 and Q_2 are close to zero) and $n_n \approx 0.96 \cdot 10^{-4}$ with a full orientation of nanoparticles along the field. The nanocomposite described in work (Vituxnovski et. al. 2008) was based upon the immersion of quantum dotes $CdSe/ZnS$ in toluene with sizes of 1.9 nm and 2.6 nm with a concentration 10^{14} cm^{-3} approx. Similar to the previous case, laser radiation frequencies were inside the limits of the absorption bands and $(\omega - \omega_{ng}) \sim (10^{12} \div 10^{13})$ Hz. Theoretically estimated values of nonlinear addition to the refraction index were $n_n \in \left[8 \cdot 10^{-5} \div 1.3 \cdot 10^{-3}\right]$, the values defined experimentally, were described in the work and were $n_n^{ex} \in \left[8.75 \cdot 10^{-5} \div 3.5 \cdot 10^{-3}\right]$. Due to limitation of the paper we have closely considered only nonlinear refraction with no regard to light absorption and scattering, which can be investigated by using the methods discussed above.

A good qualitative and quantitative correspondence of theoretical model proposed with experimental data makes it possible to say that low threshold nonlinear optical response of dielectric nanostructures and its behavior ,with the growth of the external illumination intensity, are caused, mainly, by photoinduced dipole electric moment of nanoparticles and by dependence charge carriers population difference on light intensity as well as their orientation along the polarization vector of the external optical field.

We have to notice that we didn't take into account a mechanisms, additional to regarded ones, which consists of spatial redistribution of nanoparticles in the field of a laser beam. This mechanisms can be effective in the constant laser fields with their high irregularity along the beam profile and a high mobility of nanoparticles inside the matrix.

3. Nonlinear optical effects in the laser field. Experiment.

3.1 Experimental setup

For investigation of the influence of optical properties of the liquid phase matrix upon nonlinear optical properties of heterogeneous liquid-phase nanocomposites (HLPN) the experimental setup shown in Fig. 7 was used (Kul'chin(a) et.al.,2008). As a source of continuous illumination we used a 50 mW He-Ne laser (1) at the wavelength of 632 nm pumped by array of 40 mW continuous semiconductor YAG:Nd lasers at the wavelength of a second harmonic λ = 532 nm. The laser power was controlled by a mounted set of neutral light filters (2). In order to control the radiation power level a part of the laser beam was separated by a beam splitter (3) and was registered by a photodetector (9); a signal of the photodetector after passing through an ADC (3) was received by a PC (11). A 30 cm focal length lens (4) was used for focusing the laser beam on the plane-parallel cuvette (5) with the HLPN under study. The cuvette used was made of 160 μm optical glass and was of a length of 20 mm along the light propagation direction. An axially mounted diaphragm with the aperture of 1 mm diameter separated a central area of the laser beam to be registered by the photodetector (7). The registered signal was digitalized by ADC (7) and processed by the PC (11).

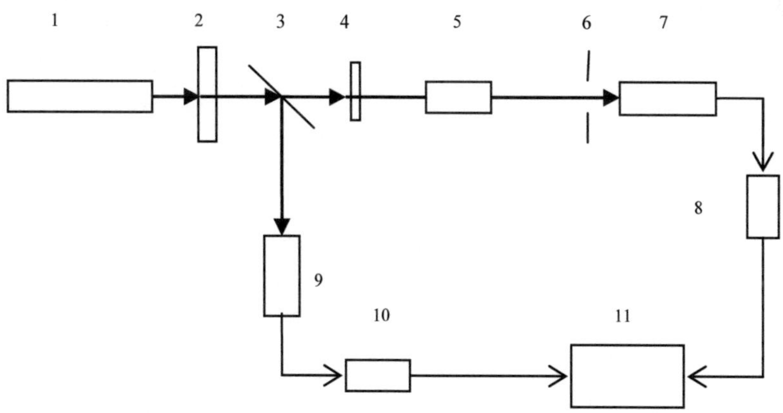

Fig. 7. The scheme of experimental setup

As liquid phase matrices of HLPN the next transparent media were used: vacuum oil VM-4 (refraction index n~1,4) and immersion oil (n~1,5) with linear and nonlinear optical properties in visible and near infrared ranges of light correspondingly. Nanoparticles in the

HLPN were of wide band dielectric Al_2O_3 (corundum) with n~ 1.7. A corundum crystal has a hexagonal symmetry and its energy gap width, according to different sources, varies from 6.26 to 8.8 eV [9,10]. The nanoparticles of a current shape were extracted from industrially prepared by the explosion method nanopowders by levigating in the acetone. Volume concentration of nanoparticles in HLPN was varied from 0.03% to 0.3%. Maximal average lateral dimension of nanoparticles was within range from 40 to 50 nm and the surface of Al_2O_3 nanoparticles isn't flat but of a complex multipeak pyramidal form (Fig. 3).

3.2 Single-frequency light-HLPN interaction.

To estimate the influence of nanoparticles upon the refraction index of HLPN experimentally, a liquid-phase matrix based upon the vacuum oil VM-4 was used. The oil, according to technical conditions, was of $dn/dT| \approx 0$ degree^{-1}. Fig.8 shows normalized dependencies of paraxial part of light radiation passed through HLPN (I_{out}/I_0) - upon input light intensity I_{in} obtained experimentally for wavelengths λ = 633 nm (curve 1) and λ = 532 nm (curve 2). Volume concentration of Al_2O_3 nanoparticles in HLPN was equal to 0.3%.

As it follows from the results obtained, the curves shown if Fig.8, starting from intensity values of 160 mW/mm^2 approx., display nonlinear portions. In case of the light with λ = 532 nm one can observe a limitation of the light passed through HLPN, for the light with λ = 633 nm we see a «bleaching» of the light transmission. When a liquid-phase matrix based upon immersion oil with $dn/dT| \approx -3 \cdot 10^{-3}$ degree^{-1} was used, a thermal mechanism of change in the refraction index was triggered. Fig.9 displays normalized dependencies of output light intensities (I_{out}/I_0) on the input light intensity obtained experimentally for both pure immersion oil (wavelength λ = 633 nm, curve 1) and HLPN (λ = 633 nm and λ = 532 nm) with volume concentration of nanoparticles equal to 0.3%.

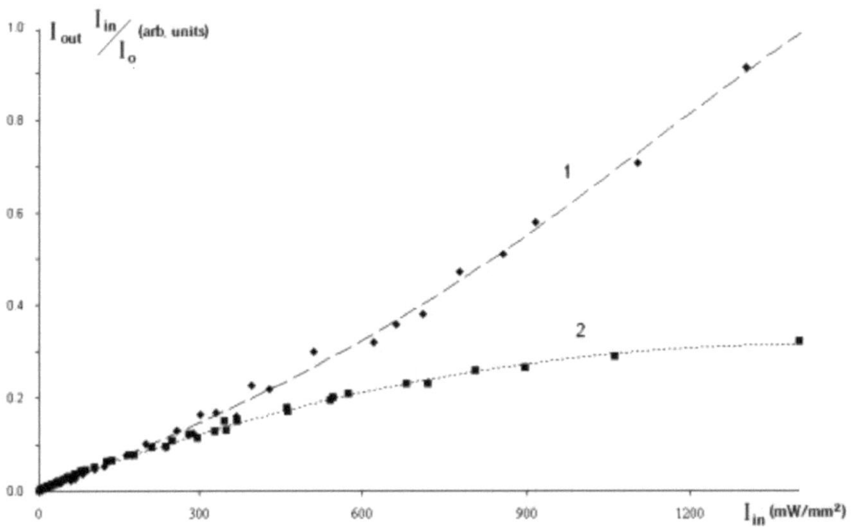

Fig. 8. Dependencies of intensity of the light passed through the VM-4 oil based HLPN on input light intensity (curve 1 - λ = 633 nm, curve 2 - λ = 532 nm, I_0= 1400 mW/mm^2).

As it follows from the Figure, in contrast to the previous case (Fig. 8), this nanocomposite demonstrates the light transmission nonlinearity when light intensities are of 40 mW/ mm² approx. The other distinguishing feature of this nanocomposite is the behavior of curves 1-3 in Fig.9 similar to that of curves 1-3 in Fig.8 for the both wavelengths at the first phase of I_{in} growth (up to 300 - 350 mW/mm²). However, as the input light intensity I_{in} (λ = 633 nm) grows up, further portion of HLPN light transmission curve oscillates, and for the light of 532 nm, after reaching a limitation level, one can observe a considerable decrease in HLPN light transmission. Thus, even the small addition (0.3% approx.) of the nanoparticles to a matrix can cause substantial alterations of optical properties of the medium. In particular, limitation level of light intensity for HLPN is proved to be 6 times smaller.

Oscillations of HLPN light transmission are caused by self-refraction and diffraction smearing of the light with λ = 633 nm along the beam cross-section after the light passing through the HLPN (Fig. 10).

As it follows from the experimental data, matrix material affects considerably upon optical properties of HLPN. Dependently on the wavelength, light transmission nonlinearity can be expressed as a light radiation limitation or as HLPN «bleaching».

The basics of the theory of nonlinear optical properties of dielectric nanocomposite are described in chapter 2. In case of immersion oil matrix one has to take into account the mechanism of formation of a thermal negative cylindrical lens.

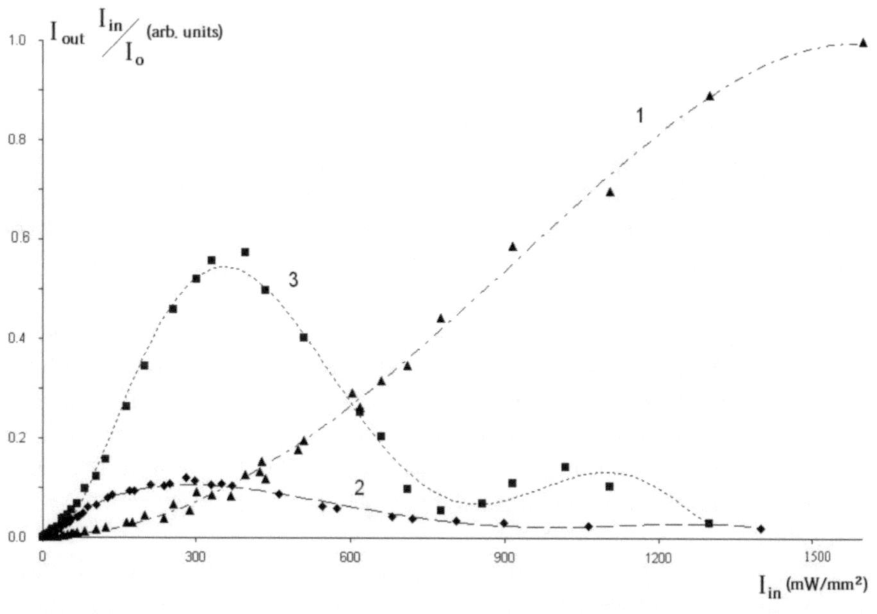

Fig. 9. Dependencies of intensities of light (λ = 633nm) passed through a pure immersion oil (curve 1) and through the HLPN based upon the immersion oil (curve 2 - λ = 532 nm, curve 3 - λ = 633 nm) upon the input light intensity, I_0 = 1600 mW/mm².

For an accurate description of mechanisms of formation of nonlinear optical properties observed, the knowledge of electronic structure of nanoparticles embedded into matrix is required. It is a complex issue with a solution rarely possible to obtain.

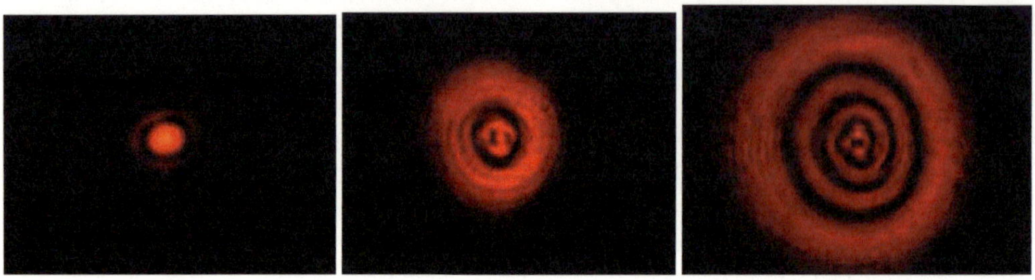

I = 0.1 W\mm2   I = 0.8 W\mm2   I = 3.12 W\mm2

Fig. 10. Light intensity distributions in the far field along the beam cross-section after the light passing through the HLPN based upon immersion oil at the different light intensities I_{in} with nanoparticles concentration equal to 0.03%.

However, it follows from the common considerations that a complex shape of a nanoparticle and a higher dielectric permittivity of the nanoparticle's material, in comparison with that of a matrix material, should cause a broadening of exciton band of charge carriers energy levels. When the nanoparticles are illuminated by a light with the wavelengths of 532 nm and 633 nm, light quanta energy isn't high enough to transmit electrons from the valence band to the conduction band ($E_{\lambda=532}$ = 2,34 eV; $E_{\lambda=633}$ = 1,97 eV). Therefore light quanta interact with electrons inside the impurity band. Quantum energy of a light with λ = 532 nm is higher than that of a photon with frequency of ω_p , which is responsible for change of sign of nonlinear addition to a negative one when being exceeded.

As a result of this processes, a nonlinear negative addition to the refraction index appears. It leads to the formation of a negative gradient lens within the HLPN. It explains the form of the curves 2 in Fig. 8 and 9, and in doing so limitation saturation, proceeding along with the growth of light intensity, is connected with a finite amount of electrons in impurity band as well as approach of photoinduced state population difference $\Delta\rho_{ng}$ of charge carriers to zero.

A frequency of the 633 nm light is lesser than ω_p and therefore nonlinear addition to refraction index has a positive sign. When the length of interaction between the light wave and the HLPN is high enough the increase in refraction index can cause a formation of a waveguide channel and additional concentration of laser radiation in the paraxial area. As a result, with the increase of light intensity one can see a «bleaching» of HLPN (curve 1 in Fig.8). Limitation of the «bleaching» for the light of 633 nm has the same reason as in case of the 532 nm light.

Utilizing a HLPN based upon immersion oil with $\partial n / \partial T < 0$, we observe an additional mechanism of formation of thermal negative cylindrical lens. It causes a faster saturation of the light limitation dependence and its consequent rapid decay with increase of input light intensity (curve 2 in Fig.9). When the light of 633 nm is passed through HLPN, the increase of refraction index at the initial phase of growth of I_{in} dominates over its thermal decay, and it leads to a medium «bleaching». Then, owing to a saturation, a process of antiwaveguide formation will be dominating, and it will cause less light transmission of a medium. The saturation process, being an obstacle for light selfrefraction in the medium, as well as subsequent increase in input light intensity, contributes largely to a formation of multilayer waveguides in the composite which it its turn causes a light intensity changes in the paraxial area. This effect allows us to explain the form of the curve 3 shown in Fig. 9.

3.3 Double-frequency interaction of collinear light laser beams in HLPN.

In order to investigate the process of collinear interaction between the laser beams of different wavelengths in heterogeneous liquid-phase nanocomposite the experimental setup shown in Fig. 11 was used (Kul'chin(b) et.al.,2008. We chose a continuous illumination source based on 50 mW He-Ne laser (1) at the wavelength $\lambda = 632$ nm pumped by array of continuous semiconductor 40 mW YAG:Nd lasers at the wavelength of a second harmonic $\lambda = 532$ nm. The laser power was controlled by a mounted set of neutral light filters (3) and (6). In order to control a radiation power level a part of the laser beam was separated by beam splitters (4) and (7) and was registered by photodetectors (5) and (8); signals from the photodetectors after digitalizing with ADC (16) and (17) were processed by the PC (18). Laser beams (1) and (2) were combined collinearly by using a beam splitter (9). A lens (10) with a high focal length (F = 30 cm) was used for focusing the laser beam upon the cuvette (11) with HLPN.

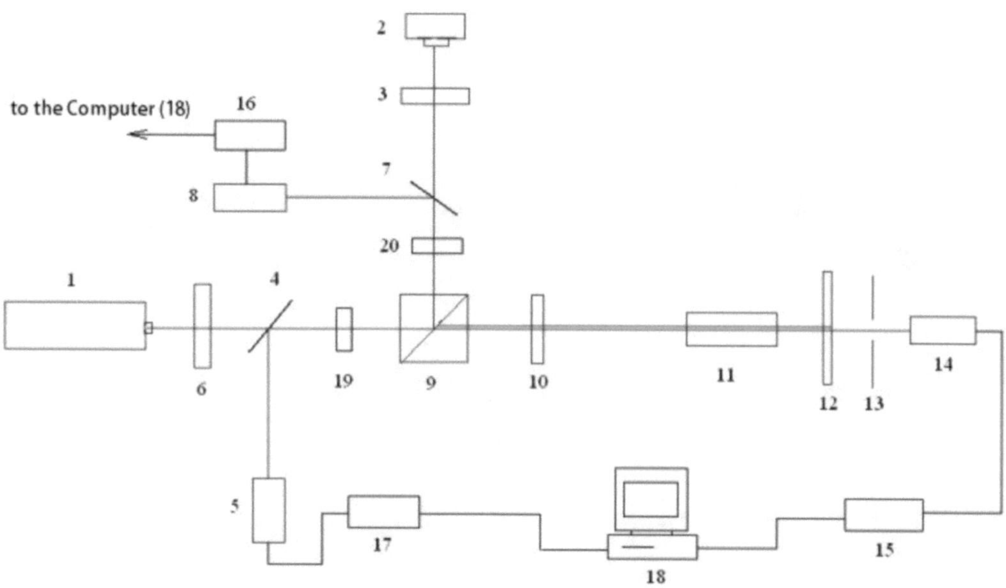

Fig. 11. The scheme of experimental setup

Laser beam diameter in the waist area was of 0.6 mm approx. The plate-parallel cuvette used in work was made of optical glass and was of a length of 20 mm along the light propagation direction. Narrow-band interferometric filter (12) was used to separate an illumination of a current wavelength. Point diaphragm (13) with a 1 mm diameter aperture filtered paraxial area of a laser beam. The passed light was registered by CCD matrix (14) and the light signal after ADC (15) was processed by a PC (18). Mechanical shutters (19) and (20) were used to control a light input into a sample cuvette. During experimental studies liquid phase matrices of HLPN based upon the immersion oil were used. The nanoparticles in HLPN were of wide band dielectric Al_2O_3 with an average size of 40 - 50 nm approx. Volume concentration of nanoparticles in HLPN was equal to 0.3%. The experiment was carried out to investigate a dependence of HLPN output light intensity in the paraxial area $I_+(I_-)$ on the intensity of input light with wavelengths of 532 nm and 633 nm in case of their collinear co-propagation.

The measurements of $I_+(I_-)$ were carried out after the extinction of transient processes and a stationary regime establishment. The period of transients occur during activation of the second wavelength (λ = 532 nm or 633 nm) light was up to 2 - 2.5 sec. Fig. 12 shows experimental dependencies of light transmission (λ = 633 nm) at the collinear (control) laser beam (λ = 532 nm) power equal to: 2 mW - curve 1, 5 mW - curve 2, 10 mW - curve 3. Fig. 13 shows experimental dependencies of light (λ = 532 nm) transmission at the power of collinear (control) laser beam (λ = 633 nm) equal to: 8 mW - curve 1, 4 mW - curve 2, 1 mW - curve 3.

It follows from the measurement results (Fig.12), that during double-frequency collinear propagation of the light in the course of input light (λ = 632 nm) increase the medium light transmission represents itself amplitude-decaying oscillations. In this case maximum values of light transmissions and intensities, associated with these maxima, decrease when the light (λ = 532 nm) intensity grows up.

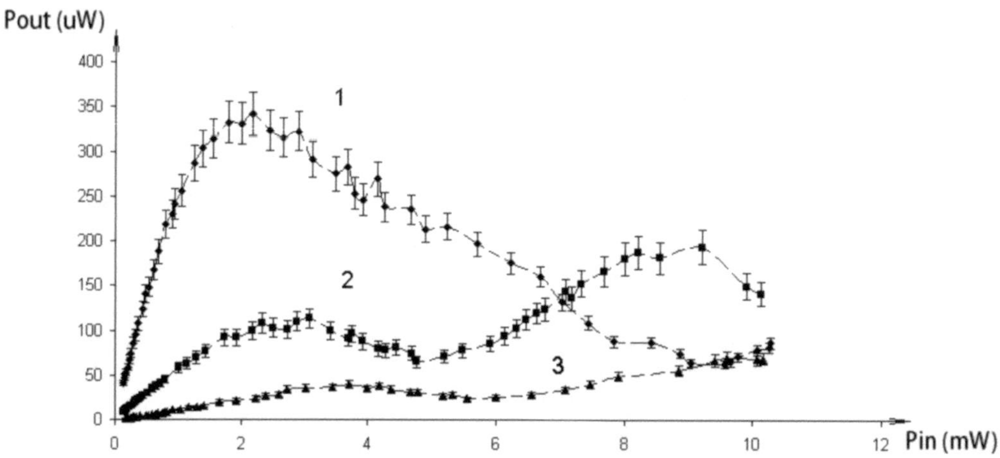

Fig. 12. The dependence of laser output power (λ = 633 nm) in the paraxial field on the input power at the different values of the power: curve 1 - P_{green} = 2mW, curve 2 - P_{green} = 5mW, curve 3 - P_{green} = 10mW.

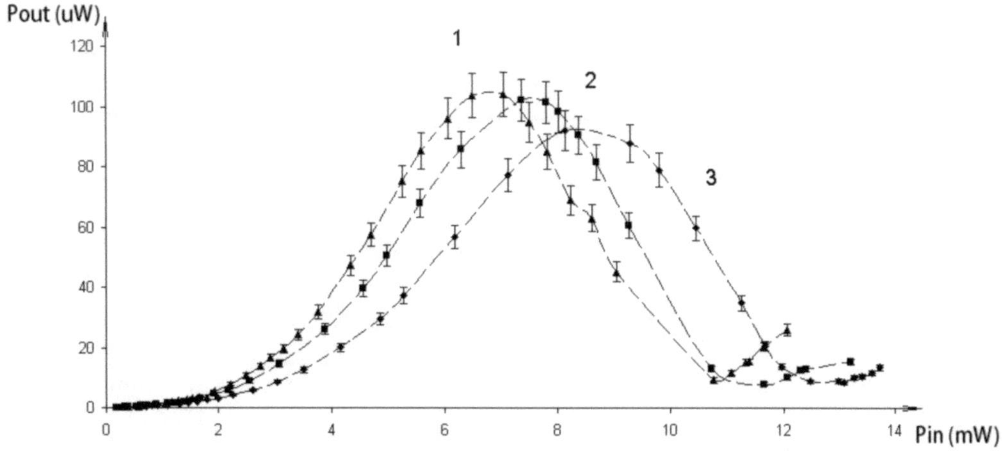

Fig. 13. The dependence of output light (λ = 532 nm) intensity at paraxial area of the beam on the input light intensity (λ = 532 nm) at the different values of the power: curve 1 - P_{red} = 8 mW, curve 2 - P_{red} = 4 mW, curve 3 - P_{red} = 1 mW.

The dependencies of light (λ = 532 nm) transmission during a co- propagation of the light (λ = 633 nm), obtained at the power range less than 6 mW, don't display such oscillations as it shown in Fig. 13. In both cases one can observe deviation from the linear dependence $I_+(I_-)$ witnessed at small power of input light (λ = 633 nm) equal to 2 mW (Fig. 12). For the 532 nm light these values depend upon control light intensity and lie within range from 0.2 mW to 2 mW (Fig. 13).

In case of interaction between collinear laser beams with different wavelengths one can witness interference between induced by these beams nonlinear mechanisms of modulation of optical properties of HLPN. In particular, for the 532 nm laser beam which interacts with the 633 nm laser beam in the HLPN on the conditions of frequency dispersion a nonlinear addition n_1 for the green light can be less than that for the red light; that, being combined with a thermal heating of the medium, cause less power of the green light passed through HLPN at smaller values of input light power. Curves 2-3 shown in Fig. 13 demonstrate this effect.

For the light with λ = 633 nm collinear interaction with the laser beam of 532 nm wavelength enhances the competition between the two processes: negative and positive lens formations. Due to these processes the maximum transmission of 633 nm light passing through HLPN is decreased as the 532 nm light intensity grows up. It is proved with curves 2 and 3 shown in Fig. 12.

As it follows from the results obtained (Fig. 12 and 13), when the light with wavelength of 532 nm is controlled by the light with wavelength of 633 nm, the size of modulation for collinear beams - the modulated beam and the control one, of current intensities can reach 10 dB. When the 633 nm light is controlled by the 532 nm light, the attenuation value can reach 5 dB approx.

4. Conclusions

The results of latest studies of nonlinear optical effects in dielectric nano-phase composites based upon dielectric nanoparticles show that, as a result of interaction between laser radiation and a nanoparticle-matrix complex, one can observe a low-threshold optical nonlinearity which depends upon a liquid-phase matrix material and a concentration of nanoparticles in HLPN.

The observed nonlinearity of the optical properties is caused by a great amount of additional energy levels and allowed energy levels within nanoparticle's forbidden band which are connected with a complex geometrical shape of nanoparticles and with their small sizes.

Being dependent of wavelength of a laser radiation, the light transmission nonlinearity can cause

Depending upon the wavelength of laser radiation, the light transmission nonlinearity can be evident as a light limitation as well as «bleaching» within a limited range of small light intensities, due to a self-refraction or a nonlinear absorption of the light.

The main reasons of this low-threshold optical nonlinearity are: a photo-induced modulation of a medium refraction index during a generation of nonequilibrium charge carriers and a medium polarization caused by laser excitation of charge carriers states, and the effect of light defocusing which, being weakly dependent of the laser radiation wavelength, depends upon liquid-phase matrix material and is reasoned by a dependence of the heterogeneous medium refraction index on the temperature.

The properties of low-threshold nonlinear optical response as well as its behavior with an increase of the external illumination are reasoned, mostly, by value of the photo-induced dipole electric moments of nanoparticles, dependence of population difference of charge carriers states on the light intensity, and the nanoparticles orientations along the external light field polarization vector. The same mechanisms are responsible for effective nonlinear interaction between collinear laser beams of different wavelengths within the HLPN.

The optical nonlinearity, observed in dielectric nanocomposites, allows one to control a nanocomposite light transmission and, correspondingly, to modulate parameters of the light passing through. As a result, one has the way to creation of optical radiation control devices based upon the self-refraction processes, light limitation or bleaching, as well as the collinear nonlinear optical interaction between the laser beams of different wavelengths within the HLPN.

The experiments conducted feature an ability to optimize the light control process by a proper selection of energetic and spectral properties of the controlling laser beam. It opens the way to development of both optically controlled filters for lightwave systems and modulators for input devices of optical processors.

5. References

V.P. Dzyuba, A.E. Krasnok, Yu.N. Kulchin. (2010). Nonlinear refractive index of the dielectric nanocomposites in weak optical fields. *Technical Physics Letters (Rus)*, Vol.36, No 21, pp. 1-9, ISSN: 0320-0116

V.P. Dzyuba, Yu.N. Kulchin. The Variation of the Nanopaticles' Shape and Quantum - Dimensional States of the Quantum Dots or Nanoparticles. *Pacific Science Review 2010*, V.12, No 1, pp. 102-104, ISSN: 1229-5450

V.P. Dzyuba, A.E. Krasnok, Yu.N. Kulchin, I.V. Dzyuba. Nonlinear Transmission of Light by Dielectric Nanocomposites. *Pacific Science Review* 2010, V.12, No 1, pp. 106-108, ISSN: 1229-5450

A.V. Federov. (2005). *Optics of nanostructure*, Nedra, ISBN: 5-94089-059-8, Saint Petersburg

Litty Irimpan, Bindu Krishnan, V.P.N. Nampoori, P. Radhakrishnan. (2008) Nonlinear optical characteristics of nanocomposites of ZnO–TiO_2–SiO_2. *Optical Materials*, Vol.31, No2, pp.361-365, ISSN: 0925-3467

Bindu Krishnana, Litty Irimpana, V.P.N. Nampoora, V. Kumarc. (2008). Synthesis and nonlinear optical studies of nano ZnO colloids. *Physica E*, Vol.40, No8, pp. 2787–2790, ISSN: 1386-9477

Yu.N. Kulchin, A.V. Scherbakov, V.P. Dzyuba, S.S. Voznesenskiy, G.T. Mikaelyan. (2008). Nonlinear-optical properties of heterogeneous liquid nanophase composites based on high-energy-gap Al_2O_3 nanoparticles. *Quantum Electronics*, Vol. 38 , No2, pp. 154 - 158, ISSN: 1063-7818

Yu. N. Kulchin, A. V. Scherbakov, V. P. Dzyuba, S. S. Voznesenskiy. (2009). Interaction of Collinear Light Beams with Different Wavelengths in a heterogeneous liquid-phase nanocomposite. *Technical Physics Letters (Rus)*, Vol.35, No. 7, pp. 640–642, ISSN: 0320-0116

Yu. N. Kulchin, V.P. Dzyuba, A.V. Scherbakov, S.S. Voznesenskiy. (2009). Nonlinear optical interaction of radiation with heterogeneous liquid-phase nanocomposites In: Nonlinear waves (2008), academician Gaponov-Grehov, pp. 361-382, ISBN: 978-5—8048-0069-8, Nizhny Novgorod

Yu.N. Kulchin, V.P. Dzyuba, A.V. Scherbakov. (2009). Optical Transmittance Spectra of Insulator nanoparticles in bulk heterocomposites. *Semiconductors (Rus)*, Vol.43, No. 3, pp. 331–339, ISSN:0015-3222

L.D. Landau, L.D. Lifshitz. (2002). *Quantum Mechanics*, Fizmatlit, ISBN: 5-9221-0057-2, Moscow

O.P. Miheev, A.I. Sidorov. (2004). Optical nonlinearity of nanoparticles of wide-gap semiconductors and insulators in visible and near infrared spectral region. *Technical Physics (Rus)*, Vol.74, No. 6, pp.77-82, ISSN: 0044-4642

Noriyuki Miyata, Masakazu Ichikawa. (2004). Scanning-probe-induced defects in thin SiO2 film on Si: Comparison with Si clusters. *Phys. Rev.* B. 2004, V.70 No7, pp. 07306-07340

S.I. Pokutny. (2006). Absorption and scattering of light by one-particle states of charge carriers in semiconductor quantum dots. *Semiconductors (Rus)*, Vol.40, No. 2, pp. 223-228, ISSN: 0015-3222

A.I. Ryasnyanskiy, B. Palpant, S. Debrus, U. Pal, A.L. Stepanov. (2009). Nonlinear Optical Properties of Gold Nanoparticles Dispersed in Different Optically Transparent Matrices. *Physics of the Solid State*, Vol.51, No. 1, pp. 55–60, ISSN: 1063-7834

Y.R. Shen. (1983). *The Principles Of Nonlinear Optics,*. John Wiley & Sons, ISBN: 5-02-014043-0

A.G. Vituhnovsky, A.A. Isaev, V.S. Lebedev. (2008). Light-induced nonlinearity of CdSe/ZNs quantum dots with millisecond relaxation tine. *Russian nanotechnologies,* 2008. Vol. 3, No. 11-12, pp. 110-117, ISSN: 1992-7223

13

Polymeric Nanoclay Composites

Hamid Dalir, Rouhollah D. Farahani,
Martin Lévesque and Daniel Therriault
*École Polytechnique de Montréal,
Canada*

1. Introduction

Traditionally, polymeric materials have been filled with synthetic or natural inorganic compounds in order to improve their properties, or simply to reduce cost. Conventional fillers are materials in the form of particles (e.g. calcium carbonate), fibers (e.g. glass fibers) or plate-shaped particles (e.g. mica). However, although conventionally filled or reinforced polymeric materials are widely used in various fields, it is often reported that the addition of these fillers imparts drawbacks to the resulting materials, such as weight increase, brittleness and opacity (Alexandre & Dubois, 2000; Fischer, 2003; Lagaly, 1999; Giannelis, 1996; Varlot et al., 2001). Nanocomposites, on the other hand, are a new class of composites, for which at least one dimension of the dispersed particles is in the nanometer range. Depending on how many dimensions are in the nanometer range, one can distinguish isodimensional nanoparticles when the three dimensions are on the order of nanometers, nanotubes or whiskers when two dimensions are on the nanometer scale and the third is larger, thus forming an elongated structure, and, finally, layered crystals or clays, present in the form of sheets of one to a few nanometers thick and hundreds to thousands nanometers in extent (Alexandre & Dubois, 2000; Fischer, 2003; Lagaly, 1999; Giannelis, 1996). Among all the potential nanocomposite precursors, those based on clay and layered silicates have been most widely investigated, probably because the starting clay materials are easily available and because their intercalation chemistry has been studied for a long time (Gorrasi et al., 2002).

Polymer-layered silicate nanocomposites, which are the subject of the present contribution, are prepared by incorporating finely dispersed layered silicate materials in a polymer matrix (Fischer, 2003). However, the nanolayers are not easily dispersed in most polymers due to their preferred face to face stacking in agglomerated tactoids. Dispersion of the tactoids into discrete monolayers is further hindered by the intrinsic incompatibility of hydrophilic layered silicates and hydrophobic engineering plastics. Therefore, layered silicates first need to be organically modified to produce polymer-compatible clay (organoclay). In fact, it has been well-demonstrated that the replacement of the inorganic exchange cations in the cavities or "galleries" of the native clay silicate structure by alkylammonium surfactants can compatibilize the surface chemistry of the clay and a hydrophobic polymer matrix (LeBaron et al., 1999).

Thereafter, different approaches can be applied to incorporate the ion-exchanged layered silicates in polymer hosts by in situ polymerization, solution intercalation or simple melt mixing. In any case, nanoparticles are added to the matrix or matrix precursors as 1-100 μm powders, containing associated nanoparticles. Engineering the correct interfacial chemistry between nanoparticles and the polymer host, as described previously, is critical but not sufficient to transform the micron-scale compositional heterogeneity of the initial powder into nanoscale homogenization of nanoparticles within a polymeric nanocomposite (Vaia & Wagner, 2004). Therefore, appropriate conditions have to be established during the nanocomposite preparation stage.

The resulting polymer-layered silicates hybrids possess unique properties - typically not shared by their more conventional microscopic counterparts - which are attributed to their nanometer size features and the extraordinarily high surface area of the dispersed clay (Alexandre & Dubois, 2000; Fischer, 2003; Lagaly, 1999; Giannelis, 1996). In fact, it is well established that dramatic improvements in physical properties, such as tensile strength and modulus, heat distortion temperature (HDT) and gas permeability, can be achieved by adding just a small fraction of clay to a polymer matrix, without impairing the optical homogeneity of the material. Most notable are the unexpected properties obtained from the addition of stiff filler to a polymer matrix, e.g. the often reported retention (or even improvement) of the impact strength. Since the weight fraction of the inorganic additive is typically below 10%, the materials are also lighter than most conventional composites (Fischer, 2003; Ginzburg et al., 2000; Osman et al., 2004; Balazs et al., 1999; Lincoln et al., 2001). These unique properties make the nanocomposites ideal materials for products ranging from high-barrier packaging for food and electronics to strong, heat-resistant automotive components (Balazs et al., 1999). Additionally, polymer-layered silicate nanocomposites have been proposed as model systems to examine polymer structure and dynamics in confined environments (Lincoln et al., 2001; Vaia & Giannelis, 2001).

However, despite the recent progress in polymer nanocomposite technology, there are many fundamental questions that have not been answered. For example, how do changes in polymer crystalline structure induced by the clay affect overall composite properties? How does one tailor organoclay chemistry to achieve high degrees of exfoliation reproducibility for a given polymer system? How do process parameters and fabrication affect composite properties? Further research is needed that addresses such issues (Fornes et al., 2001). The objective of this work is to review recent scientific and technological advances in the field of polymer-layered silicate nanocomposite materials and to develop a better understanding of how superior nanocomposites are formed.

2. Nanoclay

2.1 Geometry and structure

Layered silicates used in the synthesis of nanocomposites are natural or synthetic minerals, consisting of very thin layers that are usually bound together with counter-ions. Their basic building blocks are tetrahedral sheets in which silicon is surrounded by four oxygen atoms, and octahedral sheets in which a metal like aluminum is surrounded by eight oxygen atoms. Therefore, in 1:1 layered structures (e.g. in kaolinite) a tetrahedral sheet is fused with an octahedral sheet, whereby the oxygen atoms are shared (Miranda & Coles, 2003).

On the other hand, the crystal lattice of 2:1 layered silicates (or 2:1 phyllosilicates), consists of two-dimensional layers where a central octahedral sheet of alumina is fused to two external silica tetrahedra by the tip, so that the oxygen ions of the octahedral sheet also belong to the tetrahedral sheets, as shown in Fig. 1. The layer thickness is around 1 nm and the lateral dimensions may vary from 300 Å to several microns, and even larger, depending

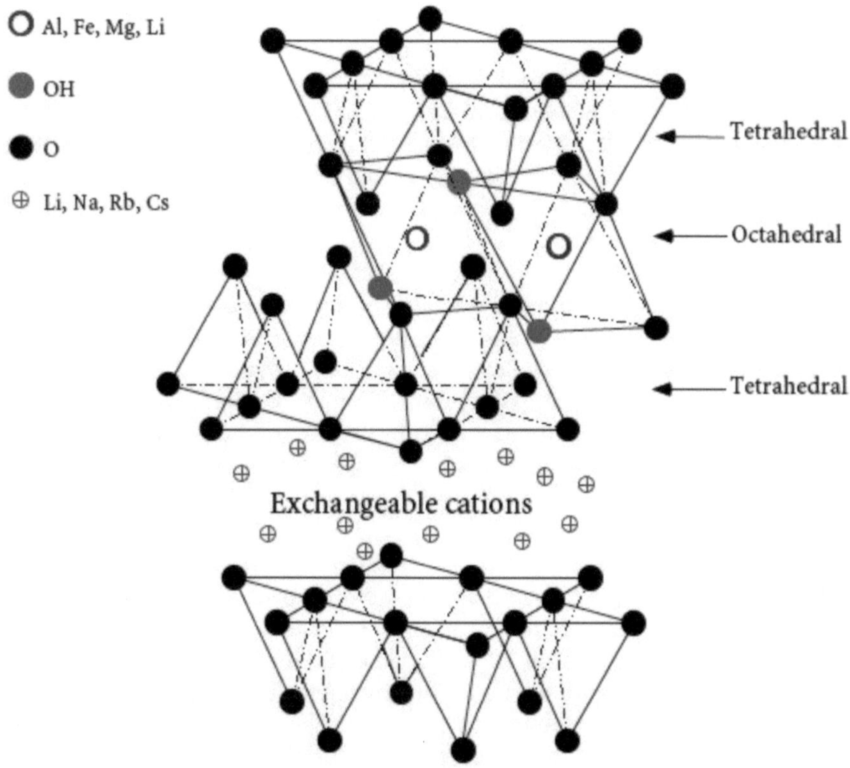

Fig. 1. The structure of a 2:1 layered silicate (Beyer et al., 2002). Reproduced from Beyer by permission of Elsevier Science Ltd., UK.

on the particulate silicate, the source of the clay and the method of preparation (e.g. clays prepared by milling typically have lateral platelet dimensions of approximately 0.1-1.0 μm). Therefore, the aspect ratio of these layers (ratio length/thickness) is particularly high, with values greater than 1000 (Beyer et al., 2002; McNally et al., 2003; Solomon et al., 2001).

Analysis of layered silicates has shown that there are several levels of organization within the clay minerals. The smallest particles, primary particles, are on the order of 10 nm and are composed of stacks of parallel lamellae. Micro-aggregates are formed by lateral joining of several primary particles, and aggregates are composed of several primary particles and micro-aggregates (Ishida et al., 2000).

2.2 Surface modification as a compatibilizer

Since, in their pristine state layered silicates are only miscible with hydrophilic polymers, such as poly(ethylene oxide) and poly(vinyl alcohol), in order to render them miscible with other polymers, one must exchange the alkali counter-ions with a cationic-organic surfactant. Alkylammonium ions are mostly used, although other "onium" salts can be used, such as sulfonium and phosphonium (Manias et al., 2001; Zanetti et al., 2000). This can be readily achieved through ion-exchange reactions that render the clay organophilic (Kornmann et al., 2001). In order to obtain the exchange of the onium ions with the cations in the galleries, water swelling of the silicate is needed. For this reason alkalications are preferred in the galleries because 2-valent and higher valent cations prevent swelling by

water. Indeed, the hydrate formation of monovalent intergallery cations is the driving force for water swelling. Natural clays may contain divalent cations such as calcium and require exchange procedures with sodium prior to further treatment with onium salts (Zanetti et al., 2000). The alkali cations, as they are not structural, can be easily replaced by other positively charged atoms or molecules, and thus are called exchangeable cations (Xie et al., 2001).
The organic cations lower the surface energy of the silicate surface and improve wetting with the polymer matrix (Giannelis, 1996; Kornmann et al., 2001). Moreover, the long organic chains of such surfactants, with positively charged ends, are tethered to the surface of the negatively charged silicate layers, resulting in an increase of the gallery height (Kim et al., 2001). It then becomes possible for organic species (i.e. polymers or prepolymers) to diffuse between the layers and eventually separate them (Kornmann et al., 2001; Zerda et al., 2001). Sometimes, the alkylammonium cations may even provide functional groups that can react with the polymer or initiate polymerization of monomers. The microchemical environment in the galleries is, therefore, appropriate to the intercalation of polymer molecules (Huang et al., 2001). Conclusively, the surface modification both increases the basal spacing of clays and serves as a compatibilizer between the hydrophilic clay and the hydrophobic polymer (Zerda et al., 2001).
There are two particular characteristics of layered silicates that are exploited in polymer-layered silicate nanocomposites. The first is the ability of the silicate particles to disperse into individual layers. Since dispersing a layered silicate can be pictured like opening a book, an aspect ratio as high as 1000 for fully dispersed individual layers can be obtained (contrast that to an aspect ratio of about 10 for undispersed or poorly dispersed particles). The second characteristic is the ability to fine-tune their surface chemistry through ion exchange reactions with organic and inorganic cations. These two characteristics are, of course, interrelated since the degree of dispersion in a given matrix that, in turn, determines aspect ratio, depends on the interlayer cation (Giannelis, 1996; Ishida et al., 2000).

3. Nanocomposite

3.1 Structural phases

Any physical mixture of a polymer and silicate (or inorganic material in general) does not necessarily form a nanocomposite. The situation is analogous to polymer blends. In most cases, separation into discrete phases normally takes place. In immiscible systems, the poor physical attraction between the organic and the inorganic components leads to relatively poor mechanical properties. Furthermore, particle agglomeration tends to reduce strength and produce weaker materials (Giannelis, 1996). Thus, when the polymer is unable to intercalate between the silicate sheets, a phase-separated composite is obtained, whose properties are in the same range as for traditional microcomposites (Alexandre & Dubois, 2000; Beyer et al., 2002).
Beyond this traditional class of polymer-filler composites, two types of nanocomposites can be obtained, depending on the preparation method and the nature of the components used, including polymer matrix, layered silicate and organic cation (Alexandre & Dubois, 2000; Beyer et al., 2002). These two types of polymer-layered silicate nanocomposites are depicted in Fig. 2 (McGlashan et al., 2003).
Intercalated structures are formed when a single (or sometimes more) extended polymer chain is intercalated between the silicate layers. The result is a well ordered multilayer structure of alternating polymeric and inorganic layers, with a repeat distance between

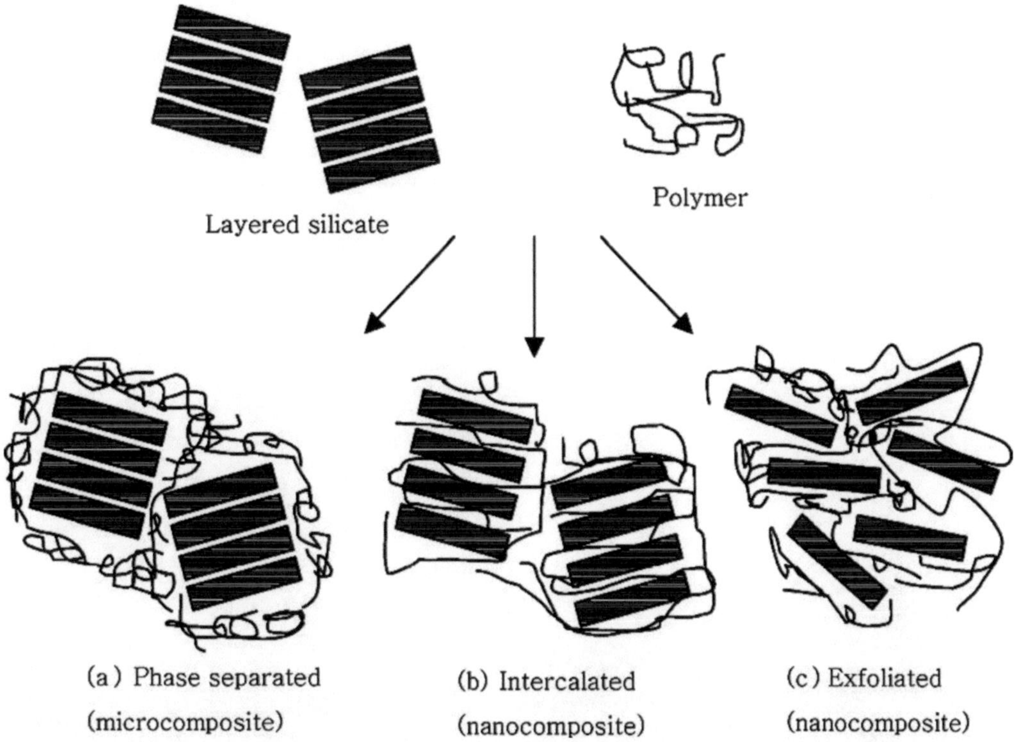

Fig. 2. Types of composite structure of polymer-layered silicate clay materials (McGlashan et al., 2003). Reproduced from McGlashan et al., by permission of John Wiley & Sons, Inc., US.

them. Intercalation causes less than 20-30 Å separation between the platelets (Beyer et al., 2002; Kim et al., 2001; Dennis et al., 2001).

On the other hand, exfoliated or delaminated structures are obtained when the clay layers are well separated from one another and individually dispersed in the continuous polymer matrix (Kim et al., 2001). In this case, the polymer separates the clay platelets by 80-100 Å or more (Dennis et al., 2001).

It is not easy to achieve complete exfoliation of clays and, indeed with few exceptions, the majority of the polymer nanocomposites reported in the literature were found to have intercalated or mixed intercalated-exfoliated nanostructures. This is because the silicate layers are highly anisotropic, with lateral dimensions ranging from 100 to 1000 nm, and even when separated by large distances (i.e. when delaminated) cannot be placed completely randomly in the sea of polymer. Furthermore, the majority of the polymer chains in the hybrids are tethered to the surface of the silicate layers. Thus, it can be expected that there are domains in these materials, even above the melting temperature of the constituent polymers, wherein some long-range order is preserved and the silicate layers are oriented in some preferred direction.

3.2 Morphological characterization

Two complementary techniques are generally used to characterize the structures of nanocomposites: X-ray differaction (XRD) and transmission electron microscopy (TEM) (Alexandre & Dubois, 2000; Huang et al., 2001; Porter et al., 2003; Ma et al., 2003).

Due to its ease of use and availability, XRD is most commonly used to probe the nanocomposite structure and occasionally to study the kinetics of polymer melt intercalation (Porter et al., 2003). This technique allows the determination of the spaces between structural layers of the silicate utilizing Bragg's law: $2d \sin\theta = n\lambda$, where λ corresponds to the wave length of the X-ray radiation used in the diffraction experiment, d the spacing between diffractional lattice planes and θ is the measured diffraction angle or glancing angle (Alexandre & Dubois, 2000; Ma et al., 2003). By monitoring the position, shape and intensity of the basal reflections from the distributed silicate layers, the nanocomposite structure may be identified (Porter et al., 2003).

(b)

(a)

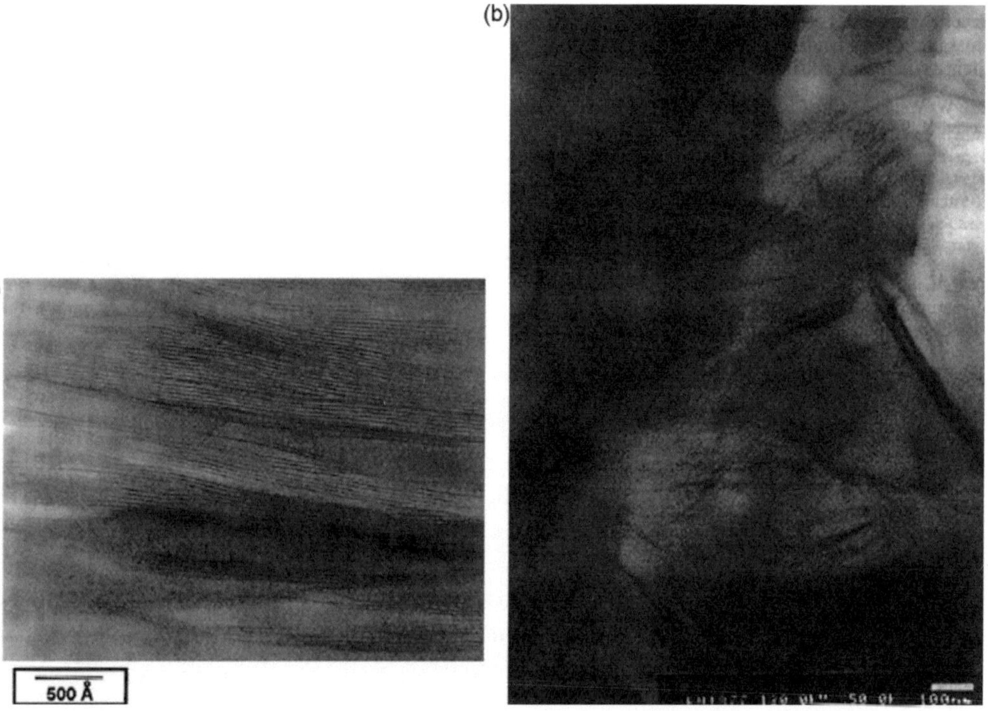

Fig. 3. TEM micrographs of poly(styrene)-based nanocomposites: (a) intercalated nanocomposite and (b) exfoliated nanocomposite (Alexandre & Dubois, 2000). Reproduced from Alexandre and Dubois by permission of Elsevier Science Ltd., UK.

Although XRD offers a conventional method to determine the interlayer spacing of the silicate layers in the original layered silicates and the intercalated nanocomposites, little can be said about the spatial distribution of the silicate layers or any structural inhomogeneities in nanocomposites. Additionally, some layered silicates initially do not exhibit well-defined basal reflections. Thus, peak broadening and intensity decreases are very difficult to study systematically. Therefore, conclusions concerning the mechanism of nanocomposite formation and structure based solely on XRD patterns are only tentative. On the other hand, TEM allows a qualitative understanding of the internal structure and can directly provide information in real space, in a localized area, on morphology and defect structures (Morgan et al., 2003; Usuki et al. (a), 1993).

Since the silicate layers are composed of heavier elements (Al, Si and O) than the interlayer and surrounding matrix (C, H and N), they appear darker in bright-field images. Therefore, when nanocomposites are formed, the intersections of the silicate sheets are seen as dark lines which are the cross sections of the silicate layers, measuring 1 nm thick. Fig. 3 shows the TEM micrographs obtained for an intercalated and an exfoliated nanocomposite.

4. Preparation of nanoclay composites

4.1 Polymer-templated nanoclay nucleation

In this technique, the clay minerals are synthesized within the polymer matrix, using an aqueous solution (or gel) containing the polymer and the silicate building blocks. As precursors for the clay silica sol, magnesium hydroxide sol and lithium fluoride are used. During the process, the polymer aids the nucleation and growth of the inorganic host crystals and gets trapped within the layers as they grow. Although theoretically this method has the potential of promoting the dispersion of the silicate layers in a one-step process, without needing the presence of the onium ion, it presents serious disadvantages. First of all, the synthesis of clay minerals generally requires high temperatures, which decompose the polymers. An exception is the synthesis of hectorite-type clay minerals which can be performed under relatively mild conditions. Another problem is the aggregation tendency of the growing silicate layers (Alexandre & Dubois, 2000; Lagaly, 1999; Zanetti et al., 2000).

4.2 Single layered nanoclay-polymer solution

Following this technique, the layered silicate is exfoliated into single layers using a solvent in which the polymer is soluble. It is well known that such layered silicates, owing to the weak forces that stack the layers together can be easily dispersed in an adequate solvent. After the organoclay has swollen in the solvent, the polymer is added to the solution and intercalates between the clay layers. The final step consists of removing the solvent, either by vaporization, usually under vacuum, or by precipitation. Upon solvent removal the sheets reassemble, sandwiching the polymer to form a nanocomposite structure. The major advantage of this method is that intercalated nanocomposites can be synthesized that are based on polymers with low or even no polarity. However, the solvent approach is difficult to apply in industry owing to problems associated with the use of large quantities of solvents (Alexandre & Dubois, 2000; Beyer et al., 2002).

4.3 Monomer polymerization migrated into layered nanoclay

In this technique, the modified layered silicate is swollen by a liquid monomer solution. The monomer migrates into the galleries of the layered silicate, so that the polymerization reaction can occur between the intercalated sheets. The reaction can be initiated either by heat or radiation, by the diffusion of a suitable initiator or by an organic initiator or catalyst fixed through cationic exchange inside the interlayer before the swelling step by the monomer. Polymerization produces long-chain polymers within the clay galleries. Under conditions in which intra- and extra-gallery polymerization rates are properly balanced, the clay layers are delaminated and the resulting material possesses a disordered structure (Alexandre & Dubois, 2000; Beyer et al., 2002; Solomon et al., 2001).

4.4 Polymer replacement of a previously intercalated solvent

Intercalation of a polymer from a solution is a two-stage process in which the polymer replaces an appropriate, previously intercalated solvent. Such a replacement requires a negative variation in the Gibbs free energy. It is thought that the diminished entropy due to the confinement of the polymer is compensated by an increase due to desorption of intercalated solvent molecules. In other words, the entropy gained by desorption of solvent molecules is the driving force for polymer intercalation from solution (Arada et al., 1992; Tunney et al., 1996; Fischer et al., 1999; Theng et al., 1979; Ogata et al., 1997; Yano et al., 1993).

Several studies have focused on the preparation of PLA-layered silicate nanocomposites using intercalation from solution. The first attempts by Ogata (Usuki et al. (b), 1993), involved dissolving the polymer in hot chloroform. However, TEM analysis revealed that only microcomposites were formed and that an intercalated morphology was not achieved.

In the case of polymeric materials that are infusible and insoluble even in organic solvents, the only possible route to produce nanocomposites with this method is to use polymeric precursors that can be intercalated in the layered silicate and then thermally or chemically converted to the desired polymer (Alexandre & Dubois, 2000; Fornes et al., 2002).

4.5 In situ intercalative polymerization

4.5.1 Thermoplastic polymers

The Toyota research group first reported the ability of α,ω-amino acid ($COOH\text{-}(CH_2)_{n1}\text{-}NH_2^+$, with n = 2, 3, 4, 5, 6, 8, 11, 12, 18) to be swollen by ε-caprolactam monomer at 100 °C and subsequently initiate ring opening polymerization to obtain PA6/MMT nanocomposites (Kojima et al. (a), 1993). The number of carbon atoms in the α,ω-amino acid was found to have a strong effect on the swelling behavior as reported in Fig. 4, indicating that the extent of intercalation of ε-caprolactam monomer is high when the number of carbon atoms in the ω-amino acid is large (Arada et al., 1990). Moreover, it was found from a comparison of different types of inorganic silicates that clays having higher CEC lead to more efficient exfoliation of the silicate platelets (Sepehr et al., 2005).

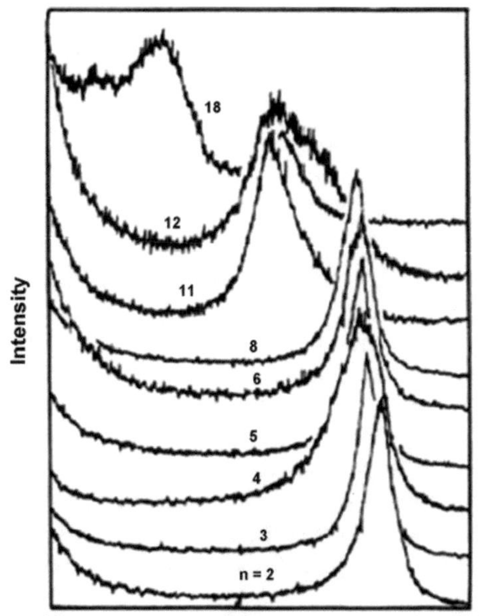

Fig. 4. XRD patterns of ω-amino acid [$NH_2(CH_2)_{n1}COOH$] modified Na^+-MMT (Arada et al., 1990). Reproduced from Usuki et al. (Usuki et al. (a), 1993), by permission of Materials Research Society, USA.

Intercalative polymerization of ε-caprolactam could be realized without modifying the MMT surface. Indeed, this monomer was able to directly intercalate the Na^+-MMT in water in the presence of hydrochloric acid, as proved by the increase in interlayer spacing from 10 to 15.1 Å. At high temperature (200 °C), in the presence of excess ε-caprolactam, the clay so modified can be swollen again, allowing the ring opening polymerization to proceed when 6-aminocaproic acid is added as an accelerator. The resulting composite does not present a diffraction peak in XRD, and TEM observation agrees with a molecular dispersion of the silicate sheets (Lan et al. (a), 1994).

At this point, it is worth mentioning that, even though in situ intercalative polymerization has proved successful in the preparation of various polymer-layered silicate nanocomposites, important drawbacks of this technique have also been pointed out: (1) it is a time-consuming preparation route (the polymerization reaction may take more than 24 h); (2) exfoliation is not always thermodynamically stable; and the platelets may re-aggregate during subsequent processing steps; and (3) the process is available only to the resin manufacturer who is able to dedicate a production line for this purpose (Kornmann et al., 1998).

4.5.2 Thermosetting polymers

Despite the aforementioned disadvantages of in situ intercalative polymerization, this is the only viable technique for the preparation of thermoset-based nanocomposites, since such nanocomposites obviously cannot be synthesized by melt intercalation, which is the other commercially important preparation method (Kornmann et al., 2001; Jiankun et al., 2001; Lan et al. (b), 1994; Liu et al., 2005).

In this case, the exfoliation ability of the organoclays is determined by their nature, including the catalytic effect on the curing reaction, the miscibility with the curing agent, etc. Since there is a curing competition between intragallery and extragallery resin, as long as the intragallery polymerization occurs at a rate comparable to the extragallery polymerization, the curing heat produced is enough to overcome the attractive forces between the silicate layers and an exfoliated nanocomposite structure can be formed. In contrast, if the extragallery polymerization is more rapid than the intragallery diffusion and polymerization or if intragallery polymerization is retarded, the extragallery resin will gel before the intragallery resin produces enough curing heat to drive the clay to exfoliate; consequently, exfoliation will not be reached. It can be inferred, therefore, that factors promoting the curing reaction of intragallery resin will facilitate the exfoliation of the clay. Such factors include the catalytic effect of organoclay on the curing reaction, the good penetrating ability of curing agent to clay, the long alkyl-chain of the organo-cation, meaning a greater amount of intragallery resin preload and a completed organization of the clay, and meaning weaker attractive forces between the silicate layers (Becker et al., 2004).

In fact, a number of research groups have studied the effect of various parameters on the exfoliation of clays in epoxy resins. Pioneering studies by Pinnavaia and coworkers (Hackman et al., 2006) on MMT/epoxy systems established the initial conceptual methodology. Interfacial modifiers, such as primary ammonium alkyls are intercalated between the MMT layers, not only to compatibilize the inorganic aluminosilicate and organic resin, but also to accelerate the crosslinking reaction between the layers through acid catalysis. That is, as the curing agent is mixed into the clay/epoxy mixture, it is thought that the modifiers introduced into the galleries of the clay sheets would promote the reaction between the epoxy in the gallery with the curing agent. This would make the intragallery curing reaction faster than the extragallery reaction, thus facilitating the expansion of the clay sheets and helping to achieve exfoliation (Liu et al., 2002).

Other researchers investigated the effect of the polymer resin. For example, Becker et al. (Vaia et al., 1997) prepared nanocomposites of three different epoxy resins: triglycidyl p-aminophenol (TGAP) and tetrafunctional tetraglycidyldiamino diphenylmethane (TGDDM), using a mixture of two diethyltoluene diamine (DETDA) isomers as the hardener and a commercially available octadecyl ammonium ion modified MMT as the clay. All epoxy resin systems intercalated the organically modified layered silicate and increased the d-spacing from 23 up to 80 Å. Similarly, Hackman and Hollaway (Vaia et al., 1993) noted that the epoxy resin component of the nanocomposite has little effect on the exfoliation of the clay layers; although it is the basic unit, the curing agent controls the rate of cure. Lower viscosity resins lead to faster pre-intercalation, but they do not seem to offer any significant long-term advantage.

4.6 Molten polymer intercalation

For most technologically important polymers, both in situ polymerization and intercalation from solution are limited because neither a suitable monomer nor a compatible polymer-silicate solvent system is always available. Moreover, they are not always compatible with current polymer processing techniques. These disadvantages drive the researchers to the direct melt intercalation method, which is the most versatile and environmentally benign among all the methods of preparing polymer-clay nanocomposites (PCNs) (Giannelis, 1996; Zheng et al., 2006).

As already mentioned, nanocomposite synthesis via polymer melt intercalation involves annealing, usually under shear, of a mixture of polymer and layered silicate above the softening point of the polymer. During annealing, polymer chains diffuse from the bulk polymer melt into the galleries between the silicate layers (Vaia & Giannelis, 2001; Fornes et al., 2003).

The advantages of forming nanocomposites by melt processing are quite appealing, rendering this technique a promising new approach that would greatly expand the commercial opportunities for nanocomposites technology (Fornes et al., 2001; Huang et al., 2001; Fornes et al., 2003). If technically possible, melt compounding would be significantly more economical and simpler than in situ polymerization. It minimizes capital costs because of its compatibility with existing processes. That is, melt processing allows nanocomposites to be formulated directly using ordinary compounding devices such as extruders or mixers, without the necessary involvement of resin production. Therefore, it shifts nanocomposite production downstream, giving end-use manufacturers many degrees of freedom with regard to final product specifications (e.g. selection of polymer grade, choice of organoclay, level of reinforcement, etc.). At the same time, melt processing is environmentally sound since no solvents are required (Fornes et al., 2001); and it enhances the specificity for the intercalation of polymer, by eliminating the competing host-solvent and polymer-solvent interactions (Shia et al., 1998).

Zheng et al. (Gorrasi et al., 2003) used an oligomerically modified clay, prepared by ion-exchange with the oligomer prepared from maleic anhydride (MA), styrene (ST) and vinylbenzyltrimethylammonium chloride (VBTACl) terpolymer, herein called MAST, to prepare PS/clay nanocomposites by melt blending. Thereafter, a portion of MAST oligomer, dissolved in acetone was added drop-wise to a dispersion of clay in distilled water and acetone. A precipitate (MAST hectorite clay) formed immediately. Nanocomposites were subsequently prepared by melt blending in a Brabender Plasticorder at 60 rpm and 190 °C for 15 min. XRD measurements indicated a mixed intercalated/delaminated structure for the MAST modified clay, whereas no peaks were observed for the PS/MAST. By combining XRD and TEM analyses the authors concluded that the hybrids formed were characterized by a mixed immiscible/intercalated/delaminated structure.

5. Characterization the properties of nanoclay composites

5.1 Mechanical properties

5.1.1 Load transfer mechanism

The first mechanism that has been put forward to explain the reinforcing action of layered silicates is one also valid for conventional reinforcements, such as fibers. That is, rigid fillers are naturally resistant to straining due to their high moduli. Therefore, when a relatively softer matrix is reinforced with such fillers, the polymer, particularly that adjacent to the filler particles, becomes highly restrained mechanically. This enables a significant portion of an applied load to be carried by the filler, assuming that the bonding between the two phases is adequate (Tortora et al. (a), 2002). From this mechanism it becomes obvious that the larger the surface of the filler in contact with the polymer, the greater the reinforcing effect will be. This could partly explain why layered silicates, having an extremely high specific surface area impart dramatic improvements of modulus even when present in very small amounts in a polymer. In fact, the low silicate loading required in nanocomposites to effect significant property improvements, is probably their most distinguishing characteristic.

In most conventionally filled polymer systems, the modulus increases linearly with the filler volume fraction, whereas for nanocomposites much lower filler concentrations increase the modulus sharply and to a much larger extent (Porter et al., 2003).

However, some authors have argued that the dramatic improvement of modulus for such extremely low clay concentrations (i.e. 2-5 wt.%) cannot be attributed simply to the introduction of the higher modulus inorganic filler layers. A proposed theoretical approach assumes a layer of affected polymer on the filler surface, with a much higher modulus than the bulk equivalent polymer. This affected polymer can be thought of as a region of the polymer matrix that is physisorbed on the silicate surface, and is thus stiffened through its affinity for and adhesion to the filler surface. Obviously, for such high aspect ratio fillers as the layered silicate layers, the surface area exposed to the polymer is huge and, therefore, the significant increases in the modulus with very low filler content are not surprising. Furthermore, beyond the percolation limit, the additional silicate layers are incorporated in polymer regions that are already affected by other silicate layers, and thus it is expected that the enhancement of modulus will become much less dramatic (Bharadwaj et al., 2002).

In order to prove the effect of degree of exfoliation on nanocomposite mechanical properties, Fornes and Paul (Fornes et al., 2003) used an analytical approach to elucidate how incomplete exfoliation influences nanocomposite stiffness. They expressed the modulus of a simple clay stack in the direction parallel to its platelets, by using the rule of mixtures:

$$E_{\text{stack}} = \emptyset_{\text{MMT}} E_{\text{MMT}} + \emptyset_{\text{gallery}} E_{\text{gallery}} \tag{1}$$

where $\emptyset_{\text{MMT}}$ is the volume fraction of silicate layers in the stack, E_{MMT} is the modulus of MMT, $\emptyset_{\text{gallery}}$ is the volume fraction of gallery space and E_{gallery} is the modulus of the material in the gallery, which is expected to be much less than E_{MMT}. The volume fraction occupied by gallery space, $\emptyset_{\text{gallery}}$ can be expressed in terms of X-ray d-spacings, as

$$\emptyset_{\text{gallery}} = \frac{(n-1)(d_{001} - t_{\text{platelet}})}{d_{001}(n-1) + t_{\text{platelet}}} \tag{2}$$

where n is the number of platelets per stack, d_{001} is the repeat spacing between silicate particles, and t_{platelet} is the thickness of a silicate platelet. Obviously, when the number of platelets in a stack is equal to one, the system represents an individual exfoliated platelet. As it can be seen, the number of platelets in a stack affects the reinforcement factor in an unexchanged, non-expandable clay ($d_{001} = 0.96\,nm$) as well as in an intercalated or organically modified clay ($d_{001} = 1.8\,nm$).

5.1.2 Modulus and strength

In general, the addition of an organically modified layered silicate in a polymer matrix results in significant improvements of Young's modulus. For example, Gorrasi et al. (Liu et al., 1999) reported an increase from 216 to 390 MPa for a PCL nanocomposite containing 10 wt.% ammonium-treated montmorillonite, while in another study (Manias (b), 2001), Young's modulus was increased from 120 to 445 MPa with addition of 8 wt.% ammonium treated clay in PCL. Similarly, in the case of nylon 6 nanocomposites obtained through the intercalative ring opening polymerization of ε-caprolactam, a large increase in the Young's modulus at rather low filler content has been reported, whatever the method of preparation: polymerization within organo-modified montmorillonite, polymerization within protonated ε-caprolactam swollen montmorillonite or polymerization within natural montmorillonite in the presence of ε-caprolactam and an acid catalyst (Zerda et al., 2001).
However, exceptions to this general trend have been reported. As shown in Fig. 5, in crosslinked polyester/OMLS nanocomposites, the modulus decreases with increasing clay content; in fact, the drop for the 2.5 wt.% nanocomposite was greater than expected. To explain this phenomenon, it was proposed that the intercalation and exfoliation of the clay in the polyester resin serve to effectively decrease the number of crosslinks from a topological perspective. The origin of the greater drop in properties of the 2.5 wt.% nanocomposites may be traced to the morphology; i.e. it was observed that the sample showed exfoliation on a global scale compared to the nanocomposite containing 10 wt.%

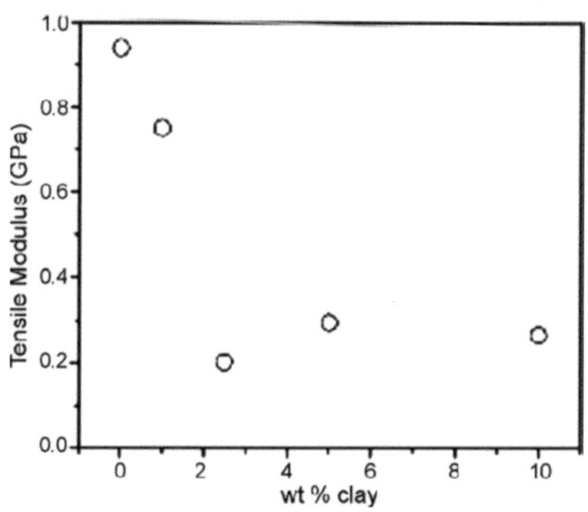

Fig. 5. Tensile modulus vs. clay concentration for crosslinked polyester nanocomposites (Manias (b), 2001). Reproduced from Manias et al., by permission of Elsevier Science Ltd., UK.

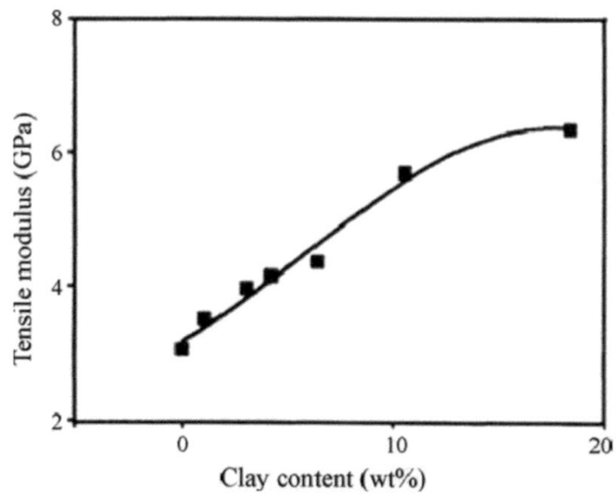

Fig. 6. Effect of clay content on tensile modulus, measured at room temperature, of organo-modified montmorillonite/nylon-6-based nanocomposite obtained by melt intercalation (Cho & Paul, 2001). Reproduced from Cho and Paul by permission of John Wiley & Sons, Inc., US.

clay, indicating that the crosslinking density is inversely proportional to the degree of exfoliation (Manias (b), 2001).

Apart from the modulus, the addition of OMLS in a polymer matrix usually also increases the tensile strength compared to that of the neat polymer material. For example, Shelley et al. (Xiong et al., 2004) reported a 175% improvement in yield stress accompanied by a 200% increase in tensile modulus for a nylon 6 nanocomposite containing 5 wt.% clay.

Most polymer-clay nanocomposite studies report tensile properties, such as modulus, as a function of clay content (Kojima et al. (b), 1993), as in Fig. 6. This plot of Young's modulus of nylon 6 nanocomposite vs. filler weight content, shows a constant large rate of increase of modulus up to ca. 10 wt.% of nanoclay, whereas above this threshold the aforementioned levelling-off of Young's modulus is observed. This change corresponds to the passage from totally exfoliated structure (below 10 wt.%) to partially exfoliated-partially intercalated structure (for 10 wt.% clay and above), as determined by XRD and TEM (Alexandre & Dubois, 2000; Porter et al., 2003).

In another study, Liu and Wu (Zheng et al., 2006) studied the mechanical performance of PA66 nanocomposites prepared via melt intercalation, using epoxy co-intercalated clay. The tensile strength increases rapidly from 78 MPa for PA66 up to 98 MPa for PA66CN5, but the increasing amplitude decreases when the clay content is above 5 wt.%. A similar phenomenon is observed in the dependence of tensile modulus of PA66CN on clay content. The smaller increase in amplitude observed with a clay loading above 5wt.%was again attributed to the inevitable aggregation of the layers at high clay content.

Similarly, other factors that influence the degree of exfoliation, apart from the clay content, also have an impact on nanocomposite modulus and strength. This explains the variations observed in moduli of PA6 nanocomposites prepared by intercalative ring opening polymerization of ε-caprolactam, with different kinds of acids to catalyze the polymerization.

Cho and Paul (Cho & Paul, 2001) studied the effect of mixing device and processing parameters on the mechanical properties of polyamide nanocomposites. In the case of composites formed by single-screw extrusion, the exfoliation of the clay platelets is not extensive. Even after a second pass through this extruder, undispersed tactoids are still easily observed with naked eye. However, the tensile strength and modulus were slightly improved by the second pass. On the other hand, nylon 6 nanocomposites with good properties can be obtained over a broad range of processing conditions in the twin screw extruder. The final nanocomposite properties are almost independent of the barrel temperature over the range of typical nylon 6 processing, but they are slightly improved by increasing the screw speed or by a second pass through the extruder. Therefore, processing conditions need to be optimized to allow greater exfoliation of the clay platelets and, thus, greater improvement in mechanical properties. Other factors that may play a crucial role in improvement of nanocomposite mechanical properties include the organic modification of the clay and the addition of compatibilizers to the polymer matrix.

The effect of clay organic modification on nanocomposite mechanical properties is also demonstrated in Fig. 7, which presents the ultimate strength of Punano composites with different contents of two organically treated montmorillonites: MO-MMT, treated with a thermally stable, aromatic amine modifier containing active groups, and C16-MMT, treated with a quaternary alkyl ammonium salt. As can be seen the ultimate strength

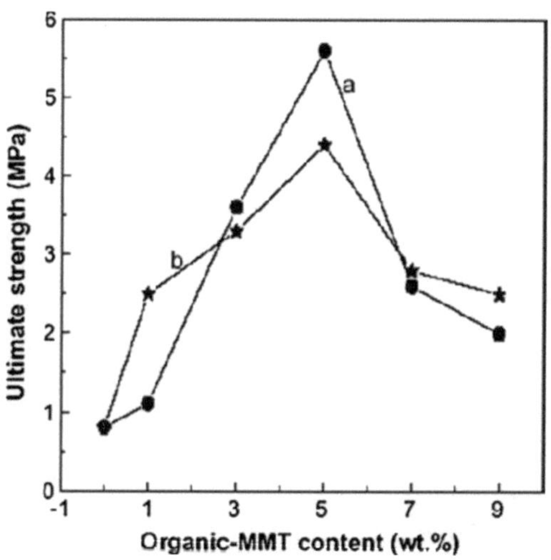

Fig. 7. Effect of organic-MMT loading on the tensile strength of (a) PU/MO-MMT and (b) PU/C16-MMT (Chaudhary et al., 2005). Reproduced from Chaudhary et al., by permission of Elsevier Science Ltd., UK.

increased dramatically with clay content and reached a maximum at 5 wt.% MMT, where the ultimate strength of the nanocomposites increased by about 450% for C16-MMT and 600% for MO-MMT, compared with that of pure PU, indicating that the improved mechanical strength depends on the characteristics of the modifier (Chaudhary et al., 2005).

The extent of improvement of nanocomposite mechanical properties will also depend directly upon the average length of the dispersed clay particles, since this determines their aspect ratio and, hence, their surface area (Porter et al., 2003; Srivastava et al., 2006). At this point we note that several authors have also pointed out factors that have an adverse effect on nanocomposite modulus and/or strength and need to be taken into consideration when preparing nanocomposite materials.

Quite interestingly, Gopakumar et al. (Gopakumar et al., 2002) found that the exfoliation of 5 and 10 wt.% clay in PE-MA, increases Young's modulus by 30 and 53%, respectively, whereas the tensile stress at yield showed only a marginal increase, up to a maximum of 15% for the 10 wt.% clay composition. The authors noted that the greatly enhanced interfacial area derived from exfoliation of the clay improves the mechanical reinforcement potential of the filler. However, given that the mechanical properties of a filled system depend on two principal factors, i.e. crystallinity of the polymer matrix and the extent of filler reinforcement, the degree of crystallinity must also be considered.

In another study dealing with the effect of matrix variations on mechanical properties of nanocomposites, Chaudhary et al. (Chaudhary et al., 2005) studied the tensile properties of nanocomposites based on EVAs with various VA contents and two alternative organoclays. Since in EVA with increasing VA content the crystallinity of the polymer decreases (and will lower the stiffness), while the polarity increases (and will increase the intercalation), the authors suggested that in their system, the stiffness and toughness responses would reflect an interplay of two factors: (a) an increase in the "rigid" amorphous phase due to polymer-clay intercalation and (b) an increase in the "mobile" amorphous phase due to the increasing VA content. Experimental results showed that the influence of increasing clay concentration on the tensile behavior of EVA matrices was significant only with a low or moderately polar EVA matrix (9 and 18% VA). Thus, a linear proportionality was found between clay concentration and tensile modulus for EVA-9 and EVA-18, a relation not observed with EVA-28. In fact, it is very difficult to compare the extent of the improvement of the mechanical properties of different EVA/clay nanocomposites reported so far, because EVAs of different vinyl acetate contents have been processed into the nanocomposites with different clays and different modifying agents by different methods (Pegoretti et al., 2004). Upon silicate addition large improvements in stiffness were observed, which however were accompanied by a decrease in tensile strength and elongation (Chang et al., 2004). Similar trends have been reported by Tortora et al. (Tortora et al. (b), 2002). Both exfoliated and intercalated PU/o-MMT nanocomposites showed an improvement in the elastic modulus upon increasing the clay content, but a decrease in the stress and strain at break.

In general, it has been argued that in the presence of polar or ionic interactions between the polymer and the silicate layers, the stress at break is usually increased, whereas when there is lack of interfacial adhesion, no or very slight tensile strength enhancement is recorded (Alexandre & Dubois, 2000). Pegoretti et al. (Pegoretti et al., 2004) found that the yield strength was not reduced by the addition of clay to recycled PET and considered this to be a sign of good interfacial adhesion; however, in the same study, a slight decrease of stress at break and a dramatic reduction of strain were reported. On the other hand, in PS intercalated nanocomposites the ultimate tensile stress was found to decrease compared to that of the PS matrix and dropped further at higher filler content. This lack of strength was attributed to the fact that only weak interactions exist at the PS/clay interface, contrary to other compositions in which polar interactions may prevail, strengthening the matrix interface (Chang et al., 2004).

An interesting study was performed by Chang et al. (Chang et al., 2004) who prepared PET-based nanocomposites through in situ intercalative polymerization, and subsequently produced nano-hybrid fibers by extrusion through the die of a capillary rheometer. The hot extrudates were stretched through the die of a capillary rheometer at 270 °C and immediately drawn to various draw ratios (DR). The tensile properties of the fibers formed increased with increasing amount of organoclay at DR = 1. When the organoclay was increased from 0 to 3 wt.% in hybrids at DR = 1, the strength linearly improved from 46 to 71 MPa, and the modulus from 2.21 to 4.10 GPa.

Finally, even though nanocomposite researchers are generally interested in the tensile properties of the final materials, there are a few reports concerning the flexural properties of PLS nanocomposites (Morawiec et al., 2005; Liu et al., 2001).

5.1.3 Toughness and strain

The brittle behavior often exhibited by nanocomposites probably originates from the formation of microvoids due to debonding of clay platelets from the polymer matrix upon failure. This has been testified through careful inspection of fracture surfaces and is also correlated to observations by in situ deformation experiments using TEM (Liu et al., 2001; Hong et al., 2005). In fact, the observation of nanocomposite fracture surfaces is quite interesting. Fig. 8(a) shows a typical fracture morphology in virgin nylon 12 and a ductile fracture as evidenced by plastic deformation. Fig. 8(b) and (c) show fracture surfaces of the nanocomposites containing 1 and 5 wt.% clay, respectively. No distinct clay agglomerates are observed by scanning electron microscopy (SEM) even at high magnification, as shown in Fig. 8(d). For 1wt.% clay addition (Fig. 8(b)), the fracture surface became smoother

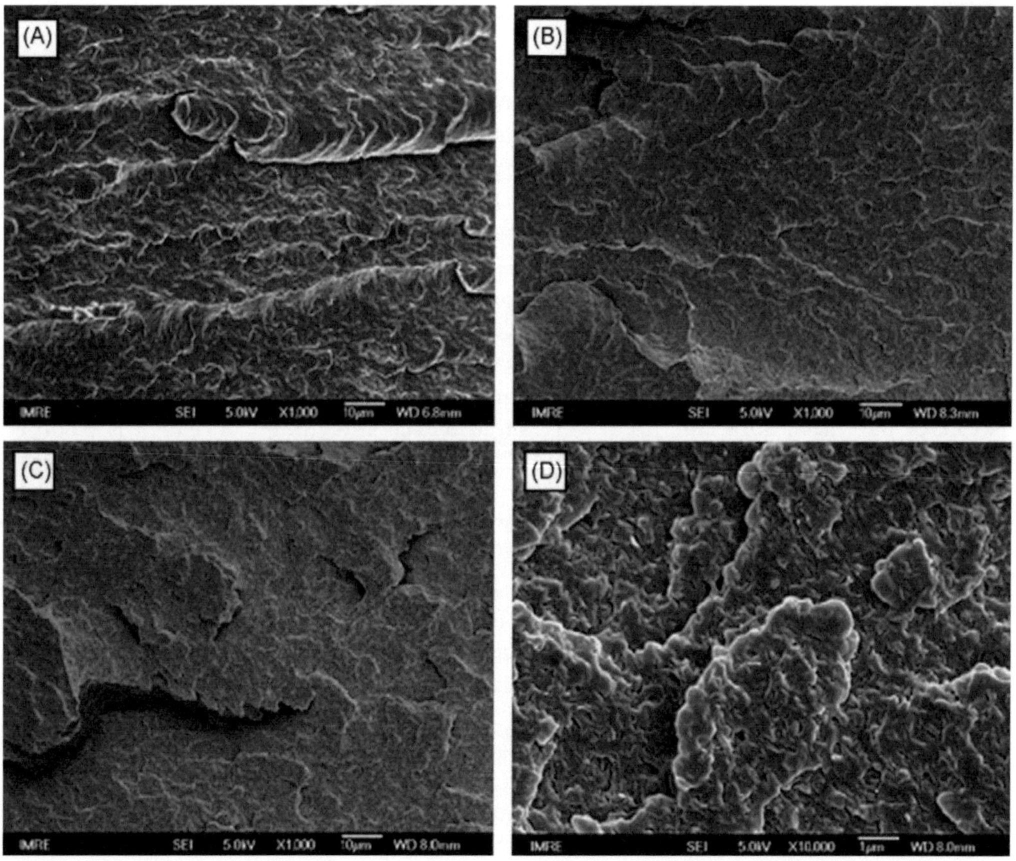

Fig. 8. SEM images showing fracture surfaces after impact tests. (a) neat PA12; (b) and (c) PA12 nanocomposites containing 1 and 5 wt.% clay, respectively; (d) high magnification of (c) Reproduced from Phang et al. (Phang et al., 2005), by permission of John Wiley & Sons Ltd., US.

compared with that of neat PA12; an even more brittle feature for clay concentration of 5 wt.% was observed in Fig. 8(c). Careful inspection of the fracture surface at higher magnification of nanocomposite with 5wt.% clay (Fig. 8(d)) verifies the formation of microvoids due to the debonding of clay platelets from the matrix. Usually, microvoids are formed around the large inhomogeneities, which become evident especially at high clay loadings. These microvoids will coalesce with formation of larger cracks causing embrittlement, ultimately resulting in reduced toughness (Liu et al., 2001).

In the case of nylon 12 nanocomposites, Fig. 9 shows that the Izod impact strength monotonically decreases as the clay concentration increases. The toughness (representing the energy absorption during the fracture process) decreases by about 25% with 5 wt.% of clay. Similar observations of reduction in impact strength are also reported in nylon 6/clay

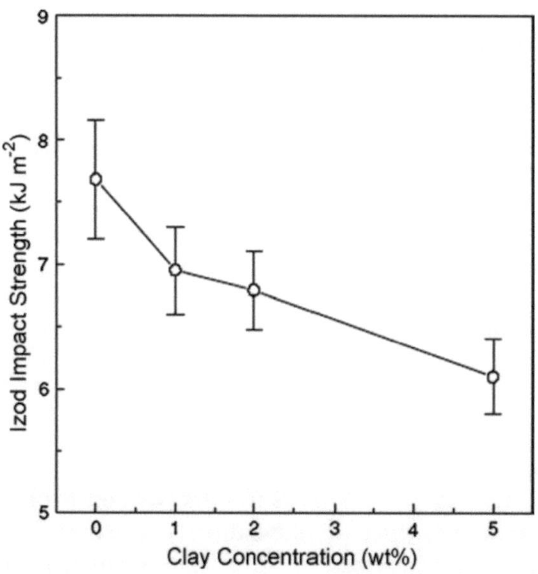

Fig. 9. Izod impact strength of PA12/clay nanocomposites as a function of clay concentration (Liu et al., 2001). Reproduced from Liu et al., by permission of John Wiley & Sons Ltd., US.

nanocomposites and PE-based nanocomposites, indicating that the incorporation of clay into semicrystalline thermoplastics usually results in toughness reduction, i.e. the aforementioned embrittlement effect from clay addition (Liu et al., 2001).

On the other hand, some studies report little or no change of toughness upon clay intercalation/exfoliation. For example, while the tensile strength and modulus of PP nanocomposites increased rapidly with increasing clay content from 0 to 5 wt.%, the notched Izod impact strength was constant, within experimental error, in the clay content range between 0 and 7 wt.% (Messersmith et al., 1994). Another study reports the impact properties for exfoliated nylon 6-based nanocomposites prepared either by in situ intercalative polymerization or by melt intercalation. In that study marginal reductions in impact properties are reported, whatever the exfoliation process used. In the case of in situ intercalative polymerization, the Izod impact strength is reduced from 20.6 to 18.1 J/m when 4.7 wt.% clay is incorporated. Charpy impact tests show similar reduction in the impact strength, with a drop from 6.21 kJ/m^2 for the filler free matrix, down to 6.06 kJ/m^2 for the 4.7 wt.% nanocomposite.

Furthermore, toughness improvements upon clay dispersion have also been reported a remarkable result, considering that conventional polymer-clay composites, containing aggregated nanolayer tactoids ordinarily improve rigidity but sacrifice toughness and elongation (LeBaron et al., 1999).
Finally, it is worth summarizing the work of Hong et al. (Nam et al., 2001) on PP-based RTPO/clay nanocomposites, prepared by using PP-MA as a compatibilizer. PP-based RTPO (or in reactor made TPO) is a blend of PP and poly(ethyleneco-propylene) (EPR), produced by the bulk polymerization of propylene, followed by gas-phase copolymerization of ethylene and propylene driven by the $TiCl_4/MgCl_2$-based catalyst system. Such materials, like the conventional blends of PP/EPR prepared by mechanical blending, exhibit improved flexibility and toughness compared to neat PP. Moreover, because the rubber phase can be dispersed uniformly and reach a high degree of dispersion in these in situ blends, it is possible to achieve more intimate interaction between the matrix and the rubber phase. The tensile moduli of the nanocomposites became higher as the clay content increases. On the other hand, the elongation at break decreases as the clay content increases, but the value of nanocomposites containing 10 wt.% clay is 437%, which is much higher than that of PP/clay nanocomposites reported elsewhere. As the authors claim, these longational properties of PP based RTPO/clay nanocomposites are unique and promising for many applications. In fact, for reasons of comparison, Hong et al. also prepared and tested nanocomposites using PP/EPR mechanical blend matrix, modified with PP-MA. For these materials, the elongation at break values were about 50%, which are much lower than those of RTPO clay nanocomposites and is not suitable for industrial application. The authors attributed this discrepancy to the difference of dispersion homogeneity and domain size of ethylene copolymer between RTPO and PP/EPR mechanical blends.

5.1.4 Dynamic analysis

Dynamic mechanical analysis (DMA) measures the response of a material to a cyclic deformation (usually tension or three-point bending type deformation) as a function of the temperature. DMA results are expressed by three main parameters: (i) the storage modulus (E′ or G′), corresponding to the elastic response to the deformation; (ii) the loss modulus (E″ or G″), corresponding to the plastic response to the deformation and (iii) tanδ, that is, the E'/E'' (or G'/G'') ratio, useful for determining the occurrence of molecular mobility transitions such as the glass transition temperature (Alexandre & Dubois, 2000).
In the case of nanocomposites, the main conclusion derived from dynamic mechanical studies is that the storage modulus increases upon dispersion of a layered silicate in a polymer. This increase is generally larger above the glass transition temperature, and for exfoliated PLS nanocomposite structures is probably due to the creation of a three-dimensional network of interconnected long silicate layers, strengthening the material through mechanical percolation (Alexandre & Dubois, 2000). Above the glass transition temperature, when materials become soft, the reinforcement effect of the clay particles becomes more prominent, due to the restricted movement of the polymer chains. This results in the observed enhancement of G′ (Porter et al., 2003). For example, an epoxy-based nanocomposite, containing 4 vol.% silicates, showed a 60% increase in G′ in the glassy region, compared to the unfilled epoxy, while the equivalent increase in the rubbery region was 450% (Jimenez et al., 1997). Similar results have also been reported in the case of PP- (Laus et al., 1997), PCL- (Okamoto et al., 2001), SBS- (Ray et al., 2002), PA- (Ray et al. (b),

2003), PLA- (Nielsen et al., 1981; Nam et al., 2001; Fornes & Paul, 2003), and epoxy-based nanocomposites (Jimenez et al., 1997).
Enhancement of the loss modulus, G′′, has also been reported for nanocomposite materials, however this aspect of dynamic mechanical performance is far less discussed in the literature.
Finally, the tanδ values are affected in different ways by nanocomposite formation, depending on the polymer matrix. For example, in PS based nanocomposites, a shift of tanδ to higher temperatures has been observed, accompanied by a broadening of this transition (Chang et al., 2004), while the opposite effect was reported in the case of PP-based nanocomposites (Nielsen, 1967). Some authors observed a decrease of tanδ peaks, and considered this indicative of a glass transition suppression by the presence of the clay. However, Fornes and Paul (Fornes et al., 2003) pointed out that this conclusion is a misinterpretation, since the low values for the nanocomposites are simply the result of dividing the relatively constant loss modulus values in the T_g region, by larger storage modulus values.
Quite surprisingly, DMA showed that above T_g, the moduli for the pure PU and the PU/o-MMT nanocomposites show no obvious difference, while below T_g, addition of o-MMT strongly influences the modulus values. Interestingly, the authors found that E′ and E′′ of the PU/o-MMT decrease in comparison with values for the PU, for unclear reasons. On the other hand, significant enhancements of E′ and E′′ were seen for the nanocomposite prepared using a particular modified clay (Kojima et al. (b), 1993). In the case of PLA-based nanocomposites, it was observed that PLACNs with a very small amount of o-PCL as a compatibilizer exhibited a very large enhancement of mechanical properties compared to that of PLACN with comparable clay loading (Nielsen et al., 1981).

5.2 Barrier properties

Generally, polymer/layered silicate nanocomposites are characterized by very strong enhancements of their barrier properties. Polymers ranging from epoxies and good sealants (like siloxanes) to semi-permeable (e.g. polyureas) and highly hydrophilic (e.g. PVA) are all improved up to an order of magnitude by low clay loadings (Manias (b), 2001).
The dramatic improvement of barrier properties can be explained by the concept of tortuous paths. That is, when impermeable nanoparticles are incorporated into a polymer, the permeating molecules are forced to wiggle around them in a random walk, and hence diffuse by a tortuous pathway (Giannelis, 1996; LeBaron et al., 1999; Dennis et al., 2001; Phang et al., 2005).
The tortuosity factor is defined as the ratio of the actual distance, d′, that the penetrant must travel to the shortest distance d that it would travel in the absence of barriers. It is expressed in terms of the length L, the width W and the volume fraction of the sheets $\emptyset_S$ as

$$\tau = \frac{d'}{d} = 1 + \frac{L}{2W}\emptyset_S$$

It becomes obvious from this expression that a sheet-like morphology is particularly efficient at maximizing the path length, due to the large length-to-width ratio, as compared to other filler shapes (Alexandre & Dubois, 2000; Porter et al., 2003).
According to the model proposed by Nielsen, the effect of tortuosity on the permeability may, in turn, be expressed as

$$\frac{P_{PCN}}{P_P} = \frac{1 - \emptyset_S}{\tau}$$

where P_{PCN} and P_P represent the permeability of the nanocomposite and the pure polymer, respectively and $\emptyset_S$ is the clay content (Porter et al., 2003; Cussler et al., 1988).

Although the above equations were developed to model the diffusion of small molecules in conventional composites, they have also been used in reproducing experimental results for the relative permeability in PLS nanocomposites. Discrepancies between the experimental data and the theoretical line may be attributed either to inadequacies of the model or to incomplete orientation of the particles within the nanocomposite film plane (Petricova et al., 2000). In fact, the key assumption of the Nielsen model is that the sheets are placed in an arrangement such that the direction of diffusion is normal to the direction of the sheets. Clearly, this arrangement results in the highest tortuosity, and any deviation from it would, in fact, lead to deterioration of the barrier properties (Porter et al., 2003).

Moreover, the tortuous path theory, including the Nielsen equation as well as other phenomenological relations (e.g. the Cussler (Cussler et al., 1988) formula, the Barrel (Petricova et al., 2000) formula and the power law equation (Fukuda & Kuwajima, 1997)), is grounded on the assumption that the presence of nanoparticles does not affect the diffusivity of the polymer matrix. However, experimental observations demonstrate that molecular mobility in a polymer matrix, which is intimately connected to the mass transport properties, diminished by clay incorporation. This reduction should be accompanied by a decrease in diffusivity of small molecules, which is not considered in the concept of tortuous paths.

Messersmith and Giannelis (Messersmith & Giannelis, 1995) studied the permeability of liquids and gases in nanocomposites and they observed that water permeability in PCL nanocomposites is dramatically reduced compared to the unfilled polymer. They also noted how the decrease in permeability is much more pronounced in the nanocomposites compared to conventionally filled polymers with much higher filler content.

Many studies reported in the literature have focused on nanocomposite barrier properties against gases and vapors. As an example, Tortora et al. (Tortora et al. (b), 2002) measured the transport properties of PU/o-MMT nanocomposites (prepared using a PCL nanocomposite "master-batch") using water vapor as hydrophilic permeant and dichloromethane as hydrophobic one. For both vapors, the sorption behavior changed in the presence of the clay, where the equilibrium concentration of water vapor is represented as a function of the vapor activity for all nanocomposites and for the o-MMT. The sorption curve of water vapor for o-MMT follows the Langmuir sorption isotherm, in which the sorption of solvent molecules occurs at specific sites; therefore, when all the sites are saturated, a plateau is reached. On the other hand, the sorption of neat PU shows a linear dependence of equilibrium concentration on activity, while nanocomposites show a dual sorption shape, that is a downward concavity, an inflection point and an upward curvature. The prevailing mechanism in the first zone is the sorption of solvent molecules on specific sites, due to interacting groups. Tortora et al. inferred that this type of sorption is due to the presence of clay in the polymers. At higher activities, the plasticization of the polymeric matrix determines a more than linear increase of vapor concentration and a transition in the curve is observed, from a dual type to a Flory-Huggins behavior. From the calculated values of the sorption parameters, defined as: $S = d(C_{eq})/dp$, and the zero-concentration diffusion coefficients for water sorption and dichloromethane vapor, the authors concluded that the sorption did not drastically change on increasing the clay content, whereas the zero-concentration diffusion coefficient D_0 strongly decreased with increasing inorganic content. The permeability calculated as the product SD_0, was largely dominated by the diffusion parameter; it showed a remarkable decrease up to 20 wt.% of clay and a levelling off at higher contents.

Summarizing: although a decrease of diffusivity is a well-established result of nanocomposite formation, contradictory results are reported concerning the saturation uptake values of various solvents or gases. Increases of the saturation uptake level are usually attributed to clustering phenomena. It is worth noticing, however, that in nanocomposites the coexistence of phases with different permeabilities can cause complex transport phenomena.

On the one hand, the organophilic clay gives rise to superficial adsorption and to specific interactions with the solvents. In turn, the polymer phase can be considered, in most cases, as a two-phase, crystalline-amorphous system, the crystalline regions being generally impermeable to penetrant molecules. The presence of the silicate layers may be expected to cause a decrease in permeability, due to the more tortuous path for the diffusing molecules that must bypass impenetrable platelets (Becker et al., 2004). Simultaneously, the influence of changes in matrix crystallinity and chain mobility, induced by the presence of the filler, should always be taken into consideration (Osman et al., 2004).

5.3 Thermal stability

The thermal stability of polymeric materials is usually studied by thermogravimetric analysis (TGA). The weight loss due to the formation of volatile products after degradation at high temperature is monitored as a function of temperature (and/or time). When heating occurs under an inert gas flow, a non-oxidative degradation occurs, while the use of air or oxygen allows oxidative degradation of the samples (Porter et al., 2003).

Generally, the incorporation of clay into the polymer matrix was found to enhance thermal stability by acting as a superior insulator and mass transport barrier to the volatile products generated during decomposition, as well as by assisting in the formation of char after thermal decomposition (Porter et al., 2003; Becker et al., 2004; Zhu et al., 2001).

Vyazovkin et al. (Vyazovkin et al., 2004) compared the thermal degradation of a PS nanocomposite with that of the virgin polymer under nitrogen and air. Both nitrogen and air the decomposition temperature of nanocomposites increased by 30-40 °C. The authors also observed that the virgin polymer degrades without forming any residue, whereas the nanocomposite (as expected) leaves some residue.

Zanetti et al. (Zanetti et al., 2004) reported TGA curves of a nanocompsite PE/EVA/o-MMT and the corresponding matrix PE/EVA. Under nitrogen, these samples do not show great differences of stability. However, in air, the PE/EVA blend is subject to a marked weight loss above 350 °C, to form a 5 wt.% residue at 450 °C, which is completely oxidized to volatile products between 470 and 550 °C. The nanocomposite, on the other hand, displays a different pattern. The presence of 5 wt.% o-MMT is enough to change the polymer's thermo-oxidative behavior and between 350 and 480 °C the amount of residue is higher to that observed in a nitrogen flow. According to the authors, the organoclay shields the polymer from the action of oxygen, dramatically increasing the thermal stability under oxidative conditions.

Bandyopadhyay et al. (Bandyopadhyay et al., 1999) reported the first improved thermal stability of biodegradable nanocomposites that combined PLA and organically modified fluorohectorite or montmorillonite. They showed that the PLA intercalated between the galleries of FH or MMT clay resisted the thermal degradation under conditions that would otherwise completely degrade pure PLA. This conclusion has been verified by a number of researchers in subsequent studies. Thellen et al. (Thellen et al., 2005) presented TGA curves for the neat polymer and corresponding nanocomposites and reported that the onset of thermal degradation was approximately 9 ^{o}C higher for the nanocomposite than for the neat PLA.

The thermal stability of PCL-based nanocomposites has also been studied by TGA. Generally, the degradation of PCL fits a two-step mechanism. First, random chain scission through pyrolysis of the ester groups, with the release of CO_2, H_2O and hexanoic acid, and in the second step, ε-caprolactone (cyclic monomer) formation as a result of an unzipping depolymerization process. It has been reported that the thermal stability of PCL/o-MMT nanocomposites systematically increases with increasing clay, up to a loading of 5 wt.% (Alexandre & Dubois, 2000; Thellen et al., 2005).

In fact, despite the general improvement of thermal stability, decreases in the thermal stability of polymers upon nanocomposite formation have also been reported, and various mechanisms have been put forward to explain the results. It has been argued, for example, that after the early stages of thermal decomposition the stacked silicate layers could hold accumulated heat, acting as a heat source to accelerate the decomposition process, in conjunction with the heat flow supplied by the outside heat source (Porter et al., 2003). Also, the alkylammonium cations in the organoclay could suffer decomposition following the Hoffmann elimination reaction, and the product could catalyze the degradation of polymer matrices. Moreover, the clay itself can also catalyze the degradation of polymer matrices. Thus, it becomes obvious that the organoclay may have two opposing functions in thermal stability of nanocomposites: a barrier effect, which should improve the thermal stability and a catalytic effect on the degradation of the polymer matrix, which should decrease the thermal stability (Zhao et al., 2005).

As deduced from the previous examples, even though contradictory results are sometimes found in the literature concerning the thermal stability of polymeric nanocomposites, the opportunity of achieving a significant improvement in thermal stability through low filler content is particularly attractive because end-products can be made cheaper, lighter and easier to process (Beyer et al., 2002).

6. Summary

Polymer-layered silicate nanocomposites, although known for many years, have attracted recent attention due to the report of the Toyota research group on the improved properties of PA6 nanocomposites and also due to the observation by Giannelis and co-workers that their preparation is possible by simple melt-mixing of the polymer with the layered silicate. Other preparation routes include intercalation of polymer or prepolymer from solution, in situ intercalative polymerization and template synthesis. In most cases, layered silicates first need to be modified with cationic-organic surfactants, in order to become miscible with polymeric matrices. Then, whether a nanocomposite will form or not, and whether this will be intercalated or exfoliated, depends on a variety of factors. These include the type of polymer, layered silicate and organic modifier, the preparation technique and processing conditions.

In general, nanocomposite materials, particularly those with exfoliated structures present significant improvements of modulus and strength, whereas contradictory results are reported concerning their elongation and toughness. Improvements of storage and loss moduli are also reported by many authors. Other interesting characteristics of this class of materials include improved barrier properties and thermal stability. Despite some contradictory results reported in the literature and presented here, concerning certain aspects of polymer-layered silicate nanocomposite technology, we hope this review will be a useful tool for those conducting research in this field.

7. References

Alexandre, M. & Dubois, P. (2000). Polymer–layered silicate nanocomposites: preparation, properties and uses of a new class of materials. *Mater Sci Eng R,* Vol. 28, pp. 1-63.

Arada, P. & Ruiz, E. (1990). Polymer–salt intercalation complexes in layer silicates. *Adv Mater,* Vol. 2, pp. 545–547.

Arada, P. & Ruiz, E. (1992). Poly(ethylene oxide)–silicate intercalation materials. *Chem Mater,* Vol. 4, pp. 1395–1403.

Balazs, AC.; Singh, C.; Zhulina, E. & Lyatskaya, Y. (1999). Modeling the phase behavior of polymer/clay nanocomposites. *Acc Chem Res,* Vol. 32, pp. 651–657.

Bandyopadhyay, S.; Chen, R. & Giannelis, EP. (1999). Biodegradable organic–inorganic hybrids based on poly(L-lactide). *J Macromol Sci Rev,* Vol. 81, pp. 159 - 160.

Becker, O.; Varley, R. & Simon, G. (2002). Morphology, thermal relaxations and mechanical properties of layered silicate nanocomposites based upon high-functionality epoxy resins. *Polymer,* Vol. 43, pp. 4365–4373.

Becker, O.; Varley, RJ. & Simon, GP. (2004). Thermal stability and water uptake of high performance epoxy layered silicate nanocomposites. *Eur Polym J,* Vol. 40, pp. 187 - 195.

Beyer, G. (2002). Nanocomposites: a new class of flame retardants for polymers. *Plast Addit Compound,* Vol. 10, pp. 22-27.

Bharadwaj, RK.; Mehrabi, AR.; Hamilton, C.; Trujillo, C.; Murga, MF. & Chavira, A. (2002). Structure-property relationships in cross-linked polyester–clay nanocomposites. *Polymer,* Vol. 43, pp. 3699–3705.

Chang, JH.; Kim, SJ.; Joo, YL. & Im, S. (2004). Poly(ethylene terephthalate) nanocomposites by in situ interlayer polymerization: the thermomechanical properties and morphology of the hybrid fiber. *Polymer,* Vol. 45, pp. 919-926.

Chaudhary, DS .; Prasad, R.; Gupta, RK. & Bhattacharya, SN. (2005). Clay intercalation and influence on crystallinity of EVA-based clay nanocomposites. *Thermochim Acta,* Vol. 433, pp. 187-195.

Cho, JW. & Paul, DR. (2001). Nylon 6 nanocomposites by melt compounding. *Polymer,* Vol. 42, pp. 1083-1094.

Cussler, EL.; Hughes, SE.; Ward, WJ. & Aris R. (1988). Barriermembranes. *JMembr Sci* Vol. 38, pp. 161-174.

Dennis, HR.; Hunter, DL.; Chang, D.; Kim, S.; White, JL.; Cho, JW. et al. Effect of melt processing conditions on the extent of exfoliation in organoclay-based nanocomposites. *Polymer,* Vol. 42, pp. 9513-9522.

Finnigan, B.; Martin, D.; Halley, P.; Truss, R. & Campell, K. (2004). Morphology and properties of thermoplastic polyurethane nanocomposites incorporating hydrophilic layered silicates. *Polymer,* Vol. 45, pp. 2249-2260.

Fischer, HR.; Gielgens, LH. & Koster, TP. (1999). Nanocomposites from polymers and layered minerals. *Acta Polym*, Vol. 50, pp. 122-126.
Fischer, H. (2003). Polymer nanocomposites: from fundamental research to specific applications. *Mater Sci Eng C*, Vol. 23, pp. 763–772.
Fornes, TD.; Yoon, PJ.; Keskkula, H. & Paul, DR. (2001). Nylon 6 nanocomposites: the effect of matrix molecular weight. *Polymer*, Vol. 42, pp. 9929-9940.
Fornes, TD.; Yoon, PJ.; Hunter, DL.; Keskkula, H. & Paul, DR. (2002). Effect of organoclay structure on nylon 6 nanocomposite morphology and properties. *Polymer*, Vol. 43, pp. 5915-5933.
Fornes, TD. & Paul, DR. (2003). Modelling properties of nylon 6/clay nanocomposites using composite theories. *Polymer*, Vol. 44, pp. 4993-5013.
Fukuda, M. & Kuwajima, S. (1997). Molecular-dynamics simulation of moisture diffusion in polyethylene beyond 10 ns duration. *J Chem Phys*, Vol. 107, pp. 2149–2159.
Giannelis, EP. (1996). Polymer layered silicate nanocomposites. *Adv Mater*, Vol. 8, pp. 29–35.
Ginzburg, VV.; Singh, C. & Balazs, AC. (2000). Theoretical phase diagrams of polymer/clay composites: the role of grafted organic modifiers. *Macromolecules*, Vol. 33, pp. 1089-1099.
Gopakumar, TG.; Lee, JA.; Kontopoulou, M. & Parent, JS. (2002). Influence of clay exfoliation on the physical properties of montmorillonite/polyethylene composites. *Polymer*, Vol. 43, pp. 5483-5491.
Gorrasi, G.; Tortora, M.; Vittoria, V.; Galli, G. & Chiellini, E. (2002). Transport and mechanical properties of blends of poly(e-caprolactone) and a modified montmorillonite–poly(e-caprolactone) nanocomposite. *J Polym Sci Polym Phys*, Vol. 40, pp. 1118–1124.
Gorrasi, G.; Tortora, M.; Vittoria, V.; Pollet, E.; Lepoittevin, B.; Alexandre, M. et al. (2003). Vapor barrier properties of polycaprolactone montmorillonite nanocomposites: effect of clay dispersion. *Polymer*, Vol. 44, pp. 2271-2279.
Hackman, I. & Hollaway, L. (2006). Epoxy-layered silicate nanocomposites in civil engineering. *Compos Part A-Appl S*, Vol. 37, pp. 1161-1170.
Hong, HC.; Lee, YB.; Bae, JW.; Jho, YJ.; Nam, BU.; Nam, GJ. et al. (2005). Tensile and flammability properties of polypropylene-based RTPO/clay nanocomposites for cable insulating material. *J Appl Polym Sci*, Vol. 97, pp. 2375-2381.
Hotta, S. & Paul DR. (2004). Nanocomposites formed from linear low density polyethylene and organoclays. *Polymer*, Vol. 45, pp. 7639-7654.
Huang, JC.; Zhu, ZK.; Yin J.; Qian, XF. & Sun, YY. (2001). Poly(etherimide)/montmorillonite nanocomposites prepared by melt intercalation: morphology, solvent resistance properties and thermal properties. *Polymer*, Vol. 42, pp. 873-877.
Ishida, H.; Campbell, S. & Blackwell, J. (2000). General approach to nanocomposite preparation. *Chem Mater*, Vol. 12, pp. 1260-1267.
Jiankun, L.; Yucai, K.; Zongneng, Q. & Xiao, Y. (2001). Study on intercalation and exfoliation behavior of organoclays in epoxy resin. *J Polym Sci Polym Phys*, Vol. 39, pp. 115-120.
Jimenez, G.; Ogata, N.; Kawai, H. & Ogihara, T. (1997). Structure and thermal/mechanical properties of poly(ε-caprolactone)-clay blend. *J Appl Polym Sci*, Vol. 64, pp. 2211-2220.

Kim, CM.; Lee, DH.; Hoffmann, B.; Kressler, J. & Stoppelmann, G. (2001). Influence of nanofillers on the deformation process in layered silicate/polyamide 12 nanocomposites. *Polymer,* Vol. 42, pp. 1095-1100.

Kojima (a), Y.; Usuki, A.; Kawasumi, M.; Okada, A.; Kurauchi, T. & Kamigaito, O. (1993). Synthesis of nylon-6 hybrid by montmorillonite intercalated with ε-caprolactam. *J Polym Sci Polym Chem,* Vol. 31, pp. 983-986.

Kojima (b), Y.; Usuki, A.; Kawasumi, M.; Okada, A.; Fukushima, Y.; Karauchi, T. et al. (1993). Mechanical properties of nylon-6/clay hybrid. *J Mater Res,* Vol. 6, pp. 1185-1189.

Kornmann, X.; Berglund, LA. & Sterte, J. (1998). Nanocomposite based on montmorillonite and unsaturated polyester. *Polym Eng Sci,* Vol. 38, pp. 1351-1358.

Kornmann, X.; Lindberg, H. & Berglund, LA. (2001). Synthesis of epoxy-clay nanocomposites: influence of the nature of the clay on structure. *Polymer,* Vol. 42, pp. 1303-1310.

Lagaly, G. (1999). Introduction: from clay mineral-polymer interactions to clay mineral-polymer nanocomposites. *Appl Clay Sci,* Vol. 15, pp. 1-9.

Lan, T.; Kaviratna, PD. & Pinnavaia, TJ. (1994). Synthesis, characterization and mechanical properties of epoxy-clay nanocomposites. *Polym Mater Sci Eng,* Vol. 71, pp. 527-528.

Lan, T. & Pinnavaia, TJ. (1994). Clay-reinforced epoxy nanocomposites. *Chem Mater,* Vol. 6, pp. 2216-2219.

Laus, M.; Francesangeli, O. & Sandrolini, F. (1997). New hybrid nanocomposites based on an organophilic clay and poly(styrene-b-butadiene) copolymers. *J Mater Res,* Vol. 12, pp. 3134-3139.

LeBaron, PC.; Wang, Z. & Pinnavaia, TJ. (1999). Polymer–layered silicate nanocomposites: an overview. *Appl Clay Sci,* Vol. 15, pp. 11–29.

Lepoittevin, B.; Pantoustier, N.; Devalckenaere, M.; Alexandre, M.; Kubies, D. & Calderg, C. (2002). Poly(ε-caprolactone)/clay nanocomposites by in situ intercalative polymerization catalyzed by dibutyltindimethoxide. *Macromolecules,* Vol. 35, pp. 8385 - 8390.

Lincoln, DM.; Vaia, RA.; Wang, ZG. & Hsiao, BS. (2001). Secondary structure and elevated temperature crystallitemorphology of nylon 6/layered silicate nanocomposites. *Polymer,* Vol. 42, pp. 1621–1631.

Liu, LM.; Qi, ZN. & Zhu, XG. (1999). Studies on nylon-6 clay nanocomposites by melt-intercalation process. *J Appl Polym Sci,* Vol. 71, pp. 1133-1138.

Liu, X. & Wu, Q. (2001). PP/clay nanocomposites prepared by grafting melt intercalation. *Polymer,* Vol. 42, pp. 10013-10019.

Liu, X. & Wu, Q. (2002). Polyamide 66/clay nanocomposites via melt intercalation. *Macromol Mater Eng,* Vol. 287, pp. 180-186.

Liu, W.; Hoa, SV. & Pugh, M. (2005). Fracture toughness and water uptake of high-performance epoxy/nanoclays nanocomposites. *Compos Sci Technol,* Vol. 65, pp. 2364-2373.

Ma, J.; Xu, J.; Ren, JH.; Yu, ZZ. & Mai, YW. (2003). A new approach to polymer/montmorillonite nanocomposites. *Polymer,* Vol. 44, pp. 4619-4624.

Manias, E.; Touny, A.; Wu, L.; Strawhecker, K.; Lu, B. & Chung, TC. (2001). Polypropylene/montmorillonite nanocomposites: Review of the syntheticroutes and materials properties. *Chem Mater,* Vol. 13, pp. 3516-3523.

Manias, E. (2001). Origins of the materials properties enhancements in polymer/clay nanocomposites, http://raman.plmsc.psu.edu/~manias/pdfs/nano2001b.pdf.

McGlashan, SA. & Halley, PJ. (2003). Preparation and characterization of biodegrdable starch-based nanocomposite materials. *Polym Int,* Vol. 52 , pp. 1767-1773.

McNally, T.; Murphy, WR.; Lew, CY.; Turner, RJ. & Brennan, GP. (2003). Polyamide-12 layered silicate nanocomposites by melt compounding. *Polymer,* Vol. 44, pp. 2761-2772.

Messersmith, PB. & Giannelis EP. (1994). Synthesis and characterization of layered silicate-epoxy nanocomposites. *Chem Mater,* Vol. 6, pp. 1719-1725.

Messersmith, PB. & Giannelis, EP. (1995). Synthesis and barrier properties of poly(ε-caprolactone)-layered silicate nanocomposites. *J Polym Sci Polym Chem,* Vol. 33, pp. 1047 - 1057.

Morawiec, J.; Pawlak, A.; Slouf, M.; Galeski, A.; Piorkowska, E. & Kransnikowa, N. (2005). Preparation and properties of compatibilized LDPE/organo-modified montmorillonite nanocomposites. *Eur Polym J,* Vol. 41, pp. 1115-1122.

Morgan, AB. & Gilman, JW. (2003). Characterization of polymer-layered silicate (clay) nanocomposites by transmission electron microscopy and X-ray diffraction: a comparative study. *J Appl Polym Sci,* Vol. 87, pp.1329-1338.

Nam, PH.; Maiti, P.; Okamoto, M.; Kotaka, T.; Hasegawa, N. & Usuki, A. (2001). A hierarchical structure and properties of intercalated polypropylene/clay nanocomposites. *Polymer,* Vol. 42, pp. 9633-9640.

Nielsen, LE. (1967). Models for the permeability of filled polymer systems. *J Macromol Sci,* A1, pp. 929-42.

Nielsen, LE. (1981). Mechanical properties of polymer and composites, vol. 2. New York: *Marcel Dekker.*

Noh, MW. & Lee, DC. (1999). Synthesis and characterization of PS-clay nanocomposite by emulsion polymerization. *Polym Bull,* Vol. 42, pp. 619-626.

Ogata, N.; Jimenez, G.; Kawai, H. & Ogihara, T. (1997). Structure and thermal/mechanical properties of poly(l-lactide)-clay blend. *J Polym Sci Polym Phys,* Vol. 35, pp. 389-396.

Okamoto, M.; Nam, PH.; Maiti, P.; Kotaka, T.; Nakayama, T.; Takada, M. et al. (2001). Biaxial flow-induced alignment of silicate layers in polypropylene/clay nanocomposite foam. *Nano Lett,* Vol. 1(9), pp. 503-505.

Osman, MA. ; Mittal, V. & Lusti, HR. (2004). The aspect ratio and gas permeation in polymer-layered silicate nanocomposites. *Macromol Rapid Commun,* Vol. 25, pp. 1145-1149.

Pegoretti, A.; Kolarik, J.; Peronic, C. & Migliaresi, C. (2004). Recycled poly(ethyleneterephthalate) layered silicate nanocomposites: morphology and tensile mechanical properties. *Polymer,* Vol. 45, pp. 2751-2759.

Petricova, E.; Knight, R.; Schadler, LS. & Twardowski, TE. (2000). Nylon 11/silica nanocomposite coatings applied by the HVOF process. II. Mechanical and barrier properties. *J Appl PolymSci,* Vol. 78, pp. 2272-2289.

Phang, IY.; Liu, T.; Mohamed, A.; Pramoda, KP.; Chen, L.; Shen, L. et al. (2005). Morphology, thermal and mechanical properties of nylon 12/organoclay nanocomposites prepared by melt compounding. *Polym Int,* Vol. 54, pp. 456-464.

Porter, D.; Metcalfe, E. & Thomas, MJ. (2000). Nanocomposite fire retardants-a review. *Fire Mater,* Vol. 24, pp. 45-52.

Ray, SS.; Maiti, P.; Okamoto, M.; Yamada, K. & Ueda, K. (2002). New polylactide/layered silicate nanocomposites. 1. Preparation, characterization and properties. *Macromolecules,* Vol. 35, pp. 3104-3110.

Ray, SS.; Yamada, K.; Okamoto, M. & Ueda, K. (2003). New polylactide/layered silicate nanocomposites. 2. Concurrent improvements of material properties, biodegradability and melt rheology. *Polymer,* Vol. 44, pp. 857-866.

Ray, SS. & Okamoto, M. (2003). Polymer - layered silicate nanocomposite: a review from preparation to processing. *Prog Polym Sci,* Vol. 28, pp. 1539-1641.

Sepehr, M.; Utracki, LA.; Zheng, X. & Wilkie, CA. (2005). Polystyrenes with macrointercalated organoclay. Part I. Compounding and characterization. *Polymer,* Vol. 46, pp. 11557-11568.

Shelley, JS.; Mather, PT.; Devries, KL. (2002). Reinforcement and environmental degradation of nylon 6/clay nanocomposites. *Polymer,* Vol. 42, pp. 5849-5858.

Shia, D.; Hui, CY.; Burnside, SD. & Giannelis, EP. (1998). An interface model for the prediction of Young' s modulus of layered silicate-elastomer nanocomposites. *Polym Compos,* Vol. 19, pp. 608-617.

Solomon, MJ.; Almusallam, AS.; Seefeldt, KF.; Somwangthanaroj, A. & Varadan, P. (2001). Rheology of polypropylene/clayhybridmaterials. *Macromolecules,* Vol. 34, pp. 1864-1872.

Srivastava, SK.; Pramanik, M. & Acharya, H. (2006). Ethylene/vinyl acetate copolymer/clay nanocomposites. *J Polym Sci Polym Phys,* Vol. 44, pp. 471-480.

Thellen, C.; Orroth, C.; Froio, D.; Ziegler, D.; Lucciarini, J.; Farrell, R. et al. (2005). Influence of montmorillonite layered silicate on plasticized poly(llactide) blown films. *Polymer,* Vol. 46, pp. 11716-11727.

Theng, BKG. (1979). Formation and properties of clay-polymer complexes. *Amsterdam: Elsevier.*

Tortora, M.; Gorrasi, G.; Vittoria, V.; Galli, G.; Ritrovati, S. & Chiellini, E. (2002). Structural characterization and transport properties of organically modified montmorillonite/polyurethane nanocomposites. *Polymer,* Vol. 43, pp. 6147-6157.

Tortora, M.; Vittoria, V.; Galli, G.; Ritrovati, S. & Chiellini, E. (2002). Transport properties of modified montmorillonite-poly(ε-caprolactone) nanocomposites. *Macromol Mater Eng,* Vol. 287, pp. 243-249.

Trevino, JC. & Coles, CA. (2003). Kaolinite properties, structure and influence of metal retention on pH. *Appl Clay Sci,* Vol. 23, pp. 133-139.

Tunney, JJ. & Detellier, C. (1996). Poly(ethylene glycol)-kaolinite intercalates. *Chem Mater,* Vol. 8, pp. 927-935.

Usuki, A.; Kojima, Y.; Kawasumi, M.; Okada, A.; Fukushima, Y.; Kurauchi, T. et al. (1993) Synthesis of nylon 6-clay hybrid. *J Mater Res,* Vol. 8, pp. 1179-1184.

Usuki, A.; Kawasumi, M.; Kojima, Y.; Okada, A.; Kurauchi, T. & Kamigaito, O. (1993). Swelling behavior of montmorillonite cation exchanged for ω-amino acid by ε-caprolactam. *J Mater Res,* Vol. 8, pp. 1174-1178.

Vaia, RA.; Ishii, H. & Giannelis, EP. (1993). Synthesis and properties of two dimensional nanostructures by direct intercalation of polymer melts in layered silicates. *Chem Mater,* Vol. 5, pp. 1694-1696.

Vaia, RA. & Giannelis, EP. (1997). Lattice of polymer melt intercalation in organicallymodified layered silicates. *Macromolecules,* Vol. 30, pp. 7990-7999.

Vaia, RA. & Wagner HD. (2000). Framework for nanocomposites. *Mater Today,* Vol. 7, pp. 32-37.

Vaia, RA. & Giannelis, EP. (2001). Liquid crystal polymer nanocomposites: direct intercalation of thermotropic liquid crystalline polymers into layered silicates. *Polymer,* Vol. 42, pp. 1281-1285.

Varlot, K.; Reynaud, E.; Kloppfer, MH.; Vigier, G. & Varlet, J. (2001). Clay-reinforced polyamide: preferential orientation of the montmorillonite sheets and the polyamide crystalline lamellae. *J Polym Sci Polym Phys,* Vol. 39, pp. 1360–1370.

Vyazovkin, S.; Dranka, I.; Fan, X. & Advincula, R. (2004). Kinetics of the thermal and thermo-oxidative degradation of a polystyrene-clay nanocomposite. *Macromol Rapid Commun,* Vol. 25, pp. 498-503.

Xie, W.; Gao, Z.; Liu, K.; Pan, WP.; Vaia, R.; Hunter, D. et al. (2001). Thermal characterization of organicallymodifiedmontmorillonite. *Thermochim Acta,* Vol. 367, pp. 339-350.

Xiong, J.; Liu, Y.; Yang, X. & Wang, X. (2004). Thermal and mechanical properties of polyurethane/montmorillonite nanocomposites based on a novel reactive modifier. *Polym Degrad Stabil,* Vol. 86, pp. 549-555.

Yano, K.; Usuki, A.; Okada, A.; Kurauchi, T. & Kamigaito, O. (1993). Synthesis and properties of polyimide/clay hybrid. *J Polym Sci Polym Chem,* Vol. 31, pp. 2493-2498.

Zanetti, M.; Lomakin, S. & Camino, G. (2000). Polymer layered silicate nanocomposites. *Macromol Mater Eng,* Vol. 279, pp. 1-9.

Zanetti, M ; Bracco, P. & Costa, L. (2004). Thermal degradation behavior of PE/clay nanocomposites. *Polym Degrad Stabil,* Vol. 85, pp. 657-665.

Zerda, AS. & Lesser, AJ. Intercalated clay nanocomposites: morphology, mechanics and fracture behavior. *J Polym Sci Polym Phys,* Vol. 39, pp. 1137-1146.

Zhao, C.; Qin, H.; Gong, F.; Feng, M.; Zhang, S. & Yang, M. (2005). Mechanical, thermal and flammability properties of polyethylene/clay nanocomposites. *Polym Degrad Stabil,* Vol. 87, pp. 183-189.

Zheng, X.; Jiang, DD. & Wilkie, CA. (2006). Polystyrene nanocomposites based on an oligomerically-modified clay containing maleic anhydride. *Polym Degrad Stabil,* Vol. 91, pp. 108-113.

Zhu, J.; Uhl, FM.; Morgan, AB. & Wilkie, CA. (2001). Studies on the mechanism by which the formation of nanocomposites enhances thermal stability. *Chem Mater,* Vol. 13, pp. 4649-4654.

14

Structural and Electron Transport Properties of Ultrathin SiO_2 Films with Embedded Metal Nanoclusters Grown on Si

Andrei Zenkevich[1], Yuri Lebedinskii[1], Oleg Gorshkov[2], Dmitri Filatov[2] and Dmitri Antonov[2]
[1]*National Research Nuclear University "Moscow Engineering Physics Institute"*
[2]*N.I. Lobachevskii University of Nizhnii Novgorod*
Russia

1. Introduction

During the last decade, much attention has been focused on the investigation of the semiconductor and metal nanocrystals (NCs) embedded in the dielectric matrices. The interest was generated by the promising applications of the nanocomposite structures in nanoelectronics. Particularly, the semiconductor or metal NCs embedded in the dielectric layer of a metal-insulator-semiconductor field-effect transistor (MOSFET) may replace the SiN_x floating gates in the nonvolatile memory devices, allowing for thinner injection oxides, and subsequently, smaller operating voltages, longer retention times, and faster write/erase speeds (Tiwari et al., 1996). The performance of such memory devices strongly depends on the parameters of NCs arrays, such as their size, shape, spatial distribution, electronic band alignment, as well as on the possibility to make reproducibly the uniform tunnel transparent oxide films.

The charge accumulation in the NCs can be limited by the single-electron effects such as Coulomb blockade provided that the cluster has a sufficiently small size, which, in principle, allows for the single electron memory devices (Yano et al., 1994; Guo et al., 1997).

Up to the present time, the thin film nanocomposite structures have been studied extensively (Ruffino et al., 2007). The most popular methods to fabricate NCs in the dielectric matrices include low-energy ion implantation with subsequent annealing (Bonafos et al., 2000), the deposition of the non-stoichiometric oxide layers also followed by the annealing (Tiwari et al., 2000), and the deposition of the multilayered oxide/NC structures (Ruffino & Grimaldi, 2007).

The main disadvantage of the ion implantation is a considerable thickness of the layer where the NCs are nucleating, and also a rather large dispersion in the NCs' sizes. The latter fact is a direct consequence of the NCs' nucleation by Ostwald ripening. The largest NCs with the minimum density are concentrated at the mean projected ion path, while the smaller NCs with higher density are nucleated in the tails of the implanted ions depth distribution. Recent attempts to improve the ion implantation technique to address the above problems are concerned mainly about the reduction of the ion energy down to ~ 1 keV or less (Ren et al., 2009). However, it requires a considerable complication of the implanters to provide a high enough ion beam density at low ion energies.

The SiO_2 layers containing the Si or Ge NCs arranged in single sheet(s) can be obtained by deposition of the thin (3 to 7 nm) suboxidized SiO_x (GeO_x, x = 1.3 to 1.7) layers sandwiched between the SiO_2 spacers, e. g., by electron beam or magnetron sputtering followed by annealing (Zacharias et al, 2002). However, the NC size dispersion inside the layers due to their nucleation through Ostwald ripening still remains an unresolved problem. More recently, ordered multilayered arrays of the Au NCs have been fabricated using the self-assembling effects in nucleation of the Au NCs on the surface of SiO_2 spacer layer both deposited by magnetron sputtering (Cho et al., 2004).

In this chapter, the two novel approaches to the fabrication of the ultrathin SiO_2/SiO_2:NC-Me/SiO_2/Si (Me = Au, Pt) nanocomposite structures are described. One of these approaches is based on the room temperature deposition of an ultrathin mixed Si —Me amorphous film by Pulsed Laser Deposition (PLD) combined with the further oxidation of the Si-Me mixture in the glow discharge oxygen plasma in a single vacuum cycle (Zenkevich et al., 2009). Another one exploits the effect of metal segregation during the thermal oxidation of the pre-deposited ultrathin Si—Me layers. Both approaches allowed the fabrication of the single sheet two-dimensional arrays of the Me NCs sandwiched between the two ultrathin SiO_2 layers with precisely controlled thickness.

Among the other ones, the SiO_2:NC-Au material system is of a special interest. Silicon is known to have an extremely high diffusivity in Au, that promotes the phase segregation in the SiO_2—Au system even at low temperatures (Hiraki et al., 1972). Since Au and Si do not form any stable chemical compound and there is no stable gold oxide, Au can be expected to precipitate into the NCs during the oxidation of the Au—Si mixture. In this scope, it is important to have an initial uniform amorphous mixture of Au and Si atoms. PLD technique, owing to its pulsed nature and low deposition rates (0.01–0.05 monolayers (ML) per pulse) allows a precise control over both the composition and the thickness of the depositing layer, and hence is particularly suitable technique to prepare uniform mixed layers by sequential deposition from the elemental targets.

The structural properties of thus fabricated nanocomposite structures as well as the metrology of the metal NCs, including their spatial and size distribution in the dielectric matrix as a function of the initial Me/Si ratio as well as of the processing conditions will be described. The structural properties of the ultrathin SiO_2/SiO_2:NC-Me/SiO_2/Si layers will be further related to their electronic properties. The electron tunnelling through the individual metal NCs embedded in the ultrathin dielectric films have been investigated using Tunnelling Atomic Force Microscopy (AFM) technique. The Au NCs sandwiched between the ultrathin SiO_2 layers on the conductive Si substrates are visualized in the Tunnelling AFM images of the nanocomposite films as the spots of increased probe current (or the current channels) and attributed to the electron tunnelling through the individual Au NCs. The tunnelling spectra of the nanocomposite films measured in the current channels exhibit a Coulomb staircase at room temperature, while a negative differential resistance is observed in the spectra measured in the smallest current channels ascribed to the resonant electron tunnelling through the size-quantized Au NCs. Finally, a theory describing the imaging of the metal NCs embedded into a thin dielectric film on a conductive substrate is presented.

2. Growth and characterization of the SiO_2:nc-metal nanocomposite films

The investigated SiO_2/SiO_2:NC-Me/SiO_2 nanocomposite structures have been grown in an ultrahigh vacuum (UHV) setup based on Kratos® XSAM-800 electron spectrometer. The schematic of the growth setup is presented in Fig. 1. This setup allows the growth of the multilayered nanocomposite structures by PLD (including reactive PLD in various gas ambient) combined with the thermal/plasma oxidation and the analysis of the deposited layers *in situ* by the combination of surface sensitive techniques, particularly, by X-ray Photoemission Spectroscopy (XPS).

The deposition of the SiO_2:NC-Me films was carried out in the preparation chamber of the Kratos® XSAM-800 spectrometer, equipped with PLD module based on the pulsed YAG:Nd Q-switched laser. The second harmonic of the YAG:Nd laser radiation was used (the emission wavelength $\lambda \approx 532$ nm). The energy in the laser pulse was varied in the range from 0.1 to 0.3 J, the pulse duration was ≈ 15 ns, and the repetition rate was set 25 Hz.

The SiO_2/SiO_2:NC-Au/SiO_2 films for the Tunnelling AFM investigations were grown on the As doped n^+-Si(001) substrates with the resistivity of $\rho \approx 0.002$ Ohm·cm. The Si substrates were chemically cleaned to remove native oxide prior to loading into the preparation chamber using a standard Radio Corporation of America (RCA) treatment. It is worth noting that XPS revealed no traces of the oxide on the Si substrate surface after loading in UHV.

The process of SiO_2/SiO_2:NC-Au/SiO_2/Si nanocomposite structures growth using the glow discharge oxygen plasma oxidation is shown schematically in Fig. 2. All the steps of the growth process were performed at room temperature. First, a the uniform SiO_2 layers with the thickness $d_u \sim 1$ nm were formed by the glow discharge plasma oxidation of the chemically cleaned Si substrates (Fig. 2, *a*). The partial pressure of oxygen p_O in the preparation chamber during the oxidation was $p_O \sim 10^{-2}$ mbar, the discharge voltage U was maintained in the range from 500 to 800 V.

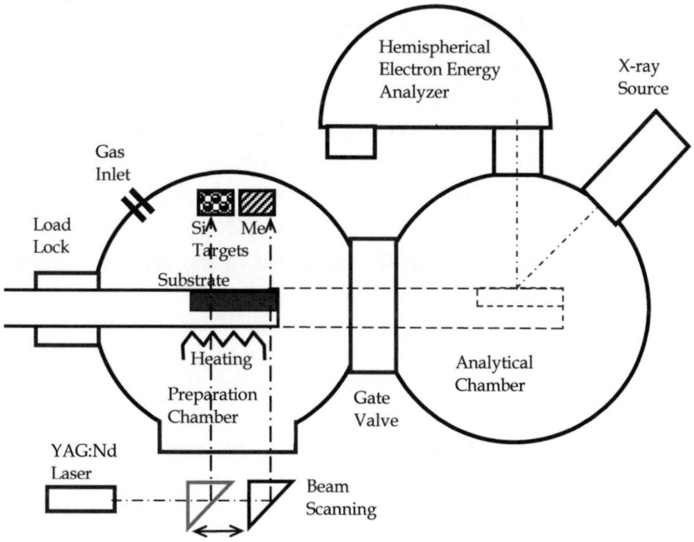

Fig. 1. The schematic of the growth setup.

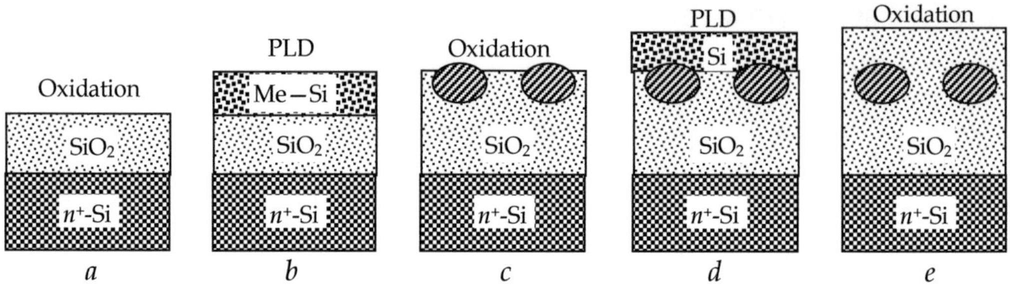

Fig. 2. The schematic of the SiO_2/SiO_2:NC-Me/SiO_2/n^+-Si structure growth using the combination of PLD and the oxidation in the glow discharge plasma.

The ultrathin mixed amorphous Au–Si layers with the thickness $d_{Au-Si} \sim 1$ nm (Fig. 2, *b*) were deposited onto the SiO_2 surface by sequential deposition of the submonolayer portions of Au and Si from the elemental targets mounted inside the preparation chamber (Fig. 1). The Au/Si composition ratio was varied form 0.15 to 0.5 by varying the number of pulses per each target. The laser beam was alternately directed onto Si and Au targets. A computerized double prism beam scanning system was used both for raster scanning of the laser beam over the targets' surface and for switching from one elemental target to another according to the deposition program. The calibration of the Au and Si deposition rates was made by Rutherford Backscattering Spectrometry (RBS) using He^+ ions with the energy of 1.5 to 2 MeV. RUMP® software was used to analyze the RBS spectra. According to RBS, the deposition rate was ~ 10^{13} atoms/cm^2 (~ 0.1 ML) per a pulse and depended on the specific target.

At the next step, the Au–Si mixture layer was oxidized in the glow discharge oxygen plasma at the same parameters as were used in growing the underlying SiO_2 layer. The oxidation has resulted in the precipitation of the Au NCs in and/or at the surface of the SiO_2 layer (Fig. 2, *c*). Finally, to cap the formed Au NCs, a ~ 1 nm thick amorphous Si layer was deposited on top and oxidized in the oxygen plasma (Fig. 2, *d* and *e*).

The process of the SiO_2/SiO_2:NC-Au/SiO_2 nanocomposite structure growth was monitored *in situ* by XPS in the analytical chamber of Kratos® XSAM-800 spectrometer (Fig. 1) as described in detail elsewhere (Lebedinskii et al., 2005). The XPS analysis provided the information on the chemical state of the film constituents as well as on the layers' thickness at each step of the growth process. The latter was calculated using the well-known procedure based on the attenuation of the respective XPS lines by the growing layers (Hochella Jr. & Carim, 1988). In particular, the thickness of the underlying SiO_2 layer (Fig. 2, *a*) was calculated as

$$d_u = \lambda_a \ln \frac{I_0}{I_1}, \tag{1}$$

where I_0 and I_1 are the intensities of the Si 2*p* XPS line recorded from the bare Si substrate and upon its oxidation, respectively (Fig. 3, spectrum 1), and $\lambda_a \approx 2.4$ nm is the free path of the photoelectrons with the energy $E \approx 100$ eV in SiO_2. For the room temperature glow discharge plasma oxidation of Si surface d_u was found to vary depending on the oxidation time, but was limited to ≈ 3 nm.

The evolution of the XPS Si 2*p* and Au 4*f* lines during SiO_2/SiO_2:NC-Au/SiO_2/n^+-Si structure growth is presented in Fig. 3. Upon deposition of the mixed Si–Au layer (Fig. 2, *b*), an additional component of the Si 2*p* line appears in the XPS spectrum (Fig. 3, spectrum 2). In addition, a shift of the Au 4*f* line with respect to the metallic Au reference spectrum (cf. spectra 2 and 6 in Fig 3) is observed. Both effects indicate the formation of a metastable Au silicide. The thickness of the deposited Si–Au layer was set to be less than 2 nm to ensure its full oxidation in the glow discharge oxygen plasma at the next step (Fig. 2, *c*). Upon oxidation (Fig. 3, spectrum 3), the Au 4*f* line is shifted close to its position in bulk metallic Au, while the silicide component of the Si 2*p* line line is converted into the Si^{4+} indicating the formation of SiO_2. From these changes the formation of the Au precipitates in or at the surface of the SiO_2 layer is evident.
Upon deposition of the amorphous Si onto the SiO_2:NC-Au layer (Fig. 2, *d*), the formation of the Au silicide bonds is again clearly observed in XPS spectra (Fig. 3, cf. spectra 4 and 2). This observation we take as an evidence that at least some of the precipitated Au NCs are not embedded in the SiO_2 layer.

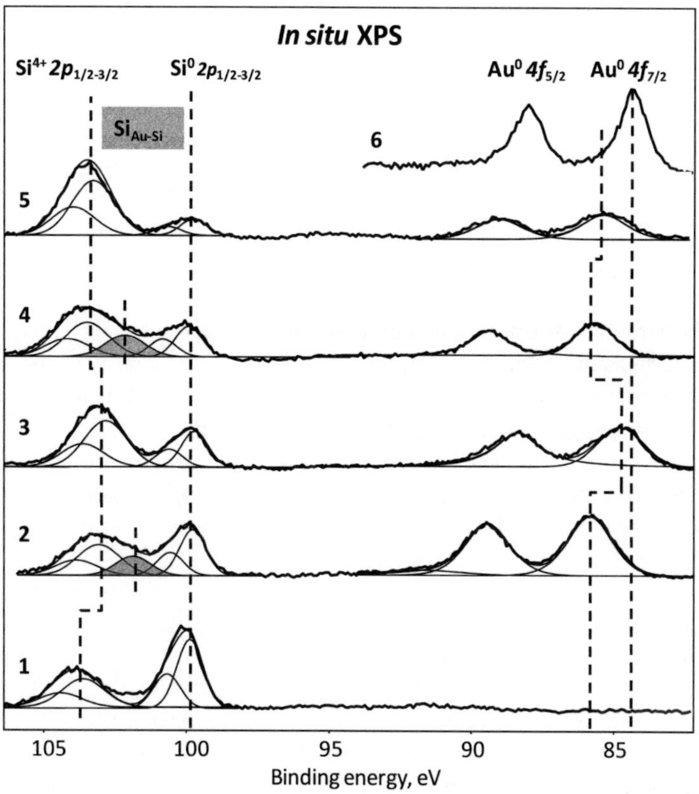

Fig. 3. The evolution of the XPS spectra during the growth of a SiO_2/SiO_2:NC-Au/SiO_2/n^+-Si(001) nanocomposite structure. *1* – upon oxidation of the Si substrate in the glow discharge oxygen plasma; *2* – upon deposition of the mixed Au–Si layer; *3* – upon oxidation of the Au–Si layer; *4* – after deposition of the amorphous Si cap layer; *5* – after oxidation of the latter; 6 – the spectrum of the bulk metallic Au (reference). Reprinted from (Zenkevich et al., 2009), with permission from ©Elsevier®.

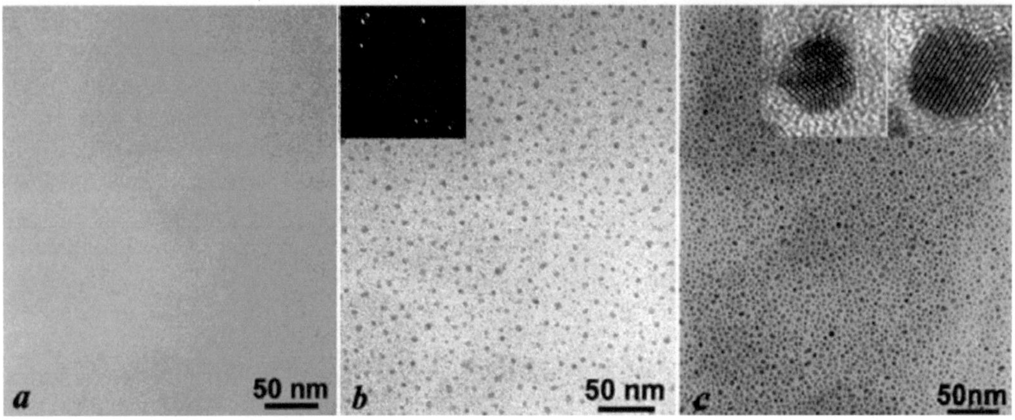

Fig. 4. The plain-view TEM bright-field images of the nanocomposite structures grown on NaCl(100): *a* – Si/Si-Au/Si (reference); *b, c* – SiO_2/SiO_2:NC-Au/SiO_2. Au/Si ratio: *a, b* – 0.15; *c* – 0.5. Insets: to Fig. *b* – dark-field image indicating the presence of crystalline NCs; to Fig. *c*– the high-resolution TEM images of the individual Au NCs. Reprinted from (Zenkevich et al., 2009), with permission from ©Elsevier®.

Finally, upon the oxidation of the capping amorphous Si layer, the Au 4*f* line is back at metallic Au position again, while all the deposited Si is in the oxidized state (SiO_2). The latter conclusion is evident from the increase of the Si^{4+} peak at the expense of the Si^0 components in the Si 2*p* line (Fig. 3, spectrum 5).The quantitative analysis of the XPS spectra presented in Fig. 3 gives the Au/Si ratio to be ≈ 0.2, the thickness of the bottom SiO_2, SiO_2:NC-Au and capping SiO_2 layers to be $d_b \approx 1.6$ nm, $d_{NC} \approx 1.6$ nm, and $d_c \approx 1.8$ nm, respectively.

The structural characterization of the nanocomposite structures was carried out with the plain-view Transmission Electron Microscopy (TEM) in the bright- and dark-field geometries using JEOL® JEM 2000EX instrument operating at the beam energy of 180 keV. The nanocomposite SiO_2/SiO_2:NC-Au/SiO_2 structures for the TEM investigations were grown on the freshly cleaved NaCl(001) substrates. Both underlying and cap SiO_2 layers were formed by oxidation of the deposited amorphous Si. In addition, a 10 nm thick amorphous carbon (graphite) film was deposited *in situ* by PLD onto the cap SiO_2 layer to ensure the integrity of the nanocomposite film after the lift-off.

To prove that the precipitation of Au NCs occurs during the oxidation of the mixed Au–Si layer, a reference sample has been grown missing the oxidation steps. Fig. 4, *a* presents a plain-view TEM image of the reference sample. No NCs are observed. In contrast, in the bright-field TEM image of an oxidized sample (Fig. 4, *b*) the Au NCs are clearly visible. The sheet density of the Au NCs N_s is clearly dependent on the Au/Si ratio (cf. Au/Si ratio 0.15 vs. 0.5 in Fig. 4, *b* and *c,* respectively). The dark-field TEM image presented on the inset in Fig. 4, *b* reveals the crystalline structure of the Au NCs as is further evident from the high-resolution TEM images of the individual Au NCs shown on the inset in Fig. 4, *c*. The detailed high resolution TEM analysis reveals both faceted and round shapes of the Au NCs, which is suggested to depend on whether the NCs precipitate near the surface or inside the SiO_2 layer during the oxidation of the Au–Si mixture, respectively. According to the detailed analysis of the plain-view TEM images, the typical values of the lateral size *D* and the areal density N_s of the Au NCs are 2 to 5 nm and (1–3) × 10^{13} cm^{-2}, respectively, subject to the Au/Si ratio. In order to characterize the spatial distribution of the Au NCs in the nanocomposite structures with different Au/Si ratios, the separations between the nearest neighbors for each NC were calculated from the plain-view TEM images. The distribution of the separations between the NCs ℓ found experimentally were then fitted with the model function derived from the Poisson distribution (valid for a small surface coverage):

$$f(\ell) = 2\pi \exp\left[-\pi N_s (\ell - R_0)^2\right] N_s (l - R_0), \tag{2}$$

for $\ell \geq R_0$. Here $R_0 = \langle D \rangle / 2$ is the mean lateral radius of the NCs and the areal density N_s were the fitting parameters. The values of $\langle D \rangle$ and N_s obtained by the fitting of the distributions of ℓ for the samples with different Au/Si ratios were in good agreement with those directly measured from the TEM images. Having determined R_0 and N_s, one can calculate the relative dispersion of the NCs' separation:

$$\frac{\Delta \ell}{\langle \ell \rangle} = \frac{\sqrt{4-\pi}}{\sqrt{\pi}\left(2R_0\sqrt{N_s}+1\right)}. \tag{3}$$

It was found that $\Delta\ell / \langle \ell \rangle$ decreased from ≈ 0.28 down to ≈ 0.15 with increasing Au/Si ratio from ≈ 0.2 to ≈ 0.5.

The kinetics of Au segregation during the plasma oxidation of the ultrathin amorphous Au–Si mixture layers at room temperature has been also analyzed. It should be noted that the oxygen plasma treatment of ultrathin Si and/or Au–Si layers, besides providing the active oxygen atoms, may also produce thermal and radiation effects in the deposited layer worth evaluated. To assess possible heating of the Si substrate surface by the glow discharge plasma with the discharge current $I = 10$ mA at $U = 500$ V during the treatment time $t = 60$ s, one can first estimate the heating expansion length x in Si at given conditions $x \sim (\kappa t/\rho c)^{1/2} \sim$ 10 cm (here κ is the thermal conductivity, c is the heat capacity, and ρ is the density of Si) that is much larger than the substrate thickness ($\approx$ 0.3 mm). Thus, the heating of the Si substrate during the plasma treatment is rather uniform in depth, and can be evaluated suggesting the fraction of the total plasma energy $Q = UIt \sim 300$ J incident onto the Si sample surface with the area of ~ 1 cm^2 while the total chamber area is $\sim 2 \times 10^3$ cm^2. So far, the heating by the plasma $\Delta T = Q/mc$ (where m is the mass of the Si substrate) appears to be limited to $\sim$ 10 K, and can thus be neglected. The radiation effects of the oxygen glow discharge plasma can certainly play an essential role in the oxidation of the Au–Si ultrathin layers as well as on the kinetics of the Au NCs' precipitation. We suggested that the oxidation starts from the upper Si atoms forming a few ML thick SiO_2. At this stage, the Au–Si chemical bonds are breaking, and due to the local strain and/or radiation heating, the Au atoms receive a sufficient energy to travel a distance of few lattice constants. Due to low solubility of Au in SiO_2, the Au atoms segregate at SiO_2 surface and further nucleate in the NCs. An alternative way is the segregation of Au NCs at the bottom interface of the SiO_2 layer through a well known mechanism of Au segregation by the Si oxidation front due to anomalously high diffusivity of the unoxidized Si atoms through the nucleated Au NCs (Hiraki et al., 1972).

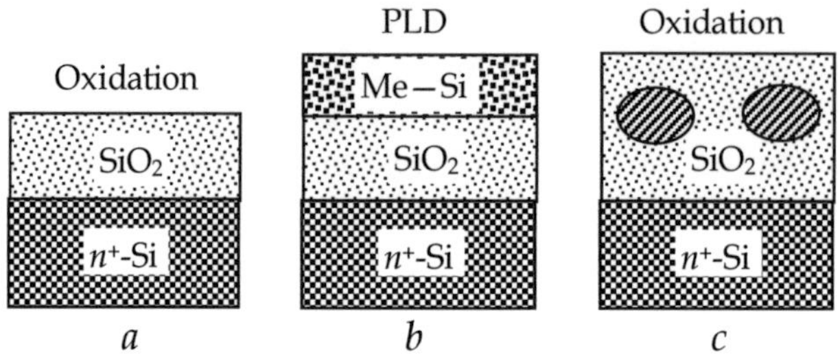

Fig. 5. The schematic of a SiO_2/SiO_2:NC-Me/SiO_2/*n*-Si structure growth using thermal oxidation.

As a result, the two types of the Au NCs precipitating at the surface of the oxidizing Au–Si layer and at the bottom interface with the underlying SiO_2 layer can be expected. The observation by TEM of two types of Au NCs corroborates this suggestion.

In Fig. 5, the schematic of the alternative growth process of the SiO_2/SiO_2:NC-Me/SiO_2/*n*-Si nanocomposite structures using the thermal oxidation of the mixed Me–Si layer is presented. The process consists of the three stages. At the first stage (Fig. 5, *a*), a SiO_2 layer 6 to 8 nm in thickness is formed by the thermal oxidation of the Si substrate.

The *n*-Si(001) substrates doped by phosphorus with the resistivity of $\rho \approx 4.5$ Ohm·cm have been used. The thermal oxidation of the Si substrates as well as that of the mixed Me–Si layers have been carried out using an industrial SDO-125/4A furnace. The temperature of the oxidation of the Me–Si layers T_A was chosen in the range T_A = 640 ÷ 725°C, the oxidation time t_A was varied from 1 to 9 hours. At the next stage, the mixed Me–Si layer (Me = Au, Pt) was deposited onto the underlying SiO_2 layer surface by the co-deposition of the noble metal and Si by PLD at the parameters described above.

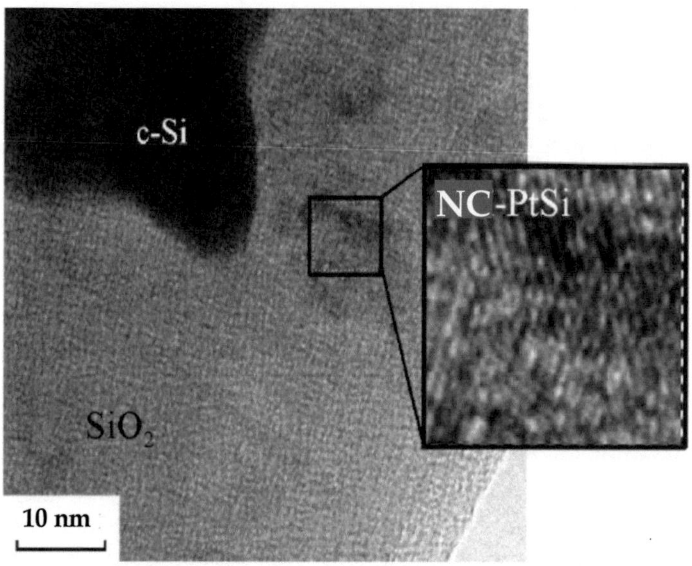

Fig. 6. A plain-view TEM image of the SiO_2/SiO_2:NC-Pt/SiO_2/*n*-Si nanocomposite structure. Reprinted from (Maksimova et al., 2009), with permission from ©Intercontact Science® Publishing.

The Me/Si ratio in the deposited Me—Si layers was varied from 1 : 20 to 1 : 30, the layer thickness was 5 to 20 nm, and the corresponding nominal thickness of the deposited Me layer was 0.5 to 0.7 nm. Finally, the thermal oxidation of the deposited Me—Si layer was performed. During the oxidation process, Me atoms effectively segregate towards the lower interface between the mixed Me—Si layer and the underlying SiO_2 one. The metal NCs segregation process is facilitated by the preferential oxidation of Si, by a low solubility of the noble metals in SiO_2, and by the anomalously high diffusivity of Si through the nucleated metallic NCs. The bottom SiO_2 layer serves as a blocking layer for the segregation process, thus strictly defining the arrangement of the Me NCs in a single sheet.

The TEM investigations of the samples prepared by the thermal oxidation have been carried out using fEI® Tecnai™ G2 30 instrument at the accelerating voltage of 300 kV. A plain view TEM image of the SiO_2/SiO_2:NC-Pt/SiO_2/*n*-Si nanocomposite structure is presented in Fig. 6 *a*. The TEM analysis reveals that besides the Me NCs in the amorphous SiO_2 layer the nanoinclusions of the non-oxidized crystalline Si of ~ 50 nm in size are also present. These inclusions are likely formed at the interface between the Me—Si and the underlying SiO_2 layers. Also the analysis of the micro diffraction patterns shows that during the thermal oxidation of the Pt—Si mixture the platinum silicide NCs (particularly, the phase Pt_2Si) are formed with the size of 5 to 10 nm.

To perform $C-V$ measurements, the Al contacts with the area of 5×10^{-5} to 8×10^{-4} cm^2 have been deposited onto the SiO_2 cap layers by PLD through a shadow mask. The back ohmic contacts to the Si substrate were made using In—Ga alloy. In Fig. 7 the $C-V$ curves of the Al/SiO_2/SiO_2:NC-Me/SiO_2/*n*-Si nanocomposite metal-oxide-semiconductor (MOS) structures grown in various conditions are presented. The high frequency $C-V$ measurements had been carried out using Keithley® 590 setup at 1 MHz. The $C-V$ curve measured from the SiO_2/SiO_2:NC-Au/SiO_2/*n*-Si sample produced at the oxidation parameters T_A = 640°C, t_A = 540 min (Fig. 7, *a*, curve *1*) exhibits a broad hysteresis loop compared to the sample oxidized during t_A = 300 min (Fig. 7 *b*, curve 2) where the hysteresis loop is much narrower.

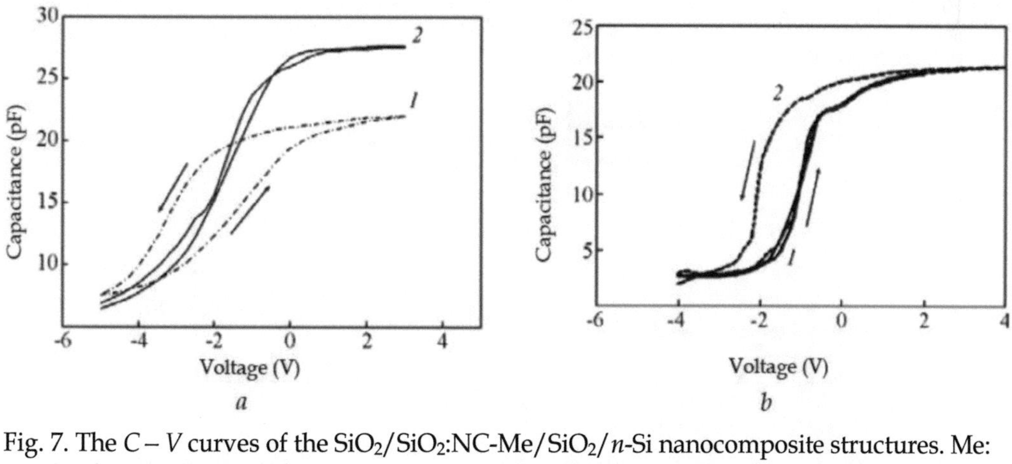

Fig. 7. The $C-V$ curves of the SiO_2/SiO_2:NC-Me/SiO_2/*n*-Si nanocomposite structures. Me: *a*— Au; *b* — Pt. T_A = 640°C; t_A, min: *1* — 540; 2 — 300. Reprinted from (Maksimova et al., 2009), with permission from ©Intercontact Science® Publishing.

The hysteresis in $C-V$ curves is ascribed to the charge accumulation in the Me NCs. The smaller value of the saturated capacitance for longer t_A is probably related to the larger total oxide thickness. It is important to note that almost no hysteresis is observed when the amplitude of the voltage sweep V is less than 2 V, however, a broad hysteresis loop appears when V is increased up to 4 V (Fig. 7, *b*). The dependence of the hysteresis in the $C-V$ curves on the sweep voltage amplitude points at the tunneling mechanism of the filling and the depletion of the NCs with electrons. The electron tunneling rate is determined by the thickness of the bottom SiO_2 layer d_u. Alternatively, the NC-Si grains observed in this sample by TEM (see Fig. 6) can also accumulate the electrons and therefore contribute to the memory effect.

3. Imaging of the metal nanoclusters embedded into the ultrathin dielectric films by Tunnelling AFM

The Tunneling AFM investigations of the SiO_2/SiO_2:NC-Au/SiO_2/n^+-Si nanocomposite structures were carried out in Omicron® MultiProbe S™ UHV system at room temperature. The measurement scheme is presented in Fig. 8. The sample surface was scanned across by a Pt coated AFM probe in the contact mode. Simultaneously, the $I-V$ curves of the probe-to-sample contact were acquired in each point of the scan.

Earlier, Tunnelling AFM had been applied mainly to study the defects in the thin dielectric films (Yanev et al., 2008). Antonov et al., 2004 have applied Tunnelling AFM to the investigation of the electron transport through the Zr NCs produced by ion implantation in the Zr(Y)O_2/p^+-Si films. Scanning Tunnelling Microscopy (STM) had been used earlier to study the electron transport through the metal NCs dispersed inside the dielectric films (Bar-Sadeh et al., 1994; Imamura et al., 2000). However, the applicability of STM to such objects is limited to the case when the NCs' concentration is high enough, i.e. when the films possess a sufficient percolation conductivity. In this case, keeping the STM feedback is provided by the continuous switching of the tunnelling electrons path through the film (hereinafter referred to as the current channel) from one chain of the NCs to another. The application of the Tunneling AFM allowed the studying of the locally nonconductive Zr(Y)O_2:nc-Zr/p^+-Si films since the feedback (AFM) channel (F_n) and the measuring one (the probe current I_t) were decoupled (Fig. 8).

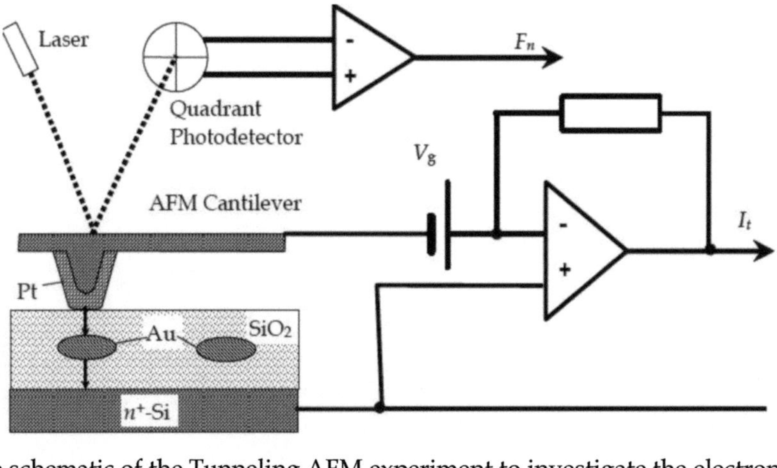

Fig. 8. The schematic of the Tunneling AFM experiment to investigate the electronic properties of a SiO_2/SiO_2:NC-Au/SiO_2/n^+-Si(001) nanocomposite structure.

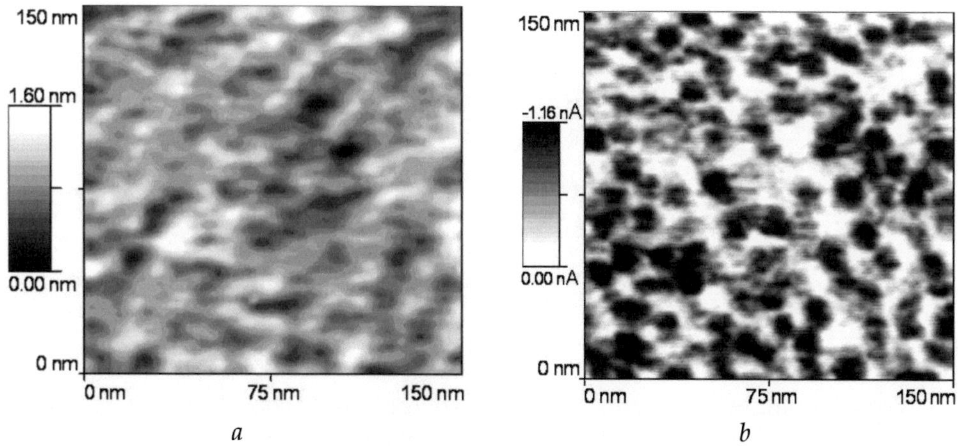

Fig. 9. The AFM (*a*) and the current (*b*) images of the SiO_2(1.5 nm)/ SiO_2:NC-Au/SiO_2(1.5 nm)/n^+-Si(001) nanocomposite structure. V_g = –2.5 V. Reprinted from (Filatov et al., 2010) under license by IoP Publishing Ltd.

In Fig. 9 the AFM and current images of the PLD/glow discharge oxidation grown SiO_2(1.5 nm)/SiO_2:NC-Au/SiO_2(1.5 nm)/n^+-Si film are presented. The inverted contrast in the current image in Fig. 9, *b* (the larger the probe current I_t, the darker the area) is related to the negative polarity of the bias voltage applied between the AFM probe and the sample V_g which, in turn, corresponds to the injection of the electrons from the Pt coated AFM tip into the n^+-Si substrate through the nanocomposite film.

The spots of increased I_t or the current channels of 3 to 10 nm in size observed in the current image (Fig. 9, *b*) are attributed to the electron tunnelling through the individual Au NCs. It is worth noting that the current image in Fig. 9, *b* weakly correlates with the AFM one presented in Fig. 9, *a* (the Pirson's correlation coefficient R_P calculated for this particular pair of images was $\approx$ 0.15).

In order to establish the relationship between the geometrical and electronic properties of the NCs in the current images of the SiO_2/SiO_2:NC-Me/SiO_2/n^+-Si nanocomposite structures , we have developed a model for the formation of the current images of the metal NCs in a dielectric film by Tunnelling AFM. The NCs were treated as the metal droplets of a spherical shape with the radius R_c. The model is based on the theory of STM (Tersoff & Hamann, 1985), which, in turn, employs the concept of the tunnel matrix element (Bardeen, 1961):

$$M_{pc} = \frac{\hbar^2}{2m} \int_S \left(\chi_p^* \nabla \chi_c - \chi_c^* \nabla \chi_p \right) d\mathbf{S}, \quad (4)$$

where χ_p and χ_c are the envelope wavefunctions of the electrons in the probe and in the NC, p and c are the generalized quantum numbers, indexing the electronic states in the probe and in the NC, respectively. Since we treat the probe coating as a bulk material, it seems reasonable to select $p = \{k_x, k_y, k_z, s\}$ where the first three elements are the respective components of the electron wavevector in the conduction band of the probe coating material and $s = \pm½$ is the spin.

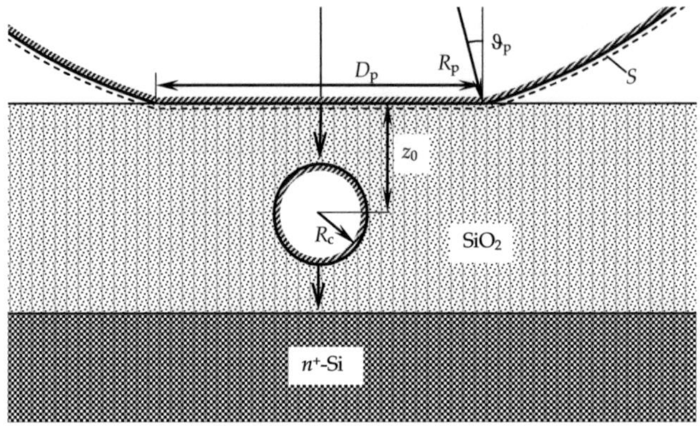

Fig. 10. The schematic of the electron tunneling between a metal coated AFM probe and a metal NC embedded inside a dielectric film on a conductive substrate.

We have considered the electron states both in the probe coating and in the NC to be twofold spin degenerate and therefore have left all the spin-related effects as well as the spin-orbit interaction induced ones beyond the present consideration. In the NC, due to a spherical symmetry of the problem, it seems reasonable to select $c = \{n, l, m, s\}$ where n, l, and m are the radial, orbital, and magnetic quantum numbers, respectively.

According to (Bardeen, 1961), the surface of integration S could be any one located entirely inside the potential barrier between the probe and the NC. We have considered a model probe shape to be a cone with a truncated spherical apex (Fig. 10). We have selected the integration surface S in a close proximity to the model probe surface but at the minimum distance $\delta > k_p^{-1} = \hbar/(2m_0A)^{1/2}$, where A is the workfunction of the probe coating material, and m_0 is the free electron mass. The integration surface S can be divided into the three parts: (i) the area of the contact between the probe and the dielectric film surface that is a circle of the diameter D_p; (ii) a spherical belt defined by the two polar angles ϑ_p and ϑ_c; and (iii) a conical side surface of the AFM probe tip. Taking into account a rapid decay of χ_c with increasing distance from the NC surface, one can neglect the contribution of the side cone surface into the integral (4). Following (Tersoff & Haman, 1985), we have selected the probe envelope wavefunction in the range $\vartheta_p < \vartheta < \vartheta_c$ in the asymptotic spherical form valid at least at the distance from the probe surface greater than the decay length of the electron wavefunction in the probe coating δ:

$$\chi_p = C_p \frac{R_p}{|\mathbf{r} - \mathbf{r}_0|\sqrt{\Omega_p}} \exp\left[k_p\left(R_p - |\mathbf{r} - \mathbf{r}_0|\right)\right], \quad (5)$$

where Ω_p is the probe volume and $C_p \sim 1$ is the normalizing constant. At $\vartheta < \vartheta_p$ (i. e. within the probe to surface contact area) the probe electron wavefunction was selected in the form:

$$\chi_{pd} = C_d \exp(k_d z) . \quad (6)$$

Here $k_d = [2m(A - X_d)]^{1/2}/\hbar$, X_d and m are the electron affinity and the effective electron mass in the dielectric, respectively. The normalizing constant C_d was adjusted to provide the continuity of the probe envelope wavefunction at $\vartheta = \vartheta_p$. Nevertheless, both the magnitude and the direction of the envelope wavefunction gradient still undergo a kink at $\vartheta = \vartheta_p$,

which is a drawback of the model. However, the calculations had shown the contribution of this "edge effect" into the total probe current to be small (< 0.3%) for the selected model parameters (R_p and D_p) so that it could be neglected. Note that at $D_p \to 0$ the probe envelope wavefunction is reduced to (5) (just this case has been considered by Tersoff & Hamann, 1985). Also, following (Tersoff & Hamann, 1985), we have assumed for simplicity that the workfunctions of the probe coating and NC materials are equal. Also, we have neglected the effect of the kinks of the effective electron mass and of the potential barrier height at the interface between the dielectric film and vacuum on the envelopes and their gradients.
The contact area size D_p can be estimated from the solution of Hertz' problem:

$$D_p = 2\sqrt[3]{\frac{F_n R_p}{K}}, \tag{7}$$

where F_n is the loading force,

$$\frac{1}{K} = \frac{3}{4}\left(\frac{1-\gamma_p^2}{Y_p} + \frac{1-\gamma_d^2}{Y_d}\right), \tag{8}$$

$Y_p(Y_d)$ are the Young's moduli and $\gamma_p(\gamma_d)$ are the Poisson coefficients of the probe coating and the dielectric film materials, respectively. For typical values of $F_n \sim 1$ nN and $R_p \approx 35$ nm, $D_p \sim 10$ nm. The tunneling rate of the electrons from the NC to the substrate Γ_u was assumed to exceed the one from the probe to the NC Γ_c, i. e. the NC was treated as a sink for the electrons with an infinite capacity. Therefore, we neglected the effect of Coulomb blockade considered in the next section.
The envelope wavefunctions of the electron in the NC were selected in the form of the radial part of a well known solution of the Schrödinger's equation for a spherical quantum dot (QD) with a finite potential barrier height U_0 in the effective mass approximation:

$$\cdot\ \chi_c = \begin{cases} C_i j_l\left(k_i\left|\mathbf{r}-\mathbf{r}_c\right|\right), \left|\mathbf{r}-\mathbf{r}_c\right| < R_c \\ C_c h_l^{(1)}\left(ik_c\left|\mathbf{r}-\mathbf{r}_c\right|\right), \left|\mathbf{r}-\mathbf{r}_c\right| > R_c \end{cases} \tag{9}$$

where $j_l(r)$ and $h_l^{(1)}(r)$ are the spherical Bessel and Hankel functions, respectively, $k_i = (2m_c E_F)^{1/2}/\hbar$, $k_c = [2m_d(A - X_d)]^{1/2}/\hbar$, where E_F is the Fermi energy in the NC (probe coating) material, $\mathbf{r}_c = \{0, 0, -z_0\}$ is the position of the NC's center beneath the dielectric film surface, C_i and C_c are the normalizing constants. We have neglected the angular dependence of χ_c because of the spherical symmetry of the NC and have normalized χ_c to unity as usual:

$$4\pi\int_0^\infty \chi_c(r) r^2 dr = 1. \tag{10}$$

The total probe current is given by (Tersoff & Hamann, 1985)

$$I_t = \frac{2\pi e}{\hbar}\sum_{c,p}\left|M_{cp}\right|^2 f(E_p)\left[1 - f(E_c + eV_g)\right]\delta\left(E_c - E_p\right). \tag{11}$$

In the limit of low voltages ($eV_g << A$) and low temperatures $T << eV_g/k_B$ where k_B is the Boltzmann constant, (11) can be expanded in the form:

$$I_t = \frac{2\pi e^2 V_g}{\hbar} \sum_c \left| M_{cp} \right|^2 \rho_p (E_F), \tag{12}$$

where $\rho_p(E_F)$ is the density of states (DOS) at the Fermi level in the probe coating material. The generalized index *c* counts the quantum confined states in the NC the energies of which E_c fall in the limits $E_F - eV_g < E_c < E_F$.

In Fig. 11 the model current images of a metal NC with R_c = 1 nm calculated for various values of D_p and of the depth of the NC's center position beneath the dielectric film surface z_0 (see Fig. 10) are presented. The two main conclusions important for the interpreting of the Tunnelling AFM results could be made from the modeling.

First, the size of the current image of an individual NC inside a dielectric film was found to be determined by the size of the contact area between the AFM probe tip to the sample surface D_p (Figs. 11 & 12). This is a particular manifestation of the convolution effect (Bukharaev et al., 1997). This effect, in turn, originates from a general principle of the theory of measurements. It claims that the result of any measurement is a convolution of the object function with the measuring instrument one. In STM, $|\chi_c(\mathbf{r})|^2$ and $|\chi_p(\mathbf{r})|^2$ play the roles of the object and measuring functions, respectively. However, in Tunnelling AFM the convolution effect manifests itself in a special way. In ordinary STM (in the constant current mode), the finite R_p results in smoothing and spreading of the features in the STM images: the less R_p, the sharper the STM image. In the limit of a point probe $[\chi_p(\mathbf{r}) \sim \delta(\mathbf{r}_0)]$, the STM

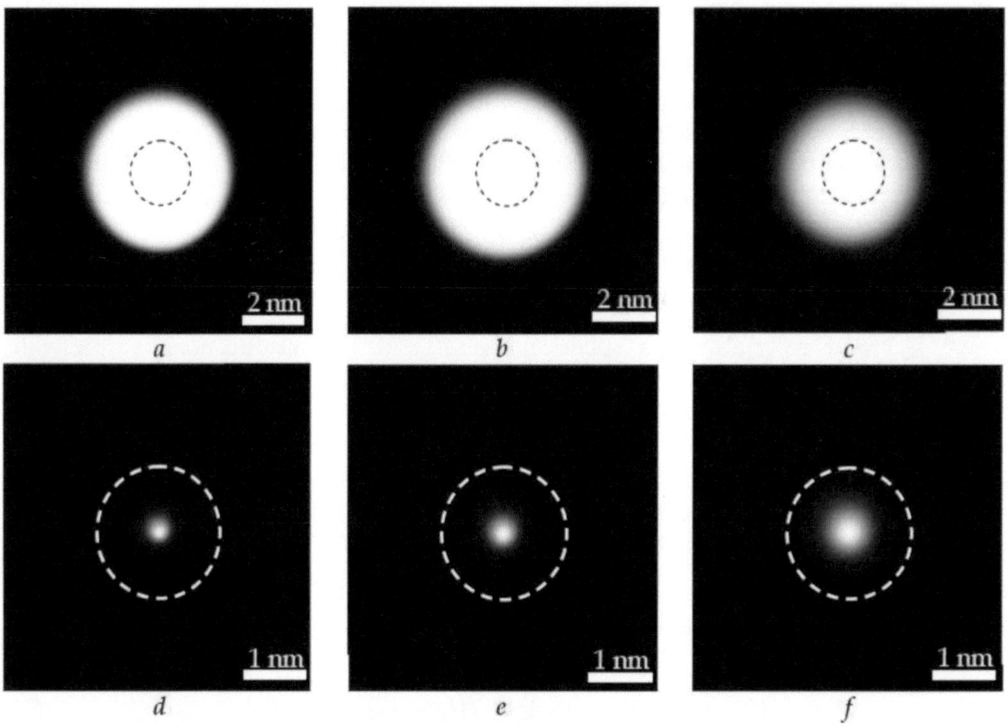

Fig. 11. The model current images $I_t(x, y)$ of a spherical Au NC in a SiO_2 film. R_c = 1 nm. R_p, nm: *a* – *c* — 35, *d* – *f* — 0.1. D_p, nm: *a* – *c* — 10, *d* – *f* — 0. z_0, nm: *a, d* — 1.2, *b, e* — 2.0, *c, f* — 5.0. The NC's contour is shown by the dashed circle.

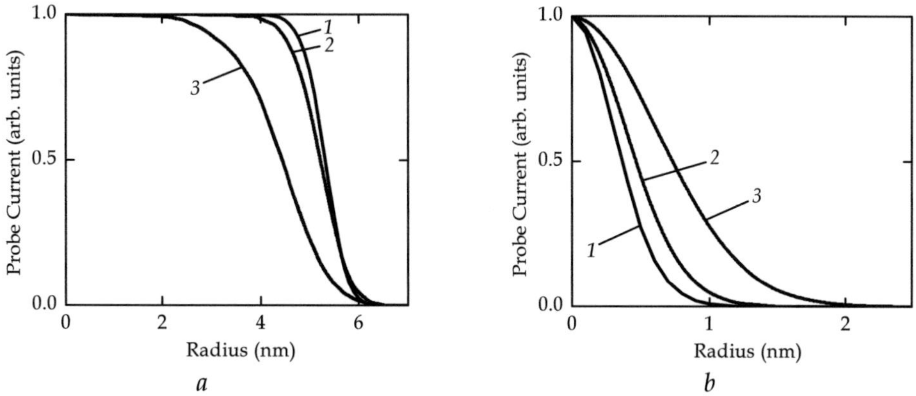

Fig. 12. The normalized model radial current image profiles $I_t(r)$ of a spherical Au NC in a SiO_2 film. $R_c = 1$ nm. R_p, nm: *a* – 35, *b* – 0.1. D_p, nm: *a* – 10, *b* – 0. z_0, nm: *1* – 1.2, *2* – 2.0, *3* – 5.0.

image reproduces the spatial distribution of the local density of states (LDOS) over the sample surface (Tersoff & Hamann, 1985). In contrary, in Tunnelling AFM the size of the current image of a NC is still determined by D_p even if $R_p << R_c$ (Fig. 11, *d–f* & Fig. 12, *b*). The latter case takes place in the STM investigation of the granular nanocomposite films (Bar-Sadeh et al., 1994; Imamura et al., 2000). The above effect originates from the fact that the AFM (or STM) probe follows the dielectric film surface in its moving (ideally, at constant height, as in constant height STM mode) and could be understood qualitatively from a simple geometrical consideration (Fig. 10). Due to the fast decay of $\chi_c(r)$ with increasing r, the major contribution into the total probe current I_t is made by the tunneling via the shortest way between the NC and the film surface (shown by an arrow in Fig. 10). The resulting current image is actually the result of the imaging of the probe by this very narrow (δ-like) current channel regardless to R_c.

The second conclusion is the dependence of the size and the contrast of the current image of a NC on the depth of its location beneath the dielectric film surface z_0. When $D_p >> R_c$ the size of the NC's current image decreases with increasing z_0 (Fig. 11 *a–c* & Fig. 12, *a*). This effect is explained by increasing the relative contribution from the near-polar segments of the NC surface into the total probe current with increasing z_0. The observation of the current channels of various sizes and contrast in the current image of a nanocomposite SiO_2/SiO_2:NC-Au/SiO_2/n^+-Si film (Fig. 9, *b*) may be considered as an evidence for the suggestion that some NCs nucleate at the upper interface of the nanocomposite layer while the others – at the lower one, as had been discussed in the previous section. When $R_p << R_c$ the size of the NC's current image increases with increasing z_0 (Fig. 11 *d–f* & Fig. 12, *b*). In both cases, the less z_0, the sharper the NC's current image. Earlier we had used a quasi classical approach to analyze the formation of the current images of the metal NCs embedded into a dielectric film on a conductive substrate by Tunneling AFM (Lapshina et al., 2008). For $R_c << D_p$ the tunnel current

$$I_t = \int_0^{2\pi}\int_0^{\rho_p} j(\rho,\varphi,V_g)\cos(\alpha)\rho d\rho d\varphi, \quad (13)$$

where $j(\rho, \varphi, V_g)$ is the tunnel current density through the element of the probe-to-sample contact area $\rho d\rho d\varphi$. In the quasi classical limit, α takes a meaning of the angle of incidence of the tunneling electrons onto the contact area. In (4) this factor is accounted for by the product $\nabla\chi\cdot d\mathbf{S}$. For $R_c >> R_p$

$$I_t = R_c^2 \int_0^{2\pi} \int_0^{\vartheta_m} j\left(\vartheta, \varphi, V_g\right) \cos(\alpha) \sin\vartheta d\vartheta d\varphi \,, \tag{14}$$

where ϑ_m is defined by the condition $\cos(\alpha) = 0$. The tunnel current density j was calculated according to Simmons' formula (Simmons, 1963). The results of the model current images calculations using the quasi classical approach agree with the ones obtained using the model described above.

The main advantage of the latter is more accurate accounting for the edge effects. On the other hand, its main drawback is the utilization of the low voltage approximation.

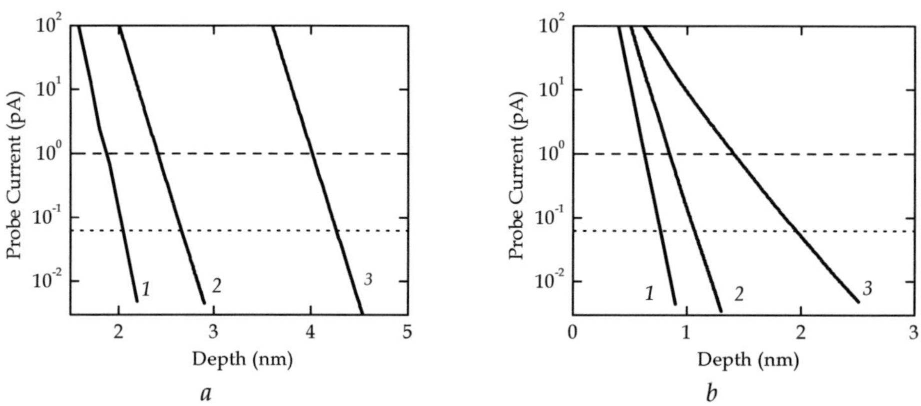

Fig. 13. The model dependences of the probe current I_t on the depth of the Au NC beneath the SiO_2 film surface z_0 calculated in the quasi classical approximation. $R_c = 1$ nm. R_p, nm: *a* – 35, *b* – 0.1. D_p, nm: *a* – 10, *b* – 0. V_g, V: *1* – 2, *2* – 6, *3* – 9. The dashed and dotted lines denote the typical noise levels of an ordinary STM preamplifier and for the low-current one, respectively.

Experimentally, a considerable tunnel current I_t (greater than the noise level of the STM preamplifier used ~ 1 pA) has been observed in the strong field regime ($eV_g \sim A$) when the electron tunneling between the probe and the NC goes through the triangle potential barrier (the Fowler-Nordheim mechanism). The expressions for the envelope wavefunctions (5, 6, and 9) are obviously not valid in this case. Using Simmons' formula allowed the evaluation of I_t in both regimes. In particular, it allowed the estimation of the maximal thickness of the dielectric layers in the structures destined to the Tunneling AFM or the STM investigations (Fig. 13, *a* & *b*, respectively).

4. Coulomb blockade of the electron tunnelling through individual Au nanoclusters in the SiO_2/Si films

The $I-V$ curves of the probe-to-sample contact measured in the current channels of larger sizes (D = 5 to 10 nm) exhibited the Coulomb staircase, more or less expressed (Fig 13, *a*).

The Coulomb staircase is more pronounced in the differential tunnel spectra of the probe-to-sample contact dI_t/dV_g *vs* V_g (Fig. 13, *b*) obtained from the measured $I-V$ curves by the numerical differentiation with nonlinear smoothing.

In order to relate the Coulomb staircase period ΔV to the NCs' morphology and to the geometrical parameters of the nanocomposite structures, we have used a model of an ellipsoidal NC (taking into account the flattened shape of the NCs discussed above in Sec. 2) between two flat electrodes (Fig. 14, *a*). The capacitance of a flattened ellipsoidal droplet with respect to a semi-infinite flat metal electrode can be calculated as

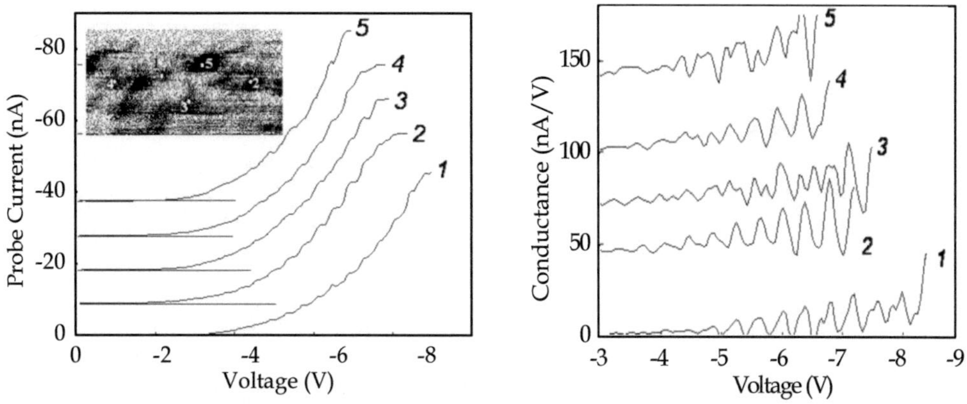

a *b*

Fig. 13. The $I-V$ curves (*a*) and the respective differential tunnel spectra dI_t/dV_g *vs* V_g (*b*) of the Pt coated Si AFM probe contact to the SiO_2/ SiO_2:NC-Au/SiO_2/n^+-Si structure measured in different current channels. Reprinted from (Filatov et al., 2010) under license by IoP Publishing Ltd.

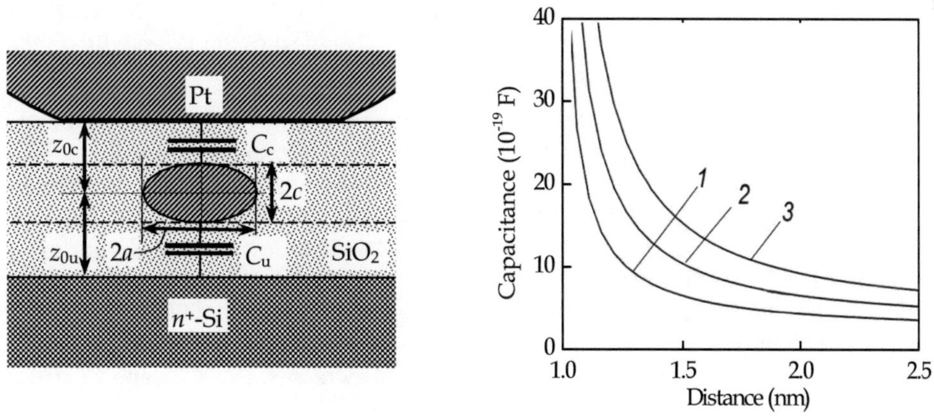

a *b*

Fig. 14. The calculated dependencies of the capacitance of an ellipsoidal metal NC with respect to a flat metal electrode (*a*) on the separation between the NC centre and the electrode z_0 (*b*). c = 1 nm. a = b, nm: *1* — 1.5; *2* — 2.0; *3* — 2.5.

$$C^{-1} = \frac{1}{2\pi\varepsilon\varepsilon_0\sqrt{a^2 - c^2}}\left[\mathrm{arctg}\left(\frac{2z_0 - c}{\sqrt{a^2 - c^2}}\right) - \mathrm{arctg}\left(\frac{c}{\sqrt{a^2 - c^2}}\right)\right], \quad (15)$$

where a and c are the bigger and smaller axes of the ellipsoid (Fig. 14, *a*), ε and ε_0 are the dielectric constants of the dielectric and vacuum, respectively. The expression (15) can be derived out easily from the well-known one for the capacitance of a lonely flattened metal ellipsoid:

$$C^{-1} = \frac{1}{8\pi\varepsilon\varepsilon_0\sqrt{a^2 - c^2}}\mathrm{arctg}\left[\frac{\sqrt{a^2 - c^2}}{c}\right], \quad (16)$$

using the image potential method. It should be stressed here that (15) gives only an approximate value of C when $z_0 \sim c$. When deriving out (15) the ellipsoidal NC surface was assumed to be an equipotential one. Obviously, this is not the case when $z_0 \sim c$ because of the polarization of the metal NC induced by the flat electrode. On the other hand, (15) gives the right asymptotic of $C(z_0)$ at $z_0 \to c$: $C \to \infty$ (Fig. 14, *b*) while (16) gives a finite value of C that is not true at $z_0 \to c$ since in this case the NC and the metal electrode contact each other.

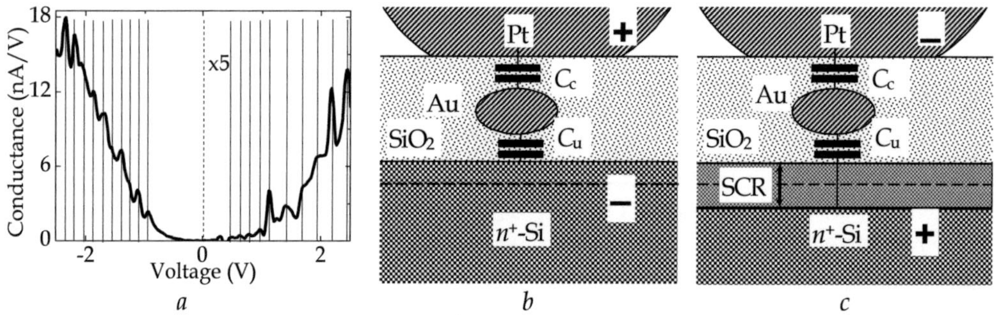

Fig. 15. A differential tunnel spectrum dI_t/dV_g *vs* V_g of a SiO_2(1.5 nm)/SiO_2:NC-Au(1 nm)/SiO_2(1.5 nm)/n^+-Si nanocomposite structure (*a*); The effect of the V_g polarity on the capacitance between the Au NC, the probe, and the substrate: *b* — $V_g < 0$; *c* — $V_g > 0$.

For the asymmetric scheme of Coulomb blockade, the period of the Coulomb staircase $\Delta V = e/C_\Sigma$, where $C_\Sigma = C_c + C_u$ (Fig. 14, *a*), C_c and C_u are the capacitances of the metal NC with respect to the probe and to the substrate, respectively (Averin et al, 1991). Assuming $c = 1$ nm, from $\Delta V = 0.4$ to 0.25 V one has $D = 2a = 1.6$ to 2.6 nm which satisfactory agrees with the TEM data.

In some cases, ΔV was different at the different polarities of V_g (Fig. 15, *a*). The effect was attributed to the modulation of the space charge region (SCR) at the SiO_2/n^+-Si interface by the probe potential (Fig. 15, *b* & *c*). At $V_g > 0$ (Fig. 15, *b*) the SCR thickness decreases, $C_u >> C_c$, and $C_\Sigma \approx C_u$. At $V_g < 0$ (Fig. 15, *c*) the SCR thickness increases, $C_u << C_c$, and $C_\Sigma \approx C_c$.

5. Resonant electron tunnelling through the individual Au nanoclusters in the SiO_2/Si films

The $I - V$ curves measured in the current channels of smaller sizes (3 to 5 nm) are shown in Fig. 16, *a*. The peaks attributed to the resonant tunnelling through the Au NCs are seen at the negative V_g. The interpretation of the tunnelling spectra was based on a double barrier $Pt/SiO_2/Au/SiO_2/Si$ structure model (Lapshina et al., 2009). The model was based on the exact solution of one-dimensional Schrödinger's equation in the effective mass approximation. The applicability of the one-dimensional model was justified by the flattened shape of the NCs so that they could be treated as thin disks which, in turn, allowed neglecting the size quantization in the film plane. The band diagrams of the model structures were calculated taking into account the band bending at the n^+-Si/SiO_2 interface. The profile of the potential $\varphi(z)$ in the SCR was calculated from the solution of the one-dimensional Poisson's equation for a uniformly doped semiconductor with the boundary condition at the Si/SiO_2 interface: $\varepsilon_{Si}F_{Si} = \varepsilon F$ where ε_{Si} and ε are the dielectric constants of Si and SiO_2, F_{Si} and F are the electric field strengths in Si and SiO_2, respectively:

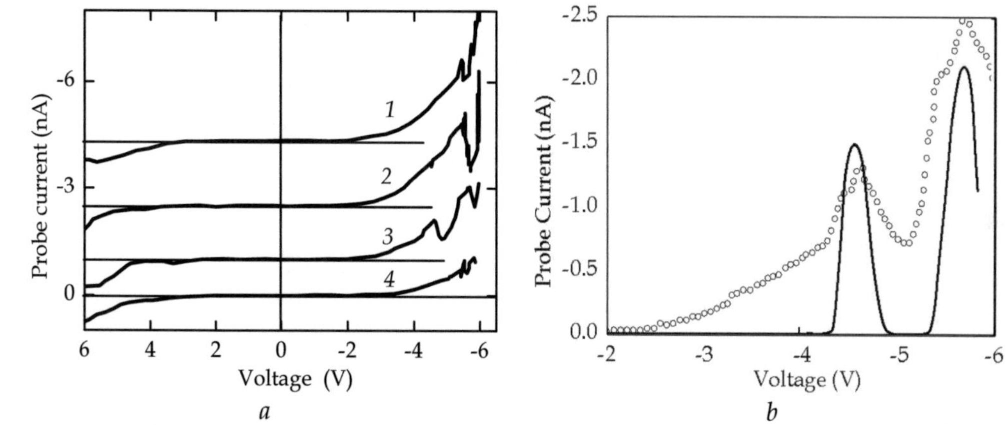

Fig. 16. The $I-V$ curves of the contact of a Pt coated Si AFM probe to the SiO_2(1.5 nm)/SiO_2:NC-Au/SiO_2(1.8 nm)/n^+-Si nanocomposite structure measured in different current channels (*a*); an example of the approximation of a part of curve *3* (circles) by a calculated one (solid line) (*b*). Reprinted from (Filatov et al., 2010) under license by IoP Publishing Ltd.

$$\varphi(z) = \frac{\varepsilon F}{2\varepsilon_{Si} l_0}(z - l_0 - z_d)^2 . \tag{17}$$

Here l_0 is the SCR thickness and z_d is the position of the SiO_2/Si interface. The NCs were assumed to be depleted completely due to the quantum size effect that allowed neglecting the screening of the electric field F by the electrons confined in the NC. Earlier, the STM/STS and AES studies revealed that the similar Au NCs on the highly oriented pyrolitic graphite (HOPG) surface could undergo the metal-to-insulator transition at $h < 1$ nm (Borman et al., 2006); an energy gap was observed in their tunnel spectra. In this case, the electron-electron scattering inside the NCs would not affect the coherent tunnelling through them. It is not the case for the NCs of larger size ($h \geq 1$ nm), which are likely filled with the electrons.
The part of V_g, which drops on the SCR

$$\Delta\varphi = \frac{\varepsilon F l_0}{2\varepsilon_{Si}} = \frac{\varepsilon_0 \varepsilon^2 F^2}{2e\varepsilon_{Si} N_D} , \tag{18}$$

where N_D is the donor concentration in Si. At $V_g \approx 5$ V, $\Delta\varphi \approx V_g/2$ (Fig. 17, *a*).
The solution of the Schrödinger's equation in the SiO_2 and Au layers for the potential shown in Fig. 17, *a* can be written as:

$$\chi(z) = C_1\mathrm{Ai}[s(z)] + C_2\mathrm{Bi}[s(z)]\,, \tag{19}$$

where Ai and Bi are the Airy functions of the I and II types, respectively, C_1 and C_2 are the normalization constants,

$$s(z) = \frac{(2m)^{1/3}}{(e\hbar F)^{2/3}}\left[eF(z - z_d) + E_c(z_d) - E\right], \tag{20}$$

E is the electron energy and $E_c(z_d)$ is the bottom of the conduction band of Au (SiO_2) at the respective position. With respect to SiO_2, taking into account the amorphous state of the material, one should rather consider the mobility gap edge than the conduction band bottom. The origin of the electron energy E was selected at the bottom of the conduction band in the probe coating material (Pt). This solution is valid for any F, including $F > A/ed_c$, i. e. in the strong electric field (or in the Fowler-Nordheim tunnelling mode) and for any E, including $E > A$, i. e. at the tunnelling over the potential barrier.
Let us consider the case of electrons injection from the AFM probe into the Si substrate (i. e. at $V_g < 0$). In this case, the eigenfunctions of the electron states in the Pt and Si layers can be written as usual:

$$\begin{aligned} \chi_e(z) &= \exp(ik_e z) + C_0\exp(-ik_e z)\,, \\ \chi_c(z) &= C_c\exp(ik_c z)\,, \end{aligned} \tag{21}$$

respectively, where $k_e = (2m_{Pt}E)^{1/2}/\hbar$ and $k_c = [2m_{Si}(E - E_{cSi})]^{1/2}/\hbar$ are the electron wave vectors of the electron eigenstates in Pt and in Si, respectively. For given selection of the normalization constants C, the tunnel transmission coefficient of the double barrier structure

$$\mathrm{T}(E) = \frac{m_{Pt}k_c(E)}{m_{Si}k_e(E)}\left|C_c(E)\right|^2, \tag{22}$$

as follows from the condition of the probability density flux continuity straightforward. Here m_{Pt} and m_{Si} are the electron effective masses in Pt and in Si, respectively. The values of the normalization constants C at given E were found from the Bastard's boundary conditions (the continuity of $\chi(z)$ and of $m^{-1}d\chi/dz$) at each interface in the structure. In Fig. 17, *b* a model tunnel transparency spectrum of the double barrier structure T(E) calculated for $V_g = 5.7$ V is presented.
The total probe current at $V_g < 0$ is given by (Esaki & Stiles, 1966):

$$I_t(V_g) = \frac{em_{Pt}k_B TD^2}{8\pi\hbar^3}\int_0^\infty \mathrm{T}(E, V_g)\ln\left[\exp\left(\frac{E_F - E}{k_B T}\right) + 1\right]dE\,. \tag{23}$$

The reverse current was neglected since I_t becomes considerable ($I_t > 1$ pA) at $V_g \geq 2$ V >> $k_B T/e$ only.
An example of the fitting of a measured $I - V$ curve of the probe-to-sample contact by a model one is presented in Fig. 16, *b*; d_u, d_c, and h were the fitting parameters. The best fit has

been achieved at d_u = 1.4 nm, d_c = 1.1 nm, and h = 1.3 nm (the band diagram in Fig. 17, *a* has been calculated just for these values) which agrees with the XPS and TEM data satisfactory although the model described above is rather rough. First of all, the effect of the image potential at the Pt/SiO_2 and Si/SiO_2 interfaces on the band profile has been neglected. Then, the electron energy distribution has been calculated in the simple parabolic band approximation, the global band minima were taken into account only while the higher band extrema as well as the band nonparabolicity were neglected. Also, the values of the band offset at the SiO_2/Au and SiO_2/Pt interfaces were calculated from the values of the workfunction and of the electron affinity for the respective materials.

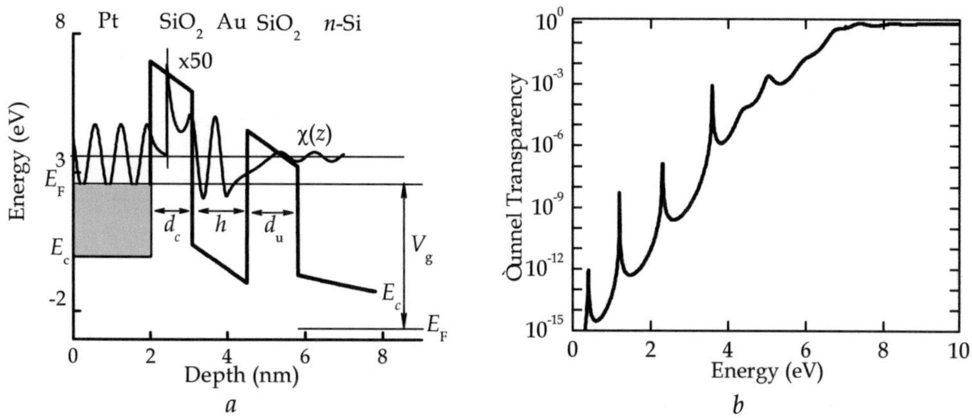

Fig. 17. The calculated band diagram (*a*) and the tunnel transparency spectrum (*b*) for a Pt/SiO_2/Au/SiO_2/n^+-Si double barrier structure. d_c = 1.1 nm, h = 1.3 nm, d_u = 1.4 nm, V_g = 5.7 V. Reprinted from (Filatov et al., 2010) under license by IoP Publishing Ltd.

In addition, the dielectric constant ε for such an exotic material as depleted Au layer was not known, so it was assumed to be equal to the one of SiO_2 to simplify the model. Nevertheless, from the comparison of the results of the modeling with the experimental data (Fig. 16, *b*) one can conclude that the present model describes the electron tunnelling through the small Au NCs in a thin SiO_2 film satisfactory and that it is likely the peaks observed in the measured $I-V$ curves originate from the resonant tunnelling effect.

In Fig. 16, *b* one can notice that the resonant peaks in the calculated $I-V$ curve are narrower compared to the measured ones that should be attributed to the inhomogeneous broadening due to the nonuniformity of the actual NC thickness. Also, one can note the splitting of the resonant peaks in the measured $I-V$ curve that could probably be ascribed to the stepwise fluctuations in the NC thickness.

6. Conclusion

The results presented in this chapter demonstrate the approaches based on the deposition of a nm-thick Me–Si (Me = Au, Pt) amorphous mixture with well-controlled Me/Si compositional ratio followed by its oxidation by the glow discharge oxygen plasma at room temperature or alternatively by its thermal oxidation to allow growing the ultrathin SiO_2 layers with the embedded Me NCs. The resulting SiO_2:NC-Me nanocomposite layers contain the Me NCs of 2 to 5 nm in the lateral size with the sheet density of ~ 10^{13} cm^{-2}, which can be

controlled by varying the Me/Si ratio in the initial Me—Si mixture. Furthermore, the methods allow growing the SiO_2/SiO_2:NC-Me/SiO_2 sandwiched nanocomposite structures on any substrate in a single vacuum cycle. It should be stressed here that the NCs formed by the proposed methods are confined in the single sheets with the precisely controlled thickness of the underlying and cap layers. The $C-V$ measurements demonstrated a clearly expressed hysteresis indicating that the nanocomposite MOS structures grown by the described methods have the potential for the nanoscale nonvolatile memory applications.
The application of the Tunneling AFM allowed the visualization of the Au NCs embedded into the ultrathin SiO_2 films on the Si substrates. The size of the current image of a NC embedded in a thin dielectric film was found to be determined primarily by the size of the contact area between the AFM probe and the film surface. The second factor affecting the image size and contrast was found to be the depth of the NC's location beneath the film surface. The application of Tunnelling AFM allowed also the observation of the Coulomb blockade and of the resonant electron tunnelling through the individual Au NCs ~ 1 nm in thickness. The observed effects can be utilized in future nanoelectronic devices based on single NCs embedded in ultrathin dielectric films.

7. References

Antonov, D.A.; Gorshkov, O.N.; Kasatkin, A.P.; Maximov, G.A.; Saveliev, D.A.; & Filatov, D.O. (2004). Investigation of the electronic properties of the Zr nanoclusters in the YSZ films by Combined STM/AFM. *Physics of Low - Dimensional Structures*, 1/2 (January 2004), 139-142, 0204-3467.

Antonov, D.A.; Filatov, D.O.; Zenkevich, A.V. & Lebedinskii Yu.Yu. (2007). Investigation of the electronic properties of Au nanoclusters in SiO_2 by Combined Scanning Tunneling/Atomic Force Microscopy. *Bulletin RAS: Physics*, 71, 1 (January 2007), 56–58, 1062-8738.

Averin, D.V.; Korotkov, A.N. & Likharev, K.K. (1991). Theory of single-electron charging of quantum wells and dots. *Phys. Rev. B*, 44, 12 (September 1991), 6199-6211, 1098-0121.

Bardeen, J. (1961). Tunnelling from a many-particle point of view. *Phys. Rev. Lett.*, 6, 2 (January 1961), 57-59, 0031-9007.

Bar-Sadeh, E.; Goldstein, Y.; Zhang, C.; Deng, H.; Abeles, B. & Millo, O. (1994). Single-electron tunneling effects in granular metal films. *Phys. Rev. B*, 50, 12 (September 1994), 8961–8964, 1098-0121.

Bonafos, C.; Garrido, B.; Lopez, M.; Perez-Rodriguez, A.; Morante, J.R.; Kihn, Y.; Ben Assayag, G. & Claverie, A. (2000). An electron microscopy study of the growth of Ge nanoparticles in SiO_2. *Appl. Phys. Lett.*, 76, 26 (June 2000), 3962-8964, 0003-6951.

Borman, V.D.; Borisyuk, P.V.; Lebid'ko, V.V.; Pushkin, M.A.; Tronin, V.N.; Troyan, V.I.; Antonov, D.A. & Filatov, D.O. (2006). A study of many-body phenomena in metal nanoclusters (Au, Cu) close to their transition to the nonmetallic state. *JETP*, 102, 2 (February 2006), 303-313, 1063-7761.

Bukharaev, A.A.; Berdunov, N.V.; Ovchinnikov, D.V. & Salikhov, K.M. (1998). Three-dimensional probe and surface reconstruction for Atomic Force Microscopy using a deconvolution algorithm. *Scanning Micros.*, 12, 1 (January 1998), 225-234, 0891-7035.

Cho, S.H.; Lee, S.; Ku, D.Y.; Lee, T.S.; Cheong, B.; Kim, W.M. & Lee, K.S. (2004). Growth behavior and optical properties of metal-nanoparticle dispersed dielectric thin films formed by alternating sputtering. *Thin Solid Films*, 447-448, 1 (January 2004), 68-73, 0040-6090.

Esaki, L. & Stiles, P.J. (1966). New type of negative resistance in barrier tunneling. *Phys. Rev. Lett.*, 16, 24 (June 1966), 1108-1110, 0031-9007.

Filatov, D.O.; Lapshina, M.A.; Antonov, D.A.; Gorshkov, O.N.; Zenkevich, A.V. & Lebedinskii, Yu.Yu. (2010). Resonant tunnelling through individual Au nanoclusters embedded in ultrathin SiO_2 films studied by Tunnelling AFM. *J. Phys. Conf. Ser.*, 245 (September 2010) 012018, 1742-6596.

Guo, L.; Leobandung, E. & Chou, S.Y. (1997). A silicon single-electron transistor memory operating at room temperature. *Science*, 275, 5300 (January 1997) 649-651, 0036-8075.

Hiraki, A.; Lugujjo, E. & Mayer, J.W. (1972). Formation of silicon oxide over gold layers on silicon substrates. *J. Appl. Phys.*, 43, 9 (September 1972), 3643-3650, 0021-8979.

Hochella Jr., M.F. & Carim, A.H. (1988). A reassessment of electron escape depths in silicon and thermally grown silicon dioxide thin films. *Surf. Sci. Lett.*, 197, 3 (March 1988), L260-L268, 0039-6028.

Imamura, H.; Chiba, J.; Mitani, S. & Takanashi, K. (2000). Coulomb staircase in STM current through granular films. *Phys. Rev. B*, 61, 1 (January 2000), 46-49, 1098-0121.

Lapshina, M.A.; Filatov, D.O. & Antonov, D.A. (2008). Current imaging in Combined STM/AFM of metal nanoclusters in dielectric films. *J. Surf. Investigation: X-ray, Synchrotron & Neutron Techn.*, 2, 4 (April 2008), 616–619, 1027-4510.

Lapshina, M.A.; Filatov, D.O.; Antonov, D.A. & Barantsev, N.S. (2009). Study of resonant tunneling in Au nanoclusters on the surface of SiO_2/Si thin films using the Combined Scanning Tunneling Microscopy and Atomic-Force Microscopy technique. *J. Surf. Investigation: X-ray, Synchrotron & Neutron Techn.*, 3, 4 (April 2009), 559–565, 1027-4510.

Lebedinskii, Yu.Yu.; Zenkevich, A.; Gusev, E.P. & Gribelyuk, M. (2005). *In situ* investigation of growth and thermal stability of ultrathin Si layers on the HfO_2/Si(100) high-κ dielectric system. *Appl. Phys. Lett.*, 86, 19 (May 2005), 1904-1906, 0003-6951.

Maksimova, K.Yu.; Matveev, Yu.A.; Zenkevich, A.V.; Nevolin, V.N.; Novikov, A.G.; Gaiduk, P.I. & Orekhov, A.C. (2010). Investigation of the SiO_2:Me nanocomposite structures formed by the metal segregation at the front of silicon oxidation in the Si:Me layers. *Perspektivnye Materialy (Promising Materials)*, 2010, 2 (April 2010), 33-38 (*in Russian*), 2075-1133.

Ren, F.; Heng, X.; Guang, X.; Cai, X.; Wang, J.B. & Jiang, C.Z. (2009). Engineering embedded metal nanoparticles with ion beam technology. *Appl. Phys. A*, 96, 2 (August 2009), 317-325, 0947-8396.

Ruffino, F.; de Bastiani, R.; Grimaldi, M.G.; Bongiorno, C.; Giannazzo, F.; Roccaforte, F.; Spinella, C. & Raineri, V. (2007). Self-organization of Au nanoclusters on the SiO_2 surface induced by 200 keV-Ar^+ irradiation. *Nucl. Instr. Meth. B*, 257, 1-2, (April 2007), 810-814, 0168-583X.

Ruffino, F. & Grimaldi, M.G. (2007). Structural and electrical characterization of gold nanoclusters in thin SiO_2 films: Realization of a nanoscale tunnel rectifier. *Microelectron. Eng.*, 84, 3 (March 2007), 532-537, 0167-9317.

Simmons, J.G. (1963). Generalized formula for the electric tunnel effect between similar electrodes separated by a thin insulting film. *J. Appl. Phys.*, 34, 6 (June 1963), 1793-1804, 0021-8979.

Tersoff, J. & Hamann, D.B. (1985). Theory of the scanning tunneling microscope. *Phys. Rev. B*, 31, 2 (January 1985), 805-813, 1098-0121.

Tiwari, S.; Rana, F.; Hanafi, H.; Hartstein, A.; Crabbe, E. F. & Chan, K. (1996). A silicon nanocrystals based memory. *Appl. Phys. Lett.*, 68, 10 (March 1996) 1377-1379, 0003-6951.

Tiwari, S.; Wahl, J.A.; Silva, H.; Rana, F. & Welser, J.J. (2000). Small silicon memories: confinement, single-electron, and interface state considerations. *Appl. Phys. A*, 71, 4 (October 2000), 403-414, 0947-8396.

Yanev, V.; Rommel, M.; Lemberger, M.; Petersen, S.; Amon, B.; Erlbacher, T.; Bauer, A.J.; Ryssel, H.; Paskaleva, A.; Weinreich, W.; Fachmann, C.; Heitmann, J. & Schroeder, U. (2008). Tunneling atomic-force microscopy as a highly sensitive mapping tool for the characterization of film morphology in thin high-*k* dielectrics. *Appl. Phys. Lett.* 92, 25 (June 2008), 2910-2912, 0003-6951.

Yano, K.; Ishii, T. Hashimoto, T.; Kobayashi, T.; Murai, F. & Seki, K. (1994). Room-temperature single-electron memory. *IEEE Trans. Electron Dev.*, 41, 9, (September 1994), 1628-1638, 0018-9383.

Zacharias, M.; Heitmann, J.; Scholz, R.; Kahler, U.; Schmidt, M. & Bläsing, J. (2002). Size-controlled highly luminescent silicon nanocrystals: a SiO/SiO_2 superlattice approach. *Appl. Phys. Lett.*, 80, 4 (January 2002), 661–664, 0003-6951.

Zenkevich, A.V.; Lebedinskii, Yu.Yu.; Timofeyev, A.A.; Isayev, I.A. & Tronin, V.N. (2009). Formation of ultrathin nanocomposite SiO_2:nc-Au structures by Pulsed Laser Deposition. *Appl. Surf. Sci.*, 255, 10 (March 2009), 5355-5358, 0169-4332.

15

From Zeolite to Host-Guest Nanocomposite Materials

Masoud Salavati-Niasari[1,2] and Fatemeh Mohandes[1]
[1]*Department of Inorganic Chemistry, Faculty of Chemistry, University of Kashan,*
[2]*Institute of Nano Science and Nano Technology, University of Kashan,*
Islamic Republic of Iran

1. Introduction

In nature, zeolites as aluminosilicate members of the family of microporous solids are often formed where volcanic rock of specific chemical composition is immersed in water so as to leach away some of the components. Zeolites are known as "molecular sieves". They have many useful applications such as: ion exchange technology, filtering, odor removal, chemical sieve and gas absorption tasks but the most well known use for zeolites is in water filtration applications. In a typical example, Figure 1 presents a cubic structure of zeolite Y with 3D system of channels.

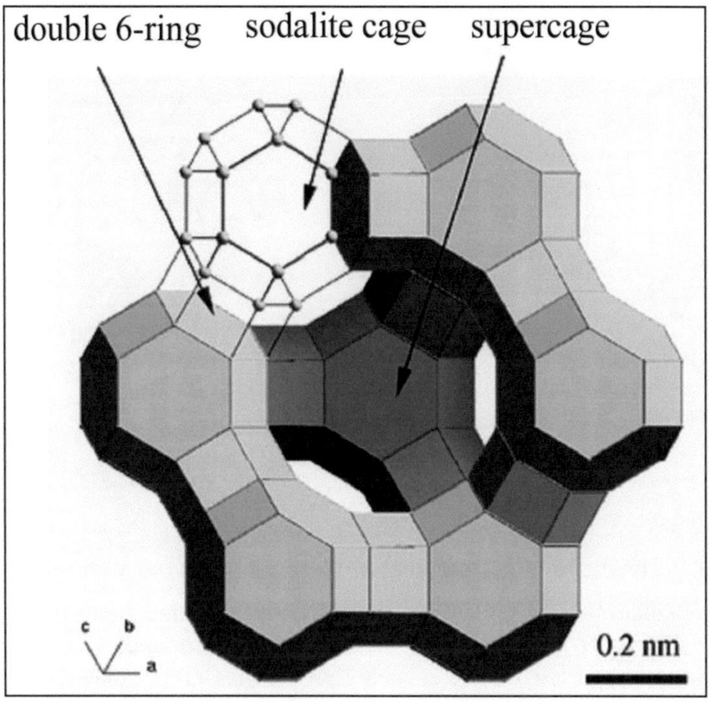

Fig. 1. The structure of zeolite Y.

Due to the uniformity of the pore dimensions, zeolites can act as "host materials" for other molecules. Only molecules of certain size are able to be absorbed by a given zeolite material, or pass through its pores, while molecules of bigger size cannot. During the last two decades, zeolites with nanoscale dimensions open a new view in host (nanopores of zeolite)–guest (transition metal complex) nanocomposite materials. The encapsulation of transition metal complexes into the framework sites of the molecular sieve has attracted considerable attentions as the preparation of novel catalysts, owing to which possess the advantages of both homogeneous catalysis as the metal ion in the solution and heterogeneous catalysis as the molecular sieve in the polyphase system. So that, the catalytic efficiency of zeolite encapsulated metal complexes (ZEMC) is much higher than that of the neat complexes and the ZEMC nanocomposites have provided the opportunity to develop catalytic process for the selective oxidation, alkylation, dehydrogenation, cyclization, amination, acylation, isomerisation and rearrangement of various substrates and are able to produce intermediates as well as most industrial.

Basically, there are three main approaches to encapsulate complexes: the zeolite synthesis (ZS), the flexible ligand (FL) and ship-in-a-bottle methods. In the ZS method, transition metal complexes, which are stable under the conditions of zeolite synthesis (high pH and elevated temperature), are included in the synthesis mixture. The resulting zeolite encapsulates the transition metal complex in its voids but in the FL method, a flexible ligand is able to diffuse freely through the zeolite pores. Scheme 1 shows a typical "flexible ligand" synthesis of zeolite-Y encapsulated metal complexes of Ni(II).

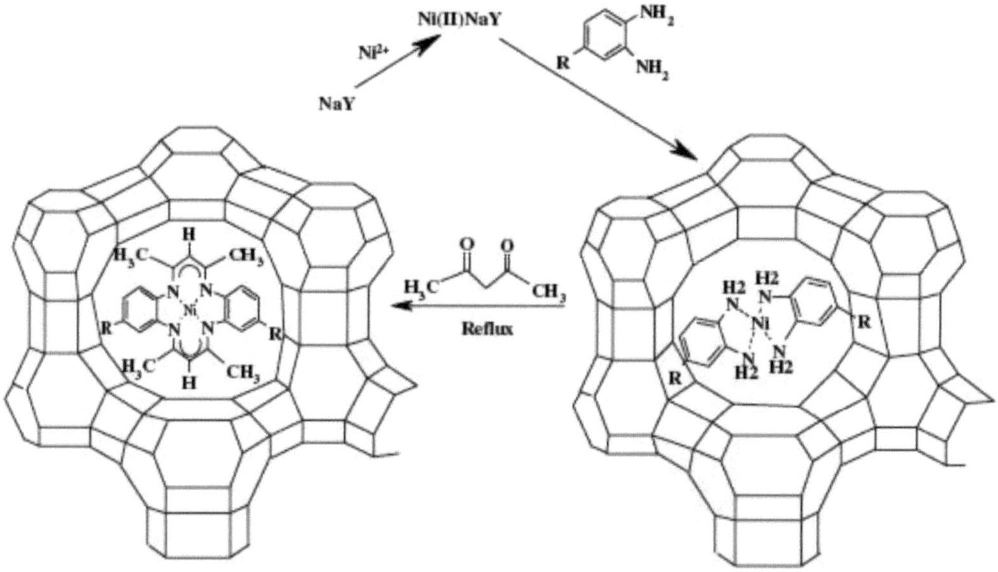

Scheme 1. "Flexible ligand" synthesis of zeolite-Y encapsulated metal complexes of Ni(II).

To ensure encapsulation, FT-IR, UV–Vis spectroscopy, powder X-ray diffraction patterns (XRD) and BET (Brunauer-Emmett-Teller) technique can be applied.

The investigation of reported results reveals that zeolites encapsulated metal complex as heterogeneous catalysts make higher selectivity and conversion percentage than that of the homogeneous catalysts.

2. Host molecules

The long history "host-guest" or "supramolecular" chemistry has come to the fore front in contemporary researches with the awarding of the 1987 Nobel Prize in chemistry to Donald J. Cram, Jean-Marie Lehn and Charles J. Pedersen. In molecular engineering, host-guest chemistry describes compounds that are composed of two or more molecules or ions that are held together in unique structural relationships by forces other than those of full covalent bonds. Host-guest chemistry encompasses the idea of molecular recognition and interactions through noncovalent bonding (hydrogen bonds, ionic bonds, van der Waals forces, and hydrophobic interactions). Although supramolecular systems are often held together by weaker interactions than covalent systems, in some cases the "strength" of supramolecular systems actually lies in the weakness of these intermolecular forces. In particular, the weakness of these interactions allows host-guest binding to become a reversible process, so that a host and guest can associate and dissociate without either of the building blocks being damaged or altered. This can be useful for systems such as molecular switches and for catalysis, and is utilized frequently in natural enzymes. Macrocyclic hosts are molecular receptors that are arranged as a ring, in which the guest, or substrate, may bind in the interior. Macrocyclic hosts can be generally classified as types of "endoreceptors", in that they bind with their guests located in the interior of the host [1]. "Exoreceptors", in contrast, bind to substrates on their exterior. Some common examples of macrocyclic hosts include:

- Cyclodextrins
- Calixarenes
- Cucurbiturils
- Crown ethers
- Cryptophanes
- Porphyrins
- Cyclotriveratrylenes
- Carcerands
- Zeolites

2.1 Cyclodextrins (CDs)

In 1891, cyclodextrin molecules made up of cyclic oligosaccharides containing α-(1,4)-linked glucopyranosyl units were described by A. Villiers [2]. Typical cyclodextrins contain a number of glucose monomers ranging from six to eight units in a ring, creating a cone shape. Thus denoting:

- α -cyclodextrin: six membered sugar ring molecule
- β -cyclodextrin: seven sugar ring molecule
- γ -cyclodextrin: eight sugar ring molecule

Chemical structure of the three main types of cyclodextrins is shown in Fig. 1.

The production of cyclodextrins (CDs) involves treatment of ordinary starch with a set of easily available enzymes [3]. Several synthetic methods for the preparation of glycosylated cyclodextrins have been reported [4, 5]. Recently, β-cyclodextrin and hydroxy propyl β-cyclodextrin spheres with a particle size less than 400 nm were prepared by Mammucari and co-workers via an Aerosol Solvent Extraction System (ASES) process [6].

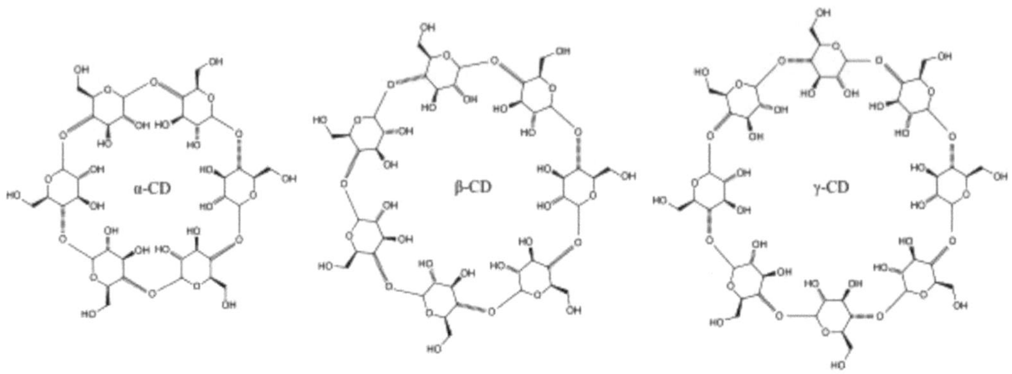

Fig. 1a. Chemical structure of cyclodextrins.

As a class of water-soluble and nontoxic cyclic oligosaccharides with a hydrophilic exterior and a hydrophobic interior, CDs have been extensively investigated in host-guest chemistry. One of the most important properties of CDs is the inclusion of guest molecules into their cavities. The cavity size of CDs increases with the increasing number of glucopyranose repeating units. The cavity diameter of CDs is about 0.44 - 0.83 nm (*α*-CD: 0.49 nm, *β*-CD: 0.62 nm, *γ*-CD: 0.80 nm) [7, 8]. Furthermore, recent novel approaches to chemical sensing systems using supramolecular structures such as CD dimers, trimers and cooperative binding systems of CDs with the other supramoleculars have been mentioned [9]. When a competitive guest is added to the aqueous solution of fluorophore-CD, the fluorophore is excluded from the inside to the outside of the CD cavity. The fluorescent CD exhibits strong emission in the self-inclusion state due to the hydrophobic environment of the CD cavity and exclusion of the fluorophore from the cavity to bulk aqueous media weakens its fluorescence intensity. Therefore, fluorescence intensity of many kinds of "turn-off" fluorescent chemical sensors decreases by complexation with guest molecules. On the other hand, in "turn-on" fluorescent chemical sensors by inclusion of a hydrophobic guest molecule into the cavity of the fluorophore-CD conjugate, the fluorophore is located in a more hydrophobic environment, thus the fluorescence intensity increases (Fig. 2a and 2b).

CDs can be used as modified surface agent in the synthesis of nanostructure materials. Zhao and Chen have synthesized multi-petals ZnO nano-structure in the presence of *β*-cyclodextrin and they found that if not using *β*-CD in the preparation process, only larger rods could be obtained [10] and according to recent report, *β*-Cyclodextrin (*β*-CD)-functionalized CdSe/ZnS quantum dots (QDs) apply for optical sensing [11]. To develop functional nanowebs, Uyar et al. have synthesized cyclodextrin functionalized polyethylene oxide (PEO) nanofibers (PEO/CD) [12]. Additionally, CDs are also used for filtration purposes [13-15]. Cyclodextrin-capped nanoparticles can apply in catalyst industry. Fig. 3 represents TEM images of *α*- Cyclodextrin capped Au nanoparticles [16].

Other important applications of CDs involve: supramolecular carrier in organometallic reactions [17] and pharmaceutical applications for drug release [18,19]. To produce HPLC columns allowing chiral enantiomers separation, *β*-cyclodextrins are used [20].

2.2 Calixarenes

Calixarenes as macrocycles or cyclic oligomers are based on a hydroxyalkylation product of a phenol and an aldehyde [21]. These macromolecules have hydrophobic cavities that can

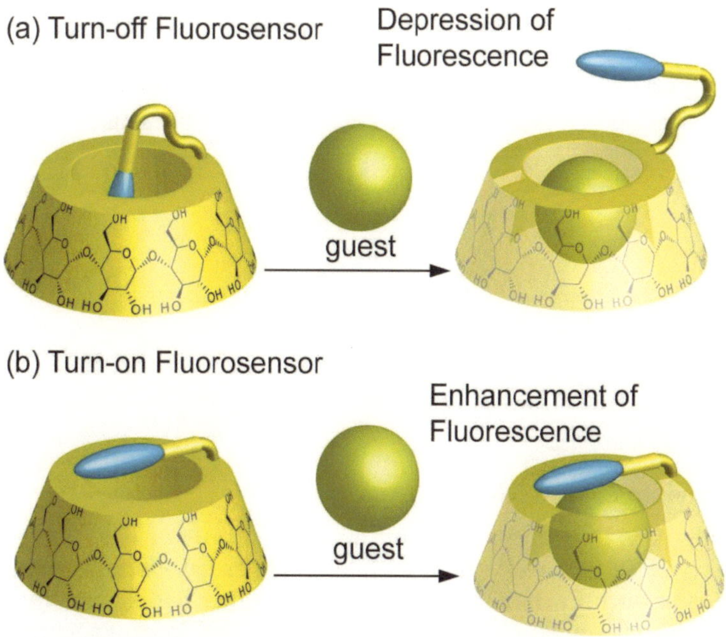

Fig. 2. (a) Turn-off and (b) Turn-on fluorosensors [9].

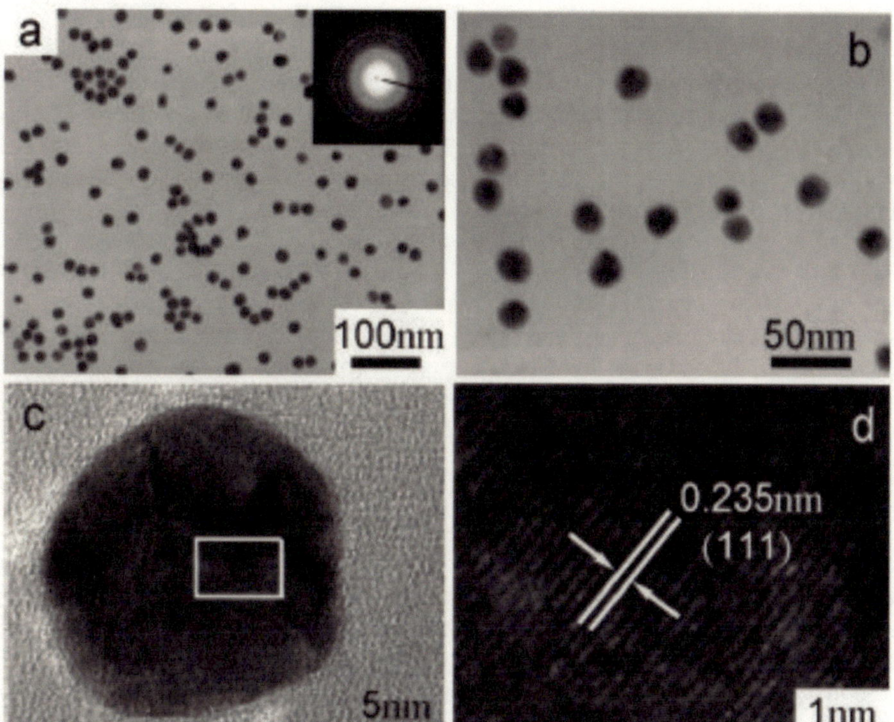

Fig. 3. TEM (a,b) and HRTEM (c,d) images of *a*- Cyclodextrin capped Au nanoparticles [16].

hold smaller molecules or ions and belong to the class of cavitands known in host-guest chemistry. Like cyclodextrins, these scaffolds are able to bind hydrophobic organic molecules and can be modified with carbohydrate ligands on one face of the molecule with control over ligand to ligand spacing. The general structure of calixarenes represents in Fig.4.

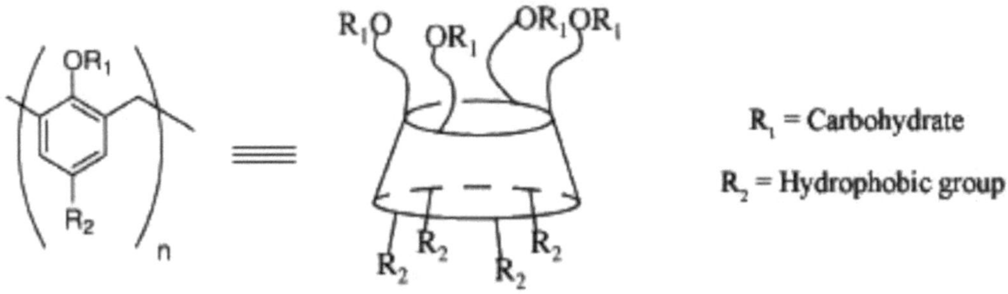

Fig. 4. The general structure of calixarenes.

These conjugates have been useful for both site-directed drug delivery [22] and for studies of water-monolayer surface interactions [23]. Calixarenes are efficient sodium ionophores and are applied as such in chemical sensors. On the other hand, these molecules exhibit great selectivity towards other cations. Calixarenes are used in commercial applications as sodium selective electrodes for the measurement of sodium levels in blood. In this case, Jin reported several new fluorescent Na^+ sensors based on calixarenes (Fig.5) [24]. Use of different types of calixarenes as sensitive layers reveals wide possibilities in control of sensitivity and selectivity of sensors [25, 26].

Ludwing and Dzung have discussed about molecular design principles of calixarene-type macrocycles for ion recognition and the relationship between their structure and selectivity [27]. Calixarenes as excellent surfactants have a remarkable capacity for molecular encapsulation, are also proficient at encapsulating nanoparticles. Therefore, these compounds can be use to control the growth of nanomaterials because of their stable cyclic structure in spite of low molecular weight and also very favorable properties for use as high-resolution resists [28]. Wei reported the synthesis of Au and Co nanoparticles by using C-undecylcalix[4]-resorcinarene (C11 resorcinarene), as shown in Fig. 6 [29].

Calixarene derivatives such as *p*-sulfonato-calix[n]arenes can form host–guest inclusion complexes in a similar way to cyclodextrins. Such water soluble calixarenes display interesting biological properties such as antiviral and anti-bacterial activity [30]. It is found that calixarene derivatives as resist materials are used for electron-beam lithography [31, 32].

2.3 Cucurbiturils (CBs)

A cucurbituril is a macrocyclic molecule consisting of several glycoluril [$=C_4H_2N_4O_2=$] repeat units, each joined to the next one by two methylene [$-CH_2-$] bridges to form a closed band. The oxygen atoms are located along the edges of the band and are tilted inwards, forming a partly enclosed cavity. Cucurbiturils were first synthesized in 1905 by Behrend, by condensing glycoluril with formaldehyde in concentrated sulfuric acid at a fairly high temperature (>110 °C) [33], but their structure was not elucidated until 1981 [34]. Cucurbiturils are commonly written as cucurbit[*n*]uril, where *n* is the number of repeat units (*n*= 5,6,7,8 and 10). A common abbreviation is CB[*n*]. Fig. 7 and Fig. 8 show X-ray crystal structures of CB[*n*] (*n* = 5-8) and synthetic method of CB[6], respectively [35, 36].

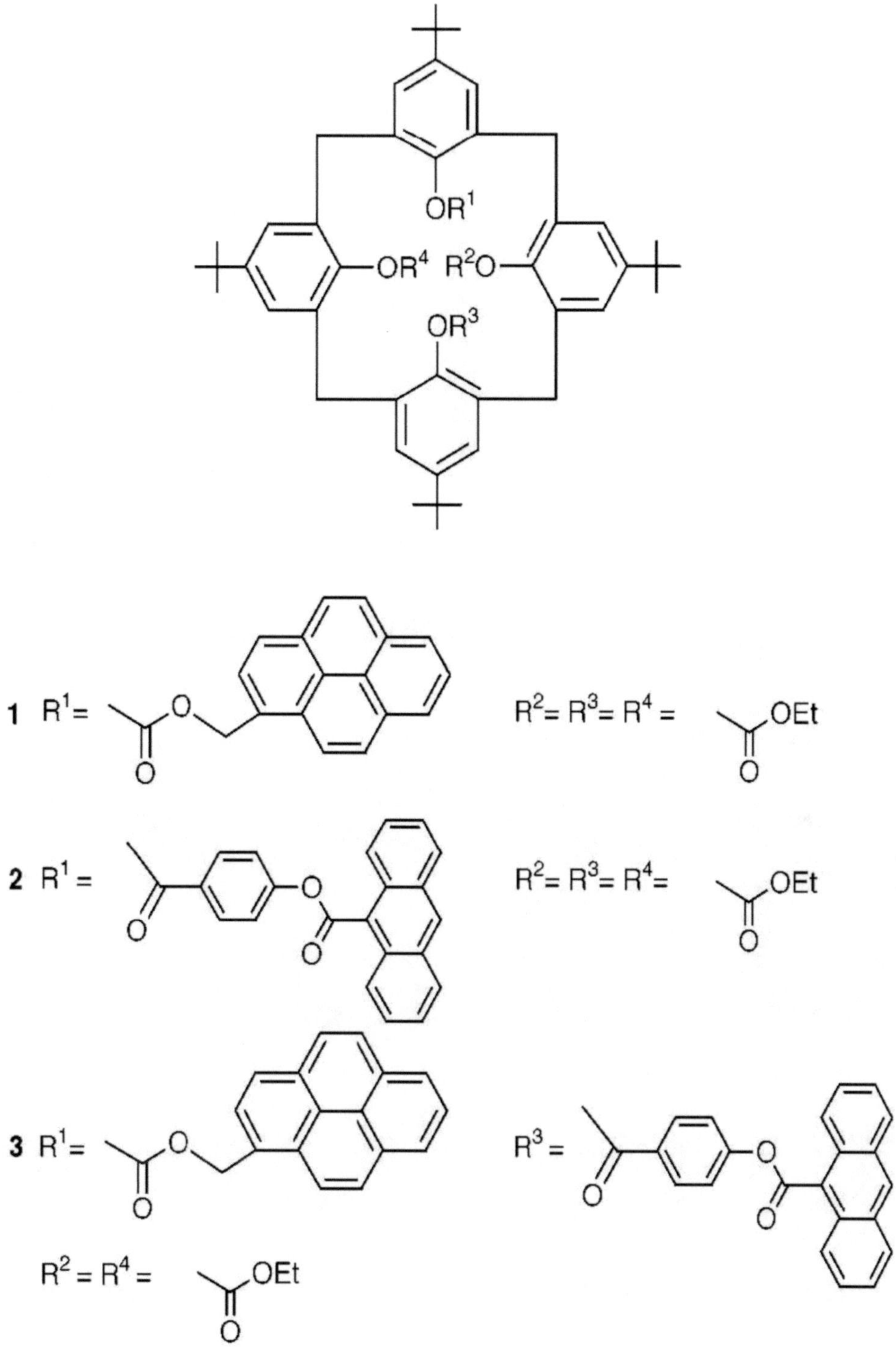

Fig. 5. New fluorescent Na^+ sensors based on Calixarenes [24].

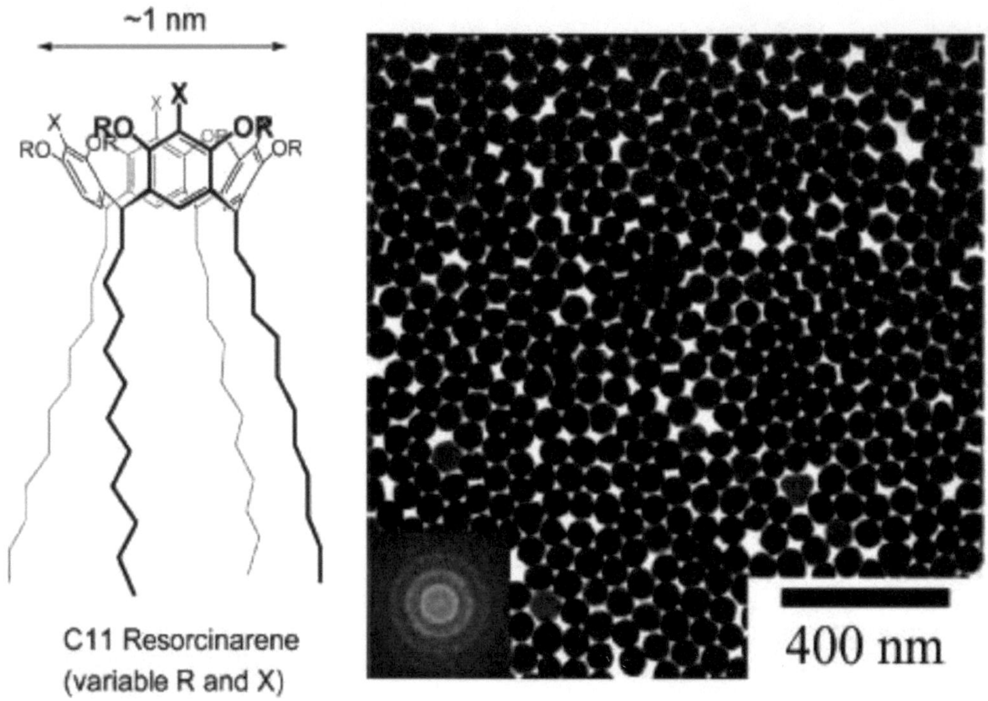

Fig. 6. TEM image of Au nanoparticles prepared by using C-undecylcalix[4]-resorcinarene (C11 resorcinarene) [29].

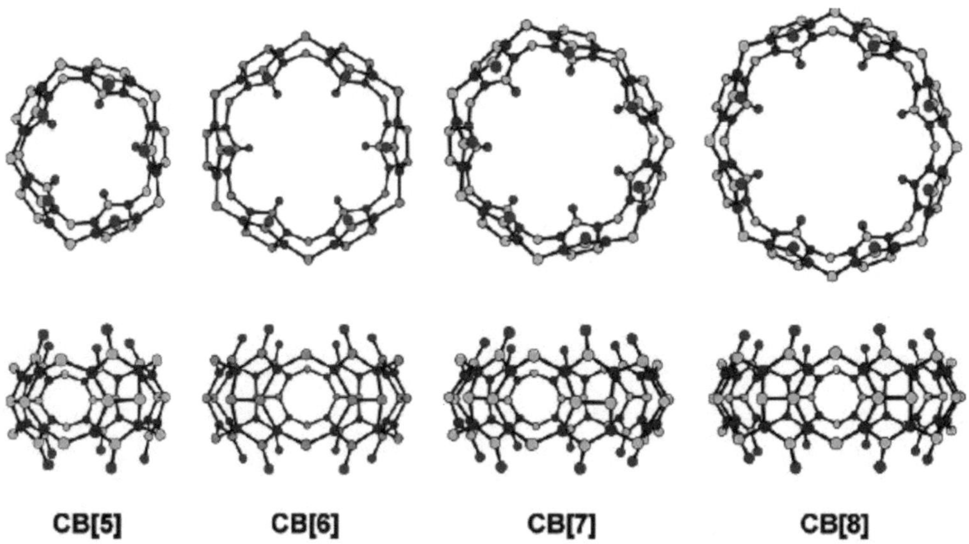

Fig. 7. X-ray crystal structures of CB[*n*] ($n = 5$-8). Color codes: carbon, gray; nitrogen, blue; oxygen, red [35].

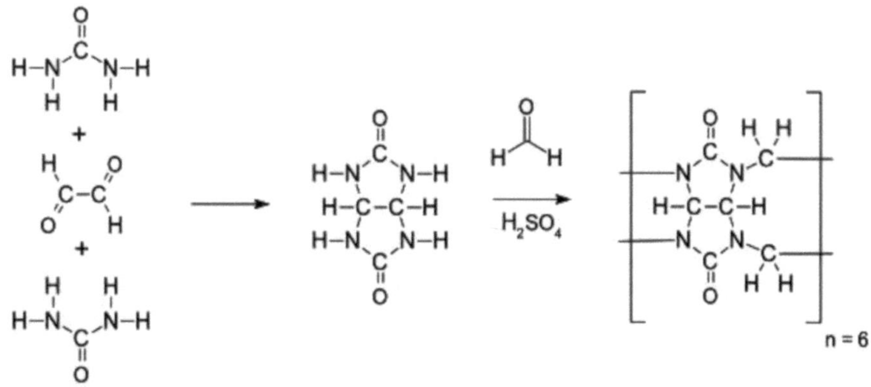

Fig. 8. Synthetic method of CB[6] [36].

The cucurbiturils are often compared to cyclodextrins (Table 1), which are considered to be their closest relatives in terms of size and shape, and the fact that both are often studied in aqueous solution [37].

Host	Portal Diameter (Å)	Interior Cavity Diameter (Å)	Height (Å)	Cavity Volume (Å)	Solubility in Water (Mm)
CB[5]	2.4	4.4	9.1	82	20-30
CB[6]	3.9	5.8	9.1	164	0.018
CB[7]	5.4	7.3	9.1	279	20-30
CB[8]	6.9	8.8	9.1	479	<0.01
CB[10]	9.5-10.6	11.3-12.4	9.1	870	Not reported
α-CD	4.7	5.3	7.9	174	297
β-CD	6.0	6.5	7.9	262	16
γ-CD	7.5	8.3	7.9	427	293

Table 1. Cavity dimensions and aqueous solubilities of CB[*n*] and CD host [37].

The cavity size of CB[6] as the first member of the CB[n] family is approximately the same as that of α-CD, and both hosts have hydrophobic cavities. As a result, binding involving CB[6] can be more selective compared to binding involving the more flexible α-CD host. Cucurbiturils have been shown to have a variety of potential applications. The ability of CB[n] to act as a catalyst for a chemical reaction was recognized by Mock relatively soon after his characterization of CB[6] [38, 39]. He realized that, since CB[6] has two electron rich portals, it had the potential to bring two cationic reactants in close proximity, as well as in specific orientations, so that the two reactants could undergo an accelerated reaction if they were simultaneously bound with the macrocycle. The larger macrocycles, such as CB[7] and especially CB[8], have seen further use for catalyzing and inhibiting reactions, as their larger cavity size allow them to form complexes, including ternary complexes, with a greater variety of guests. Considerable interest has been devoted to the use of cucurbiturils as drug delivery vehicles. As mentioned in the discussion of CB[7], Day [40-42] as well as Kim [43] have studied complexation between CB[7] and platinum-based anti-cancer drugs. Kim and co-workers have recently explored the use of nanoparticles based on functionalized CB[6] as drug delivery vehicles [44, 45]. Also, depending on the situation, they may also improve the solubility of a drug [46], reduce its undesired toxicity [47], improve its stability or activate

the drug [48]. In many cases the complexation between cucurbituril and a guest causes a change in the guest's spectroscopic properties. While this is useful for measuring the complexation between the host and guest, it can also have applications towards detecting the presence of a particular compound that the cucurbituril has an affinity for. Recent developments in the use of cucurbiturils as biosensors has been shown by Nau and co-workers, to detect enzymes such as amino acid decarboxylases [49, 50]. On the other hand, the effects of CB[7] on the activity of proteases have been studies.[51].
It is reported that the most photosensitizer dyes aggregate easily and have poor solubility in aqueous solutions. To overcome these limitations, there have been several approaches such as combination of photosensitizers with polymers or macrocycles. Because of water-soluble cyclic oligomers of glycoluril units linked by methylene bridges in cucurbiturils, CBs have emerged as versatile cyclic hosts that can form strong inclusion complexes with positively charged organic guests [52]. Additional above mentioned applications, CBs as organic capsules would be suitable hosts to improve the stability and solubility of gold nanoparticles in common solvents. Because a general problem of metal nanoparticles is their tendency to undergo agglomeration, which increases their particle size, and therefore, dramatically reduces the catalytic activity. Two general strategies have been developed to stabilize the particle size of the metal; 1) supporting the metal nanoparticles on suitable solid surfaces and 2) by using suitable ligands [53-56]. Therefore, CB[n] macrocycles can be used to stabilize the metal nanoparticles. Corma and co-workers have shown that the size of metal nanoparticles can be controlled by encapsulation in CB[7], which shows unprecedented behavior at stabilizing very small gold nanoparticles [57]. TEM images of gold nanoparticles obtained in the presence and absence of CB are shown in Fig. 9.

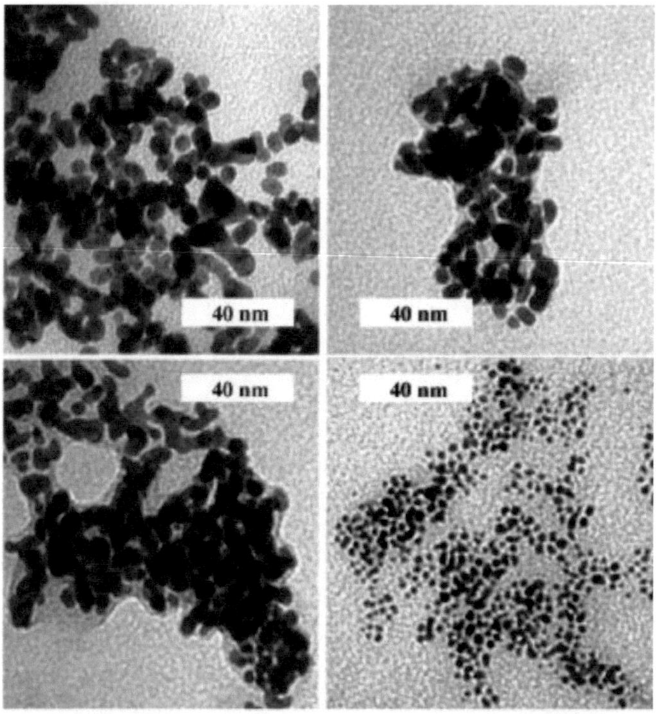

Fig. 9. TEM images of gold nanoparticles obtained in the absence of CB (top, left) and in the presence of CB[5] (top, right), CB[6] (bottom, left), and CB[7] (bottom, right) [57].

2.4 Crown ethers

In 1967, Charles Pedersen discovered a simple method of synthesizing a crown ether when he was trying to prepare a complexing agent for divalent cations [58, 59]. His strategy entailed linking two catecholate groups through one hydroxyl on each molecule. This linking defines a polydentate ligand that could partially envelop the cation and, by ionization of the phenolic hydroxyls, neutralize the bound dication. He was surprised to isolate a by-product that strongly complexed potassium cation. He realized that the cyclic polyethers represented a new class of complexing agents that were capable of binding alkali metal cations. He proceeded to report systematic studies of the synthesis and binding properties of crown ethers in a seminal series of papers. The fields of organic synthesis, phase transfer catalysts, and other emerging disciplines benefited from the discovery of crown ethers. Pedersen shared the 1987 Nobel Prize in Chemistry for the discovery of crown ethers.

Crown ethers are named as x-crown-y where x denotes the total number of atoms in the cyclic backbone and y denotes the number of oxygen atoms (Fig. 10).

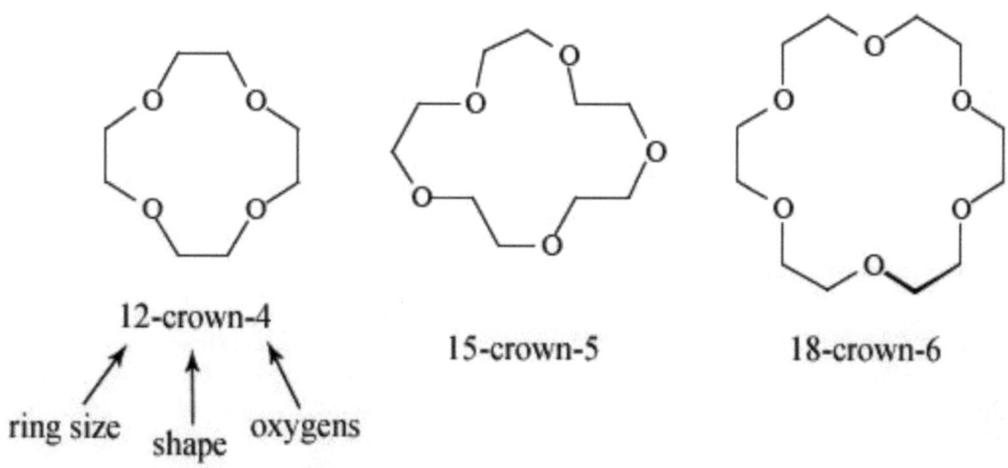

Fig. 10. General structure of crown ethers.

Crown ethers may be used to vary the solubility of the metal salts in non-polar and dipolar aprotic solvents [60]. They may be also used to catalyze reactions between reagents contained in two different phases (either liquid/liquid or solid/liquid). The polyether is present only in "catalytic" amounts and the process is named "phase transfer catalyst" (TPC). Because of their "open" structure, crown ethers can extract cations from a crystalline solid. Depending on the catalyst used cations can be taken to a solid or liquid phase [61, 62]. According to Table 2, crown ethers display selectivity in complexation based on cavity and ion size.

Cation	Ionic Radius	Crown Ether	Cavity Diameter	Strongest Binding
Li^+	1.39Å	12-crown-4	1.2-1.4Å	Li^+
Na^+	1.94Å	15-crown-5	1.7-2.2Å	Na^+
K^+	2.66Å	18-crown-6	2.6-3.2Å	K^+

Table 2. The selectivity of crown ethers based on ion size.

Other related macripolycyclic ligands called "cryptates" (or cryptands or simply crypts) are more potent, selective and even stronger complexing agents for alkali metals. Besides, they show high complexing ability with alkaline earth cations, few transition and lanthanide ions also. The crypts are considered to be the three-dimensional equivalents of crown ethers but incorporate N as well as O atoms (sometimes S or P atoms also) to show higher denticity than the crown ethers. The molecules are appropriately cross-linked with donor atoms correctly positioned in the bridging group to encapsulate metal ions in cage-like structures. A typical crypt is the molecule $N(CH_2CH_2OCH_2CH_2OCH_2CH_2)_3N$, called Crypt-222 (222 denotes the no. of ethereal O atoms in each N-N bridge) having the basic structure given in Fig. 11. Cryptates (meaning hidden) are so called because they wrap around and hide the cation. All these ligands from complexes with very high formation constants.

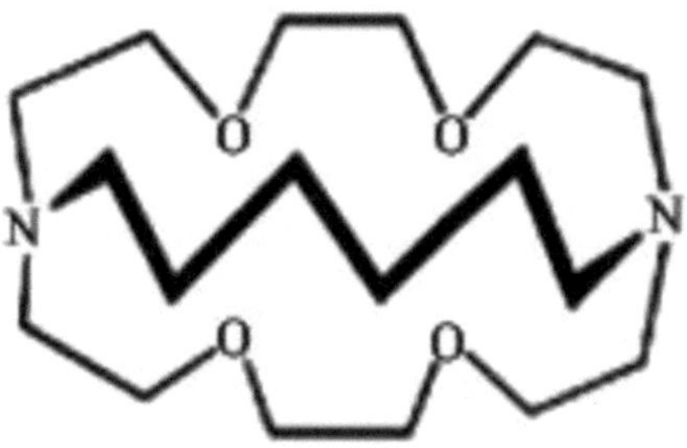

222 cryptand

Fig. 11. Structure of 222 cryptand.

The detection and elimination of heavy metal ions are important for life, environment and medicine sciences. Up to now, many analytical methods have been developed for detection of heavy metals [63, 64]. Very recentiy, smart responsive microcapsules capable of recognizing heavy metal ions based on crown ethers due to the "host–guest" complexation between the receptors and ions have been successfully developed by Pi et al [65]. Besides supramolecules like calixarenes and cyclodextrins, Schiff's bases (azomethines or imines) and crown ethers can be used as sensing material in the construction of potentiometric ion selective electrodes or catalytic applications [66-68]. Ganjali and co-workers studied different supramolecules based on Schiff's bases and crown ethers as potentiometric membrane sensors for cations and anions [69].

2.5 Cryptophanes

For the first time, cryptophanes were discovered by Collet and co-workers in 1981 [70]. Cryptophanes are molecular hosts having high binding affinities for small, neutral molecules, consisting of two cup-shaped, C3-symmetric, [1.1.1]orthocyclophane units, connected by three bridges (Y). These host–guest complexes are supramolecular assemblies mainly stabilized by van der Waals forces and specific π-interactions between the aromatic molecular cavity and the encapsulated guest. Most cryptophanes exhibit two diastereomeric forms (*syn* and *anti*), distinguished by their symmetry type. This general scheme offers a variety of choices (Y, R_1, R_2, and symmetry type) by which the shape, the volume, and the chemical properties of the generally hydrophobic pocket inside the cage can be modified, making cryptophanes suitable for encapsulating many types of small molecules and even chemical reactions. Fig. 12 represents the general structure of cryptophanes.

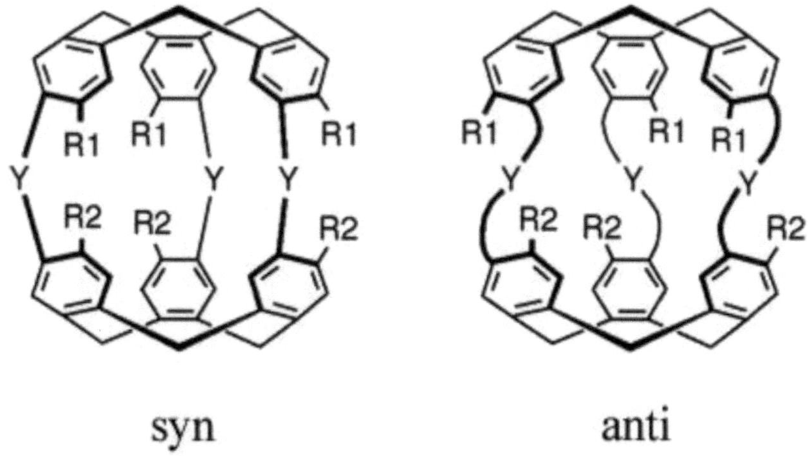

Fig. 12. General structure of cryptophanes.

Cryptophanes can be synthesized through any of three methods: (i) the "template method", (ii) "two-step method", or (iii) a "capping method". "template method" was the first employed by Collet for the synthesis of cryptophanes [70]. A suitably substituted benzyl alcohol is first cyclized to form a trimeric cup. Additional benzyl alcohol substituents are then appended via a bridging *Y* group, and these groups are then subjected to an intramolecular cyclization to yield the corresponding cryptophanes. The initially formed C_3-cup thus serves as a template for the second cup-forming reaction, with appreciable diastereoselectivity. This approach, for instance, allows the synthesis of cryptophanes that are asymmetric with respect to the peripheral substituents attached to the top and the bottom cups (R_1 and R_2 are different). Second, provided that the initially formed trimer can be resolved, this route can be employed to provide optically active *anti* cryptophanes of known absolute configuration. Method (ii), known as Collet's "two step" synthesis, yields suitable cryptophanes in only two synthetic steps from readily available vanillyl alcohol (R = OCH_3, X = OH) [71]. In this approach, a monomer possessing two vanillyl alcohol residues is synthesized and then cyclized directly, under acidic conditions, to the correspondingcryptophanes. Notably, this reaction tends to display diastereoselectivity in favor of the chiral *anti* cryptophanes, making available some chiral *anti* cryptophanes that give low yields by the "template method". Cram and co-workers demonstrated "capping method" (iii), the "capping" of one preformed *C3*-trimer with another, is also viable [72]. Though Cram's synthesis involved the oxidative coupling of two identical cups possessing terminal acetylenic groups (R = CH_3; X = OCH_2C CH), "capping" one *C3*-trimer with a dissimilar trimer can also be envisaged (R_1 and R_2, X_1 and X_2 are different). Though the yields of cryptophanes synthesized by the "capping" method were low (< 15%), lessons learned from similar approaches toward the syntheses of (hemi)carcerand molecular containers suggest that it may be possible to improve yields by employing suitable guest templates. Depending on their structure, cryptophane cages are classified according to the following Table 3 [73].

Cryptophanes are interesting and versatile molecular receptors, whose complexation properties have been widely studied. For instance, cryptophanes are adapted for the recognition of dichloromethane [74, 75], xenon or biosensors based on xenon [76-79] and methane [80, 81].

Structure			Name	
Bridges Y	R_1	R_2	anti	syn
3 × $O(CX_2)_2O$ (where X is H or D)	OCX_3	OCX_3	A	
3 × $O(CX_2)_2O$	OCH_2CO_2H	OCH_2CO_2H	A3	
3 × $O(CX_2)_2O$	OCH_3	H	C	D
3 × $O(CH_2)_3O$	OCH_3	OCH_3	E	F
3 × $O(CH_2)_3O$	OCH_2CO_2H	OCH_2CO_2H	E3	
3 × $O(CH_2)_5O$	OCH_3	OCH_3	O	P
3 × $O(CH_2)_5O$	OCH_2CO_2H	OCH_2CO_2H	O3	
3 × $OCH_2C{\equiv}CC{\equiv}CH_2O$	CH_3	CH_3	γ (gamma)	δ (delta)
2 × $O(CH_2)_2O$, 1 × $O(CH_2)_3O$	OCH_3	OCH_3	223	
2 × $O(CH_2)_3O$, 1 × $O(CH_2)_2O$	OCH_3	OCH_3	233	
2 × $O(CH_2)_2O$, 1 × $O(CH_2)_4O$	OCH_3	OCH_3	224	

Table 3. Classification of known cryptophane structures [73].

2.6 Porphyrins

Porphyrins are heterocyclic macrocycles composed of four modified pyrrole subunits interconnected at their *a* carbon atoms via methine bridges and they obey Hückel's rule for aromaticity, possessing 4n+2 = π electrons that are delocalized over the macrocycle. The synthetic chemistry of porphyrins is extremely rich, as its history began in the mid-1930s. The general synthetic route of porphyrins is shown in Fig. 13.

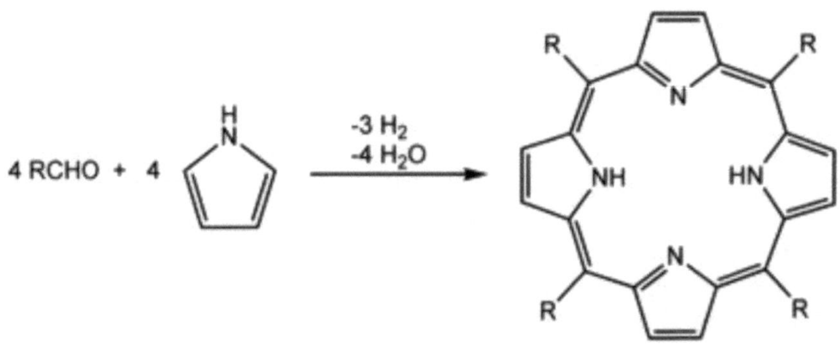

Fig. 13. General synthesize of porphyrins.

Porphyrins are also capable of being metallated and demetallated. Almost every metal in the periodic chart has been inserted into the porphyrin macrocycle but most typically are: Fe, Zn, Cu, and Ni, which can be inserted into the porphyrin cavity by using simple metal salts. Demetallation can usually be achieved by treatment with acids of various strengths. Porphyrins are often used to construct structures in supramolecular chemistry. These systems take advantage of the Lewis acidity of the metal, typically zinc. An example of a host-guest complex that was constructed from a macrocycle composed of four porphyrins

[82]. A guest-free base porphyrin is bound to the center by coordination with its four pyridine substituents. On the other hand, it is reported that multi-porphyrinic arrays attract more and more attention for the elaboration of photonic and electronic wires [83, 84]. In particular, metal ion coordination utilising ligation of the metallated porphyrin nucleus has proven efficient for positioning photoactive components. Because fullerenes are known as one type of acceptor with strong п-electron accepting ability, fullerene-porphyrin supramolecular assemblies have been studied extensively to generate photocurrents as well as to elucidate their unique photophysical and photochemical properties [85-89]. Supramolecular complexes were elaborated using these porphyrin-peptide oligomers as host and fullerenes as guest molecules, in collaboration with Professor Fukuzumi from Osaka University in Japan (Fig. 14).These supramolecular devices behave as organic photovoltaic cells and demonstrate remarkable enhancement in the photoelectrochemical performance as well as broader photoresponse in the visible and near-infrared regions with increasing the number of porphyrin units in a-polypeptide structures. These results clearly show that the formation of molecular assembly between fullerene and multi-porphyrin arrays with a polypeptide backbone controls the electron-transfer efficiency in the supramolecular complex, which is essential for the light-energy conversion [90].

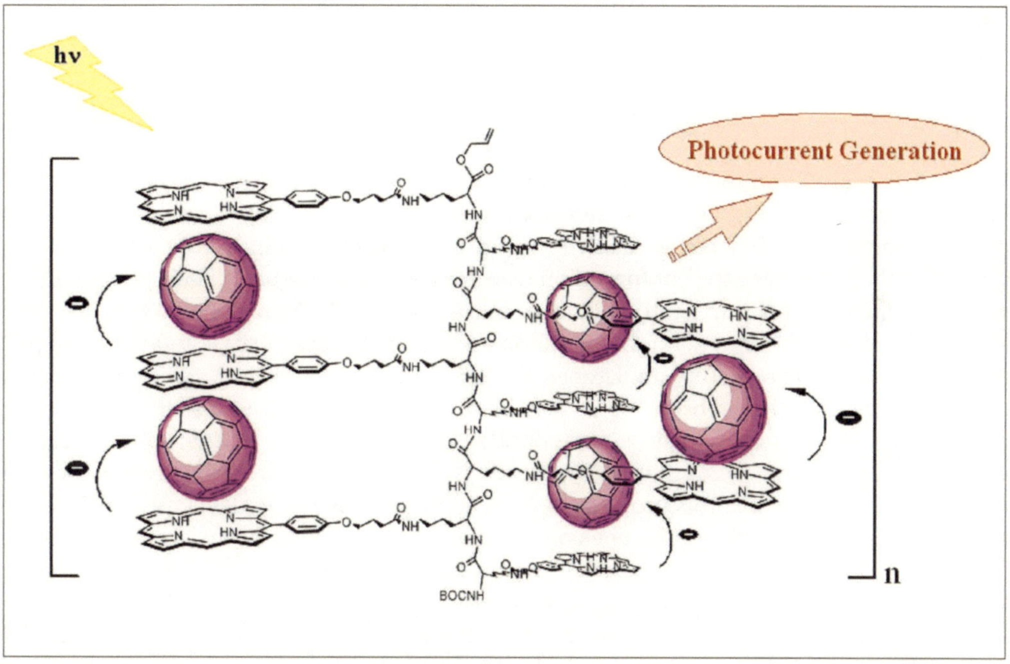

Fig. 14. A supramolecular consist of fullerene and multi-porphyrin arrays for the light-energy conversion [90].

Porphyrins with other guest molecules such as urea can provide a guideline to improve the structural homogeneity of the photonic wire [91]. It is possible that porphyrins are applied as guest molecules for special purposes. For example, a supramolecular system composed of photoswitchable β-cyclodextrin dimmers (host) and tetrakis-sulfonatophenyl porphyrin (guest) is a good candidate for using as photocontrollable (drug) delivery systems and might find application in, for example, photodynamic cancer therapy [92]. Besides, porphyrin nanostructures are attractive building blocks because of their electronic, optical, and catalytic properties [93-95].

2.7 Cyclotriveratrylenes

Cyclotriveratrylene macrocycles are used in host-guest chemistry as molecular hosts. These compounds can be synthesised from veratrole alcohol by addition of a suitable acid which can be perchloric acid in methanol or formic acid or sulfuric acid in acetic acid [96].

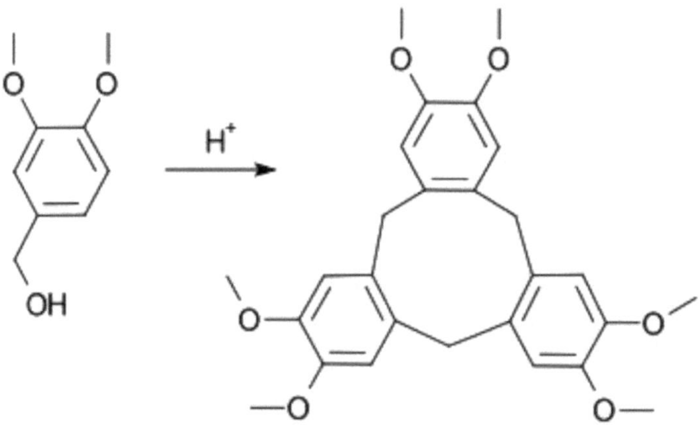

Fig. 15. General method for the preparation of cyclotriveratrylene.

For the first time, Cyclotriveratrylene (CTV) was synthesised by Gertrude Robinson in 1915 [97] and in 1965, Lindsey identified the correct structure [98]. The host molecule cyclotriveratrylene (CTV) has a relatively rigid bowl shape, creating a molecular cavity, much like those seen in cone conformation calixarenes. The host-guest chemistry of CTV is, however, quite distinct from that of the calixarenes. CTV forms balland-socket type supermolecules with large neutral and cationic guest molecules such as fullerene-C_{60} [99] and $\{Na\text{-}[2.2.2]cryptate\}^+$ [100]. Fullerenes are able to generate efficiently cytotoxic singlet oxygen by energy transfer from the metastable fullerene triplet excited-state to molecular oxygen making them promising compounds for antitumor activities (photodynamic therapy). But, the low solubility of fullerenes in aqueous media limits their biological applications. To overcome this problem, several strategies have been developed such as the preparation of water soluble fullerene complexes with macrocyclic host systems [101, 102]. Therefore, Rio and Nierengarten synthesized water soluble supramolecular cyclotriveratrylene–[60]fullerene complexes for biological applications [103]. On the other hand, small organic solvent molecules have been encapsulated within the CTV molecular cavity [104, 105]. CTV has also proved to be a valuable building-block for the synthesis of more sophisticated host molecules such as cryptophanes and extended arm cavitands, where intracavity complexation of small and large guest molecules is common place. CTV has been employed as a host in coordination complexes or organometallic complex [106-108]. Recently, Nuriman and co-workers reported a novel chemosensor, based on tris[2-(4-phenyldiazenyl)phenylaminoethoxy]cyclotriveratrylene (TPPECTV) for the colorimetric determination and visual detection of Hg(II) ions. The preparation of TPPECTV and its structure are shown in Fig. 16 [109]. Schematic representation of heteroditopic hosts (CTVs) for ion-pair recognition is shown in Fig. 17 [110].

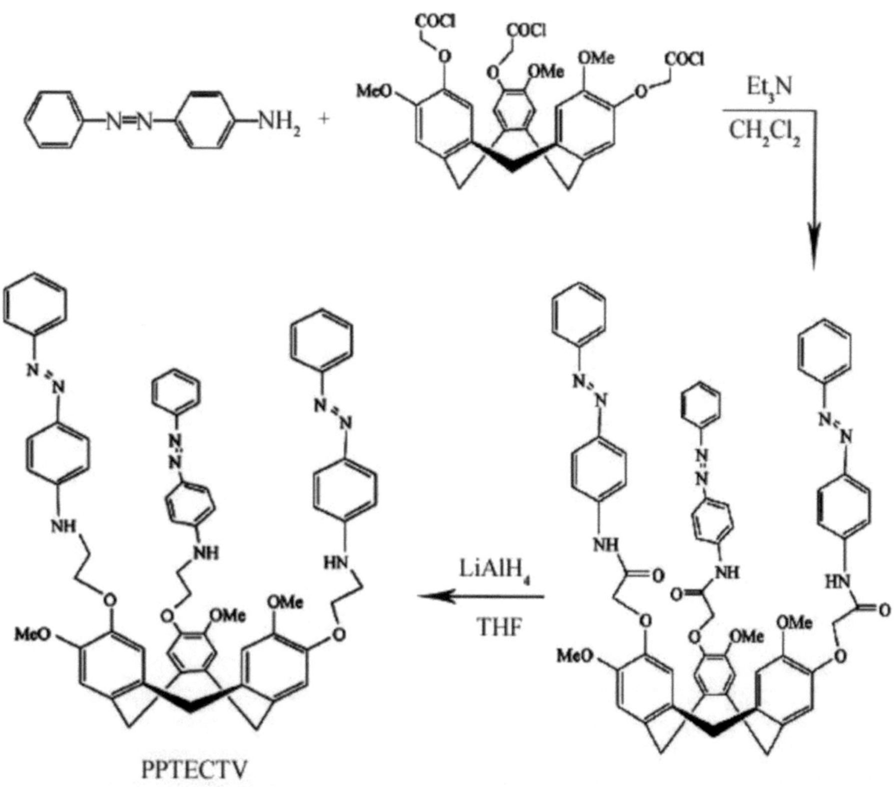

Fig. 16. The preparation of TPPECTV and its structure [109].

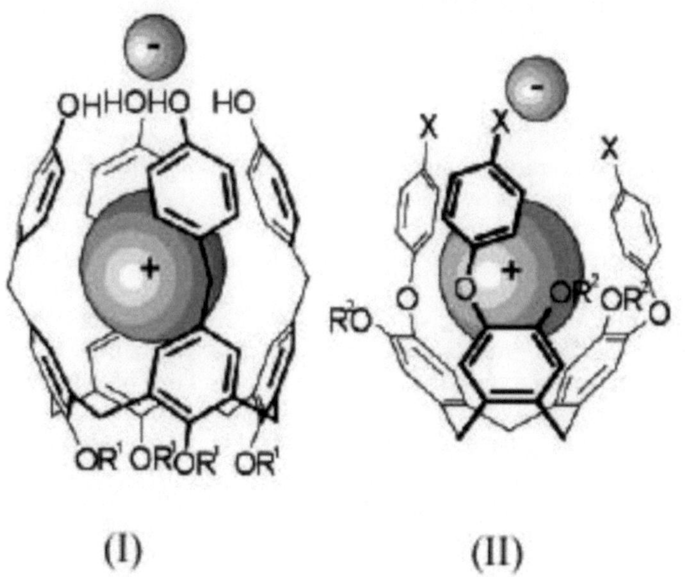

Fig. 17. Schematic representation of heteroditopic hosts (CTVs) for ion-pair recognition [110].

2.8 Carcerands

In 1983, Donald J. Cram introduced the concept of carcerand as a closed-surface binding host that can imprison simple organic molecules as guest [111], and in 1985, Cram and co-workers synthesized a typical carcerand by multiply binding cavitand (**1**) to (**2**) (Fig. 18) [112]. A carcerand as a synthetic host molecule completely entraps its guest so that it will not escape even at high temperatures. The generation of carcerands is based on calixarenes [113].

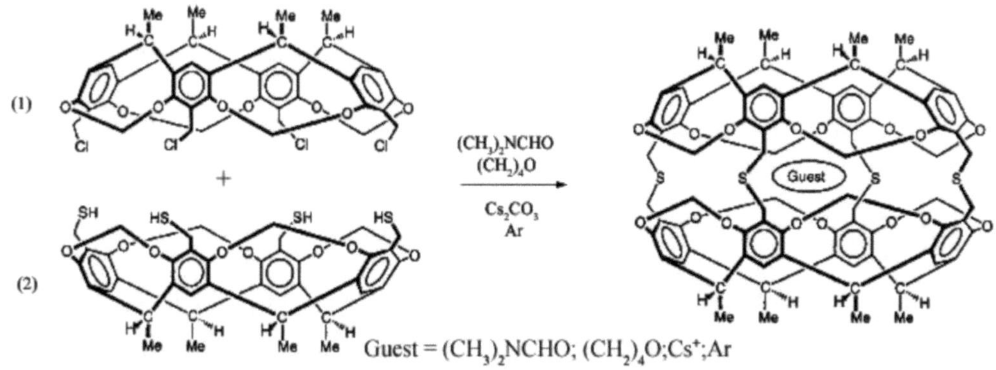

Fig. 18. Synthesized carcerand from (1) and (2) molecules [112].

The name carcerand is derived from the Latin word *carcer*, meaning "prison". In each carcerand, the incarcerated guests did not leave their prison, even at high temperatures. Complexes with permanently imprisoned guests are called "carceplexes" [114, 115]. In contrast, "hemicarcerands" incarcerate and liberate guests at elevated temperatures but form stable hemicarceplexes at ambient temperature [116]. It is reported that water-soluble hemicarcerands can be useful for drug delivery systems [117]. Piotrowiak and co-worker investigated novel water-soluble hemicarcerands to develop unique hybrid materials composed of semiconductor nanoparticles (TiO_2) and host-guest assemblies [118].

2.9 Zeolites

Zeolites as microporous aluminosilicate minerals are found in nature, and the zeolite mineral stilbite was first discovered in 1756 by the Swedish mineralogist A. F. Cronstedt, who observed that upon rapidly heating the material stilbite, it produced large amounts of steam from water that had been adsorbed by the material. Zeolites with microporous structure are known as a member of the family solids called "molecular sieves". The term "molecular sieve" refers to a particular property of these materials, the ability to selectively sort molecules based primarily on a size exclusion process. This is due to a very regular pore structure of molecular dimensions. Natural zeolites form where volcanic rocks and ash layers react with alkaline groundwater. Zeolites also crystallize in post-depositional environments over periods ranging from thousands to millions of years in shallow marine basins. Naturally occurring zeolites are rarely pure and are contaminated to varying degrees by other minerals, metals, quartz, or other zeolites. For this reason, naturally occurring zeolites are excluded from many important commercial applications where uniformity and purity are essential. Besides natural zeolites, synthetic zeolites can be obtained through hydrothermal or sol-gel method. Zeolite synthesis via a hydrothermal method occurs with reagents being a silica source, an alumina source, a mineralizing agent such as OH^- or F^-, and for higher Si/Al ratio zeolites, organic molecules as structure-directing agents. A schematic of the zeolite growth process is shown in Fig. 19 [119].

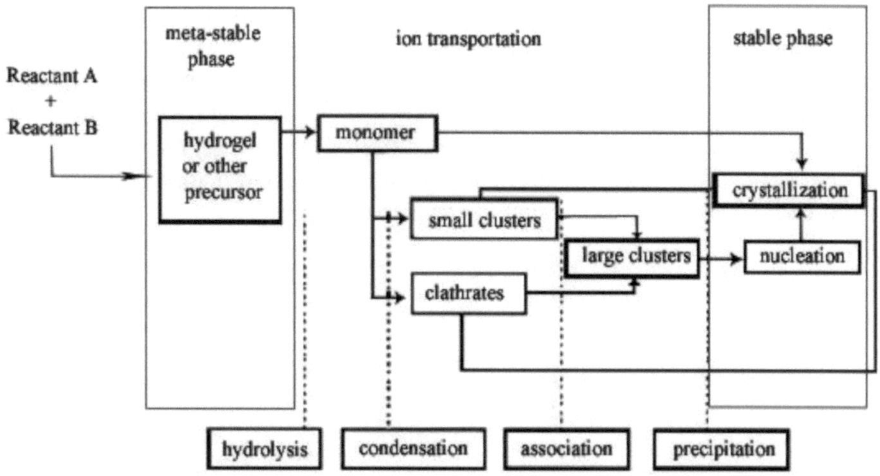

Fig. 19. Schematic representation of zeolite crystallization process [119].

Several independent processes are occurring in the medium, including nucleation of various structures, crystallization as well as dissolution of meta-stable phases. It is commonly observed that the conversion of the composition (gel or solution) to crystals is quite rapid once the crystallization process gets started. This suggests that nucleation is the rate-limiting step and is consistent with studies that report addition of seed crystals decrease the induction time [120]. Tezak suggested that rather than viewing the synthesis process as nucleation and crystallization, at least four subsystems be considered: (a) formation of simple and polymeric aluminosilicates, (b) aggregation of these complexes to form embryo, (c) nucleation as aggregate formation with a well ordered core and micelle formation (primary particles), and (d) aggregation of primary particles via oriented aggregation [121]. In sol-gel process, other elements (metals, metal oxides) can be easily incorporated but silicalite sol formed by the hydrothermal method is very stable.

2.9.1 Structure of zeolites

"Tectosilicates" (framework silicates) are inorganic 3D polymers based on SiO_4 tetrahedral units some of which can be replaced with AlO_4 tetrahedral units. Tectosilicates form 3/4 of the Earth's crust and zeolites are a sub-group of tectosilicates. Liebau and co-workers have proposed a classification for porous tectosilicates that is summarized in Fig. 20 [122].

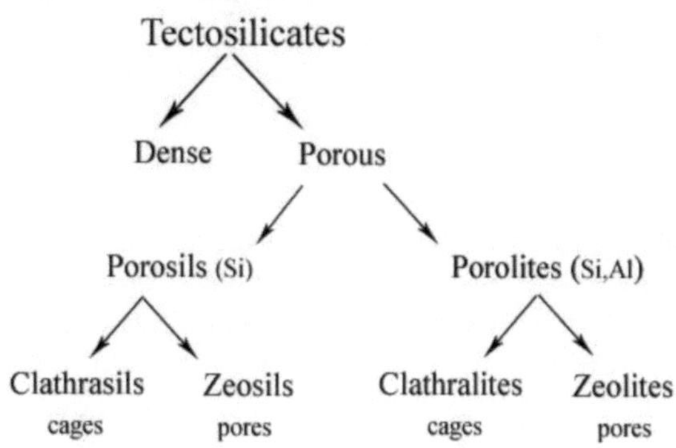

Fig. 20. Classification of porous tectosilicates [122].

Zeolites are composed of TO_4 tetrahedra (T = Si, Al) with O atoms connecting neighboring tetrahedra. For a completely siliceous structure, combination of TO_4 (T = Si) units in this fashion leads to silica (SiO_2), which is an uncharged solid. Upon incorporation of Al into the silica framework, the +3 charge on the Al makes the framework negatively charged, and requires the presence of extraframework cations (inorganic and organic cations can satisfy this requirement) within the structure to keep the overall framework neutral. The zeolite composition can be best described as having three components:

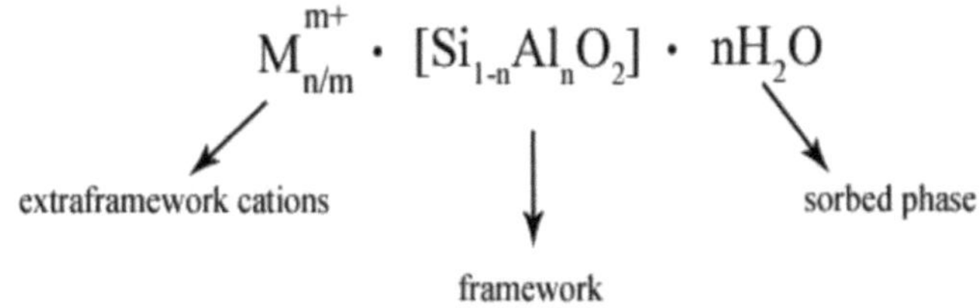

Fig. 21. Chemical composition of zeolites.

The extraframework cations are ion exchangeable and give rise to the rich ion-exchange chemistry of these materials. The novelty of zeolites stems from their microporosity and is a result of the topology of the framework. Other elements, such and B, Ge, Zn, P, and transition elements, can also be incorporated into the framework and are referred to as crystalline molecular sieves. Aluminophosphates (AlPOs) have strictly alternating AlO_2^- and PO_2^+ units, and the framework is neutral, organophilic and nonacidic. The alternation of Al or P leads to structures lacking in odd-numbered rings. Substitution of P by Si leads to silicoaluminophosphates (SAPOs), with cation-exchange abilities. Metal cations can also be introduced into the framework, including transition metal ions such as Co, Fe, Mn, and Zn. Discovery of these solids has led to the development of several new structures [123]. The framework composition depends on the synthesis conditions. As the Si/Al ratio of the framework increases, the hydrothermal stability as well as the hydrophobicity increases.

Typically, in as-synthesized zeolites, water present during synthesis occupies the internal voids of the zeolite. The sorbed phase and organic non-framework cations can be removed by thermal treatment/oxidation, making the intracrystalline space available. The fact that zeolites retain their structural integrity upon loss of water makes them different from other porous hydrates, such as $CaSO_4$. The crystalline nature of the framework ensures that the pore openings are uniform throughout the crystal and can readily discriminate against molecules with dimensional differences less than 1 Å, giving rise to the name molecular sieves. According to the molar ratio of Si/Al, zeolites are classified to three types:

1. Low-silica or Al-rich zeolites
2. Intermediate silica zeolites
3. High-silica zeolites

About 40 natural zeolites are known. Most zeolites known to occur in nature are of lower Si/Al ratios, since organic structure–directing agents necessary for formation of siliceous zeolites are absent. Table 4 indicates the natural zeolites [124]. Sometimes natural zeolites are found as large single crystals, though it is very difficult to make large crystals in the laboratory. High-porosity zeolites such as faujasite (X,Y) are known as intermediate silica. Two natural zeolites that find extensive use are clinoptilolite (HEU) and mordenite (MOR) for ion-exchange (radioactive) agricultural uses and as sorbents. The catalytic activity of natural zeolites is limited by their impurities and low surface areas. Another natural zeolite, erionite (ERI) has toxicity comparable to or even worse than some of the most potent forms of asbestos, especially in causing a form of lung mesothelioma.

Zeolites A and X (as low silica) were discovered by Milton and Breck at Union Carbide in 1959. These zeolites have the highest cation contents and are excellent ion-exchange agents. Breck reported the synthesis of zeolite Yin 1964, which spans a Si/Al ratio of 1.5–3.8 and with framework topology similar to that of zeolite X and the mineral faujasite. Decreasing the Al content led to both thermal and acid stabilities and paved the way for development of zeolite Y–based processes in hydrocarbon transformations. Large–port mordenite, also with a Si/Al ratio of 5, was reported by Sand [125]. Zeolites with Si/Al ratios of 10–100 (or higher) such as ZSM-5 were reported by Mobil Research and Development Laboratories in the 1960s and 1970s [126]. Fig. 22 shows how the sodalite unit can be assembled to form common zeolitic frameworks. For example, in LTA, each sodalite(sod)-cage is connected to six nearest neighbouring sod-cages through double T4-rings (Fig. 22a) but in sodalite (SOD), each sod-cage is connected to six nearest neighbouring sod-cages through common T4-rings (Fig. 22b) [127].

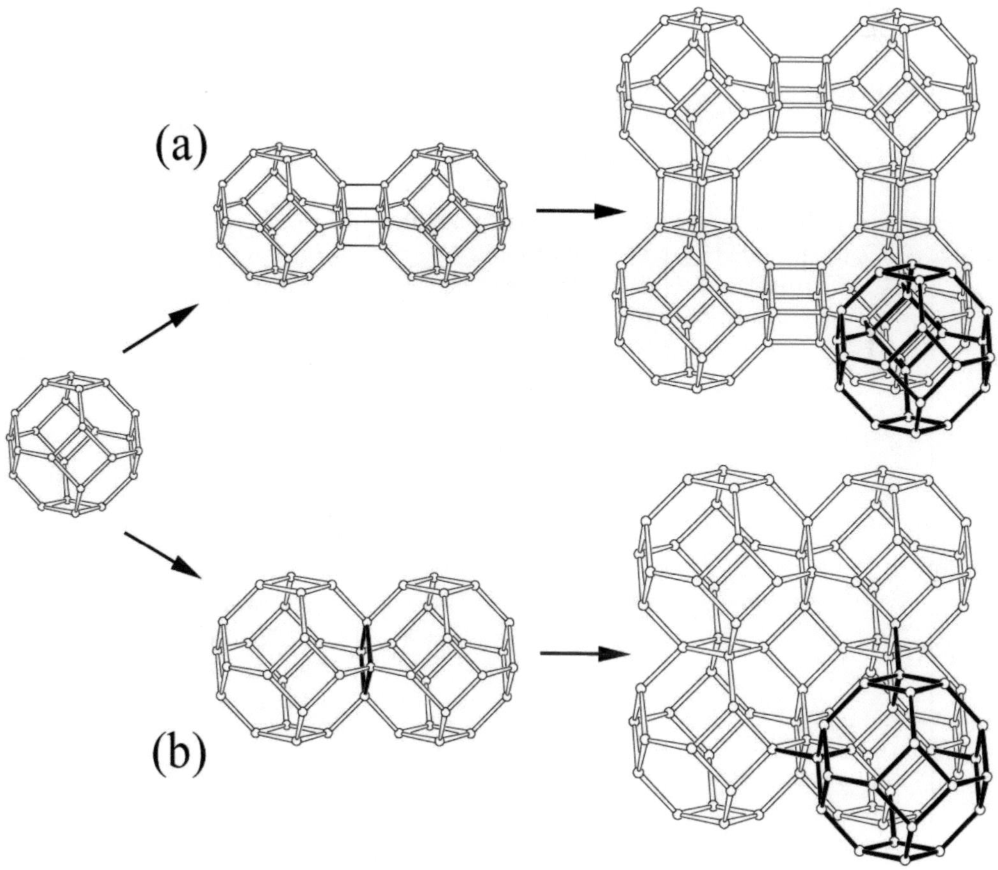

Fig. 22. Structure of LTA (a) and sodalite (b) [127].

Low silica Si/Al ≤ 2	Intermediate silica 2 < Si/Al ≤ 5	High silica 5< Si/Al
ABW, Li-A(BW)	BHP, linde Q	ASV, ASU-7
AFG, afghanite [a]	BOG, boggsite [a]	BEA, zeolite β
ANA, analcime [a]	BRE, brewsterite [a]	CFI, CIT-5
BIK, bikitaite [a]	CAS, Cs-aluminosilicate	CON, CIT-1
CAN, cancrinite [a]	CHA, chabazite [a]	DDR, decadodelcasil 3R
EDI, edingtonite [a]	CHI, chiavennite [b]	DOH, dodecasil 1H
FAU, NaX	DAC, dachiardite [a]	DON, UTD-1F
FRA, franzinite	EAB, EAB	ESV, ERS-7
GIS, gismondine [a]	EMT, hexagonal faujasite	EUO, EU-1
GME, gmelinite [a]	EPI, epistilbite [a]	FER, ferrierite [a]
JBW, NaJ	ERI, erionitea	GON, GUS-1
LAU, laumonite [a]	FAU, faujasite [a], NaY	IFR, ITQ-4
LEV, levyne [a]	FER, ferrierite [a]	ISV, ITQ-7
LIO, liottite [a]	GOO, goosecreekite [a]	ITE, ITQ-3
LOS, losod	HEU, heulandite [a]	LEV, NU-3
LTA, linde Type A	KFI, ZK-5	MEL, ZSM-11
LTN, NaZ-21	LOV, lovdarite [b]	MEP, melanopholgite [a]
NAT, natrolite [a]	LTA, ZK-4	MFI, ZSM-5
PAR, partheite [a]	LTL, linde L	MFS, ZSM-57
PHI, phillipsite [a]	MAZ, mazzite [a]	MSO, MCM-61
ROG, roggianite [a]	MEI, ZSM-18	MTF, MCM-35
SOD, sodalite	MER, merlinoite [a]	MTN, dodecasil 3C
WEN, wenkite [a]	MON, montasommaite [a]	MTT, ZSM-23
THO, thomsonite [a]	MOR, mordenite [a]	MTW, ZSM-12
TSC, tschortnerite	OFF, offretite [a]	MWW, MCM-22
	PAU, paulingite [a]	NON, nonasil
	RHO, rho	NES, NU-87
	SOD, sodalite	RSN, RUB-17
	STI, stilbite [a]	RTE, RUB-3
	YUG, yugawaralite [a]	RTH, RUB-13
		RUT, RUB-10
		SFE, SSZ-48
		SFF, SSZ-44
		SGT, sigma-2
		SOD, sodalite
		STF, SSZ-35
		STT, SSZ-23
		TER, terranovaite
		TON, theta-1
		ZSM-48
		VET, VPI-8
		VNI, VPI-9
		VSV, VPI-7

[a] Natural materials; [b] Beryllosilicates (natural).

Table 4. Classification of zeolites [124].

2.9.2 Zeolite properties

The decomposition temperature for Low-silica zeolites is ~700 °C, whereas completely siliceous zeolite, such as silicalite, is stable up to 1300 °C. Low-silica zeolites are unstable in acid, whereas high-silica zeolites are stable in boiling mineral acids, though unstable in basic solution. Cation concentration, siting, and exchange selectivity vary significantly with Si/Al ratios and play an important role in adsorption, catalysis, and ion-exchange applications. Though acid site concentration decreases with increase in Si/Al ratio, the acid strength and proton activity coefficients increase with decreasing aluminum content. The internal volume of zeolites consists of interconnected cages or channels, which can have dimensionalities of one to three. Pore sizes can vary from 0.2 to 0.8 nm, and pore volumes from 0.10 to 0.35 cm^3/g. The framework can exhibit some flexibility with changes in temperature and via guest molecule-host interaction, as noted for the orthorhombic-monoclinic transformations in ZSM-5 [128].

2.9.3 Zeolite applications

Zeolites are widely used in primarily three applications, including:

1. Ion-exchange beds
2. Catalysts
3. Adsorbents

The major use of zeolites as ion-exchange agents is for water softening applications in the detergent industry and substitute use of phosphates. Natural zeolites find considerable use for removal of Cs^+ and Sr^{2+} radioisotopes by ion exchange from radioactive waste streams [129]. Synthetic zeolites are extensively used as catalysts in the petrochemical industry. Zeolites confine molecules in small spaces, which cause changes in their structure and reactivity. The hydrogens formed of zeolites (prepared by ion-exchange) are powerful solid-state acids, and can facilitate a host of acid-catalyzed reactions, such as isomerisation, alkylation and cracking [130]. On the other hand, zeolites have the potential of purification and separation of gases, including the removal of H_2O, CO_2 and SO_2 from low-grade natural gas streams. Other separations include noble gases, N_2, O_2, freon and formaldehyde [131]. Besides, zeolites because of their microporous and nanoporous structure are used to form host-guest nanocomposite materials as host molecules for special applications, which discuses in future.

3. Host-guest chemistry

Commonly, there are four essential fields to study host-guest chemistry, including:

- Inclusion Compounds
- Intercalation Compounds
- Clathrates
- Molecular Tweezers

The definition of "inclusion compounds" is very broad, extending to channels formed between molecules in a crystal lattice in which guest molecules can fit. In host-guest chemistry, an inclusion compound is a complex in which one chemical compound (host) forms a cavity in which molecules of a second (guest) compound are located. In 1978, different cryptands with three-dimensional, two-dimensional, macrobicyclic, macrotricyclic and spheroidal topologies were introduced to form inclusion compounds [132]. Traditionally, inclusion compounds have been synthesized by crystallizing the host and guest compounds from a solution [133-135]. Whereas the choice of components available for the construction of such systems is limited by the solubilities of the host and the guest, the limitations become more severe in multicomponent systems that require balancing the solubilities of several molecular species. A possible way to overcome solubility issues is through the use of mechanochemical methods. So that, Braga and co-workers have used the mechanochemical synthesis for the preparation of metal-organic intercalation compounds [136]. The design of synthetic strategies for creating novel host-guest inclusion compounds with zeolites provides the applications of these compounds in diverse areas of materials science, such as size /shape selective catalysis, nonlinear optics, quantum electronics, and photonics [137]. For example, cyclodextrins (CDs) with drugs can improve the solubility, stability, and bioavailability of drugs. The enhancement of drug activity and selective transfer or the reduction of side effects can also be achieved by inclusion compound formation [138]. Harris reported that thiourea inclusion compounds exhibit a wide range of interesting and important fundamental physicochemical properties [139].

In chemistry, "intercalation" is the reversible inclusion of a molecule (or group) between two other molecules (or groups). Many layered solids intercalate guest molecules. A famous example is the intercalation of potassium into graphite. Intercalation expands the "van der Waals gap" between sheets, which requires energy. Usually this energy is supplied by charge transfer between the guest and the host solid, i.e. redox.

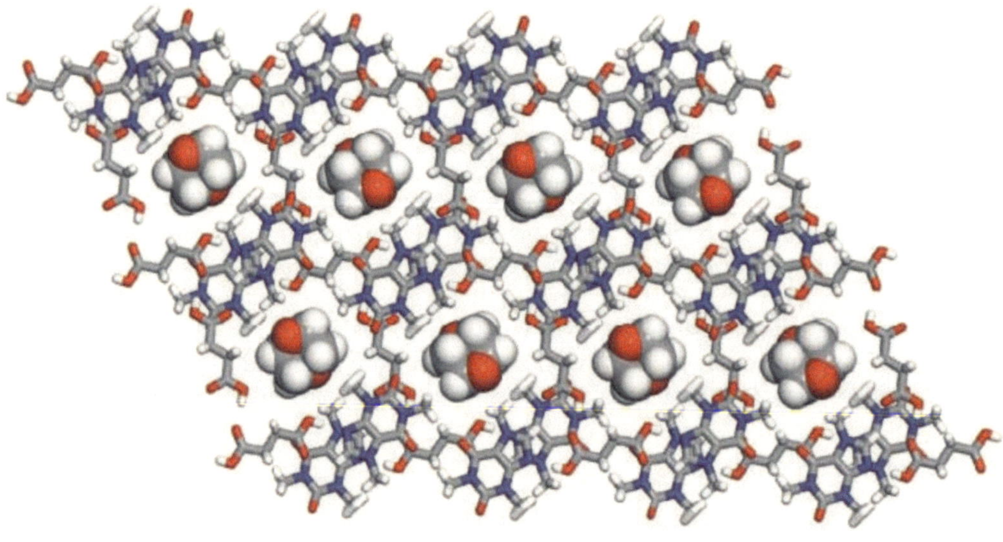

Fig. 23. An inclusion compounds consisting of a caffeine and succinic acid in a 1:1 stoichiometric ratio bound within 1,4-dioxane molecules reported by Jones [140]

Graphite intercalation compounds are formed by the insertion of atomic or molecular layers of a different chemical species called the intercalant between layers in a graphite host material, as shown in Fig. 24. The intercalation compounds occur in highly anisotropic layered structures where the intraplanar binding forces are large in comparison with the interplanar binding forces [141].

Aside from graphite, well known intercalation hosts are the layered dichalcogenides such as tantalum disulfide, molybdenum disulfide and iron oxychloride. Layered nanocomposites derived from two-dimensional MoS_2 host and coordination complexes exhibit good catalytic activities in the oxidization of sulfide ions into thiosulfate ions, due to their large surface areas [142]. It is reported that there are three ways to intercalate guest species in layered inorganic hosts (inorganic-organic hybrid materials or nanocomposites): i) intercalation in layered host solids, ii) entrapping into inorganic matrices generated by sol-gel, and iii) grafting of macrocycles on inorganic surfaces. So that, the resulting nanocomposites have useful applications as solid electrolytes, ion-selective membranes and electrochemical sensors [143].

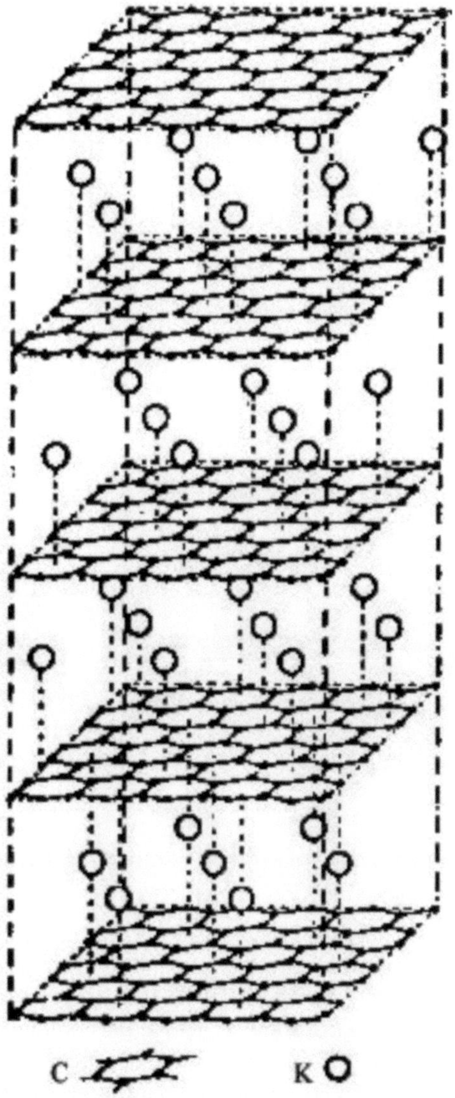

Fig. 23. The intercalation of potassium (guest) into graphite layers (host) [141].

"Clathrate compounds" or cage compounds are consisting of a lattice of one type of molecule trapping and containing a second type of molecule. In 1810, a clathrate hydrate discovered by Humphry Davy is a special type of gas hydrate in which a lattice of water molecules encloses molecules of a trapped gas. Very recently, Lee et al. studied ionic clathrate hydrates [144]. Clathrate hydrates are inclusion compounds where the host is a polyhedral framework built of hydrogen-bonded water molecules in which guest molecules are trapped. Host-guest interactions are usually van der Waals in nature. Structures of clathrate hydrates are classified by the type of water framework, which is considered to be a three-dimensional packing of polyhedral cages with tetra-, penta-, and hexagonal faces, oxygen atoms as vertices, and hydrogen bonds as edges. The cages can act as hosts to a variety of guest molecules, from inert gases to large organic molecules (including those with hydrophilic groups). The shape and size of the guest molecules define the structural type of hydrate formed.

According to available investigations, ionic clathrate hydrates involve acid and base clathrate hydrates. In some strong acids, anion guests are enclathrated into the protonated host lattice to form ionic clathrate hydrates such as HPF_6 hydrate with a cubic *Im3m* structure [145, 146] but, some types of ionic clathrate hydrates are composed to cationic guests and a host lattice balanced by anions such as OH^-, F^- and Br^-, showing strong or weak basicity in many cases [147-149].

Fig. 24. The truncated-octahedral cage of HPF_6^- hydrate and PF_6^- is encaged into the water cage [145].

A number of hydrates of quaternary ammonium salts (QAS) can form ionic clathrate hydrates. The most usual types of QAS hydrates are the tetra-nbutyl ammonium ($(nBu)_4N^+$) and tetra-iso-amyl ammnonium ($(i\text{-}amyl)_4N^+$) hydrates. So that, these types of ionic clathrate hydrates are also called "semi-clathrate" because some cages are partially broken in order to encage the large cations [150]. Recently, ionic clathrate hydrates have received increased attention arising from their potential applications in various fields, including solid electrolytes, gas separation, and gas storage [151-153]. Kamata et al. [154, 155] examined the ability of the $(nBu)_4NBr$ clathrate hydrate to separate the gas mixtures. Small gas molecules such as CH_4, N_2 and H_2S were selectively encaged in the cages of this hydrate during clathrate formation, whereas larger gas molecules such as C_2H_6 and C_3H_8 were not incorporated into the $(nBu)_4NBr$ double clathrate hydrates. Among the various gas molecules, H_2S was found to be the most efficient guest for the $(nBu)_4NBr$ double clathrate hydrate owing to its molecular size and solubility in water.

Among the various types of artificial receptors molecular "tweezers" has attracted more and more attention in molecular recognition, mimic enzyme catalysis, the resolution of racemates as well as molecular devices [156-159]. Related studies by Whitlock [160, 161] and Zimmerman and their co-workers [162-164] enabled the development of a subset of this receptor class that is referred to as "molecular tweezers". The open cavity of the molecular tweezers may bind guests using non-covalent bonding which includes hydrogen bonding, metal coordination, hydrophobic forces, van der Waals forces, π-π interactions, and/or electrostatic effects. These complexes are a subset of macrocyclic molecular receptors and their structure is that the two "arms" that bind the guest molecule between them are only connected at one end. The reduction of high cholesterol levels in humans which result from the consumption of dairy products is very essential for human health. Several studies have focused on receptors for cholesterol in order to extract the sterol from the food source [165]. Molecular tweezer type receptors have been developed for this purpose [166-168].

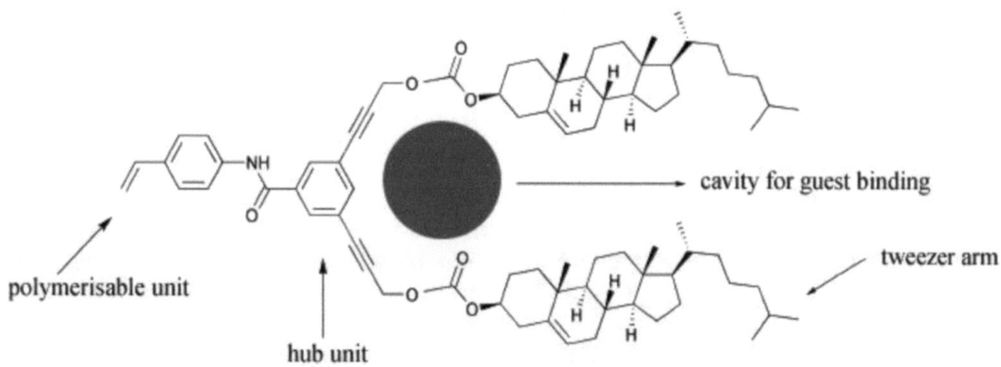

Fig. 25. Molecular tweezer type receptor for cholesterol [168].

4. Zeolite encapsulation and host-guest nanocomposite materials

A "nanocomposite" consists of multiphase solid material where one of the phases has one, two or three dimensions of less than 100 nanometers (nm), or structures having nano-scale repeat distances between the different phases that make up the material [169]. The mechanical, electrical, thermal, optical, electrochemical, catalytic properties of the nanocomposite will differ markedly from that of the conventional composite materials due to the exceptionally high surface to volume ratio. Some zeolites (pore size from 0.2 to 0.8 nm) because of their nanoporous structure are used to form host-guest nanocomposite materials as host molecules for special applications. Recently, encapsulation of different guest molecules into zeolites has attracted much interest due to the possibility of obtaining stable structures with potential applications in different fields [170-178]. Herein, several zeolite encapsulation systems and their applications are studied.

4.1 Host (zeolite)/guest (transition metal complex) nanocomposites: catalytic applications

Zeolites and zeotypes (molecular sieves), owing to their varied intrinsic properties (e.g., acidity, basicity, redox behavior, etc.), channel sizes, high surface areas, thermal and chemical stabilities and channel structures (shape selectivity) have been extensively used to form heterogeneous catalysts with transition metal complexes. The encapsulation of transition metal complexes within small voids in zeolites or other microporous solids can be used to contain and protect active structures and to select reactants, products, or transition states, thus providing diverse strategies for the design and synthesis of more effective catalysts. The small and uniform size of channels and voids within crystalline microporous solids allows selective access of molecules to intracrystalline active sites based on size or shape.

Although homogeneous metal complex catalysts exhibit good activities and selectivities, heterogenization of homogeneous metal complex catalysts by encapsulating them inside zeolites offers the advantages of both homogeneous and heterogeneous counterparts such as: (1) easy separation of the catalysts from the reagents and reaction products; (2) simplification of methods to recycle expensive catalysts; (3) nonvolatile and nontoxic characteristics to high molecular weight zeolite backbones; (4) minimization of certain catalyst deactivation pathways by site isolation [179]. There are three approaches to synthesize zeolite encapsulated metal complexes, including:

1. Zeolite Synthesis Method (ZS)
2. Flexible Ligand Method (FL)
3. "Ship-in-a-Bottle"

In zeolite synthesis method (ZS), transition metal complexes, which are stable under the condition of zeolite synthesis such as high pH and high temperature, are added to the synthesis mixture. The resulting zeolite encapsulates the metal complexes in its cavity. In flexible ligand method (FL), if the size of ligand is smaller than the diameter of zeolite channels, the ligand is diffused freely through the zeolite channels. The ligand reacts easily with the desired metal ions which have been previously exchanged in the supercages of zeolite and makes a stable metal complex. When the ligand size is larger than the diameter of zeolite channels, then template synthesis (TS) method is used. In this method, the molecules of the ligand species are diffused freely into the zeolite cavity where they assemble around the resident metal ions (fixed by ion exchange) in the zeolite cavity. The size of the synthesized metal complex will be too large to escape out from the supercages of zeolite. These encapsulated metal complexes in the cavity of zeolite have been called by several research groups as "ship-in-a-bottle" [180, 181].

4.1.1 Cyclohexane oxidation reactions

To produce more valuable organic compounds such as alcohols, aldehydes, and ketones, the oxyfunctionalization of inexpensive hydrocarbons requires the selective oxidation of strong C–H bonds. Because alcohols and ketones are known as important intermediate materials for the manufacture of many important products such as: fiber, drugs and fragrance. Amongst heterogeneous catalysts, zeolite encapsulated transition metal complexes catalysed oxidation of cyclohexane into a variety of products such as cyclohexanol and cyclohexanone has received much attention [182-184]. Maurya and co-workers introduced a new catalyst based on oxovanadium(IV) and copper(II) exchanged zeolite-Y for the oxidation of styrene, cyclohexene and cyclohexane [182]. As shown in Table 5, oxidation of cyclohexane catalysed by [Cu(sal-dach)]-Y and [VO(sal-dach)]-Y gave cyclohexanone, cyclohexanol and cyclohexane-1,2-diol with different conversion and selectivity in the presence of H_2O_2. Amongst the various products formed, the selectivity of cyclohexanol was found to be highest (91.2 % for [Cu(sal-dach)]-Y and 82.2 % for [VO(sal-dach)]-Y) while selectivities of other two products were much less. On the other hand, catalytic potential of neat complexes has been compared. These results revealed that a maximum of 78.1% conversion of cyclohexane catalysed by [VO(sal-dach)]-Y and only 21.0% conversion by [Cu(sal-dach)]-Y with major reaction products of cyclohexanone, cyclohexanol and cyclohexane-1,2-diol have been obtained. The structure of [VO(sal-dach)] and [Cu(sal-dach)] complexes are shown in Fig. 26.

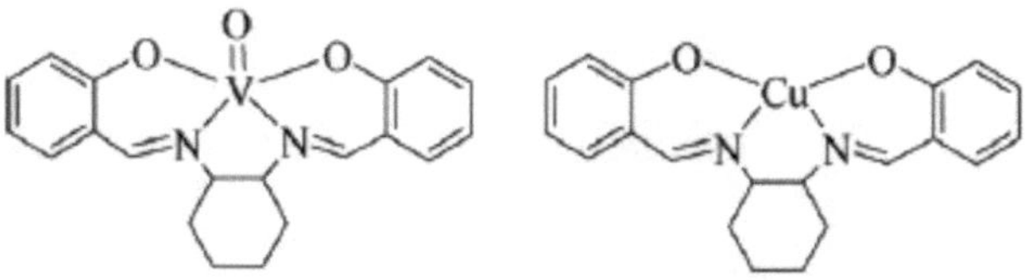

Fig. 26. The structure of [VO(sal-dach)] and [Cu(sal-dach)] complexes [182].

Catalyst	Conversion (%)	Selectivity (%)	
		Cyclohexanol	Cyclohexanone
[VO(sal-dach)]-Y	78.1	82.2	4.8
[Cu(sal-dach)]-Y	21.0	91.2	2.0
[VO(sal-dach)]	36.8	93.1	2.4
[Cu(sal-dach)]	40.9	90.3	1.8

Table 5. Effect of different catalysts on the oxidation of cyclohexene and product selectivity [182].

Besides zeolite encapsulated transition metal complexes, metal complexes with Schiff-base ligands can be entrapped in the nanocavity of zeolite-Y. Recently, two novel host (zeolite-Y)/guest (Schiff base metal complex) nanocomposites were synthesized for the oxidation of cyclohexane by our group [183, 184]. Metal complexes of Mn(II), Co(II), Cu(II) and Ni(II) with tetradendate Schiff-base ligand, (salicylaldehyde)oxaloyldihydrazone, were entrapped in the nanocavity of zeolite-Y via a "ship-in-a-bottle" approach (Fig. 27).

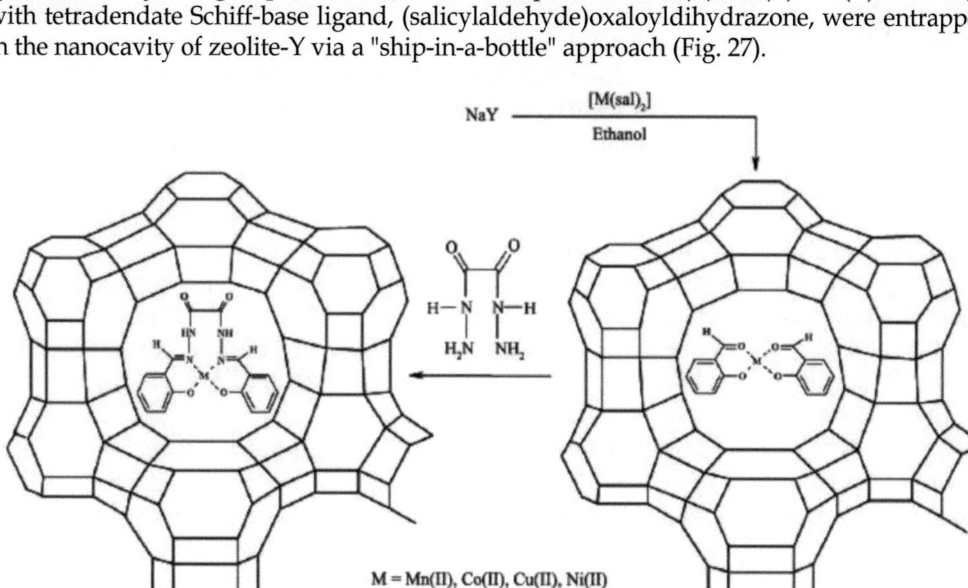

Fig. 27. Schiff-base ligand entrapped in the nanocavity of zeolite-Y [183].

The catalytic effects of Schiff base metal complexes encapsulated in zeolite-Y and neat complexes were studied on the oxidation of cyclohexane in the presence of hydrogen peroxide in CH_3CN and the results are shown in Table 6 [183]. Fig. 29 and Fig. 30 show the catalytic effects of octahydro-Schiff base metal complexes and octahydro-Schiff base metal complexes encapsulated in zeolite-Y, respectively.

Catalyst	Conversion (%)	Selectivity (%)	
		Cyclohexanol	Cyclohexanone
[MnL]	12.6	35.9	64.1
[CoL]	31.2	31.6	68.4
[NiL]	5.7	29.5	70.5
[CuL]	46.8	28.4	71.6
[MnL]-NaY	8.4	25.9	74.1
[CoL]-NaY	28.6	22.4	77.6
[NiL]-NaY	2.1	20.3	79.7
[CuL]-NaY	42.9	17.5	82.5
Cu(II)-NaY	10.6	11.7	88.3

Table 6. Oxidation of cyclohexane with H_2O_2 catalyzed by host-guest nanocomposites and neat complexes [183].

According to Table 6, it is found that cyclohexanone is selectively formed in the presence of all catalysts and the catalytic potential of Schiff base metal complexes encapsulated in zeolite-Y decreases in the series [CuL]-NaY > [CoL]-NaY > [MnL]-NaY > [NiL]-NaY.

For the first time, transition metal (M = Mn(II), Co(II), Ni(II) and Cu(II)) complexes with octahydro-Schiff base (H_4-N_4O_4) have been encapsulated in nanopores of zeolite-Y; [M([H]$_8$-N_4O_4)]@NaY; through Flexible Ligand Method (FLM) and utilized as oxidation catalysts [184]. The formation of these host-guest nanocomposite materials was illustrated in Fig. 28. The results clearly suggest that [Cu_2([H]$_8$–N_4O_4)]@NaY efficiently catalyses the conversion of cyclohexane to cyclohexanol and cyclohexanone with 59.3% and 40.7% selectivity, respectively.

4.1.2 Cyclohexene oxidation reactions

The encapsulated complexes catalyse the cyclohexene oxidation reactions in good yield. Under the optimized conditions, the oxidation of cyclohexene gave cyclohexeneoxide, 2-cyclohexene-1-ol, cyclohexane-1,2-diol and 2-cyclohexene-1-one as major products (Fig. 31). Various high reactive zeolite encapsulated complex catalysts used for the oxidation of cyclohexene were illustrated in Table 7.

Besides the type of catalyst, temperature and amount of catalyst effect on the catalytic activity and product selectivity [193-195]. Nanoscale microreactor containing (5,7,7,12,14,14-hexamethyl-1,4,8,11-tetraazacyclotetradecane-4,11-diene)nickel(II) were entrapped in the supercage of zeolite Y and cyclohexene was catalytically oxidized in the presence of molecular oxygen and $[Ni(Me_6[14]aneN_4)]^{2+}$-NaY. Effects of temperature and amount of catalyst on the reactivity and product selectivity were investigated. These results were illustrated in Table 8 and Table 9 [194].

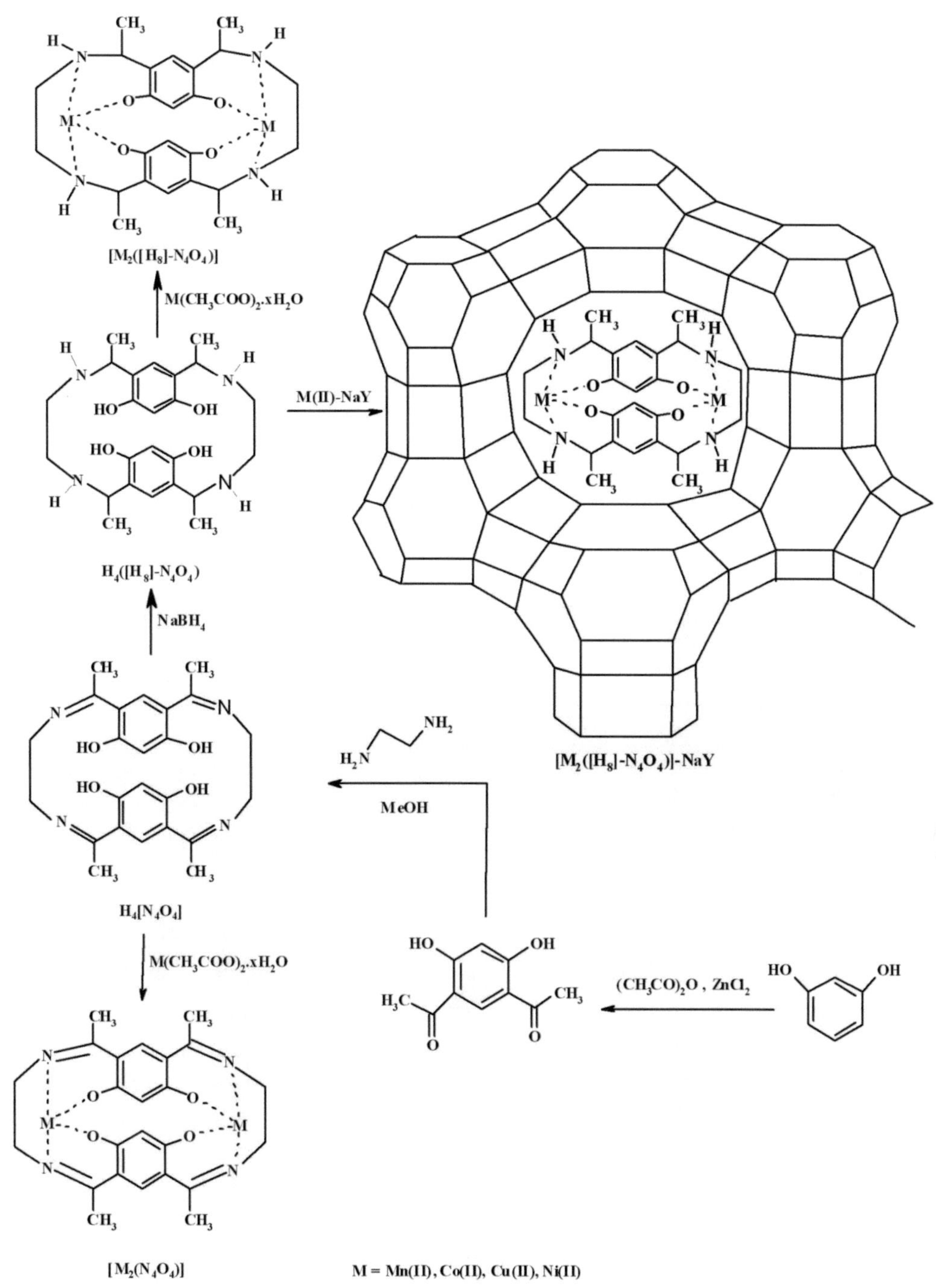

Fig. 28. Octahydro-Schiff base (H_4-N_4O_4) encapsulated in nanopores of zeolite-Y through Flexible Ligand Method [184].

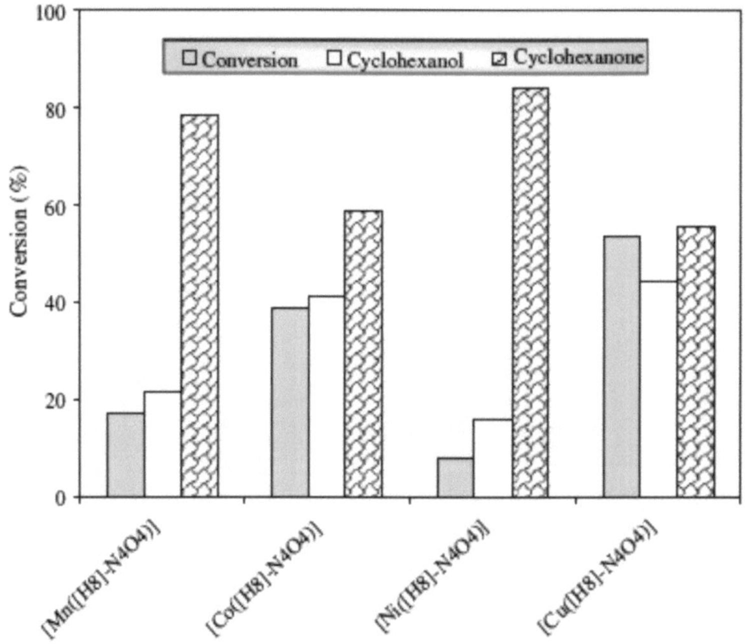

Fig. 29. Conversion and oxidation products distribution in CH_3CN with neat octahydro-Schiff base complexes in the oxidation of cyclohexane with H_2O_2 [184].

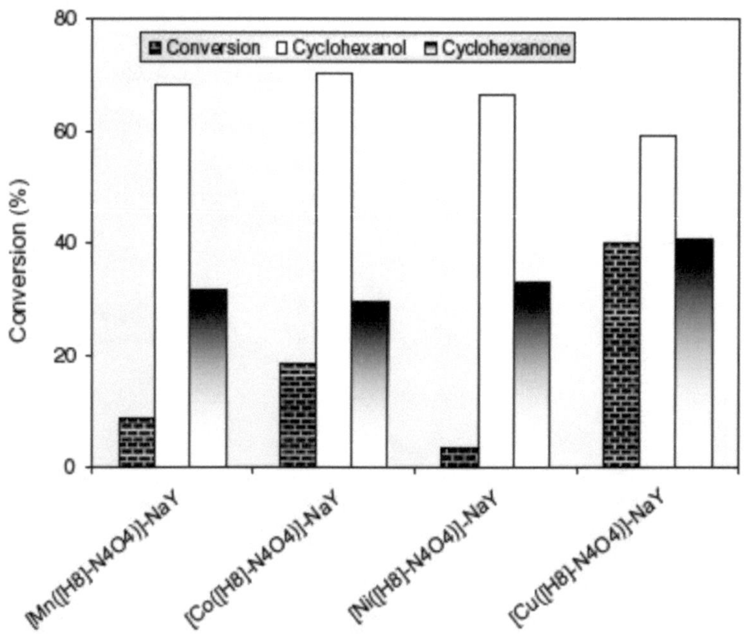

Fig. 30. Conversion and oxidation products distribution in CH_3CN with octahydro-Schiff base (H_4-N_4O_4) encapsulated in nanopores of zeolite-Y in the oxidation of cyclohexane with H_2O_2 [184].

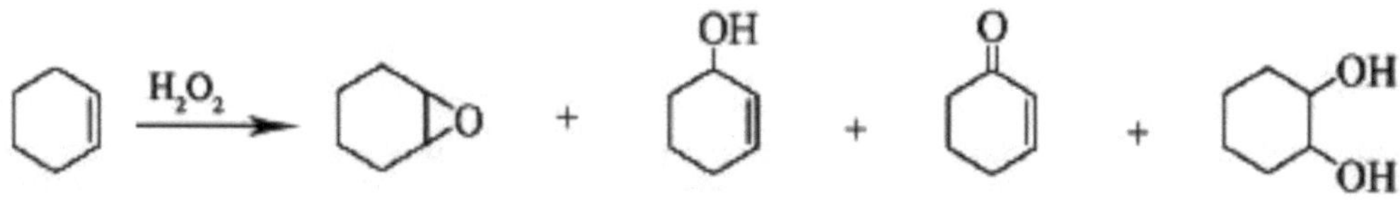

Fig. 31. Oxidation of cyclohexene.

Catalyst	Conversion (%)	Major Product	Ref.
[VO(sal-dach)]-Y	86.6	2-cyclohexene-1-one	182
$[Ni(Bzo_2[14]aneN_4)]^{+2}$-NaY	49.6	2-cyclohexene-1-ol	185
$[Mn(H_4C_6N_6S_2)]$-NaY	90.3	2-cyclohexene-1-one	186
$[Mn(Bzo_2)[12]aneN_4)]^{+2}$-NaY	80.34	di-2-cyclohexenylether	187
$[Ni((Benzyl)_2[16]aneN_6)]^{+2}$-NaY	59.7	2-cyclohexene-1-ol	188
$[Ni((Benzyl)_2Bzo_2[14]aneN_6)]^{+2}$-NaY	68.7	2-cyclohexene-1-ol	189
[Mn(sal-2,6-py)]-NaY	92.5	2-cyclohexene-1-one	190
$[Ni((C_6H_5)_2[12]1,3\text{-diene}N_2O_2)]^{+2}$-NaY	67.5	2-cyclohexene-1-ol	191
$[Ni([H]_2\text{-}N_4)]^{+2}$-NaY	70.8	2-cyclohexene-1-ol	192

Table 7. High reactive zeolite encapsulated complex catalysts used for the oxidation of cyclohexene.

Temperature (oC)	Conversion (%)	Selectivity (%)	
		2-Cyclohexene-1-ol	2-Cyclohexene-1-one
50	4.8	60.2	39.8
60	10.6	62.6	37.4
70	61.6	65.5	34.5
80	20.6	68.6	31.4

Table 8. Effect of temperature on the reactivity and product selectivity [194].

Amount of catalyst (mg)	Conversion (%)	Selectivity (%)	
		2-Cyclohexene-1-ol	2-Cyclohexene-1-one
5	53.2	53.6	46.4
6	54.5	55.4	44.6
7	56.3	57.6	42.4
8	57.4	59.4	40.6
9	59.6	61.3	38.7
10	61.6	65.5	34.5
11	59.1	60.6	39.4
12	58.3	51.4	48.6

Table 9. Effect of amount of catalyst on the reactivity and product selectivity [194].

It is found that the reactivity and selectivity to 2-cyclohexen-1-ol increase in the range of 50–70 ·C. On the other hand, the optimized amount of catalyst is 10 mg with the highest conversion and selectivity to 2-cyclohexen-1-ol.

4.1.3 Phenol and alcohol oxidation reactions

Host (zeolite)/guest(metal complex) nanocomposite materials (HGNMs) can be used as suitable catalysts for oxidation of aromatic and aliphatic alcohols. Catalytic activity of [Cu(salpn)]-Y in the oxidation of phenol to a mixture of catechol and hydroquinone using H_2O_2 as an oxidant has been studied by Maurya and co-workers and the best suited reaction conditions have been optimized by considering the effect of solvents used, concentration of substrate, temperature, reaction time, amount of catalyst and oxidant [196]. Under the best suited conditions, the selectivity towards the formation of catechol and hydroquinone is about 80 and 20%, respectively. Very recently, solvent free catalyzed oxidation of benzyl alcohol by 7, 16-diacetyl[Cu(Me_4($Bzo)_2$[14]tetraeneN_4)]-NaY was reported [197]. Saha et al. showed that the complex Cu(salen) [salen = N,N0-(ethylene)bis(salicylaldiimine)] encapsulated in NaY zeolite exhibits remarkable catalytic activity for oxidizing of 1-naphthol and norbornene [198]. Herein, we summarize several effective HGNM catalysts for the oxidation of phenol, benzyl alcohol and cyclohexanol (Table 10).
Functionalization of carbon nanotubes with Schiff base complexes is an effective way to enhance their physical and chemical properties, and improve solubility. Our experimental results showed that functionalized multi-wall carbon nanotubes (MWNTs) by Schiff-base complexes can catalyze the oxidation of aliphatic and aromatic alcohols into the corresponding carboxylic acids and ketones in the presence of H_2O_2 in good yields [206, 207].

4.1.4 Other catalytic applications

Besides above mentioned catalytic application of HGNMs, these novel materials can catalyze other important reactions in good yield such as: hydroxylation [208-214], oxidation of sulfides and ethers [215-219] and epoxidation [220-222].

Catalyst	Substrate	Conversion (%)	Major Product	Ref.
[Cu(($C_6H_5)_2$[13]1,4dieneN_4O_2)]$^{+2}$-NaY	Cyclohexanol	88.6	Cyclohexanone	199
[Cu((Benzyl)$_2$Bzo$_2$[14]aneN_6)]$^{+2}$-NaY	Benzyl alcohol	75.7	Benzaldehyde	200
[Cu(Me_4($NO_2)_2$Bzo[14]tetraeneN_4]$^{+2}$-NaY	Benzyl alcohol	83.1	Benzaldehyde	201
[Cu(Me_4[14]aneN_8]$^{+2}$-NaY	Benzyl alcohol	73.2	Benzaldehyde	202
[VO_2(sal-ambmz)]Y	Phenol	43.9	Catechol	203
[Cu(Eto-salen)-NaY	Phenol	44	Catechol	204
[Ni(Me_4($NO_2)_2$Bzo[14]tetraeneN_4]$^{+2}$-NaY	Phenol	60.3	Catechol	205

Table 10. Several effective HGNM catalysts for the oxidation of phenol and alcohols.

5. Conclusions

A topic of great interest in recent years is the synthesis of molecules (hosts) capable of encapsulating smaller molecules (guests) within open interior cavities containing portals allowing the smaller guest species to enter and depart. Such molecules can serve as models of molecular recognition in biological systems and have provided better understanding of the nature of hydrophobic and van der Waals interactions in aqueous solutions. Besides catalytic applications of host-guest nanocomposite materials, these materials because of their nano-size dimension can be used as gas adsorption and seperation [223-225] sunscreen additives for reducing adverse effects from UV [226] and for water splitting [227].

6. References

[1] K. Ariga, T. Kunitake, *Supramolecular Chemistry–Fundamentals and Applications*, Springer-Verlag: Berlin, 2006.
[2] A. Villiers, *Compt. Rend. Fr. Acad. Sci.* 435 (1891) 8.
[3] A. Biwer, G. Antranikian, E. Heinzle, *Appl. Microbiol. Biotechnol.* 59 (2002) 17.
[4] D.A. Fulton, J.F. Stoddart, *Org. Lett.* 2 (2000) 1113.
[5] J.M. Garcia-Fernandez, C. Ortiz-Mellet, *Adv. Carbo. Chem. Biochem.* 55 (2000) 35.
[6] R. Mammucari, F. Dehghani, N. R. Foster, *Pharmaceutical Research* 23 (2006) 429.
[7] M.V. Rekharsky, Y. Inoue, *Chem. Rev.* 98 (1998) 1875.
[8] K.A. Connors, *Chem. Rev.* 97 (1997) 1325.
[9] T. Ogoshi, A. Harada, *Sensors* 8 (2008) 4961.
[10] B. Zhao, H. Chen, *Mater. Lett.* 61 (2007) 4890.
[11] R. Freeman, T. Finder, L. Bahshi, I. Willner, *Nano Lett.* 9 (2009) 2073.
[12] T. Uyar, F. Besenbacher, *Eur. Polym. J.* 45 (2009) 1032.
[13] J. Olah, T. Cserhati, J. Szejtli, *Water Res.* 22 (1988) 1345.
[14] B. Martel, P. L. Thuaut, S. Bertini, G. Crini, M. Bacquet, G. Torri, M. Morcellt, *J. Appl. Polym. Sci.* 85 (2002) 1771.
[15] P. Blach, S. Fourmentin, D. Landy, F. Cazier, G. Surpateanu, *Chemosphere* 70 (2008) 374.
[16] T. Huang, F. Meng, L. Qi, *J. Phys. Chem. C* 119 (2009) 13636.
[17] L. Leclercq, H. Bricout, S. Tilloy, E. Monflier, *J. Colloid Interface Sci.* 307 (2007) 481.
[18] G. Becket, L.J. Schep, M.Y. Tan, *Int. J. Pharm.* 179 (1999) 65.
[19] C.-S. Lu, C.-J. Hu, Y. Yu, Q.-J. Meng, *Chem. Pharm. Bull.* 48 (2000) 56.
[20] A. Motoyama, A. Suzuki, O. Shirota, R. Namba, *J. Pharm Bio. Anal.* 28 (2002) 97.
[21] C. D. Gutsche, *Calixarenes*. Cambridge: Royal Society of Chemistry, (1989).
[22] W. Spevak, J.O. Nagy, D.H. Charych, M.E. Schaefer, J.H. Gilbert, M.D. Bednarski, *J. Am. Chem. Soc.* 115 (1993) 1146.
[23] N.V. Bovin, J.-H. Gabius, *Chem. Soc. Rev.* 24 (1995) 413.
[24] T. Jin, *Chem. Commun.* (1999) 2491.
[25] V.I. Kalchenko, I.A. Koshets , E.P. Matsas, O.N. Kopylov, A. Solovyov, Z.I. Kazantseva, Y. M. Shirshov, *Mater. Science-Poland* 20 (2002) 73.
[26] S. F. Chin, M. Makha, C. L. Raston, M. Saunders, *Chem. Commun.* (2007) 1948.
[27] R. Ludwig, N. T. K. Dzung, *Sensors* 2 (2002) 397.
[28] J. Fujita, Y. Ohnishi, Y. Ochiai, S. Matsui, *Appl. Phys. Lett.* 68 (1996) 1297.
[29] A. Wei, *Chem. Commun.* (2006) 1581.
[30] F. Perret, A. N. Lazar and A. W. Coleman, *Chem. Commun.* (2006) 2425.
[31] M. Aktary, K. L. Westra, M. R. Freeman, Y. Tanaka, *J. Vac. Sci. Technol. B* 24 (2006) 267
[32] J.-I. Fujita, Y. Ohnishi, S. Manako, Y. Ochiai, E. Nomura, T. Sakamoto, S. Matsui, *Jpn. J. Appl. Phys.* 36 (1997) 7769.
[33] R. Behrend, E. Meyer, F. Rusche, *Justus Liebigs Ann. Chem.* 339 (1905) 1.
[34] W.A. Freeman, W.L. Mock, N.Y. Shih, *J. Am. Chem. Soc.* 103 (1981) 7367.
[35] J. W. Lee, S. Samal, N. Selvapalam, H.-J. Kim, K. Kim, *Acc. Chem. Res.* 36 (2003) 621.
[36] S. Liu, K. Kim, L. Isaacs, *J. Org. Chem.* 72 (2007) 6840.
[37] J. Lagona, P. Mukhophadhyay, S. Chakrabarti, L. Isaacs, *Angew. Chem. Int. Ed.* 44 (2005) 4844.

[38] W.L. Mock, T.A. Irra, J.P. Wepsiec, T.L. Manimaran, *J. Org. Chem.* 48 (1983) 3619.
[39] W.L. Mock, T.A. Irra, J.P. Wepsiec, M. Adhya, *J. Org. Chem.* 54 (1989) 5302.
[40] N.J. Wheate, A.I. Day, R.J. Blanch, A.P. Arnold, C. Cullinane, J.G. Colline, *Chem. Commun.* (2004) 1424.
[41] M.S. Bali, D.P. Buck, A.J. Coe, A.I. Day, J.G. Collins, *Dalton Trans.* (2006) 5337.
[42] S. Kemp, N.J. Wheate, S. Wang, J.G. Collins, S.F. Ralph, A.I. Day, V.J. Hughs, J.R. Aldrich-Wright, *J. Biol. Inorg. Chem.* 12 (2007) 969.
[43] Y.J. Jeon, S.-Y. Kim, Y.H. Ko, S. Sakamoto, K. Yamaguchi, K. Kim, *Org. Biomol. Chem.* 3 (2005) 2122.
[44] K.M. Park, K. Suh, H. Jung, D.-W. Lee, Y. Ahn, J. Kim, K. Baek, K. Kim, *Chem. Commun.* (2009) 71.
[45] D. Kim, E. Kim, J. Kim, K.M. Park, K. Baek, M. Jung, Y.H. Ko, W. Sung, H.S. Kim, J.H. Suh, C.G. Park, O.S. Na, D. Li, K.E. Lee, S.S. Han, K. Kim, *Angew. Chem. Int. Ed.* 46 (2007) 3471.
[46] Y. Zhao, D.P. Buck, D.L Morris, M.H. Pourgholami, A.I. Day, J.G. Collins, *Org. Biomol. Chem.* 6 (2008) 4509.
[47] N.J. Wheate, A.I. Day, R.J. Blanch, A.P. Arnold, C. Cullinane, and J.G. Colline, *Chem. Commun.* (2004) 1424.
[48] N. Saleh, A.L. Koner, W.M. Nau, Angew. Chem., Int. Ed. 47 (2008) 5398.
[49] A. Hennig, H. Bakirci, W.M. Nau, *Nat. Methods* 4 (2007) 629.
[50] D.M. Bailey, A. Hennig, V.D. Uzunova, W.M. Nau, *Chem. Eur. J.* 14 (2008) 6069.
[51] A. Hennig, G.Ghale,W. M. Nau, *Chem. Commun.* (2007) 1614.
[52] M. González-Béjar, P.Montes-Navajas, H.García, J.C.Scaiano, *Langmuir* 25(2009) 10490.
[53] M. Haruta, M. Date, *Appl. Catal. A* 222 (2001) 427.
[54] M. Haruta, *Stud. Surf. Sci. Catal.* 145 (2003) 31.
[55] S. Praharaj, S. K. Ghosh, S. Nath, S. Kundu, S. Panigrahi, S. Basu, T. Pal, *J. Phys. Chem. B* 109 (2005) 13166.
[56] G. Budroni, A. Corma, *Angew. Chem.* 118 (2006) 3406.
[57] A. Corma, H. Garc, P. Montes-Navajas, A. Primo, J. J. Calvino, S. Trasobares, *Chem. Eur. J.* 13 (2007) 6359.
[58] C. J. Pedersen, *J. Am. Chem. Soc.* 89 (1967) 7017.
[59] C. J. Pedersen, *J. Am. Chem. Soc.* 89 (1967) 2495.
[60] E. Weber, J.L. Toner, I. Golberg, F. V6gtle, D.A. Laidler, J.F. Stoddart, R.A. Bartsch, C.L. Liotta, *Crown Ethers and Analogs* ,Wiley, New York, 1989.
[61] L.F. Lindoy, *The Chemistry of Macrocyclic Ligand Complexes*, Cambridge University, Cambridge, 1989.
[62] E. Cordoncillo, G. Monrós, M.A. Tena, P. Escribano, J. Carda, *J. Non-Cryst. Solids* 171 (1994) 105.
[63] M. Ghaedi, K. Niknam, A. Shokrollahi, E. Niknam, H.R. Rajabi, M. Soylak, *J. Hazard. Mater.* 155 (2008) 121.
[64] K. Uysal, Y. Emre, E. Kose, *Microchem. J.* 90 (2008) 67.
[65] S.-W. Pi, X.-J. Ju, H.-G. Wu, R. Xie, L.-Y. Chu, *J.Colloid Interface Sci.* 349 (2010) 512.
[66] W. Zeng, Z. Mao, X. Wei, J. Li, Z. Hong, S. Y. Qin, *J. Supramolecular Chem.* 2 (2002) 501.
[67] R. M. Wang, Z. F. Duan, Y. F. He, *J. Macromol. Sci. A: Pure & Appl. Chem.* 42 (2005) 231.
[68] R.M. Wang, X. Dong LI, Y. Feng HE, *Chinese Chem. Lett.* 17 (2006) 265.
[69] F. Faridbod, M. R. Ganjali, R. Dinarvand, P. Norouzi, S, Riahi, *Sensors* 8 (2008) 1645.
[70] J. Gabard, A. J. Collet, *Chem. Soc., Chem. Commun.* (1981) 1137.
[71] J. Canceill, A. Collet, *J. Chem. Soc., Chem. Commun.* (1988) 582.
[72] D.J. Cram, M.E. Tanner, S.J. Keipert, C.B. Knobler, *J. Am. Chem. Soc.* 113 (1991) 8909.

[73] J. L. Atwood, J. W. Steed, *"Encyclopedia of Supramolecular Chemistry."*, ISBN: 978-0-8247-4725-1 (electronic), 2004, Publisher: Taylor & Francis.
[74] J. Canceill, M. Cesario, A. Collet, J. Guilhem, L. Lacombe, B. Lozach, C. Pascard, *Angew. Chem. Int. Ed. Engl.* 28 (1989) 1246.
[75] Z. Tošner, J. Lang, D. Sandström, O. Petrov, J. Kowalewski, *J. Phys. Chem. A 106* (2002) 8870.
[76] K. Bartik, M. Luhmer, J.-P. Dutasta, A. Collet, J. Reisse, *J. Am. Chem. Soc.* 120 (1998) 784.
[77] H. A. Fogarty, P. Berthault, T. Brotin, G. Huber, H. Desvaux, J.-P. Dutasta, *J. Am. Chem. Soc.* 129 (2007) 10332.
[78] V. Roy, T. Brotin, J.-P. Dutasta, M.-H. Charles, T. Delair, F. Mallet, G Huber, H.Desvaux, Y. Boulard, P. Berthault, *ChemPhysChem.* 8 (2007) 2082.
[79] J. M. Chambers, P. A. Hill, J. A. Aaron, Z. Han, D. W. Christianson, N. N. Kuzma, I. J. Dmochowski, *J. Am. Chem. Soc.* 131 (2009) 563.
[80] L. Garel, J.-P. Dutasta, A. Collet, *Angew. Chem. Int. Ed. Engl.* 32 (1993) 1169.
[81] M. Benounis, N. Jaffrezic-Renault, J.-P. Dutasta, K. Cherif, A. Abdelghani, *Sensors Actuators B* 107 (2005) 32.
[82] S. Anderson, H. L. Anderson, A. Bashall, M. McPartlin, J. K. M. Sanders, *Angew. Chem. Int. Ed. Engl.* 34 (1995) 1096.
[83] R. W. Wagner, J. S. Lindsey, J. Seth, V. Palaniappan, D. F. Bocian, *J. Am. Chem. Soc.* 118 (1996) 3996.
[84] A. Osuka, H. Shimidzu, *Angew. Chem. Int. Ed. Engl.* 36 (1997) 135.
[85] D. Gust, T.A. Moore, A.L. Moore, *Acc. Chem. Res.* 34 (2001) 40.
[86] P.D.W. Boyd, C.A. Reed, *Acc. Chem. Res.* 38 (2005) 235.
[87] H. Imahori, S. Fukuzumi, *Adv. Funct. Mater.* 14 (2004) 525
[88] T. Hasobe, S. Fukuzumi, P.V. Kamart, *Interface* 15 (2006) 47.
[89] S. Yoshimoto, K. Itaya, *J. Porphyrins Phthalocyanines* 11 (2007) 313.
[90] T. Hasobe, P. V. Kamat, V. Troiani, N. Solladié, T. K. Ahn, S. K. Kim, D. Kim, A. Kongkanand, S. Kuwabata, S. Fukuzumi, *J. Phys. Chem B.* 109 (2005) 19.
[91] Z. S. Y., S. Easwaramoorthi, D. Kim, *Bull. Korean Chem. Soc.* 29 (2008)197.
[92] A. Mulder, A. Juković, F. W. B. van Leeuwen, H. Kooijman, A. L. Spek, J. Huskens, D. N. Reinhoudt, *Chem. Eur. J.* 10 (2004) 1114.
[93] C. M. Drain, A. Varotto, I. Radivojevic, *Chem. Rev.* 109 (2009) 1630.
[94] C. J. Medforth, Z. Wang, K. E. Martin, Y. Song, J. L. Jacobsenc, J. A. Shelnutt, *Chem. Commun.* (2009) 7261.
[95] Z. Wang, C. J. Medforth, J. A. Shelnutt, *J. Am. Chem. Soc.* 126 (2004) 16720.
[96] M. Hardie, *Chem. So. Rev.* 39 (2010) 516.
[97] G. M. Robinson, *J. Chem. Soc. Transactions* 107 (1915) 267.
[98] A. S. Lindsey, *J. Chem. Soc. (Resumed)* (1965) 1685.
[99] J. L. Atwood, M. J. Barnes, M. G. Gardiner, C. L. Raston, *Chem. Commun.* (1996) 1449.
[100] M. J. Hardie, C. L. Raston, *Chem. Commun.* (2001) 905.
[101] C. N. Murthy, K. E. Geckeler, *Chem. Commun.* (2001) 1194.
[102] S.-I. Takekuma, H. Takekuma, T. Matsumoto, Z.-I. Yoshida, *Tetrahedron Lett.* 41 (2000) 4909.
[103] Y. Rio, J.-F. Nierengarten, *Tetrahedron Lett.* 43 (2002) 4321.
[104] M. J. Hardie, C. L. Raston, *Cryst. Growth Des.* 1 (2000) 53.
[105] M. J. Hardie, C. L. Raston, A. Salinas, *Chem. Commun.* (2001) 1850.
[106] Z. Zhong, A. Ikeda, S. Shinkai, S. Sakamoto, K. Yamaguchi, *Org. Lett.* 3 (2001) 1085.
[107] R. Ahmad, M. J. Hardie, *Cryst. Growth Des.* 3 (2003) 493.
[108] M. J. Hardie, R. M. Mills, C. J. Sumby, *Org. Biomol. Chem.* 2 (2004) 2958.

[109] Nurimana, B. Kuswandia, W. Verboomb, *Anal. Chim. Acta* 655 (2009) 75.
[110] A. Arduini, F. Calzavacca, D. Demuru, A. Pochini, A. Secchi, *J. Org. Chem.* 69 (2004) 1386.
[111] D. J. Cram, *Science* 219 (1983) 1177.
[112] D. J. Cram, S. Karbach, Y. H. Kim, L. Baczynskyj, G. W. Kalleymeyn, *J. Am. Chem. Soc.* 107 (1985) 2575.
[113] L. R. MacGillivray, P. R. Diamente, J. L. Reid, J. A. Ripmeester, *Chem. Commun.* (2000) 359.
[114] D. J. Cram, S. Karbach, Y. H. Kim, L. Baczynskyj, K. Marti, R. M. Sampson, G.W. Kalleymeyn, *Ibid.* 110 (1988) 2554.
[115] J. C. Sherman, C. B. Knobler, D. J. Cram, *J. Am. Chem. Soc.* 113 (1991) 2194.
[116] R. Warmuth, J. Yoon, *Acc. Chem. Res.* 34 (2001) 95.
[117] J. Yoon, D. J. Cram, *Chem. Commun.* (1997) 497.
[118] P. Piotrowiak, K. Deshayes, Z. S. Romanova, C. Pagba, S. Hore, G. Zordan, I. Place,
A. Farrán, *Pure Appl. Chem.* 75 (2003) 1061.
[119] E.N. Coker, J.C. Jasen, J.A. Martens, P.A. Jacobs, F. DiRenzo, F. Fajula, A. Sacco Jr, *Micropor Mesopor Mater.* 23 (1998) 119.
[120] D.W Breck, *Zeolite Molecular Sieves*, New York: Wiley, 1974.
[121] B. Tezak, *Disc. Faraday Soc.* 42 (1966) 175.
[122] F. Liebau, H. Gies, R.P. Gunawardne, B. Marles, *Zeolites* 6 (1986) 373.
[123] C.K. Cheetham, G. Ferey, T. Loiseau. *Angew. Chem. Int. Ed.* 38 (1999) 3268.
[124] Ch Baerlocher, W.M. Meier, D. Holson, *Atlas of Zeolite Framework Types*. Amsterdam: Elsevier, 2001.
[125] W.L. Kranich, Y.H. Ma, L.B. Sand, A.H. Weiss, I. Zwiebel. *Adv. Chem. Ser.* 101 (1971) 502.
[126] D.H. Olson, G.T. Kokotaillo, S.L. Lawton, W.M. Meier. *J. Phys. Chem.* 85 (1981) 2238.
[127] P. A. Jacobs, J. A. Martens, J. Weitkamp, H. K. Beyer, Faraday Disc. Chem. Sci. 72 (1981) 353.
[128] H. van Koningsveld, J.C. Jansen, H. van Bekkum, *Zeolites* 10 (1990) 235.
[129] E. M. Flanigen. *Pure Appl. Chem.* 52 (1980) 2191.
[130] A. Galarneau, F. Di Renzo, F. Fajula, J. Vedrine, Zeolites and Mesoporous Materials at the dawn of the 21st century, Proceedings of the 13th International Zeolite Conference. *Stud. Surf. Sci. Catal.* 135 (2001) v-vi .
[131]. D. Barthomeuf, *Catal. Rev. Sci. Eng.* 38 (1996) 521.
[132] J.-M. Lehn, *Acc. Chem. Res.* 11 (1978) 49.
[133] V. S. S. Kumar, F. C. Pigge, N. P. Rath, *Cryst. Growth Des.* 4 (2004) 1217.
[134] L. R. Nassimbeni, *Acc. Chem. Res.* 36 (2003) 631.
[135] S. Hirano, S. Toyota, M. Kato, F. Toda, *Chem.Commun.* (2005) 3646.
[136] D. Braga, S. L. Giaffreda, F. Grepioni, A. Pettersen, L. Maini, M.Curzi, M. Polito, *Dalton Trans.* (2006) 1249.
[137] G. A. Ozin, S. Ozkart , *Chem. Mater.* 4 (1992) 511.
[138] K. Uekama, F. Hirayama, T. Irie, *Chem. Rev.* 98 (1998) 2045.
[139] K. D. M. Harris, J. Mol, *Structure* 374 (1996) 241.
[140] T. Friŝĉiĉ, A. V. Trask, W. Jones, W. D. Samuel Motherwell, *Angew. Chem. Int. Ed.* 45 (2006) 7546.
[141] M. S. Dresselhaus, G. Dresselhaus, *Adv. Phys.* 51 (2002) 1.
[142] B.-Z. Lin, X.-K. Pei, J.-F. Zhang, G.-H. Han, Z. Li, P. -D. Liu, J.-H. Wu, *J. Mater. Chem.* 14 (2004) 2001.
[143] E. Ruiz-Hitzky, B. Casal, P. Aranda, J.C. Galván, *Reviews in Inorganic Chemistry*, 21 (2001) 125.
[144] K. Shin, J.-H. Cha, Y. Seo, H. Lee, *Chem. Asian J.* 5 (2010) 22.

[145] H. Bode, G. Teufer, *Acta Crystallogr.* 8 (1955) 611.
[146] D. W. Davidson, S. K. Garg, *Can. J. Chem.* 50 (1972) 3515.
[147] D. L. Fowler, W. V. Loebenstein, D. B. Pall, C. A. Kraus, *J. Am. Chem. Soc.* 62 (1940) 1140.
[148] R. K. McMullan, T. C. W. Mak, G. A. Jeffrey, *J. Chem. Phys.* 44 (1966) 2338.
[149] D. Mootz, R. Seidel, *J. Inclusion Phenom.* 8 (1990) 139.
[150] W. Shimada, M. Shiro, H. Kondo, S. Takeya, H. Oyama, T. Ebinuma, H. Narita, *Acta Crystallogr.* C61 (2005) 065.
[151] S. Hashimoto, S. Murayama, T. Sugahara, H. Sato, K. Ohgaki, *Chem.Eng.Sci.*61 (2006) 7884.
[152] S. Hashimoto, T.Sugahara, M.Moritoki, H. Sato, K. Ohgaki, *Chem. Eng. Sci.* 63 (2008) 1092.
[153] J. Sakamoto, S. Hashimoto, T. Tsuda, T. Sugahara, Y. Inoue, K. Ohgaki, *Chem. Eng. Sci.* 63 (2008) 5789.
[154] Y. Kamata, H. Oyama, W. Shimada, T. Ebinuma, S. Takeya, T. Uchida, J. Nagao, H. Narita, *Jpn. J. Appl. Phys.* 43 (2004) 362.
[155] Y. Kamata, Y. Yamakoshi, T. Ebinuma, H. Oyama, W. Shimada, H. Narita, *Energy Fuels* 19 (2005) 1717.
[156] Y. Habata, J. S. Bradshaw, X. X. Zhang, R. M. Izatt, *J. Am. Chem. Soc.* 119 (1997) 7145.
[157] Y. N. Ito, T. katsuki, *Tetrahedron Lett.* 39 (1998) 4325.
[158] K. Manabe, *Tetrahedron Lett.* 39 (1998) 5807.
[159] H. Dugas, *Bioorganic Chemistry,* Springer-Verlag, New York, 1996, P:593.
[160] C. W. Chen, H. W. Whitlock, *J. Am. Chem. Soc.* 100 (1978) 7921.
[161] K. M. Nedar, H. W. Whitlock, *J. Am. Chem. Soc.* 112 (1990) 7269.
[162] S. C. Zimmerman, Z. Zeng, W. Wu, D. E. Reichert, *J. Am. Chem. Soc.*113 (1991) 183.
[163] S. C. Zimmerman, W. Wu, Z. Zeng, *J. Am. Chem. Soc.* 113 (1991) 196.
[164] S. C. Zimmerman, *Bioorganic Chemistry Frontiers,* Vol. 2, Springer-Verlag, New York, 1991
[165] M. J. Whitcombe, M. E. Rodriguez, P. Villar, E. Vulfson, *J. Am. Chem. Soc.* 117 (1995) 7105.
[166] U. Maitra, L. J. D'Souza, *J. Chem. Soc. Chem. Commun.* (1994) 2793.
[167] L. J. D'Souza, U. Maitra, *J. Org. Chem.* 61 (1996) 9494.
[168] L. Davidson, A. Blencowe, M. G. B. Drew, K. W. Freebairnb, W. Hayes, *J. Mater. Chem.* 13 (2003) 758.
[169] P.M. Ajayan, L.S. Schadler, P.V. Braun (2003). *Nanocomposite science and technology.* Wiley. ISBN 3527303596.
[170] M. Salavati-Niasari, F. Davar, K. Saberian, *Polyhedron* 29 (2010) 2149.
[171] M. Salavati-Niasari, *Inorg. Chem.Commun.* 13 (2010) 266.
[172] M. Salavati-Niasari, *Inorg. Chem. Acta* 362 (2009) 3738.
[173] M. Salavati-Niasari, *Inorg. Chem.Commun.* 12 (2009) 359.
[174] M. Salavati-Niasari, *Polydedron* 27 (2008) 3207.
[175] M. Salavati-Niasari, *Inorg. Chem.Commun.* 8 (2005) 174.
[176] M. Salavati-Niasari, S. N. Mirsattari, K. Saberyan, *Bull. Korean Chem. Soc.* 30 (2009) 348.
[177] M. Salavati-Niasari, *Chem. Lett.* 34 (2005) 1444.
[178] M. Salavati-Niasari, *J. Incl. Phenom. Macrocycl. Chem.* 62 (2008) 65.
[179] M. R. Maurya, S. J. J. Titinchi, S. Chand, I. M. Mishra, *J. Mol.Catal. A: Chem.* 180 (2002) 201.
[180] N. Herron, *Inorg. Chem.* 25 (1986) 4714.
[181] A. Corma, H. Garcia, *Eur. J. Inorg. Chem.* (2004) 1143.
[182] M. R. Maurya, A. K. Chandrakar, S. Chand, *J. Mol. Catal. A: Chem.* 270 (2007) 225.
[183] M. Salavati-Niasari, A. Sobhani, *J. Mol. Catal. A: Chem.* 285 (2008) 58.
[184] M. Salavati-Niasari, Z.Salimi, M. Bazarganipour, F. Davar, *Inorg. Chim. Acta* 362 (2009) 3715.
[185] M. Salavati-Niasari, *Polyhedron* 27 (2008) 3132.

[186] M. Salavati-Niasari, *J. Mol. Catal. A: Chem.* 283 (2008) 120.
[187] M. Salavati-Niasari, *J. Mol. Catal. A: Chem.* 272 (2007) 249.
[188] M. Salavati-Niasari, *Micropor. Mesopor. Mater.* 92 (2006) 173.
[189] M. Salavati-Niasari, F. Davar, *Inorg. Chem.Commun.* 9 (2006) 263.
[190] M. Salavati-Niasari, M. Shaterian, *J. Porous Mater.* 15 (2008) 581.
[191] M. Salavati-Niasari, *Transition Met. Chem.* 31 (2006) 964.
[192] M. Salavati-Niasari, *J. Coord. Chem.* 62 (2009) 980.
[193] A. Günes, O. Bayraktar, S. Yılmaz, *Ind. Eng. Chem. Res.* 45 (2006) 54.
[194] M. Salavati-Niasari, *J. Mol. Catal. A: Chem.* 229 (2005) 159.
[195] M. Salavati-Niasari, M. Bazarganipour, *Transition Met. Chem.* 32 (2007) 9.
[196] M. R. Maurya, S. J. J. Titinchi, S. Chand, *Appl. Catal. A: General* 228 (2002) 177.
[197] V. K. Bansal, P. P. Thankachan, R. Prasad, *Appl. Catal. A: General* 381 (2010) 8.
[198] P. K. Saha, S. Banerjee, S. Saha, A. K. Mukherjee, S. Sivasanker, S. Koner, *Bull. Chem. Soc. Jpn.*, 77 (2004) 709.
[199] M. Salavati-Niasari, E.Zamani, M.R. Ganjali, P. Norouzi, *J. Mol. Catal. A: Chem.* 261 (2007) 196.
[200] M. Salavati-Niasari, F. Davar, *Inorg. Chem. Commun.* 9 (2006) 304.
[201] M. Salavati-Niasari, *J. Mol. Catal. A: Chem.* 245 (2006) 192.
[202] M. Salavati-Niasari, M. R. Ganjali, P. Norouzi, *Transition Met. Chem.* 32 (2007) 1.
[203] M. R. Maurya, A. K. Chandrakar, S. Chand, *J. Mol. Catal. A: Chem.* 263 (2007) 227.
[204] P.K. Saha, S. Koner, *Inorg. Chem. Commun.* 7 (2004) 1164.
[205] M. Salavati-Niasari, M. Bazarganipour, *Catal. Commun.* 7 (2006).
[206] M. Salavati-Niasari, M. Bazarganipour, *Transition Met. Chem.* 34 (2009) 605.
[207] M. Salavati-Niasari, M. Bazarganipour, *Appl. Surf. Sci.* 255 (2008) 2963
[208] M. Salavati-Niasari, *Inorg. Chim. Acta* 362 (2009) 2159.
[209] M. Salavati-Niasari, *J. Incl. Phenom. Macrocycl. Chem.* 65 (2009) 349.
[210] M. Salavati-Niasari, M. R. Ganjali, P. Norouzi, *J. Porous Mater.* 14 (2007) 14.
[211] L. Wang, A. Kong, B. Chen, H. Ding, Y. Shan, M. He, *J. Mol.Catal. A: Chem.* 230 (2005) 143.
[212] M. R. Maurya, S. J. J. Titinchi, S. Chand, *J. Mol. Catal. A: Chem.* 214 (2004) 257.
[213] P. K. Saha, S. Koner, *Inorg. Chem.Commun.* 7 (2004) 1164.
[214] R. Dimitrova, M. Spassova, *Catal. Commun.* 8 (2007) 693.
[215] S. Bunce, R. J. Cross, L. J. Farrugia, S, Kunchandy, L. L. Meason, K. W. Muir, M. ÓDonnell, R. D. Peacock, D. Stirling, S. J. Teat, *Polyhedron* 17 (1998) 4179.
[216] M.J. Alcón, A. Corma, M. Iglesias, F. Sánchez, *J. Mol. Catal. A: Chem.* 178 (2002) 253.
[217] M. Salavati-Niasari, *Inorg. Chem.Commun.* 9 (2006) 628.
[218] M. Salavati-Niasari, *J. Mol. Catal. A: Chem.* 217 (2004) 87.
[219] M. Salavati-Niasari, N. Mir, *J. Incl. Phenom. Macrocycl. Chem.* 59 (2007) 223.
[220] M. Salavati-Niasari, *J. Mol. Catal. A: Chem.* 278 (2007) 22.
[221] M. Salavati-Niasari, E. Zamani, M. Bazarganipour, *Appl. Clay Sci.* 38 (2007) 9.
[222] M. Salavati-Niasari, *J. Mol. Catal. A: Chem.* 310 (2009) 51.
[223] A. H. Ahmed, Z. M. El-Bahy, T. M. Salama, *J. Mol. Structure* 969 (2010) 9.
[224] I. O. Ali, *Mater. Sci. Eng. A* 459 (2007) 294.
[225] A. H. Ahmed, *J. Appl. Sci. Res.* 3 (2007) 1663.
[226] M. Sathish, B.Viswanathan, R.P. Viswanath, *Int. J. Hydrogen Energy* 31 (2006) 891.
[227] M. N. Chrétien, E. Heafey, J. C. Scaiano, *Photochem. Photobiol.* 86 (2010) 153.

16

Morphology Development of Polymer Nanocomposites: Utilizing Interstratified Clay Minerals from Natural Systems

Kenji Tamura and Hirohisa Yamada
National Institute for Materials Science, Japan

1. Introduction

Various nano-scale materials and an abundance of formulas that offer many potential benefits in our life and contribute to creating a sustainable society are latent in nature. For example, ubiquitous substances such as efflorescent soil minerals and rock contain various polymorphous nanomaterials like nano-balls, nano-tubes, and nano-sheets, which each have unique physicochemical properties including ion exchange, adsorption, colloidal characteristics, and swelling capabilities (van Olphen, 1977; Suzuki at al., 2001; Tamura & Nakazawa, 1996; Tamura et al., 1999, 2000). We benefit unknowingly from these materials in many ways. We can create innovative materials by learning their properties and mineral genesis.

Novel functional composite materials are being studied in our laboratory as a means to develop new engineering technologies for engineering ends. To improve various material properties such as strength, stiffness, corrosion resistance, surface finish, weight, and fatigue life, composites combine two or more substances on micro- and macroscopic scales to form useful materials. Of particular interest are the recently developed nanocomposites that consist of a polymer with a layered silicate, as they often exhibit remarkably improved mechanical and other properties when compared with pure polymers or conventional micro- and macro-composites (Kurauchi at al., 1991; Usuki et al., 1993; Maiti et al., 2001; Messersmith & Giannelis, 1994; Wang & Pinnavaia, 1998; Haraguchi & Takehisa, 2002; Yano, et al., 1997).

Generally, layered silicate-polymer nanocomposites are classified according to their levels of intercalation and exfoliation (Giannelis, 1992). At one end of this classification scheme are well ordered and stacked multilayers. These consist of intercalated polymer chains resident between host layers. At the other end are exfoliated silicate layers in a continuous polymer matrix. Some exfoliated clay-polymer nanocomposites have already been used in practical applications because of they are suitably rigid, strong, and exhibit barrier properties, whilst having far less silicate content than conventional composites that are filled with other minerals. Nanocomposite properties vary widely between the two extremes. The best results with respect to rigidity or barrier properties are obtained if the layered silicate is fully exfoliated into single layers with a thickness of about 1 nm. In spite of numerous attempts reported during the past couple of decades, there has been no report of a nanocomposite with a rigorously controlled morphology.

Here, *morphology* is defined as the dispersion state of the clay components, called particles hereinafter, in the polymer matrix. Precise control of morphology is one of the most important factors in production of nanocomposites with desired properties. Although the control of morphology is still difficult, it can be a pivotal technique in studying how morphology, such as particle shape and size, affects the properties of nanocomposites. In this chapter, with respect to control of morphology, our attention was directed at the clay components to be dispersed and the number of layers of clay particles.

The following section highlights the various species and structures of layered silicates that are important when controlling the number of layers of dispersed silicate platelets. In particular, we focus strongly on interstratified clay minerals where two or more different individual component layers are stacked in various ways to make up a new structure, different from those of its constituents. These interstratified structures result from the strong similarity that exists between the layers of the different clay minerals, which are all composed of tetrahedral and octahedral sheets of hexagonal arrays of atoms, and from the distinct differences in the thickness of clay mineral layers. Later sections will explore specific types of nanocomposites that have been prepared using several types of interstratified clay minerals.

2. Structure of layered silicates

2.1 Tetrahedral sheet and octahedral sheet

The layers that form all phyllosilicates, and therefore clay minerals, are characterized by two-dimensional sheets with octahedral **(O)** or tetrahedral **(T)** configurations (Fig. 1). The octahedral (O) sheets are called dioctahedral if they contain two M^{3+} cations in six-fold coordination with O^{2-} or OH^-, in which case one of three cation sites remain vacant. If the octahedral sheets are made up of three M^{2+} cations coordinated with six O^{2-} or OH units$^-$, then all the cation sites are occupied by M^{2+} ions. The layers are then termed trioctahedral.

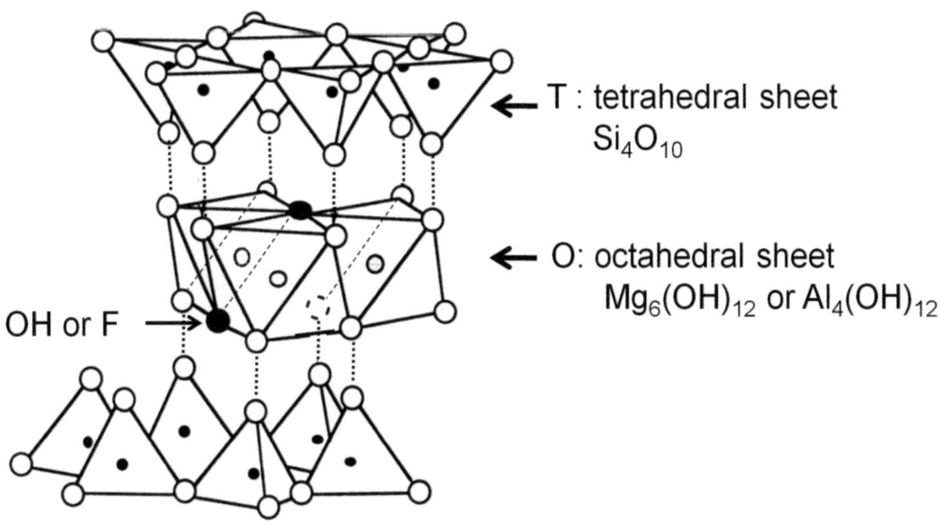

Fig. 1. Two-dimensional octahedral (O) and tetrahedral (T) sheets..

Tetrahedral (T) sheets are composed of cations organized in fourfold coordination with O^{2-} or OH^{-}. The dominant cation in such layers is usually Si^{4+}, but may also be Al^{3+}. The atomic arrangements of octahedral and tetrahedral layers are shown schematically in Fig. 1. The junction plane between the tetrahedral and octahedral sheets consists of shared apical oxygen atoms of the tetrahedrons and unshared hydroxyls (or fluorides) (Fig. 1). Typical layered clay minerals are listed in Table 1.

2.2 1:1 type layered silicates

Clays can be categorized depending on the way that tetrahedral and octahedral sheets are packaged into layers. If there is only one tetrahedral and one octahedral group in each layer the clay is known as a 1:1 TO layered silicate. The 1:1 TO layer type (kaolin-serpentine group) has one tetrahedral sheet, which shares corners with an octahedral sheet. The thickness of this two-sheet unit is about 0.7 nm (Fig. 2a). The kaolin group consists of 1:1 layer structures with the general composition of $Al_2Si_2O_5(OH)_5$. Kaolinite, dickite and nacrite are polytypes. Halloysite is a hydrated polymorph of kaolinite with curved layers and a basal spacing of about 1 nm due to hydration.

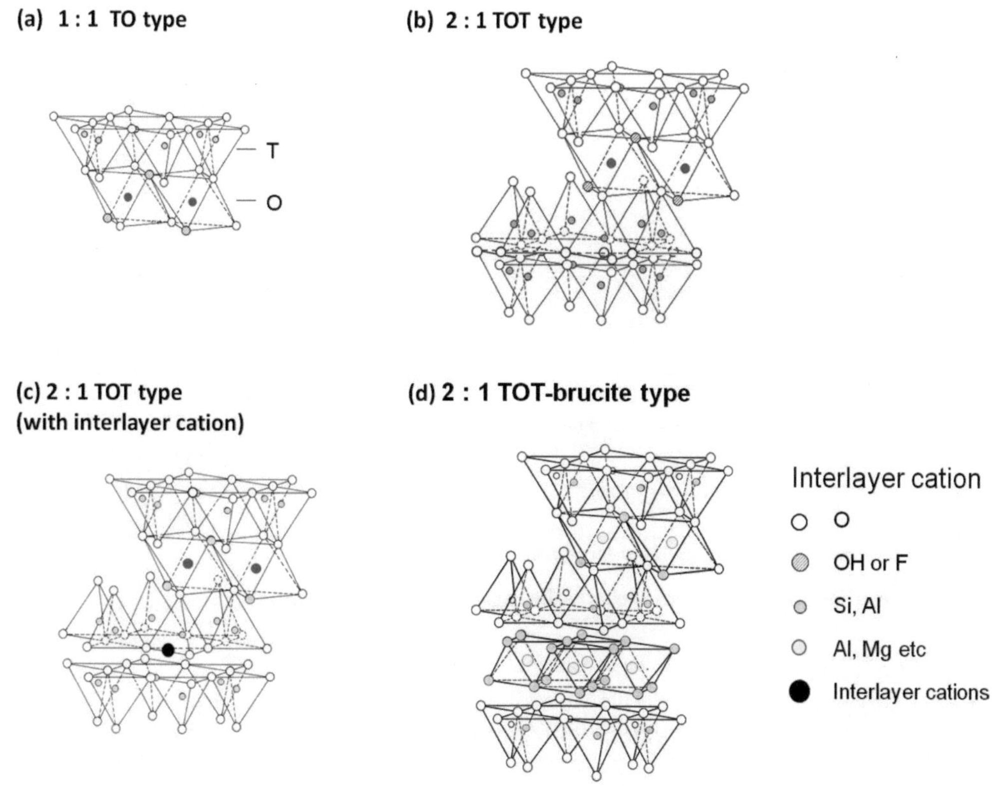

Fig. 2. Typical structures of layered silicates. (a) 1:1 layer type, kaolin-serpentine group; (b) 2:1 layered type, e.g. pyrophyllite or talc; (c) 2:1 layer type with interlayer cation (primarily K, Na, Ca, and Mg); and (d) 2:1 TOT layer type with brucite-like layer.

Layer Type	Group	Sub-group	Species	Chemical composition
1 : 1	kaolin · serpentine	kaolin	kaolinite	$Al_2Si_2O_5(OH)_4$
	(x~0)		dickite	↑
			nacrite	↑
			halloisite	$Al_2Si_2O_5(OH)_4 \cdot 2H_2O$
		serpentine	chrysotile	$Mg_3Si_2O_5(OH)_{4,}$
			lizardite	↑
			antigorite	$(Mg, Fe^{2+})_{2.0}Al(Si_{1.0}Al_{1.0})O_5(OH)_4 \cdot 2H_2O$
2 : 1	pyrophillite · talc	pyrophillite	pyrophillite	$Al_2Si_4O_{10}(OH)_2$
	(x~0)	talc	talc	$Mg_3Si_4O_{10}(OH)_2$
	smectite	dioctahedral smectite	montmorillonite	$E_{0.33}(Al_{1.67}Mg_{0.33})\ Si_4O_{10}(OH)_2.nH_2O$
	(x=0.2~0.6)		beidellite	$E_{0.33}Al_2(Si_{3.67}Al_{0.33})O_{10}(OH)_2.nH_2O$
		trioctahedral smectite	saponite	$E_{0.33}Mg_3(Si_{3.67}Al_{0.33})O_{10}(OH)_2.nH_2O$
			hectorite	$E_{0.33}(Mg_{2.67}Li_{0.33})Si_4O_{10}(OH)_2 \cdot nH_2O$
			stevensite	$E_{0.33}Mg_{2.92}Si_4O_{10}(OH)_2 \cdot nH_2O$
			sauconite	$E_{0.33}Zn_3(Si_{3.67}Al_{0.33})O_{10}(OH)_2.nH_2O$
	vermiculite	dioctahedral verimiculite	dioctahedral verimiculite	$Mg_{2/x}Mg_3(Si_{4-x}Al)O_{10}(OH)_2 \cdot nH_2O$
	(x=0.6~0.9)	trioctahedral vermiculite	trioctahedral vermiculite	$Mg_{2/x}Al_2(Si_{4-x}Al)O_{10}(OH)_2 \cdot nH_2O$
	mica	dioctahedral mica	muscobite	$KAl_2(Si_3Al)O_{10}(OH)_2$
	(x=0.6~1.0)		paragonite	$NaAl_2(Si_3Al)O_{10}(OH)_2$
			illite	$K_{0.75}Al_{1.75}R_{0.1}{}^{2+}(Si_{3.5}Al_{0.5})O_{10}(OH)_2$
			celadonite	$KFe^{3+}(Mg,Fe^{2+})\square Si_4O_{10}(OH)_2$
			tobelite	$(NH_4)_{0.6}K_{0.2}Al_2(Si_{3.2}Al_{0.8})O_{10}(OH)_2$
		trioctahedral mica	phlogopite	$KMg_3(Si_3Al)O_{10}(OH)_2$
			annite	$KFe^{2+}{}_3(Si_3Al)O_{10}(OH)_2$
			lepidolite	$K(Li,Al)_3(Si,Al)_4O_{10}(F,OH)_2$
	brittle mica	dioctahedral brittle mica	margarite	$CaAl_2(Si_2Al_2)O_{10}(OH)_2$
	(x~2.0)	trioctahedral brittle mica	clintonite	$Ca(Mg,Al_3)(SiAl3)O_{10}(OH)_2$
			anandite	$Ba(Fe,Mg)_3(Si_3Fe^{3+})O_{10}(OH)_2$
2 : 1: 1	chlorite	dioctahedral chlorite	donbassite	$Al_{4.27}(Si_{3.2}Al_{0.8})O_{10}(OH)_8$
	(x valiable)	di-·tri-octahedral chlorite	cookeite	$(LiAl_4)(Si_3Al)O_{10}(OH)_8$
			sudoite	$(Al_{2.7}Mg_{2.3})(Si_{3.3}Al_{0.7})O_{10}(OH)_8$
		trioctahedral chlorite	clinoclore	$(Mg,Fe^{2+})_5Al(Si_3Al)O_{10}(OH)_8$
			chamocite	$(Fe^{2+},Mg,Fe^{3+})_5Al(Si_3Al)O_{10}(OH)_8$

x : net layer charge per formula unit. E : exchangeable cation. □ : vacant.

Table 1. Classification of typical layered silicates related to clay minerals.

2.2 2:1 type layered silicate

The alternative, known as a 2:1 TOT layered silicate, has two tetrahedral sheets with the unshared vertices of each sheet pointing towards each other and forming each side of the octahedral sheet. The 2:1 TOT layer type with no interlayer cation, which includes pyrophyllite and talc, consists of two sheets of tetrahedra with an octahedral sheet sandwiched in between. These 2:1 layers are electrostatically neutral in the ideal case, with no interlayer ion present (Fig. 2b). The thickness of this 2:1 unit is about 0.91 nm to 0.94 nm. The 2:1 TOT layer type with interlayer cations consists mainly of the smectite, vermiculite, illite, and mica groups. It also contains 2:1 TOT layers, similar to Fig. 2b, which corresponds to pyrophyllite and talc, but differs in that there is more significant isomorphous substitution, sufficient to raise the charge per unit cell. The net negative charge that arises from this is compensated for by interlayer cations (primarily K, Na, Ca and Mg) (Fig. 2c). The thickness of this platelet is about 1 nm.

2.3 2:1 type layered silicates (TOT-brucite)

Unlike other 2:1 clay minerals, in chlorites, the interlayer spaces between each 2:1 sandwich contain (Mg^{2+}, Fe^{3+})$(OH)_6$. This (Mg^{2+}, Fe^{3+})$(OH)_6$ unit is more commonly referred to as the brucite-like layer (an octahedral sheet), due to its close resemblance to the mineral brucite ($Mg(OH)_2$). Thus, chlorite's structure is as follows: T-O-T-brucite-...(Fig. 2d). The typical general formula is: $(Mg,Fe)_3(Si,Al)_4O_{10}(OH)_2 \cdot (Mg,Fe)_3(OH)_6$ (see Table 1).

3. Interstratified clay minerals

3.1 Natural systems

Some natural clays consist of particles in which different types of silicate layers are stacked together (interstratification), and each type of silicate layer has swelling or non-swelling properties. These silicates are called interstratified clay minerals (Reynolds, 1980).

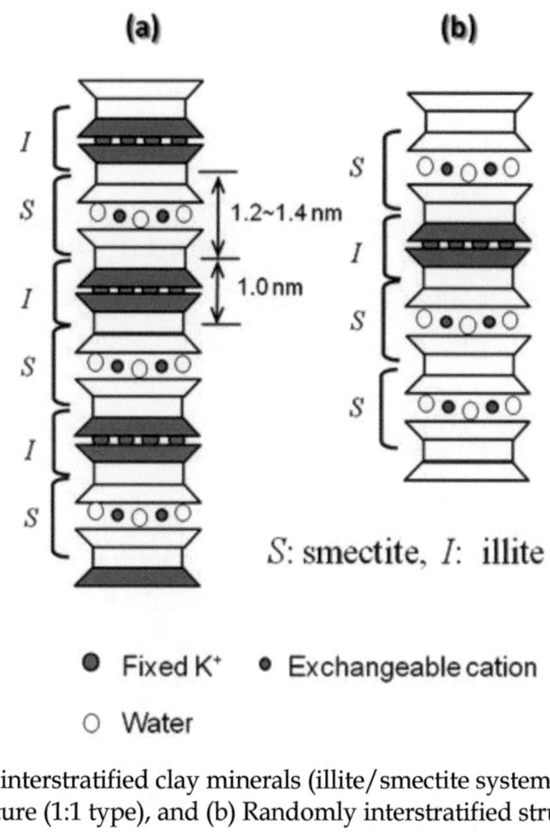

Fig. 3. Structures of interstratified clay minerals (illite/smectite system): (a) Regularly interstratified structure (1:1 type), and (b) Randomly interstratified structure.

Interstratified clay minerals can have ordered (or regular) mixed layer structures if different layers alternate along the c* direction in a periodic pattern (e.g. the stacking of generic type

A and type B layers can be ...ABABAB... or ...AABAABAA...etc.). They can also have disordered (or irregular) mixed layer structures, wherein the stacking of type A and type B layers in the c* direction is random (e.g. ...ABAAABABB...). These orderings are called regular or random, and are described by a Reichweite (R) ordering parameter (Jagodzinski, 1949a, 1949b, 1949c). For example, the regularly interstratified sequence of illite (I) and smectite (S) layers with a ratio of 1:1 refer to an ordering parameter R1. This type would be ordered in an ISISIS fashion (Fig. 3a). On the other hand, Figure 3b shows random ordering which refer to R0. Figure 3 shows an example of interstratification between illite (*I*): anhydrous layers, with a periodicity of about 1.0 nm, and smectite (*S*): hydrated layers with a periodicity of about 1.2 nm to 1.4 nm. Interstratified clay minerals which are perfect R1 types often have individual names (Table 2). R1-ordered chlorite-smectite is known as corrensite, and R1 mica-smectite is rectorite. Special names are assigned to regularly alternating sequences of components present in a fixed ratio. Irregularly stacked structures are identified using the names of the two components, such as illite-smectite, smectite-chlorite and kaolinite-smectite. Typical interstratified clay minerals are listed in Table 3.

Mineral Name	Layer components	Regular alternation of layer types
Rectorite	Mica (di)/Smectite (di)	1 : 1
Tarasovite	Illite/Smectite (di)	3 : 1
Tosudite	Chlorite (di)/Smectite(di)	1 : 1
Brinrobertsite	Pyrophillite/ Smectite (di)	1 : 1
Hydrobiotite	Mica(tri, Biotite) /Vermiculite(tri)	1 : 1
Corrensite	Chlorite(tri) / Vermiculite (tri)	1 : 1
	Chlorite(tri) / Smectite (tri)	1 : 1
Aliettite	Talc / Smectite (tri)	1 : 1
Kulkeite	Chlorite(tri) /Talc	1 : 1
Dozyite	Serpentine / Chlorite(tri)	1 : 1

Table 2. Regularly stacked, mixed clay minerals. di: dioctahedral, tri: trioctahedral.

In table 3, which lists the combinations between dioctahedral type layers and between trioctahedral type layers, there is only one combination between dioctahedral and trioctahedral type layers.

As shown in table 3, natural interstratified clay minerals mainly comprise combinations of two types of dioctahedral layers, or combinations of two types of trioctahedral layers. A combination of dioctahedral-type layer and trioctahedral-type layer is rarely found in nature. This is due to a larger misfit between the dioctahedral-type layer with the trioctahedral-type layer along the b* axis. Each basic building block (layer) of interstratified clay minerals is generally assumed to possess the properties which are the same as the corresponding naturally-occurring mineral.

In order to define the basic building block (layer) of interstratified clay minerals, two possible structure models, which either have or do not have a center of symmetry placed between illite layers or smectite layers, are provided for regularly interstratified R1 I/S clay minerals. In the model which does not have the center of symmetry, an individual 2 : 1 layer in R1 I/S is always chemically homogeneous, consisting of either smectite or illite layer (Fig.4a). This model is referred to as non-polar 2:1 model. On the other hand, in the model having the center of symmetry (the symmetry is across the interlayer), an individual 2 : 1 layer in R1 I/S can be chemically homogeneous, containing a low charge smectite layer-like layer on one side and high charge illite-like layer on the other (Fig. 4b) which is referred to as polar 2:1 model. The polar model is supported by a number of NMR and XRD results (Altaner et al., 1988; Barron at al., 1985; Jakobsen at al., 1995; Plançon, 2004), and the first principle calculation (energy calculations) further reinforces the polar model (Olives et al., 2000; Stixrude & Peacor, 2002). The polar model is also consistent with the fundamental particle model by Nadeau et al.(1984) and the latest high resolution TEM result (Murakami et al., 2005).

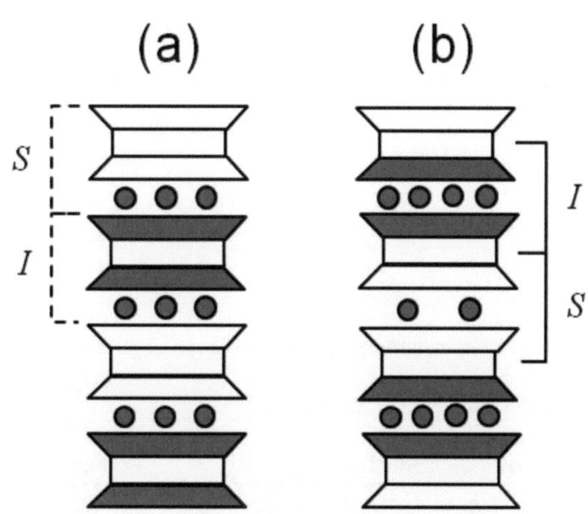

Fig. 4. Schematic of (a) non-polar and (b) polar models of the R1 I/S structure.

In the natural process of diagenesis, the transition of smectite to illite is an established axiom within the field of clay mineralogy. The formation of I/S minerals is actually affected not only by temperature, but also by other factors such as fluid chemistry, time, fluid/ rock (W/R) ratio during the formation of I/S minerals, and the chemical composition of the precursor materials (Altaner & Ylagan, 1997). Three of the most apparent variables driving the smectite-to-illite transition, called by some the illitization of smectite, are the time involved, the temperature, and the availability of interlayer K ions. As with all chemical reactions, time and temperature are inversely related (S´rodon´ & Eberl, 1984).

3.2 Synthesis of interstratified clay minerals

Many kinds of regularly interstratified clay minerals such as illite/smectite, mica/smectite, and chlorite/smectite have been found in nature. They are not, however, generally used as industrial materials because they are rare. Because of their scarcity, it is important to attempt to synthesize some interstratified clay minerals. Examples of previous research on synthesizing regularly interstratified clay minerals include studies on lizardite/saponite (smectite) and serpentine/smectite (Torii et al., 1998; Nagase et al., 2000).

	Combination	Layer components
dioctahedral type - dioctahedral type	non-expandable layer/expandable layer	Illite - Smectite (di)
		Illite - Vermiculite (di)
		Mica (di) - Vermiculite (di)
		Mica (di) - Smectite (di)
		Glauconite - Smectite (di)
		Chlorite (di) - Smectite (di)
		Pyrophillite - Smectite (di)
		Pyrophillite - Vermiculite (di)
		Kaoline - Smectite (di)
		Halloisite (7Å) - Halloisite (10Å)
	non-expandable layer/non-expandable layer	Chlorite (di) - Kaoline
		Mica (di) - Pyrophilite
		Mica (di) - Chlorite (di)
	expandable layer/expandable layer	Vermiculite (di) - Smectite (di)
		Smectite (di) - Smectite (di)
trioctahedral type - trioctahedral type	non-expandable layer/expandable layer	Mica(tri, Biotite) - Vermiculite(tri)
		Mica(Biotite) - Smectite(tri)
		Chlorite(tri) - Vermiculite(tri)
		Chlorite(tri) - Smectite(tri)
		Talc - Smectite(tri)
	non-expandable layer/non-expandable layer	Chlorite(tri) - Talc
		Serpentine - Chlorite(tri)
		Chlorite(tri) - Mica(Biotite)
	expandable layer/expandable layer	Vermiculite (tri) - Vermiculite (tri)
dioctahedral type - trioctahedral type	non-expandable layer/expandable layer	Chlorite(tri) - Smecite (di)

Table 3. Combinations of silicate layers in reported interstratified clay minerals.

In a previous study, we performed a series of hydrothermal experiments to determine the phase relationships of montmorillonite – stevensite and beidellite - saponite pseudo-binary joins (Yamada et al., 1999, 2010). Fig. 5 shows a three-component phase diagram of smectite with Si, Mg, and Al. Na is constant, and the figure shows the proportion of each element present in the smectites. Hydrothermal experiments were conducted using a rapid-quench type hydrothermal apparatus running at temperatures from 250 °C to 500 °C, with a constant pressure of 100 MPa and varying duration times. The time-temperature-phase change diagram in the both pseudo-binary join shows that immiscibility occurs between dioctahedral smectite (ex. montmorillonite-like smectite) and trioctahedral smectite (ex. stevensite-like smectite) below 400°C. It should be noted that interstratified clay minerals including rectorite-like interstratified layered silicate (composition I); a regularly interstratified chlorite/smectite (composition II); and a regularly interstratified talc/talc/smectite (composition III) were obtained at 350 °C (composition III), 450°C (composition I, II) in the intermediate chemical regions of these joins (Yamada et al., 1999, 2010). Of particular note was the montmorillonite-stevensite pseudo-binary join. On the

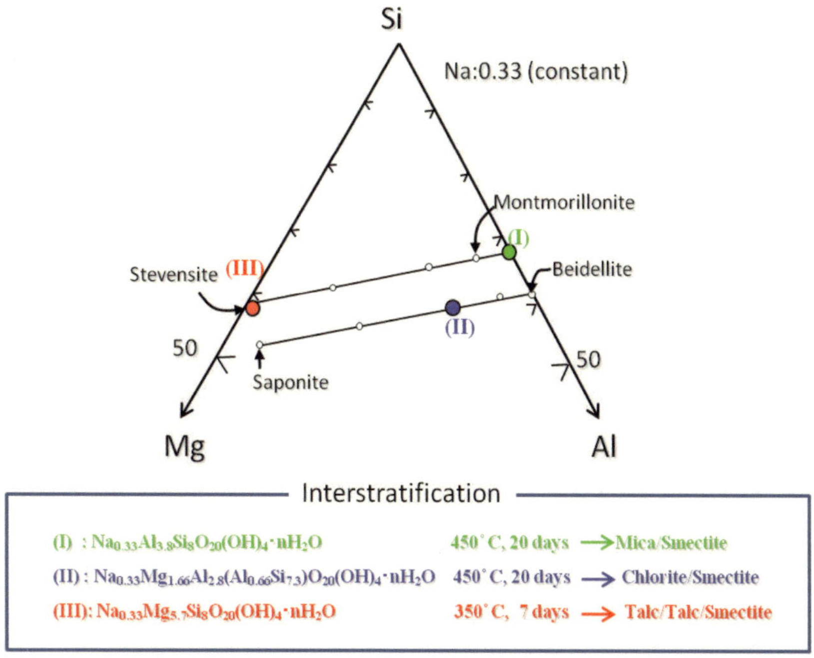

Fig. 5. A three-component phase diagram of smectites with Si, Mg, and Al. Na is constant, showing the proportion of each element present in the smectites.

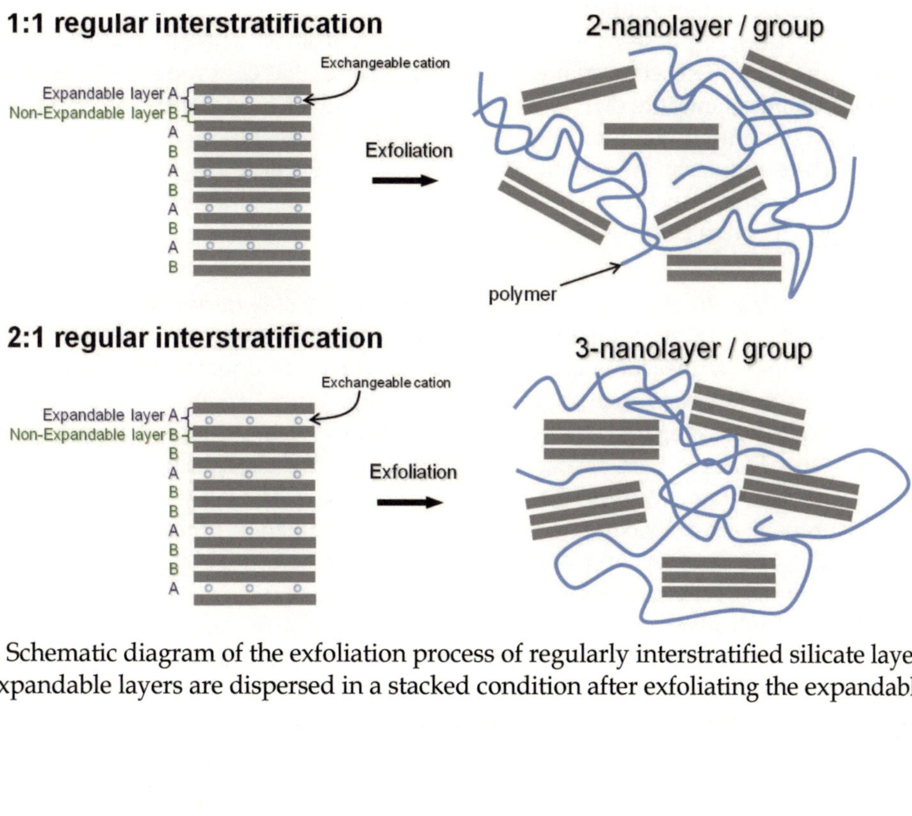

Fig. 6. Schematic diagram of the exfoliation process of regularly interstratified silicate layers: non-expandable layers are dispersed in a stacked condition after exfoliating the expandable layer.

stevensite side of this region, the trioctahedral smectite was found below 300°C, which transformed to stevensite and quartz with aging. Above 350°C, however, in the composition III, a regular interstratification was recognized, which was the regularly interstratified sequence of talc-like and smectite-like layers with a ratio of 2:1 (talc/talc/smectite) and a Reichweite (R) ordering parameter R2. The assemblage of regularly interstratified clay minerals was recognized in the intermediate chemical composition region of their joins. The following section explains the characterization of the synthesized interstratified clay minerals in more detail.

4. Interstratified clay minerals: polymer nanocomposites

4.1 Control of the number of layers

In order to achieve control of the number of layers, a layered silicate with two antagonistic properties, expandable and non-expandable characteristics, was prepared in each particle. Using these interstratified clay minerals with regularly alternating sequences of components, exfoliated nanolayers consisting of two or three layers per group could be dispersed in a polymer matrix (Fig. 6). The hydrophilic nature of these pristine clay minerals impedes their homogeneous dispersion in the polymer matrix. In most cases, the interlayer that has a expandable property must be organically modified. Unfortunately, it is difficult to obtain favorable materials having interstratified structures in ordered sequences because natural sources are impure and non-homogeneous, so it is important to create better-designed interstratified materials for inclusion in nanocomposites.

4.2 Morphology development of polymer nanocomposites

4.2.1 Synthesized interstratified clay minerals

The time-temperature-phase change diagram described above showed that three types of regularly interstratified clay minerals occur above 350°C; product (I), a regularly interstratified mica/smectite (M/S), with a regular 1:1 layer type alternation, product (II), a regularly interstratified chlorite/smectite (C/S), which also has a 1 : 1 alternation, and product (III), a regularly interstratified talc/talc/smectite (T/T/S) with a 2:1 alternation (Fig. 5). The stacking structures were determined by examining different treatment samples using X-ray diffraction (XRD) measurement.

4.2.2 Characterization of the synthesized interstratified clay minerals

To evaluate the layer stacking of the resulting material, two quenched samples were examined using XRD measurements. (1) The samples were oriented by sedimentation in water onto glass slides and then drying at room temperature. (2) An ethylene glycol (EG)-treated sample was prepared as follows: the oriented sample was placed in a sealed container with EG and then heated overnight at 60 °C. The phases encountered were identified using previously described criteria (Yamada et al., 1999, Tamura at al., 2008). Then, to prepare the nanocomposites, organically modified synthesized materials were made utilizing the ion-exchange reaction between exchangeable interlayer cations and octadecylamine (ODA) hydrochloride.

Product (I) By hydrothermal treatment of quenched glass with composition (I) at 450°C for 20 days, a layered silicate with a regularly interstratified structure was obtained (Tamura at al., 2009). Figure 7 shows the XRD patterns of (a) the pristine product (I), (b) an EG-treated sample and (c) an ODA treated sample. In the pristine product (I), a shoulder is observed at

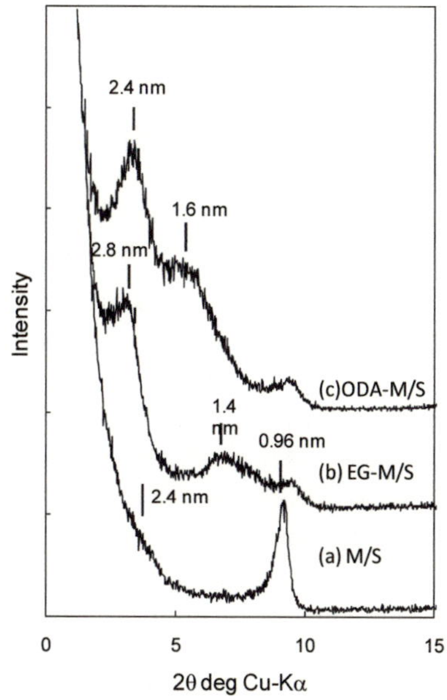

Fig. 7. XRD patterns of (a) product (I), (b) ethylene glycol-treated product (I) (EG-MS), and (c) ODA-modified product (I). XRD data were collected using Cu-Kα radiation.

2θ = 3.6°, corresponding to 2.4 nm. After treatment with EG, the 001 reflection shifted to a lower angle, from 2.4 nm to 2.8 nm and a 002 reflection appeared. Curve (c) is the XRD pattern of the ODA treated sample. Three broad reflections at 2.4 nm, 1.6 nm, and 0.94 nm were observed, which correspond to the 002, 003 and mica phases, respectively. The basal spacing of the ODA-product (I) is estimated to be 4.8 nm. All XRD results indicate that the layer structure is a 1:1 regularly interstratified structure consisting of the mica/smectite sequence we refer to as M/S. This is a "Rectorite-like type interstratified silicate" (Fig. 8).

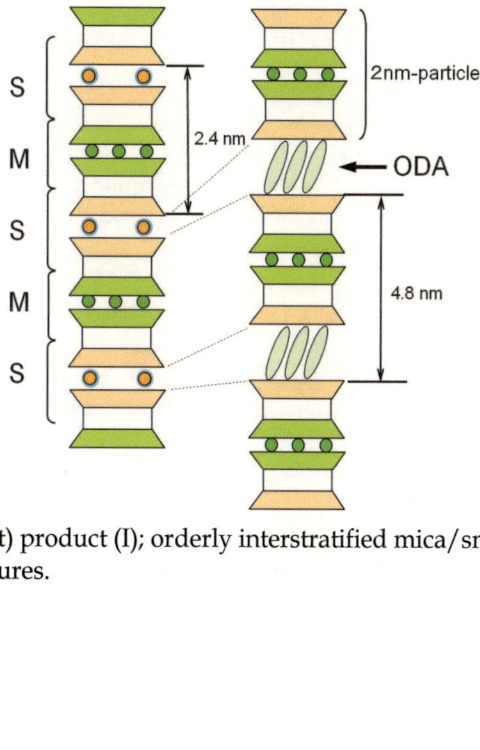

Fig. 8. Schematics of (left) product (I); orderly interstratified mica/smectites (M/S) and (right) ODA-M/S structures.

Product (II) By hydrothermal treatment of quenched glass with composition (II) at 450°C for 20 days, a regularly interstratified structure was recognized (Tamura at al., 2009). The pristine product shows three broad reflections of 2.7 nm, 1.35 nm and 1 nm, respectively (Fig. 9, curve (a)). For the EG treated sample, curve (b), two reflections shift to lower angles. These correspond to 001, and 002, respectively (curve (b)). The line profiles suggest that the layer structure of product (II) is a 1:1 regularly interstratified structure consisting of the chlorite(C)/smectite (S) sequence we refer to as C/S. Thus, we have a "corrensite"-like interstratified silicate (Fig. 10). Curve (c) in Fig. 9 is an XRD pattern of the ODA treated sample. Two broad reflections of 3.2 nm and 1.6 nm were observed, which correspond to the 001 and 002 reflections, respectively.

Product (III) By hydrothermal treatment of quenched glass with composition (III) at 350°C for 7 to 21 days, a unique regularly interstratified clay mineral composed of talc/talc/smectite (T/T/S) was recognized (Tamura et al., 2008). For the pristine product, three broad reflections of 3.5 nm, 1.7 nm, and 1.0 nm were observed, which correspond to the 001, 002, and 003 reflections, respectively (Fig. 11, curve (a)). After treatment with EG, all three reflections shifted to lower angles of 3.7 nm, 1.8 nm, and 1.2 nm (Fig. 11, curve (b)). These results suggest that the layer structure of product (III) is a 2:1 regularly interstratified structure consisting of the talc/talc/smectite sequence that we call T/T/S (Reichweite ordering parameter: R2). This is a new regularly interstratified clay mineral in an ordered sequence.

The d_{001} phase at 5.2 nm is a super-lattice of the compound formed by the intercalation of ODA^{+} molecules (Fig. 11, curve(c)). Figure 12 schematically depicts the ordered interstratified T/T/S and organically modified T/T/S (ODA-T/T/S) structures.

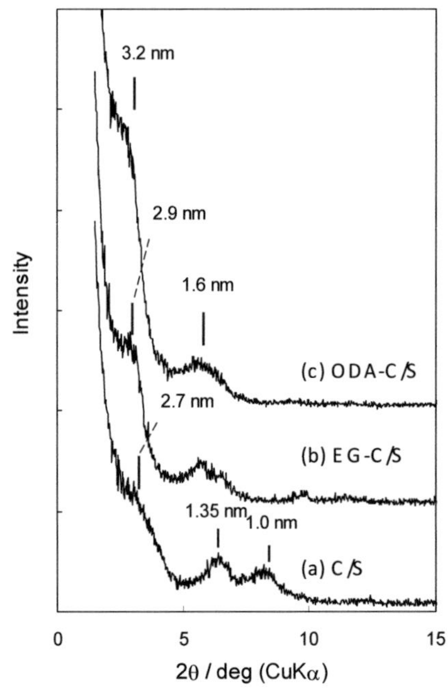

Fig. 9. XRD patterns of (a) product (II), (b) ethylene glycol-treated product (II) (EG-C/S), and (c) ODA-modified product (II). XRD data were collected using Cu-Kα radiation.

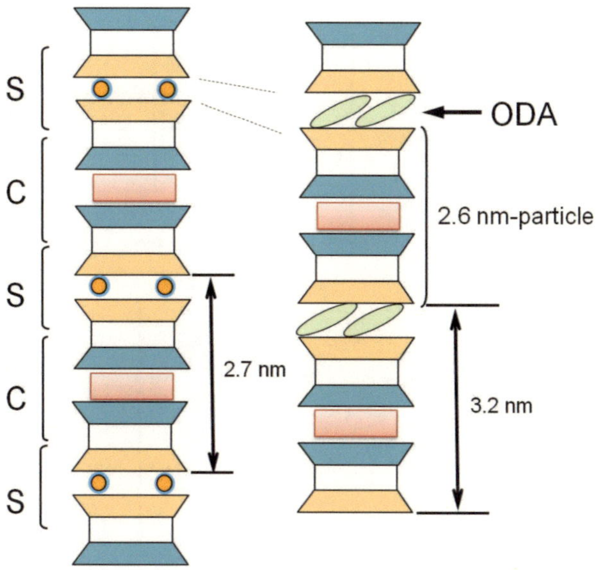

Fig. 10. Schematics of (left) product (II); orderly interstratified chlorite/smectites (C/S) and (right) ODA-C/S structures.

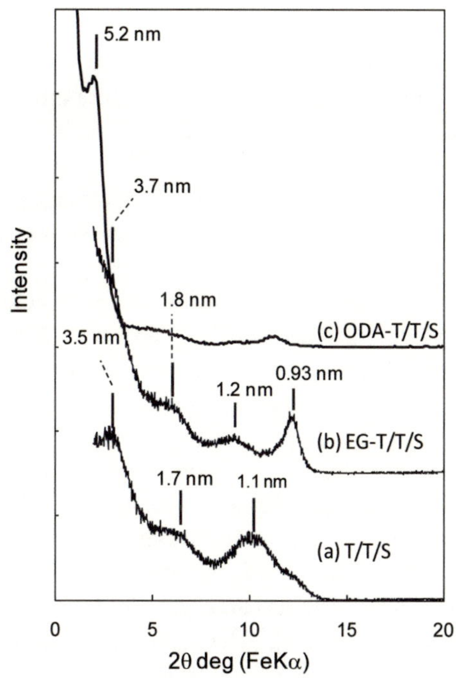

Fig. 11. XRD patterns of (a) product (III), (b) ethylene glycol-treated product (III) (EG-T/T/S), and (c) ODA-modified product (III). XRD data were collected using Fe-Kα radiation.

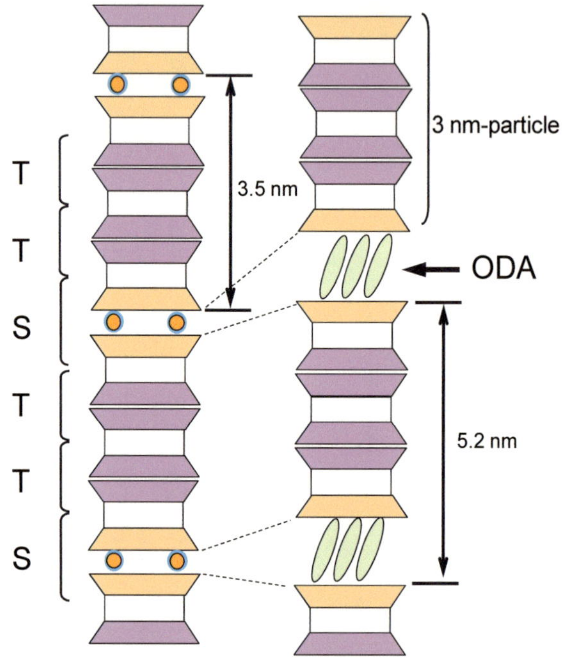

Fig. 12. Schematics of (left) product (III); orderly interstratified talc/talc/smectites (T/T/S) and (right) ODA-T/T/S structures.

4.2.3 Preparation of nanocomposite

The ODA-modified silicate powder was added to diglycidyl ether of bisphenol A (DGEBA) and cured with methyl nadic anhydride (MNA). Benzyl dimethylamine (BDMA) was used as a catalyst. Samples were cured at 150°C for 6 h.

Figure 13 shows transmission electron microscopy (TEM) images of (a) [ODA-M/S]-epoxy, (b) [ODA-C/S]-epoxy, and (c) [ODA-T/T/S]-epoxy, respectively. The dark areas represent the silicate layers and the gray/white areas represent the epoxy matrix. It would seem that the dispersed platelets (units) are thicker than those of a completely exfoliated (monolayer dispersed) clay-polymer nanocomposite. In the TEM image of the [ODA-M/S]-epoxy composite, partially exfoliated double layers (2-nanolayer/group) can be seen (Fig. 13a). In the [ODA-C/S]-epoxy composite, flocculated silicate layers ranging from about 50 nm to 200 nm in thickness were observed (Fig. 13b). It seems, however, that fine particles might be partially exfoliated in the epoxy matrix. Closer observation of the nanocomposite at high magnification confirms a rather rigid platelet (cross section), in which the length of the cross section is a few hundred nanometers and its thickness is approximately 3 nm. Most of the nanolayers appear to be 3 layers per unit (C/S: the thickness is a ca. 2.5 nm).

The [ODA-T/T/S]-epoxy nanocomposite has an especially homogeneous dispersion of ODA-TTS platelets in the epoxy matrix (Fig. 13c). The TEM image confirms a rather rigid platelet (cross section), in which the length of the cross section is a few hundred nanometers and it is approximately 3 nm to 5 nm thick. Most of the dispersed nanolayers shown in Fig. 13c are actually 3 nanolayers grouped together. The 3-layer group thickness of approximately 5 nm, deduced from the TEM images, is thicker than that of the 3 silicate layers (minimum thickness

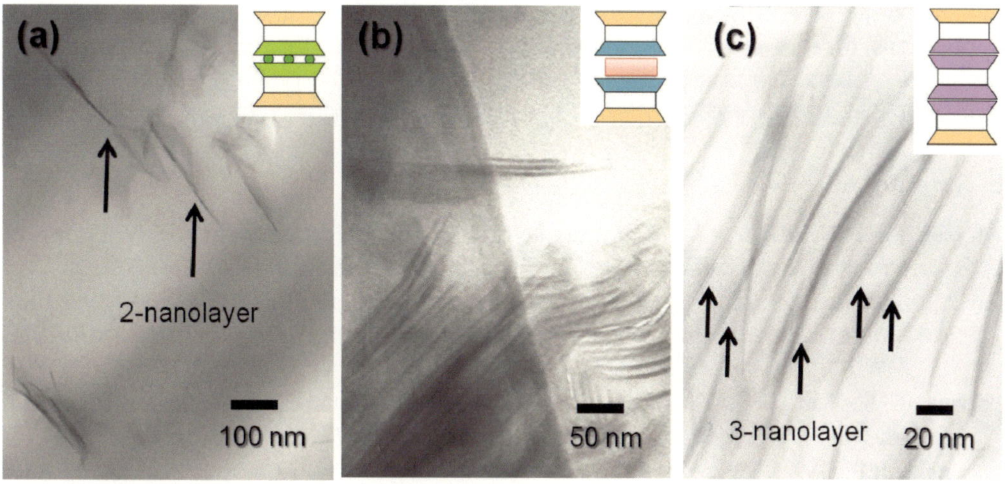

Fig. 13. TEM images of thin sections of glassy epoxy composites cured at 150°C for 6 h: (a) [ODA-M/S]-epoxy system, (b) [ODA-C/S]-epoxy system, (c) [ODA-T/T/S]-epoxy nanocomposite. The weight fraction of each loading was 2.5% silicate. The insets show supposed unit layers .

of about 3.5 nm) that could be caused by the tilting of layers that are inclined obliquely to the sectioned plane. Despite the thicker dimension, the image clearly shows that perfect exfoliation of a 3-layer unit (T/T/S) has taken place, showing that we succeeded in dispersing silicate nanolayers consisting of several layers in one group in a polymer.

5. Summary

Following the pioneering work of Toyota researchers who demonstrated the first practical application of a clay-nylon 6 nanocomposite in the automobile industry (Kurauchi et al., 1991), numerous researchers have reported the preparation of exfoliated clay-polymer nanocomposites in various polymer systems. A range of factors that influence not only the morphology but also the final properties of composites have been identified, including the nature of the polymer, the nature of the clay minerals, interfacial interactions between the clay minerals and the polymer, the processing methodologies, and the amount of clay. In order to optimize the properties of nanocomposites so that they can be used effectively in practical applications, we need to take a comprehensive view of all these factors when designing the material. We strongly believe that broadening our scope in order to utilize various types of clay minerals and methods will be essential to achieve that goal.

In this chapter, several examples of polymer nanocomposites using different types of regularly interstratified clay minerals have been introduced. Generally, smectites and other expandable layered silicates are versatile materials that, in a practical sense, are well adapted to yield a large variety of nanocomposites due to their ion exchange, expandability, and colloidal properties. On the other hand, no one has attempted to use regularly interstratified clay minerals, the so-called "mixed clay minerals", despite their unique structure of highly ordered silicate nanolayers (an ordered sequence along the c* direction) to create new nanocomposites due to their unavailability and metastability. Peculiar clay minerals could potentially extend the field of high performance materials beyond traditional applications to encompass new unexpected functions. For the future, to design high-performance nanocomposites, highly precise control of morphology ranging from the microscopic to the macroscopic scale will be probed using various interstratified minerals.

6. References

Altaner, S.P., Weiss, C. A. & Kirkpatrick, R.J., (1988). Evidence from ^{29}Si NMR for the structure of mixed-layer illite/smectite clay minerals. *Nature*, 331, 699-702.

Altaner, S. P. & Ylagan, R. F., (1997). Comparison of Structural models of Mixed-Layer Illite-Smectite and reaction Mechanisms of Smectite Illitization. *Clays. Clay. Miner.*, 45, 517-533.

Barron, R., F., Slade, P. & Frost, R.L., (1985). Solid-state silicon-29 spin-lattice relaxation in several 2:1 phyllosilicate minerals. *J. Phys. Chem.*, 89, 3880-3885.

Giannelis, E.P. (1992). A New Strategy For Synthesizing Polymer Ceramic Nanocomposites. *JOM*, 44, 28-30.

Haraguchi, K. & Takehisa, T. (2002). Nanocomposite hydrogels: a unique organic-inorganic network structure with extraordinary mechanical, optical, swelling/de-swelling properties. *Adv. Mater.*,14, 1120-1124.

Jagodzinski, H. (1949a). Eindimensionale Fehlordnung in Kristallen und ihr Einfluss auf die Röntgeninterferenzen. I. Berechnung des Fehlordnungsgrades aus den Röntgenintensitäten. *Acta. Cryst.*, 2., 201-207.

Jagodzinski, H. (1949b). Eindimensionale Fehlordnung in Kristallen und ihr Einfluss auf die Röntgeninterferenzen. II. Berechnung der fehlgeordneten dichtesten Kugelpackungen mit Wechselwirkungen der Reichweite 3. *Acta. Cryst.*, 2., 208-214.

Jagodzinski, H. (1949c). Eindimensionale Fehlordnung in Kristallen und ihr Einfluss auf die Röntgeninterferenzen. III. Vergleich der Berechnungen mit experimentellen Ergebnissen. *Acta. Cryst.*, 2., 298-304.

Jakobsen, H.J., Neilsen, N. C. & Lindgreen, H. (1995). Sequences of Charged Sheets in Rectorite. *Am. Mineral.*, 80, 247-252.

Kurauchi, T., Okada, A., Nomura, T., Nishio,T., Saegusa, S. & Deguchi, R. (1991). Nylon 6-Clay Hybrid-Synthesis, Properties and Application to Automotive Timing Belt Cover. *SAE Technical Paper*, 910584, 1-7.

Maiti, P., Yamada, K., Okamoto, M., Ueda, K. & Okamoto, K.(2002). New polylactide /layered silicate nanocomposites: Role of organoclays. *Chem. Mater.*, 14, 4654-4661.

Messersmith, P.B. & Giannelis, E.P. (1994). Synthesis and characterization of layered silicate-epoxy nanocomposites. *Chem. Mater.*, 6, 1719-1725.

Murakami, T., Inoue, A., Lanson, B., Meunir, A., Beaufort, D. (2005). Illite-smectite mixed-layer minerals in the hydrothermal alteration of volcanic rocks: II. One-dimensional HRTEM structure images and formation mechanisms. *Clays. Clay. Miner.*, 53, 440-451.

Nadeau, P. H., Tait, J. M. McHardy, W. J. & Wilson, M. J. (1984). Interstratified XRD Characteristics of Physical Mixtures of Elementaly Clay Particles. *Clay. Miner.*, 19, 67-76.

Nagase, T., Ebina, T., Torii, K., Iwasaki,T., Hayashi, H., Onodera, Y. & M. Chatterjee, M. (2000). TEM Observation of Interstratified Ni-Serpentine/Smectite Compounds. *Chem. Lett.*, 344-345.

Olives, J., Amouric, M., Perbost., R. (2000). Mixed Layering of Illite-Smectites: Results from High-Resolution Transmission Electron Microscopy. *Clay. Clay. Miner.*, 48, 282-289.

Plançon, A., (2004). Consistent Modeling of the XRD Patterns of Mixed-Layer Phyllosilicates. *Clays. Clay. Miner.*, 52, 47-54.

Reynolds, R. C. (1980). Interstratified Clay Minerals, In: *Crystal Structures of Clay Minerals and Their X-ray Identification*, Brindley, G. W. & Brown, G., (Ed.), 249-303, Mineralogical Society, London.

Stixrude, L. & Peacor, D. R., (2002) First-Principles Study of Illite–Smectite and Implications for Clay Mineral Systems. *Nature,* 420, 165-168.

Suzuki, M., Ohashi, F., Inukai, K., Maeda, M., Tomura, S. & Mizota, T. (2001). Hydration enthalpy measurement and evaluation as heat exchangers of allophane and imogolite. *J. Ceram. Soc. Jpn.*, 109, 681-685.

S´rodon´, J. & Eberl, D. D. (1984). Illite, In: *Micas, Vol. 13 in Reviews in Mineralogy,* Bailey, S. W., (ed), 495-544, Mineralogical Society of America, Washington, D.C.

Tamura, K. & Nakazawa, H. (1996). Intercalation of n-alkyltrimethylammonium into swelling fluoro-mica. *Clays. Clay. Miner.*, 44, 501-505.

Tamura, K., Yamada, H. & Nakazawa, H. (2000) Stepwise hydration of high-quality synthetic smectite with various cations. *Clays. Clay. Miner.*, 48, 400-404.

Tamura, K., Sasaki, T., Yamada, H. & Nakazawa, H. (1999). Laue function analysis of colloidal lithium taeniolite. *Langmuir,* 15, 5509-5512.

Tamura, K., Yamada, H., Yokoyama, S. & Kurashima, K. (2008). Regularly Interstratified Layered Silicate-Polymer Nanocomposite, *J. Am. Ceram. Soc.*, 91, 3668-3672.

Tamura, K., Yamada, H., Yokoyama, S., Kurashima, K. (2009). *Abstracts of 14th International Clay Conference (ICC 2009)*, pp , Preparation of Regularly Interstratified Layered Silicate-Polymer Nanocomposite. Castellaneta Marina (Italy), June, 2009.

Torii, K., Onodera,Y., Hayashi, H., Nagase, T. & Iwasaki,T. (1998). Hydrothermal Synthesis of Interstratified Lizardite/Saponite. *J. Am. Ceram. Soc.*, 81, 447-449.

Usuki, A., Kawasumi, M., Kojima, Y., Fukushima, Y., Okada, A., Kurauchi, T. & Kamigaito, O. (1993). Synthesis of Nylon 6-Clay Hybrid. *J. Mater. Res.*, 8, 1179-1184.

van Olphen, H. (1977). Clay Mineralogy, In: *An introduction to clay colloid chemistry: for clay technologists, geologists, and soil scientists.*, 57-82, John Wiley & Sons Inc., New York.

Wang, Z. & Pinnavaia, T.J. (1998). Nanolayer reinforcement of elastomeric polyurethane. *Chem. Mater.*, 10, 3769-3771.

Yamada,H., Yoshioka, K., Tamura, K., Fujii, K. & Nakazawa, H. (1999). Compositional gap in dioctahedral-trioctahedral smectite system: beidellite-saponaite pseudo-binary join. *Clays. Clay. Miner.*, 47, 803-810.

Yamada, H., Tamura, K., Morimoto, K., Yokoyama, S. & Hatta, T. (2010). *Abstracts of Mid European Clay Conference (MECC2010),* pp. -, Synthesis and Characterization of Phyllosilicates in Montmorillonite - Stevensite Compositional Series. Budapest (Hungary), Aug, 2010.

Yano, K. Usuki, A. & Okada, A. (1997). Synthesis and Properties of Polyimide-Clay Hybrid Films. *J. Polym. Sci. A. Polym. Chem.*, 35, 2289-2294.

17

Nanocomposite and Nanostructured Carbon-based Films as Growth Substrates for Bone Cells

Lucie Bacakova[1], Lubica Grausova[1], Jiri Vacik[2], Alexander Kromka[3], Hynek Biederman[4], Andrei Choukourov[4] and Vladimir Stary[5]
[1]*Department of Growth and Differentiation of Cell Populations, Institute of Physiology, Academy of Sciences of the Czech Republic*, Prague,
[2]*Nuclear Physics Institute, Academy of Sciences of the Czech Republic & Research Center Rez*,
[3]*Institute of Physics, Academy of Sciences of the Czech Republic*, Prague,
[4]*Department of Macromolecular Physics, Faculty of Mathematics and Physics, Charles University*, Prague,
[5]*Faculty of Mechanical Engineering, Czech Technical University*, Prague,
Czech Republic

1. Introduction

In recent years, the need to substitute and repair damaged hard tissues (bone, joints, teeth) has been increasing due to diseases and injuries of the locomotory system, which are relatively widespread in the civilized world. The most frequently used implants are based on metals; for example, the metallic part of the socket and the metallic stem of a total hip prosthesis, anchored to the hip-bone and femur, respectively, or dental implants anchored to the jawbone. In bone implants, primary stability is determined mainly by the shape of the implant and by the quality of the bone preparation. However, the secondary stability depends strongly on the attractiveness of the implant surface for the adhesion, growth and phenotypic maturation of osteoblasts, and the formation of mineralized bone tissue at the bone-implant interface. The attractiveness of an implant for colonization with bone cells can be markedly enhanced by various surface modifications of the implant, e.g., by deposition of nanocomposite and nanostructured films.

In this chapter, we summarize more than ten years of experience in our laboratory (in the context of investigations by other authors) of studying the interaction of bone cells *in vitro* with various types of nanocomposite layers based on carbon and carbon enriched with metalloid or metal atoms. These layers include nanocomposite hydrocarbon plasma polymer (ppCH) films with various concentrations of Ti or Ag, titanium-enriched amorphous carbon (C:Ti), pyrolytic graphite and also novel boron-doped nanocrystalline diamond films, nanostructured hybrid fullerene C_{60}/Ti composite films and carbon nanotube-based substrates.

The nanocomposite films are expected to support their colonization with cells particularly by their nanoscale surface roughness, i.e., the presence of irregularities smaller than 100 nm.

The surface nano-roughness mimics, at least to some degree, the nanoarchitecture of the natural extracellular matrix (ECM) as well as of the cell membrane, such as the size of some ECM molecules, their folding and branching, or prominences on the cell membrane, e.g., extracellular parts of the cell adhesion receptors. The nanostructure of a material also improves the adsorption of cell adhesion-mediating ECM molecules, present in biological fluids, such as the serum supplement of cell culture media, blood or interstitial fluid, or synthesized and deposited by cells contacting the material. These molecules involve, e.g., vitronectin, fibronectin, collagen, laminin and also fibrin, a molecule of the provisional matrix formed during tissue healing. On nanostructured materials, the cell adhesion-mediating molecules are adsorbed in advantageous geometrical conformations, allowing for good accessibility of the active sites in these molecules by the cell adhesion receptors (Webster *et al.*, 2000; Price *et al.*, 2004). The active sites include specific amino acid sequences, e.g., Arg-Gly-Asp (RGD), bound by many cells types, or cell-specific sequences, such as Lys-Arg-Ser-Arg (KRSR), Arg-Glu-Asp-Val (REDV) and Val-Ala-Pro-Gly (VAPG), bound by osteoblasts, vascular endothelial cells and vascular smooth muscle cells, respectively. The cells bind the active sites in the matrix molecules by their cell adhesion receptors. The most known and systemized are receptors of the integrin superfamily, but non-integrin receptors, usually of the proteoglycan nature, also take an important place in the cell adhesion. The non-integrin receptors can bind either specific amino acid sequences, e.g., VAPG, or other specific saccharide ligands also present in the cell adhesion-mediating ECM molecules (for a review, see Bacakova *et al.*, 2004; 2008a; Bacakova and Svorcik 2008).

In addition, nanostructured surfaces are believed to preferentially adsorb vitronectin, due to its relatively small and linear molecule compared to the larger and more complicated, e.g., branched, molecules of other ECM proteins (Webster *et al.*, 2000; Price *et al.*, 2004). Vitronectin is then preferentially recognized by osteoblasts rather than by other cell types, thus nanostructured surfaces are considered to be particularly suitable for bone tissue engineering.

In supporting the protein adsorption and cell adhesion, the nanoscale-roughness acts synergistically with the polarity and wettability of the material surface, which is often due to the presence of oxygen-containing chemical functional groups on the surface. On wettable surfaces, the cell adhesion-mediating molecules are adsorbed in a flexible and reorganisable form, enabling good accessibility of specific sites on these molecules by the cell adhesion receptors. In our earlier studies on ion-irradiated polymers, the creation of oxygen-containing groups on the polymer surface and the increased material wettability were associated with increased adsorption of collagen IV and enhanced adhesion, growth and phenotypic maturation of vascular endothelial and smooth muscle cells in cultures on the polymers (Bacakova *et al.*, 2000, 2001a). In comparison with hydrophobic surfaces, the wettable surfaces also adsorb a lower amount of albumin, i.e., a protein non-adhesive for cells. However, the cell adhesion is optimal only on moderately wettable surfaces. On highly hydrophilic surfaces, e.g., on oxygen-terminated nanocrystalline diamond (water drop contact angle less than 2°; Clem *et al.*, 2008) or surfaces with PEO chains, particularly in the form of brush, mobile in water environment (Bacakova *et al.*, 2007a), the cell adhesion-mediating molecules were not adsorbed at all or very weakly, which disabled or significantly limited the cell adhesion.

Another important property of some carbon-based materials, particularly nanocrystalline diamond films doped with boron or substrates with carbon nanotubes, is the electroactivity of these substrates, e.g., their electrical charge, electrical potential and electrical conductivity, which enable the electrical stimulation of cells (Supronowicz *et al.*, 2002;

Kromka *et al.*, 2010). Interestingly, the adhesion, growth, maturation and function of cells on electroactive surfaces are improved even without active stimulation of cells with an electrical current. The underlying mechanism probably includes enhanced adsorption of cell adhesion-mediating proteins, a more advantageous geometrical conformation of these proteins for their accessibility by cell adhesion receptors and facilitation of cellular processes, such as activation of ion channels in the cell membrane, movement of charged molecules inside and outside the cell, upregulated mitochondrial activity and enhanced proteosynthesis (for a review, see Schmidt *et al.*, 1997; Gomez and Schmidt, 2007; Khang *et al.*, 2008; Shi *et al.*, 2008).

Last but not least, the nanocomposite films can significantly increase the mechanical and chemical resistance of the implant surface. In other words, these coatings can prevent the release of ions and material particles from the bulk material. For example, NCD films impermeably covered the underlying silicon substrates, which in their uncoated state acted as highly cytotoxic for cells, and carbon-titanium layers deposited on carbon fiber-reinforced carbon composites significantly attenuated the release of carbon particles from these materials.

Thus, all nanocomposite layers mentioned in this chapter are promising for coating biomaterials designed for hard tissue implantation in order to enhance their integration with the surrounding tissue. The potential use of our nanocomposite films for constructing biosensors and biostimulators (boron-doped NCD films) or micropatterned surfaces for regionally-selective cell adhesion and directed cell growth (fullerene C_{60}/Ti composite films) are other promising applications, which are discussed in this chapter.

2. Hydrocarbon plasma polymers

Plasma polymers (Tab. 1) are materials created from small organic molecules (e.g., methane, n-hexane, acrylic acid, octadiene, tetrafluoroethylene, acetonitrile) by glow discharge, which is usually carried out at low temperatures, low pressures and without solvents. Glow discharge as a source of highly energetic electrons and UV radiation causes ionization, excitation and fragmentation of the monomer molecules employed, creating a number of reactive chemical species, which further participate in crosslinking and formation of quazi-polymeric network. Thus, plasma polymers differ from conventional polymers, i.e., materials with chains containing regularly repeated monomer units, by short and randomly branched chains. Plasma polymers usually exist in the form of thin films which can be deposited on various substrates, such as glass, metals, silicon, polystyrene cell culture dishes etc. These films have a broad spectrum of applications, including use in medicine and various biotechnologies (for a review, see Biederman 2004). For example, plasma polymers have been used for the creation of antibacterial coatings of body implants and other medical devices (Vasilev *et al.*, 2010), surfaces for studying protein adsorption (Zhang and Feng 2007), protein interaction (Menzies *et al.*, 2010), protein separation (Tsai *et al.*, 2003), and also for immobilization of proteins (Kim *et al.*, 2003), e.g., enzymes, such as lysozyme (Thierry *et al.*, 2008) or trypsin (Abbas *et al.*, 2009). Plasma polymer films with immobilized heparin can serve as hemocompatible surfaces preventing blood coagulation (Yang *et al.*, 2010). Another broad and advanced application of plasma polymers is the construction of biosensors (Seyama *et al.*, 2004; Chu *et al.*, 2009).

As for interaction with anchorage-dependent cells, plasma polymer films can be constructed as repulsive or attractive for cells. For example, a plasma polymer of methane was constructed to prevent cell and tissue adhesion on contact lenses (Ho and Yasuda 1988), while plasma polymers of acrylic acid supported the adhesion of epithelial cells on special therapeutic contact lenses designed to transfer these cells onto the denuded rabbit cornea *in vitro* (Deshpande *et al.*, 2009). Acrylic acid-based surfaces were used for the creation of cell sheets for tissue engineering (Majani *et al.*, 2010). Cell-attractive and cell-repulsive plasma polymers have also been combined in order to create micropatterned surfaces for regionally-selective adhesion and directed growth of cells. For example, surfaces containing cell-adhesive microdomains of tetrafluoroethylene, placed on a cell-repulsive background of tetraethylene glycol dimethyl ether, were used for controlling the shape, degree of spreading, assembly of actin cytoskeleton, phenotypic modulation and proliferation of vascular smooth muscle cells (Goessl *et al.*, 2001). Similarly, cell-repellent domains of poly(ethylene) oxide-like plasma-deposited films, combined with cell-adhesive poly-L-lysine domains, were used for controlled adhesion, arrangement and neurite outgrowth of human neural stem cells derived from umbilical cord blood (Ruiz *et al.*, 2008). In our earlier study (Filova *et al.*, 2009), micropatterned surfaces were prepared by the successive plasma polymerization of acrylic acid (AA) and 1,7-octadiene (OD) through a metallic mask on the inner surface of 24-well polystyrene multidishes. A reactor with a cooper radio frequency power source for initiating and sustaining the plasma was used for this procedure. Rat vascular smooth muscle cells (VSMC), bovine endothelial cells (EC), porcine mesenchymal stem cells (MSC) or human skeletal muscle cells (HSKMC) seeded on these surfaces adhered and grew preferentially on hydrophilic strip-like AA domains rather than on hydrophobic OD domains. In addition, both VSMC and EC on AA domains displayed a higher degree of phenotypic maturation. In VSMC, this was manifested by a higher concentration of alpha-actin (i.e., an important marker of VSMC differentiation towards the contractile phenotype), measured per mg of protein in cell homogenates by an enzyme-linked immunosorbent assay. In EC on AA domains, immunofluorescence staining showed more apparent Weibel-Palade bodies, containing the von Willebrand factor. Also MSC growing on AA domains had a better developed beta-actin cytoskeleton, and also contained a higher concentration of beta-actin per mg of protein (Filova *et al.*, 2009).

Our other earlier studies were focused on composites of hydrocarbon plasma polymers with metals, namely titanium or silver (Bacakova *et al.*, 2008b; Grinevich *et al.*, 2009). The benefits of composite materials arise from combining materials with very different properties to produce structural or functional properties not present in any individual component. The type and concentration of the metal component in the hydrocarbon plasma polymers, such as silver, molybdenum, nickel, germanium or titanium, significantly influence the optical, electrical and mechanical properties of these composites. Metal/hard carbon plasma polymer thin films have been a special issue among the carbon-based materials, because of their favorable properties such as high hardness, low friction coefficient, high wear resistance, optical transparency and chemical inertness. The advantageous physicochemical properties of these films are important not only for their industrial applications but also for their use in medicine and in various biotechnologies.

In our earlier study (Grinevich *et al.*, 2009), nanocomposite Ti/hydrocarbon plasma polymer (Ti/ppCH) films were deposited by DC magnetron sputtering of a titanium target in n-hexane, argon or a mixture of these two gases. The resultant films were heterogeneous with inorganic regions of nanometer scale distributed within a plasma polymer matrix. The titanium content was controlled by adjusting the argon/n-hexane ratio in the working gas. In the pure n-hexane atmosphere, the Ti concentration was found to be below 1 atomic %, whereas in pure argon, it reached 20 atomic %, as revealed by Rutherford Backscattering/Elastic Recoil Detection Analysis (RBS/ERDA). A high level of titanium oxidation was detected with TiO_2, sub-stoichiometric titania and titanium carbide composing an inorganic phase of the composite films. In addition, high hydrogen content was detected in films rich with titanium.

As for the interaction with cells, Ti-deficient and Ti-rich films proved to be equally good substrates for adhesion and growth of cultured human osteoblast-like MG 63 cells. In these cells, the population densities on days 1, 3 and 7 after seeding, spreading area on day 1, formation of talin-containing focal adhesion plaques, as well as concentrations of talin and osteocalcin (per mg of protein), were comparable to the parameters obtained in cells on the reference cell culture materials, represented by microscopic glass coverslips or polystyrene dishes. In contrast to our results, the number of rat bone marrow cells in primary or low-passaged cultures on a similar material, i.e., amorphous hydrogenated carbon with 0 to 13 atomic % of Ti, was significantly increased on Ti-containing samples compared to cells on glass (Schroeder *et al.*, 2000). In addition, the activity of alkaline phosphatase, an important marker of osteoblastic cell differentiation, tended to be higher in the cells on films with higher Ti concentrations. This disproportion between our results and results of the study by Schroeder *et al.*, (2000) may be explained by a higher sensitivity of primocultured or low-passaged bone marrow cells to physical and chemical properties of the material surface in comparison with the line MG 63 used in our study, which is tumor-derived, highly passaged and well-adapted to long-term cultivation *in vitro* conditions. Similar higher sensitivity to the roughness, topography and other physical and chemical surface properties were observed in our earlier study in rat vascular smooth muscle cells (VSMC) in low-passaged cultures on carbon fiber-reinforced carbon composites (CFRC) coated by a carbon-titanium layer. The increase in cell adhesion and growth on Ti-containing surfaces was more pronounced in VSMC than in MG 63 cells (Bacakova *et al.*, 2001b).

However, another cell line, namely bovine pulmonary artery endothelial CPAE cells, reacted sensitively to the presence of Ti in the ppCH films. The cell population densities, cell spreading area and also the concentration of von Willebrand factor, a marker of endothelial cell maturation, were significantly higher in CPAE cells on Ti-rich than on Ti-deficient films. On Ti-rich films, these parameters were also higher or similar in comparison with the reference cell culture materials. The main underlying mechanism of this cell response is probably an increased content of oxygen in the Ti-rich films, which is present in the form of TiO, TiO_2 and oxygen-containing chemical functional groups, such as C-O, C=O and O-C=O. The increased oxygen content leads to a higher wettability of the material surface and improved adsorption of cell adhesion-mediating molecules in a flexible form, enabling good accessibility of specific sites on these molecules by cell adhesion receptor (Bacakova *et al.*, 2000, 2001a; for a review, see Bacakova and Svorcik 2008).

Another important factor improving the protein adsorption and cell adhesion is the nanoscale roughness of the material surface, which resembles the nanoarchitecture of the natural extracellular matrix (ECM). However, both Ti-deficient and Ti-rich films in our study (Grinevich *et al.*, 2009) were of similar nano-roughness (root mean square roughness, RMS, was in the range of 1-6 nm), as measured by Atomic Force Microscopy (AFM) on the 1x1 μm images.

Thus, both Ti-deficient and Ti-rich hydrocarbon plasma polymer films could be used for coating bone implants, the Ti-rich film also being effective in enhancing the endothelialization of blood contacting artificial materials (Grinevich *et al.*, 2009).

	Carbon films														
Designation	**1 Plasma-polymer films**	**2 Amorphous carbon films (diamond-like-carbon films / DLC)**							**3 Crystalline carbon films**						
									Diamond films						Graphite films
Thin film / thick film	Thin film	Thin film							Thin film			Thick film (free standing)			Thin film
Doping, additional elements		hydrogen-free			hydrogenated				undoped		doped	undoped	doped	undoped	
				modified			modified								
				with metal			with metal	with non-metal							
Crystal size on the growth side	./.	(amorphous)							1 to 500 nm, nano-crystalline	0.5 to 10 µm, mikro-crystalline	0.1 to 5 µm	(5 µm to) 80 to 500 µm	80 to 500 µm		
Predominating C-C-bond type	sp^2 or sp^3, linear bond	sp^2	sp^3	sp^2	sp^2 or sp^3	sp^3	sp^2	sp^2	sp^3	sp^3	sp^3	sp^3	sp^3	sp^2	
Film No.	**1**	**2.1**	**2.2**	**2.3**	**2.4**	**2.5**	**2.6**	**2.7**	**3.1**	**3.2**	**3.3**	**3.4**	**3.5**	**3.6**	
Designation	**Plasma-polymer film**	**Hydrogen-free amorphous carbon film**	**Tetrahedral hydrogen-free amorphous carbon film**	**Metal-containing hydrogen-free amorphous carbon film**	**Hydrogenated amorphous carbon film**	**Tetrahedral hydrogenated amorphous carbon film**	**Metal-containing hydrogenated amorphous carbon film**	**Modified hydrogenated amorphous carbon film**	**nano-crystalline CVD diamond film**	**micro-crystalline CVD diamond film**	**doped CVD diamond film**	**CVD diamond**	**doped CVD diamond**	**graphite film**	
Recommended abbreviation	./.	a-C	ta-C	a-C:Me	a-C:H	ta-C:H	a-C:H:Me (Me = W, Ti, ...)	a-C:H:X (X = Si, O, N, F, B, ...)	./.	./.	./.	./.	./.	./.	

Table 1. Classification of carbon films by Fraunhofer Institute, © 2009. Electronic Name Index of Carbon Coatings at http://www.ist.fraunhofer.de/english/c-products/tab/complete.html

Hydrocarbon plasma polymers (Bacakova *et al.*, 2008b) or amine plasma polymers (Vasilev *et al.*, 2010) have also been combined with silver in order to create antimicrobial surfaces. Silver atoms can bind bacterial DNA, inactivate bacterial proteins, inhibit a number of important transport processes of bacterial cells and interact with cellular oxidation processes, as well as the respiratory chain. Silver has been used as an important component of bioactive wound dressings, eye drops, urinary, vascular and blood dialysis catheters, artificial vascular grafts and heart valves, devices for bone fixation, orthopaedic and dental implants, bone cements, three-dimensional scaffolds designed for engineering various tissues, e.g., bone, and sponges for the potential construction of bioartificial skin (for a review, see Bacakova *et al.*, 2008b, Vasilev *et al.*, 2010). In comparison with synthetic antibiotics, the advantage of silver is to kill a much wider spectrum of microbes and not to develop resistance. However, the disadvantage is cytotoxicity of silver for mammalian cells, which is based on the inhibitory action of silver on DNA synthesis, damage and loss of cellular proteins, mitochondrial dysfunction, inhibition of adenosine triphosphate synthesis and increased generation of oxygen radicals. Therefore, an ideal silver-loaded material coating should completely inhibit microbial colonization, but allow for the normal adhesion and spreading of mammalian cells. For this purpose, it is necessary to have control over the concentration and release of the silver atoms from the coating.

In our earlier study (Bacakova *et al.*, 2008b), the composite Ag/hydrocarbon plasma polymer films were deposited on microscopic glass slides using an unbalanced planar magnetron with an Ag target (78 mm in diameter) operated in the dc mode in a working gas mixture of n-hexane and argon (pressure 2 Pa, deposition time 10 minutes, magnetron current 0.1 A). The concentration of silver was controlled by the ratio of n-hexane and argon in the working gas mixture. The concentration of silver in the films increased proportionally to the amount of Ar in the working gas mixture, ranging from 0 atomic % to 39 atomic % (i.e., 0, 3, 10, 30 and 39 atomic %), as estimated by RBS/ERDA.

On day 1 after seeding, the CPAE cells adhered to the films with 3 and 10 atomic % of Ag in similar numbers, and by similar cell spreading areas as on the control pure hydrocarbon plasma polymer films and glass slides. On day 3 after seeding, the cell numbers on pure hydrocarbon films, films with 3 atomic % of Ag and control glass coverslips reached similar values, only the cell number on films with 10 atomic % of Ag was significantly lower than on the pure hydrocarbon plasma polymer films. Nevertheless, on day 7 after seeding, the cells on pure hydrocarbon films, films with 10 atomic % of Ag and glass coverslips reached similar population densities, and on the films with 3 atomic % of Ag, this value was even significantly higher than on Ag-free films. The CPAE cells on films with 3 atomic % of Ag developed confluent layers of a cobblestone-like pattern, typical for mature endothelium, and contained well-apparent Weibel-Palade bodies with the von Willebrand factor, a marker of endothelial cell maturation. Moreover, immunofluorescence staining of the cells on films with 3 atomic % of Ag showed numerous and bright dot-like vinculin-containing focal adhesion plaques, which were even more apparent than on the control pure hydrocarbon plasma polymer films and glass (Fig. 1). Similar results were obtained in a rat mast cell line RBL-2H3 exposed to subtoxic concentrations of Ag^+ ions ranging from 10 to 100 μM, which induced degranulation and the release of histamine and leukotrienes via activation of focal adhesion kinase (Suzuki *et al.*, 2001). Adhesion of vascular endothelial cells was also enhanced on polystyrene implanted with Ag^- ions (Sato *et al.*, 1999). However, this effect seemed to be due to the degradation of the polymer by the ion bombardment followed by the polymer oxidation and the increase in its wettability, rather than due to a direct action of Ag^- ions entrapped in the polymer structure. Nevertheless, a moderate surface wettability of our surfaces with 3 atomic % of Ag (static water drop contact angle ~67°) could also

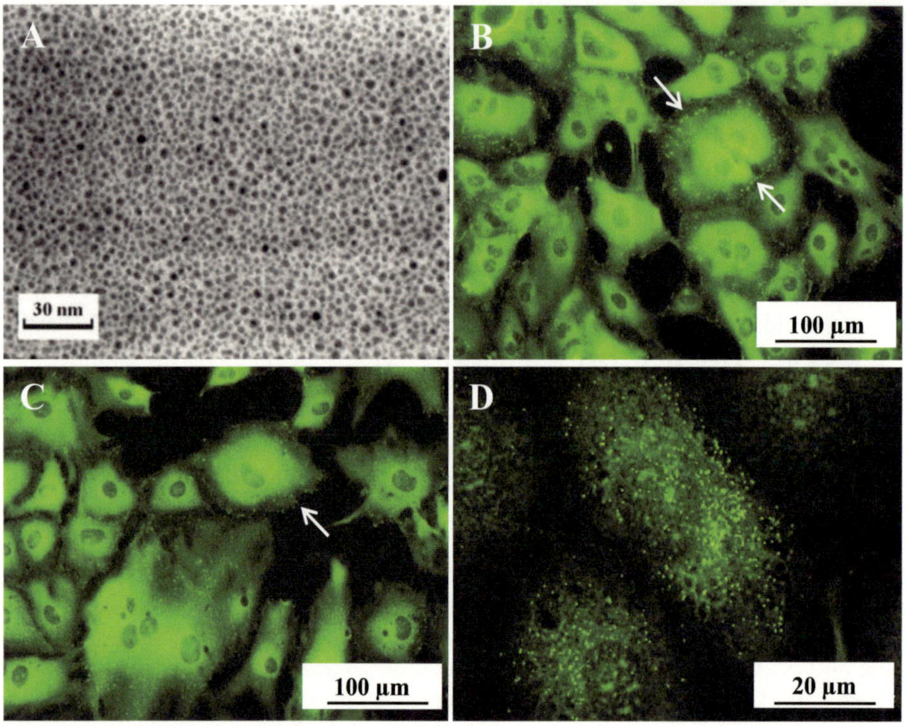

Fig. 1. Morphology of hydrocarbon plasma polymer composite films and their colonization with bovine pulmonary artery endothelial CPAE cells. A: Transmission electron micrograph of a film with 3 at.% of Ag. B, C: Immunofluorescence staining of vinculin, a protein of focal adhesion plaques (arrows), in CPAE cells on day 3 after seeding on films with 3 at.% of Ag (B) and 0 at.% of Ag (C). D: Immunofluorescence staining of von Willebrand factor, a marker of endothelial cell maturation in CPAE cells on day 3 after seeding on a film with 3 at.% of Ag. Olympus IX 51 microscope, DP 70 digital camera, obj. 20x and bar 100 µm (B, C), obj. 100x and bar 20 µm (D).

contribute to the good cell adhesion on these surfaces in our study (Bacakova *et al.,* 2008b). Another supportive factor was the nanoscale roughness of the Ag-containing surfaces. Transmission Electron Microscopy (TEM) revealed that Ag formed dark nanoclusters randomly dispersed in a lighter plasma polymer matrix (Fig. 1), and their diameter increased from 4 nm to 20 nm. However, at higher silver concentrations (30 and 39 atomic %), the supportive effects of the substrate nanostructure on its colonization with CPAE cells was disabled by the cytotoxicity of silver. On these films, the cells adhered at very low initial numbers, did not spread and usually died before day 7 of cultivation. As for the antibacterial effects of the silver-containing films, all these films, including those with the lowest Ag concentration (i.e., 3 atomic %), attenuated the growth of *E. coli.* The number of bacteria in wells (with glass coated with Ag-containing films) did not change significantly during 6 hours after inoculation, whereas the values on pure polystyrene dishes or on pure hydrocarbon polymer films increased by at least one order of magnitude after a 4-hour cultivation (Fig. 2). Similar results were obtained in a study by Vasilev *et al.,* (2010), performed on n-heptylamine plasma polymer (HApp) loaded with silver nanoparticles (at a concentration of 500 ng/cm^2) and coated with additional HApp films of various thicknesses, controlling the rate of release of silver atoms from the material. These films allowed for the

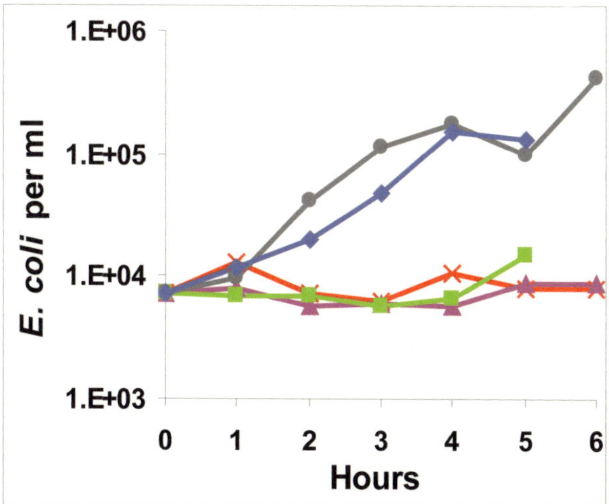

Fig. 2. Growth of *E. coli* on polystyrene dishes (blue), hydrocarbon plasma polymer films without Ag (gray), with 3 at.% Ag (green), with 30 at.% Ag (pink) or with 39 at.% Ag (red).

adhesion and growth of human SaOs-2 osteoblastic cells, while the colonization of the films with *Staphylococcus epidermidis* was markedly attenuated and even disabled.

However, it should be taken into account that for cultivation of mammalian cells, such as endothelial cells or osteoblasts, media supplemented with foetal bovine serum are used, while bacteria are usually incubated in a medium without serum. For example, in our study (Bacakova *et al.*, 2008b) an LB medium, containing bacto-tryptone (10 g/l), bacto-yeast extract (5g/l) and NaCl (10g/l) in H_2O was used. It has been reported that the serum supplement of the culture media has protective effects against the cytotoxic action of silver (Hidalgo *et al.*, 1998; Sun *et al.*, 2006). Thus, the promising results obtained in both studies by Bacakova *et al.* (2008b) and Vasilev *et al.* (2010) should be further investigated, particularly using the culture media of the same chemical composition for both mammalian and bacterial cells.

3. Amorphous carbon

Amorphous carbon, also referred to as diamond-like carbon (DLC; Tab. 1), possesses a number of favourable properties, such as high hardness, a low friction coefficient, chemical inertness and high corrosion resistance, which is due to its particular structure, i.e., cohabitation of the sp^2 and sp^3 phases (for a review, see Chai *et al.*, 2008). These properties make DLC, especially in the form of thin films, attractive for various biomedical applications, particularly for hard tissue surgery, such as orthopaedics and stomatology. For example, DLC has been deposited on the articular surfaces of artificial joint replacements (Santavirta 2003), dental prostheses or orthodontic archwires (Kobayashi *et al.*, 2007) in order to improve the resistance of these devices against wear, corrosion, debris formation and release of metallic ions, which can act as cytotoxic, immunogenic or even carcinogenic materials. In addition, due to their bioinertness, DLC films were also found to be suitable for cardiovascular applications, mainly as protective coatings of blood-contacting devices (intravascular stents, mechanical heart valves, pumps) in order to prevent hemocoagulation, thrombosis and inflammatory reaction on these surfaces (Monties *et al.*, 1994; Tran *et al.*, 1999; for a review, see Roy and Lee 2007).

However, DLC films can also act as bioactive substrates for the adhesion and subsequent growth of osteogenic cells, and thus they could be used as a protective coating for the bone-anchoring parts of joint prostheses, in which strong and quick integration with the surrounding bone tissue is required. In a study by Chai *et al.* (2008), DLC films were deposited on silicon substrates by Plasma-Enhanced Chemical Vapor Deposition (PECVD) using methane (CH_4) as a precursor gas. In order to improve their mechanical properties, some DLC films were deuterated, i.e., deposited using deuterated methane (CD_4) as a precursor gas. All surface treatments were performed under two different self-bias voltages (V_{sb}): -400 and -600V. DLC coatings prepared under the V_{sb} of -600V in pure methane (600CH_4) or in pure deuterated methane (600CD_4), significantly increased the proliferation rate of mouse osteogenic MC3T3-E1 cells compared to a pure Si substrate. Scanning Electron Microscopy (SEM) observations confirmed that the optimal cell adhesion behavior also occurred on the surface of the 600CH_4 and 600CD_4 groups of DLC, where the cells showed increased amounts of filopodia and microvilli, allowing a higher cell-environment exchange. Surprisingly, the mentioned DLC surfaces were relatively highly hydrophobic, having total surface energy and its polar component significantly lower than the values obtained on bare silicon substrates. It is generally accepted that the cell proliferation is optimal on moderately hydrophilic and polar surfaces, which enable adsorption of cell adhesion-mediating extracellular matrix molecules (such as vitronectin, fibronectin, collagen, laminin) in appropriate geometrical conformations, allowing accessibility of specific sites on these molecules to cell adhesion receptors (Chai *et al.*, 2008; Coelho *et al.*, 2010). In accordance with this proposition, the number and spreading of MG 63 cells was higher on micropatterned silicon substrates (grooves of 4, 8 and 10 μm in width) and coated with Ti, which were more hydrophilic and of a higher surface energy than the corresponding micropatterned substrates coated with DLC (Ismail *et al.*, 2007).

Also the addition of Ti into DLC films enhanced their attractiveness for cell colonization. In our earlier study (Bacakova *et al.*, 2001b), an amorphous carbon film enriched with Ti was used for coating carbon fiber-reinforced carbon composites (CFRC), i.e., a material which has been considered as promising for the construction of bone and dental implants. However, the surface roughness of CFRC is relatively high to enable optimal adhesion and growth of cells. As revealed by SEM, this was due to carbon fibers prominent over the carbon matrix. Another important problem is a tendency of CFRC to release particles, which is due to the relative brittle nature of the carbon matrix as well as carbon fibers. A possible solution of these problems may consist in covering the material surface with a mechanically resistant and highly biocompatible layer.

Therefore, CFRC were covered with a carbon-titanium layer, prepared by the PECVD method using unbalanced d.c. planar magnetron sputtering (2 x 4 kW, substrate temperature about 260°C) of a carbon poisoned titanium target in an acetylene-argon gas mixture at a total pressure of roughly 4 x 10^{-3} mbar. First, a metallic film (about 1 μm in thickness) was deposited by sputtering the metal target in pure argon and then C_2H_2 gas was added successively, giving a C:Ti film of at least 3.3 μm thickness. In this way, a layer consisting of a mixture of carbon and titanium atoms was created with an amorphous or very fine crystalline ("nanocrystalline") structure and with a gradually decreasing content of metal atoms from substrate to the surface. Clearly, the layer (and especially the layer surface) contains a large number of free bonds. These bonds cause strong adhesion of the layer to a substrate with a chemically very active surface.

The CFRC were then seeded with human osteoblast-like cells of the MG 63 line (passage 100) and vascular smooth muscle cells (i.e., another cell type present in the bone tissue), isolated from the rat aorta (passage 9). On day 1 after seeding, both cell types adhered at higher numbers. From days 1 to 4 the cells exhibited shorter population doubling times. As a result, on day 4 after seeding, the cells attained higher population densities, volume and protein content. Since the surface roughness of CFRC was not significantly changed by the coating with the carbon-titanium layer, the beneficial effects of this coating on cell colonization could be explained by spontaneous oxidation of Ti and C in the cell culture system, including the formation of titanium oxides and oxygen-containing chemical functional groups. These features are known to enhance cell adhesion and growth (He *et al.*, 2008; Sawase *et al.*, 2008) and attract phosphate and calcium ions (Toworfe *et al.*, 2006; Rakngarm *et al.*, 2008), which facilitate bone tissue formation and bonding between the artificial implant and bone. Interestingly, the improvement of cell colonization by the C:Ti film was more pronounced in VSMC than in MG 63 cells, which was probably due to a different sensitivity of low-passaged cells and cell line to the physicochemical properties of the material surface. In addition, the release of carbon particles from the CFRC composites was significantly decreased (24 times) in the coated samples, and this effect was further enhanced by polishing the CFRC with colloidal SiO_2 (the particle release was decreased 42 times in polished + covered samples). Moreover, polishing the CFRC markedly improved the cell spreading on the material surface (Fig. 3). These results show that both polishing and carbon-titanium covering significantly improve the biocompatibility of CFRC composites *in vitro*, especially when these two modifications are combined (Bacakova *et al.*, 2001b).

Similar positive results were obtained in a study by Schroeder *et al.* (2000), where the hardness and inertness of a-C:H films were also combined with the biological acceptance of titanium. Different amounts of titanium of various concentrations, ranging from 7 to 24 atomic %, were incorporated into a-C:H films by a combined radio frequency (rf) and magnetron sputtering set-up. The X-ray Photoelectron Spectroscopy (XPS) of air-exposed a-C:H/Ti films revealed that the films were composed of TiO_2 and TiC embedded in and connected to an a-C:H matrix. Cell culture tests using primary bone marrow cells (BMC), derived from adult rat femora, revealed that the relative cell number, expressed as total DNA content per culture sample, was significantly increased on a-C:H with 10 and 13 atomic % of Ti compared with BMC cultures grown on glass. On all tested materials, i.e., a-C:H, a-C:H/Ti and control glass coverslips, BMC were able to differentiate to osteoblasts and osteoclasts, as revealed by measurements of the activity of alkaline phosphatase (ALP) and tartrate resistant acid phosphatase (TRAP), respectively. No significant differences in osteoblast differentiation on a-C:H, a-C:H/Ti and control glass coverslips were detected. However, the TRAP activity of osteoclast was lower on a-C:H with Ti than on the pure a-C:H. This decrease of osteoclast activity can be considered as a favourable result, i.e., an indicator for reduced inflammatory cell development (since osteoclast-like cells are derived from the monocyte family). Even in the form of particles added to BMC cultures, a-C:H and a-C:H/Ti did not significantly stimulate the osteoclast-related TRAP activity (Bruinink *et al.*, 2005). In addition, the Inductively Coupled Plasma Optical Emission Spectrometry (ICP-OES) measurements of culture media exposed to titanium containing a-C:H films showed that no titanium was dissolved from the films after one week under cell culture conditions (the detection limit for the ICP-OES method is 0.05 mg l^{-1}, i.e., 1.043 μmol l^{-1}; (Schroeder *et al.*, 2000). Thus, it can be concluded that a-C:H/Ti could be a valuable coating for bone implants by supporting bone cell proliferation, reducing osteoclast-like cell activation and bone resorption, and providing chemical stability to the implant surface.

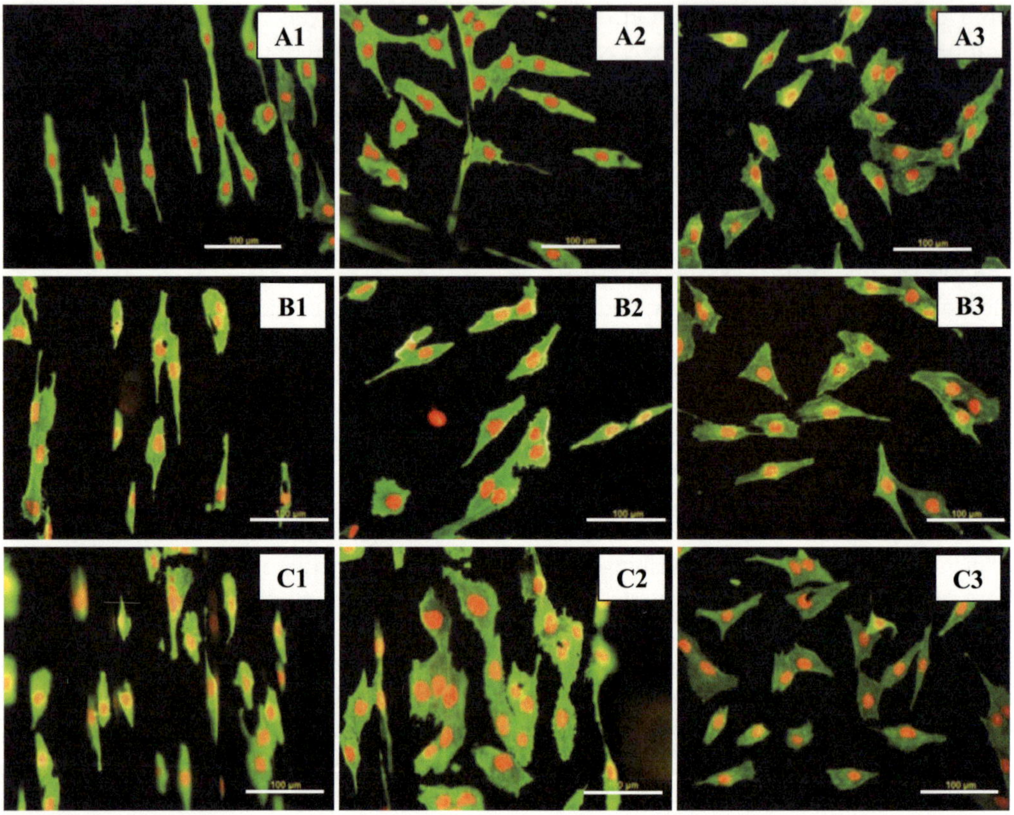

Fig. 3. Morphology of human osteoblast-like MG 63 cells on day 1 after seeding on control untreated CFRC composites (**A1**), CFRC composites ground by metallographic paper of 4000 grade (**A2**), CFRC composites ground with metallographic paper of 4000 grade, covered with pyrolytic graphite, ground by metallographic paper of 4000 grade and polished by diamond paste of 3/2 grade (**A3**). **B1, B2, B3** represent A1, A2, A3 covered with a-C:H, and **C1, C2, C3** are A1, A2, A3 covered with Ti-C:H layer. The cells were visualized by immunofluorescence staining of a cytoskeletal protein beta-actin, the nuclei were counterstained with propidium iodide. Olympus IX 51 microscope, obj. 20x, DP 70 digital camera, bar = 100 μm.

Also other C- and Ti-containing multicomponent coatings for load-bearing medical applications have been developed, such as the systems Ti-Ca-C-O-(N), Ti-Zr-C-O-(N), Ti-Si-Zr-O-(N) and Ti-Nb-C-(N). These films were deposited on various substrates, such as single crystal silicon (100), stainless steel, titanium alloy, glass coverslips and Teflon, by DC magnetron sputtering (atmosphere of Ar or mixture of Ar and N) from composite targets, which were manufactured by means of self-propagating high-temperature synthesis (Shtansky *et al.*, 2005). The films showed high hardness in the range of 30-37 GPa, significantly reduced Young's Modulus, low friction coefficient down to 0.1-0.2 and low wear rate in comparison with conventional magnetron-sputtered TiC and TiN films. The adhesion and proliferation of cultured cells (i.e., rat liver epithelial IAR-2 cells and Rat-1 embryonic fibroblasts), measured by cell numbers, were similar on all films and uncoated substrata. Only the cells cultivated on the Ti-Zr-C-O-(N) and Ti-Nb-C-(N) films disturbed their actin cytoskeleton, which was manifested by the disappearance of circumferential actin bundles in IAR-2 cells and irregular short and thin actin bundles in the cytoplasma of Rat-1 fibroblasts. Nevertheless, the assessment of the population of cells covering the Teflon plates coated with the Ti-Ca-C-O-(N) and Ti-Zr-C-O-(N) films after 16 weeks of subcutaneous implantation into mice revealed a high biocompatibility level of tested films and the absence of inflammatory reactions (Shtansky *et al.*, 2005).

Other related materials promising for protective, tribological and load-bearing coatings of medical devices include nanostructured and nanocomposite binary (TiC-a:C), quaternary (Ti-B-C-N) and quinternary (Ti-Si-B-C-N) multicomponent films, prepared using unbalanced magnetron sputtering and closed field unbalanced magnetron sputtering from both elemental and composite targets (Lin *et al.*, 2009). Other nanocomposite films were based on nanocrystalline phases of metals and ceramics (Mo, Si, Al, Ti), embedded in an amorphous carbon matrix, and carbide-forming metal/carbon (Me/C) composite films (Me = Mo, W or Ti) deposited by a hybrid technique using Physical Vapor Deposition (PVD; magnetron sputtering) and PECVD, by the use of CH_4 gas (Teixeira *et al.*, 2009). Ti-TiC-TiC/DLC gradient films have been prepared by plasma immersion ion implantation and deposition (PIIID) combined with PECVD (Zheng *et al.*, 2008). All the mentioned films were usually characterized by excellent mechanical properties, particularly high hardness, low friction coefficient, high wear-resistance, improved adherence to the substrate and fracture toughness, corrosion resistance, and were constructed in order to overcome the limitations of pure DLC films.

4. Nanodiamond

From all nano-sized carbon allotropes, diamond (Tab. 1) can be considered as the most advantageous material for advanced biomedical and biosensoric applications, which is mainly due to the absence of its cytotoxicity, immunogenicity and other adverse reactions (Schrand *et al.*, 2007; Huang *et al.*, 2008; Grausova *et al.*, 2009a,b; Rezek *et al.*, 2009; for a review, see Bacakova *et al.*, 2008a). Other remarkable properties of nanodiamond, enabling its application in biotechnologies and medicine (particularly in hard tissue surgery), are high hardness, a low friction, and also high chemical, thermal and wear resistance. In our earlier studies and in studies by other authors, nanodiamond has proven itself as an excellent substrate for the adhesion, growth, metabolic activity and phenotypic maturation of several cell types *in vitro*, including osteogenic cells (Schrand *et al.*, 2007; Amaral *et al.*, 2008; Grausova 2008, 2009a,b; Kalbacova *et al.*, 2009). The beneficial effects of the nanodiamond layers on cell colonization lie in their nanoscale surface roughness (i.e., the size of irregularities less than 100 nm), which support the adsorption of cell-adhesion-mediating molecules in the appropriate geometrical conformation, enabling exposure of bioactive sites in these molecules (e.g., specific amino acid sequences such as RGD, KRSR etc.) to the cell adhesion receptors. In contrast, the adhesion of human osteoblasts was attenuated on microcrystalline diamond layers (Yang *et al.*, 2009).

The nanodiamond layers also acted as suitable platforms for the adhesion of mammalian neurons, neurite outgrowth and the formation of excitable neuronal networks *in vitro*, which have great potential for chronic medical implants (Thalhammer *et al.*, 2010). In studies *in vivo*, diamond layers, deposited on Ti6Al4V probes and implanted into a rabbit femur, showed very high bonding strength to the metal base as well as to the surrounding bone tissue, and prevented material corrosion (Rupprecht *et al.*, 2002). Nanocrystalline and multilayer diamond thin films were also used for coating the heads and cups of an artificial temporomanibular joint made of Ti6Al4V alloy (Papo *et al.*, 2004). Another advantageous property of diamond and/or nanodiamond is its ability to bind various biological molecules, which can be utilized for sensing, detecting, separating and purifying these molecules (Bondar *et al.*, 2004; Kong *et al.*, 2005; Rezek *et al.*, 2007).

Nanocrystalline diamond (NCD) films used in our earlier studies also proved to be an impermeable coating preventing adverse effects of the underlying substrate on cell colonization. For deposition of NCD films (Kromka *et al.*, 2008), we used (100) oriented silicon substrates. Uncoated substrates acted cytotoxically on human osteoblast-like MG 63 cells and bovine pulmonary artery endothelial CPAE cells. On day 1 after seeding, both cell types adhered to silicon substrates in significantly lower numbers than on NCD films, control polystyrene dishes and microscopic glass coverslips. On day 5 after seeding, no MG 63 cells and few CPAE cells were detected on these substrates, while the cells on NCD films grew to similar cell population densities as on the reference polystyrene dishes and glass (Grausova *et al.*, 2008).

An interesting issue is doping of NCD films with boron. This doping renders the NCD films electroconductive (Gajewski *et al.*, 2009). Boron-doped NCD films have been applied in electronics and sensorics, e.g., for the construction of sensors for DNA hybridization (Nebel *et al.*, 2007), bacteria (Majid *et al.*, 2008) or glucose (Zhao *et al.*, 2009). However, little is known about the influence of boron-doped NCD films on the adhesion, growth, differentiation and function of osteogenic cells, and thus on their potential use as a substrate for bone tissue regeneration.

In our earlier studies, the nanocrystalline diamond films (NCD) were deposited on glass (Kopecek *et al.*, 2008) or silicon substrates (Kromka *et al.*, 2010) by a microwave plasma-enhanced CVD process. Boron doping was achieved by adding trimethylboron (TMB) to the gas mixture, and the B:C ratio varied from 100 to 6700 ppm. The study by Kromka *et al.* (2010) revealed that the increase in the number of human osteoblast-like MG 63 cells in 7-day-old cultures was most apparent on NCD doped with 133 and 1000 ppm of B than on films with 6700 ppm of B. At the same time, the cells on NCD with 6700 ppm of B developed large and numerous focal adhesion plaques containing talin (Fig. 4), and contained the highest concentration of focal adhesion protein vinculin. Both features are indicators of firm cell-substrate adhesion, which was probably enabled by the flattest appearance of the nanodiamond crystals on films doped with 6700 ppm of B. It has been reported that cell proliferation activity is the highest at intermediate cell-substrate adhesion strength (for a review, see Bacakova *et al.*, 2004; Bacakova and Svorcik 2008). If this strength is relatively high, the cells slow down their proliferation activity and enter the differentiation program. In accordance with this, the cells on NCD films doped with 6700 ppm of B contained one of the highest concentrations of osteocalcin (i.e., a calcium-binding non-collagenous ECM protein and an important marker of osteogenic cell differentiation), which was significantly higher than the value on non-doped NCD and on the control polystyrene cell culture dishes. Similar results were also obtained in the study by Kopecek *et al.* (2008) performed on NCD films doped with 3000 ppm of B, where MG 63 cells were less numerous on boron-doped NCD films than on non-doped NCD films, but they were more intensely immunofluorescently stained against osteocalcin.

The beneficial effects of boron-doping on cell colonization of NCD films can be attributed to the electroactivity of these films. The electrical resistivity of the films decreased from >100 MΩ (non-doped films) to 100, 2, and 0.4 kΩ in films doped with 133, 1000 and 6700 ppm of

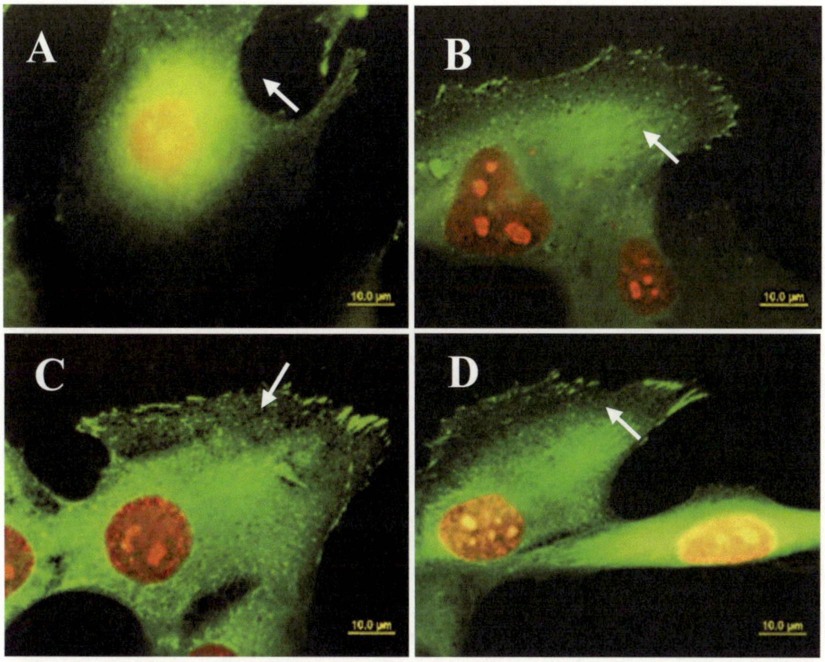

Fig. 4. Immunofluorescence staining of talin in MG 63 cells on day 3 after seeding on non-doped NCD (**A**), NCD films doped with boron in concentrations of 133 ppm (**B**), 1000 ppm (**C**) and 6700 ppm (**D**). The cell nuclei were counterstained with propidium iodide. Olympus IX 51 epifluorescence microscope, DP 70 digital camera, obj. 100x, bar = 10 μm.

B. In addition, Kelvin Force Microscopy revealed that the surface potential of NCD films increased proportionally to the boron concentration, i.e., from +19 ± 10 mV in non-doped films to +27 ± 26 mV, +112 ± 19 mV and +97 ± 34 mV in films with 133, 1000 and 6700 ppm of B, respectively (Kromka *et al.*, 2010). It is known that electroactive surfaces stimulate the adhesion, growth and specific functions of cells even without their active stimulation with electrical current. For example, Schwann cells cultured on electrically conductive melanin films accelerated their proliferation, and rat pheochromocytoma PC12 cells enhanced extension of their neurites (Bettinger *et al.*, 2009). Similarly, composite nanofibers made of electrically conductive polyaniline blended with poly(L-lactide-co-epsilon-caprolactone) enhanced the adhesion and proliferation of C2C12 murine skeletal muscle myoblasts, as well as their differentiation towards myotubes (Jeong *et al.*, 2008; Jun *et al.*, 2009).

The mechanisms of the positive effects of electroactive materials on cell colonization and function (which can be further enhanced by additional electrical stimulation of cells through the materials) have not yet been fully elucidated and systemized. These mechanisms probably include enhanced adsorption of cell adhesion-mediating ECM proteins from biological environments, a more advantageous geometrical conformation of these proteins for their accessibility by cell adhesion receptors, redistribution of cell membrane growth factors and adhesion receptors or cytoskeletal proteins such as actin, activation of ion channels in the cell membrane, followed by cell depolarization, hyperpolarization or generation of action potential, movement of charged molecules inside and outside the cell, upregulated mitochondrial activity and enhanced protein synthesis (for a review, see (Schmidt *et al.*, 1997; Gomez *et al.*, 2007; Khang *et al.*, 2008; Shi *et al.*, 2008).

Another factor contributing to better performance of MG 63 cells on B-doped films could be the slightly increased RMS surface roughness of these films (RMS from 25 to 30 nm versus 19 nm on non-doped films; Kromka *et al.*, 2010). Similarly, in a study by Kalbacova *et al.* (2007), the metabolic activity of SaOs-2 osteoblasts, measured by the activity of dehydrogenases in these cells, increased with increasing surface roughness of O-terminated NCD films (RMS from 11 to 39 nm). Direct action of boron on the cell metabolism is less probable, because boron atoms are strongly bound in the NCD lattice or around the grain boundaries, and their release is very limited, even at very high temperatures combined with a vacuum and a strong electrical field, i.e., conditions not occurring in biological environments (for a review, see Kopecek *et al.*, 2008).

All these results indicate that the NCD films provided very good supports for colonization with osteogenic cells. The potential of these films for bone tissue regeneration can be further enhanced by boron-doping, which can be explained by the electroactivity (i.e., electrical conductivity and changes of surface electrical potential) of boron-doped NCD films.

Boron-doped diamond films have also been used for the construction of patterned surfaces for regionally-selective adhesion and guided growth of cells. In a recent study by Marcon *et al.* (2010), the boron-doped polycrystalline diamond was prepared from a $B_2H_6/CH_4/H_2$ source gas mixture by Hot Filament-Assisted Chemical Vapour Deposition (HFCVD). These substrates were then patterned with stripe-like hydrophilic microdomains with oxygen containing chemical functional groups or amine groups (water drop contact angle of 4° and 53°, respectively) and hydrophobic microdomains terminated with hydrogen or methyl-, trifluoromethyl- and vinyl- groups (contact angle from 92° to 114°). Human osteosarcoma U2OS and mouse fibroblast L929 cells preferentially colonized the hydrophilic domains, forming confluent arrays with distinguishable edges separating the hydrophobic alkyl regions. Similar results were obtained with human osteosarcoma SaOs-2 cells in cultures on NCD films patterned with hydrophilic O-terminated and hydrophobic H-terminated domains (Kalbacova *et al.*, 2008). The micropatterned diamond-based surfaces have a great potential for the development of biosensors, biostimulators, construction of cell microarrays for advanced genomics and proteomics and also for tissue engineering. In addition, the size and shape of the adhesive domains can be utilized for regulation of the cell spreading, which has a strong influence on further cell behaviour, e.g., switching between apoptosis and proliferation or proliferation and differentiation programs (for a review, see Bacakova *et al.*, 2004; Bacakova and Svorcik 2008).

5. Graphite

Graphite is one of the most common allotropes of carbon (named by Abraham Gottlob Werner in 1789 after the Greek word graphein, i.e., "to draw/write", for its use in pencils). Graphite is the most stable form of carbon under standard conditions; it is an electrical conductor, and can be used, for instance, as the material in the electrodes of an electrical discharge. However, despite its electrical conductivity, which is usually associated with the stimulatory effects on cell colonization and functioning, unmodified graphite is rather bioinert, i.e., less adhesive for cells (Watari *et al.*, 2009). It is due to a relatively low ability of graphite to adsorb cell adhesion-mediating proteins from the serum supplement of the culture medium (Aoki *et al.*, 2007; Li *et al.*, 2009a) and also bone morphogenetic proteins (BMP), such as BMP-2, a factor promoting the osteogenic cell differentiation (Li *et al.*, 2009b). As a result, graphite compacts allowed the attachment of mouse myoblastic C2C12 cells in significantly lower numbers compared to carbon nanotube compacts, which showed a significantly high capability to adsorb proteins. The C2C12 cells then proliferated more slowly, and contained a lower amount of alkaline phosphatase and total protein (Li *et al.,* 2009a). A similar response was found in human osteosarcoma SaOs-2 cells, which attached and proliferated much worse on graphite than on single- and multi-walled carbon nanotube films (Aoki *et al.,* 2007). In addition, SaOs-2 cells cultured on graphite showed a lower expression of osteonectin, osteopontin and osteocalcin, i.e., markers of osteogenic cell differentiation, and a lower content of alkaline phosphatase and total protein after pre-soaking both graphite and nanotube compacts in the culture medium with BMP-2 (Li *et al.,* 2009b).

In our earlier studies, graphite was prepared by pyrolysis, i.e., by a process defined as thermochemical decomposition of organic material at elevated temperatures (above 430 °C) in the absence of oxygen. However, in practice it is not possible to achieve a completely oxygen-free atmosphere. Because some oxygen is present in any pyrolysis system, a small amount of oxidation occurs. In our experiments, carbon fiber-reinforced carbon composites (CFRC) were coated with pyrolytic graphite by decomposition of butane in a mixture with N_2 at a pressure of 4 Pa, a temperature of 1900 °C and for 300–400 minutes. This coating increased the number of human osteoblast-like MG 63 cells on days 1 and 3 after seeding, and shortened the cell population doubling time compared to uncoated composites (Stary *et al.,* 2002). These results could be attributed to the presence of oxygen-containing chemical functional groups (e.g., hydroxyl, carbonyl, carboxyl) on the pyrolytic graphite films (Stary *et al.,* 2003). It is known that these groups promote cell adhesion and growth by increasing the polarity, and wettability of the material and thus its attractiveness for adsorption of cell-adhesion mediating molecules (Bacakova *et al.,* 2001a). At the same time, the surface roughness and topography, measured by the parameters R_a (departures of the roughness profile from the mean line), R_q (root mean square deviation of the assessed profile) and S (the mean spacing of the adjacent local peaks), was not significantly changed by coating CFRC by pyrolytic graphite (Stary *et al.,* 2002, 2003).

Graphite has also been used in three-dimensional systems as a component of various composite materials used for construction of bone implants. For example, carbon/graphite fibers were used for reinforcing poly(methylacrylate) implants designed for reconstruction of the maxilla damaged by malignant tumors (Ekstrand and Hirsch 2008). Graphite fibers were also applied for reinforcing of polyetheretherketone (PEEK) for construction of composite hip joint prostheses (Yildiz *et al.,* 1998a,b).

Unlike some other carbon allotropes, namely fullerenes and nanotubes, graphite is not cytotoxic. The cell viability tests, performed on human epithelial L132 cells (colony-forming method) and mouse osteoblast-like MC3T3-E1 cells (staining with Blue Alamar dye) showed excellent cytocompatibility of graphite powder (Hornez *et al.,* 2007). Only pyrolytic carbon (i.e., carbon more or less similar to graphite prepared by pyrolysis, which has lost its lubrication properties due to a large number of crystallographic defects), showed some tendency to induce inflammatory or hyperplastic reactions. Coating polyethylene terephthalate (PET) with pyrolytic carbon induced expression of mRNAs specific for platelet-derived growth factors (PDGF-A, PDGF-B) and transforming growth factors (TGF-β1 and TGF-β2) in human umbilical vein endothelial cells (HUVEC) cultured on this substrate, but did not change expression of mRNA for interleukin-6 (Cenni *et al.,* 2000). As determined by flow cytometry, pyrolytic carbon coating on PET did not increase the percentage of cells positive for platelet endothelial cell adhesion molecule-1 (PECAM-1), endothelial leukocyte adhesion molecule-1 (ELAM-1), intercellular adhesion molecule-1

(ICAM-1) and vascular cell adhesion molecule-1 (VCAM-1), nor the intensity of fluorescence of these molecules (Cenni *et al.*, 1995). However, in experiments *in vivo*, pyrolytic carbon and pyrolytic graphite/silicon-carbide, implanted into rabbit mandibles, induced fibrous capsule formation and infiltration with multinucleated phagocytic cells (Maropis *et al.*, 1977). On the other hand, similarly as the amorphous hydrogenated carbon, pyrolytic carbon (particularly in the form of so-called Low Temperature Isotropic pyrolytic carbon, LTI) has been used as a coating for commercially available blood contacting devices, such as artificial heart valves (Kwok *et al.*, 2004; Jackson *et al.*, 2006), in order to prevent hemocoagulation and thrombus formation on these devices, although its hemocompatibility was not ideal. LTI has been also used in orthopedic applications, namely for construction of joint replacements, because it has been reported to reduce the cartilage wear (Bernasek *et al.*, 2009). In addition, when deposited on CFRC composites, pyrolytic carbon supported proliferation of embryonal human lung fibroblasts of the line LEP, particularly if the composites were polished after coating (Pesakova *et al.*, 2000). Also in our studies performed on CFRC, coating with pyrolytic graphite combined with polishing the material surface increased spreading and subsequent growth of human osteoblast-like MG 63 cells (Stary *et al.*, 2002, 2003; Figs. 3, 5)

6. Fullerenes

Fullerenes were discovered, prepared and systemized by Kroto *et al.* (1985). These spheroidal molecules are made exclusively of carbon atoms (e.g., C_{60}, C_{70}) and display a diverse range of biological activity. Their unique hollow cage-like shape and structural analogy with clathrin-coated vesicles in cells support the idea of the potential use of fullerenes as drug or gene delivery agents. For example, a fullerene derivative, namely bisphosphonate fullerene $C_{60}(OH)_{16}$AMBP, i.e., (4,4-bisphosphono-2-(polyhydroxyl-1,2-dihydro-1,2-methanfullerene$_{60}$-61-carboxamido) butyric acid) has been shown to have a strong affinity for the bone tissue, which is due to binding sites of this compound for hydroxyapatite. Thus, this bone-vectored fullerene derivative could be used as a carrier for radionuclides or other drugs for selective therapy of bone tumors (Gonzales *et al.*, 2002).

Other biological activities of fullerenes arise from their reactivity, due to the presence of double bonds and bending of sp^2-hybridized carbon atoms, which produces angle strain. Fullerenes can act either as acceptors or donors of electrons. When irradiated with ultraviolet or visible light, fullerenes can convert molecular oxygen into highly reactive singlet oxygen. Thus, they have the potential to inflict photodynamic damage on biological systems, including damage to cellular membranes, inhibition of various enzymes or DNA cleavage. This harmful effect can be exploited for photodynamic therapy against tumors (Liu and Tabata 2010), viruses including HIV-1 (Marchesan *et al.*, 2005) and broad spectrum of bacteria and fungi (Huang *et al.*, 2010). On the other hand, C_{60} is considered to be the world's most efficient radical scavenger. This is due to the relatively large number of conjugated double bonds in the fullerene molecule, which can be attacked by radical species. Thus, fullerenes would be suitable for applications in quenching oxygen radicals, and thus preventing inflammatory and allergic reactions (Dellinger *et al.*, 2009) and damage of various tissues and organs, including the lung (Chen *et al.*, 2004), blood vessels (Maeda *et al.*, 2008) and brain (Tykhomyrov *et al.*, 2008; Lao *et al.*, 2009). Fullerenes also protected epithelial cells *in vitro* from anoikis, i.e., apoptosis due to adhesion deprivation, by a mechanism supporting the formation of focal adhesion plaques, assembly of the actin cytoskeleton and cell spreading, which was also attributed to the antioxidative action of fullerenes (Straface *et al.*, 1999).

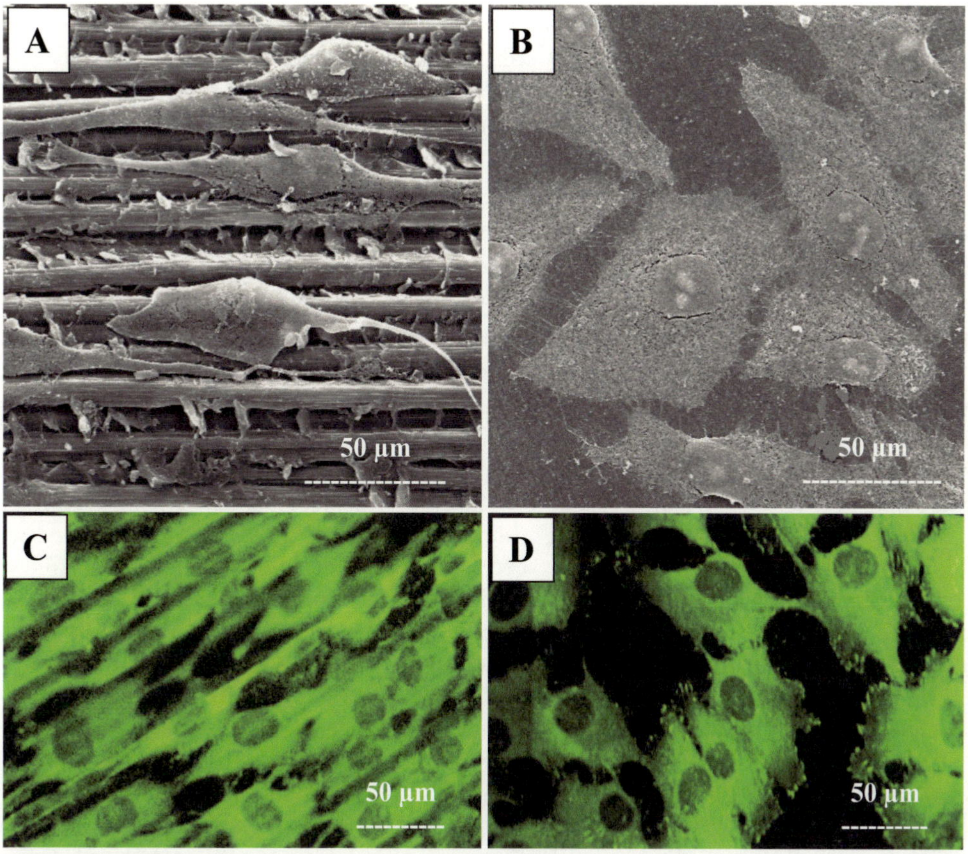

Fig. 5. Morphology of human osteoblast-like MG 63 cells on pristine CFRC composites (**A, C**) and composites ground with metallographic paper, coated with pyrolytic graphite, again ground with metallographic paper and polished with diamond paste 3/2 (**B, D**). Note that on pristine unmodified composites, the cells are spindle-shaped, arranged in parallel to carbon fibers prominent on the surface (**A**), and do not contain distinguishable vinculin-containing focal adhesion plaques (**C**), while on modified composites, the cells are well-spread, polygonal (**B**) and with clearly developed vinculin-containing focal adhesion plaques (**D**). **A, B**: scanning electron microscope JSM5410, JEOL, Japan, day 1 after seeding; **C, D**: Immunofluorescence of vinculin, an integrin-associated focal adhesion protein, in cells on day 4 after seeding, Bio-Rad MRC600 confocal laser scanning microscope, oil immersion 60× objective, (numerical aperture = 1.4), excitation wavelength 488 nm.

Recently, several derivatives of fullerenes C_{60} were synthesized, e.g., glutathione C_{60}, beta-alanine C_{60}, cystine C_{60} or folacin C_{60} (Hu *et al.*, 2007a,b,c; 2010), and were proven to have a protective activity against oxidative stress in rat pheochromocytoma PC12 cells, induced by treatment of these cells with hydrogen peroxide and leading to apoptotic death of unprotected cells.

In addition, fullerenes emit photoluminescence which could be utilized in advanced imaging technologies for diagnostics of various diseases, e.g., cancer (Levi *et al.*, 2006). For a more detailed review of biological effect of fullerenes and their derivatives, see our earlier studies (Bacakova *et al.*, 2008a, Grausova *et al.*, 2009c).

Our earlier studies were focused on fullerenes deposited in the form of thin films for cell cultivation, because relatively little is still known about the behaviour of cells when fullerenes act as their growth supports. For this purpose, fullerenes C_{60} (purity 99.5 %, SES Research, USA) were deposited onto microscopic glass coverslips by evaporation of C_{60} in the Univex-300 vacuum system (Leybold, Germany) in the following conditions: room temperature of the substrates, C_{60} deposition rate $\preccurlyeq 1$ Å/s, temperature of C_{60} evaporation in the Knudsen cells about 450° C. The thickness of the layers increased proportionally to the temperature in the Knudsen cell and the time of deposition. As revealed by AFM, the layer thickness was 505 ± 43 nm and 1090 ± 8 nm. The Raman spectroscopy showed that the fullerene films were prepared with high quality, i.e., without fragmentation and graphitization. After sterilization of the fullerene films with ethanol, which was necessary for cell cultivation, some of the C_{60} molecules reacted with oxygen or polymerized (Grausova *et al.*, 2009c,d).

The fullerene-coated glass coverslips were then inserted into 24-well polystyrene multidishes (TPP, Switzerland), seeded with human osteoblast-like MG 63 cells and incubated in Dulbecco's modified Eagle's Minimum Essential Medium supplemented with 10% fetal bovine serum. On day 1 after seeding, the cells on both thin and thick fullerene layers adhered at similar numbers, comparable to the values found on standard cell culture substrates, represented by the tissue culture polystyrene dishes and microscopic glass coverslips. However, the cell spreading area (i.e., the cell area projected on the material) was significantly smaller on both thin and thick fullerene films, which could be attributed to a relatively high hydrophobicity of these films (sessile water drop contact angle from 97 ± 2° to 101 ± 7°). Nevertheless, between days 1 to 5 after seeding, the cells on both thin and thick C_{60} films proliferated with similar and relatively short cell population doubling times (about 19 hours) as the cells on polystyrene and glass, and on day 5, the cells on all tested substrates reached similar cell population densities. The cells on the fullerene layers were of normal polygonal or spindle-like shape, they were homogeneously distributed on the material surface and on day 7 after seeding, they practically reached confluence (Grausova *et al.*, 2009c; Fig. 6).

Thus, the fullerene C_{60} layers in our study were colonized with human osteoblast-like MG 63 cells to a similar extent as standard cell culture polystyrene dishes. We considered this result rather surprising, because many studies have reported cytotoxic and even genotoxic effects of fullerenes (Xu *et al.*, 2009; Johnson-Lyles *et al.*, 2010; for a review, see Bacakova *et al.*, 2008a). However, the cytotoxic action of fullerenes was usually associated with suspending fullerenes in the cell culture medium (Yamawaki *et al.*, 2006; Gelderman *et al.*, 2008) or UV-light irradiation (Prylutska *et al.*, 2010), while our fullerene layers were resistant to be dissolved in a water environment, and the cells were cultured in the dark. In addition, the cytotoxic effects of fullerenes are based mainly on the reactivity of fullerenes, which may weaken with time due to the oxidation and polymerization of fullerenes in an air atmosphere, ethanol or cell culture medium, as revealed by Raman spectroscopy.

Similarly as in other carbon-based thin films described in this review, the supportive effect of fullerene layers on cell colonization could also be explained by their surface nanostructure, mimicking the nanoarchitecture of the natural ECM, i.e., a physiological substrate for cell adhesion. In addition, the cell colonization could be supported by the presence of oxygen, revealed by Raman spectroscopy. Both the nanostructure and the oxygen content in our C_{60} layers may act synergetically by improving the adsorption of cell

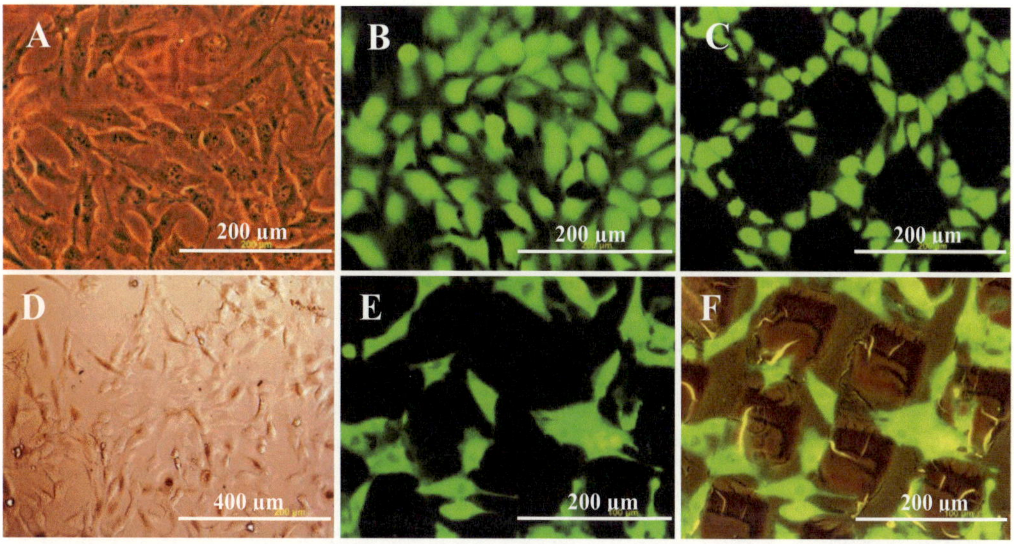

Fig. 6. Human osteoblast-like MG 63 cells in cultures on continuous and micropatterned C_{60} and C_{60}/Ti films. **A-C**: Cells on day 7 after seeding on a continuous fullerene C_{60} layer (505 ± 43 nm in thickness, **A**), and micropatterned C_{60} layers with prominences 128 ± 8 nm in height (**B**) or 1043 ± 57 nm in height (**C**). **D-F**: Cells on day 3 after seeding on a continuous C_{60}/Ti film (274 ± 11 nm in thickness, **D**) and micropatterned C_{60}/Ti layer with prominences 351 ± 18 nm in height (**E, F**). **A, D**: Native cells without staining; **B, C**: Cells stained with LIVE/DEAD viability/cytotoxicity kit; **E, F**: Cells visualized by immunofluorescence of β-actin. **E**: Fluorescence only, **F**: Combination of fluorescence and conventional light in order to visualize the grooves among the prominences. Olympus IX 51 microscope, DP 70 digital camera, obj. 20x and bar = 200 µm (except **D** where obj. is 10x and bar = 400 µm).

adhesion-mediating ECM molecules, such as vitronectin, fibronectin and collagen, provided by the serum of the culture medium or synthesized by the cells. The spatial conformation of these molecules on our films may also be more appropriate for the accessibility of specific sites on these molecules by cell adhesion receptors.

In our studies, fullerenes C_{60} were also deposited on microscopic glass coverslips in the form of micropatterned layers, i.e., through a metallic mask with rectangular holes (Grausova *et al.*, 2009c,d). Proportionally to the increasing temperature and the time of deposition, the thickness of the layers on sites underlying the openings of the grid was 128 ± 8 nm, 238 ± 3 nm, 326 ± 5 nm and 1043 ± 57 nm (data presented as mean ± standard deviation). Fullerene layers were also formed below the metallic part of the grid, where their thickness was usually within the size of the standard deviation. Thus, the fullerene layers contained bulge-like prominences separated by grooves.

From day 1 to 7 after seeding, the cells on the layers with the fullerene prominences of 128 ± 8 nm, 238 ± 3 nm and 326 ± 5 nm were distributed almost homogeneously over the entire material surface. Thus, the cell population densities on the prominences and in the grooves, and also the total cell population densities on all three surfaces, were similar. However, the cells on the layers with prominences of 1043 ± 57 nm were found preferentially in the grooves (Fig. 6). These grooves contained from 75.0 ± 13.4 % to 99.6 ± 0.3 % of cells, although they occupied only 44 ± 3 % of the material surface. The cell population density in the grooves was about 5 to 192 times higher than on the bulges, and these differences increased with time of cultivation.

On various polymeric, metallic or carbon-based materials, the cells were usually able to colonize the prominences on these surfaces, although these prominences amounted to several micrometers or even tens of μm (for a review, see Bacakova *et al.*, 2008a; Grausova *et al.*, 2009c,d). Surprisingly, on our fullerene C_{60} layers, the MG 63 cells were not able to "climb up" relatively low prominences (i.e., only about 1 μm in height), even at a relatively late culture interval of 7 days after seeding. This may be due to a synergistic action of certain physical and chemical properties of the fullerene bulges less appropriate for cell adhesion, such as their hydrophobicity, a steep rise as well as the tendency of spherical ball-like fullerene C_{60} molecules to diffuse out of the prominences (Guo *et al.*, 2007).

In our studies, hybrid C_{60}/Ti layers were also formed by co-deposition of C_{60} molecules and Ti atoms in an ultra-high vacuum (UHV) chamber (background pressure of ~10^{-7} Torr) using the Knudsen cell and electron gun for vaporization of C_{60} and Ti, respectively. Ti was chosen as a highly biocompatible metal often used for the construction of bone implants. Its concentration in the layers varied from about 25% to 75%, i.e., from 25 to 75 Ti atoms per 75 to 25 C_{60} molecules. Similarly to pure C_{60} films, hybrid C_{60}/Ti layers also promoted the adhesion and growth of MG 63 cells in a similar or even higher extent compared to standard polystyrene culture dishes and microscopic glass coverslips, and when deposited in micropatterned form, they induced regionally-selective cell adhesion and growth of these cells in grooves among the prominences (Vandrovcova *et al.*, 2008; Vacik *et al.*, 2010; Fig. 6).

At the same time, potential DNA damage by the hybrid C_{60}/Ti films was analyzed by immunofluorescence staining of markers of DNA damage response, such as phosphorylation of histone H2AX (Bekker-Jensen *et al.*, 2006) and focal recruitment of p53-binding protein (Schultz *et al.*, 2000). H2AX is a member of the histone H2A family. The formation of nuclear double-strand breaks (DSBs) triggers phosphorylation of H2AX at Ser139, referred as gamma-H2AX. Phosphorylation of H2AX is the first step in recruiting and localizing DNA repair proteins. The combination of phosphospecific antibodies, which recognize the phosphorylated Ser139 residue of gamma-H2AX, with immunofluorescence microscopy document the local phosphorylation through the formation of distinct foci in the vicinity of DSBs and allow for monitoring of their induction and repair. Another important protein involved in DNA repair is the p53 binding protein 1 (53BP1). Efficient 53BP1 focal recruitment depends on a number of upstream factors, including phosphorylation of H2AX at Ser139, recruitment of other mediator proteins (such as MDC1 and the E3 ubiquitin ligase RNF8), methylation of histones H3 and H4 and Tip60 HAT (histone acetyltransferase) activity, which can acetylate histone residues and therefore may facilitate 53BP1 recruitment through the formation of a more open chromatin structure. However, it is important to note that gamma-H2AX is not required for the initial recruitment of 53BP1 immediately following DNA damage.

As a positive control to the markers of DNA damage response, a 7-day-long treatment with 2.5 mM thymidine was used. The proliferation activity and morphological changes of the cells were also monitored. After 7 days of cultivation, we observed no cytotoxic morphological changes, such as enlarged cells or cytosolic vacuole formation, which are signs of cell senescence, and no increased induction of cell death. In addition, there was no increased level of DNA damage response on the C_{60}/Ti composites. We also found no significant differences in cell population densities and no increased level of DNA damage among various Ti concentrations (Kopova *et al.*, 2010). These results suggest that fullerenes, in combination with Ti, do not cause cytotoxic injury, and thus this material could be used for coating bone implants. However, Raman spectroscopy revealed that most fullerene molecules were disintegrated by their co-deposition with Ti, and converted to amorphous carbon (a-C). In addition, the hybrid C_{60}/Ti films also contained polymerized and oxidized structures (Vandrovcová *et al.*, 2008; Vacík *et al.*, 2010).

7. Carbon nanotubes

Carbon nanotubes are, in comparison with other carbon allotropes, relatively newly discovered carbon nanoparticles, which were first described, synthesized and systemized by Iijima (Iijima and Ichihashi 1993; Iijima 2002). These tubular structures are formed by a single cylindrically-shaped graphene sheet (single-wall carbon nanotubes, usually referred to as SWNT or SWCNT) or several graphene sheets arranged concentrically (multi-wall carbon nanotubes, referred to as MWNT or MWCNT). Carbon nanotubes have excellent mechanical properties, mainly due to sp^2 bonds. The tensile strength of single-walled nanotubes is about one hundred times higher than that of the steel, while their specific weight is about six times lower (Iijima and Ichihashi 1993; Yakobson *et al.*, 1997; Iijima 2002). Thus, carbon nanotubes could be utilized in hard tissue surgery, e.g., to reinforce artificial bone implants, particularly scaffolds for bone tissue engineering made of relatively soft synthetic or natural polymers. Carbon nanotubes have been used in combination with poly(carbonate) urethane (Khang *et al.*, 2007, 2008), biodegradable polymers such as polylactic acid (Supronowicz *et al.*, 2002), propylene fumarate (Shi *et al.*, 2006), poly(3-hydroxybutyrate) (Misra *et al.*, 2010), a copolymer of polylactide-caprolactone (Lahiri *et al.*, 2009) or a copolymer of polypyrrole-hyaluronic acid (Pelto *et al.*, 2010). Also hydroxyapatite (HAp), i.e., a ceramic material widely used in bone tissue engineering, but known for its high brittleness, has been reinforced with carbon nanotubes (Balani *et al.*, 2007; Hahn *et al.*, 2009). Carbon nanotubes not only improved the mechanical properties of the mentioned materials, such as tensile (Young's) modulus, compressive and flexural moduli and compressive, flexural and tensile strength in the polymeric materials (Shi *et al.*, 2006; Lahiri *et al.*, 2009; Misra *et al.*, 2010), and fracture toughness, hardness, elastic modulus and adhesion to the underlying substrate in HAp coatings (Balani *et al.*, 2007; Hahn *et al.*, 2009), but also increased the attractiveness of these materials for the adhesion, growth, differentiation and phenotypic maturation of cells, such as osteoblasts, chondrocytes and stem cells. One of the mechanisms of the improved cell colonization was an increased adsorption of fibronectin, i.e., an important cell-adhesion mediating ECM protein, to these composites, which has been explained by creating a nanoscale surface roughness of the material by the addition of nanotubes, and also by an increased material surface hydrophilia due to the presence of the polymeric component (pure carbon nanotube surfaces were highly hydrophobic, Khang *et al.*, 2007, 2008). Another important mechanism is the electroactivity of carbon nanotubes, i.e., their electrochemical activity, electrical charge and conductivity, which enable electrical stimulation of cells (Supronowicz *et al.*, 2002; Zanello *et al.*, 2006; Khang *et al.*, 2008; Pelto *et al.*, 2010). For example, osteoblasts cultured on nanocomposites consisting of polylactic acid and carbon nanotubes and exposed to electrical stimulation (current 10 μA, frequency 10 Hz), increased their proliferation activity, concentration of extracellular calcium and mRNA expression for collagen type I (Supronowicz *et al.*, 2002). Similarly, human chondrocytes cultured on MWCNT/poly(carbonate) urethane composites (weight ratio 1:2) and subjected to electrical stimulation (current 10 μA, frequency 10 Hz) adhered to these substrates in higher numbers and reached higher population densities after 2-day-cultivation than the cells on non-stimulated composites (Khang *et al.*, 2008).

In our earlier studies (Bacakova *et al.*, 2007b; 2008a), carbon nanotubes were used for the creation of composite materials by their mixing with terpolymer of polytetrafluoroethylene, polyvinyldifluoride and polypropylene (PTFE/PVDF/PP) or polysulfone (PSU), i.e., polymers which have been considered to be promising for the construction of bone implants. For example, PTFE/PVDF/PP has been tested as a material for septal nasal reconstruction (Scierski *et al.*, 2007). PSU was used for the creation of composites with bioactive glass particles for potential bone and cartilage tissue engineering (Zhang et al., 2002; Orefice et al., 2007) or for fabrication of bone joint screws (Jan & Grzegorz 2005).

In our experiments, the carbon nanotubes were expected to reinforce the relatively soft and elastic polymeric materials, and thus to improve their mechanical properties for hard tissue surgery. The other desired function of nanotubes was the formation of prominences in nanoscale on the material surface, which could promote its colonization with bone cells.

The PTFE/PVDF/PP-nanotube composites were prepared from a commercially available PTFE/PVDF/PP terpolymer (density of 1600 g/dm^3, Aldrich Chemical Co., U.S.A.) and single-wall carbon nanohorns (SWCNH; belonging to the single-wall nanotube family) or high crystalline electric arc multi-wall nanotubes (MWCNT). Both types of nanoparticles were purchased from NanoCraft Inc., Renton, U.S.A. SWCNH were 2 to 3 nm in diameter, 30 to 50 nm in length, and 19° closed end (referred to as a horn). The diameter of MWCNT ranged from 5 to 20 nm and the length from 300 to 2000 nm. PTFE/PVDF/PP was dissolved in acetone to a concentration of 0.1 g/ml, and mixed with 2, 4, 6 or 8 wt. % of SWCNH or MWCNT. The suspensions were then exposed to ultrasound in a PARMER INSTRUMENTS sonicator (model CP 130PB, power 130W, frequency 20 kHz) for 10 minutes at 20° C in order to prevent clustering of the nanoparticles. Finally, the suspensions were poured on to Petri dishes and left to evaporate the solvent (Bacakova *et al.*, 2007b, 2008a).

The addition of SWCNH or MWCNT to the PTFE/PVDF/PP terpolymer markedly improved several parameters of the adhesion and growth of human osteoblast-like MG 63 cells (Bacakova *et al.*, 2007b, 2008a). They were well-spread, polygonal, and contained distinct beta-actin filament bundles, whereas most cells on the pure terpolymer were less spread or even round and clustered into aggregates (Fig. 7). The enzyme-linked immunosorbent assay (ELISA) revealed that the cells on the material with 4 wt. % of SWCNH contained a higher concentration of vinculin (by 64 % and 69 %), a component of focal adhesion plaques, in comparison with the values in cells on the pure terpolymer and tissue culture polystyrene, respectively. The concentration of talin, another important integrin-associated focal adhesion protein, was also higher in cells grown on terpolymer samples with 4 and 8 wt.% of SWCNH (by 35 % and 28 %, respectively). In addition, the cells on carbon nanotube-terpolymer composites were more active in proliferation, which was most apparent on samples with 4 wt. % of MWCNT. On day 7, the cells on this composite reached a population density 4.5 times higher (228 000 ± 10 050 $cells/cm^2$) than the density on the unmodified PTFE/PVDF/PP (50 300 ± 5 400 $cells/cm^2$).

After the addition of SWCNH or MWCNT, the wettability of the composite material was relatively low (sessile water drop contact angle from 99 ± 7° to 105 ± 2°) and unchanged in comparison with the non-modified terpolymer (contact angle 100 ± 4°). However, the addition of carbon nanotubes significantly increased the nano- and submicron-scale surface roughness. The R_a parameter (i.e., departures of the roughness profile from the mean line) measured by AFM, ranged from 101 ± 15 nm to 150 ± 23 nm on the nanotube-modified samples, whereas on the pure terpolymer, it was only 30 ± 5 nm. Thus, the increased nano- and submicron-scale surface roughness could explain the beneficial effects of the terpolymer-nanotube composites on the adhesion and growth of MG 63 cells.

On the other hand, the terpolymer-carbon nanotube composites also displayed microscale surface roughness, ranging from 0.59 ± 0.1 µm to 2.22 ± 0.36 µm in the composites with 2 to 8 wt. % of SWCNT, and from 0.41 ± 0.01 µm to 1.63 ± 0.3 µm in the material with 2 to 8 wt. % of MWCNT, as measured by surface profilometry (T500 Hommel Tester, Hommelwerke Co., Germany). The microscale roughness was most probably due to clustering of carbon nanotubes and to prominences of these microclusters on the material surface (Bacakova *et al.,* 2007b, 2008a). An increase in the microscale surface roughness in polymer-carbon nanotube composites has been considered as a factor decreasing the colonization of the composites with cells. For example, on poly(3-hydroxybutyrate)/bioactive glass composites with MWCNT, the cell population density and growth rate of MG 63 cells decreased proportionally to the increasing concentrations of MWCNT (2, 4, 7 wt. %) and increasing microscale surface roughness (RMS from 6 ± 2 to 10 ± 2 µm) of the material (Misra *et al.,* 2010). Also in our study, the cell adhesion and growth of MG 63 cells (evaluated by the concentration of integrin-associated proteins talin and vinculin and the cell population density) was, in most cases, the highest on composites with 4 wt. % of SWCNT or MWCNT (R_a in both cases about 1 µm) and then decreased (Bacakova *et al.,* 2008a).
The polysulfone-nanotube composites were prepared by a similar manner from a commercially available polysulfone (Aldrich Chemical Co., U.S.A.) and SWCNH or MWCNT referred to above. PSU was dissolved in dichloromethane to a concentration of 0.1 g/ml and mixed with 0.5, 1.0 or 2.0 wt.% of SWCNH or MWCNT. The suspensions were then sonicated, poured on to Petri dishes and left to evaporate the solvent (Bacakova *et al.,* 2008a).
The addition on carbon nanotubes to PSU usually did not significantly influence the adhesion, spreading, morphology and growth of MG 63 cells in cultures on these materials (Fig. 7). From days 1 to 5 after seeding, the cell numbers on these composites were usually similar to the values found on the pure PSU. Only the cell adhesion area, measured on day 3 after seeding, was significantly larger on PSU with 2 wt. % of MWCNT (1 770 ± 90 μm^2) than on pristine PSU (1 330 ± 70 μm^2) and cell culture polystyrene dishes (1 400 ± 110 μm^2). These results could be explained by a relatively higher surface hydrophilicity of the pure PSU (water drop contact angle 85 ± 5°) than of the terpolymer PTFE/PVDF/PP (contact angle 100 ± 4°), which might mask the supportive effects of the surface nano-roughness, created by the addition of carbon nanotubes (R_a from 11 ± 4 nm to 27 ± 8 nm *versus* 4 ± 2 nm on the pure PSU, as measured by AFM), on the cell colonization. After the addition of carbon nanotubes, the surface hydrophilicity of the materials slightly decreased on average (i.e., the water drop contact angle increased from 85.0 ± 5° to 93 ± 4°), but these differences were not statistically significant.
Also in the form of thin films, carbon nanotubes acted as good substrates for cell colonization. Carbon nanotubes deposited on various substrates, such as polycarbonate membranes (Aoki *et al.,* 2007), glass coverslips (Holy *et al.,* 2009), titanium (Terada *et al.,* 2009), silicone rubber (Matsuoka *et al.,* 2009), or pressed in compacts (Li *et al.,* 2009a) markedly improved the adhesion, proliferation and differentiation of cells in cultures on these films in comparison with the non-coated materials or pure graphite. Similarly as in the composites of carbon nanotubes with polymers and ceramics, these favorable effects were mediated by an increased adsorption of specific proteins promoting cell adhesion, growth and differentiation, such as cell adhesion-mediating proteins from the serum of the culture medium (Aoki *et al.;* 2007, Li *et al.,* 2009a) and bone morphogenetic protein-2 (Li *et al.,* 2009b), and particularly by the electrical activity of carbon nanotubes (Zanello *et al.,* 2006). The electroactivity of carbon nanotube films makes these materials advantageous for tissue engineering applications involving electrically excitable tissues, such as muscles and nerves.

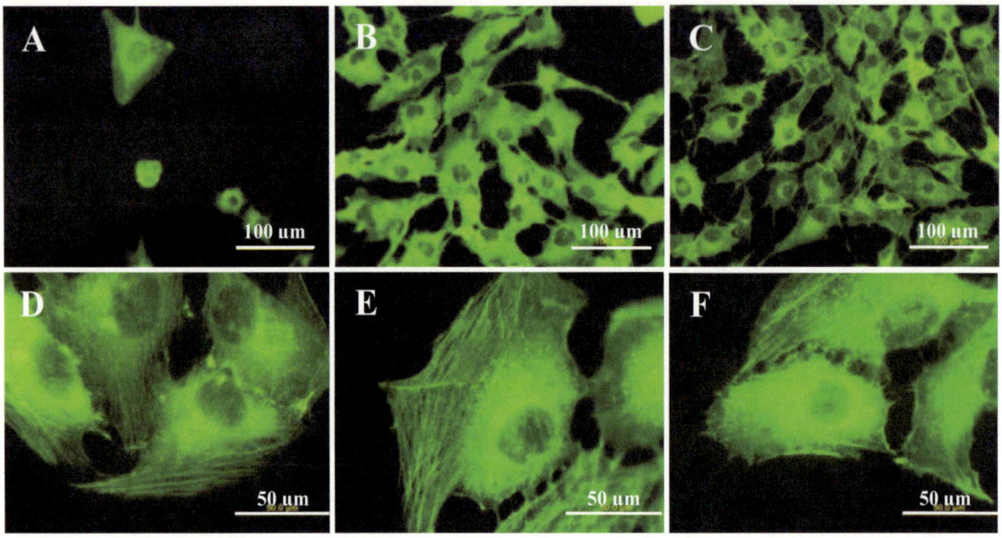

Fig. 7. Immunofluorescence staining of β-actin in human osteoblast-like MG 63 cells on day 3 after seeding on a terpolymer of polytetrafluoroethylene, polyvinyldifluoride and polypropylene (**A**), terpolymer mixed with 4 wt.% of SWCNH (**B**), terpolymer mixed with 4 wt.% of MWCNT (**C**), pure polysulfone (**D**), polysulfone with 2 wt.% of SWCNH (**E**) and polysulfone with 2 wt.% MWCNT (**F**). Olympus IX 51 epifluorescence microscope, DP 70 digital camera. Obj. 20, bar=100 μm (**A-C**), obj. 100x, bar = 20 μm (**D-F**).

For example, SWCNT films sprayed on glass coverslips promoted the outgrowth of neurites and enlargement of the neuron cell body area in hippocampal neurons isolated from newborn rats (Malarkey *et al.*, 2009). The hybrid neuroblastoma*glioma cell line NG108, a model of neuronal cells, and also primary rat peripheral neurons, were electrically coupled to conductive SWCNT films and showed robust voltage-activated currents when electrically stimulated through these substrates (Liopo *et al.*, 2006). Also mouse embryonic neural stem cells, isolated from the brain cortex and seeded on layer-by-layer assembled SWCNT-polyelectrolyte multilayer thin films, were successfully differentiated to neurons, astrocytes, and oligodendrocytes with clear formation of neurites. The neurite outgrowth and expression of specific cell type markers, such as nestin (a marker of neuronal stem cells), microtubule-associated protein 2 (MAP2; a marker of neurons), glial fibrillary acidic protein (anti-GFAP; a marker of astrocytes) and oligodendrocyte marker (O4) in cells cultured on the SWCNT films were similar as on poly-L-ornithine, i.e., one of the most widely used growth substrata for neural stem cells (Jan and Kotov 2007).

Not only continuous films, but also micropatterned carbon nanotube films have been applied for directed adhesion and growth of osteoblasts and neurons. For example, MWCNT were deposited by a hot embossing imprint lithography process on polycarbonate substrates, where they formed 28 different patterns of microscale lanes and circles. The feature diameters and spacing ranged from 9 to 76 μm. Osteoblast-like MC3T3-E1 cells seeded on these surfaces showed a maximum alignment (55 ± 6% of cells) to the microlane patterns with the lane diameter of 16 μm and spacing of 14 μm, and no significant alignment to microcircle patterns (Eliason *et al.*, 2008). Microdomains of carbon nanotubes (CNT) patterned on non-adhesive substrates, such as clean glass or quartz surfaces, also functioned as effective anchoring sites for neurons and glial cells *in vitro*, and thus provided a means to form complex, engineered, interconnected neuronal networks with pre-designed geometry via utilizing the self-assembly process of neurons. Depositing these CNT islands onto a multielectrode array chip can also facilitate electrical interfacing between the electrodes and the neurons (David-Pur *et al.*, 2010). In addition, the communication between neurites of neurons attached to different nanotube-containing domains induce a mechanical tension in the neuronal network, which serves as a signal for survival of the axonal branch and perhaps for the subsequent formation of synapses (Eliason *et al.*, 2008). Neuronal networks developed on surfaces micropatterned with CNT domains can be utilized not only for neural tissue engineering, but also as advanced neuro-chips for bio-sensing applications, e.g., drug and toxin detection (Sorkin *et al.*, 2006).

Carbon nanotube films can be deposited not only on planar 2D substrates, but also on three-dimensional matrices, i.e., on the walls of pores inside sponge-like collagenous scaffolds for bone tissue engineering. MWCNT deposited on these scaffolds improved the ingrowth of mouse osteoblast MC3T3-E1 cells and human osteosarcoma SaOs-2 cells inside the scaffolds, increased the strength of the cell adhesion and cell number (measured by DNA content), when compared to the scaffolds without nanotubes (Hirata *et al.*, 2009, 2010).

Despite all of these encouraging results, the potential cytotoxicity of carbon nanotubes should also be taken into account. Cytotoxicity of carbon nanotubes were mainly reported in cases when the nanotubes were dispersed in a liquid environment and could penentrate through the cell membrane inside the cells. For example, multi-wall carbon nanotubes (MWCNT) administered intrascrotally behaved as a carcinogen and induced intraperitoneally-disseminated mesothelioma in male Fischer 344 rats (Sakamoto *et al.*, 2009). Carboxylated SWCNT and MWCNT, suspended in the culture medium (concentrations 3, 6 and 30 μg/ml) and added to multipotent mesenchymal stem cells decreased the proliferation, viability, osteogenic differentiation (measured by the activity of alkaline phosphatase), matrix mineralization and also adipogenic differentiation (determined by staining intracytoplasmic lipids with Oil Red O) in these cells. SEM and TEM images revealed that carbon nanotubes might interact with proteins located on the cell membrane or in the cytoplasm, which has a further impact on subsequent cellular signaling pathways. Q-PCR results and Western blot analysis together verified that the inhibition of proliferation and osteogenic differentiation of MSCs may be modulated through a Smad protein-dependent bone morphogenetic protein signaling pathway (Liu *et al.*, 2010).

In addition, carbon nanotubes were revealed to act as cytotoxic not only in suspension, but also in the form of films. Thin films consisting of a SWCNT network, deposited on glass coverslips, were inhibitory to the proliferation, viability, and neuritegenesis of rat pheochromocytoma PC12 cells, and also to the proliferation of human fetal osteoblasts cultured on the top of these substrates, whereas these effects were not observed in cells grown on reduced graphene oxide films. As shown by MTT assay, based on the reduction of a tetrazolium bromide to a purple formazan in living cells, the SWCNT network decreased the activity of mitochondrial enzymes. However, at the same time, similar conductive carbon nanotube networks have been used to electrically stimulate neurons, facilitated propagation of action potentials, boosted neuronal electrical signaling and have been considered to be promising for neural tissue engineering (Agarwal *et al.*, 2010).

8. Conclusion and further perspectives

It can be concluded that all carbon-based materials discussed in this chapter, i.e., hydrocarbon plasma polymers and amorphous carbon (especially when enriched with Ti), nanocrystalline diamond films (particularly those doped with boron), pyrolytic graphite/carbon, fullerenes C_{60}, binary C_{60}/Ti composites and carbon nanotube-containing substrates provided good supports for the adhesion and growth of cell types present in the bone tissue, such as osteoblasts, endothelial cells or vascular smooth muscle cells. In the form of thin films, all these materials could be used for the coating of bone implants, in which a firm and fast integration with the surrounding bone tissue is desirable, e.g., bone-anchoring parts of joint prostheses or dental replacements. For this purpose, nanocrystalline diamond films are particularly suitable, because they are mechanically and chemically resistant, non-cytotoxic and provide excellent support for the adhesion, spreading, viability, growth and maturation of osteogenic cells. Their beneficial effect on cell colonization can be further enhanced by boron doping, which renders these films electrically conductive and thus enabling electrical stimulation of cells. Electrical conductivity and increased adsorption of molecules mediating cell adhesion or stimulating osteogenic cell differentiation is also characteristic for carbon nanotubes; however, their potential cytotoxicity should be taken into account. Cytotoxicity is also a main risk of fullerene-based films, although in our studies, these films supported the colonization with osteogenic cells to a similar extent as on standard cell culture substrates, such as polystyrene dishes and microscopic glass coverslips. Moreover, binary C_{60}/Ti composite films did not cause DNA damage, as revealed by immunoflurescence of markers of DNA damage response, such as gamma-H2AX and 53BP1. In addition, fullerenes, C_{60}/Ti composites, carbon nanotubes and nanocrystalline diamond can be deposited in the form of micropatterned surfaces for regionally-selective adhesion, directed growth and controlled function of cells. Micropatterned and electroconductive substrates also have great potential to be applied for the construction of biosensors and biostimulators.

9. Acknowledgements

This study was supported by the Academy of Sciences of the Czech Republic (grants No. KAN101120701, KAN400480701, KAN400100701 and IAAX00100902), and the Grant Agency of the Czech Republic (grant No. P108/10/1858). We also wish to thank to Dr. Ken Haenen (Institute for Materials Research, Hasselt University & Division IMOMEC, Belgium) for the growth of boron-doped diamond thin films, Dr. Bohuslav Rezek (Institute of Physics, Acad. Sci. CR) for his characterization of diamond thin films, and Mrs. Jana Vobornikova (Inst. Physiol., Acad. Sci. CR) for her excellent technical assistance in the manuscript preparation. Mrs. Sherryl Ann Vacik (Pangrac & Associates, Angel Fire, New Mexico, U.S.A.) is gratefully acknowledged for her language revision of the manuscript.

10. References

Abbas, A.; Vercaigne-Marko, D.; Supiot, P.; Bocquet, B.; Vivien, C. & Guillochon, D. (2009). Covalent attachment of trypsin on plasma polymerized allylamine. *Colloids Surf B Biointerfaces* 73: 315-324

Agarwal, S.; Zhou, X.; Ye, F.; He, Q.; Chen, G.C.; Soo, J.; Boey, F.; Zhang, H. & Chen, P. (2010). Interfacing live cells with nanocarbon substrates. *Langmuir* 26: 2244-2247

Amaral, M.; Dias, A.G.; Gomes, P.S.; Lopes, M.A.; Silva, R.F.; Santos, J.D. & Fernandes, M.H. (2008). Nanocrystalline diamond: In vitro biocompatibility assessment by MG 63 and human bone marrow cells cultures. *J Biomater Res A* 87: 91-99

Aoki, N.; Akasaka, T.; Watari, F. & Yokoyama, A. (2007). Carbon nanotubes as scaffolds for cell culture and effect on cellular functions. *Dent Mater J* 26: 178-185

Bacakova, L.; Mares, V.; Lisa, V. & Svorcik, V. (2000). Molecular mechanisms of improved adhesion and growth of an endothelial cell line cultured on polystyrene implanted with fluorine ions. *Biomaterials* 21: 1173-1179

Bacakova, L.; Walachova, K.; Svorcik, V. & Hnatowicz, V. (2001a). Adhesion and proliferation of rat vascular smooth muscle cells (VSMC) on polyethylene implanted with O^+ and C^+ ions. *J Biomater Sci Polym Ed* 12: 817-834

Bacakova, L.; Stary, V.; Kofronova, O. & Lisa, V. (2001b). Polishing and coating carbon fiber-reinforced carbon composites with a carbon-titanium layer enhances adhesion and growth of osteoblast-like MG63 cells and vascular smooth muscle cells in vitro. *J Biomed Mater Res* 54: 567-578

Bacakova, L.; Filova, E.; Rypacek, F.; Svorcik, V. & Stary, V. (2004). Cell adhesion on artificial materials for tissue engineering. *Physiol Res* 53 Suppl. 1: S35-S45

Bacakova, L.; Filova, E.; Kubies, D.; Machova, L.; Proks, V.; Malinova, V.; Lisa, V. & Rypacek, F. (2007a). Adhesion and growth of vascular smooth muscle cells in cultures on bioactive RGD peptide-carrying polylactides. *J Mater Sci Mater Med* 18: 1317-1323

Bacakova, L.; Grausova, L.; Vacik, J.; Fraczek, A.; Blazewicz, S.; Kromka, A.; Vanecek, M. & Svorcik, V. (2007b). Improved adhesion and growth of human osteoblast-like MG 63 cells on biomaterials modified with carbon nanoparticles. *Diamond Relat Mater* 16: 2133-2140

Bacakova, L.; Grausova, L.; Vandrovcova, M.; Vacik, J.; Frazcek, A.; Blazewicz, S.; Kromka, A.; Vanecek, M.; Nesladek, M.; Svorcik, V. & Kopecek, M. (2008a). Carbon nanoparticles as substrates for cell adhesion and growth. In: *Nanoparticles: New Research.* Simone Luca Lombardi, (Ed.), pp. 39-107, Nova Science Publishers, Inc., ISBN: 978-1-60456-704-5 Hauppauge, New York

Bacakova, L.; Koshelyev, H.; Noskova, L.; Choukourov, A.; Benada, O.; Mackova, A.; Lisa, V. & Biederman, H. (2008b). Vascular endothelial cells in cultures on nanocomposite silver/hydrocarbon plasma polymer films with antimicrobial activity. *J Optoelectron Adv Mater* 10: 2082-2087

Bacakova, L. & Svorcik, V. (2008). Cell colonization control by physical and chemical modification of materials, In: *Cell Growth Processes: New Research,* Daiki Kimura, (Ed.), pp. 5-56, Nova Science Publishers, Inc., ISBN: 978-1-60456-123-6, Hauppauge, New York

Balani, K.; Anderson, R.; Laha, T.; Andara, M.; Tercero, J.; Crumpler, E. & Agarwal, A. (2007). Plasma-sprayed carbon nanotube reinforced hydroxyapatite coatings and their interaction with human osteoblasts in vitro. *Biomaterials* 28: 618-624

Bekker-Jensen, S.; Lukas, C.; Kitagawa, R.; Melander, F.; Kastan, M.B.; Bartek, J. & Lukas, J. (2006). Spatial organization of the mammalian genome surveillance machinery in response to DNA strand breaks. *J Cell Biol* 173: 195-206

Bernasek, T.L.; Stahl, J.L. & Pupello, D. (2009). Pyrolytic carbon endoprosthetic replacement for osteonecrosis and femoral fracture of the hip: a pilot study. *Clin Orthop Relat Res* 467: 1826-1832

Bettinger, C.J.; Bruggeman, J.P.; Misra, A.; Borenstein, J.T. & Langer, R. (2009). Biocompatibility of biodegradable semiconducting melanin films for nerve tissue engineering. *Biomaterials* 30: 3050-3057

Biederman, H. (2004). *Plasma polymer films.* Biederman, H., (Ed.), Imperial College Press, ISBN: 1-86094-467-1, London

Bondar, V.S.; Pozdnyakova, I.O. & Puzyr, A.P. (2004). Applications of nanodiamonds for separation and purification of proteins. *Phys Solid State* 46: 758-760

Bruinink, A.; Schroeder, A.; Francz, G. & Hauert, R. (2005). In vitro studies on the effect of delaminated a-C:H film fragments on bone marrow cell cultures. *Biomaterials* 26: 3487-3494

Cenni, E.; Granchi, D.; Arciola, C.R.; Ciapetti, G.; Savarino, L.; Stea, S.; Cavedagna, D.; Di Leo, A. & Pizzoferrato, A. (1995). Adhesive protein expression on endothelial cells after contact in vitro with polyethylene terephthalate coated with pyrolytic carbon. *Biomaterials* 16: 1223-1227

Cenni, E.; Granchi, D.; Ciapetti, G.; Savarino, L.; Corradini, A. & Di Leo, A. (2000). Cytokine expression in vitro by cultured human endothelial cells in contact with polyethylene terephthalate coated with pyrolytic carbon and collagen. *J Biomed Mater Res* 50: 483-489

Chai, F.; Mathis, N.; Blanchemain, N.; Meunier, C. & Hildebrand, H.F. (2008) Osteoblast interaction with DLC-coated Si substrates. *Acta Biomater* 4: 1369-1381

Chen, Y.W., Hwang, K.C., Yen, C.C. & Lai, Y.L. (2004). Fullerene derivatives protect against oxidative stress in RAW 264.7 cells and ischemia-reperfused lungs. *Am J Physiol Regul Integr Comp Physiol* 287: R21-R26

Chu, L.Q.: Knoll, W. & Förch, R. (2009). Plasma polymerized non-fouling thin films for DNA immobilization. *Biosens Bioelectron* 25: 519-522

Clem, W.C.; Chowdhury, S.; Catledge, S.A.; Weimer, J.J.; Shaikh, F.M.; Hennessy, K.M.; Konovalov, V.V.; Hill, M.R.; Waterfeld, A.; Bellis, S.L. & Vohra, Y.K. (2008). Mesenchymal stem cell interaction with ultra-smooth nanostructured diamond for wear-resistant orthopaedic implants. *Biomaterials* 29: 3461-3468

Coelho, N.M.; González-García, C.; Planell, J.A.; Salmerón-Sánchez, M. & Altankov, G. (2010). Different assembly of type IV collagen on hydrophilic and hydrophobic substrata alters endothelial cells interaction. *Eur Cell Mater* 19: 262-272

David-Pur, M.; Shein, M. & Hanein, Y. (2010). Carbon nanotube-based neurochips. *Methods Mol Biol* 625: 171-177

Dellinger, A.; Zhou, Z.; Lenk, R.; MacFarland, D. & Kepley, C.L. (2009) Fullerene nanomaterials inhibit phorbol myristate acetate-induced inflammation. *Exp Dermatol* 18: 1079-1081

Deshpande, P.; Notara, M.; Bullett, N.; Daniels, J.T.; Haddow, D.B. & MacNeil, S. (2009). Development of a surface-modified contact lens for the transfer of cultured limbal epithelial cells to the cornea for ocular surface diseases. *Tissue Eng Part A* 15: 2889-2902

Eliason, M.T.; Sunden, E.O.; Cannon, A.H.; Graham, S.; García, A.J. & King, W.P. (2008). Polymer cell culture substrates with micropatterned carbon nanotubes. *J Biomed Mater Res A* 86: 996-1001

Ekstrand, K. & Hirsch, J.M. (2008). Malignant tumors of the maxilla: virtual planning and real-time rehabilitation with custom-made R-zygoma fixtures and carbon-graphite fiber-reinforced polymer prosthesis. *Clin Implant Dent Relat Res* 10: 23-29

Filova, E.; Bullett, N.A.; Bacakova, L.; Grausova, L.; Haycock, J.W.; Hlucilova, J.; Klima, J. & Shard, A. (2009). Regionally-selective cell colonization of micropatterned surfaces prepared by plasma polymerization of acrylic acid and 1,7-octadiene. *Physiol Res* 58: 669-684

Gajewski, W.; Achatz, P.; Williams, O.; Haenen, K.; Bustarret, E.; Stutzmann, M. & Garrido, J. (2009). Electronic and optical properties of boron-doped nanocrystalline diamond films. *Physical Review B* 79: 045206

Gelderman, M.P.; Simakova, O.; Clogston, J.D.; Patri, A.K.; Siddiqui, S.F.; Vostal, A.C. & Simak, J. (2008). Adverse effects of fullerenes on endothelial cells: fullerenol C60(OH)24 induced tissue factor and ICAM-I membrane expression and apoptosis in vitro. *Int J Nanomedicine* 3: 59-68

Goessl, A.; Garrison MD.; Lhoest, JB. & Hoffman, A.S. (2001). Plasma lithography--thin-film patterning of polymeric biomaterials by RF plasma polymerization I: Surface preparation and analysis. *J Biomater Sci Polym Ed* 12: 721-738

Gomez, N. & Schmidt, C.E. (2007). Nerve growth factor-immobilized polypyrrole: bioactive electrically conducting polymer for enhanced neurite extension. *J Biomed Mater Res A* 81: 135-149

Gonzales, K.A.; Wilson, L.J.; Wu, W. & Nancollas, G.H. (2002). Synthesis and in vitro characterization of tissue-selective fullerene: vectoring C60(OH)16AMBP to mineralized bone. *Bioorg Med Chem* 10: 1991-1997

Grausova, L.; Kromka, A.; Bacakova, L.; Potocky, S.; Vanecek, M. & Lisa, V. (2008). Bone and vascular endothelial cells in cultures on nanocrystalline diamond films. *Diamond Relat Mater* 17: 1405–1409

Grausova, L.; Bacakova, L.; Kromka, A.; Vanecek, M. & Lisa, V. (2009a). Molecular markers of adhesion, maturation and immune activation of human osteoblast-like MG 63 cells on nanocrystalline diamond films. *Diamond Relat Mater* 18: 258–263

Grausova, L.; Bacakova, L.; Kromka, A.; Potocky, S.; Vanecek, M.; Nesladek, M. & Lisa, V. (2009b). Nanodiamond as a promising material for bone tissue engineering. *J Nanosci Nanotechnol* 9: 3524-3534

Grausova, L.; Vacik, J.; Vorlicek, V.; Svorcik, V.; Slepicka, P.; Bilkova, P.; Vandrovcova, M.; Lisa, V. & Bacakova, L. (2009c). Fullerene C_{60} films of continuous and micropatterned morphology as substrates for adhesion and growth of bone cells. *Diamond Relat Mater* 18: 578–586

Grausova, L.; Vacik, J.; Bilkova, P.; Vorlicek, V.; Svorcik, V.; Soukup, D.; Bacakova, M.; Lisa, V. & Bacakova, L. (2009d) Regionally-selective adhesion and growth of human osteoblast-like MG 63 cells on micropatterned fullerene C60 layers. *J Optoelectron Adv Mater* 10: 2071-2076

Grinevich, A.; Bacakova, L.; Choukourov, A.; Boldyryeva, H.; Pihosh, Y.; Slavinska, D.; Noskova, L.; Skuciova, M.; Lisa, V. & Biederman, H. (2009). Nanocomposite Ti/hydrocarbon plasma polymer films from reactive magnetron sputtering as growth supports for osteoblast-like and endothelial cells. *J Biomed Mater Res* 88A: 952–966

Guo, S.; Fogarty, D.P.; Nagel, P.M. & Kandel, S.A. (2007). Scanning tunneling microscopy of surface-adsorbed fullerenes: C_{60}, C_{70}, and C_{84}. *Surf Sci* 601: 994-1000

Hahn, B.D.; Lee, J.M.; Park, D.S.; Choi, J.J.; Ryu, J.; Yoon, W.H.; Lee, B.K.; Shin, D.S. & Kim, H.E. (2009). Mechanical and in vitro biological performances of hydroxyapatite-carbon nanotube composite coatings deposited on Ti by aerosol deposition. *Acta Biomater* 5: 3205-3214

He, J.; Zhou, W.; Zhou, X.; Zhong, X.; Zhang, X.; Wan, P.; Zhu, B. & Chen, W. (2008). The anatase phase of nanotopography titania plays an important role on osteoblast cell morphology and proliferation. *J Mater Sci Mater Med* 19: 3465-3472

Hidalgo, E.; Bartolome, R.; Barroso, C.; Moreno, A. & Domínguez, C. (1998). Silver nitrate: antimicrobial activity related to cytotoxicity in cultured human fibroblasts. *Skin Pharmacol Appl Skin Physiol* 11: 140-151

Hirata, E.; Uo, M.; Takita, H.; Akasaka, T.; Watari, F. & Yokoyama, A. (2009). Development of a 3D collagen scaffold coated with multiwalled carbon nanotubes. *J Biomed Mater Res B Appl Biomater* 90: 629-634

Hirata, E.; Uo, M.; Nodasaka, Y.; Takita, H.; Ushijima, N.; Akasaka, T.; Watari, F. & Yokoyama, A. (2010). 3D collagen scaffolds coated with multiwalled carbon nanotubes: initial cell attachment to internal surface. *J Biomed Mater Res B Appl Biomater* 93: 544-550

Ho, C.P. & Yasuda, H. (1988). Ultrathin coating of plasma polymer of methane applied on the surface of silicone contact lenses. *J Biomed Mater Res* 22: 919-937

Holy, J.; Perkins, E. & Yu, X. (2009). Differentiation of pluripotent stem cells on multiwalled carbon nanotubes. *Conf Proc IEEE Eng Med Biol Soc* 2009: 6022-6025

Hornez, J.C.; Chai, F.; Monchau, F.; Blanchemain, N.; Descamps, M. & Hildebrand, H.F. (2007). Biological and physico-chemical assessment of hydroxyapatite (HA) with different porosity. *Biomol Eng* 24: 505-509

Hu, Z.; Liu, S.; Wei, Y.; Tong, E.; Cao, F. & Guan, W. (2007a). Synthesis of glutathione C60 derivative and its protective effect on hydrogen peroxide-induced apoptosis in rat pheochromocytoma cells. *Neurosci Lett* 429: 81-86

Hu, Z.; Guan, W.; Wang, W.; Huang, L.; Xing, H. & Zhu, Z. (2007b). Synthesis of beta-alanine C60 derivative and its protective effect on hydrogen peroxide-induced apoptosis in rat pheochromocytoma cells. *Cell Biol Int* 31: 798-804

Hu, Z.; Guan, W.; Wang, W.; Huang, L.; Xing, H. & Zhu, Z. (2007c). Protective effect of a novel cystine C(60) derivative on hydrogen peroxide-induced apoptosis in rat pheochromocytoma PC12 cells. *Chem Biol Interact* 167: 135-144

Hu, Z.; Guan, W.; Wang, W.; Zhu, Z. & Wang, Y. (2010). Folacin C60 derivative exerts a protective activity against oxidative stress-induced apoptosis in rat pheochromocytoma cells. *Bioorg Med Chem Lett* 20: 4159-4162

Huang, H.; Pierstorff, E.; Osawa, E. & Ho, D. (2008). Protein-mediated assembly of nanodiamond hydrogels into a biocompatible and biofunctional multilayer nanofilm. *ACS Nano* 2: 203-212

Huang, L.; Terakawa, M.; Zhiyentayev, T.; Huang, Y.Y.; Sawayama, Y.; Jahnke, A.; Tegos, G.P.; Wharton, T. & Hamblin, M.R. (2010). Innovative cationic fullerenes as broad-spectrum light-activated antimicrobials. *Nanomedicine* 6: 442-452

Iijima, S. & Ichihashi, T. (1993). Cobalt-catalysed growth of carbon nanotubes with single-atomic-layer walls. *Nature* 363: 603-605

Iijima, S. (2002). Carbon nanotubes: past, present, and future. *Physica B* 323: 1–5

Ismail, F.S.; Rohanizadeh, R.; Atwa, S.; Mason, R.S.; Ruys, A.J.; Martin, P.J. & Bendavid, A. (2007). The influence of surface chemistry and topography on the contact guidance of MG63 osteoblast cells. *J Mater Sci Mater Med* 18: 705-714

Jackson, M.J.; Robinson, G.M.; Ali, N.; Kousar, Y.; Mei, S.; Gracio, J.; Taylor, H. & Ahmed, W. (2006). Surface engineering of artificial heart valve disks using nanostructured thin films deposited by chemical vapour deposition and sol-gel methods. *J Med Eng Technol* 30: 323-329

Jan, C. & Grzegorz, K. (2005). The study of lifetime of polymer and composite bone joint screws under cyclical loads and in vitro conditions. *J Mater Sci Mater Med* 16: 1051-1060

Jan, E. & Kotov, N.A. (2007). Successful differentiation of mouse neural stem cells on layer-by-layer assembled single-walled carbon nanotube composite. *Nano Lett* 7: 1123-1138

Jeong, S.I.; Jun, I.D.; Choi, M.J.; Nho, Y.C.; Lee, Y.M. & Shin, H. (2008). Development of electroactive and elastic nanofibers that contain polyaniline and poly(L-lactide-co-epsilon-caprolactone) for the control of cell adhesion. *Macromol Biosci* 8: 627-637

Johnson-Lyles, D.N.; Peifley, K.; Lockett, S.; Neun, B.W.; Hansen, M.; Clogston, J.; Stern, S.T. & McNeil, S.E. (2010). Fullerenol cytotoxicity in kidney cells is associated with cytoskeleton disruption, autophagic vacuole accumulation, and mitochondrial dysfunction. *Toxicol Appl Pharmacol*, in press

Jun, I.; Jeong, S. & Shin, H. (2009). The stimulation of myoblast differentiation by electrically conductive sub-micron fibers. *Biomaterials* 30: 2038-2047

Kalbacova, M.; Kalbac, M.; Dunsch, L.; Kromka, A.; Vanecek, M.; Rezek, B.; Hempel, U. & Kmoch, S. (2007). The effect of SWCNT and nano-diamond films on human osteoblast cells. *Phys Stat Sol (b)* 244: 4356-4359

Kalbacova, M.; Michalikova, L.; Baresova, V.; Kromka, A.; Rezek, B. & Kmoch, S. (2008). Adhesion of osteoblasts on chemically patterned nanocrystalline diamonds. *Phys Stat Sol (b)* 245: 2124-2127

Kalbacova, M.; Rezek, B.; Baresova, V.; Wolf-Brandstetter, C. & Kromka, A. (2009). Nanoscale topography of nanocrystalline diamonds promotes differentiation of osteoblasts. *Acta Biomater* 5: 3076-3085

Khang, D.; Kim, S.Y.; Liu-Snyder, P.; Palmore, G.T.; Durbin, S.M. & Webster, T.J. (2007). Enhanced fibronectin adsorption on carbon nanotube/poly(carbonate) urethane: independent role of surface nano-roughness and associated surface energy. *Biomaterials* 28: 4756-4768

Khang, D.; Park, G.E. & Webster, T.J. (2008). Enhanced chondrocyte densities on carbon nanotube composites: the combined role of nanosurface roughness and electrical stimulation. *J Biomed Mater Res A* 86: 253-260

Kim, J.; Park, H.; Jung, D. & Kim, S. (2003). Protein immobilization on plasma-polymerized ethylenediamine-coated glass slides. *Anal Biochem* 313: 41-45

Kobayashi, S.; Ohgoe, Y.; Ozeki, K.; Hirakuri, K. & Aoki, H. (2007). Dissolution effect and cytotoxicity of diamond-like carbon coatings on orthodontic archwires. *J Mater Sci Mater Med* 18: 2263-2268

Kong, X.L.; Huang, L.C.L.; Hsu, C.-M.; Chen, W.-H.; Han, C.-C. & Chang, H.-C. (2005). High-affinity capture of proteins by diamond nanoparticles for mass spectrometric analysis. *Anal Chem* 77: 259-265

Kopecek, M.; Bacakova, L.; Vacik, J.; Fendrych, F.; Vorlicek, V.; Kratochvilova, I.; Lisa, V.; Van Hove, E.; Mer, C.; Bergonzo, P. & Nesladek, M. (2008). Improved adhesion, growth and maturation of human bone-derived cells on nanocrystalline diamond films. *Phys Stat Sol (a)* 205: 2146-2153

Kopova, I.; Bacakova, L.; Vacik, J. & Lavrentiev, V. (2010). No evidence for DNA damage by fullerene (C_{60})-transitional metal (Ti) composites developed for potential bone tissue engineering. The 2nd International Conference on Cellular and Molecular Bioengineering (ICCMB2), Book of Abstracts, p. 21, abstract No. C1013, Nanyang Technological University, School of Chemical & Biomedical Engineering, Singapore, August 2 - 4, 2010

Kromka, A.; Rezek, B.; Remes, Z.; Michalka, M.; Ledinsky, M.; Zemek, J.; Potmesil, J. & Vanecek M. (2008). Formation of continuous nanocrystalline diamond layer on glass and silicon at low temperatures. *Chem Vap Deposition* 14: 181-186

Kromka, A.; Grausova, L.; Bacakova, L.; Vacik, J.; Rezek, B.; Vanecek, M.; Williams, O.A. & Haenen, K. (2010). Semiconducting to metallic-like boron doping of nanocrystalline diamond films and its effect on osteoblastic cells. *Diamond Relat Mater* 19: 190-195

Kroto, H.W.; Heath, J.R.; O'Brien, S.C.; Curl, R.F. & Smalley, R.E. (1985). C_{60}: Buckminsterfullerene. *Nature* 318: 162-163

Kwok, S.C.; Yang, P.; Wang, J.; Liu, X. & Chu, P.K. (2004). Hemocompatibility of nitrogen-doped, hydrogen-free diamond-like carbon prepared by nitrogen plasma immersion ion implantation-deposition. *J Biomed Mater Res A* 70: 107-114

Lahiri, D.; Rouzaud, F.; Namin, S.; Keshri, A.K.; Valdes, J.J.; Kos, L.; Tsoukias, N.; & Agarwal, A. (2009). Carbon nanotube reinforced polylactide-caprolactone copolymer: mechanical strengthening and interaction with human osteoblasts in vitro. *ACS Appl Mater Interfaces* 1: 2470-2476

Lao, F.; Chen, L.; Li, W.; Ge, C.; Qu, Y.; Sun, Q.; Zhao, Y.; Han, D. & Chen, C. (2009). Fullerene nanoparticles selectively enter oxidation-damaged cerebral microvessel endothelial cells and inhibit JNK-related apoptosis. *ACS Nano* 3: 3358-3368

Levi, N.; Hantgan, R.R.; Lively, M.O.; Carroll, D.L. & Prasad, G.L. (2006). C_{60}-fullerenes: detection of intracellular photoluminescence and lack of cytotoxic effects. *J Nanobiotechnol* 4:14

Li, X.; Gao, H.; Uo, M.; Sato, Y.; Akasaka, T.; Feng, Q.; Cui, F.; Liu, X. & Watari, F. (2009a). Effect of carbon nanotubes on cellular functions in vitro. *J Biomed Mater Res A* 91: 132-139

Li, X.; Gao, H.; Uo, M.; Sato, Y.; Akasaka, T.; Abe, S.; Feng, Q.; Cui, F. & Watari, F. (2009b). Maturation of osteoblast-like SaoS2 induced by carbon nanotubes. *Biomed Mater* 4: 015005

Lin, J.; Park, I.W.; Mishra, B.; Pinkas, M.; Moore, J.J.; Anton, J.M.; Kim, K.H.; Voevodin, A.A. & Levashov, E.A. (2009). Processing, structure, and properties of nanostructured multifunctional tribological coatings. *J Nanosci Nanotechnol* 9: 4073-4084

Liopo, A.V.; Stewart, M.P.; Hudson, J.; Tour, J.M. & Pappas, T.C. (2006). Biocompatibility of native and functionalized single-walled carbon nanotubes for neuronal interface. *J Nanosci Nanotechnol* 6: 1365-1374

Liu, D.; Yi, C.; Zhang, D.; Zhang, J. & Yang, M. (2010). Inhibition of proliferation and differentiation of mesenchymal stem cells by carboxylated carbon nanotubes. *ACS Nano* 4: 2185-2195

Liu, J. & Tabata, Y. (2010). Photodynamic therapy of fullerene modified with pullulan on hepatoma cells. *J Drug Target* 18: 602-610

Maeda, R.; Noiri, E.; Isobe, H.; Homma, T.; Tanaka, T.; Negishi, K.; Doi, K.; Fujita, T. & Nakamura, E. (2008). A water-soluble fullerene vesicle alleviates angiotensin II-induced oxidative stress in human umbilical venous endothelial cells. *Hypertens Res* 31: 141-151

Majani, R.; Zelzer, M.; Gadegaard, N.; Rose, F.R. & Alexander, M.R. (2010). Preparation of Caco-2 cell sheets using plasma polymerised acrylic acid as a weak boundary layer. *Biomaterials* 31: 6764-6771

Majid, E.; Male, K.B. & Luong, J.H. (2008). Boron doped diamond biosensor for detection of Escherichia coli. *J Agric Food Chem* 56: 7691-7695

Malarkey, E.B.; Fisher, K.A.; Bekyarova, E.; Liu, W.; Haddon, R.C. & Parpura, V. (2009). Conductive single-walled carbon nanotube substrates modulate neuronal growth. *Nano Lett* 9: 264-268

Marchesan, S.; Da Ros, T.; Spalluto, G.; Balzarini, J. & Prato M. (2005). Anti-HIV properties of cationic fullerene derivatives. *Bioorg Med Chem Lett* 15: 3615-3618

Marcon, L.; Spriet, C.; Coffinier, Y.; Galopin, E.; Rosnoblet, C.; Szunerits, S.; Heliot, L.; Angrand, P.O. & Boukherroub, R. (2010). Cell Adhesion properties on chemically micropatterned boron-doped diamond surfaces. *Langmuir*, in press

Maropis, P.S.; Molinari, J.A.; Appel, B.N. & Baumhammers, A. (1977). Comparative study of vitreous carbon, pyrolytic carbon, pyrolytic graphite/silicon-carbide, and titanium implants in rabbit mandibles. *Oral Surg Oral Med Oral Pathol* 43: 506-512

Matsuoka, M.; Akasaka, T.; Hashimoto, T.; Totsuka, Y. & Watari, F. (2009). Improvement in cell proliferation on silicone rubber by carbon nanotube coating. *Biomed Mater Eng* 19: 155-162

Menzies, D.J.; Cowie, B.; Fong, C.; Forsythe, J.S.; Gengenbach, T.R.; McLean, K.Mm; Puskar, L.; Textor, M.; Thomsen, L.; Tobin, M. & Muir, B.W. (2010). One-step method for generating PEG-like plasma polymer gradients: chemical characterization and analysis of protein interactions. *Langmuir*, in press

Misra, S.K.; Ohashi, F.; Valappil, S.P.; Knowles, J.C.; Roy, I.; Silva, S.R.; Salih, V. & Boccaccini, A.R. (2010). Characterization of carbon nanotube (MWCNT) containing P(3HB)/bioactive glass composites for tissue engineering applications. *Acta Biomater* 6: 735-742

Monties, J.R.; Havlik, P.; Mesana, T.; Trinkl, J.; Tourres, J.L. & Demunck, J.L. (1994). Development of the Marseilles pulsatile rotary blood pump for permanent implantable left ventricular assistance. *Artif Organs* 18: 506-511

Nebel, C.E.; Shin, D.; Rezek, B.; Tokuda, N.; Uetsuka, H. & Watanabe, H. (2007). Diamond and biology. *J R Soc Interface* 4: 439-461

Orefice, R.; Clark, A.; West, J.; Brennan, A. & Hench, L. (2007). Processing, properties, and in vitro bioactivity of polysulfone-bioactive glass composites. *J Biomed Mater Res A* 80: 565-580

Papo, M.J.; Catledge, S.A.; Vohra, Y.K. & Machado, C. (2004). Mechanical wear behavior of nanocrystalline and multilayer diamond coatings on temporomandibular joint implants. *J Mater Sci: Mater Med* 15: 773-777

Pelto, J.; Haimi, S.; Puukilainen, E.; Whitten, P.G.; Spinks, G.M.; Bahrami-Samani, M.; Ritala, M. & Vuorinen, T. (2010). Electroactivity and biocompatibility of polypyrrole-hyaluronic acid multi-walled carbon nanotube composite. *J Biomed Mater Res A* 93: 1056-1067

Pesakova, V.; Klezl, Z.; Balik, K. & Adam, M. (2000). Biomechanical and biological properties of the implant material carbon-carbon composite covered with pyrolytic carbon. *J Mater Sci Mater Med* 11: 793-798

Price, R.L.; Ellison, K.; Haberstroh, K.M. & Webster, T.J. (2004). Nanometer surface roughness increases select osteoblast adhesion on carbon nanofiber compacts. *J Biomed Mater Res A* 70: 129-138

Prylutska, S.V.; Grynyuk, I.I.; Palyvoda, K.O. & Matyshevska, O.P. (2010). Photoinduced cytotoxic effect of fullerenes C60 on transformed T-lymphocytes. *Exp Oncol* 32: 29-32

Rakngarm, A.; Miyashita, Y. & Mutoh, Y. (2008). Formation of hydroxyapatite layer on bioactive Ti and Ti-6Al-4V by simple chemical technique. *J Mater Sci Mater Med* 19: 1953-1961

Rezek, B.; Shin, D. & Nebel, C.E. (2007). Properties of hybridized DNA arrays on single-crystalline undoped and boron-doped (100) diamonds studied by atomic force microscopy in electrolytes. *Langmuir* 23: 7626-7633

Rezek, B.; Michalikova, L.E.; Ukraintsev, E.; Kromka, A. & Kalbacova, M. (2009). Micro-pattern guided adhesion of osteoblasts on diamond surfaces. *Sensors* 9: 3549-3562

Roy, R.K. & Lee, K.R. (2007). Biomedical applications of diamond-like carbon coatings: a review. *J Biomed Mater Res B Appl Biomater* 83: 72-84

Ruiz, A.; Buzanska, L.; Gilliland, D.; Rauscher, H.; Sirghi, L.; Sobanski, T.; Zychowicz, M.; Ceriotti, L.; Bretagnol, F.; Coecke, S.; Colpo, P. & Rossi, F. (2008). Micro-stamped surfaces for the patterned growth of neural stem cells. *Biomaterials* 29: 4766-4774

Rupprecht, S.; Bloch, A.; Rosiwal, S.; Neukam, F.W. & Wiltfang, J. (2002). Examination of the bone-metal interface of titanium implants coated by the microwave plasma chemical vapor deposition method. *Int J Oral Maxillofac Implants* 17: 778-785

Sakamoto, Y.; Nakae, D.; Fukumori, N.; Tayama, K.; Maekawa, A.; Imai, K.; Hirose, A.; Nishimura, T.; Ohashi, N. & Ogata, A. (2009). Induction of mesothelioma by a single intrascrotal administration of multi-wall carbon nanotube in intact male Fischer 344 rats. *J Toxicol Sci* 34: 65-76

Santavirta, S. (2003). Compatibility of the totally replaced hip. Reduction of wear by amorphous diamond coating. *Acta Orthop Scand* Suppl 74: 1-19

Sato, H.; Tsuji, H.; Ikeda, S.; Ikemoto, N.; Ishikawa, J. & Nishimoto, S. (1999). Enhanced growth of human vascular endothelial cells on negative ion (Ag-)-implanted hydrophobic surfaces. *J Biomed Mater Res* 44: 22-30

Sawase, T.; Jimbo, R.; Baba, K.; Shibata, Y.; Ikeda, T. & Atsuta, M. (2008). Photo-induced hydrophilicity enhances initial cell behavior and early bone apposition. *Clin Oral Implants Res* 19: 491-496

Schmidt, C.E.; Shastri, V.R.; Vacanti, J.P. & Langer, R. (1997). Stimulation of neurite outgrowth using an electrically conducting polymer. *Proc Natl Acad Sci U S A* 94: 8948-8953

Schrand, A.M.; Huang, H.; Carlson, C.; Schlager, J.J.; Osawa, E.; Hussain, S.M. & Dai, L. (2007). Are diamond nanoparticles cytotoxic? *J Phys Chem* 111: 2-7

Schroeder, A.; Francz, G.; Bruinink, A.; Hauert, R.; Mayer, J. & Wintermantel, E. (2000). Titanium containing amorphous hydrogenated carbon films (a-C: H/Ti): surface analysis and evaluation of cellular reactions using bone marrow cell cultures in vitro. *Biomaterials* 21: 449-456

Schultz, L.B.; Chehab, N.H.; Malikzay, A. & Halazonetis, T.D. (2000). p53 binding protein 1 (53BP1) is an early participant in the cellular response to DNA double-strand breaks. *J Cell Biol* 151: 1381-1390

Scierski, W.; Polok, A.; Namyslowski, G.; Błazewicz, M.; Pamula, E.; Stodolak, E.; Nozynski, J.; Zwirska-Korczala, K.; Szwarc, K.; Misiolek, M.; Czecior, E.; Turecka, L.; Lisowska, G. & Orecka, B. (2007). Study of selected biomaterials for reconstruction of septal nasal perforation. *Otolaryngol Pol* 61: 842-846

Seyama, M.; Sugimoto, I. & Nakamura, M. (2004). Aroma sensing and indoor air monitoring by quartz crystal resonators with sensory films prepared by sputtering of biomaterials and sintered polymers. *Biosens Bioelectron* 20: 814-824

Shi, X.; Hudson, J.L.; Spicer, P.P.; Tour, J.M.; Krishnamoorti, R. & Mikos, A.G. (2006). Injectable nanocomposites of single-walled carbon nanotubes and biodegradable polymers for bone tissue engineering. *Biomacromolecules* 7: 2237-2242

Shi, G.; Rouabhia, M.; Meng, S. & Zhang, Z. (2008). Electrical stimulation enhances viability of human cutaneous fibroblasts on conductive biodegradable substrates. *J Biomed Mater Res A* 84: 1026-1037

Shtansky, D.V.; Gloushankova, N.A.; Sheveiko, A.N.; Kharitonova, M.A.; Moizhess, T.G.; Levashov, E.A. & Rossi, F. (2005). Design, characterization and testing of Ti-based multicomponent coatings for load-bearing medical applications. *Biomaterials* 26: 2909-2924

Sorkin, R.; Gabay, T.; Blinder, P.; Baranes, D.; Ben-Jacob, E. & Hanein, Y. (2006). Compact self-wiring in cultured neural networks. *J Neural Eng* 3: 95-101

Stary, V.; Bacakova, L.; Hornik, J. & Chmelik, V. (2002). Bio-compatibility of the surface layer of pyrolytic graphite. *Thin Solid Films* 433: 191-198

Stary, V.; Glogar, P.; Bacakova, L.; Hnilica, F.; Chmelik, V.; Korinek, Z.; Gregor, J.; Mares, V. & Lisa, V. (2003). A study of surface properties of composite materials and their influence on the biocompatibility. *Acta Montana* AB 11: 19-36

Straface, E.; Natalini, B.; Monti, D.; Franceschi, C.; Schettini, G.; Bisaglia, M.; Fumelli, C.; Pincelli, C.; Pellicciari, R. & Malorni, W. (1999). C3-fullero-tris-methanodicarboxylic acid protects epithelial cells from radiation-induced anoikia by influencing cell adhesion ability. *FEBS Lett* 454: 335-340

Sun, T.; Jackson, S.; Haycock, J.W. & MacNeil, S. (2006). Culture of skin cells in 3D rather than 2D improves their ability to survive exposure to cytotoxic agents. *J Biotechnol* 122: 372-381

Supronowicz, P.R.; Ajayan, P.M.; Ullmann, K.R.; Arulanandam, B.P.; Metzger, D.W. & Bizios, R. (2002). Novel current-conducting composite substrates for exposing osteoblasts to alternating current stimulation. *J Biomed Mater Res* 59: 499-506

Suzuki, Y.; Yoshimaru, T.; Yamashita, K.; Matsui, T.; Yamaki, M. & Shimizu, K. (2001). Exposure of RBL-2H3 mast cells to Ag(+) induces cell degranulation and mediator release. *Biochem Biophys Res Commun* 283: 707-714

Teixeira, V.; Soares, P.; Martins, A.J.; Carneiro, J. & Cerqueira, F. (2009). Nanocomposite metal amorphous-carbon thin films deposited by hybrid PVD and PECVD technique. *J Nanosci Nanotechnol* 9: 4061-4066

Terada, M.; Abe, S.; Akasaka, T.; Uo, M.; Kitagawa, Y. & Watari, F. (2009). Multiwalled carbon nanotube coating on titanium. *Biomed Mater Eng* 19: 45-52

Thalhammer, A.; Edgington, R.J.; Cingolani, L.A.; Schoepfer, R. & Jackman, R.B. (2010). The use of nanodiamond monolayer coatings to promote the formation of functional neuronal networks. *Biomaterials* 31: 2097-2104

Thierry, B.; Jasieniak, M.; de Smet, L.C.; Vasilev, K. & Griesser, H.J. (2008). Reactive epoxy-functionalized thin films by a pulsed plasma polymerization process. *Langmuir* 24: 10187-10195

Toworfe, G.K.; Composto, R.J.; Shapiro, I.M. & Ducheyne, P. (2006). Nucleation and growth of calcium phosphate on amine-, carboxyl- and hydroxyl-silane self-assembled monolayers. *Biomaterials* 27: 631-642

Tran, H.S.; Puc, M.M.; Hewitt, C.W.; Soll, D.B.; Marra, S.W.; Simonetti, V.A.; Cilley, J.H. & DelRossi, A.J. (1999). Diamond-like carbon coating and plasma or glow discharge treatment of mechanical heart valves. *J Invest Surg* 12: 133-140

Tsai, S.W.; Loughran, M.; Hiratsuka, A.; Yano, K. & Karube, I. (2003). Application of plasma-polymerized films for isoelectric focusing of proteins in a capillary electrophoresis chip. *Analyst* 128: 237-244

Tykhomyrov, A.A.; Nedzvetsky, V.S.; Klochkov, V.K. & Andrievsky, G.V. (2008). Nanostructures of hydrated C60 fullerene (C60HyFn) protect rat brain against alcohol impact and attenuate behavioral impairments of alcoholized animals. *Toxicology* 246: 158-165

Vacik, J.; Lavrentiev, V.; Novotna, K.; Bacakova, L.; Lisa, V.; Vorlicek, V. & Fajgar, R. (2010). Fullerene (C60)–transitional metal (Ti) composites: Structural and biological properties of the thin films. *Diamond Relat Mater* 19: 242-246

Vandrovcova, M.; Vacik, J.; Svorcik, V.; Slepicka, P.; Kasalkova, N.; Vorlicek, V.; Lavrentiev, V.; Vosecek, V.; Grausova, L.; Lisa, V. & Bacakova, L. (2008). Fullerene C60 and hybrid C60/Ti films as substrates for adhesion and growth of bone cells. *Phys Stat Sol (a)* 205: 2252-2261

Vasilev, K.; Sah, V.; Anselme, K,; Ndi, C,; Mateescu, M,; Dollmann, B.; Martinek, P.; Ys, H.; Ploux, L. & Griesser, H.J. (2010). Tunable antibacterial coatings that support mammalian cell growth. *Nano Lett* 10: 202-207

Watari, F.; Takashi, N.; Yokoyama, A.; Uo, M.; Akasaka, T.; Sato, Y.; Abe, S.; Totsuka, Y. & Tohji, K. (2009). Material nanosizing effect on living organisms: non-specific, biointeractive, physical size effects. *J R Soc Interface* 6 Suppl 3: S371-388

Webster, T.J.; Ergun, C.; Doremus, R.H.; Siegel, R.W. & Bizios, R. (2000). Specific proteins mediate enhanced osteoblast adhesion on nanophase ceramics. *J Biomed Mater Res* 51: 475-483

Xu, A.; Chai, Y.; Nohmi, T. & Hei, T.K. (2009). Genotoxic responses to titanium dioxide nanoparticles and fullerene in gpt delta transgenic MEF cells. *Part Fibre Toxicol* 6: 3

Yakobson, B. & Smalley, R.E. (1997). Fullerene nanotubes: C[sub1,000,000) and beyond. *Am Sci* 85, 324-333

Yamawaki, H. & Iwai, N. (2006). Cytotoxicity of water-soluble fullerene in vascular endothelial cells. *Am J Physiol Cell Physiol* 290: C1495-C1502

Yang, L.; Sheldon, B.W. & Webster, T.J. (2009). Orthopedic nano diamond coatings: control of surface properties and their impact on osteoblast adhesion and proliferation. *J Biomed Mater Res A* 91: 548-556

Yang, Z.; Wang, J.; Luo, R.; Maitz, M.F.; Jing, F.; Sun, H. & Huang, N. (2010). The covalent immobilization of heparin to pulsed-plasma polymeric allylamine films on 316L stainless steel and the resulting effects on hemocompatibility. *Biomaterials* 31: 2072-2083

Yildiz, H.; Ha, S.K. & Chang, F.K. (1998a). Composite hip prosthesis design. I. Analysis. *J Biomed Mater Res* 39: 92-101

Yildiz, H.; Chang, F.K. & Goodman, S. (1998b) Composite hip prosthesis design. II. Simulation. *J Biomed Mater Res* 39: 102-119

Zanello, L.P.; Zhao, B.; Hu, H. & Haddon, R.C. (2006). Bone cell proliferation on carbon nanotubes. *Nano Lett* 6: 562-567

Zhang, K.; Ma, Y. & Francis, L.F. (2002). Porous polymer/bioactive glass composites for soft-to-hard tissue interfaces. *J Biomed Mater Res* 61: 551-563

Zhang, Z.H. & Feng, C.L. (2007). The investigation of protein adsorption behaviors on different functionalized polymers films. *Biotechnol J* 2: 743-751

Zhao, J.; Wu, L. & Zhi, J. (2009). Non-enzymatic glucose detection using as-prepared boron-doped diamond thin-film electrodes. *Analyst* 134: 794-799

Zheng, Y.; Liu, D.; Liu, X. & Li, L. (2008). Ti-TiC-TiC/DLC gradient nano-composite film on a biomedical NiTi alloy. *Biomed Mater* 3: 044103

18

Multiscale Manufacturing of Three-Dimensional Polymer-Based Nanocomposite Structures

Louis Laberge Lebel and Daniel Therriault
École Polytechnique of Montreal,
Canada

1. Introduction

There is currently a worldwide effort for advances in micro and nanotechnologies due to their high potential for technological applications in fields such as microelectromechanical systems (MEMS), organic electronics and high-performance structures for aerospace. In these fields, nanoparticle-filled composites, i.e. nanocomposites, represent an interesting material option compared to conventional resins due to their enhanced properties and multi-functional potential. However, several significant scientific and technological challenges must be first overcome in order to fabricate nanocomposite-based structures and devices rapidly and cost-effectively. Various fabrication techniques of one or two-dimensional (1D/2D) nanocomposite structures have been developed, but few techniques are available for three-dimensional (3D) nanocomposite structures. The capability to manufacture complex 3D structures will greatly expand the practical applications of nanocomposite materials and enable the development of novel devices such as a 3D nanocomposite micro-coil spring. This chapter will present an overview of the challenges in multiscale fabrication and the current manufacturing techniques available to create 3D structures of polymer-based nanocomposites.

2. A multiscale approach

Due to the several orders of magnitude involved in the fabrication of nanocomposite devices, an efficient manufacturing technique must address the challenges at the nano-, micro- and macroscales. Figure 1 shows this multiscale concept for the creation of a 3D scaffold structure using a single-walled carbon nanotube (SWNT) and a polymer nanocomposite.

At the nanoscale, the dispersion of SWNTs should respond to the targeted usage of the nano-reinforcement. Individualization of the nanoparticles, so the particles are in contact with the matrix only, might be desirable when nanoscale properties are to be present in the final product. Conversely, slight contact between the nanoparticles is needed when the percolation phenomena through the entire domain are needed. In both cases, the interaction with the host polymer must be controlled. At the microscale, the production of nanocomposite structures must allow a control over the orientation of high aspect-ratio nanoparticles such as SWNTs. The arrangement of the nanocomposite microscale structures in 3D permits the localization and orientation of the nano-reinforcement in a macroscale product.

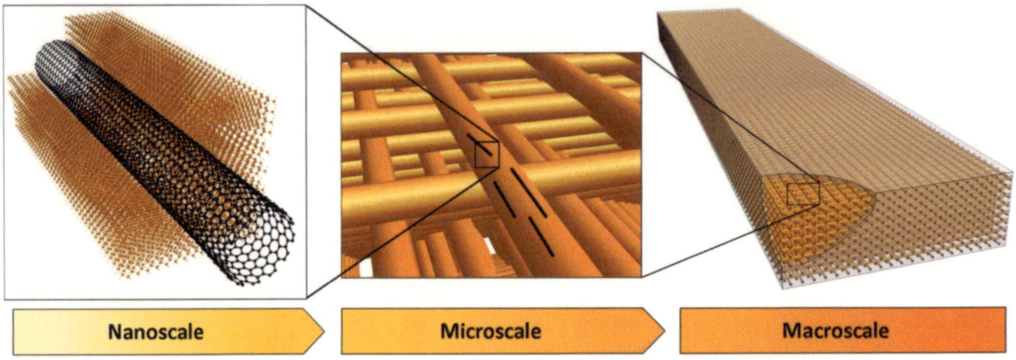

Fig. 1. Nano, micro and macro scales have to be considered for the successful manufacturing of a SWNT/Polymer nanocomposite.

3. Nanoscale

The nanoscale poses important manufacturing challenges such as the dispersion of the nanoparticles and the close interaction of the nanoparticles with the host polymer matrix (Byrne & Guin'Ko, 2010). In a dispersed state, the distance between particles in the matrix must be controlled to achieve the targeted properties. For mechanical reinforcement, every nanoparticle should be separated from each other to maximize the interface between the matrix and the nano-reinforcement, thus enhancing the available area for stress transfer to the nano-reinforcement. Aggregated nanoparticles are often responsible for underperformances. For example, carbon nanotube (CNT) aggregates in polymer nanocomposites can act as stress concentrators that trigger the fracture of the nanocomposite when submitted to moderate stress (Wong, Paramsothy et al., 2003). In carbon/magnetite nanocomposite produced by electro-spinning, the particle growth due to agglomeration during pyrolysis drastically reduces the paramagnetic property of the magnetite nanoparticle (Bayat, Yang et al., 2010). Conversely, contact between particles can be desirable to achieve an electrically percolated network in nanocomposites using CNTs. It has been observed that polymer coating around CNTs can lower the electrical conductivity of a nanocomposite due to the separation of the CNTs (Kilbride, Coleman et al., 2002).
Several studies theorized the quantity of particles that can be added to a polymer while keeping the continuity of the matrix phase (Coleman, Khan et al., 2006; Sandler, Pegel et al., 2004). The maximum concentration of SWNTs that can be added to a polymer without having discontinuities in the polymer phase was calculated using a hexagonal-packed array of aligned SWNTs and an interface zone of 5 nm. The maximum volume/volume (vol%) concentration was calculated at 1 vol%. Equation 1 can be used to understand this conclusion.

$$V_{f,\max} = \frac{\pi r^2}{2\sqrt{3}(1+r)^2} \quad with\ r = \frac{d}{c} \tag{1}$$

The maximum volume concentration ($V_{f,max}$) of cylindrical particles in a hexagonal packed arrangement is expressed as a function of the ratio (r) of the diameter of the cylinders (d) over the separation distance (c) between them. Figure 2 plots this relation and two examples are given: (1) the ratio of 0.12 for SWNTs of 1.2 nm diameter and a separation distance of 10

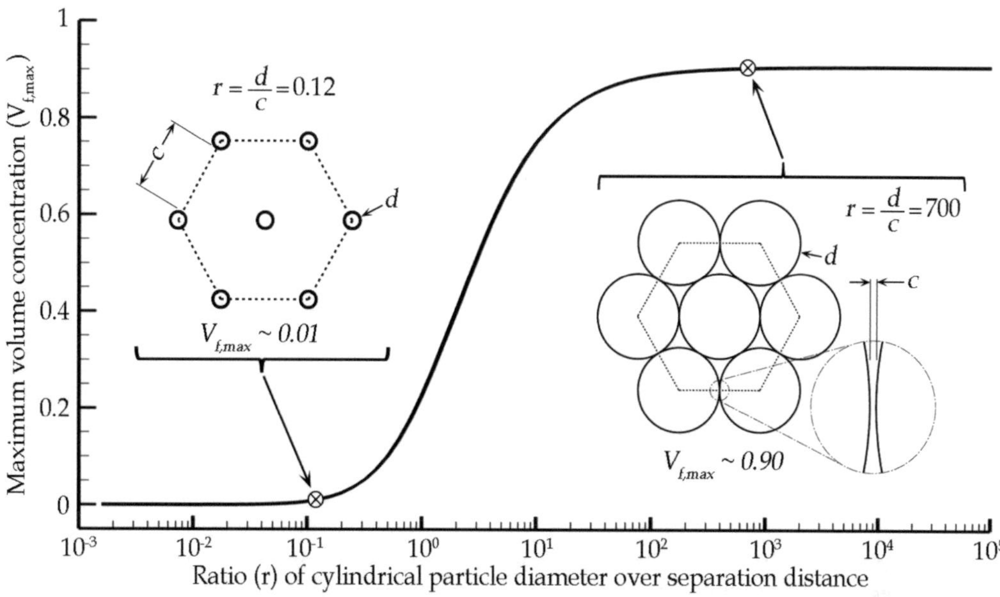

Fig. 2. Relation between the maximum volume concentration of cylindrical particles in hexagonal packing and the ratio (*r*) of the diameter (*d*) over the separation distance (*c*). Two examples are given for ratios of 0.12 and 700.

nm, which gives a maximum concentration of ~ 1 vol%; and (2) the ratio of 700 for carbon fibers of 7 μm with also a separation distance of 10 nanometers, which gives maximum concentration of ~ 90 vol%.

The interface zone is a certain thickness around nanoparticles where the polymer conformation is influenced by the nanoparticle. The value of the polymer radius of gyration is commonly employed to gage this distance. The radius of gyration of a polymer is the quadratic average of the distance between every monomer of the chain and the polymer mass center (Bicerano, 2002). It generally varies from 3 to 30 nm (Winey & Vaia, 2007). Using simulations, it has been shown that the surface of the nanoparticle provides a preferential orientation for the polymers also for a distance equal to the radius of gyration of the polymer (Starr, Schrder et al., 2002). However, special polymer conformation might as well exist in the case of higher strength or longer ranged interaction between the polymer and the nanoparticle. For example, several polymers having aromatic cycles can adopt a wrapping configuration around carbon nanotubes. This configuration has been observed with polyaniline (Sainz, Benito et al., 2005) and polyvinyl pyrolidone (O'Connell, Boul et al., 2001).

Moreover, nanoparticles, i.e. particles with at least one dimension in the order of the nanometer, possess a large area available for interaction with the host matrix compared to traditional composite reinforcements due to their high surface/volume ratio. For small diameter nanoparticles, the interface-affected volume might be larger than the volume of all the dispersed nanoparticles (Winey & Vaia, 2007). This interaction volume with the polymer can be tailored using covalent functionalization. For example, heating CNTs in inorganic acids (sulfuric and nitric) oxidize the CNTs mainly from the ends, grafting carboxylic groups. Such carboxylic groups may interact with the matrix to improve the stress transfer in mechanical applications. This carboxylic group is often used to further functionalize the CNTs by covalently attaching DNA (Baker, Cai et al., 2002) or ionically attaching octadecylamine (Chen, Rao et al., 2001), for example. However, the chemical modification of the nanoparticle has to be done carefully to prevent the creation of defects that would lower the mechanical properties or change the electronic signature. Frankland *et al.* simulated a SWNT in a polyethylene matrix having a low density (< 1%) of covalent bounds between the matrix and the SWNT sidewalls. This configuration contributed to an augmentation of one order of magnitude of the shear stress transfer between the polymer and the SWNT without lowering significantly the tensile strength of the SWNT. However, the mechanical properties of the modified SWNT significantly dropped when 10% of the carbon atoms where modified (Frankland, Caglar et al., 2002). Moreover, extensive modification of the SWNT sidewall may destroy the delocalized π-electron system responsible for their electrical conductivity (Choi & Ihm, 1999).

Another approach is the usage of non-covalent functionalization of the carbon nanotube. The molecules used to perform this functionalization should have two functional ends: one that will preferably interact with the nanoparticle and another one that will interact with the host polymer matrix. Surfactants such as the sodium dodecyl sulfate, sodium dodecylbenzene sulfonate, or Triton X-100 are efficient to help the dispersion of carbon nanotubes in water (Islam, Rojas et al., 2003; Jiang, Gao et al., 2003). These molecules have a hydrophobic extremity that interacts with the CNT and a hydrophilic extremity that helps the dispersion in polar mediums like water. As stated previously, polymers can also wrap around CNTs. The wrapping process is attributed to the thermodynamic forces that tend to eliminate the hydrophobic interface between the CNT and the water (O'Connell, Boul et al., 2001). However, these wrapping techniques are not easily transferable to non-polar solutions such as the organic polymer matrices. Other successful non-covalent functionalizations are performed with molecules having polycyclic aromatic rings such as pyrenes (Chen, Zhang et al., 2001) and porphyrins (Satake, Miyajima et al., 2005). These groups can interact strongly with graphitic basal plane through π-π interactions.

Nanoscale dispersion is of paramount importance to benefit from the nanoparticle mechanical properties in a polymer nanocomposite. Nanoparticles have a tendency to agglomerate because, at such a small scale, the van der Waals forces are significant. This type of interaction comes from the deformation of the electron cloud around the atom, which creates an instant polarity that can be attracted by a nearby atom. In any case, a proper modification of the nanoparticle surface through covalent or non-covalent functionalization will prevent the re-agglomeration of the particles through steric or electrostatic forces.

Moreover, nanoparticles such as CNTs may have irregular shapes and, therefore, can be aggregated in their as-synthesized form. These aggregates greatly reduce the aspect-ratio of the nanoparticle or the available surface for interaction with the polymer (Thostenson & Chou, 2002). In this case, it might be difficult to achieve a nanoscale dispersion throughout the matrix (Stoeffler, Perrin-Sarazin et al., 2010).

Polymer nanocomposite with high aspect ratio nanoparticles can be highly viscous. Thus, the ultrasonic or mechanical mixing operations might generate undesirable heat, especially when integrating the nanoparticles to a thermoset polymer with a cross-linking reaction accelerated by temperature. The usage of a proper solvent is then favorable to reduce the viscosity during the integration into the matrix. Solvent-based processing also prevents the emission of air-born particles that might cause respiratory problems. An appropriate solvent must be able to both maintain the nanoparticles in suspension and dissolve the polymer. For CNT nanocomposites, tetrahydrofuran or toluene has been employed with polystryrene

(Thostenson & Chou, 2002; Wong, Paramsothy et al., 2003), dimethylformamide with poly-methyl methacrylate (Haggenmueller, Gommans et al., 2000), and chloroform with epoxy (Xu, Thwe et al., 2002). The nanocomposite is then recovered by evaporating the solvent in a vacuum oven or simply under a fume hood.

Ultrasonication is a common treatment employed to disperse nanoparticles in solvents and matrices. However, this technique does not individualize the CNTs and might just provide a good dispersion at the microscale. Also, re-agglomeration of CNTs have been observed when the ultrasonic treatment is stopped (Haggenmueller, Gommans et al., 2000).

Fluidic shear forces can help the dispersion of CNTs in a polymer matrix. Several types of mechanical mixers and homogenizers provide these forces. For example, the three-roll mill is composed of three cylinders rotating at different speeds and opposite directions. When the material passes through the small gap between to counter-rotating cylinders, the extreme shear forces generated are strong enough to disperse the nanoparticles at the nanoscale level (Gojny, Wichmann et al., 2004). When a thermoplastic polymer matrix is used, the fluidic shear forces in the extruder can help the dispersion of the nanoparticles. The shear forces can also come from the manufacturing process of the nanocomposite. For example, electro-spinning is a process in which a fiber is ejected from a spinneret with the help of a voltage applied between the spinneret and the substrate. The extreme shear forces involved during the ejection of the fiber have been considered responsible for the exfoliation of CNTs from their bundled nature (Ayutsede, Gandhi et al., 2006).

4. Microscale

The arrangement of nanocomposite material in structures typically at the micron to several hundred micron range offers several advantages. First, the material needed is reduced when the cost is an issue. In addition, the microstructures manufacturing techniques allow a better control on the nanoparticle disposition due to their microscale confinement. For example, high aspect ratio nanoparticles, such as CNTs, can align themselves along the flow direction with the help of the high shear achievable in small-scale manufacturing. Several techniques are used to produce nanocomposite at the microscale.

Microinjection molding (MIM) is an emerging method to manufacture microscale devices from polymer nanocomposites (Huang, Chen et al., 2006). Polycarbonate and multi-walled carbon nanotube (MWNT) nanocomposite microscale dog bone samples were manufactured by MIM. The high shear flow in the cavity while molding resulted in the alignment of the MWNTs along the flow direction (Abbasi, Carreau et al., 2010).

The process of fiber production, i.e. the "spinning", works by forcing a viscous material through a small orifice, called a «spinneret». There are several variations of this process and each of them rely on specific viscoelastic properties of the spun material (Katayama & Tsuji, 1994).

The "melt spinning" is the most common method of producing fibers and has been used to produce CNT nanocomposite in thermoplastic matrices. After the integration of the nanoparticle in the polymer, the nanocomposite is extruded through a small orifice in a die, and the produced fiber is allowed to cool. During spinning, the fiber can be drawn. This will provide an extensional flow that orients the polymer molecules along the fiber direction. Several studies showed the alignment of CNTs along the flow direction and the positive effect on the mechanical properties of the fiber. (Sandler, Pegel et al., 2004; Thostenson & Chou, 2002).

"Solution spinning" was also used to produce carbon-nanotube fibers. This process extrudes a solution of nanocomposite in a non-solvent (Tsurumi, 1994). When the material is in contact with the non-solvent, a coagulation of the material occurs to form a fiber that can be collected. MWNT/polyvinyl alcohol (PVA) fibers were produced by solution spinning of water-based MWNT dispersion in an aqueous solution of PVA. The contact between the two solutions induced a polymer bridging coagulation of the MWNTs (Miaudet, Bartholome et al., 2007).

"Gel spinning" has been developed for the production of ultra-high molecular weight polyethylene fibers having ultra-high strength. The material has first to be treated to form a gel-like state that can be spun. This technique allows a superior control onto the molecular chains arrangement to produce a defect-free highly crystalline polymer fiber. Gel-spinning has also been used to produce SWNT/PVA nanocomposite fibers. The high alignment of the nano-reinforcement along the fiber axis improved the mechanical properties compared with pure PVA fibers (Minus, Chae et al., 2009).

Electro-spinning is another method widely used to produce nanocomposite fibers. In this method, the driving force for the spinning of the fiber is a voltage applied between the spinneret and the substrate. Highly charged polymer is ejected from the spinneret towards the substrate in the form of a fiber. The diameter of the produced fibers is typically in the range of 50 to 800 nm. SWNT reinforced polylactic acid and polyacrilonitrile nano-fibers have been produced using electro-spinning (Ko, Gogotsi et al., 2003).

5. Macroscale

Different techniques exist to manufacture nanocomposite products at the macroscale. A polymer nanocomposite can be simply molded in a shape before hardening either by cooling or by the effect of curing reaction. This relatively simple technique could find applications in traditional fiber reinforced composites by modifying the matrix-dominated properties. While the micro-fibers (e.g., glass, carbon, Kevlar) still serve as the main structural reinforcement, the nanoparticle enhanced polymer matrix could provide better fracture toughness and thermal stability, for example. More precisely, electrically conductive CNT/polymer matrix could give other functionalities to the composite such as electrostatic discharge protection, electromagnetic shielding and even strain and damage sensing (Thostenson & Chou, 2006). Although molding techniques are simple, they do not offer the possibility of controlling the position and orientation of the nanoparticles locally. Hence, it is challenging to position the nanoparticles where they are needed in a macroscopic product. Moreover, high-aspect ratio nanoparticles cannot be oriented along different directions corresponding to design requirements.

If the nanocomposite is manufactured in the form of continuous micro-fibers, weaving and braiding techniques can be used to dispose the nanocomposite in 3D. However textile-processing methods are labor-intensive and manufacture structures in repeating units only. Nanocomposite fibers produced by electro-spinning are usually randomly deposited on a substrate producing non-woven mats. Oriented electro-spun fiber layers can be assembled by employing a special collector electrode arrangement (Li & Xia, 2004). The obtained uni- or multi-directional fiber films can provide a control of the orientation of nanoparticles in 1D or 2D. Techniques that produce films – that is, solvent casting (Xu, Zhang et al., 2006), spin coating (Xu, Thwe et al., 2002), layer-by-layer (Mamedov, Kotov et al., 2002), extrusion and stretching (Thostenson & Chou, 2002), or compression (Haggenmueller, Gommans et al., 2000) - are known to produce random alignment of CNTs in 2D or even alignment in 1D. Stereo-lithography can also be used to build 3D macroscale products from UV-curing polymers. A thin layer of nanocomposite is cured on the surface of a resin bath using localized UV exposure. Layers of material are successively hardened on top of each other to form a 3D structure. Stereo-lithography has been used to manufacture 3D structures with CNT nanocomposite (Varadan, Xie et al., 2001), or ceramic nanoparticle nanocomposites (Varadan & Varadan, 2001). However, this technique cannot orient axisymmetric nanoparticles during manufacturing because no force is applied on the particles randomly dispersed in the resin bath. Also, due to the layer-by-layer curing, curved surfaces on the vertical plane will be formed by a succession of discrete right angles.

Microfibers can be assembled in structures using the direct-write fabrication method. This manufacturing technique consists of the motion of an extrusion micro-nozzle that generates defined patterns in 3D with material filaments (Lewis & Gratson, 2004). The deposited material, generally called "ink", is extruded and allowed to lie on a substrate along a defined pattern. Then, the extrusion nozzle is raised by an increment and other filaments can be deposited on the previous layer. The stacking of the filaments produces a 3D structure. This method is relatively simple and the resolution of products can be down to the micrometer range (Gratson, Garcia-Santamaria et al., 2006). During direct writing, the displacement speed of the extrusion device should match the linear extrusion rate of material from the nozzle. However, a slightly higher displacement speed will stretch the filament, thereby providing an extensional flow, enhancing the alignment of the nanoparticles.

The ink must have tailored viscoelastic properties to enable 3D manufacturing. That is, it must be fluid enough to be extruded through a micro-nozzle. Additionally, its rigidity must increase after extrusion to form a filament that can be deposited between support points. Several materials offer this type of behavior. Organic fugitive inks with superior stiffness have been used as self-supporting spanning filaments. This material also exhibited moderate shear yield stress to facilitate extrusion through a deposition nozzle. The development of this fugitive ink enable the direct-write manufacturing of filamentary scaffold structures having more than a hundred layers (Therriault, Shepherd et al., 2005). Spanning filaments of silver nanoparticle reinforced polyacrylic acid were deposited by direct-write between two electrodes with out-of-plane filament curvature (Ahn, Duoss et al., 2009). After material sintering, the filament insured the electrical connectivity between individual micro-solar cells of a solar panel. Figure 3 illustrates the deposition of filaments using the direct-write technique.

The various direct-write techniques have been limited mainly to supported structures in a layer-by-layer building sequence and straight spanning filaments between support points. This limitation comes from the fact that for shear-thinning inks, the ink rigidity increases at the same moment as the ink exits the extrusion nozzle, that is, when the shear strain applied to the material returns to a near zero value. Therefore, the extruded filament has the same rigidity from its extrusion point to the previous support point. Given these conditions, the application of a side force due to the extrusion nozzle changing direction results in a bending moment that reaches a maximum at the support point, hence creating a deflection of the whole deposited filament (see Figure 3b).

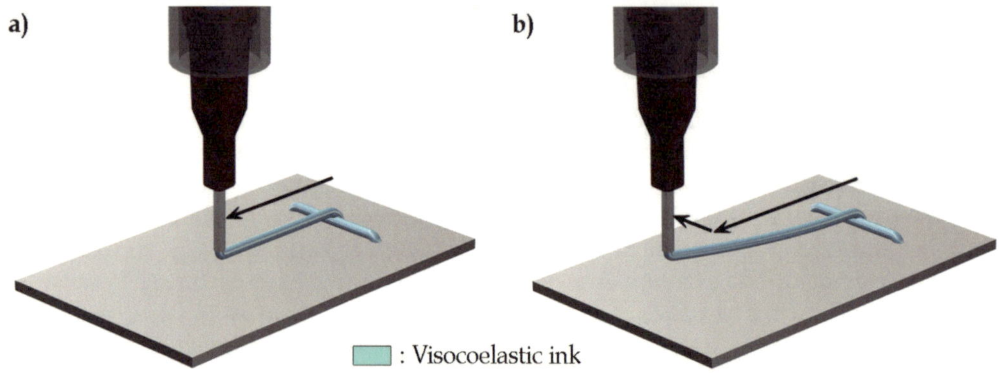

Fig. 3. Direct-write deposition of a viscoelastic filament: a) Extrusion of a filament having the same rigidity from the extrusion point to its anchoring point. b) Bending of the whole filament when the extrusion nozzle changes direction.

6. Multi-scale fabrication of SWNT/polyurethane nanocomposite structures

The following sections present an example of multi-scale manufacturing of nanocomposite structures using SWNTs and a UV-curable polyurethane. The challenges involved at the nano-, micro-, and macroscales are addressed step-by-step. First, the SWNTs are dispersed in the polyurethane using functionalization and shear mixing. Then, micro-fibers are deposited using the direct-write technique. Finally, two methods involving the direct-write assembly technique are employed to manufacture tridimensional nanocomposite structures.

6.1 Nanoscale: integration of SWNTs to a polyurethane polymer

A SWNT/polyurethane compound can be prepared according to the procedure schematized in Figure 4. Before integration to the polymer, a reflux of five hours in a nitric acid solution was used to treat the as-grown SWNTs. This treatment purified the as-grown material and also attached carboxylic chemical groups onto the nanotube sidewall (see Figure 4-1). The nanotube purity and covalent modification was verified by transmission electron microscope, Raman microscopy, thermo--gravimetric analysis, and X-ray photoelectron microscopy. A non-covalent modification was also performed by sonicating a specific amount of the purified SWNTs in a 0.1 mM solution of Zinc Protoporphyrin IX (ZnPP) in the dichloromethane (DCM) solvent (see Figure 4-2). As stated earlier, the porphyrin end of the ZnPP molecule have polycyclic aromatic hydrocarbon end that adhere to the nanotube sidewall through $\pi - \pi$ interaction. The other end of the molecule is composed of carboxylic groups, which offer possible interaction sites with the matrix.

A commercially available UV-curable polyurethane (NEA123MB, Norland Products) was then slowly added to the SWNT solution in DCM while stirring with a magnetic stirrer (see Figure 4-3 and 4-4). The polyurethane has -CONH- groups on the polymer backbone capable of interacting with the carboxylic groups on the covalently and non-covalently modified SWNTs (Sahoo, Jung et al., 2006). After evaporation of the DCM solvent (see Figure 4-5), the nanocomposite mixture was passed several times in a three-roll mixer mill where the gap between the rolls and the speed of the apron roll were controlled (Figure 4-6). Fumed silica nanoparticles were also added to increase the viscosity and to yield a shear-thinning behavior (Raghavan, Walls et al., 2000). The nanocomposite was slowly added to a

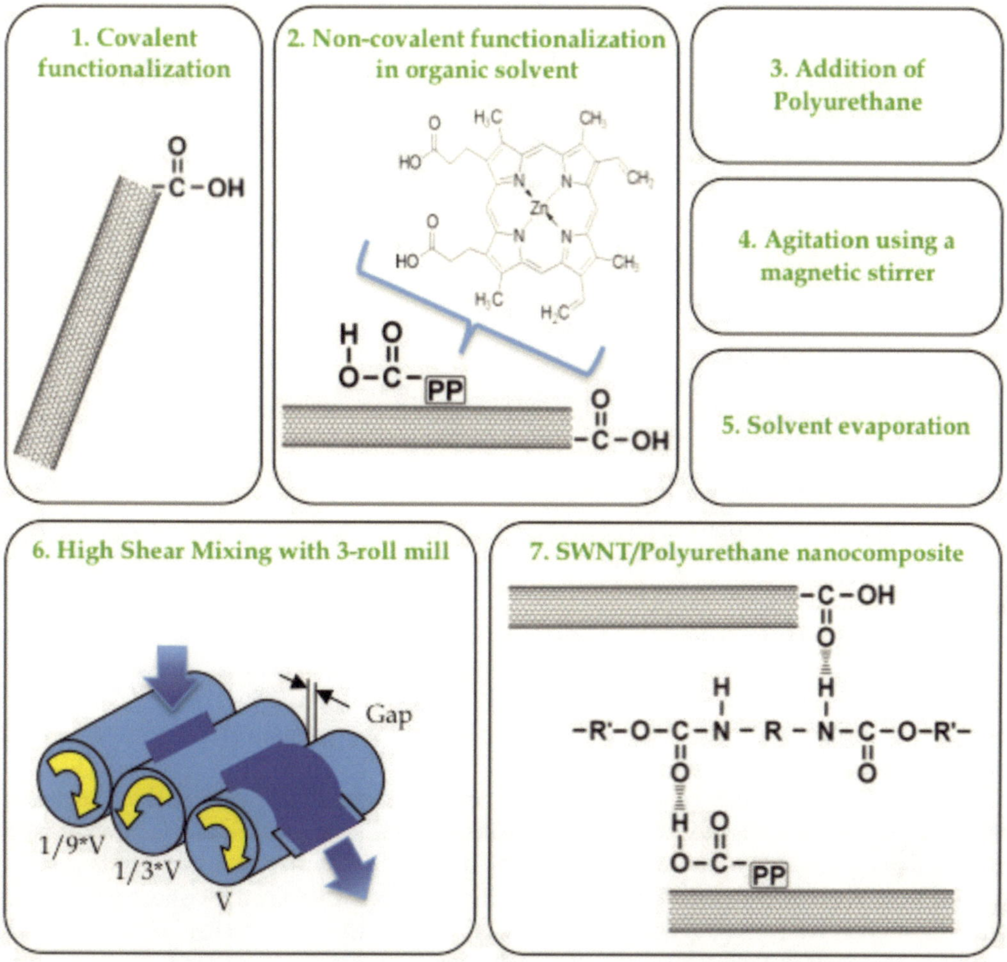

Fig. 4. Preparation steps for the integration of SWNTs to a polyurethane matrix

fumed silica nanoparticle solution in DCM while mixing. Finally, the nanocomposite was poured into syringe barrels, where the DCM solvent was fully evaporated. The whole procedure produced UV-curable nanocomposites containing 0.5 wt% of purified and functionalized SWNTs and 5 wt% of fumed silica (see complete experimental details in (Lebel, Aissa et al., 2009)).

Figure 5a shows the microscopic scale dispersion of the final nanocomposite. Dark spots, of average size of ~ 1.3 µm, are observed in the nanocomposite blend. These spots are believed to be aggregates of carbonaceous materials or SWNTs entangled around larger carbon particles. From the observation of the distribution of the visible dark spots, one can assume that the SWNTs, along with their associated agglomerates, were uniformly dispersed in the polyurethane matrix. However, it is difficult to ascertain that all the SWNTs are exfoliated and dispersed at the nanoscale level. Such distribution was also often obtained in other studies (Bose, Khare et al., 2010). This trend poses a serious question about the possibility of de-agglomeration of SWNT when they are aggregated or bundled in their as-grown form by traditional means. In case the raw material is highly aggregated, an ultra-centrifugation

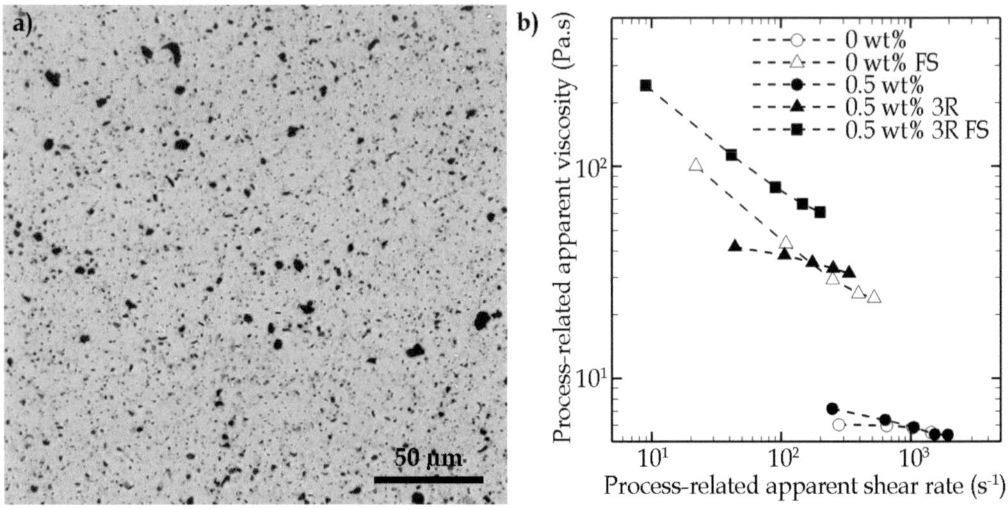

Fig. 5. a) Optical microscope image of SWNT/Polyurethane nanocomposite illustrating its dispersion at the microscale. b) Apparent viscosity measurements of the nanocomposite at different stages during the integration process.

procedure should be performed after the solubilization of the nanoparticle in the organic solvent with the aid of functionalization (see figure 4, between steps 2 and 3). Ultracentrifugation is a well-known technique to separate the SWNT material from the aggregates (Kim, Nepal et al., 2005) and can also be used to segregate SWNTs based on their electronic properties (Green & Hersam, 2007).

The apparent viscosity is a good indicator of the dispersion of the SWNTs. Well-dispersed SWNTs increase the viscosity and may create a shear thinning behavior. It has been observed that the shear-thinning behavior of a SWNT/polymer nanocomposite can be caused by a rheological percolation network that is created when the critical percolation concentration is reached (Du, Scogna et al., 2004). The latter is more sensitive to the aspect ratio (i.e. length over diameter) of the reinforcing nano-material rather than to its actual size (Balberg, Binenbaum et al., 1984).

Figure 5b shows the process-related viscosity at different shear rates measured by capillary viscometry at different steps during the integration procedure. The indication 0 wt% stands for pure polyurethane; 0.5 wt% stands for the polyurethane having a 0.5 wt% concentration of SWNTs; 0.5 wt% 3R stands for the nanocomposite mix after the three-roll mixing; and 0.5 wt% 3R FS stands for the nanocomposite after the addition of fumed silica. These results where compared with polyurethane also having a 5 wt% fumed silica charge (0 wt% FS). The process-related apparent viscosity (η) values were plotted on a log-log graph as a function of the shear rate ($\dot{\gamma}$) and fitted with a power-law fluid relationship according to the following equation:

$$\eta = m\dot{\gamma}^{(1-n)} \tag{2}$$

Table 1 lists the calculated values of the power-law indexes (n) as an indication of the viscosity behavior with respect to the strain rate. The sample 0 wt% has an index near 1 showing a Newtonian behavior, as the value of viscosity does not vary across the strain rates investigated. The first mixing stage consisted of the incorporation of the SWNTs in the polyurethane resin. After the nanotube incorporation, a slight shear-thinning behavior was observed (0.5 wt%, $n \approx 0.85$) due to nanotubes that formed a viscosity-percolated network. After the three-roll mixing, a drastic viscosity increase was observed, indicating that the three-roll mill reduced the size of aggregates and successfully dispersed the added material. However, the calculated viscosity indexes after this mixing stage remained unchanged. Thus, it can be argued that the three-roll mixing step did not generate higher aspect-ratio entities (i.e., individual nanotubes). The incorporation of fumed silica particles further increased the viscosity and changed the rheological behavior, as indicated by the viscosity index reaching a value of 0.5. A similar response is observed at different intensities for the 0 wt%-FS, whose power-law index value is comparable. This effect is the result of a network formation of hydrogen bonded fumed silica particles that impart a gel-like rheological behavior to the mixture at rest. The apparent viscosity diminishes under the application of a moderate shear force destroying the weakly bounded network (Raghavan, Walls et al., 2000).

Sample	Viscosity index (n)
0 wt%	0.95
0 wt% FS	0.53
0.5 wt%	0.85
0.5 wt% 3R	0.86
0.5 wt% 3R FS	0.55

Table 1. Viscosity indexes of nanocomposite blends.

The preparation procedure resulted in two nanocomposite mixtures. The low viscosity nanocomposite, that is, the 0.5 wt% 3R, can be used in casting or molding situations, while the high viscosity nanocomposite, that is, the 0.5 wt% 3R FS, has a gel-like rheological behavior that is readily spinnable in the form of fibers.

6.2 Microscale: UV-assisted direct-write fabrication of nanocomposite fibers.

The UV-curable nanocomposite (0.5 wt% 3R FS) prepared according to the procedure in section 6.1 can be placed in a syringe barrel equipped with a micro-nozzle tip. Under the application pressure, the gel-like nanocomposite is extruded through the micro-nozzle to form a fiber. Also, the syringe barrel can be displaced in the 3D space using a micro-positioning robot. Then, using a UV source that follows the extrusion point, the material is cured right after extrusion to form a fiber that can span between two support points. Figure 6a describes the ultraviolet-assisted direct-write (UV-DW) fabrication of nanocomposite fibers. A fiber is extruded between two polymer pads by robotic displacement of an extrusion nozzle. The fiber is exposed right after extrusion to a UV-source to cure the nanocomposite. Figure 6b shows the actual fiber produced and Figure 6c shows the stress-strain curve measured using a dynamic mechanical analyzer while testing the nanocomposite (0.5 wt% 3R FS) and polyurethane (0 wt% FS) fibers under tensile loading.
The stress-strain curves shown in Figure 6c indicate a clear change in the mechanical behavior between the unloaded polyurethane fibers and the nanocomposite ones. The unloaded material (0 wt% FS) exhibits a non-linear response of the stress under strain and experiences a large elongation before rupture. In contrast, the nanocomposite (0.5 wt% 3R FS) response is rather elastic and linear until breakage, which is reached at rather shorter

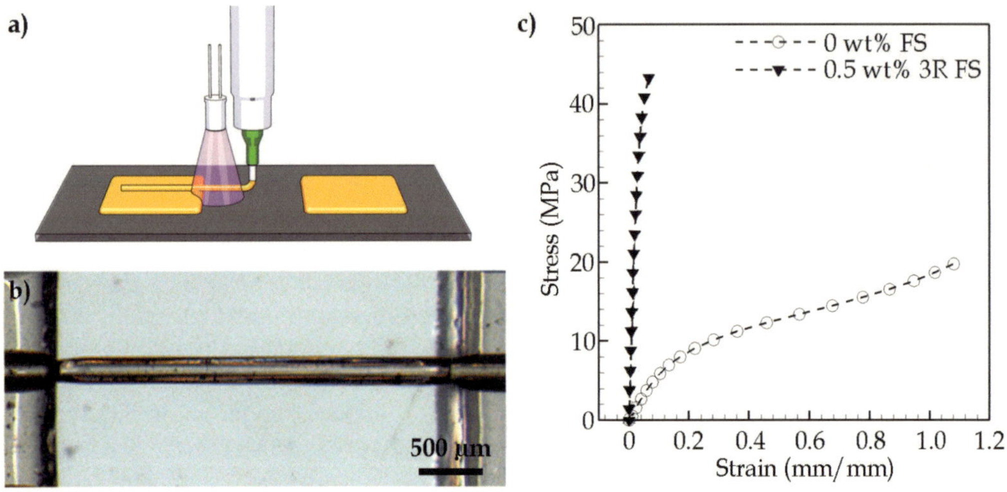

Fig. 6. Production of a SWNT/Polyurethane nanocomposite fiber by ultraviolet-assisted direct writing. a) The fiber is extruded using a moving micro-nozzle between two polymer pads. b) Optical microscope image of the produced fiber between two pads. c) Mechanical characterization of the nanocomposite fiber.

elongations. This radically different behavior in comparison with the unloaded polyurethane is attributed to the SWNTs incorporation and a significant stress transfer between the host matrix and the nano-reinforcement. The mechanical characterization of the microfibers revealed a significant increase in both the strength (by ~64%) and the stiffness (by more than 15 times). These mechanical enhancements were attributed to the presence of SWNT, and both the covalent and the non-covalent functionalizations of the SWNTs (Lebel, Aissa et al., 2009).

6.3 Macroscale: fabrication of 3D SWNT/polyurethane nanocomposite structures using direct-write techniques

The direct-write techniques offer the possibility of arranging fibers in three dimensions. Two approaches have been developed using the nanocomposite prepared according to the procedure in section 6.1. The low viscosity nanocomposite (0.5 wt% 3R) was used for the infiltration of 3D micro-fluidic networks. The more viscous and shear thinning nanocomposite (0.5 wt% 3R FS) was deposited in 3D using the UV-DW approach. The following sections briefly present these two approaches.

6.3.1 Nanocomposite infiltration in 3D microfluidic network

This approach permits the fabrication of a 3D-reinforced product through the directed and localized infiltration of SWNT/polymer nanocomposites into a 3D microfluidic network. These microfluidic networks are created using the flexible direct-write assembly method of organic fugitive ink. Figure 7 illustrates the manufacturing steps to manufacture nanocomposites beams by the microfluidic infiltration approach. The direct-write assembly method is first used to generate a scaffold of fugitive ink filaments (Figure 7a). The rigidity of this type of ink at room temperature is high enough to resist deformation when the filament is disposed between two support points. The filaments can also be layered without the collapse of the scaffold structure (Therriault, White et al., 2003). The architecture of the scaffold can be tailored at will by controlling the number and orientation of filaments in every layer (Figure 7b). After completion of the deposition, an epoxy matrix is used to encapsulate the scaffold (Figure 7c). The resin is then allowed to harden around the fugitive ink filaments. Heat generation during curing reaction of the resin is kept to a minimum to prevent deformation of the fugitive ink scaffold. In the fourth step, the encapsulated ink filaments are heated to liquid state and then drained to obtain a 3D microfluidic network, which is reproducing the ink filament scaffold architecture (Figure 7d). In the fifth step, the microfluidic network is infiltrated with a SWNT/polymer nanocomposite fluid (Figure 7e). The final step is to cure the nanocomposite to obtain an epoxy beam reinforced by a 3D network of nanocomposite fibers (Figure 7f). This technique allows the designer to place nanocomposite filaments were they are needed in a macroscopic product. Moreover, at the structural level, stress concentration is low due to the circular cross-section of the produced microfluidic channels.

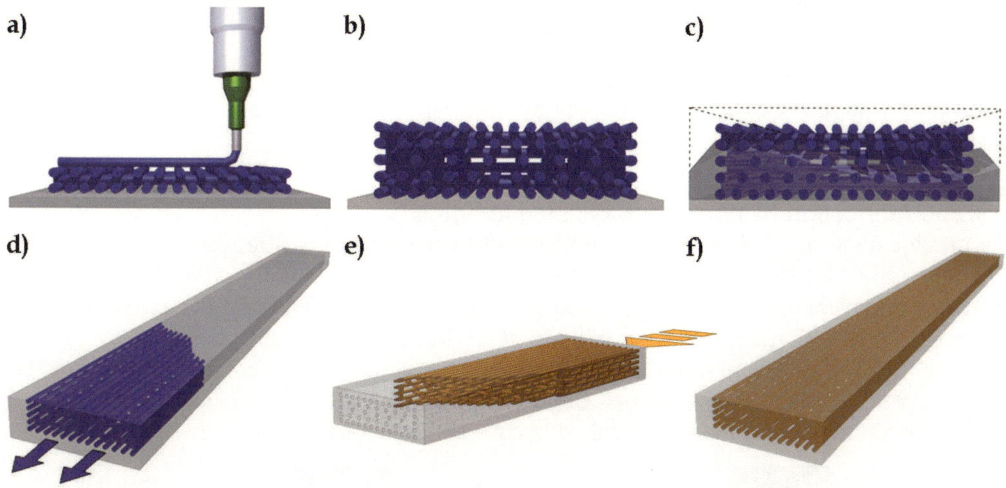

Fig. 7. Production of a macroscale nanocomposite structure by the infiltration of a microfluidic network.

By using the technique depicted in Figure 7, nanocomposite-reinforced rectangular beams were manufactured to provide good bending rigidity. Flexural mechanical solicitation generates an axial stress distribution that linearly increases from the neutral mid-plane until reaching a maximum value at the outer surfaces. Figure 8a shows the as-produced beam reinforced by a nanocomposite micro-fiber skeleton. Figure 8b shows a top view image of the beam where perpendicular (in the lateral direction) microfluidic channels were required to support and interconnect the different levels of axial fibers to facilitate the complete filling of the 3D microstructure. Figure 8c shows a beam cross-section where the number of axial reinforcing infiltrated microfibers was increased in higher stress regions of the cross-section. Dynamic mechanical analysis in bending showed an increase of 12.5% in the storage modulus at temperatures below 35°C compared to the neat resin infiltrated beams (Lebel, Aissa et al., 2009). This technique can be compared to microinjection molding with the particularity that both the mold and the injected structure are part of the same product. No particular rheological behavior is needed as long as the viscosity is sufficiently low. High viscosity nanocomposite need high injection pressure that could damage the microfluidic network during nanocomposite injection.

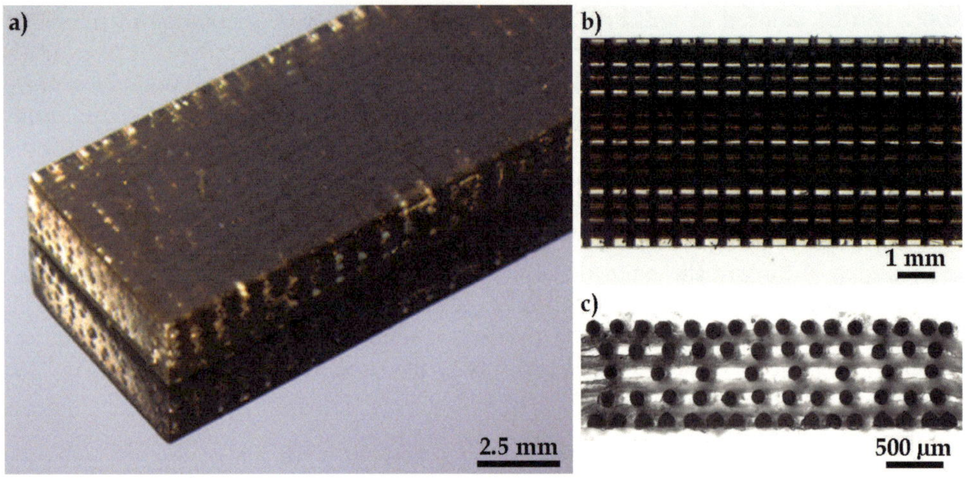

Fig. 8. a) Nanocomposite reinforced beam. b) Top view of the beam showing the injected nanocomposite fibers. c) Cross-section of the beam. The black circles are the axial fiber cross-sections.

6.3.2 3D fabrication by ultraviolet-assisted direct-write

As previously seen, DW techniques are limited to deposition of straight or simply curved filaments. In order to create a curved shape by changing the moving path of the extrusion nozzle, another increase in rigidity has to occur slightly distanced from the extrusion point. As a result, the filament bending will occur at the transition zone between to the low bending rigidity of the newly deposited material and the higher bending rigidity of the cured spanning filament previously deposited. This principle is illustrated in Figure 9, where the filament bends before and during cure of the material under exposure to UV.

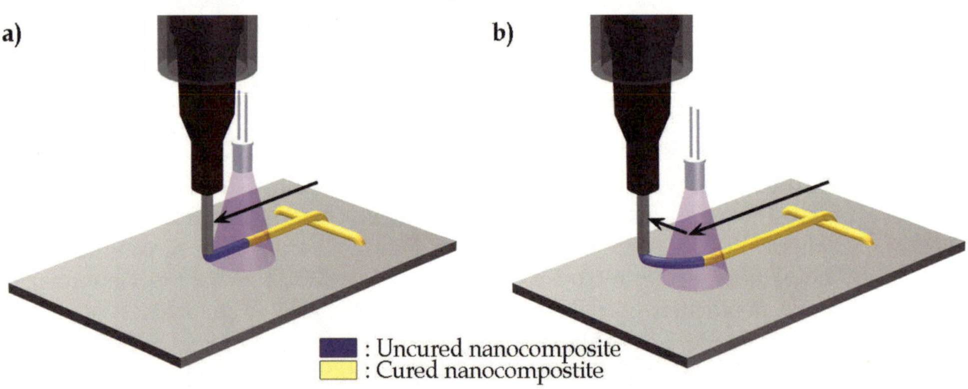

Fig. 9. Concept of ultraviolet-assisted direct-write technique a) Deposition and curing of a nanocomposite filament using an UV source following the extrusion nozzle. b) Filament curvature initiated at the curing transition zone, following the extrusion nozzle path.

The ultraviolet-assisted direct-write method extends the manufacturing space for nanocomposite structures and opens up new prospects for nanocomposite devices in 3D. For successful fabrication using the UV-DW technique, the nanocomposite curing speed under radiation must match closely or be faster than the moving speed of the extrusion device.

Moreover, the distance between the UV radiation exposure and the filament extrusion point has to be carefully calibrated. If the curing reaction is triggered in the extrusion nozzle, there is a risk of clogging. If the curing reaction is triggered far from the extrusion point, the nanocomposite filament will not reproduce the moving path of the extrusion point. The ability to cure the nanocomposite along a certain extrusion path allows the designer to create curve-shaped structures in space such as the ones presented in Figure 10.

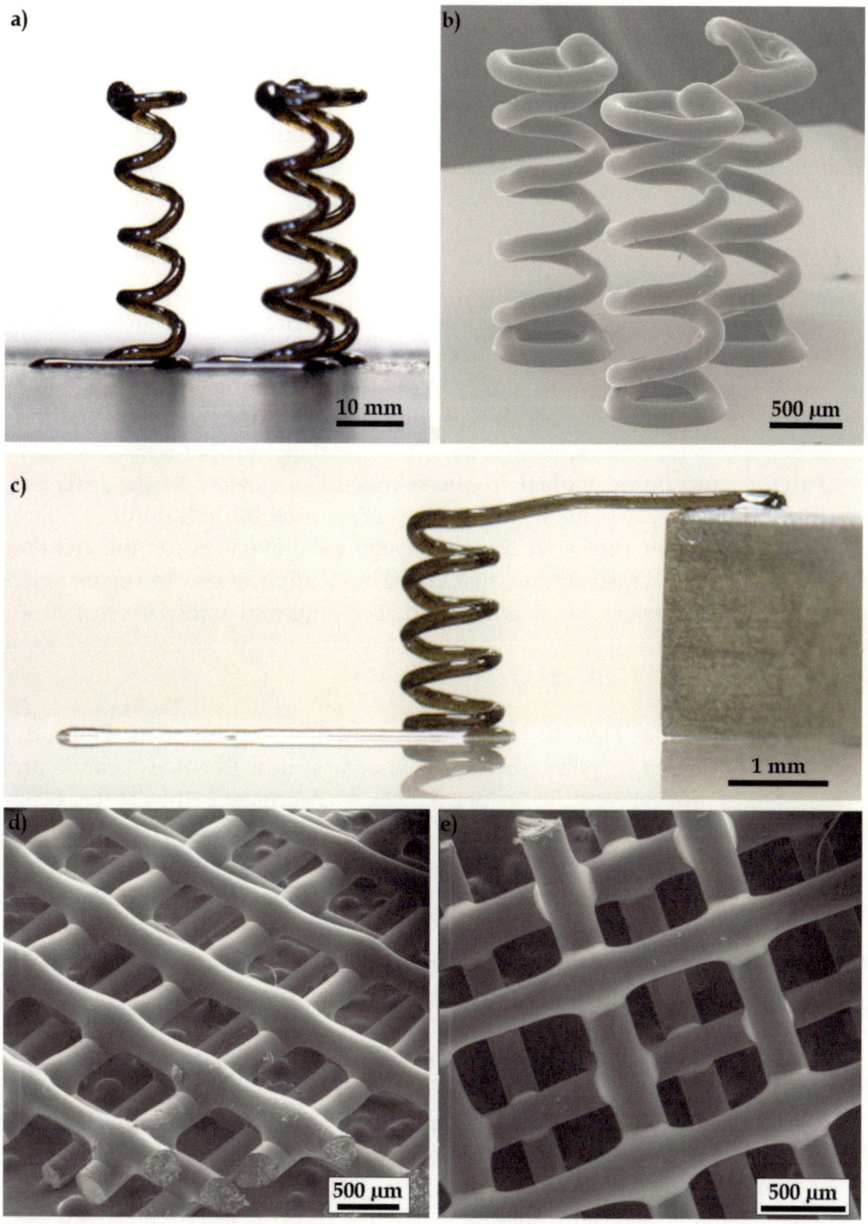

Fig. 10. Several nanocomposite structures manufactured by UV-DW: a-b) Coiled micro-spring network. c) Coiled wire between a substrate and an electrode. d-e) Scaffold structure.

Nanocomposite spring network composed of three micro-coils (Figure 10a) were fabricated using the UV-DW technique. Each of the springs had a filament diameter of ~ 150 μm, a coil diameter of ~ 1 mm and a height of ~ 4 mm. The networks were mechanically tested under compression and showed a rigidity of ~ 11.5 mN/mm. A smaller network of coils was also fabricated (Figure 10b). In this case, the coils had a ~100 μm diameter filament and a ~ 500 μm coil diameter. A micro-coil using a ~ 200 μm nanocomposite filament was also deposited to bridge the gap of ~2.1 mm between two uneven substrates (Figure 10c). This demonstration of an organic inductance device had a 10^{-6} S/cm electrical conductivity (Lebel, Aissa et al., 2009). Nanocomposite scaffold structures (Figure 10d-e) were also deposited using the UV-DW technique. The filaments were deposited in a layer-by layer sequence with the orientation of the filament being perpendicular from a layer to the other. The filament diameter were ~ 200 μm and the spacing between filaments among the same layer was ~ 1 mm.

7. Conclusion

The fabrication of high-performance nanocomposite materials and complex 3D structures must overcome the different challenges at the nano-, micro-, and macroscale. Dispersion and interaction with the polymer matrix are of paramount importance at the nanoscale. The microscale manufacturing techniques should provide a control over the orientation of high aspect-ratio nanoparticles such as carbon nanotubes. Finally, proper assembly technique of microstructures should be developed to create functional devices at the macroscale. The manufacturing techniques explained in this chapter, i.e. the infiltration of 3D microfluidic networks and UV-assisted direct writing, represent new avenues for the creation of 3D reinforced micro- and macrostructures that could find applications in organic electronics, polymer-based MEMS, sensors, tissue engineering scaffolds and aerospace structures.

8. References

Abbasi, S.; Carreau, P. J. & Derdouri, A. (2010). Flow induced orientation of multiwalled carbon nanotubes in polycarbonate nanocomposites: Rheology, conductivity and mechanical properties. *Polymer*, Vol. 51, No. 4, (February 2010) 922-935, 0032-3861

Ahn, B. Y.; Duoss, E. B.; Motala, M. J.; Guo, X.; Park, S.-I.; Xiong, Y.; Yoon, J.; Nuzzo, R. G.; Rogers, J. A. & Lewis, J. A. (2009). Omnidirectional Printing of Flexible, Stretchable, and Spanning Silver Microelectrodes. *Science*, Vol. 323, No. 5921, (March 2009) 1590-1593, 0036-8075

Ayutsede, J.; Gandhi, M.; Sukiraga, S.; Ye, H.; Hsu, C.-M.; Gogotsi, Y. & Ko, F. (2006). Carbon nanotube reinforced Bombyx mori silk nanofibers by the electrospinning process. *Biomacromolecules*, Vol. 7, No. 1, (January 2006) 208-214, 1525-7797

Baker, S. E.; Cai, W.; Lasseter, T. L.; Weidkamp, K. P. & Hamers, R. J. (2002). Covalently Bonded Adducts of Deoxyribonucleic Acid (DNA) Oligonucleotides with Single-Wall Carbon Nanotubes: Synthesis and Hybridization. *Nano Letters*, Vol. 2, No. 12, (October 2002) 1413-1417, 1530-6984

Balberg, I.; Binenbaum, N. & Wagner, N. (1984). Percolation thresholds in the three-dimensional sticks system. *Physical Review Letters*, Vol. 52, No. 17, (April 1984) 1465-8, 0031-9007

Bayat, M.; Yang, H. & Ko, F. (2010). Structure and Properties of Superparamagnetic Composite Nanofiber, *Proceedings of 8th Joint Canada-Japan Workshop on Composites*, Montreal, Canada, July 26-29, 2010, DEStech Publications Inc, In press

Bicerano, J. (2002). *Prediction of polymer properties*, Marcel Dekker, 0-8247-0821-0, New York

Bose, S.; Khare, R. A. & Moldenaers, P. (2010). Assessing the strengths and weaknesses of various types of pre-treatments of carbon nanotubes on the properties of polymer/carbon nanotubes composites: A critical review. *Polymer*, Vol. 51, No. 5, (March 2010) 975-993, 0032-3861

Byrne, M. T. & Guin'Ko, Y. K. (2010). Recent advances in research on carbon nanotube - polymer composites. *Advanced Materials*, Vol. 22, No. 15, (April 2010) 1672-1688, 0935-9648

Chen, J.; Rao, A. M.; Lyuksyutov, S.; Itkis, M. E.; Hamon, M. A.; Hu, H.; Cohn, R. W.; Eklund, P. C.; Colbert, D. T.; Smalley, R. E. & Haddon, R. C. (2001). Dissolution of full-length single-walled carbon nanotubes. *Journal of Physical Chemistry B*, Vol. 105, No. 13, (April 2010) 2525-2528, 1089-5647

Chen, R. J.; Zhang, Y.; Wang, D. & Dai, H. (2001). Noncovalent sidewall functionalization of single-walled carbon nanotubes for protein immobilization. *Journal of the American Chemical Society*, Vol. 123, No. 16, (April 2001) 3838-3839, 0002-7863

Choi, H. J. & Ihm, J. (1999). Exact solutions to the tight-binding model for the conductance of carbon nanotubes. *Solid State Communications*, Vol. 111, No. 7, (July 1999) 385-90, 0038-1098

Coleman, J. N.; Khan, U.; Blau, W. J. & Gun'ko, Y. K. (2006). Small but strong: A review of the mechanical properties of carbon nanotube-polymer composites. *Carbon*, Vol. 44, No. 9, (August 2010) 1624-1652, 0008-6223

Du, F.; Scogna, R. C.; Zhou, W.; Brand, S.; Fischer, J. E. & Winey, K. I. (2004). Nanotube networks in polymer nanocomposites: rheology and electrical conductivity. *Macromolecules*, Vol. 37, No. 24, (November 2004) 9048-9055, 0024-9297

Frankland, S. J. V.; Caglar, A.; Brenner, D. W. & Griebel, M. (2002). Molecular simulation of the influence of chemical cross-links on the shear strength of carbon nanotube-polymer interfaces. *Journal of Physical Chemistry B*, Vol. 106, No. 12, (March 2002) 3046-3048, 1089-5647

Gojny, F. H.; Wichmann, M. H. G.; Kopke, U.; Fiedler, B. & Schulte, K. (2004). Carbon nanotube-reinforced epoxy-composites: Enhanced stiffness and fracture toughness at low nanotube content. *Composites Science and Technology*, Vol. 64, No. 15 SPEC ISS, (November 2004) 2363-2371, 0266-3538

Gratson, G. M.; Garcia-Santamaria, F.; Lousse, V.; Xu, M.; Fan, S.; Lewis, J. A. & Braun, P. V. (2006). Direct-write assembly of three-dimensional photonic crystals: Conversion of polymer scaffolds to silicon hollow-woodpile structures. *Advanced Materials*, Vol. 18, No. 4, (Febuary 2006) 461-465, 0935-9648

Green, A. A. & Hersam, M. C. (2007). Ultracentrifugation of single-walled nanotubes. *Materials Today*, Vol. 10, No. 12, (December 2007) 59-60, 1369-7021

Haggenmueller, R.; Gommans, H. H.; Rinzler, A. G.; Fischer, J. E. & Winey, K. I. (2000). Aligned Single-Wall Carbon Nanotubes in Composites by Melt Processing Methods. *Chemical Physics Letters*, Vol. 330, No. 3-4, (November 2000) 219-225, 0009-2614

Huang, C. K.; Chen, S. W. & Wei, W. C. J. (2006). Processing and property improvement of polymeric composites with added ZnO nanoparticles through microinjection molding. *Journal of Applied Polymer Science*, Vol. 102, No. 6, (December 2006) 6009-6016, 0021-8995

Islam, M. F.; Rojas, E.; Bergey, D. M.; Johnson, A. T. & Yodh, A. G. (2003). High Weight Fraction Surfactant Solubilization of Single-Wall Carbon Nanotubes in Water. *Nano Letters*, Vol. 3, No. 2, (February 2003) 269-273, 1530-6984

Jiang, L.; Gao, L. & Sun, J. (2003). Production of aqueous colloidal dispersions of carbon nanotubes. *Journal of Colloid and Interface Science*, Vol. 260, No. 1, (April 2003) 89-94, 0021-9797

Katayama, K.-i. & Tsuji, M. (1994). Fundamentals of spinning. In: *Advanced fiber spinning technology*. Nakajima, T., Kajiwara, K. & McIntyre , J. E. (Ed.). 1-24. Woodhead Publishing Ltd., 1 85573 182 7, Cambridge, England

Kilbride, B. E.; Coleman, J. N.; Fraysse, J.; Fournet, P.; Cadek, M.; Drury, A.; Hutzler, S.; Roth, S. & Blau, W. J. (2002). Experimental observation of scaling laws for alternating current and direct current conductivity in polymer-carbon nanotube composite thin films. *Journal of Applied Physics*, Vol. 92, No. 7, (October 2002) 4024-30, 0021-8979

Kim, D. S.; Nepal, D. & Geckeler, K. E. (2005). Individualization of single-walled carbon nanotubes: Is the solvent important? *Small*, Vol. 1, No. 11, (November 2005) 1117-1124, 1613-6810

Ko, F.; Gogotsi, Y.; Ali, A.; Naguib, N.; Ye, H.; Yang, G.; Li, C. & Willis, P. (2003). Electrospinning of continuous carbon nanotube-filled nanofiber yarns. *Advanced Materials*, Vol. 15, No. 14, (July 2003) 1161-1165, 0935-9648

Lebel, L. L.; Aissa, B.; El Khakani, M. A. & Therriault, D. (2009). Ultraviolet-assisted direct-write fabrication of carbon nanotube/polymer nanocomposite microcoils. *Advanced Materials*, Vol. 22, No. 5, (February 2010) 592-596, 0935-9648

Lebel, L. L.; Aissa, B.; Khakani, M. A. E. & Therriault, D. (2009). Preparation and mechanical characterization of laser ablated single-walled carbon-nanotubes/polyurethane nanocomposite microbeams. *Composites Science and Technology*, Vol. 70, No. 3, (March 2010) 518-524, 0266-3538

Lebel, L. L.; Aissa, B.; Paez, O. A.; El Khakani, M. A. & Therriault, D. (2009). Three-dimensional micro structured nanocomposite beams by microfluidic infiltration. *Journal of Micromechanics and Microengineering*, Vol. 19, No. 12, (December 2009) 125009 (7 pp.), 0960-1317

Lewis, J. A. & Gratson, G. M. (2004). Direct writing in three dimensions. *Materials Today*, Vol. 7, No. 7, (July/August 2004) 32-39, 1369-7021

Li, D. & Xia, Y. (2004). Electrospinning of Nanofibers: Reinventing the Wheel? *Advanced Materials*, Vol. 16, No. 14, (July 2004) 1151-1170, 1521-4095

Mamedov, A. A.; Kotov, N. A.; Prato, M.; Guldi, D. M.; Wicksted, J. P. & Hirsch, A. (2002). Molecular design of strong single-wall carbon nanotube/polyelectrolyte multilayer composites. *Nature Materials*, Vol. 1, No. 3, (November 2002) 190-194, 1476-1122

Miaudet, P.; Bartholome, C.; DerrÈ, A.; Maugey, M.; Sigaud, G.; Zakri, C. & Poulin, P. (2007). Thermo-electrical properties of PVA-nanotube composite fibers. *Polymer*, Vol. 48, No. 14, (June 2007) 4068-4074, 0032-3861

Minus, M. L.; Chae, H. G. & Kumar, S. (2009). Interfacial crystallization in gel-spun poly (vinyl alcohol)/single-wall carbon nanotube composite fibers. *Macromolecular Chemistry and Physics*, Vol. 210, No. 21, (November 2009) 1799-1808, 1022-1352

O'Connell, M. J.; Boul, P.; Ericson, L. M.; Huffman, C.; Wang, Y.; Haroz, E.; Kuper, C.; Tour, J.; Ausman, K. D. & Smalley, R. E. (2001). Reversible water-solubilization of single-walled carbon nanotubes by polymer wrapping. *Chemical Physics Letters*, Vol. 342, No. 3-4, (July 2001) 265-271, 0009-2614

Raghavan, S. R.; Walls, H. J. & Khan, S. A. (2000). Rheology of silica dispersions in organic liquids: New evidence for solvation forces dictated by hydrogen bonding. *Langmuir*, Vol. 16, No. 21, (October 2000) 7920-7930, 0743-7463

Sahoo, N. G.; Jung, Y. C.; Yoo, H. J. & Cho, J. W. (2006). Effect of functionalized carbon nanotubes on molecular interaction and properties of polyurethane composites. *Macromolecular Chemistry and Physics*, Vol. 207, No. 19, (October 2006) 1773-1780, 1022-1352

Sainz, R.; Benito, A. M.; Martinez, M. T.; Galindo, J. F.; Sotres, J.; Baro, A. M.; Corraze, B.; Chauvet, O. & Maser, W. K. (2005). Soluble self-aligned carbon nanotube/polyaniline composites. *Advanced Materials*, Vol. 17, No. 3, (Fabruary 2005) 278-281, 0935-9648

Sandler, J. K. W.; Pegel, S.; Cadek, M.; Gojny, F.; Van Es, M.; Lohmar, J.; Blau, W. J.; Schulte, K.; Windle, A. H. & Shaffer, M. S. P. (2004). A comparative study of melt spun polyamide-12 fibres reinforced with carbon nanotubes and nanofibres. *Polymer*, Vol. 45, No. 6, (March 2004) 2001-2015, 0032-3861

Satake, A.; Miyajima, Y. & Kobuke, Y. (2005). Porphyrin-carbon nanotube composites formed by noncovalent polymer wrapping. *Chemistry of Materials*, Vol. 17, No. 4, (February 2005) 716-724, 0897-4756

Starr, F. W.; Schrder, T. B. & Glotzer, S. C. (2002). Molecular dynamics simulation of a polymer melt with a nanoscopic particle. *Macromolecules*, Vol. 35, No. 11, (May 2002) 4481-4492, 0024-9297

Stoeffler, K.; Perrin-Sarazin, F.; That, M. T. T. & Denault, J. (2010). Polyethylene/Carbon Nanotubes Nanocomposites, *Proceedings of 8th Joint Canada-Japan Workshop on Composites*, Montreal, Canada, July 26-29, 2010, DEStech Publications Inc, In Press

Therriault, D.; Shepherd, R. F.; White, S. R. & Lewis, J. A. (2005). Fugitive inks for direct-write assembly of three-dimensional microvascular networks. *Advanced Materials*, Vol. 17, No. 4, (February 2005) 395-399, 0935-9648

Therriault, D.; White, S. R. & Lewis, J. A. (2003). Chaotic mixing in three-dimensional microvascular networks fabricated by direct-write assembly. *Nature Materials*, Vol. 2, No. 4, (April 2003) 265-271, 1476-1122

Thostenson, E. T. & Chou, T.-W. (2002). Aligned multi-walled carbon nanotube-reinforced composites: Processing and mechanical characterization. *Journal of Physics D: Applied Physics*, Vol. 35, No. 16, (August 2002) 77-80, 0022-3727

Thostenson, E. T. & Chou, T.-W. (2006). Carbon nanotube networks: Sensing of distributed strain and damage for life prediction and self healing. *Advanced Materials*, Vol. 18, No. 21, (November 2006) 2837-2841, 0935-9648

Tsurumi, T. (1994). Solution Spinning. In: *Advanced fiber spinning technology*. Nakajima, T., Kajiwara, K. & McIntyre , J. E. (Ed.). 1-24. Woodhead Publishing Ltd., 1-85573-182-7, Cambridge, England

Varadan, V. K. & Varadan, V. V. (2001). Micro stereo lithography for fabrication of 3D polymeric and ceramic MEMS, *Proceedings of The International Society for Optical Engineering - MEMS Design, Fabrication, Characterization, and Packaging*, 978-0-8194-4108-9, Edinburgh, United kingdom, May/June 2001, SPIE,

Varadan, V. K.; Xie, J. & Ji, T. (2001). Three dimensional MEMS devices with functionalized carbon nanotubes, *Proceedings of The International Society for Optical Engineering, Micromachining and Microfabrication Process Technology VIII*, 978-0-8194-4779-1, Adelaide, Australia, January 2003, SPIE,

Winey, K. I. & Vaia, R. A. (2007). Polymer nanocomposites. *MRS Bulletin*, Vol. 32, No. 4, (April 2007) 314-319, 0883-7694

Wong, M.; Paramsothy, M.; Xu, X. J.; Ren, Y.; Li, S. & Liao, K. (2003). Physical interactions at carbon nanotube-polymer interface. *Polymer*, Vol. 44, No. 25, (December 2003) 7757-7764, 0032-3861

Xu, M.; Zhang, T.; Gu, H.; Wu, J. & Chen, Q. (2006). Synthesis and properties of novel polyurethane - Urea/multiwalled carbon nanotube composites. *Macromolecules*, Vol. 39, No. 10, (May 2006) 3540-3545, 0024-9297

Xu, X.; Thwe, M. M.; Shearwood, C. & Liao, K. (2002). Mechanical properties and interfacial characteristics of carbon-nanotube-reinforced epoxy thin films. *Applied Physics Letters*, Vol. 81, No. 15, (October 2002) 2833-5, 0003-6951

19

The Role of Elongational Flow in Morphology Modification of Polyethylene/OMMt Nanocomposite System

N. Tz. Dintcheva[1,2*] and F. P. La Mantia[1,2]
[1]*Università di Palermo, Dipartimento di Ingegneria Chimica dei Processi e dei Materiali,*
[2]*Consorzio Interuniversitario Nazionale "La Chimica per l'Ambiente",*
Italy

1. Introduction

The formulation of high-performance polymer based nanocomposites depend on many factors, such as polymeric matrix type, nanoparticle type, loading and morphology, affinity between the polymeric matrix and nanoparticles, presence of compatibilizer and processing conditions [1-4].

During the industrial processing the polymer based systems are subjected to two different processing flows, i.e. the shear and elongational flow. The shear flow plays a significant role in polyethylene/OMMt manufacture processing but it is not able to change the system morphology, while, the elongational flow, involved in spinning and film-blowing processing operations, can induce considerably clay morphology variations [5-13]. In order to evaluate the effect of the elongational flow on the polyethylene/OMMt system morphology, the affinity between the matrix and the OMMt particles can be considered. In particular, the presence of some compatibilizer, as maleic anhydride grafted polyethylene can modify the system affinity and subsequently, the clay morphology changes are different than the uncompatibilized system also upon the extensional flow. Nevertheless, the obtained morphology changes upon the elongational flow in the polyethylene/OMMt system, without and with good system affinity, lead to significant mechanical improvements than the unfilled systems, more larger than the simple macromolecular orientation [10, 12].

If considering, from "flow point of view", the polyethylene/OMMt system as a biphasic incompatible mixture, composed by an inorganic phase dispersed in a polymeric matrix, the applied extensional flow can be able to change strongly the clay morphology. In particular, the clay nanoparticles can be broken and/or fragmentized, dispersed and oriented along the flow direction, giving rise to flow induced intercalation/exfoliation morphology transition. Indeed, the OMMt particles can be considered as hard but breakable particles, i.e. polymeric particles in a polymeric blends, while, the conventional filler particles are elastic but unbreakable. The elongational flow leads to exfoliation of intercalated OMMt tactoids and/or to some more intercalation of the same tactoids. For the systems with good affinity

* Corresponding author: N.Tz. Dintcheva, Tel: +3909123863704, e-mail: dintcheva@dicpm.unipa.it

between the components, i.e. the exfoliated clay particles are formed due to the compatibilizer loading, perpendicularly to the flow direction, some clay re-aggregation occurs, also due to action of the radial forces.

In this book chapter, will be tried to answer to two main questions, in particular: "Does the flow influence the morphology of polyethylene/OMMt nanocomposite system?" and "Which is the role of the affinity, in terms of compatibility, between the two phases?"

2. Effect of the elongational flow on the morphology modification of polyethylene/OMMt nanocomposite system

The morphology of the polyethylene/OMMTs nanocomposite system, i.e. system without any affinity between the components, upon the elongational flow, considerably changes, while, upon the shear flow, no change occurs. Indeed, as well known, the morphology of blends composed of two incompatible components is determined from the flow type. The extensional flow is able to break-up, disperse and orientate the of the dispersed system, if the viscosities of the two systems are considerably different [14-16].

In order to obtain an improvement of the OMMt morphology, also with consequent improvement in the final material performances, the study of the polyethylene/OMMt system behaviour upon the elongational flow is a critical issue. The unfilled polyethylene and OMMt filled polyethylene are subjected to the filament formation using apparatus reported in Figure 1, considering a capillary with die diameter of 1 mm and length-to-diameter, L/D=40. Before the filament formation, the polyethylene (linear low-density polyethylene, LLDPE, by Polimeri Europa, Italy) is compounded with 5 % wt./wt. organo-modified montmorillonite, OMMt (Cloisite®15A, by Southern Clay Products, USA) in a co-rotating twin screw extruder; obviously, the unfilled polyethylene is subjected to the some processing history. The formulated filaments are drawn at room temperature and the amount of drawing is characterized by the draw ratio (DRc=Lf/Lo, were Lf is the final filament length and Lo the initial length of the fibres.

The spinnability and stretchability of the polymer based systems depend on the resistance to break in the melt during the hot drawing and the melt strength, MS, and breaking stretching ratio, BSR, values can give useful information about these system capabilities. In Table 1, the values of the MS and BSR at an apparent shear rate = 60 s^{-1} are reported for the unfilled polyethylene and the OMMt filled polyethylene.

The polyethylene/OMMt system shows slightly a higher value of the melt strength and a slightly lower value of the breaking stretching ratio with respect to the unfilled polyethylene. However, the MS improvement and the relatively high value of the BSR, even if lower than the unfilled PE, indicate that the nanocomposite sample is able to filament formation as the linear low-density polyethylene sample.

In order to evaluate the mechanical behaviour as a function of the cold drawing, the filaments are subjected to the extensional flow and the dimensionless main mechanical properties, namely, elastic modulus, E, tensile strength, TS, and elongational at break, EB, are calculated and the trends in Figure 2 are reported. The dimensionless elastic modulus and tensile strength of the polyethylene/OMMt fibres are significantly higher than the unfilled polyethylene filaments. By the OMMt loading, in the isotropic state, i.e. DRc=0, compression-moulded sheets, no differences are observed in the mechanical behaviour. Some time, the dimensionless elongation at break for both systems show similar trend.

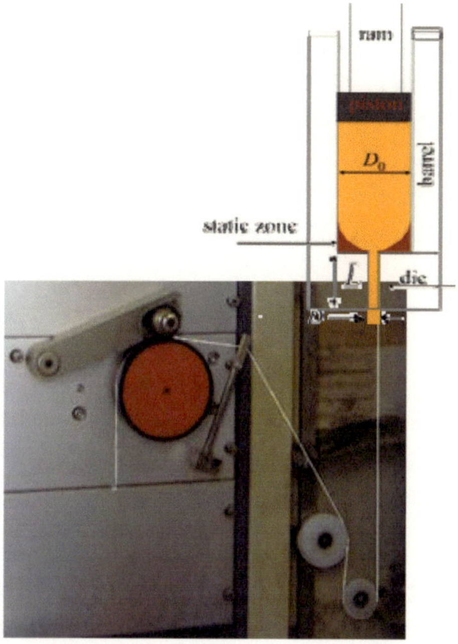

Fig. 1. Scheme of apparatus for filament formation

Sample	MS, cN	BSR
Polyethylene	3.6	39
Polythylene/OMMt nanocomposite system	4.0	30

Table 1. Melt strength, MS, and breaking stretching ratio, BSR, of unfilled Polyethylene and OMMt filled Polyethylene at apparent shear rate = 60 1/sec

The thermal analysis performed by DSC and the measurements of the total birefringence of the PE and PE/OMMt fibres at the lowest and highest draw ratio are very similar, see Table 2, suggesting similar orientation of the macromolecules for both system. However, the improvement in the mechanical behaviour of the polyethylene/OMMt system cannot be attributed only to the macromolecular orientation but can be understand and explain considering the clay morphology modification achieved upon the elongational flow.
The accurate clay morphology analysis, performed by x-ray diffraction and TEM, indicate the formation of single clay platelets and some more intercalated tactoids upon extensional flow, in particular, this phenomenon is pronounced for the highest drawn PE/OMMt fibre, see Table 3 and Figure 3. The elongational flow is able to change significantly the clay morphology, in the parallel fibre direction the interlayer distance increases from 3.30 nm for the compounded sample to 3.81 for the drawn PE/OMMt fibre at DRc=6, while, in the radial direction the morphology modification is particular, the interlayer distance slowly decreases maybe due to the some clay re-aggregation [17], due to the action of the radial fibre forces.
With the clay morphology modification upon the extensional flow, i.e. the increasing the number of the exfoliated layers and of the intercalated tactoids as a function of the draw ratio, the interface area between the polyethylene matrix and the OMMt nanoparticles increase and the mechanical improvements become much more evident than the unfilled polyethylene fibres at the same level of the macromolecular orientation.

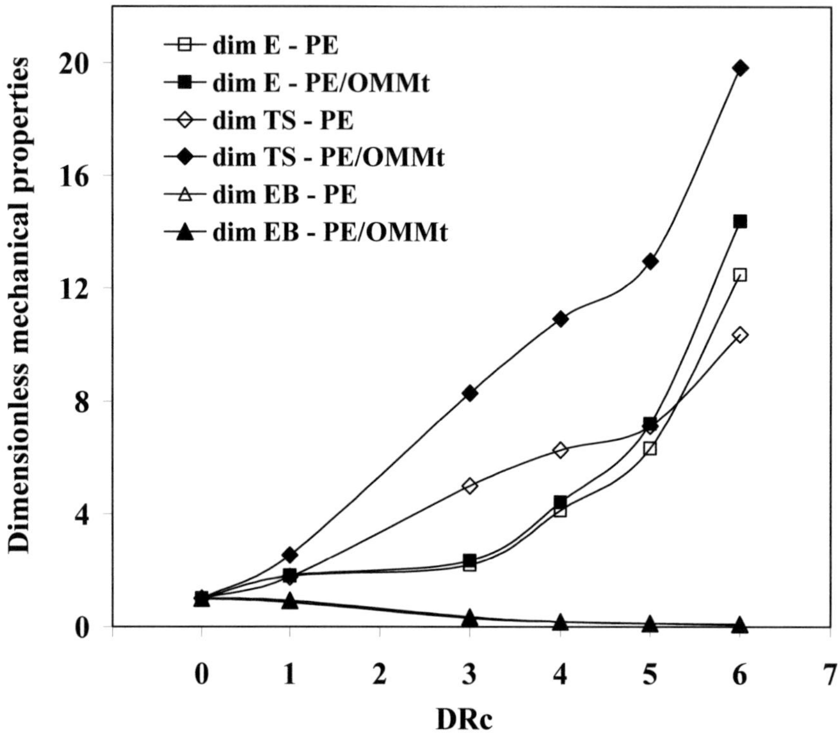

Fig. 2. Dimensionless mechanical properties as a function of the draw ratio. The dimensionless values have been calculated by dividing the values of the fibres by those of the compression-moulded sheets. *La Mantia F.P, Dintcheva N.Tz, Scaffaro R, Marino R: Morphology and Properties of Polethylene/Clay Nanocomposite Drawn Fibre. Marcomolecular Matererial and Engineering. 2008, 293, 83-91, Copyright Wiley-VCH Verlag GmbH & Co. KGaA. Reproduced with permission.*

Sample	Crystalline degree(*), %	Total birefringence x 10^{-3} (**)
PE fibre at DRc=1	43.4	4.1
PE/OMMt fibre at DRc=1	42.2	4.5
PE fibre at DRc=6	45.5	12.1
PE/OMMt fibre at DRc=6	44.5	12.6

(*) Crystalline degree is calculated by the thermal analyses (fusion enthalpy), using the formula:

$\alpha = \Delta H_f / \Delta H_{f100\%}$

(**) The total birefringence is calculated using the formula: $\Delta n = \Delta n_c f_c x_c + \Delta n_a f_a (1-x_c) + \Delta n_f$, where Δn_c and Δn_a are the intrinsic values of the birefringence of the crystalline and amorphous phase, f_c and f_a are the orientation factors of the two phases and x_c is the crystalline degree. The form birefringence, Δn_f, is usually considered negligible with respect to the other terms.

Table 2. Crystalline degree and total birefringence of PE and PE/OMMt fibres at lowest and highest draw ratio. La Mantia F.P, Dintcheva N.Tz, Scaffaro R, Marino R: Morphology and Properties of Polethylene/Clay Nanocomposite Drawn Fibre. Marcomolecular Matererial and Engineering. 2008, 293, 83-91, Copyright Wiley-VCH Verlag GmbH & Co. KGaA. Reproduced with permission.

Sample	Interlayer distance, d_{001}, nm
OMMt	3.15
Parallel direction	
compounded PE/OMMt	3.30
PE/OMMt at DRc=1	3.48
PE/OMMt at DRc=6	3.81
Radial direction	
compounded PE/OMMt	3.30
PE/OMMt at DRc=1	2.85
PE/OMMt at DRc=6	--- (*)

(*) This x-ray peak is not clearly evident

Table 3. Interlayer distances of different samples calculated by x-ray peak analysis using the Bragg's formula, $d_{001} = n\,\lambda\,/\,(2\sin\theta)$, where n is an integer, θ is the angle in incidence of x-ray beam. *La Mantia F.P, Dintcheva N.Tz, Scaffaro R, Marino R: Morphology and Properties of Polethylene/Clay Nanocomposite Drawn Fibre. Marcomolecular Matererial and Engineering. 2008, 293, 83-91, Copyright Wiley-VCH Verlag GmbH & Co. KGaA. Reproduced with permission.*

4. Effect of the elongational flow on the morphology modification of polyethylene/OMMt nanocomposite system, using capillaries with different diameters and length-to-diameter ratios

In order to evaluated quantitatively the effect of the convergent extensional flow on the morphology modification of polyethylene/OMMt system, i.e. system without affinity between the components, the capillaries with different diameters and length-to-diameter ratios are used, in particular, the die diameters are Do = 1 mm, 2 mm and 3 mm, while the length-to-diameter ratios are for Do =1 mm, L/Do = 1, 4, 10, 20, 40; for Do = 2 mm, L/Do = 1, 2, 4, 40; for Do = 3 mm, L/Do = 1, 2, 4. The values of the convergent extensional stress are calculated using the Cogswell's formula [18]:

$$\sigma_{el} = 3/8\,(n+1)\;*\;\Delta P_{ent} \qquad (1)$$

where ΔP_{ent} is the pressure drop at the entrance of the capillary and n is the flow power index.
The pressure drop was calculated adopting the followed formula:

$$\Delta P_{measured} = \Delta P_{ent} + \Delta P_{cap} \qquad (2)$$

In particular:

$$\Delta P_{measured} = \Delta P_{ent} + (\,\Delta P\,/\,L\,)_{cap}\;*\;L \qquad (3)$$

where $(\,\Delta P\,/\,L\,)_{cap}$ is the pressure drop for unit capillary length.
The calculated results are reported in the Table 4. Also, in order to make a correct rheological characterization, the flow curves of PE and PE/OMMt samples are measured (not reported) and the experimental data, taken with different capillary geometries, are same within the experimental error. This means that is not slip at the wall and that the Bagley correction are well evaluated.

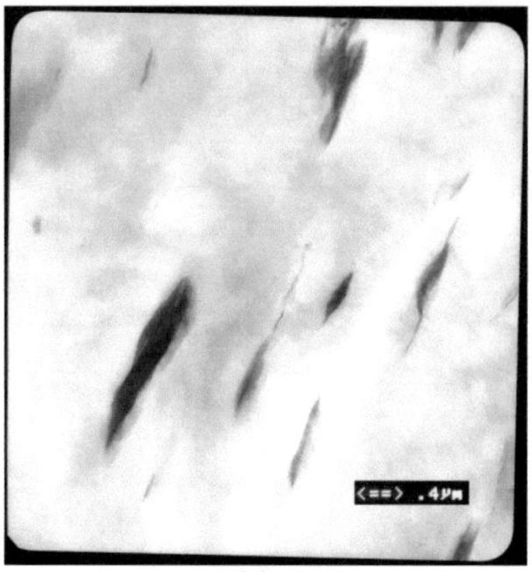

(a)

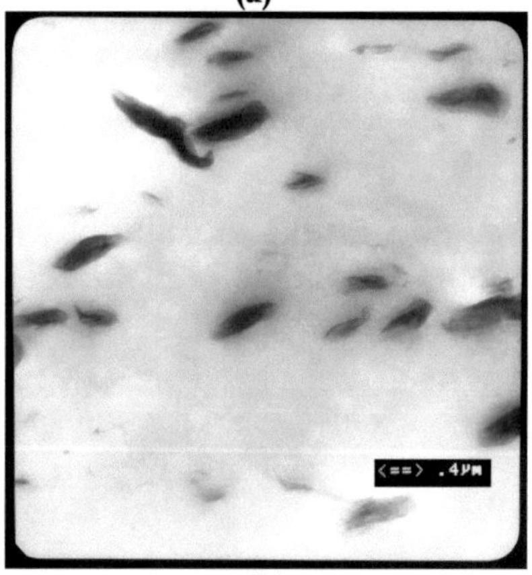

(b)

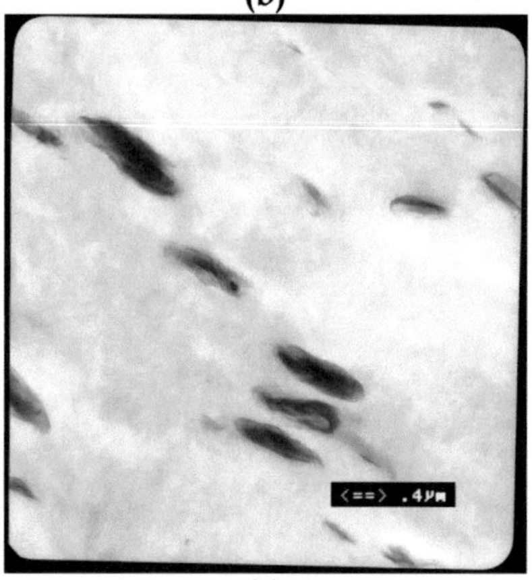

(c)

Fig. 3. TEM micrographs of (a) compounded PE/OMMt sample, (b) PE/OMMt fibre at DRc=1, (c) PE/OMMt fibre at DRc=6. La Mantia F.P, Dintcheva N.Tz, Scaffaro R, Marino R: Morphology and Properties of Polethylene/Clay Nanocomposite Drawn Fibre. Marcomolecular Matererial and Engineering. 2008, 293, 83-91, Copyright Wiley-VCH Verlag GmbH & Co. KGaA. Reproduced with permission.

The convergent extensional stress, generated at the entrance of the capillary, significantly decreases with increasing of the capillary diameter and it is almost intensive to the variation of the L/D ratios. Furthermore, the values of the convergent extensional stress for the PE/OMMt sample are significantly higher than the values of the PE one. The last can be explain considering two important phenomenon, occur at the capillary entrance, in particular, some frictional interaction between the OMMt particles and some physical interaction between the OMMt nanoparticles and the matrix macromolecules. The frictional effect and the physical interaction, due to the OMMt loading, are more pronounced by the die diameter reduction and the nanoparticle morphology modification is favoured at these conditions.
The mechanical behaviour reflects the trends of the calculated convergent extensional stress, in particular, the mechanical performances increase with decrease of the die diameter due to the action of the more severe elongational flow. Also, the mechanical performances decrease with increasing of the L/D ratio, due to different macromolecular orientation.

Do = 3 mm	55	55	---
Do = 2 mm	153	161	153
Do = 1 mm	437	446	440
	L/D = 1	**L/D = 2**	**L/D = 40**

(a)

Do = 3 mm	83	84	---
Do = 2 mm	201	205	204
Do = 1 mm	737	727	727
	L/D = 1	**L/D = 2**	**L/D = 40**

(b)

Table 4. Extensional stress values, calculated at the entrance of the capillaries of PE (a) and of PE/OMMt (b) for capillaries with different geometries. *La Mantia F.P, Marino R, Dintcheva N.Tz: Morphology Modification of Polyethylene/Clay Nanocomposite Samples under Convergent Flow. Marcomolecular Matererial and Engineering. 2009, 294, 575-581, Copyright Wiley-VCH Verlag GmbH & Co. KGaA. Reproduced with permission.*

The SEM morphology observations clearly indicate the formation of OMMt particles with smaller dimensions by increasing intensity of the applied convergent flow (lower values of the die diameter and L/D ratio), see Figure 4 (magnification is 10000x). The OMMt particles for the PE/OMMt prepared using capillary with Do = 1 mm and L/D= 1 are not visible at this magnification, but some time, using capillary with Do = 3 mm and L/D= 1, the particles are clearly evident, and, their dimensions are similar to the particle dimension for the compounded PE/OMMt sample. Also, no uniform OMMt particle dispersion can be noted when the sample is subjected to the weak convergent extensional stress.
The OMMt morphology modification is confirmed by x-ray analysis on the radial filament surfaces. The interlayer distances increases with the reduction of the die diameter at same L/D ratio but some time small reduction of the interlayer distances can be noted with increasing of the L/D, see Table 5. The last could be seen a contradiction (not clearly visible by SEM micrographs) but can be understand considering some re-aggregation of the OMMt particles, according to the reported in literature by Okamoto [17], due to the long residence times of the samples in the capillary, also, related to the macromolecule relaxation. It is interesting to highlight that the die diameter plays a significant role in the OMMt morphology modification, and, it is more pronounced that the role of the length-to diameter ratio.

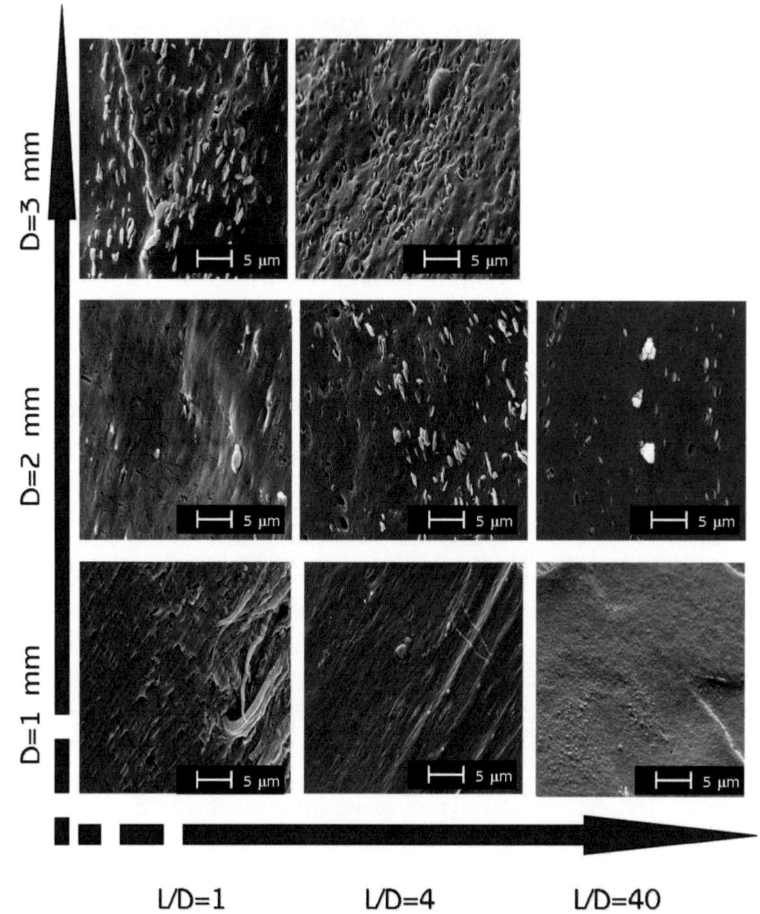

(a)

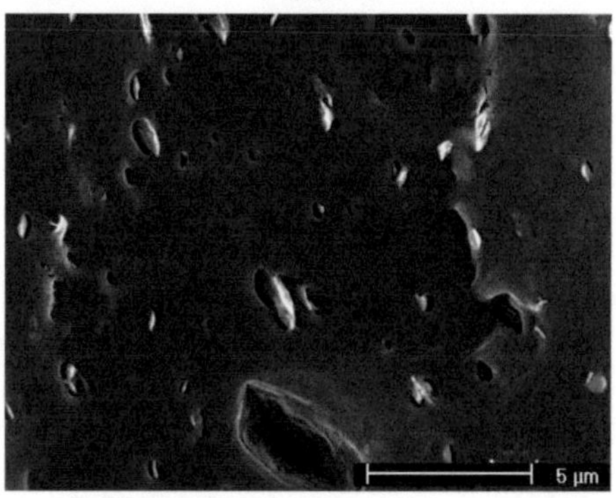

(b)

Fig. 4. SEM micrograohs of PE/OMMt fibres (a) considering capillaries with different geometries and compounded PE/OMMt sample (b). *La Mantia F.P, Marino R, Dintcheva N.Tz: Morphology Modification of Polyethylene/Clay Nanocomposite Samples under Convergent Flow. Marcomolecular Matererial and Engineering. 2009, 294, 575-581, Copyright Wiley-VCH Verlag GmbH & Co. KGaA. Reproduced with permission.*

Sample	Do, mm	L/D	Interlayer distance, d_{001}, nm
OMMt	---	---	3.15
Compounded PE/OMMt	---	---	3.30
PE/OMMt	3	1	3.40
PE/OMMt	3	4	3.34
PE/OMMt	2	1	3.54
PE/OMMt	2	4	3.44
PE/OMMt	2	40	3.40
PE/OMMt	1	1	3.84
PE/OMMt	1	4	3.42
PE/OMMt	1	40	2.85

Table 5. Interlayer distances of PE/OMMt sample using capillaries with different geometries, calculated by x-ray peak analysis using the Bragg's formula, $d_{001} = n\lambda / (2\sin\theta)$, where n is an integer, θ is the angle in incidence of x-ray beam. *La Mantia F.P, Marino R, Dintcheva N.Tz: Morphology Modification of Polyethylene/Clay Nanocomposite Samples under Convergent Flow. Marcomolecular Matererial and Engineering. 2009, 294, 575-581, Copyright Wiley-VCH Verlag GmbH & Co. KGaA. Reproduced with permission.*

Finally, the morphology of the PE/OMMt systems changes significantly upon the applied elongational flow. In particular, the morphology modification for the system with limited affinity between the components is more pronounced by increasing of the applied elongational stress (lower values of die diameter and L/D ratio), and, some reduction of the OMMt particle dimensions, according to the variation of the applied extensional stress can be noted.

5. Effect of the elongational flow on the morphology modification of polyethylene/compatibilizer/OMMt nanocomposite system

The loading of a commercial maleic anhydride grafted polyethylene is able to change considerably the OMMt morphology in polyethylene/compatibilizer/OMMt system due to the variation of the affinity between the polymeric matrix and the nanoparticles, as well known in literature [5-6, 19-20]. Obviously, the different started OMMt morphology by the PEgMA presence leads to different morphology modification, in particular, an almost compact delaminated OMMt structure can be obtained.

The tensile strength, see Figure 5, significantly increases, as discussed before, upon the extensional flow and this rise is more pronounced by the compatibilized loading. Also, no influences of the OMMt and campotibilizer loadings on the macromolecular orientations are observed, i.e. the thermal and total birefringence measurements show similar values for all investigated samples.

For system with greater affinity between the matrix and OMMt nanoparticles, i.e. polyethylene/compatibilizer/OMMt, the elongational flow is more efficient than the system with reduced system affinity, i.e. polyethylene/OMMt. In particular, the morphology of the OMMt nanoparticles of the PE/PEgMA/OMMt system, upon the extensional flow, shows higher level of intercalation and some exfoliation, more and more pronounced with the draw ratio. The initial intercalated OMMt morphology changes to some more intercalated and finally, at the highest anisotropic state evolves to delaminated OMMt structure. These remarks are on the basis of the morphological studies performed by x-ray diffraction and

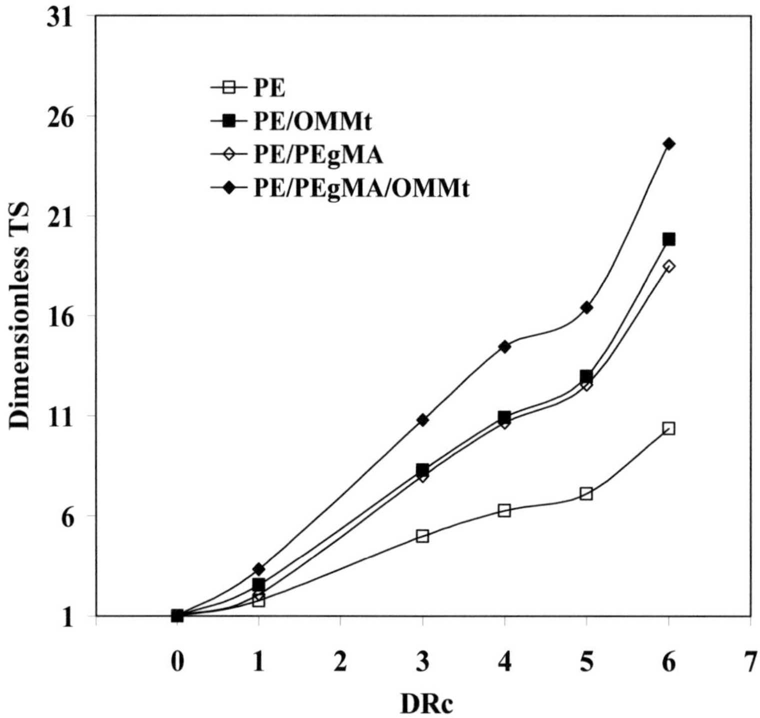

Fig. 5. Dimensionless tensile strength as a function of the draw ratio. The dimensionless values have been calculated by dividing the values of the fibres by those of the compression-moulded sheets. *Dintcheva N.Tz, Marino R, La Mantia F.P: The role of the matrix–filler affinity on morphology and properties of polyethylene/clay and polyethylene/compatibilizer/clay nanocomposite drawn fibres. e-Polymers. 2009, no. 054. Copyright e-Polymers. Reproduced with permission.*

TEM analysis. The interlayer distances for PE/PEgMA/OMMt sample are 3.42 nm for the compounded sample, 4.09 nm for the fibre at lowest draw ratio and no diffraction peak for the fibre at the highest drawing is noted. It is clear that the elongational flow, in this case, is able to change the OMMt morphology and it evolves to delaminated structured formations, confirmed also by TEM micrographs, reported in Figure 6.

6. Effect of the extensional flow on the properties of polyethylene/OMMt nanocomposite films for twist wrapping

On the basis of all the results, a practical case study for twist wrapping of PE/OMMt nanocomposite is made. As discussed before, the OMMt loading leads to more pronounced increase of the elastic modulus and tensile strength than the unfilled polyethylene sample one upon the elongational flow; together, the elongation at break and the yield strain decrease. The last means that the oriented PE/OMMt film shows improved ability in maintenance (keeping) the induced torsion deformation than the oriented unfilled PE film.

The enhanced performances of PE/OMMt film, from twist wrapping applicative point of view, are due to the significant obtained morphology modification upon the extensional flow. Some benefit synergic effect of the OMMt presence and the extensional flow on the morphology and subsequently on the applicative performances is noted.

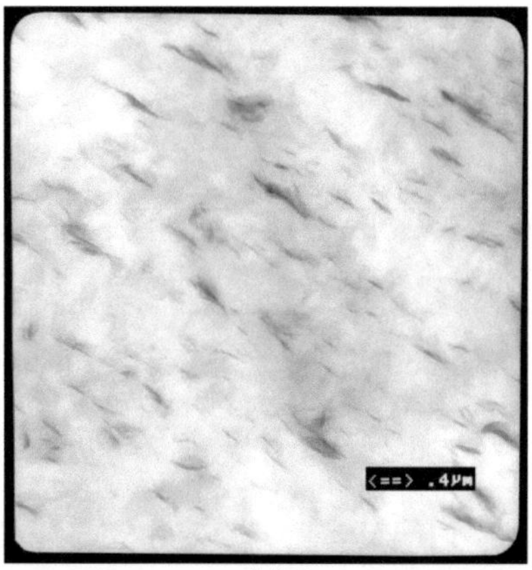

(a)

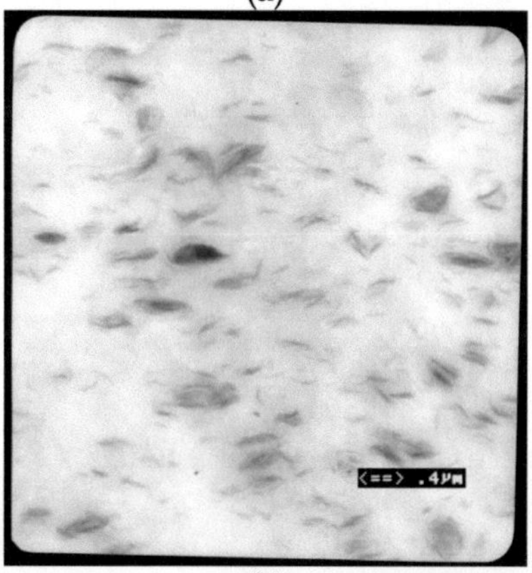

(b)

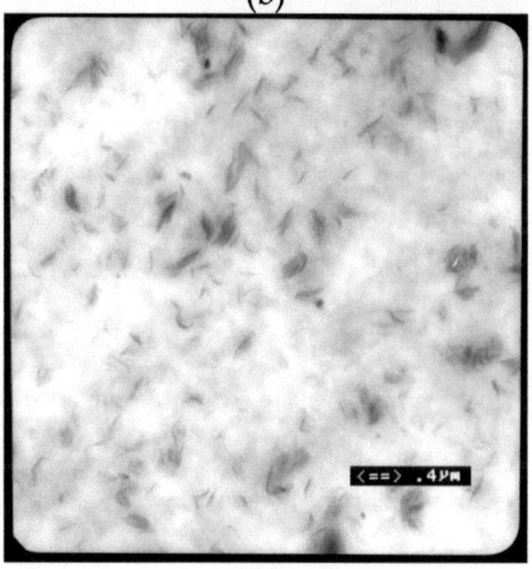

(c)

Fig. 6. TEM micrographs of (a) compounded PE/PEgMA/OMMt sample, (b) PE/PEgMA/OMMt fibre at DRc=1, (c) PE/PEgMA/OMMt fibre at DRc=6. *Dintcheva N.Tz, Marino R, La Mantia F.P: The role of the matrix-filler affinity on morphology and properties of polyethylene/clay and polyethylene/compatibilizer/clay nanocomposite drawn fibres. e-Polymers. 2009, no. 054. Copyright e-Polymers. Reproduced with permission.*

7. Conclusions

All the results agree that the elongational flow is able to change considerably the OMMt morphology. In particular, the nanoparticles can be deformed, broken and oriented along the flow direction and consequently the significant improvement in the mechanical final performances can be obtained. The improvement of the macroscopical performances cannot be attributed only to the simple macromolecular orientation but it can be explained considering the OMMt morphology modification due to two phenomena: interaction between the OMMt particles and between the OMMt particles and macromolecular matrix subjected to the elongational flow.
From an application point of view, the ability of the elongational flow in OMMt morphology modification is very important because this type of flow is involved in essential industrial processing as spinning and film-blowing. Also, the polyethylene/OMMt oriented nanocomposite film is more suitable for twist wrapping application than the unfilled polyethylene one.

8. References

[1] M. Alexandre, P. Dubois; *Mat. Sci. Eng.*, 2000, 28, 1-63.
[2] T. D. Fornes, P. J. Yoon, H. Keskkula and D. R. Paul, *Polymer*, 2001, 42, 9929-9940.
[3] R. Krishnamoorti, J. Ren, A. S. Silva,. *J Chem Phys*, 2001, 114, 4968.
[4] S. S. Ray, M. Okamoto, *Prog. Polym. Sci.*, 2003, 28, 1539–1641.
[5] S. Pavlikova, R. Thomann, P. Reichert, R. Mulhaupt, A. Marcincin, E. Borsig, *J. Appl. Polym. Sci.*, 2003, 89, 604-611.
[6] Z. Mlynarcikova, D. Kaempfer, R. Thomann, R. Mulhaupt, E. Borsig, *Polym. Adv. Technol.*, 2005, 16, 362-369.
[7] L.A. Utracki, *Indian J. Fibre and Textile Research*, 2006, 31, 15-28.
[8] M. Tokihisa, K. Yakemoto, T. Sakai, L.A. Utracki, M. Sepehr, J. Li, Y. Simard, *Polym. Eng. Sci.*, 2006, 46, 1040-1050.
[9] L.A. Utracki, M. Sepehr, J. Li, Intern. Polym. Process. 2006, 1, 3-16.
[10] F.P. La Mantia, N.Tz. Dintcheva, R. Scaffaro, R. Marino, Marcomol. Mater. Eng. 2008, 293, 83-91.
[11] F.P. La Mantia, R. Marino, N.Tz. Dintcheva, Marcomol. Mater. Eng. 2009, 294, 575-581.
[12] N.Tz. Dintcheva, R. Marino, F.P. La Mantia, e-Polymers, 2009, no. 054.
[13] N.Tz. Dintcheva, R. Marino, F.P. La Mantia, "Effect of the extensional flow on the properties of oriented nanocomposite films for twist wrapping", accepted for publication in J. Appl. Polym. Sci.
[14] J. Lyngaae-Jorgensen, Rheology of Polymer Blends, in Polymer Blends and Alloys, M.J. Folkes, P.S. Hope, Eds.; Blackie Academic & Professional: London, UK; 1993.
[15] L.A. Utracki, Commercial Polymer Blends, Chapman and Hall, London, 1998.
[16] Ch. Harrats, S. Thomas, G. Groeninckx, Eds. "Micro- and Nanostructured Multiphase Polymer Blned Systems", CRC Press, Taylor & Francis, Boca Raton FL, 2006.
[17] M. Okamoto, P.H. Nam, P. Maiti, T. Kotaka, N. Hasegawa, A. Usuki, Nano Letters, 2001, 1 (6), 295-298.
[18] F.N. Cogswell, Polym. Eng. Sci. 1972, 12, n.1, 64.
[19] M. Joshi, V. Viswnathan, J. Appl. Polym. Sci., 2006, 102, 2164.
[20] M. Zhang, U. Sundararaj, Macromol. Mater. Eng. 2006, 291, 697.

20

Dental Nanomaterials

Seyed Shahabeddin Mirsasaani[1],
Maedeh Hajipour Manjili[2] and Nafiseh Baheiraei[1]
[1]Faculty of medical Engineering, Amirkabir University of Technology,
[2]Gorgan University of Agricultural Sciences and Natural Resources,
Iran

1. Introduction

Humankind has always been plagued by the problem of restoring parts of the body lost due to an accident or disease. The replacement of lost teeth is desired for two primary reasons: esthetics and restoration of function (partial or complete).

Restorative dental materials include items such as noble and base metals, amalgam alloys, cements, composites, glass ionomers, ceramics, denture base resins, and other materials used in restorative dental operations. To describe these materials, comparisons are usually made on the basis of physical and chemical characteristics.

Most restorative materials are measured by a set of physical, chemical, or mechanical tests that lend themselves to duplicate, and efforts are being made to control the quality of materials by noting the result of these tests. This approach has led to a number of gradual improvements in the materials available to the professions. As improvements in properties have occurred, refinements in the technique of application have become necessary [1].

An examination of the history of dentistry shows that the various materials available in any given period have always been important to contemporary restorative dental operations. Improvements came slowly and steadily over the centuries at about the same rate as related developments in other fields of science. It becomes evident that many accepted techniques, materials, and practices have resulted from systematic evaluation and development, so that the branch of restorative materials has become an accepted part of the science of dentistry.

With the beginning of the twentieth century came many refinements and improvements in the quality of various materials and processes used in restorative dentistry. Physical and mechanical tests and the fundamentals of engineering practice were applied to structural designs and restorative materials [2].

The development of resin composite, glass ionomer and compomer restorative materials, new and modified polymers for restorations and impressions, new phenolic and resin cements, improved base metal alloys and amalgams have contributed to the service and function of restorative materials.

Many of the advances in biomaterials during the twentieth century occurred after 1950. These advances include chemical and light-cured composites, acid etching of enamel, glass ionomers, high copper amalgams, and cements, to list only a few of the 100 or more major advances.

A variety of cements have been used in dentistry through the years for two primary purposes: as restorative filling materials, either alone or with other materials, and to retain restorations or appliances in a fixed position within the mouth. In addition, certain cements are used for specialized purposes in the restorative, endodontic, orthodontic, periodontic, and surgical fields of dentistry.

Zinc phosphate, glass ionomer, and zinc oxide-eugenol (ZOE) cements can be applied as a base in deep cavities to insulate the pulp from possible chemical and thermal trauma. A metallic, ceramic, or composite filling material may then be placed over the cement base in sufficient bulk and in proper adaptation to the cavity walls to form the final restoration. The sedative effect of ZOE mixtures has made them valuable for a variety of applications. The ability of the glass and hybrid ionomer and compomer cements to release fluoride and to bond chemically to tooth structure has resulted in their uses as bases and for cementation. Resin cements are used for retention of orthodontic brackets, all-ceramic veneers, crowns and inlays, and resin-bonded bridges because of their strength and ability to bond to acid-etched enamel and dentin treated with a dentin bonding agent [1].

This chapter addresses the ten major classes of dental cements used today based on their chief chemical ingredients and application: Glass and Hybrid Ionomers, Zinc Polyacrilate, Zinc Phosphate, Zinc Oxide-Eugenol, Non-Eugenol-Zinc Oxide, Calcium Hydroxide, Compomers, Composites and Adhesive Resin.

2. Glass ionomer cements

As with other dental materials, the chemistry of dental cements has undergone continual evolution. The glass ionomer cements combine the technologies and chemistries from silicate and zinc polycarboxylate cements in order to incorporate the desirable characteristics of both. Thus they contain the ion-leachable fluoroaluminosilicate (FAS) glass of silicate cements but avoid their susceptibility to dissolution by substituting the phosphoric acid and its salts with the polymeric carboxylic acids of the zinc polycarboxylate cements [2]. Like their predecessors, the zinc polycarboxylates, these materials are primarily acid–base cements.

2.1 Composition

Glass ionomer cements are supplied as a powder and a liquid or as a powder that is mixed with water. Several products are encapsulated. The liquid typically is a 47.5% solution of 2:1 polyacrylic acid/itaconic acid copolymer (average molecular weight 10,000) in water. The itaconic acid reduces the viscosity of the liquid and inhibits gelation caused by intermolecular hydrogen bonding; D(+) tartaric acid (5%, the optically active isomer) in the liquid serves as an accelerator by facilitating the extraction of ions from the glass powder. The powder of glass ionomer cement is a calcium fluoroaluminosilicate glass with a formula of SiO_2-Al_2O_3-CaF_2-Na_3AlF_6-$AlPO_4$

The powder is described as an ion-leachable glass that is susceptible to acid attack when the Si/A1 atomic ratio is less than 2:1. Barium glass or zinc oxide may be added to some powders to provide radiopacity. In some products the polyacrylic acid is coated on the powder. The liquids of these products may be water or a dilute solution of tartaric acid in water [3].

2.2 Setting reaction

The setting reaction is an acid-base reaction between the acidic polyelectrolyte and the aluminosilicate glass. The polyacid attacks the glass to release cations and fluoride ions. These ions, probably metal fluoride complexes, react with the polyanions to form a salt gel matrix. The Al^{3+} ions appear to be site bound, resulting in a matrix resistant to flow, unlike

the zinc polyacrylate matrix. During the initial setting reaction in the first 3 hours, calcium ions react with the polycarboxylate chains. Subsequently, the trivalent aluminum ions react for at least 48 hours. Between 20% and 30% of the glass is decomposed by the proton attack. The fluoride and phosphate ions form insoluble salts and complexes. The sodium ions form a silica gel. The structure of the fully set cement is a composite of glass particles surrounded by silica gel in a matrix of polyanions cross-linked by ionic bridges. Within the matrix are small particles of silica gel containing fluorite crystallites.
Glass ionomer cements bond chemically to enamel and dentin during the setting process. The mechanism of bonding appears to involve an ionic interaction with calcium and/or phosphate ions from the surface of the enamel or dentin. Bonding is more effective with a cleaned surface provided the cleansing process does not remove an excessive amount of calcium ions. Treating dentin with an acidic conditioner followed by a dilute solution of ferric chloride improves the bonding. The cleansing agent removes the smear layer of dentin while the Fe^{3+} ions are deposited and increase the ionic interaction between the cement and dentin [2, 3].

2.3 Role of tartaric acid

Tartaric acid is added to prolong the working time of the cement mix, improve the manipulative characteristics and narrow the range of the setting time. In the absence of this component, the mix becomes rubbery within a few seconds and is rather difficult to work with. The mechanism by which tartaric acid operates is believed to be a temporary suppression of the ionization of the polyacid as well as a preferential extraction of the cations from the glass so that polyacrylate complexes cannot form immediately, thus leading to an increased working time. Later, it sharpens the set and accelerates the hardening of the glass ionomer mix [1].

2.4 Properties

2.4.1 Setting time

Glass ionomer cements set within 6 to 8 minutes from the start of mixing. The setting can be slowed when the cement is mixed on a cold slab, but this technique has an adverse effect on the strength [3].

2.4.2 Physical properties

The glass ionomers are rather brittle materials with high modulus of elasticity, low diametral tensile strength and low fracture toughness. They are susceptible to desiccation and hence must be protected with a varnish or a resin bonding agent during the setting process. Compressive strengths are similar to zinc phosphate cements [1, 2].

2.4.3 Solubility and erosion

The glass ionomers are susceptible to erosion in the oral environment, particularly under acidic conditions. The maximum acid erosion allowable by the lactic acid jet method is 0.05μm per hour. The solubility of this class of cements is higher than most other types of permanent luting cements [2].

2.4.4 Fluoride release

Glass ionomer materials exhibit a sustained release of fluoride over a long period of time. The uptake of the released fluoride ion in human saliva [4] and its incorporation into human enamel have been reported [5]. Although considerable debate exists about the 'clinical proof' of the benefits of fluoride, occurrence of recurrent caries in the teeth where these cements have been used is reported to be rare. Studies have shown that glass ionomers inhibit demineralization of the surrounding tooth structures in vitro [3] and in situ [1] and provide protection against recurrent caries under clinical conditions for patients with high caries risk. This can be attributed to the ability of glass ionomer cements to inhibit demineralization and enhance remineralization through release of fluoride to the adjacent tissue and surrounding fluid. The rate of fluoride release depends on a particular product brand. However, as a general class, glass ionomers release more fluoride than other types of fluoride-releasing materials [6].

2.4.5 Adhesion

Like the polycarboxylate cements, the glass ionomers can bond to the calcium of hydroxyapatite of tooth tissue [6].

2.4.6 Post-operative sensitivity

In the past, one of the common complaints against conventional glassionomer luting cements has been the high incidence of post-operative sensitivity [7] with greater incidence reported for the 'anhydrous' type [8]. The most frequent reasons cited for post-operative dentinal hypersensitivity are as given below.

- Desiccation of the tooth. The glass ionomers require water for their setting and may absorb water from the dentinal tubules during this process. This phenomenon is thought to be more prevalent in the anhydrous cements where the dried polycarboxylic acid also needs water for rehydration.
- Initial low pH of the material (although it increases rapidly during the setting process) may irritate the pulp.
- Moisture contamination leading to poor sealing of the tooth and improper margins, hence leakage to bacteria and their by-products.
- Too low a viscosity of the mix causing fluid to be forced down the dentinal tubules during seating of the prosthesis. The lower values of pH (higher acidity) encountered for prolonged periods during setting of conventional glass ionomers compared with other types of cements was cited as a possible reason for pulp sensitivity [9]. The diffusion of acid was observed through 0.25mm dentine disks [10]. However, it has been demonstrated that by following proper clinical procedures during cementation the operator can minimize or eliminate the occurrence of post-operative sensitivity [11, 12].

2.5 Applications

Glass ionomer cements are used primarily for permanent cement, as a base, and as a Class 5 filling material. The cement has been evaluated as a pit and fissure sealant and as an endodontic sealer. The sensitivity of the cement to moisture and desiccation may minimize its use in these latter applications. Glass ionomer cements are being used clinically for cementation of orthodontic bands because of their ability to minimize decalcification of enamel by means of fluoride release during orthodontic treatment [9].

3. Resin-modified glass ionomer cements

RMGIs, developed in the late 1980s [13], are more recent entrants into the dental cement arena, having been first introduced commercially as a luting cement in 1994. This class of cements is less technique-sensitive than the conventional glass ionomer materials and possesses some very favorable physicomechanical properties compared with conventional GI materials yet releases similar levels of fluoride. Since the RMGI luting cements also allow easy removal of excess cement, show minimal post-operative sensitivity and, so far, have exhibited good clinical performance and durability, this class of material has rapidly become one of the most popular materials for routine crown and bridge applications. Improvements in delivery format have now consolidated their use [14, 15].

3.1 Composition

The first-generation materials were supplied as powder–liquid configurations and are available in hand-mix or encapsulated versions. Vast differences do exist between products from different manufacturers; hence care has to be taken in choosing a particular commercial product for clinical use. The essential components of a true RMGI are [16]:

- polycarboxylic acid copolymer often modified with pendant methacrylate groups
- FAS glass
- Water
- Water-compatible methacrylate monomers
- Free-radical initiators

3.1.1 Polymeric component

The RMGIs contain some methacrylate components common in resin composites. There are two ways in which the methacrylate component can be introduced. In the first RMGIs introduced to the market, the polycarboxylic acid copolymer chain is modified to contain a pendant methacrylate group. This was achieved by treating some of the carboxylic acid groups of the polycarboxylic acid with isocyanatoethyl methacrylate to provide pendant methacrylate groups connected through the hydrolytically stable amide linkages. The first commercial glass ionomer material (Vitrebond™ RMGI liner, 3M ESPE) was introduced in 1988 using this type of chemistry. The RMGI luting cements from this manufacturer also use the same chemistry. In addition to the methacrylate-modifi ed carboxylic acid, the liquid portion contains a water-miscible methacrylate monomer, e.g. hydroxyethyl methacrylate (HEMA), glycerol dimethacrylate (GDMA), etc. In the second type of RMGI system, the polymer is unmodified polycarboxylic acid. In this case, the liquid is formulated with a mixture of hydrophilic methacrylate monomers and water. Generally, the water content of these materials is lower and the monomer content higher than for the first type. As a result, the coefficient of thermal expansion of these glass ionomers is high. Free radical initiators are added to trigger the curing of the methacrylate groups. Visible light initiators and/or self-cure redox initiators are employed to effect this curing and covalent cross-linking reaction [17, 18].

3.1.2 Nature of the glass

The FAS glass of the RMGI systems is similar in composition to the glasses described for conventional glass ionomers, although some variations are made in order to match the refractive index of the glass with that of the matrix. It is also common to treat the surface of the glass with an organic modifier, such as a methacryloyl functional alkoxysilane, to allow better binding with the methacrylate matrix [19, 20].

3.2 Setting reactions
Two distinct types of curing reactions take place in a true light-cured glass ionomer, the traditional acid-base glass ionomer cure and the free radical methacrylate polymerization [14]. In the laboratory, the former can be followed by infrared spectroscopy through the appearance of carboxylate ion peaks. The methacrylate reaction, being a chain polymerization, proceeds at a rate that is several orders of magnitude higher than the acid-base reaction. In practice, the extent to which each of these two reactions occurs is very dependent on a particular system. If the system is low in water and high in the methacrylate components, the ionization of the polycarboxylic acid will be severely suppressed resulting in little acid-base reaction. In one system, the redox initiators are microencapsulated in inert components and mixed with the powder [16]. When spatulated with the liquid, the initiators are triggered and free radical polymerization of the methacrylate groups initiates. In some commercial RMGI luting cements an additional light-curing initiation is provided. However, light will polymerize only the cement exposed at the margins leading to a false sense of security since the bulk of the cement under the crown typically would take longer to undergo a clinical set.

3.3 Role of water and tartaric acid
As with conventional systems, water is an integral part of all true RMGIs, being essential for the acid-base glass ionomer reaction to proceed to any appreciable extent and for the release of fluoride from the FAS glass by this mechanism. Tartaric acid may be optionally added to modulate the working time and acid-base glass ionomer setting reaction [1].

3.4 Properties
As with conventional glass ionomers, the properties of RMGI luting cements are regulated under the ISO 9917-1:2003 standard for dental water-based cements and self-curing resin-modified cements [3].

3.4.1 Setting and working time
The working time is typically greater than 2.5min from the start of mix at ambient temperature of 23°C. Higher temperatures and vigorous spatulation shorten the working time while lower temperatures prolong it. Excess material can be removed when the cement reaches a waxy stage after placement in the mouth (2-3min at 37°C) using a suitable instrument. The restoration should be finished and the occlusion checked when the material has completely set (about 5min from placement, depending on brand) [6].

3.4.2 Solubility and erosion
RMGI luting cements have greatly improved resistance to dissolution when compared with zinc phosphate, polycarboxylate or conventional glass ionomer luting cements [21]. The ISO 9917-1 erosion test for lactic acid shows the solubility to be virtually zero. This is one of the key advantages of this type of cement [22].

3.4.3 Fluoride release
Although differences exist between manufacturers' brands, in general, the fluoride release and rechargability of RMGI luting cements was found to be similar to that of the conventional glass ionomer cements [23]. The beneficial effect of fluoride from RMGIs has

been shown in vitro [17, 18] and in vivo. The dentinal caries inhibition of these materials due to sustained fluoride release was also reported to be similar to the conventional glass ionomers [24-26].

3.4.4 Physical and mechanical properties

Both the early mechanical properties, as well as the 'matured' properties, of the RMGIs are much improved over the conventional glass ionomer luting systems. The additional covalent cross-linking in the matrix due to the polymerization of the methacrylate groups contributes towards this effect. Thus, these materials are considerably less brittle and have significantly higher tensile strengths, flexural strengths and fracture toughness values than the corresponding conventional systems [27]. The properties of one paste–paste system were found to be similar to those of the corresponding powder–liquid type [28]. The modulus values are in general lower for the RMGIs [29] and fracture toughness is higher, contributing to the favorable clinical outcomes in high stress-bearing areas [30]. The combinations of these unique characteristics make the RMGIs particularly attractive when used as luting cements under crowns, since they can provide relief of stress created by masticatory forces. All the ultimate properties build up rapidly during the first 24h and then increase slightly during the next 7 days. As with conventional systems, most resin-modified systems do not undergo significant decrease in mechanical properties over time due to fluoride release [31].

3.4.5 Adhesion

Several RMGI systems (e.g. RelyX™ Luting Cement and RelyX™ Luting Plus Cement, 3M ESPE) do not require any etching, priming or conditioning of the tooth and thus can be considered as self-adhesive cements. Others, such as Advance™ (Caulk/Dentsply) and Fuji™ CEM or Fuji™ Plus (GC Dental), suggest the use of additional conditioning agents. Thus Fuji plus luting cement is used with a 10:2 citric acid–ferric chloride etchant due to the more resinous characteristic of the cement [32]. Clinical retention of crowns with the RMGI luting cements has also been found to be excellent in long-term clinical studies. The measured in vitro bond strengths of RMGIs are considerably higher than their conventional counterparts, although still not as high as resin adhesive–composite combinations. The failure is usually cohesive in the glass ionomer; hence, as with other glass ionomer cements, the bond strength measurements are not true reflections of interfacial adhesion. The mechanism of adhesion is a combination of two factors:

- The modification of the dentinal smear layer and interpenetration of the dentinal tubules by the fluid cement followed by polymerization and entanglement with collagen fibers
- Ionic reaction of the polycarboxylate with the calcium ions of hydroxyapatite. The first mechanism has been observed by confocal microscopy studies. The second reaction has been verified by fourier transformed infrared spectroscopic studies. The additional feature of increased stress relief afforded by these materials contributes to the long-term durability of the prosthesis. A restoration adhesively cemented with an RMGI luting agent is more resistant to microleakage when compared with a non-adherent cement like zinc phosphate [33]. The RMGIs appear to exhibit minimal microleakage similar to the performance of adhesive resin cements although the latter yield higher in vitro bond strength values [34].

3.5 Applications

RMGI luting cements are indicated for routine cementation of: PFM crowns and bridges to tooth structure, amalgam, composite or glass ionomer core buildups; metal inlays, onlays or crowns; prefabricated and cast posts. They are also indicated for cementation of copings made with all-alumina or all zirconia cores and provide retention similar to adhesive luting cements. Retention of the all-ceramic crowns with RMGI cement was reported to be excellent and similar to that achieved with adhesive resin luting cements in vitro and with zinc phosphate cements in a split-mouth randomized clinical trial [35].

4. Hybrid ionomer cement

Self-cured and light-cured hybrid ionomers (or resin-modified glass ionomers) are available for cementation. Hybrid ionomers are used for restorations in low stress bearing areas and are recommended for patients with high caries risk. These restorations are more esthetic than glass ionomers because of their resin content [36].

4.1 Composition

The powder of hybrid ionomers is similar to that of glass ionomers. The liquid contains monomers, polyacids and water. Hybrid ionomers set by a combined acid-base ionomer reaction and light-cured resin polymerization of 2-hydroxyethyl methacrylate. Placing a dentin bonding agent before inserting a hybrid ionomer is contraindicated, because it decreases fluoride release.

One self-cured hybrid ionomer cement powder contains a radiopaque, fluoroaluminosilicate glass and a micro-encapsulated potassium persulfate and ascorbic acid catalyst system. The liquid is an aqueous solution of polycarboxylic acid modified with pendant methacrylate groups. It also contains 2-hydroxyethylmethacrylate (HEMA) and tartaric acid. Another self-cured cement contains a mixture of fluoroaluminosilicate and borosilicate glasses in the powder. Its liquid is a complex monomer containing carboxylic acid groups that can undergo an acid-base reaction with glass and vinyl groups that can polymerize when chemically activated. A light-cured hybrid ionomer cement contains fluoroaluminosilicate glass in the powder and a copolymer of acrylic and maleic acids, HEMA, water, camphorquinone, and an activator in the liquid [37].

4.2 Setting reaction

Setting of hybrid ionomer cements generally results from an acid-base glass ionomer reaction and self-cured or light-cured polymerization of the pendant methacrylate groups. Some cements, however, are light-cured only [37].

4.3 Properties

Hybrid ionomers bond to tooth structure without the use of a dentin bonding agent. Typically, the tooth is conditioned (etched) with polyacrylic acid or a primer before placing the hybrid ionomer. The transverse strength of a hybrid ionomer is almost double that of a standard glass ionomer. Hybrid ionomers release more fluoride than compomers and composites but almost the same as glass ionomers. Glass ionomers and hybrid ionomers recharge when exposed to fluoride treatments or fluoride dentifrices.

Requirements for light-activated cements, which are water-based and set by multiple reactions, including an acid-base reaction and polymerization (Type I), and by cements that set only after light-activation (Type III, are described by ANSI/ADA Specification No. 96 (IS0 9917, Part 2). The compressive and tensile strengths of hybrid ionomer cements are similar to those of glass ionomer cements. The fracture toughness is higher than that of other water based cements but lower than composite cements. The bond strength to moist dentin ranges from 10 to 14 MPa and is much higher than that of most water-based cements. Hybrid ionomer cements have very low solubility when tested by lactic acid erosion. Water sorption is higher than for resin cements. Delayed fracture of ceramic restorations cemented with hybrid ionomer cements has been reported. Recently, some hybrid ionomer cements have been modified to have less water sorption. Fluoride release and rechargeability are similar to glass ionomer cements. The early pH is about 3.5 and gradually rises. Clinical experience indicates minimal postoperative sensitivity [36].

4.4 Applications

Self-cured hybrid ionomer cements are indicated for permanent cementation of porcelain-fused-to-metal crowns; bridges; metal inlays, onlays, and crowns; post cementation; and luting of orthodontic appliances. Additional uses include adhesive liners for amalgam, bases, provisional restorations, and cementation of specific ceramic restorations. Light-cured hybrid ionomer cements are used primarily for liners and bases. One light-cured product is recommended for direct bonding of orthodontic brackets and bands [35].

5. Zinc phosphate cement

Zinc phosphate cement has been a mainstay in dentistry for crown and bridge applications for well over a century and has undergone many refinements in formulation and compounding. However, with the advent of newer chemical technologies its use has gradually declined over the last decade. This type of cement is supplied as a powder–liquid formulation [38].

5.1 Composition

The powder consists of mainly amorphous zinc oxide as the major ingredient with small amounts of oxides of magnesium and bismuth added mainly to facilitate the calcining process for manufacture of the powder. Small quantities of silica (typically fumed) are often added to aid in providing the right viscosity and yield stress. Minor amounts of barium and calcium compounds may be added by various manufacturers to provide a smooth, creamy mix, which is desirable for easy flow during cementation of the restoration. Some products also contain tannin fluoride although it is doubtful if the amount of fluoride released is of clinical significance.

The liquid consists of an aqueous mixture of orthophosphoric acid with small amounts of aluminum and zinc phosphates which act as buffers in reducing the reactivity of the free acid. The amount of water used in the formulation is quite critical since it dictates the setting time - too little prolongs the setting time whereas too much shortens it [39].

5.2 Setting reaction

When the alkaline zinc oxide powder is incorporated in the acidic liquid an exothermic acid–base reaction ensues. Initially, the mix becomes fluid and creamy which aids in the placement of the prosthodontic device. Within a few minutes the mass hardens to a set cement which may contain varying amounts of crystalline phosphates of zinc including the formation of hopeite, $Zn_3(PO_4)_2 \cdot 4H_2O$, in the presence of excess moisture [22, 39].

5.3 Properties

The clinically significant properties of zinc phosphate cements include setting time, solubility and film thickness. The mechanical properties of importance are strength, modulus of elasticity and hardness. Most commercial materials meet the American National Standards Institute/American Dental Association (ANSI/ADA) Specification 96 (ISO 9917-1: 2003). In comparison with the more recent types of cements, the modulus of elasticity is quite high, which causes them to be rather brittle materials that are prone to fracture. The solubility of the set cement is relatively high at about 0.2% in 24h compared with some other classes of cements. Furthermore, these materials have virtually no adhesion to the dentinal core of the tooth preparation [38].

5.4 Applications

Zinc phosphate cements have had a long history of clinical usage for routine luting of metal-supported crowns and bridges [22]. Other uses include cementation of orthodontic bands, as a basing material and as temporary restorations. Occasional post-operative sensitivity has been reported by patients with the use of this type of cement [38]. The exact cause of this is not known and has been linked to acidity of the un-neutralized orthophosphoric acid residues in the cement and/or the movement of dentinal fluid across the tubules in excessively desiccated dentine due to an improper seal. The use of resin desensitizing agents has been advocated to counter post-operative sensitivity without compromising crown retention [39, 40].

6. Zinc polyacrylate cements

Zinc polycarboxylate cements, sometimes referred to simply as 'polycarboxylate' or 'polyacrylate' cements, are based on the reaction of zinc oxide with polycarboxylic (same as polyalkenoic) acid in water [41]. Like the zinc phosphate cements these are also supplied as powder–liquid compositions.

6.1 Composition

Zinc polyacrylate cements are supplied as a powder and a liquid or as a powder that is mixed with water. The powder is quite similar to that of zinc phosphate cements and consists mainly of sintered zinc oxide ground to a fine particle size. The sintering process reduces the reactivity of the amorphous zinc oxide and helps in the manipulation of the cement. Small amounts of magnesium oxide (1–5%) are added to aid the sintering process while the incorporation of fumed silica helps in the mixing and flow of the cement. Fluoride salts, e.g. stannous fluoride, may also be incorporated in small amounts to improve mechanical strength and to serve as a source of leachable fluoride. In some commercial embodiments the powder is coated with 5–20% of anhydrous polyacrylic acid to make it less technique-sensitive during the mixing process.

The liquid consists of an aqueous solution of a polycarboxylic acid, generally a homopolymer of polyacrylic acid or a copolymer of acrylic acid with itaconic or maleic acids. The average molecular weight (weight average) is usually in the range 20000–50000.

The viscosity of the solution may be controlled by the addition of small amounts of tartaric acid. For products where the powder is coated with polycarboxylic acid, the liquid is either a dilute solution of the polyacid or simply water [42].

6.2 Setting reaction

The polycarboxylic acid reacts with the basic zinc oxide in a neutralization reaction forming a zinc polycarboxylate complex salt [42]. This results in the formation of a cross-linked polycarboxylate hydrogel reinforced by the oxide particles. The cross-linked gel is bound to the polyanion chains by electrostatic interaction. The setting reaction has been studied by infrared spectroscopy which shows that the carboxylic acid groups (COOH) are progressively converted to carboxylate (COO−) groups as the cement hardens. Water plays several important roles in controlling the chemistry and properties of all acid–base cements. These are outlined below.

1. As a diffusion medium. Water is needed for the acids to ionize so that the protons can be dissociated and solvated and the acidic property can be manifested. It is also needed for the diffusion of the metallic ions of the bases (e.g. zinc oxide) so that these can enter the liquid phases and thus react with the acid. In addition, it is essential for the diffusion of fl uoride ions, where present, out of the set cement.
2. As a stabilizer of the carboxylate complexes. A portion of the water coordinates with the zinc carboxylate complexes by electron donation to stabilize them. This is known as bound water and cannot be easily removed from the stabilized matrix.
3. As a plasticizer. The residual water helps to plasticize the set cement and makes it more resilient and less prone to failure by fracture.

6.3 Properties

6.3.1 Setting and working time

Setting and working time are two important clinically relevant parameters for any cement. The working time is the time available for the manipulation of the unset cement while the setting time is the time required by the material to set or harden from a fluid or plastic state to a rigid one. While at the end of the setting time the hardening reaction may not be complete it generally progresses far enough for the dentist to proceed to successive steps in the procedure without fear of premature dislodgement of the prosthesis. There is no standard value for the working time, but it should be reasonably long so that adequate time is available for mixing and placing the cement. Since both the working time and setting time are dependent on the same chemical reaction the former should not be too long to ensure the setting of the cement within a reasonable time. The rate of setting is affected by the following parameters:

- Reactivity of the zinc oxide
- Particle size of the zinc oxide
- Presence of additives
- Molecular weight and concentration of the polyacrylic acid
- Powder/liquid ratio

The first four parameters are controlled by the manufacturer. The last one is dependent on the operator. Hence, accurate dispensing of the powder and liquid are essential if consistent setting is to be obtained. For commercial cements the working times are 2-4min and set times are between 5 and 8 min [41, 42].

6.3.2 Strength

In general, zinc polycarboxylate cements are lower in compressive strength than zinc phosphates but, due to the presence of a polymeric matrix, are tougher and less brittle. Plastic deformation is exhibited at higher load values, which contributes to the increased toughness of these cements. Compressive and diametral tensile strengths of the cement develop rapidly within the first hour to about 60–80% of its strength at maturity, while it takes about 24h for maximum strength to be attained [41].

6.3.3 Solubility and erosion

The solubility of the cements when tested in water alone is low (0.1–0.5%) compared with the zinc phosphate cements, although solubility in distilled water does not always correlate to solubility in oral conditions. These cements are rather susceptible to erosion in an acidic environment, however. A quantitative measure of this property is defined in the ISO 9917-1 specification using the impinging lactic acid jet method devised by Beech and Bandyopadhyay [43].

6.4 Applications

Zinc polycarboxylate cements have a fairly long record of clinical success when used for specified indications. The most frequent use is for cementation of cast alloy- and metal-supported inlays, onlays and single-unit crowns. Because of the lower values of compressive modulus these materials should not be used for permanent cementation of long-span bridges. Less frequently they are also used as cavity liners and orthodontic band cementation but in such cases a higher powder/liquid ratio is recommended by manufacturers. The comparatively high long-term solubility and low hardness limit the utility of this class of materials for permanent cementation, resulting in a decline in their use in recent times [42].

7. Zinc oxide-eugenol and non-eugenol cements

When certain types of zinc oxide are mixed with eugenol, the mix sets to a hard cement that is compatible with both the hard and soft tissues of the mouth. Cements of this type have been used extensively since the 1890s. Simple mixtures of these two materials do not have great strength when compared with zinc phosphate cements, and their use has been limited to situations in which strength is not important. Quite early it was found that they had a sedative effect on exposed dentin. For many years, ZOE cements have been used as provisional restorations, soft tissue packs in oral surgery and periodontics, and root canal sealers. Because eugenol acts as an inhibitor for free-radical polymerized materials, select other materials for provisional restorations when bonding of the permanent restoration is anticipated. Non-eugenol-zinc oxide cements are also available for temporary cementation. These cements are suitable for patients sensitive to eugenol [26, 29, 36].

7.1 Composition

The powder is mainly zinc oxide, with added white rosin to reduce the brittleness of the set cement, zinc stearate as a plasticizer, and zinc acetate to improve the strength of the cement. The liquid is eugenol with olive oil as a plasticizer. Two compositional changes have been used to increase the strength of the cement for luting purposes. In one, methyl methacrylate polymer is added to the powder, and in the other, alumina ($A1_2O_3$) is added to the powder and ethoxybenzoic acid (EBA) to the liquid. A typical polymer-reinforced cement has 80% zinc oxide and 20% poly(methy1 methacrylate) in the powder and eugenol in the liquid [14]. These cements are sufficiently strong for final cementation of fixed prostheses and are used also as cement bases and provisional restorations. A typical EBA-alumina-reinforced ZOE cement contains 70% zinc oxide and 30% alumina by weight in the powder. In some cases, rosin and copolymers may be added to reduce the brittleness and film thickness and improve the mixing qualities. The liquid of the EBA-alumina reinforced cements contains 62.5% ortho-EBA by weight and 37.5% eugenol by weight. The non-eugenol-zinc oxide cements typically contain an aromatic oil and zinc oxide. Other ingredients may include olive oil, petroleum jelly, oleic acid, and beeswax [17, 31].

7.2 Setting reaction

The setting of ZOE cements is a chelation reaction in which an amorphous, zinc eugenolate is formed. The setting reaction is shown below, where two molecules of eugenol react with ZnO in the presence of water to form the chelate, zinc eugenolate. Excess zinc oxide is always used, so the set material consists of a matrix of amorphous zinc eugenolate that binds the unreacted zinc oxide particles together. The setting reaction is accelerated by increases in temperature or humidity. EBA also forms a chelate with zinc oxide, and its presence allows some crystalline zinc eugenolate to form, which provides additional strength. The reaction is not measurably exothermic, and the presence of moisture is essential for setting to occur [32].

7.3 Properties

The variety of compositions of the ZOE cements and the many uses to which they are applied make it difficult to write a specification for these cements. ANSVADA Specification No. 30 (IS0 3107) for dental zinc oxide-eugenol cements and zinc oxide non-eugenol cements gives standards for temporary cements, permanent cements, filling materials and bases, and cavity liners. It sets requirements for the general characteristics of the powders, liquids, and pastes used in these cements and for the important physical properties of setting time, compressive strength, disintegration, film thickness, and acid-soluble arsenic content, where these are applicable [38].

7.4 Applications

A range of ZOE and modified ZOE cements are suitable for many uses in restorative dentistry, and the practitioner should become familiar with each type and its application. Base Materials having a compressive strength of 5.5 to 39 MPa are used as a cement base, and the strength reaches a maximum in about 12 to 15 minutes. They are normally used under zinc phosphate cement, which acquires about three times the strength in the same time. The ZOE cements have the advantage that the thermal insulating properties of the cements are excellent and are approximately the same as those for human dentin [33].

8. Calcium hydroxide

8.1 Composition

The base paste of a typical product contains calcium tungstate, tribasic calcium phosphate, and zinc oxide in glycol salicylate. The catalyst paste contains calcium hydroxide, zinc oxide, and zinc stearate in ethylene toluene sulfonamide. The ingredients responsible for setting are calcium hydroxide and a salicylate, which react to form an amorphous calcium disalicylate. Fillers such as calcium tungstate or barium sulfate provide radiopacity. A light-cured calcium hydroxide base consists of calcium hydroxide and barium sulfate dispersed in a urethane dimethacrylate resin [11].

8.2 Properties

Calcium hydroxide cements are used for lining deep cavities or for direct pulp capping. The antibacterial action of calcium hydroxide makes these cements useful in indirect pulp-capping procedures involving carious dentin. ZOE cements are used in deep cavities to retard penetration of acids and reduce possible discomfort to the pulp. Both calcium hydroxide and ZOE low strength bases are often used with a high strength base in restoring a tooth. Root canal sealers containing calcium hydroxide have been developed.

The important properties of these bases are mechanical and thermal properties, solubility, and pH. Calcium hydroxide (self-cured bases have low values of tensile strength and compressive strength, or elastic modulus, compared with high-strength bases. Although setting times vary between 2.5 and 5.5 minutes, compressive strengths of these cements continue to increase over a 24-hour period. For a group of five commercial products, compressive strengths ranged from 6.5 to 14.3 MPa at 10 minutes to from 9.8 to 26.8 MPa at 24 hours. The low elastic modulus of calcium hydroxide bases restricts their usage to areas not critical to the support of restorations [40].

Mechanical support should be provided by sound dentin or by a high-strength base. Calcium hydroxide bases are, however, considered strong enough to support the forces of condensation of amalgam [10].

8.2.1 Thermal properties

Calcium hydroxide bases may provide some thermal insulation to the pulp if used in sufficiently thick layers. A thickness greater than 0.5 mm is not suggested. Practically, thermal protection should be provided by the overlying high-strength base [38].

8.2.2 Solubility

The solubility of calcium hydroxide bases has been measured in several solvents for various periods of immersion. For a group of five commercial products, values ranged from 0.4% to 7.8% in distilled water at 37°C for 24 hours, from 0.1% to 6.2% in 35% phosphoric acid for 60 seconds, and from 0.3% to 1% in ether for 10 seconds [18]. One product that was resistant to dissolution in water and in acid disintegrated when exposed to ether. Some solubility of the calcium hydroxide is necessary to achieve its therapeutic properties, although an optimum value is not known. Clearly the use of acid-etching procedures and varnish in the presence of calcium hydroxide bases must be done with care. Over a long term, some calcium hydroxide products seem to "disappear" from the cavity. The cause of this dissolution is unclear, but some products have been reformulated in an attempt to minimize the problem [25].

8.2.3 pH

The pH of commercial products has been measured at between 9.2 and 11.7. Free calcium hydroxide in excess of that necessary to form the calcium disalicylate stimulates secondary dentin in proximity to the pulp and shows antibacterial activity [31].

9. Compomers

Compomer cement is the newest resin-based cement indicated for cementation of cast alloy crowns and bridges, porcelain-fused-to-metal crowns and bridges, and gold cast inlays and onlays. Cementation of all-ceramic crowns, inlays, onlays, and veneers, with some exceptions, is contraindicated. The cement should not be used as a core or filling material. Compomers are also known as *poly acid-modified composites.* A compomer cement was recently introduced for orthodontic bonding [41].

9.1 Composition

The cement powder contains strontium aluminum fluorosilicate glass, sodium fluoride, and self and light-cured initiators. The liquid contains polymerizable methacrylate/carboxylic acid monomer, multifunctional acrylate/phosphate monomer, diacrylate monomer, and water [40].

9.2 Setting reaction

Setting is the result of self- and light-cured polymerization. Once the cement comes into contact with oral fluids, an acid-base reaction may occur. The carboxylic acid groups contribute to the adhesive capability of the cement [42].

9.3 Properties

Compomer cement has high values of retention, bond strength, compressive strength, flexural strength, and fracture toughness. The cement has low solubility and sustained fluoride release [22].

10. Composites and adhesive resins

Cements based on resin composites have been used for cementation of crowns, conventional bridges, and resin-bonded bridges; for bonding of esthetic ceramic and laboratory-processed composite restorations to teeth; and for direct bonding of orthodontic brackets to acid-etched enamel. Recently, composite cements have been developed for cementation of provisional restorations. IS0 *4049* for polymer-based filling, restorative, and luting materials (ANSI/ADA No. 27) describes the following three classes of composite cements [33, 29]:

Class 1-self-cured materials

Class 2-light-cured materials

Class 3-dual-cured materials

Property requirements based on IS0 4049 can be summarized as follows:

Class 1, 2, 3: film thickness, max.—50 pm

Class 1, 3: working time, min.—60 seconds

Class 1, 3: setting time, max.—10 minutes

Class 2: depth of cure, min.—0.5 mm

(Opaque), 1.5 mm (others)

Class 1, 2, 3: water sorption, max.—40μg/mm^3

Class 1, 2, 3: solubility, max.—7.5 μg/mm^3

10.1 Cementation of alloy crowns and bridges, resin-bonded bridges, and provisional restorations

Synthetic resin cements based on methyl methacrylate have been available since 1952 for use in cementation of inlays, crowns, and appliances. In the early 1970s, a resin composite was introduced as a crown and bridge cement [1].

10.2 Composition

Self-cured composite cements are typically two paste systems. One major component is a diacrylate oligomer diluted with lower-molecular weight dimethacrylate monomers. The other major component is silanated silica or glass. The initiator-accelerator system is peroxide-amine [27].

10.3 Setting reaction

One adhesive resin cement is a self-cured, powder-liquid system formulated with methacryloxyethylphenyl phosphate or 4-methacryloxyethyl-trimellitic anhydride (4-META). The 4-META cement is formulated with methyl methacrylate monomer and acrylic resin filler and is catalyzed by tri-butyl-borane. Another adhesive resin cement is a phosphonate cement supplied as a two-paste system, containing Bis-GMA resin and silanated quartz filler. The phosphonate molecule is very sensitive to oxygen, so a gel is provided to coat the margins of a restoration until setting has occurred. The phosphate end of the phosphonate reacts with calcium of the tooth or with a metal oxide. The double-bonded ends of both 4-META and phosphonate cements react with other double bonds when available. Setting of resin cements results from self or light-cured polymerization of carbon-carbon double bonds [39].

10.4 Properties

Composite cements used for cementation of provisional restorations (25 to 70 MPa) have a substantially lower compressive strength than composite cements used for permanent cementation (180 to 265 MPa) [22].

10.5 Applications

Adhesive resin cements and composite cements in conjunction with bonding agents are being used as cements for posts and cores. Bond strengths of 14 MPa have been reported for silica-treated posts cemented with 4-META resin cement in extracted teeth. The use of resin-bonded bridges declined dramatically in the late 1980s [37].

11. Nanocomposites

Resin composites are used to replace missing tooth structure and modify tooth color and contour, thus enhancing facial esthetics. The development of composites about 1960 has resulted in higher mechanical properties, lower thermal coefficient of expansion, lower dimensional change on setting, and higher resistance to wear, thereby improving clinical performance. Later development of bonding agents for bonding composites to tooth structure has also improved the quality of composite restorations [44].

Composites were initially developed for anterior Class 3 to Class 5 restorations, in which esthetics was crucial, and for Class 1 restorations, in which moderate occlusal stresses occur.

In the 1990s, modifications of materials and techniques extended their application to Class 2 and Class 6 posterior restorations. Laboratory- processed composites are used for crowns and even bridges, when reinforced with fibers, and can be bonded to an alloy substructure.

A resin composite is composed of four major components: organic polymer matrix, inorganic filler particles, coupling agent, and the initiator-accelerator system [45].

11.1 Polymer matrix

The organic polymer matrix in most composites is either an aromatic or urethane diacrylate oligomer. Oligomers are viscous liquids, the viscosity of which is reduced to a useful clinical level by the addition of a diluent monomer [45].

The two most common oligomers that have been used in dental composites are dimethacrylates (Bis-GMA) 2,2-bis[4(2- hydroxy-3 methacryloyloxy-propy1oxy)-phenyll propane and urethane dimethacrylate (UDMA). Both contain reactive carbon double bonds at each end that can undergo addition polymerization. A few products use both Bis-GMA and UDMA oligomers. The viscosity of the oligomers, especially Bis-GMA, is so high that diluents must be added, so a clinical consistency can be reached when they are compounded with the filler. The most widely used resin in dental composites is based on the copolymer prepared from a combination of Bis-GMA and triethylene glycol dimethacrylate (TEGDMA) (Fig. 1). TEGDMA is usually added to Bis-GMA in order to achieve workable viscosity limits since the latter monomer possesses very high viscosity ($>10^6$cP) due to the intermolecular hydrogen bonding. In Bis-GMA/TEGDMA dental resin systems, Bis-GMA functions to limit the polymerization induced volumetric shrinkage and to enhance resin reactivity, while TEGDMA provides for increased vinyl double bond conversion [46].

Although Bis-GMA has widely been used as the main monomer in most resin composite systems due to its superior aesthetic quality, simple operation technique, enhanced mechanical strength, less shrinkage, higher modulus, and reduced toxicity because of its lower volatility and diffusivity into the tissue, and the composites have undergone significant development since their advent, they still have shortcomings limiting their application. Lack of good adhesion to the tooth structure and polymerization shrinkage are the most important problems. Considerable interest has been devoted to synthesizing new monomers to provide alternative monomers to overcome the problems [47].

BTDMA as a dimethacrylate monomer containing carboxylic acid groups in its structure has been shown that can interact with the Ca^{2+} ions of the tooth structure so it has the potential to provide better adhesion properties in dental composites (Fig. 1) [48].

11.2 Fillers

As dental composites cannot withstand heavy occlusal forces, many ways have been introduced to reinforce them, such as using fibers and whiskers as reinforcing agents. Some of dental resins are reinforced with inorganic fillers [49]. A wide variety of fillers have been employed in composites to improve the properties and developments in filler technology are responsible for many improvements in composites which are used today [45]. Despite many improvements in this field, dental composites do not have enough toughness, strength and durability in order to be used in stress bearing areas. There are numerous types of inorganic fillers. Fillers such as SiO_2, ZrO_2 and Al_2O_3 of micron or nano particle size are usually used. Most of the fillers which are used to reinforce dental composites are silicate glasses [50].

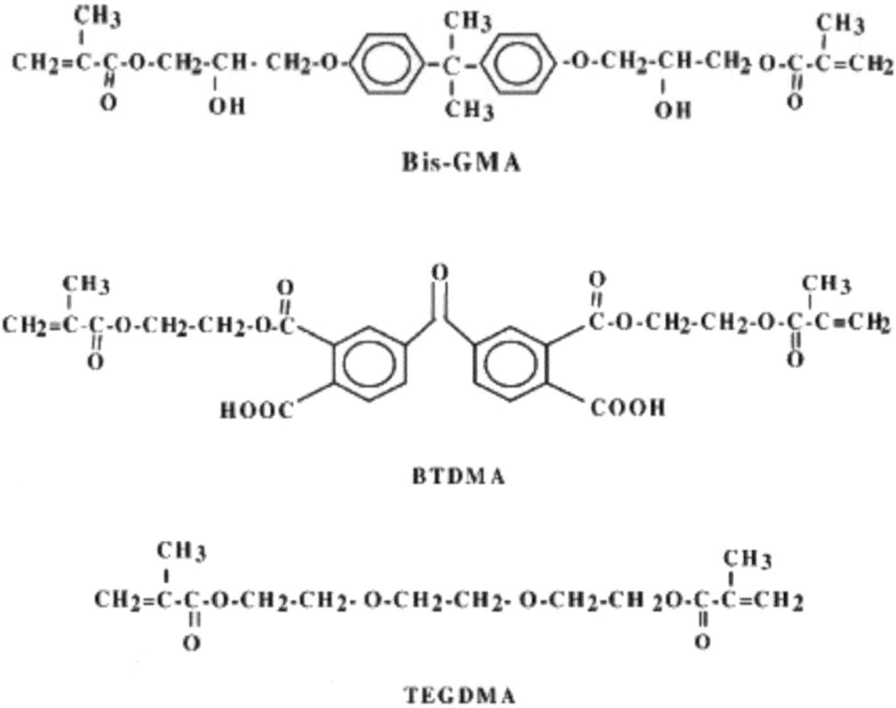

Fig. 1. Chemical structures of Bis-GMA, BTDMA and TEGDMA

The relatively low strength and durability of the composites, however, have limited their uses. The strength of the inorganic filler reinforced dental composites is usually in the range from 80 to 120 MPa, and the average life time is 5 years or less. By comparison, dental amalgams have strength over 400 MPa and have a life time of more than 15 years. Investigations of the reasons for failure revealed that, among other things, inorganic filler was a major contributor. Ironically, the inorganic fillers which are added for the purpose of fortifying the dental composites are actually responsible, at least in part, for the demise. Stresses are transmitted onto the filler particles projecting from the occlusal surfaces through the boluses of foods during chewing [51].

Glass–ceramics are polycrystalline materials which consist of a glass matrix and one or more crystalline phases. The glass fillers are not strong enough and exhibit cracks that either cut through the glass fillers or propagate around the filler particles. To overcome the problem, much effort has been made into the use of glass fibers, nanoporous fillers, branched fibers or even ceramic whiskers [45].

Since the inorganic filler particles are considerably harder than the dental resin matrices, the stresses are transmitted through the filler to the resin. Wherever the submerged portions of the filler particles are angulated or irregular in shape, the stress concentration may become excessively high. Such a condition tends to generate small cracks around the filler particles, thereby weakening the matrices locally. Reinforcement with nanofibers was shown substantial improvements on mechanical properties of dental composites, such as flexural strength, elastic modulus and the work-of-fracture. The small diameter of nanofibers also provides for a large ratio of surface area to volume, which can enhance the intermolecular hydrogen bonding between nanofiber filler and the resin matrix. Furthermore, nanofibers

are continuous. If a micro-crack is initiated in a matrix under contact wear and/or other stresses, the nanofibers remain intact across the crack planes and support the applied load. Therefore, crack opening is resisted by the nanofibers and the matrix is reinforced [52].
The uniform distribution of nanofibers improved the strength and modulus of the resulting nanocomposite nanofiber. The silanized single crystals on the surface of nanofibers also enhance the intermolecular interaction/bonding between the nanofiber filler and the resin matrix [49].
Apart from fillers, a good bond between fillers and the resin matrix is essential in dental composites. Silane coupling agents provide the bond between two components in dental composites, but this bond can be degraded by water absorbed by the composites. The idea of increasing the micro-mechanical retention between fillers and resin in order to reinforce the coupling agent was first described by Bowen et al. in 1976. Their strategy was to use multi-phase glasses which can be etched and produce porous fillers [53].
Recently, nanosized particles dispersed in the organic matrix to give high strength, hardness and toughness.
A helpful method of classifying dental composites is by the particle size, shape, and distribution of filler. Early composites contained large (20 to 30μm) spherical particles, followed by products containing large irregularly shaped particles, microfine particles (0.04 to 0.2μm), fine particles (0.4 to 3μm), and finally blends (microhybrids) containing mostly fine particles with some microfine particles. Based on the type of filler particles, composites are currently classified as *microhybrid* and *microfilled* products. Microhybrid composites contain irregularly shaped glass (borosilicate glass; lithium or barium aluminum silicate; strontium or zinc glass) or quartz particles of fairly uniform diameter. Typically, composites have a distribution of two or more sizes of fine particles plus microfine filler (5% to 15%). This distribution permits more efficient packing, whereby the smaller particles fill the spaces between the larger particles. Microhybrid composites may contain 60% to 70% filler by volume, which, depending on the density of the filler, translates into 77% to 84% by weight in the composite. Most manufacturers report filler concentration in weight percent (wt %) [50].
Microfilled composites contain silica with a very high surface area (100 to 300 m^2/g) having particle diameters of 0.04 to 0.2 μm. Because of the high surface area, only 25% by volume or 38% by weight can be added to the oligomer to keep the consistency of the paste sufficiently low for clinical applications. Fillers consisting of microfine silica in polymerized oligomers are prepared and ground into particles 10 to 20 pm in diameter. These reinforced fillers may be added to the oligomer in concentrations, so the inorganic content can be increased to 32% to 50% by volume or about 50% to 60% by weight. A variation of this modification is used in which most of the filler is reinforced filler, with smaller amounts of silica added to the oligomer. Another modification (homogeneous microfill) has no reinforced filler, but rather microfine silica dispersed in the oligomer [48].

11.3 Coupling agents

The coupling agent, an organosilane (silane), is applied to the inorganic particles by the manufacturer before being mixed with the unreacted oligomer. Silanes are called *coupling agents,* because they form a bond between the inorganic and organic phases of the composite. The silane contains functional groups (such as methoxy), which hydrolyze and react with the inorganic filler, as well as unsaturated organic groups that react with the oligomer during polymerization. During the deposition of the silane on the filler, the methoxy groups hydrolyze to hydroxyl groups that react with adsorbed moisture or -OH groups on the filler. They can also condense with -OH groups on an adjacent hydrolyzed silane to form a homopolymer film on the surface of the filler. During the setting reaction of the oligomer, the carbon double bonds of the silane react with the oligomer, thus forming a bond from the filler through the coupling agent to the polymer matrix. This coupling reaction binds the filler and the oligomer, so when a stress is applied to a composite, the stress can be transferred from one string filler particle to another through the rather low-strength polymer. As a result, the strength of the composite is microfine in polymer filler, intermediate to that of the filler and the polymer separately. This bond can be degraded by dater absorbed by the composite during clinical use [51, 53].

11.4 Initiators and accelerators

Composites are light cured or self-cured, with the former being more common. Light activation is accomplished with blue light with a peak wavelength of about 470 nm, which is absorbed usually by a photo-activator, such as camphoroquinone (Fig. 2), added by the manufacturer in amounts varying from 0.2% to 1.0%. The reaction is accelerated by the presence of an organic amine containing a carbon double bond. The amine and the camphoroquinone are stable in the presence of the oligomer at room temperature, as long as the composite is not exposed to light [54].

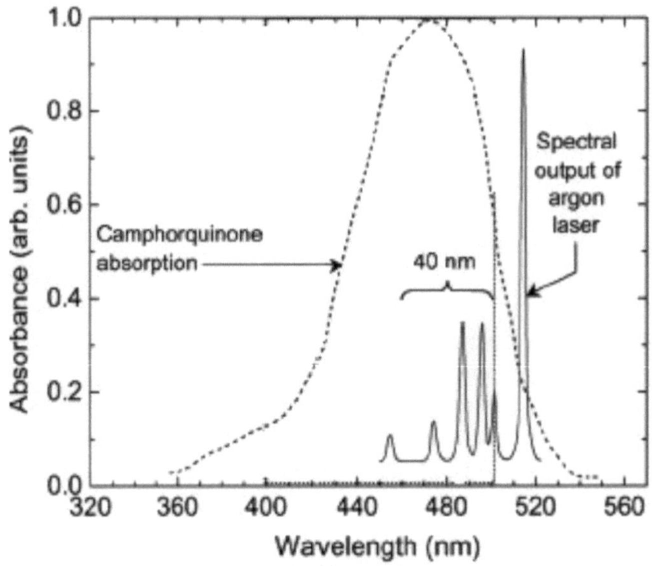

Fig. 2. Absorption spectrum of camphorquinone [55]

Although camphoroquinone is the most common photo-activator, others are sometimes used to accommodate special curing conditions, such as the use of plasma-arc lights with rapid-cured composites.

Chemical activation is accomplished at room temperature by an organic amine (catalyst paste) reacting with an organic peroxide (universal paste) to produce free radicals, which in turn attack the carbon double bonds, causing polymerization. Once the two pastes are mixed, the polymerization reaction proceeds rapidly.

Some composites are dual cured. These formulations contain initiators and accelerators that allow light activation followed by self-curing or self curing alone [55].

11.5 Pigments and other components

Inorganic oxides are usually added in small amounts to provide shades that match the majority of tooth shades. Numerous shades are supplied, ranging from white bleaching shades to yellow to gray. An ultraviolet (W) absorber may be added to minimize color changes caused by oxidation.

Composites are formulated to contain accelerators and initiators that allow self-curing, light curing, and dual curing. IS0 4049 for polymerbased filling, restorative, and luting materials (ANSVADA No. 27) describes two types and three classes of composites, as shown by the following [53]:

Type 1: Polymer-based materials suitable for restorations involving occlusal surfaces
Type 2: Other polymer-based materials
Class 1: Self-cured materials
Class 2: Light-cured materials
Group 1: Energy applied intra-orally
Group 2: Energy applied extra-orally
Class 3: Dual-cured materials

12. Types of composites

12.1 Light-cured dental nanocomposites

Light-cured composites are discussed in this chapter, because most composites and compomers are light activated. These composites are supplied in various shades in spills, syringes, and compules. The syringes are made of opaque plastic to protect the material from exposure to light and thus provide adequate shelf life. If packaged as a compule, the compule is placed on the end of a syringe, and the paste is extruded after removal of the protective tip. The advantages of compules are ease of placement of the composite paste, decrease in cross infection, and protection of the paste from exposure to ambient light.

Polymerization of light-cured composites is initiated by visible blue light [53].

The polymerized resin is highly cross-linked because of the presence of the difunctional carbon double bonds. The degree of polymerization varies, depending on whether it is in the bulk or in the air-inhibited layer of the restoration [50].

The polymerization of light-cured composites varies according to the distance from the composite to the light and the duration of light exposure. The percentage of the double bonds that react may vary from *35%* to *80%*. The degree of polymerization is higher for laboratory composites that are post-cured at elevated temperatures [51, 53].

12.2 Self-cured/dual-cured dental nanocomposites

Self- and dual-cured composites are typically packaged in syringes or tubs of paste and catalyst and require mixing. The polymerization reaction of self-cured composites is chemically initiated with a peroxide initiator and an amine accelerator.

Dual-cured products use a combination of chemical and light activation to carry out the polymerization reaction [52].

12.3 Polymerization

Exposure times for polymerization vary depending on the type of light-curing unit and the type, depth, and shade of the composite. Times may vary from 20 to 60 seconds for a restoration 2mm thick. Microfilled composites require longer exposure than microhybrid composites because the small filler particles scatter the light more. Darker shades or more opaque composites require longer exposure times (up to 60 seconds longer) than lighter shades or more translucent composites. In deep restorations, add and polymerize the composite in layers. One layer bonds to another without any loss of strength. The setting time of light-cured composite and the depth of cure within a given mass depend on the intensity and penetration of the light. A material with a low absorption coefficient cures to the greatest depth. The presence of ultraviolet absorbers for color stabilization, fluorescent dyes for esthetics, or excessive initiator concentration has a detrimental effect on completeness of cure [55].

13. Properties of dental nanocomposites

The chemical and physico-mechanical properties of composites depend on their resin compositions, filler type, size and content. Important properties of the nanocomposites include degree of conversion, flexural strength, flexural modulus, working and setting times, polymerization shrinkage, thermal properties, water sorption and solubility, color stability and radiopacity. These parameters were studied systematically. For dental nanocomposite, an improvement in physical properties is expected due to the increased interfacial interactions between resin and fillers [56].
Among the other factors such as light intensity, initiator system, light irradiation method and polymerization temperature, varying the relative amounts of the matrix monomers has a significant effect on the mechanical properties of the resin composites [57].

13.1 Degree of conversion

The most important dental composite characteristics are the degree of conversion (DC) and the depth of polymerization. These are great importance for the clinical longevity of the restorations. The DC of resins is a major factor influencing their bulk physical properties. In general, the higher conversion of double bonds, the greater mechanical strength. The DC is the measurement of the percentage of consumed double bonds. Conversion of the monomer to the polymer in light-activated composites is dependant on several factors, such as light source, power density, light wavelength, resin composition, transmission of light through the material and amount of activator-initiator and inhibitor present [58].
Low DC of dental composite may give inadequate wear resistance and a low bonding stability to tooth surface. This clinical problem can result in marginal shrinkage, subsequent loss of anatomic form and fractures in the restorations. Several methods have been used to investigate the effectiveness of irradiation source on the polymerization of dental composite, such as micro hardness, optimal microscopy and vibration methods including infrared spectroscopy (FTIR) and Raman spectroscopy (RS) [59].
Vibration methods allow precise assessment of the depth of polymerization and DC (i.e., the percentage of vinyl group converted to aliphatic functions) of methacrylate composite resins.

13.1.1 Method of test

To measure the degree of conversion, the uncured composite is placed between two polyethylene films, pressed to form a very thin film and absorbance peak obtained by transmission mode of FTIR. Then the sample is cured and the absorbance peak recorded for the cured sample. The quantity of the remaining double bonds is determined by a method described by Ruyter and Gyorosi. The percentage of unreacted carbon-carbon double bonds is determined from the ratio of absorbance intensities of aliphatic C=C (peak at 1638 cm^{-1})/aromatic C....C (peak at 1608 cm^{-1}), which is used as an internal standard, since it doesn't change during the polymerization reaction, before and after curing of the specimen. The percentage degree of conversion is obtained as follows [60]:

$$DC\% = (1 - \frac{\left[\frac{abs(aliphatic: C=C)_{1638cm^{-1}}}{abs(aromatic: C.....C)_{1608cm^{-1}}}\right] Cured}{\left[\frac{abs(aliphatic: C=C)_{1638cm^{-1}}}{abs(aromatic: C.....C)_{1608cm^{-1}}}\right] Uncured}) \times 100 \quad (1)$$

Commercial dental restorative materials are usually polymerized in less than 60 s with a degree of conversion from 40 to 75%. Our previous research on preparation of dental nanocomposites based on Bis-GMA, TEGDMA and nano SiO_2 indicated that higher resin concentration and power density resulted in a higher degree of conversion [61].

13.2 Mechanical properties

13.2.1 Flexural strength and modulus

The flexural strengths of the various composites are similar. The flexural modulus of microfilled and flowable composites are about 50% lower than values for all-purpose and packable composites, which reflects the lower volume percent filler present in the microfilled and flowable composites.
If the amalgams are to be replaced by composites as restorative materials, they must possess some acceptable mechanical characteristics such as high strength and high wear resistance to stand a normal biting or chewing force [61].

13.2.1.1 Method of test

The flexural strength and flexural modulus of the polymerized composites are measured with three-point bending test according to the ISO4049:2000 (Fig. 3). The samples for measuring both flexural strength and flexural modulus are prepared in a rectangular brass mold (length= 25 mm, width= 2 mm, height= 2 mm) (Fig. 4). Five specimens for each experimental group are required and their mean value is determined.

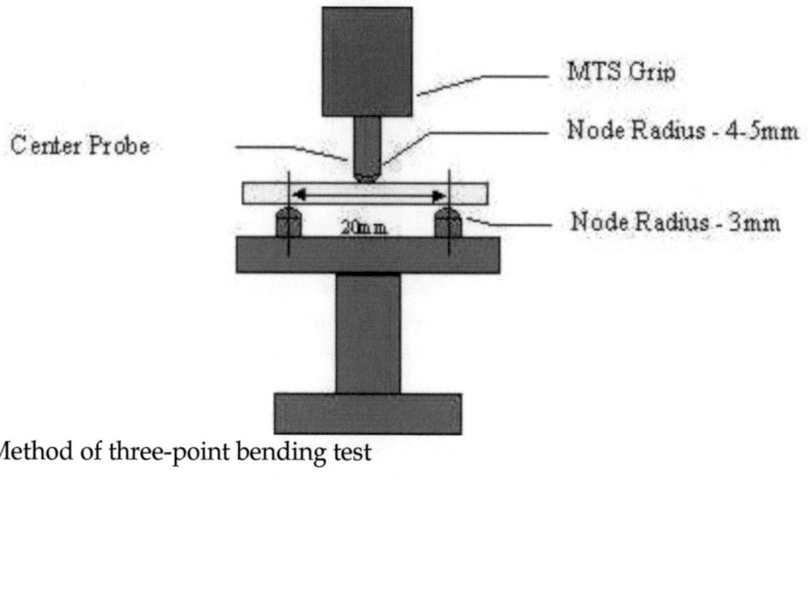

Fig. 3. Method of three-point bending test

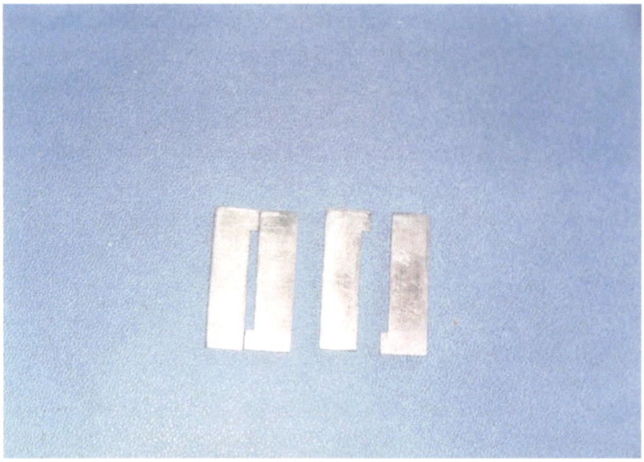

Fig. 4. Mold of flexural strength and flexural modulus tests

The flexural strength and flexural modulus were calculated from the following equation [60]:

$$FS = \frac{3fl}{2bh^2} \tag{2}$$

$$FM = \frac{f\,l^3}{4bh^3 d} \tag{3}$$

where f is the maximum force (N), l the distance between the supports (mm), b the specimen width, h the height of the bar (mm) and d rate of curvature under load (mm).
Investigations show that maximum flexural strength and flexural modulus of the nanocomposites can be achieved by increasing the intensity and percentage of filler [61].

13.2.2 Compressive strength and modulus

The compressive strengths of the various composites are similar. The flexural modulus of microfilled and flowable composites are typically lower than values for all-purpose and packable composites, which reflects the lower volume percent filler present in the microfilled and flowable composites. For comparison, the modulus of elasticity in compression is 62 GPa for amalgam, 19 GPa for dentin, and 83 GPa for enamel [22].

13.2.3 Knoop hardness

Values of Knoop hardness for composites (22 to 80 kg/mm^2) are low compared with values of 343 kg/mm^2 for human enamel and 110 kg/mm^2 for dental amalgam. The Knoop hardness of composites with fine particles is somewhat greater than values for composites with microfine particles because of the hardness and volume fraction of the filler particles. These values indicate a moderate resistance to indentation under functional stresses for the more highly filled composites, but this difference does not appear to be a major factor in resisting functional wear [59].
A microhardness measurement such as Knoop can be misleading on composites with large filler particles (>10 pm in diameter), in which the small indentation could be made solely on

the organic or the inorganic phase. However, with most current products, filler particle sizes have become much smaller (<1 pm), and the microhardness values appear more reliable [62].

13.2.4 Bond strength to dental substrates

The bond strength of composites to etched enamel and dentin is typically between 20 and 30 MPa. Bonding is principally a result of micromechanical retention of the bonding agent into the etched surfxes of enamel and dentin. In dentin, a hybrid layer of bonding resin and collagen is often formed, and the bonding adhesive penetrates the dentinal tubules [48].

Composite can be bonded to existing composite restorations, ceramics, and alloys when the substrate is roughened and primed appropriately. In general, the surface to be bonded is sandblasted (microetched) with 50 pm alumina and then treated with a resin-silane primer for composite, a silane primer for ceramic, or a special alloy primer. Bond strengths to treated surfaces are typically greater than 20 MPa [51].

13.2.5 Depth of cure (light-cured composites)

Maximum intensity of the light radiation beam is concentrated near the surface of a light-cured composite. As the light penetrates the material, it is scattered and reflected and loses intensity [49].

A number of factors influence the degree of polymerization at given depths from the surface after light curing. The concentration of photoinitiator or light absorber in the composite must be such that it will react at the proper wavelength and be present in sufficient concentration. Both filler content and particle size are critical to dispersion of the light beam. For this reason, microfilled composites with smaller and more numerous particles scatter more light than microhybrid composites with larger and fewer glass particles. Longer exposure times are needed to obtain adequate depth of cure of microfilled composites [55].

The light intensity at the resin surface is a critical factor in completeness of cure at the surface and within the material. The tip of the light source must be held within 1 mm of the surface to gain optimum penetration. More-opaque shades reduce light transmission and cure only to minimal depths (1 mm). A standard exposure time using most visible lights is 20 seconds. In general, this is sufficient to cure a light shade of resin to a depth of 2 or 2.5 mm. A 40-second exposure improves the degree of cure at all depths, but it is required to obtain sufficient cure with the darker shades. Application of the light beam through 1 mm or less thickness of tooth structure produces a sufficient cure at shallower depths, but the hardness values obtained are not consistent. Because the light beam does not spread sufficiently beyond the diameter of the tip at the emitting surface, it is necessary to "step" the light across the surface of large restorations so the entire surface receives a complete exposure [53, 56].

Larger tips have been manufactured for placement on most light-curing units. However, as the light beam is distributed over a larger surface area, the intensity at a given point is reduced. Use a longer exposure time of up to 60 seconds when larger emitting tips are used.

To evaluate the effective depth of cure of a specific light-curing unit, cut a small section of 5 to 10 mm from a clear straw and place it on a glass slide. Pack the section with composite. Apply the light directly to the top surface for 20 to 40 seconds according to the recommended technique. Cut off the straw, and scrape uncured composite from the bottom of the specimen with a sharp knife. Measure the length of the apparently cured specimen and divide in half to estimate the effective depth of cure [47, 50].

13.3 Physical properties

13.3.1 Working and setting times

For light-cured composites, initiation of polymerization is related specifically to the application of the light beam to the material; about 75% of the polymerization takes place during the first 10 minutes. The curing reaction continues for a period of 24 hours. Not all of the available unsaturated carbon double bonds react; studies report that about 25% remain unreacted in the bulk of the restoration. If the surface of the restoration is not protected from air by a transparent matrix, polymerization is inhibited; the number of unreacted carbon double bonds may be as high as 75% in the tacky surface layer. Although the restoration can be finished with abrasives and is functional after 10 minutes, the optimum physical properties are not reached until about 24 hours after the reaction is initiated [29, 32].

For most composites that are initiated by visible light, there is a critical time period after dispensing of the paste onto a paper pad during which fresh composite flows against tooth structure at an optimum level. Within 60 to 90 seconds after exposure to ambient light, the surface of the composite may lose its capability to flow readily against tooth structure, and further work with the material becomes difficult. Fluorescent lights labeled "gold" can be substituted to provide unlimited working time for light-cured composites. The setting times for chemically activated composites range from 3 to 5 minutes. These short setting times have been accomplished by controlling the concentration of initiator and accelerator [33].

A controlled setting time allows for the individual polymerization of small increments of composite, thus permitting the use of multiple shades of composite within a single restoration and accommodating polymerization shrinkage within each increment as opposed to the total shrinkage in a bulk-cure method [46].

13.3.2 Polymerization shrinkage

Free volumetric polymerization shrinkage is a direct function of the amount of oligomer and diluent, and thus microhybrid composites shrink only 0.6% to 1.4%, compared with shrinkage of microfilled composites of 2% to 3%. This shrinkage creates polymerization stresses as high as 13 MPa between the composite and tooth structure. These stresses severely strain the interfacial bond between the composite and the tooth, leading to a very small gap that can allow marginal leakage of saliva. This stress can exceed the tensile strength of enamel and result in stress cracking and enamel fractures along the interfaces. The potential for this type of failure is even greater with microfilled composites, in which there is a much higher volume percent of polymer present, and polymerization shrinkage is greater. The net effect of polymerization shrinkage can be reduced by incrementally adding a light-cured composite and polymerizing each increment independently, which allows for some contraction within each increment before successive additions [56, 57, 63, 64].

Composites with epoxy resin and nanosilica fillers provide materials with high mechanical properties and low polymerization shrinkage.

13.3.3 Thermal properties

The thermal expansion coefficient of composites ranges from 25 to 38 x 10^{-6}/°C for composites with fine particles to 55 to 68 x 10^{-6}/°C for composites with microfine particles. The values for composites are considerably less than the average of the values for the polymer matrix and the inorganic phase, however, the values are higher than those for dentin (8.3 x 10^{-6}/°C) and enamel (*11.4* x 10^{-6}/°C). The higher values for the microfilled composites are related mostly to the greater amount of polymer present. Certain glasses may be more effective in reducing the effect of thermal change than are others, and some resins have more than one type of filler to compensate for differential rates [58].

Thermal stresses place an additional strain on the bond to tooth structure, which further compounds the detrimental effect of the polymerization shrinkage. Thermal changes are also cyclic in nature, and although the entire restoration may never reach thermal equilibrium during the application of either hot or cold stimuli, the cyclic effect can lead to material fatigue and early bond failure. If a gap were formed, the difference between the thermal coefficient of expansion of composites and teeth could allow for the percolation of oral fluids [62].

The thermal conductivity of composites with fine particles (25 to 30 x 10^{-4} cal/sec/cm^2 [°C /cm]) is greater than that of composites with microfine particles (12 to 15 x 10^{-4} cal/sec/cm^2 [°C /cm]) because of the higher conductivity of the inorganic fillers compared with the polymer matrix. However, for highly transient temperatures the composites do not change temperature as fast as tooth structure and this difference does not present a clinical problem [61].

13.3.4 Water sorption

The water sorption of composites with fine particles (0.3 to 0.6 mg/cm^2) is greater than that of composites with microfine particles (1.2 to 2.2 mg/cm^2), because of the lower volume fraction of polymer in the composites with fine particles. The quality and stability of the silane coupling agent are important in minimizing the deterioration of the bond between the filler and polymer and the amount of water sorption. It has been postulated that the corresponding expansion associated with the uptake of water from oral fluids could relieve polymerization stresses. In the measurement of hygroscopic expansion starting 15 minutes after the initial polymerization, most resins required 7 days to reach equilibrium and about 4 days to show the majority of expansion. Because composites with fine particles have lower values of water sorption than composites with microfine particles, they exhibit less expansion when exposed to water [58, 60].

13.3.5 Solubility

The water solubility of composites varies from 0.01 to 0.06 mg/cm^2. Adequate exposure to the light source is critical in light-cured composites. Inadequate polymerization can readily occur at a depth from the surface if insufficient light penetrates. Inadequately polymerized resin has greater water sorption and solubility, possibly manifested clinically with early color instability [61].

During the storage of microhybrid composites in water, the leaching of inorganic ions can be detected; such ions are associated with a breakdown in interfacial bonding. Silicon leaches into the water bath in the greatest quantity (15 to 17μg/ml) during the first 30 days of storage and decreases with time of exposure. Microfilled composites leach silicon more slowly and show a 100% increase in amount during the second 30-day period (14.2pg/ml). Boron, barium, strontium, and lead, which are present in glass fillers, are leached to various degrees (6 to 17μg/ml) from the various resin-filler systems. Breakdown and leakage can be a contributing factor to the reduced resistance to wear and abrasion of composites [59, 63].

13.3.5.1 Method of test

Solubility and water sorption are measured according to ISO 4049. Composite is inserted into stainless steel mold, with 6mm diameter and 1mm thickness, between two glass slides.

The composite then cured on each side to form disk shape specimen. Five specimens are required for each test. After curing, the weight of the specimen (m_1) was measured. Then, the disc is immersed in distilled water for a week at room temperature so that unreacted monomers would be eliminated.

After that, the specimen is removed, dried and stored in an oven at 37°C until a constant mass (m_2) had been achieved. The solubility is determined from the following equation [61]:

$$SL\% = \left(\left(m_1 - m_2 \right) / m_2 \right) \times 100 \quad (4)$$

The specimen is dipped again in distilled water at room temperature so that the water sorption could be measured and after different time periods, it is removed and blotted dry according to the ISO 4049 standard to remove excess water and is then weighed. Its weight is recorded until there is no significant change in weight.

The value of water sorption is calculated for each specimen from the following equation [61]:

$$WS_t\% = \left(\left(m_t - m_2 \right) / m_2 \right) \times 100 \quad (5)$$

Where, WS_t denotes water sorption at time t, m_t weight at time t and m_2 initial weight.

Meanwhile, there is another method for calculation of SL and WS. The diameter and the thickness of the specimen is measured at five points and the volume (V) is calculated in cubic millimeters. The values of solubility and water sorption are obtained for each disc using the following formulae [61]:

$$SL = \left(m_1 - m_2 \right) / V \quad (6)$$

$$WS_t = \left(m_t - m_2 \right) / V \quad (7)$$

13.3.6 Color and color stability

The color and blending of shades for the clinical match of esthetic restorations are important. Universal shades vary in color among currently marketed products.

Change of color and loss of shade-match with surrounding tooth structure are reasons for replacing restorations. Stress cracks within the polymer matrix and partial debonding of the filler to the resin as a result of hydrolysis tend to increase opacity and alter appearance. Discoloration can also occur by oxidation and result from water exchange within the polymer matrix and its interaction with unreacted polymer sites and unused initiator or accelerator [55].

Color stability of current composites has been studied by artificial aging in a weathering chamber (exposure to UV light and elevated temperatures of 70 °C) and by immersion in various stains (coffee/tea, cranberry/grape juice, red wine, sesame oil). Composites are resistant to color changes caused by oxidation but are susceptible to staining [59].

13.3.7 Radiopacity

Modern composites include glasses having atoms with high atomic numbers, such as barium, strontium, and zirconium. Some fillers, such as quartz, lithium-aluminum glasses, and silica, are not radiopaque and must be blended with other fillers to produce a radiopaque composite. Even at their highest volume fraction of filler, the amount of radiopacity seen in composites is noticeably less than that exhibited by a metallic restorative like amalgam. The microhybrid composites achieve some radiopacity by incorporating very finely divided heavymetal glass particles [63].

Aluminum is used as a standard reference for radiopacity. A 2-mm thickness of dentin is equivalent in radiopacity to 2.5 mm of aluminum, and enamel is equivalent to 4 mm of aluminum. To be effective, a composite should exceed the radiopacity of enamel, but international standards accept radiopacity equivalent to 2 mm of aluminum.

Amalgam has a radiopacity greater than 10 mm of aluminum, which exceeds all the composite materials available [61].

13.3.8 Wear rates

Clinical studies have shown that composites are superior materials for anterior restorations in which esthetics is essential and occlusal forces are low. One problem with composites is the loss of surface contour of composite restorations in the mouth, which results from a combination of abrasive wear from chewing and toothbrushing and erosive wear from degradation of the composite in the oral environment [39, 57].

Wear of posterior composite restorations is observed at the contact area where stresses are the highest. Interproximal wear has also been observed. Ditching at the margins within the composite is observed for posterior composites, probably resulting from inadequate bonding and polymerization stresses. Currently accepted composites for posterior applications require clinical studies that demonstrate, over a 5-year period, a loss of surface contour less than 250 pm or an average of 50 pm per year of clinical service. Products developed as packable or laboratory composites usually have better wear resistance than microfilled or flowable composites [28, 34].

13.3.9 Biocompatibility

Nearly all of the major components of composites (Bis-GMA, TEGDMA, and UDMA, among others) have been found cytotoxic *in vitro* if used in pure form, but the biological liability of a cured composite depends on the release of these components from the composite. Although composites release some levels of components for weeks after curing, there is considerable controversy about the biological effects of these components. The amount of release depends on the type of composite and the method and efficiency of the cure of the composite. A dentin barrier markedly reduces the ability of components to reach pulpal tissues, but these components can traverse dentin barriers, albeit at reduced concentrations. The effects of low-dose, long-term exposures of cells to resin components are not generally known. On the other hand, the use of composite materials as direct pulp-capping agents poses a higher risk for adverse biological responses, because no dentin barrier exists to limit exposure of the pulp to the released components. The effects of released components from composites on oral or other tissues is not known with certainty, although no studies have documented any adverse biological effects. The tissue at highest risk from this type of release would appear to be gingiva in close, long-term contact with composites [1, 8, 43].

Components of composites are known allergens, and there has been some documentation of contact allergy to composites. Most of these reactions occur with dentists or dental personnel who regularly handle uncured composite and, therefore, have the greatest exposure. There are no good studies documenting the frequency of allergy to composites in the general population [61].

Finally, there has been some controversy about the ability of components of composites to act as xenoestrogens. Studies have proven that Bis-phenol A and its dimethacrylate are estrogenic in vitro tests that measure this effect using breast cancer cell growth. Trace levels of these components have been identified in some commercial composites; however, estrogencity from cured commercial composites has not been demonstrated. Furthermore, there is considerable controversy about the accuracy and utility of in vitro tests using breast cancer cells to measure a true estrogenic effect. An early study in this area, which claimed that dental sealants and composites were estrogenic in children, has since been largely discredited [53, 58].

14. Composites for special applications

14.1 Microfilled composites

These composites are recommended for use in Class 3 and Class 5 restorations, where a high polish and esthetics are most important. One product has been used successfully in posterior restorations. They are composed of light-activated, dimethacrylate resins with 0.04-µm colloidal silica fillers with a filler loading of 32-50% by volume. Because they are less highly filled, microfilled composites have higher values of polymerization shrinkage, water sorption and thermal expansion as compared with microhybrid composites [56].

14.2 Packable composites

These composites are recommended for use in Classes 1, 2, and 6 (MOD) cavity preparations. They are composed of light- activated, dimethacrylate resins with fillers (fibers or porous or irregular particles) that have filler loading of 66% to 70% by volume. The interaction of the filler particles and modifications of the resin cause these composites to be packable [63].

Important properties of packable composites include high depth of cure, low polymerization shrinkage, radiopacity, and low wear rate (3.5 pm/year), which is similar to that of amalgam. Several packable composites are packaged in unit-dose compules. A bulk-fill technique is recommended by manufacturers but has not yet been demonstrated effective in clinical studies. Use singlebottle bonding agents with these composites [28].

14.3 Flowable composites

These light-cured, low-viscosity composites are recommended for cervical lesions, pediatric restorations, and other small, low stress-bearing restorations. They contain dimethacrylate resin and inorganic fillers with a particle size of 0.7 to 3.0 pm and filler loading of 42% to 53% by volume. Flowable composites have a low modulus of elasticity, which may make them useful in cervical abfraction areas. Because of their lower filler content, they exhibit higher polymerization shrinkage and lower wear resistance than microhybrid composites. The viscosity of these composites allows them to be dispensed by a syringe for easy handling [44, 51, 61].

14.4 Laboratory composites

Crowns, inlays, veneers bonded to metal substructures, and metal-free bridges are prepared indirectly on dies from composites processed in the laboratory using various combinations of light, heat, pressure, and vacuum to increase the degree of polymerization and the wear resistance.

For increased strength and rigidity, laboratory composites can be combined with fiber reinforcement. Restorations are usually bonded with composite cements. Cavity preparations for indirect composites must be nonretentive rather than retentive, as typically prepared for direct placement [61].

14.5 Core composites

At times, so much tooth structure is lost from caries that the crown of the tooth must be built up to receive a crown. Amalgam is the most common core material, but composite is becoming popular. Composite core materials are typically two-paste, self-cured composites, although light-cured and dual-cured products are available. Core composites are usually tinted (blue, white, or opaque) to provide a contrasting color with the tooth structure. Some products release fluoride [63].

Composite cores have the following advantages as compared with amalgam: can be bonded to dentin, can be finished immediately, are easy to contour, have high rigidity, and have good color under ceramic restorations. Composite cores are bonded to remaining enamel and dentin using bonding agents. Be careful to use a bonding agent recommended by the manufacturer of the core material, because some self-cured composite core materials are incompatible with some light-cured bonding agents [64].

15. Summary

It is fortunate that many types of dental nanomaterials are available as restorative materials. The requirements may vary for different patients, and the processing facilities can dictate which material to use.

There are also significant differences between dental materials. Amalgam is a highly successful material clinically and is very cost effective, but alternatives such as esthetic restorative materials are now very competitive in terms of frequency of use. The ultimate goal of advanced dental nanomaterials research is to produce a material that can be used in all circumstances as an amalgam replacement material.

As a result, contemporary dentists have many options available for consideration when choosing a dental material. The choice of the composite or cement, whether old or new, should depend on the particular clinical situation and specific criteria. While improved physical properties or in vitro adhesion strengths may be considered advantageous, they have not automatically resulted in improved clinical performance. Ease of manipulation and handling, the materials clean-up and minimal technique sensitivity are important factors in ensuring clinical success. These factors –coupled with results of favorable long-term clinical studies including absence of post-operative sensitivity, long-term retention of the prosthodontic device and marginal integrity– should be the deciding factors that clinicians take into consideration while selecting the materials for a particular procedure.

Dental materials such as composites and cements are used to replace missing tooth structure and modify the color and contour of the teeth in order to enhance esthetics and nano particles are used in the dental materials to give excellent physical and mechanical properties due to high strength, hardness and toughness of this component.

16. References

[1] Wilson AD, Kent BE (1971) The glass-ionomer cement: a new translucent dental filling material. J Appl Chem Biotechnol, 21: 313.

[2] Rezk-lega F, Ogaard B, Rolla G (1991) Availability of fluoride from glass-ionomer luting cements in human saliva. Scand J Dent Res, 99: 60–63.
[3] Scoville RK, foreman F, Burgess JO (1990) In vitro fluoride uptake by enamel adjacent to a glass ionomer luting cement. ASDC J Dent Child, 57: 352–355.
[4] Hicks MJ, Flaitz CM, Silverstone LM (1986) Secondary caries formation in vitro around glass ionomer restorations. Quintessence Int, 17(9): 527–532.
[5] Ten Cate JM, Van Duinen RN (1995) Hypermineralization of dentinal lesions adjacent to glass ionomer cement and amalgam restorations. J Dent Res, 74: 1266–1271.
[6] Tyas MJ (1991) Cariostatic effect of glass ionomer cement: a five year clinical study. Aust Dent J, 36: 236–239.
[7] Wozniakw (1984) Reported sensitivity to glass ionomer luting cements. J Am Dent Assoc, 109: 476.
[8] Simmons JJ (1986) Post-cementation sensitivity commonly associated with the "anhydrous" forms of glass ionomer luting agents: a theory. Texas Dent J, 103: 7–8.
[9] Smith DC, Ruse ND (1986) Acidity of glass ionomer cements during setting and its relation to pulp sensitivity. J Am Dent Assoc, 112(5): 654–657.
[10] Hiraishi N, Kitasako Y, Nikaido T, Foxton RM, Tagami J, Nomura S (2003) Acidity and conventional luting cements and their diffusion through bovine dentin. Int Endontic J, 36(9): 622–628.
[11] Pameijer CH, Nilner K (1994) Long term clinical evaluation of three luting cements. Swed Dent J, 18: 59–67.
[12] Kern M, Kleimeier B, Schaller HG, Strub JR (1996) Clinical comparison of postoperative sensitivity of a glass ionomer and a zinc phosphate luting cement. Prosthet Dent, 75: 159–162.
[13] Mitra SB (1989) Photocurable ionomer cement system. European Patent Application, EP 0323120 A2.
[14] Mitra SB (1994) Curing reactions of glass ionomer materials. Proceedings of the 2nd International Conference on Glass Ionomers: 13–22 Philadelphia, USA.
[15] Uno S, Finger WJ, Fritz U (1996) Long term mechanical characteristics of resinmodified glass-ionomer restorative materials. Dent Mater, 12(1): 64–69.
[16] Mitra SB (1991) Adhesion to dentin and physical properties of a light-cured glass ionomer liner/base. J Dent Res, 70(1): 72–74.
[17] Mitra SB (1991) In vitro fl uoride release from a light-cured glass ionomer liner/base. J Dent Res, 70(1): 75–78.
[18] Nagamine M, Toshiyuki I, Irie M, Staninec M, Inoue K (1997) Effect of resin-modified glass-ionomer cements on secondary caries. J Am Dent, 10: 173–178.
[19] Tantbirojn D, Falsafi A, Mitra SB, Delong R, Douglas WH, ton t (2005) Fluoride release and dentin caries inhibition adjacent to resin-modified glass-ionomer luting cements. Chin Dent J, 24(3): 127–133.
[20] Christensen GJ (1995) A promising new category of dental cements. J Am Dent Assoc, 126: 781–782.
[21] Cho E, Kopel H, White SN (1995) Moisture susceptibility of resin-modified glass ionomer materials. Quintessence Int, 26: 351–358.
[22] Donovan TE, Cho GC (1999) Contemporary evaluation of dental cements. Compend Contin Educ Dent, 20(3): 197–219
[23] Falsafi A, Mitra SB, Ton T, Ren CD (2003) Expansion and acid erosion of different luting cements. J Dent Res, 82(Special Issue B), abstract 1990.
[24] Ton A, Falsafi A, Mitra SB (2006) Fluoride recharge and release of resin modified glass ionomer cements. ADEA/AADR/CADR Meeting, Orlando, FL, USA, abstract 1887.

[25] Tantbirojn D, Douglas WH, Versluis A (1997) Inhibitive effect of a resin-modified glass ionomer cement on remote enamel artificial caries. Caries Res, 31(4), 275–280.
[26] Donly KJ, Segura A, Kanellis M, Erickson RL (1999) Clinical performance and caries inhibition of resin-modified glass ionomer cement and amalgam restorations. J Am Dent Assoc, 130: 1459–1466.
[27] Mccomb D, Erickson RL, Maxymiw WG, Wood RE (2002) A clinical comparison of glass ionomer, resin-modified glass ionomer and resin composite restorations in the treatment of cervical caries in xerostomic head and neck radiation patients. Oper Dent, 27(5): 430–437.
[28] Haveman CW, Summitt JB, Burgess JO, Carlson K (2003) Three restorative materials and topical fluoride gel used in xerostomic patients: a clinical comparison. J Am Dent Assoc, 134(2): 177–184.
[29] Li J, Von Beetzen M, Sundstrom F (1995) Strength and setting behavior of resinmodified glass ionomer cements. Acta Odontl Scand, 53: 311–317.
[30] Falsafi A, Mitra SB, Winters EJ, Carufel RJ, Ton T (2003b) Comparison of mechanical properties and adhesion of RMGI luting cements. J Dent Res, 82 (Special Issue A), abstract 234.
[31] Smith DC (1999) Development of glass-ionomer systems. Biomaterials, 19: 467–478.
[32] Akira M, Koji H, Terenaka T (2003) Fracture toughness of resin-modified glass ionomer materials: effect of powder/liquid ratio and particle size reduction on fracture toughness. Dent Mater, 19(8): 747–757.
[33] Mitra SB, Kedrowski BL (1994) Long-term mechanical properties of glassionomers. Dent Mater, 10(2): 78–82.
[34] Mccomb D, Nathanson D (1999) Glass-ionomer luting cements. In Davidson CL and Mjor IA eds, Advances in Glass-Ionomer Cements, Quintessence Publishing Co., Chicago, USA, pp. 149–170.
[35] Lindquist TJ, Connolly J (2001) In vitro microleakage of luting cements and crown foundation material. J Prosthet Dent, 85(3): 292–298.
[36] Thonemann B, Federlin M, Schmalz G, Hiller KA (1995) Resin-modified glass ionomers for luting posterior ceramic restorations. Dent Mater, 11(3): 161–168.
[37] Palacios RP, Johnson GH, Phillips KM, Raigrodski AJ (2006) Retention of zirconium oxide ceramic crowns with three types of cements. J Prosthet Dent, 96(2): 104–114.
[38] Ernst CP, Cohnen U, Stender E, Willershausen B (2005) In vitro retentive strength of zirconium oxide ceramic crowns using different luting agents. J Prosthet Dent, 93(6): 551–558.
[39]Jokstad A (2004) A split-mouth randomized clinical trial of single crowns retained with resin-modified glass-ionomer and zinc phosphate luting cements. Int J Prosthodont, 17(4): 411–416.
[40] Johnson GH, Powell LV, Derouen TA (1993) Evaluation and control of postcementation pulpal sensitivity: zinc phosphate and glass ionomer luting cements. J Am Dent Assoc, 124(11): 38–46.
[41] Smith DC (1968) A new dental cement. Br Dent J, 125: 381–384.
[42] Smith DC(1983) Dental cements - current status and future prospects. Dent Clin North Am, 27: 763–792.
[43] Beech DR, Bandyopadhyay S (1983) A new laboratory method for evaluating the relative solubility and erosion of dental cements. J Oral Rehab, 10(1): 57–63.
[44] Craig RG, Editor (1997) Restorative dental materials. 10th ed. St. Louis, MO: C.V. Mosby Company

[45] Roberson TM, Heyman H, Swift EJ (2002) Sturtevant's art and science of operative dentistry. 4th ed. St Louis: Mosby.
[46] Bowen RL (1965) Method of preparing a monomer having phenoxy and methacrylate groups linked by hydroxy glyceryl groups. Patent US 3179623
[47] Sibold N, Madec P-J, Masson S, Pham T-N (2002) Synthesis and characterization of copolymers containing a phosphonate function for use in dental composites. Polymer, 43: 7257–67
[48] Atai M, Nekoomanesh M, Hashemi SA, Yeganeh H (2002) Synthesis and characterization of a BTDA-based dental adhesive monomer and its interaction with Ca2 ions. J Appl Polym Sci, 86: 3246–49
[49] Antonucci JM, Stansbury JW (1997) Molecular designed dental polymer. In: Arshady R, editor. Desk reference of functional polymers: synthesis and application. American Chemical Society Publication, 719–38
[50] Chen M, Chen C, Hsu S, Sun SH, Su W (2005) Low shrinkage light curable nanocomposite for dental restorative material. Dent Mater, 1–8
[51] Corbin SB, Kohn WG (1994) The benefits and risks of dental amalgam. J Am Dent Assoc, 125: 381–88
[52] Fong H (2004) Electrospun nylon6 nanofiber reinforced Bis-GMA/TEGDMA dental restorative composite resins. Polymer, 45: 2427–32
[53] Bowen RL, Reed LE (1976) Semi porous reinforcing filler for composite resins. Preparation of provisional glass formulations. J Dent Res, 55: 738–47
[54] Kobayashi M, Shinzato S, Kawanabe K, Neo M, Matsushita M, Kokubo T, Kikutani T, Nakamura T (2000) Alumina powder/Bis-GMA composite: effect of filler content on mechanical properties and osteoconductivity. Biomed Mater Res, 49: 319–27
[55] Pradhan. R.D, Melikechi N, Eichmiller F (2002) The effect of irradiaton wavelength bandwidth and spot size on the scraping depth and temperature rise in composite exposed to an argon laser or a conventional quartztungsten-halogen source. 18: 221–26
[56] Rosin M, Urban AD, Gartner C, Bernhardt O, Splieth C, Meyer G (2002) Polymerization shrinkage-strain and microleakage in dentinbordered cavities of chemically and light-cured restorative materials. Dent Mater, 18: 521–28
[57] Asmussen E, Peutzfeldt A (1998) Influence of UEDMA, BisGMA and TEGDMA on selected mechanical properties of experimental resin composites. Dent Mater, 14: 51–6
[58] Anusavic K.J (1996) Philips Science of Dental Materials.10th ed, Philadelphia: W.B. Saundrs Company
[59] Rueggeberg F.A. Hasginger D.T, Fairhust C.W (1990) Calibration of FTIR conversion analaysis of contemporary dental resin composite. Dental Mater, 6: 241 49
[60] Craig R (1997) Restorative dental materials. 10th ed. Mosby, St. Louis, MO
[61] Mirsasaani S.SH, Atai M, Hasani-Sadrabadi M.M. Photopolymerization of a dental nanocomposite as restorative material using the argon laser, Lasers in Medical Science, DOI 10.1007/s10103-009-0699-7
[62] Khosroshahi ME, Atai M, Nourbakhsh MS (2007) Photopolymerization of dental resin as restorative material using an argon laser. Lasers Med Sci. doi: 10.1007/s10103-007-0487-1
[63] Atai M, Watts DC, Atai Z (2005) Shrinkage strain-rates of dental resin-monomer and composite systems. Biomaterials, 26: 5015–20
[64] Craig GC, Powers JM (2002) Restorative Dental Materials Mosby, Inc. 11830 Westline Industrial Drive, St Louis, Missouri 64146 USA 11th Ed, 238–39

21

Influence of Nanocomposite Materials for Next Generation Nano Lithography

Scott Lewis and Lucio Piccirillo
The University of Manchester, United Kingdom

1. Introduction

The trend of Moore's law has been maintained using the current technology of the ultra large scale integration which produces circuits of sub 100 nm (Gonsalves et al., 2001). This has been made possible by sub 100nm lithography, by the design of next generation resists operating at extreme ultra violet wavelengths. However, it is well known that electron beam lithography can achieve similar dimensions using positive tone e-beam resists such as PolyMethylMethAcrylate (PMMA), Polybutene - 1 - Sulfone (PBS) and ZEP520 (Dean et al., 1994) resists. Upon exposure, the molecular weight of the polymer chains, are reduced by the chain scission process. This increases the solubility of the polymer in organic solvents. Figure 1 illustrates the chain scission process with a resist called PMMA. The chain scission process occurs when the radiolysis of the main chain carbon to carbonyl bond is initiated by the electron. The main chain is cleaved and produces unstable products.

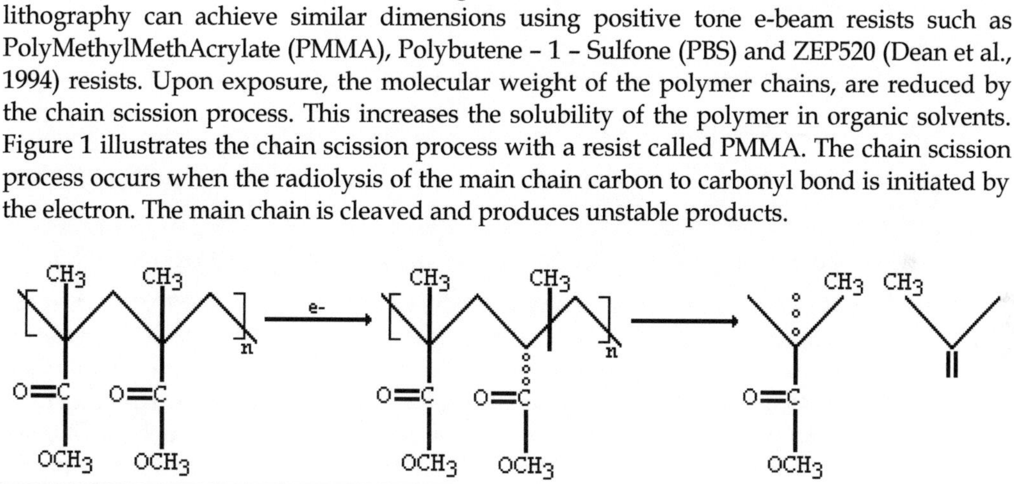

Fig. 1. E-beam chain scission process with PMMA e-beam resist.

Therefore, the dissolution rate is increased in an organic developer like Methyl - IsoButyl Ketone (MIBK): IsoPropAnol (IPA) (1:3 ratio) as the main - chain scissions reduces molecular weight of the PMMA. Unfortunately, there is a limit to the negative scaling factor, called the proximity effect (T. Ishii et al 1997). Therefore, it is essential to optimize these resists to perform with a higher sensitivity for high throughput and a higher contrast to achieve sub 100nm resolution.

As a figure of merit, the smallest feature size possible with PMMA is approximately 100nm when the resist thickness is approximately 400nm. Hence, the aspect ratio is 4:1. This is due to the fact that a minority of the incident electrons when penetrating the surface of the PMMA resist undergo small angle forward scattering while the majority of them participate in large angle scattering events that contribute to backscattering. Therefore, this leads to additional exposure in the bulk resist (the proximity effect), outside of the intended 'write volume'. The bulk of the resist is exposed by primary electrons. As their range in the resist is only a few nanometers their contributions to the proximity effect are negligible. However, secondary electrons (SE) may be generated from the primary electrons and may have energies of the order of 1 to 5KeV. This is significant, because these fast electrons can have a range of a few tenths of a micron, thus contributing to the proximity effect. This has been illustrated by the Monte Carlo simulation of electron trajectories in figure 2. Figure 2 illustrates the proximity effect which is indicated by the red line, it can be seen that the trajectories of the secondary electrons can interfere with the structure adjacent to it, thus to achieve ever decreasing features the overall resist thickness must be decreased to compensate for this.

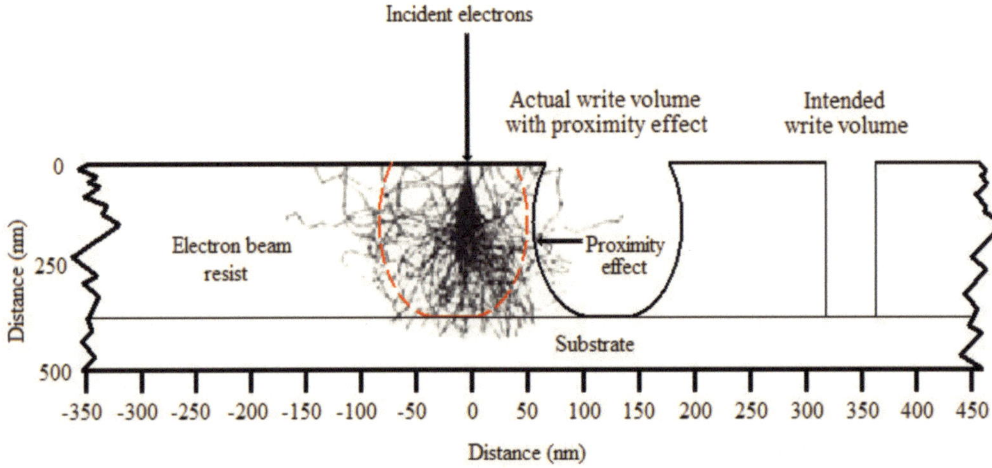

Fig. 2. Schematic of a typical Monte Carlo simulation demonstrating the proximity effect.

The limit of the smallest feature size that can be produced in PMMA is about 10nm, when the resist thickness is approximately 40nm. At these geometries, the resist cannot be used for lift off, because, if the user wanted to produce a gate structure for a transistor for example, then the power applied to the gate must be extremely low to avoid burning out the structure (due to resistive heating) and cause an open circuit. For high power devices such as Gallium Nitride (GaN) HEMTs used as power amplifier in the next generation mobile phone handsets, lithography via this method is not suitable.

In order to address this lithographic challenge, the current trend is toward the use of nanocomposite resists (Ishii et al., 2000, Ishii et al 1997). These materials can have a large impact on current lithography fabrication technologies thanks to the properties of maintaining high throughput while increasing the aspect ratio of the resist. Nanoparticles can have an enormous effect on important properties of the resist. This has been extensively studied using carbon fullerenes and silica nanoparticles (Ishii et al 2001, Merhari et al 2002).

The first requirement for the nanocomposite hybrid material is that the dispersion of the nanoparticles must be well defined. It has often proved difficult to form a stable dispersion of nanoparticles in polymer matrices, as the particles tend to aggregate. The strength of the inter-nanoparticle coupling in liquids typically ranges from weak van der Waals interactions to relatively strong hydrogen bonding. The geometrical form of the nanoparticle aggregates can be string, network, or globular structures.

The surface nature of these oxide nanoparticle species is hydrophilic due to the existence of hydroxyl groups (Sidorenko et al 1999). In order to embed the nanoparticle in a hydrophobic polymer matrix, the interfacial interaction must be improved. The modification of the nanoparticle surface with different interfacial methods has been applied. Such techniques incorporated the encapsulation of the nanoparticle by an emulsion polymerization in PMMA (Erdem et al., 2000).

The physical properties like electrical, mechanical and optical properties are enhanced by these materials. The type of nanoparticle species can affect the polymer nanocomposite properties, these are determined by its nanostructure, size, shape and concentration and by interactions with the polymer matrix (Yuwono et al., 2005, Lu et al., 2005). These hybrid materials allow the designer to manipulate their physical and chemical properties, are processed with ease and have a relatively low cost (Judeinstein & Sanchez 1996, Potyrailo 2006). The goal here was that the nanoparticles must not effect or interfere with the process of e-beam lithography.

It is expected that proximity effects can be greatly reduced by nanocomposite resists. This is achieved through the addition of nanoparticles throughout the PMMA polymer reducing the lateral distance of the electrons travelling through the PMMA polymer. This is due to the fact that the density of the nanoparticles can be substantially larger than that of the organic polymer, which reduces the lateral energy spread of secondary electrons. Hence, the nanoparticles have a larger electron energy stopping power. Reducing the proximity effect provides a means of achieving a higher aspect ratio as the overall resist thickness can be increased (Gonsalve et al 2001).

The addition of the nanoparticles will also have a direct effect on the high aspect ratios by increasing the rigidity of the film, as the mechanical properties of the PMMA resist are improved when the concentration of nanoparticles is increased. The nanoparticles will prevent the nanometer features from collapsing, giving rise to a higher achievable aspect ratio. Subsequently, this can lead to a reduction of line broadening and improve the critical dimension of feature sizes. Achieving these improvements will be the subject of the subsequent sections in this chapter.

2. Oxide nanoparticle stability and DVLO theory

Three of the fundamental states of matter are solids, liquids and gases. If one of these states is finely dispersed in another then we have a 'colloidal system'. These materials have special properties that are of great practical importance. There are various examples of colloidal systems that include aerosols, emulsions and colloidal suspensions.

An initially formed aggregate is called a cluster and the process of its formation is called clustering. The cluster may or may not sediment or phase separate. If the aggregate changes to a much denser form, it is said to undergo coagulation. An aggregate usually separates out either by sedimentation (if it is more dense than the medium) or by creaming (if it less dense than the medium). The terms clustering and coagulation have often been used interchangeably. Usually coagulation is irreversible whereas clustering can be reversed by the process of declustering. Figure 3 schematically represents some of these processes.

The quality of the nanoparticle dispersion inside the nanocomposite material can be determined by measuring its surface roughness. If the surface roughness is poor then this

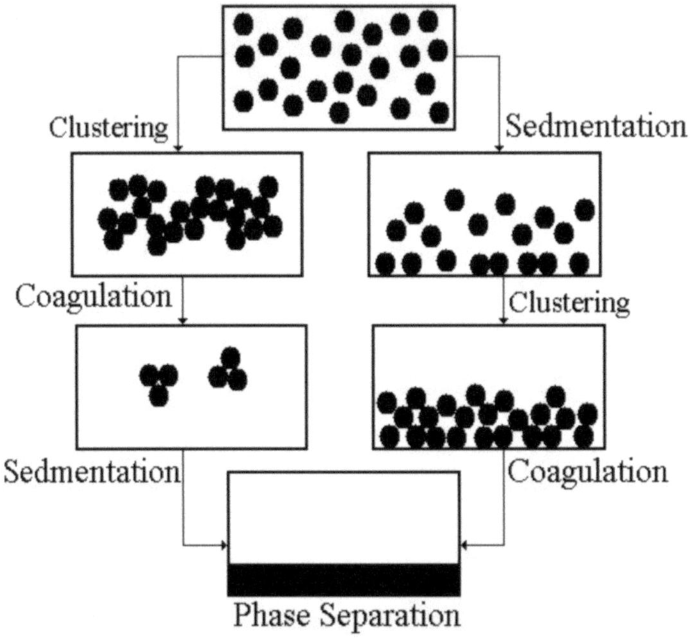

Fig. 3. Schematic diagram showing various mechanisms where stability may be lost in a colloidal dispersion.

indicates that the nanoparticles have clustered together, thus a poor suspension has occurred. This is insufficient for electron beam lithography as the nano structures written in the nanocomposite resist when developed will result with poor edge definition. The value of the surface roughness will depend on the size of the features that are to be fabricated. As a figure of merit the size of the features is below 100nm then the surface roughness must be below 50nm. However, there is a limit to the smallest nano structure that can be produced, that limit is defined by the size of the nanoparticle.

The surface roughness can be measured using a surface probe microscope topometrix explorer in non-contact mode. The Atomic Force Mircoscope (AFM) tips were cantilevers operating at a frequency range of 279-313 kHz. The length of the cantileverwas 134 µm, width 28 µmand the thickness was 3.7 µm. The pyramidal tip was 3 µm with a base of 10 µm high, the tip radius was <20 nm and the force applied was 25-38 Nm^{-1}. To determine the surface roughness of all the samples, the average was taken of the peaks and troughs, measured in 10 µm squares.

It can be clearly seen that the AFM image of figure 4 shows the surface of the nanocomposite film (loaded with 200mg of nanoparticles) is not homogenous when the oxide nanoparticles are directly dispersed into the PMMA polymer. The rms surface roughness of the Al_2O_3/PMMA nanocomposite surface was 402nm. This shows that the surface of the nanocomposite resist indicates that the nanoparticles have clustered together. This will affect the Line Edge Roughness (LER) of the nano structures directly written into the nanocomposite resist when developed. Therefore, the nanocomposite resist surface must be

Fig. 4. AFM image of 200mg of Al_2O_3 nanoparticles embedded directly into 1g of PMMA/Anisole (7%).

improved, but first a good level of understanding of the mechanism of nanoparticle/nanoparticle interactions must be gained in order to achieve this goal.

In the 1940s, scientists Derjaguin, Verwey, Landau and Overbeek (DVLO) developed a theory for the stability of colloidal systems (Derjagin & Landau 1941, Verway & Overbeek 1948)). DVLO theory proposed that the stability of a particle in a solution was dependent upon its total potential energy function V_T and is given by

$$V_T = V_A + V_R + V_S \tag{1}$$

V_S is the potential energy. As the separation occurs near interface at a distance of a few nanometers, this contribution is a small proportion of the energy of the system to the total potential energy. The terms V_A and V_R, are the van der waals attractive and the electrical double layer repulsive forces, these have a larger impact on the particles and function at larger distances. The attraction potential is given by

$$V_A = \frac{-A}{\left(12 \cdot \pi \cdot D^2\right)} \tag{2}$$

where A is the Hamaker constant and D is the particle separation. The repulsive potential V_R is given by

$$V_R = 2\pi\varepsilon r\xi^2 \exp(-\kappa D) \tag{3}$$

where κ is a function of the ionic composition, ζ is the zeta potential, r is the particle radius and π is the solvent permeability.

The theory indicates that in order to stabilize the particles in the colloidal system as they advance towards each other due to the Brownian motion, contributions of the attractive van der Waals forces (V_A) and repulsive electrical double layer forces (V_R) and this is also known as a depletion zone, which can be determined by the total potential energy. Therefore, due to this theory the oxide nanoparticles are suspected to unbalance the depletion effect and osmotic pressures in the PMMA/Anisole system. This is illustrated in figure 6. It depicts the three tier system consisting of the PMMA molecule chain indicated in green. To achieve its chain form, the PMMA is dissolved in a solvent called Anisole, shown in red. It can be seen from the illustration that each nanoparticle is surrounded by a depletion zone; this is the region immediately next to each nanoparticle surface (indicated by the dotted line).
The depletion of polymer molecules from the region between closely spaced oxide nanoparticles leads to an effective internanoparticle / nanoparticle attraction. When the nanoparticles are introduced into the PMMA polymer matrix, the unbalanced osmotic pressure pushes the nanoparticles towards one another via Brownian motion. Hence nanoparticle clustering occurs when the depletion regions of two nanoparticles intersect, due to the van der waals attractive forces indicated by the black area in the depletion zones in figure 6. This can be expressed in the form of an attractive pair potential, the depletion potential, U_{dep}

$$U_{dep} = \begin{cases} +\infty & for \quad r \le a \\ -\prod_p V_{overlap} & for\ \sigma < r \le \sigma + 2r_g \\ 0 & for \quad r > \sigma + 2r_g \end{cases} \tag{4}$$

where $\sigma = 2a$ is the particle diameter and $\prod_p$ is the osmotic pressure of the polymer. $V_{overlap}$ is the volume of the overlapping

$$V_{\text{intersect}} = \left(1 - \frac{3r}{2\sigma(1+\xi)} + \frac{1}{2}\left[\frac{r}{\sigma(1+\xi)}\right]^3\right)\frac{\pi}{6}\sigma^3(1+\xi)^3 \tag{5}$$

where $\xi = {}^{r_g}\!/_{a}$ is the size asymmetry ratio. The polymer osmotic pressure given here is

$$\prod_p = np^{(R)} k_B T \tag{6}$$

where $n_p^{(R)}$ is the polymer number density.

3. Dispersion mechanisms and the oxide nanoparticle surface properties

To improve the dispersion of the oxide nanoparticles, their surface must be encapsulated, so that they will repel each other (hence meeting the requirement of the DVLO theory) and then be homogenously dispersed into the PMMA. Therefore, the nanoparticles must be suspended in a solvent first. It is this parameter that governs its viscosity and overall thickness of the nanocomposite resist.

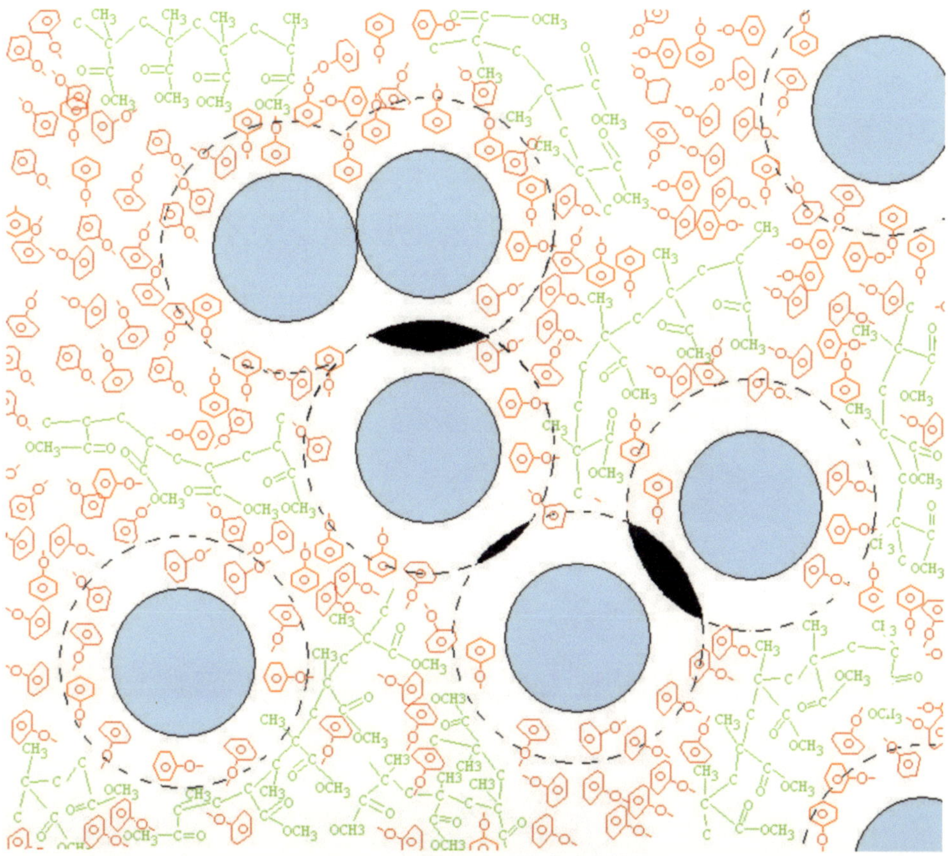

Fig. 6. Schematic representation of the nanoparticle / Anisole / PMMA system.

The stability of the oxide nanoparticle dispersion in the PMMA/anisole can be determined by understanding the behaviour of the bonding mechanism between the various solvents used to disperse the oxide nanoparticles. This will identify which solvents are most appropriate for this task of repelling the nanoparticles. The solvents considered initially were toluene, xylene, anisole, isopropanol (IPA), methanol, acetic acid and deionized water and their properties can be found in table 1. It should be noted that acetone was not used, because, the PMMA molecule breaks down in acetone (acetone is used the strip PMMA from the substrate). The interaction between the surface of the oxide nanoparticles and the encapsulating solvents will determine the effectiveness of the dispersion of the oxide nanoparticles in the solvents. This interaction can be predicted by means of contact angle measurements and a suitable hydrophilic surface. This was achieved by substituting the oxide nanoparticles with borosilicate glass, as the oxide nanoparticles are also hydrophilic.

Each solvent was dropped onto the surface of the borosilicate glass and the contact angle was measured. Due to the nature of the borosilicate glass the solvents with a relatively high polarity index should have a smaller contact angle due to the molecule having OH groups attached to them. This will lead to hydrogen bonding. The results are shown in figure 7a – g and in table 1.

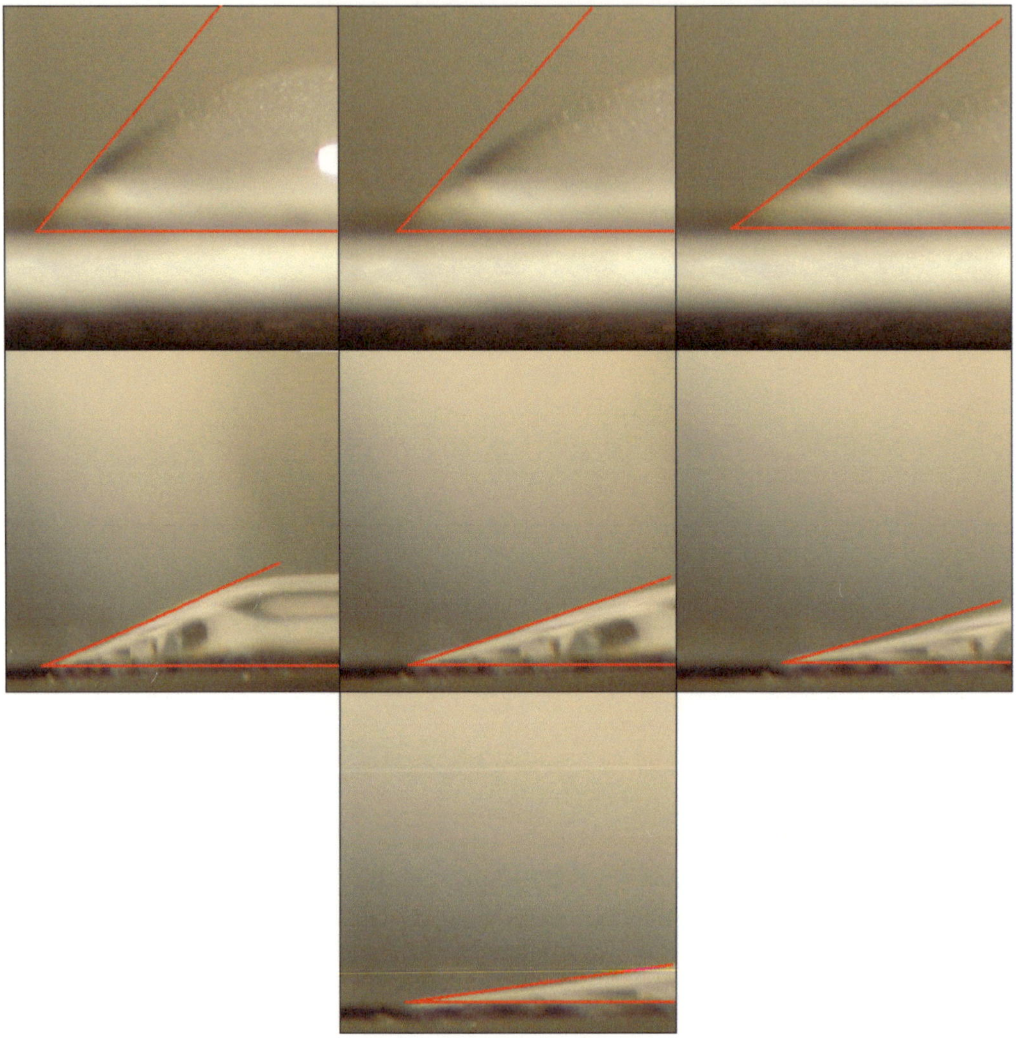

Fig. 7. Contact angles of solvents. a) Toluene, b) Xylene, c) Anisole, d) IPA, e) Methanol, f) Acetic Acid, g) Water.

Table 1 shows the molecule structure of each solvent incorporated in this experiment. It is evident that the solvents (Toluene, Xylene and Anisole) with a polarity index less than four do not have any OH groups, where as the solvents with a higher polarity index more than four have OH groups attached to the molecule, hence, hydrogen bonding can take place. Where, the electrical double layer repulsive force will have a larger impact.

It was found from the data shown in figure 8 that in order to achieve a stable oxide nanoparticle suspension, solvent molecules with a polarity index higher than 4.0 are extremely hydrophilic. Due to the nature of these solvents, hydroxyl (OH) groups are attached to them. These OH groups have a dipole, where the hydrogen atom is bonded with oxygen atom. The electron field is directed from the hydrogen atom towards the oxygen atom pulling electrons away from the hydrogen atom on to the oxygen atom. This causes the hydrogen to have a slight positive charge and the oxygen will become negatively charged. A saturated

Chemical	Molecule Structure	Polarity Index	Contact Angle
Toluene	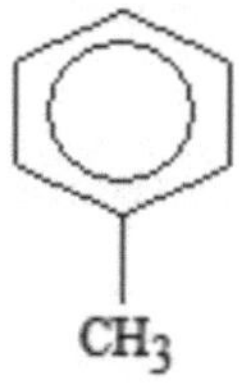	2.4	52°
Xylene	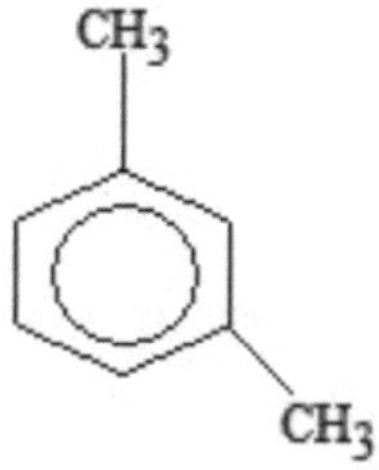	2.5	49°
Anisole	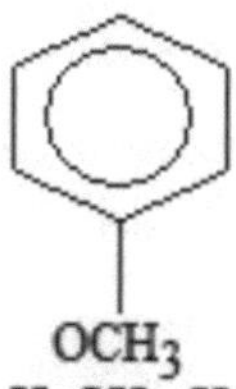	3.8	38.5°
IPA	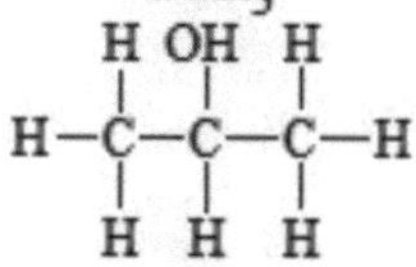	4.0	22°
Methanol	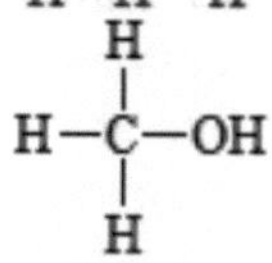	5.1	18°
Acetic Acid	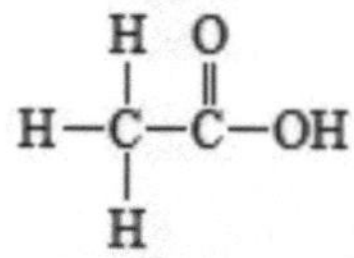	6.2	14°
Water	H_2O	9	7.5°

Table 1. Solvent molecules and contact angle.

oxygen atom in a covalent compound can have two single bonds and be bonded to other atoms; these use two of the oxygen outer shell electrons. The other four electrons form a couple of lone pair orbitals, these lone pairs and the single two bonds are arranged in a tetrahedral lattice around the oxygen. A positively charged atom can form an extra bond to another molecule (oxygen needs to be in a bonded state) by inserting itself into a lone pair orbital on the hydrogen atom.

Therefore, hydrogen bonding between the solvent (polarity index higher than 4.0) will stabilize the oxide nanoparticle suspension, (this can be seen from table 1). These solvents, with a polarity index lower than 4.0 are aromatic. They are formed around a conjugated ring, which do not have hydroxyl groups attached to them. Hence, these substances can not bond to the oxide nanoparticle via the hydrogen bonding method, as illustrated in figure 9.

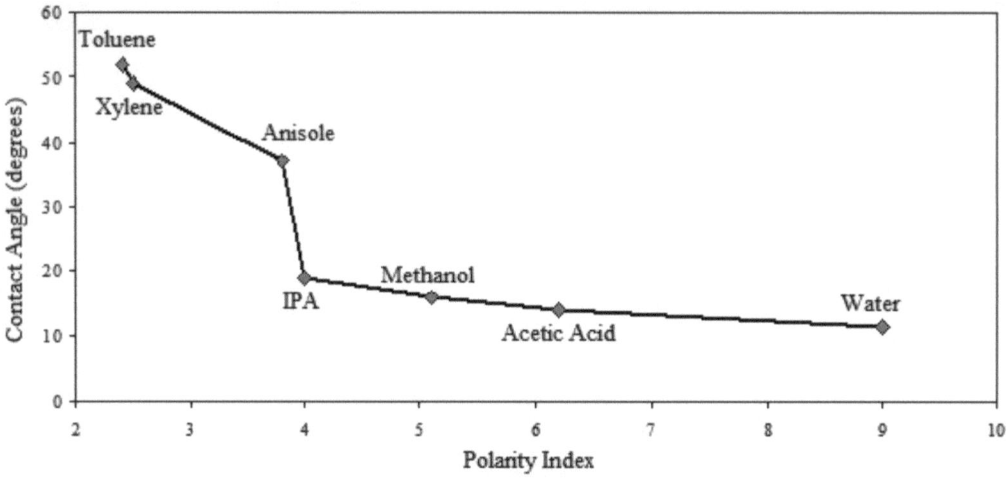

Fig. 8. Contact angles of the solvents shown in the previous figure (Lewis et al., 2010).

Figure 9 gives a schematic representation of the mechanism of hydrogen bonding, it can be seen clearly that the Toluene, Xylene and Anisole cannot suspend the oxide nanoparticle because there is no mechanism for hydrogen bonding process to occur. This is because the dipoles within the hydroxyl groups are not present. There is no barrier immediate to the surface of the oxide nanoparticle. This allows the depletion region around each oxide nanoparticle to intersect and therefore, the oxide nanoparticles are attracted to one another, hence, clustering.

Clustering of the oxide nanoparticles can be greatly reduced by using IPA, Methanol, Acetic Acid or Water. Figure 9 illustrates that these molecules have hydroxyl groups within them and are attracted to the hydroxyl groups at the surface of the oxide nanoparticles via their dipoles. Therefore, they surround the oxide nanoparticle and form a physical barrier with a thickness greater than that of the depletion region which would surround the oxide nanoparticle (see figure 6). The depletion regions of the oxide nanoparticles can never intersect. Hence, the oxide nanoparticle attraction is greatly reduced and the oxide nanoparticle suspension is stable.

Due to the fact that the PMMA polymer is dissolved in Anisole, it is evident that when the oxide nanoparticles were dispersed directly into the PMMA polymer matrix / Anisole system they would never be suspended homogeneously. The solvents used were Methanol and Toluene, the reason for this choice was that Methanol was more hydrophilic (contact angle of 18°) than Toluene (contact angle of 52°) as they have an index polarities of 5.1 and 2.4 respectively. As the homogeneous dispersal of the oxide nanoparticles could be ambiguous, the performance of the suspension was measured using the zeta potential measurement system.

4. Oxide nanoparticle stability

The Zeta potential is a physical property that is generally applied to evaluate the stability of a colloidal system. The solvent that surrounds the particle comprises of an inner region which is called the stern layer and an outer region called the diffuse layer (this makes up the electrical double layer). The first component has strong ions bound to it, while the strength

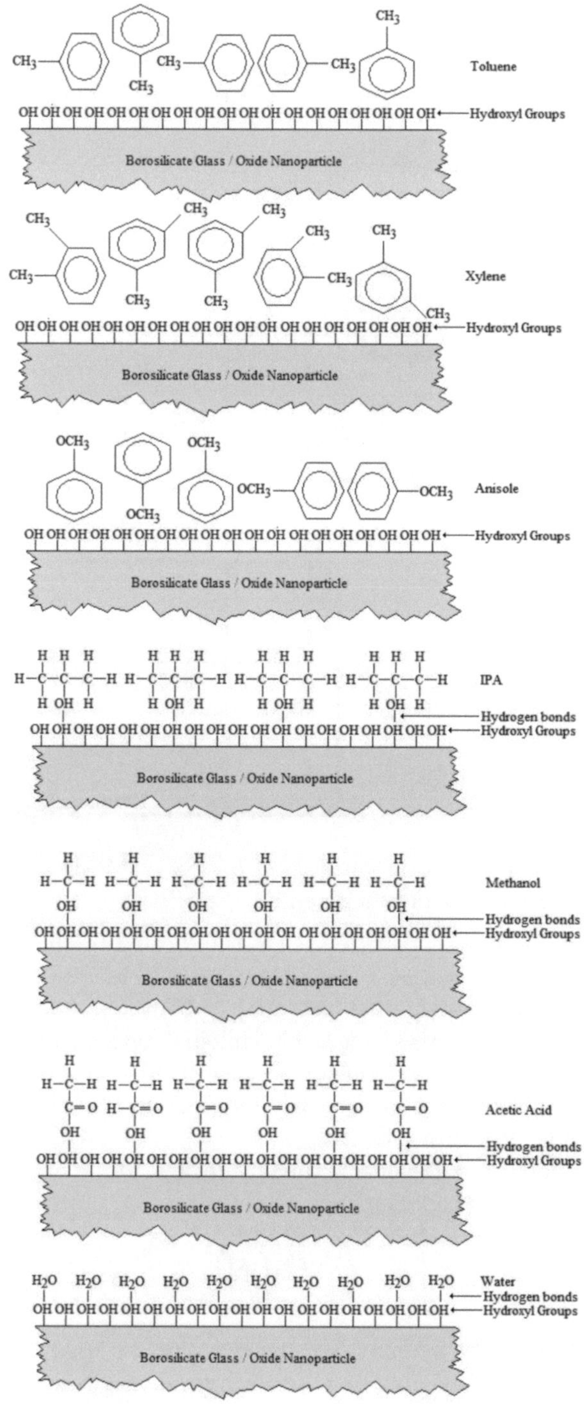

Fig. 9. Hydrogen bonding mechanism between the borosilicate glass / oxide nanoparticle surface and the various solvents.

of the ions of the secondary component are weakly associated. However, ions and particles can be stabilized due to the theoretical boundary within the diffuse layer. However, ions and particles can be stabilized due to the theoretical boundary within the diffuse layer. The potential at this boundary is called the zeta potential and this is illustrated in figure 10.

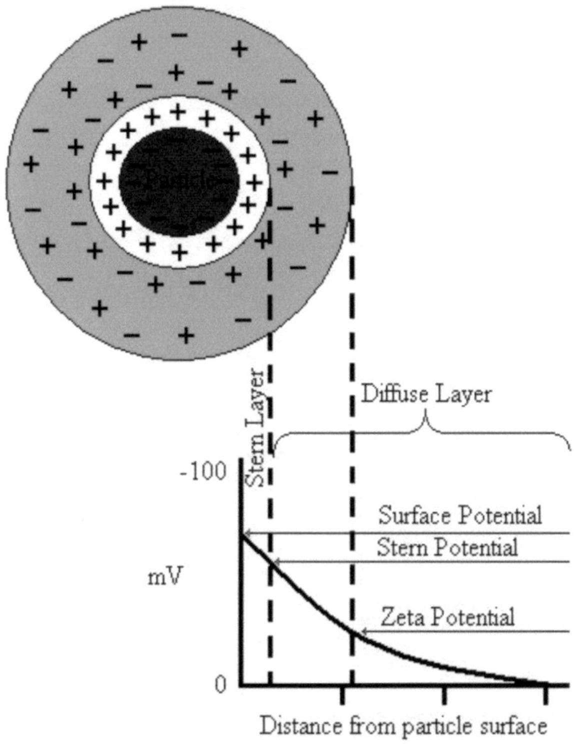

Fig. 10. Schematic representation of zeta potential.

The potential stability of the nanoparticle system is indicated by the magnitude of the zeta potential parameter. All the particles in the suspension will tend to repel each other if they have a large positive or negative zeta potential of 30mV or -30mV, therefore, particle clustering will not occur given rise to a stable colloidal system. But, if the zeta potential value is low, then the particles will cluster because there is no force to repel them. The knowledge of the zeta potential can reduce the time taken to manufacture trial formulations and assist in forecasting long term stability.

The electrophoretic mobility parameter is directly related to the zeta potential parameter, and this is given by the Henry equation

$$U_E = \frac{2\varepsilon\xi f(\kappa r)}{3\eta} \tag{7}$$

where U_E is the electrophoretic mobility, η is the viscosity and f(κa) is the Henry's function. The Debye length is termed in units of κ. Its reciprocal length κ^{-1} is the measure of the thickness of the electrical double layer. Due to the fact that the term r is the radius of the particle, then κr measures the ratio of the particle radius to the electrical double layer thickness.

To measure the electrophoretic mobility, charged particles suspended in the electrolyte are attracted towards the electrode of opposite charge under the influence of an electric field. The forces acting on these particles are likely to resist this motion. When the particles travel with a constant velocity, the two resisting force have reached an equilibrium condition. This is dependent upon the strength of the electric field, the dielectric constant of the suspending medium and its viscosity and the zeta potential. Therefore, the electrophoretic mobility is the velocity of a particle in a unit electric field.

From the Aluminium oxide (Al_2O_3) nanoparticles directly suspended in Methanol and Toluene, figure 11 shows, that Methanol suspends the oxide nanoparticles successfully. The zeta potential peaks shows that Al_2O_3 nanoparticles are both positive. From the data here, it can be seen that the nanoparticles suspended in Methanol have a zeta potential above the 30mV threshold, therefore the nanoparticle suspension was stable, however, both nanoparticle species suspended in Toluene are below the zeta potential threshold, hence nanoparticle instability, where the zeta potential value was 19.4mV. It was observed that Methanol was effective in suspending the oxide nanoparticles, prior to dispersal within the PMMA polymer. It was found that 200mg of nanoparticles aggregated together and dropped out of suspension under the force of gravity, on average in less than one minute in 0.5ml of Toluene, whereas the methanol suspended the same concentration of nanoparticles for many days. Methanol encapsulates the surface of nanoparticles and acts as a barrier so the depletion regions do not intersect. Therefore, the Figure 13 shows the Al_2O_3 nanoparticles treated with methanol and suspended directly in the PMMA polymer the rms roughness was 63nm. Thus, as more nanoparticles are embedded in the PMMA, then the surface roughness is increased. However, the surface of the nanocomposite film has been improved by a factor of 6.34 times. This is more suitable for the electron beam lithographic application, because the surface roughness is considerably reduced, hence the pattern remains intact.

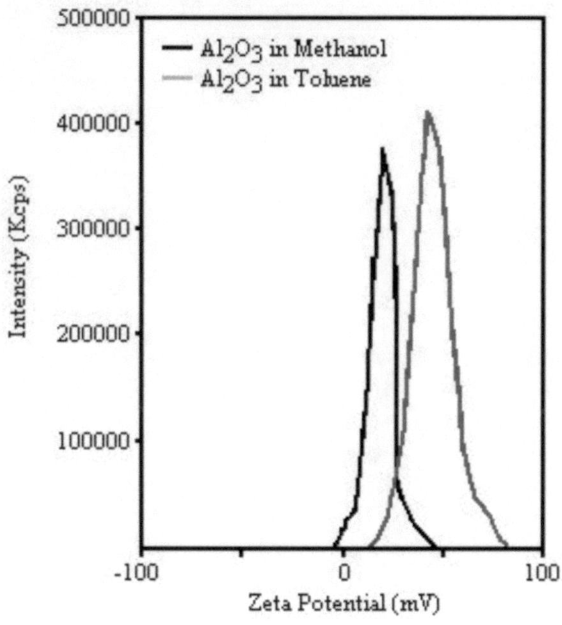

Fig. 11. Zeta potential of the two test cases – Al_2O_3 suspended in methanol and toluene respectively (Lewis et al., 2010).

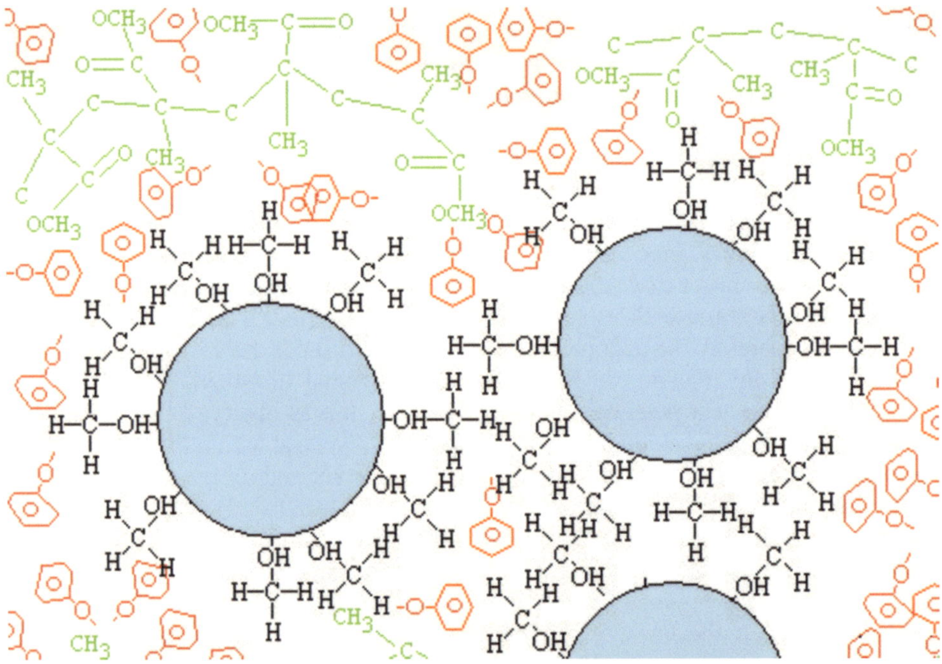

Fig. 12. Schematic representation of the nanoparticle /Methanol /Anisole /PMMA system.

Fig. 13. AFM image of Al_2O_3 nanoparticles embedded in the PMMA matrix that was suspended in Methanol (Lewis et al., 2010).

5. Characterization of the nanocomposite resist using Electron Beam Lithography

Since the early 1970's it has been recognized that by accelerating a focused beam of electrons at the surface of a resist, a very high resolution pattern can be produced, where the wavelength of electrons are smaller than the photons of ultra violet radiation by more than three orders of magnitude (Livesay & Fritz 1972).
For electron beam lithography, it is most important to determine the clearing dose of the resist material. This is defined by the exposure parameters that allows the electron to penetrate the resist to achieve excellent pattern definition and this is governed by the area dose and is given by

$$Area\ Dose = \frac{I * t}{S^2}, \tag{8}$$

Where I is the current of the incident beam, t is the dwell time, which is the time taken for the electrons to penetrate the resist and S is the step size. This is the parameter that governs the position that the beam moves from one position to its next position. Figure 14 shows a schematic representation of the step size parameter to demonstrate its impact to the proximity effect. As the electron beam is scanned over the surface of the resist from position a to position b, the electrons penetrate the resist. It can be seen that the foot print is larger at the bottom of the resist than at the top. Therefore, if the electron beam is placed too close to the previous position (as indicated by position a) that was already exposed, then the new exposure will intersect (illustrated by the red line) with the last exposure and this contributes to the proximity effect. Thus has been over exposed. This will have direct effect on the definition of the pattern. However, if the electron beam is placed too far away from the previous exposure as shown by position d and e in figure 14. The result of the pattern will become under exposed. Therefore, the step size must be balanced as seen by c and d. Here both exposures are placed by enough distance part from each other, so that they intersect at the bottom of the resist and this is illustrated by the red line.

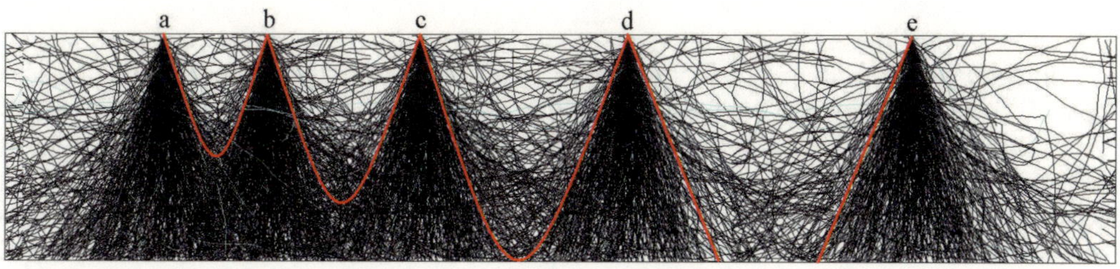

Fig. 14. Schematic representation of the step size parameter.

Over exposure may also result if the current I is too large as more SE can be generated as they scatter from the atoms at an angle of 80. Figure 15 shows an SEM image of examples of resist that has been over exposed. As the resist profile has an under cut. If the current density is reduced then this may result in under exposure, where the electrons have not completely penetrated through the resist.
Of course, in reality this is dependent upon thickness. The current and step size can be reduced, if the thickness was decreased then the scattering events inside the resist would

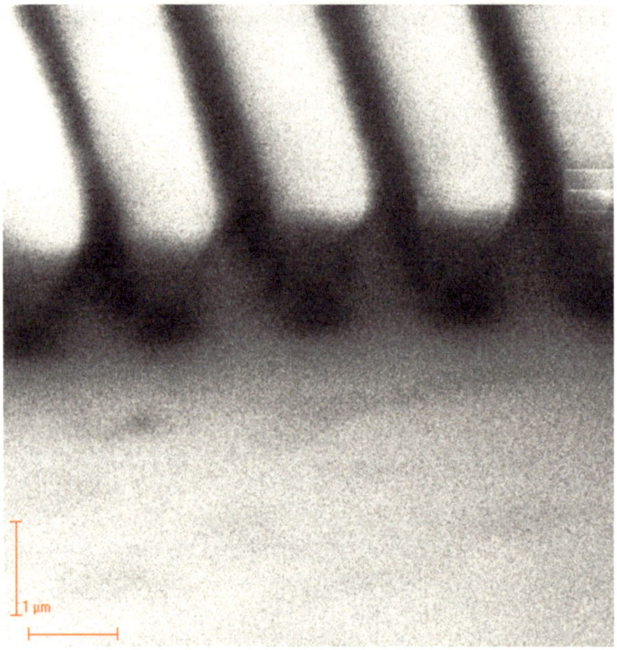

Fig. 15. An SEM image of over exposed resist.

also be decreased. This reduces the proximity effect because the number of electrons (current density) must be enough to overcome the stopping power of the atoms in the resist. This is due to as an electron collides with the atoms in the resist molecule, they will scatter in arbitrary directions. As this occurs, a fraction of this energy associated to that electron will be lost. This process will be repeated more frequently from this point onward until all its energy has been lost. If the resist thickness is large and associated energy is small (e.g. 10KeV) the then more collisions will occur and the electrons will scatter and generate SE which will damage the resist in a lateral direction as well as a vertical direction, thus, contributing to the proximity effect.

Both standard PMMA and nanocomposite resists were exposed to the electron beam, which had an acceleration voltage of 25kV and beam current of 30pA at a working distance of 7mm. This was achieved using a Scanning Electron Microscope (SEM), Cambridge S360, which was driven by Elphy quantum pattern generator. To determine the thickness and the appropriate exposure clearing dose, the exposure pattern consisted of a matrix of a hundred 20µm squares, this allowed for accurate AFM measurements of the remaining resist thickness for each dose. These squares were exposed with a dose scaled in incremental steps of $30\mu C/cm^2$, up to a highest dose of $600\mu C/cm^2$. Hence, the clearing dose could be accurately determined of each resist. The exposure pattern was developed using solution of MIBK (Methyl IsoButyl Ketone) and IPA (IsoProPanol), in the ratio of 1:3, for 30s followed by a 40s rinse in IPA.

It was found that the standard PMMA 950K 7% had a thickness of 407nm. However, when 150mg of Al_2O_3 nanoparticles was introduced into the PMMA 950K 7% polymer matrix, the overall resist thickness was increased to 562nm. Figure 16 indicates that the exposure clearing doses of PMMA 950K 7% resists increased as the density of Al_2O_3 nanoparticles embedded into the PMMA was increased. The combination of the increased resist thickness

and the Al_2O_3 nanoparticles resulted in higher exposure doses being required in order for the chain scission process to occur throughout the nanocomposite resist.
Looking in more detail at Figure 16, the exposure clearing doses of the standard PMMA 950K 7% were in the range of 180 - 210μC/cm², while when it was loaded with 150mg of Al_2O_3 nanoparticles the clearing doses were in the range of 240 - 270μC/cm², as determined from the dose scale. It should be noted that this exposure matrix method only gives a dosage range (e.g 240 - 270μC/cm²) within which complete resist exposure occurs. However, an estimate of the exact clearance dose can be obtained by extrapolation from the data set up to - but not including - the final point (where the resist thickness is zero). This method gives a clearance dose of 200μC/cm² for the standard PMMA case, while the clearance doses was 265μC/cm² for PMMA 950K 7% loaded with 150mg. These exposure doses found here are only valid for this resist system and that particular thickness. As the concentration of Al_2O_3 nanoparticles increases, the clearing dose had to be increased because the nanoparticles act as electron scattering centres and consequently decrease the electron momentum (Ishii et al 2001).

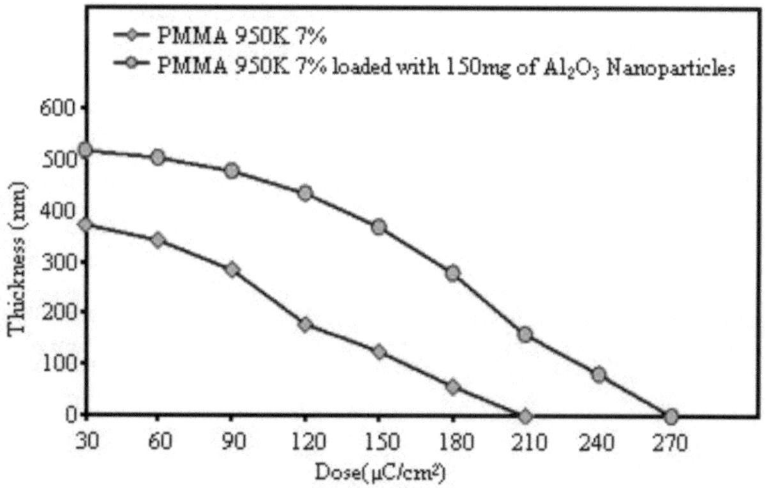

Fig. 16. Exposure doses of nanocomposite resist.

The Al_2O_3 nanoparticles also increased the overall resist thickness and therefore the electron exposure dose must be increased to overcome this.

6. Modeling the effects of the nanocomposite resist using the Monte Carlo Method

From the results of the previous section it is obvious that the introduction of Al_2O_3 nanoparticles in an electron beam resist effects the interaction between the impinging electrons and the atoms in the resist film. The Al_2O_3 nanoparticles exhibit different scattering behaviour than the traditional atoms contained in the PMMA electron beam resist, such as Carbon, Oxygen and Hydrogen. Here, the influence of the Al_2O_3 nanoparticles in the PMMA electron beam resist was studied through Monte Carlo simulations. This was determined by characterizing the scattering behaviour of the point electron distributions of the PMMA based nanocomposite resist.

The effect of the Al_2O_3 nanoparticle density was studied by comparing the standard PMMA with the PMMA based nanocomposite resist in a constant unit volume of PMMA. Therefore, the effects of nanoparticle density on the interaction with incident electrons can be fully examined.

The Monte Carlo simulation that is presented here calculates electron scattering trajectories in materials, in this case PMMA based nanocomposite resist. The physical properties of the resist were calculated using the weighted percent of nanoparticles to PMMA. PMMA has a density of 1.19g/cm^3 and Al_2O_3 has a density of 4g/cm^3. The total density of the nanocomposite resist was calculated using,

$$\rho_{resist} = \omega\rho_{Al_2O_3} + (1-\omega)\rho_{PMMA} \tag{9}$$

Where ρ is the density of each material and ω was the weighted percent of Al_2O_3 to PMMA. The atomic weight of PMMA here was 950 000 g/mol and the atomic weight of Al_2O_3 is 101.96 g/mol.

The nanocomposite resist was considered to have a homogeneous dispersion of Al_2O_3 nanoparticles throughout. As the electron travels through the resist it collides either elastically or inelastically with a molecule of PMMA or with a nanoparticle. When the electron collides with an atom, it was decided using the weighted percent of Al_2O_3 to PMMA and comparing it with a randomly generated number between 0 and 1. Whether it is an elastic or inelastic collision was decided by comparing a randomly generated number with the probability of getting an inelastic collision, the ratio between the inelastic cross section and the total cross section. Depending on the type of collision, the scattering cross sections and the mean free path are calculated.

For both elastic and inelastic scattering, the total scattering cross section is calculated again using the weighted percent of Al_2O_3 to PMMA,

$$\sigma_{total} = \omega\sigma_{Al_2O_3} + (1-\omega)\sigma_{PMMA} \tag{10}$$

To decide the outcome of the collisions as either elastic or inelastic a random number was generated. This was dependent upon the ratio of the elastic and inelastic scattering cross sections to the total scattering cross section. The next random number decides if the electron has collided with an Al_2O_3 nanoparticle or an atom in the PMMA molecule and compares this event to the nanoparticle loading in the polymer film. When this has been achieved, the Monte Carlo procedure calculated energy losses are calculated, and the scattering angles for the collision event.

The elastic scattering cross section is calculated using the screened Rutherford elastic cross section,

$$\sigma_{elastic} = \frac{Z^2}{E^2}\frac{4\pi}{\alpha(1+\alpha)}\left(\frac{E+511}{E+1024}\right)^2 cm^2/atom \tag{11}$$

Where E is the electron energy in KeV, Z is the atomic number of the material and α is the screening factor, this compensates for the fact that the electron does not see the all of the atom's charge as it is surrounded by a cloud of electrons.

For the case of inelastic scattering from the nanoparticles, the scattering cross section is calculated by considering only the coulomb interaction between the electron and the atom. In this case the inelastic scattering cross section per electron is,

$$\frac{d\sigma_{inelastic}}{d\Omega} = \frac{\pi e^4}{E^2}\left(\frac{1}{\Omega^2 + (1-\Omega)^2}\right) \tag{12}$$

Where ΩE is the energy of the secondary electron produced. The inelastic scattering event causes the primary electron to be deflected by an angle α given by,

$$\sin^2 \alpha = \frac{2\Omega}{2 + t - t\Omega} \tag{13}$$

Where t is the kinetic energy of the electron in units of its rest mass. However the secondary electron created exits the collision at an angle γ given by,

$$\sin^2 \gamma = \frac{2(1-\Omega)}{2 + t\Omega} \tag{14}$$

For an impact that transfers approximately 500eV to the secondary electron, the primary scatters at an angle of about 1° and the secondary electron exits at an angle of about 80°. This is why the secondary electrons play a major role in producing the so called 'proximity effect'.
Once the scattering cross sections are calculated, the mean free path of the electron must be calculated. This is the sum of the inelastic mean free path and elastic mean free path of the electron.

$$\frac{1}{\lambda_{total}} = \frac{1}{\lambda_{elastic}} + \frac{1}{\lambda_{inelastic}} \tag{15}$$

The elastic and inelastic mean free paths are calculated from the scattering cross sections,

$$\lambda_{elastic} = \frac{A}{N_a \rho \sigma_{elastic}} \tag{16}$$

$$\lambda_{inelastic} = \frac{A}{N_a Z \rho \sigma_{inelastic}} \tag{17}$$

where A is the atomic weight of the material, Na is Avogadro's number, ρ is the density of the resist and σ is the respective scattering cross section. The total elastic and inelastic mean free path is calculated using weighted percentages,

$$\frac{1}{\lambda} = \frac{\omega}{\lambda_{Al_2O_3}} + \frac{1-\omega}{\lambda_{PMMA}} \tag{18}$$

From the value of the mean free path, the distance the electron will travel before it collides again can be calculated. This is done using the step size equation given by,

$$s = -\lambda \ln(RND) \tag{19}$$

where λ is the total mean free path and RND is a random number between 0 and 1. This gives a distribution of step sizes with an average step size of λ. Therefore, the probability of obtaining a step size near the mean free path is larger than obtaining a step size much larger or smaller than the mean free path.

The final step of the Monte Carlo simulation was to calculate the energy lost by the electron during the scattering event. This was done using the modified Bethe equation for the stopping power of a material and is given by (Newbury & Myklebust 1981, Luo & Joy 1989),

$$\frac{dE}{dS} = 78500\frac{Z}{AE}\ln\left(\frac{1.166(E+0.85J)}{J}\right) \tag{20}$$

where J is the mean ionization potential of the material, which was 145 and 74eV for Al_2O_3 and PMMA respectively [6]. Every time the electron scatters, this value is calculated and was subtracted from the current energy of the electron. Once the electron's energy falls below 0.5 keV, the electron was no longer tracked as the distance it travels in the material is very small.

The model was used to determine the influence of the Al_2O_3 nanoparticles when a primary beam of electrons was incident upon it. The resist films considered here consisted of standard PMMA and PMMA loaded with 150mg of Al_2O_3 nanoparticles, the diameter of the nanoparticles were 25nm and the Films were uniform. The incident energy is 25KeV and to avoid over exposure the exposure dose was 265μC/cm².

From the scattering trajectories seen in figure 17, it shows that the incident electrons experience a number of collisions with the atoms in the PMMA molecule in its flight, as this process occurs; the electrons diverge away from the incident beam. Hence, as the incident electron travels through the PMMA resist then the number of collisions increases and therefore dramatically increases the proximity effect.

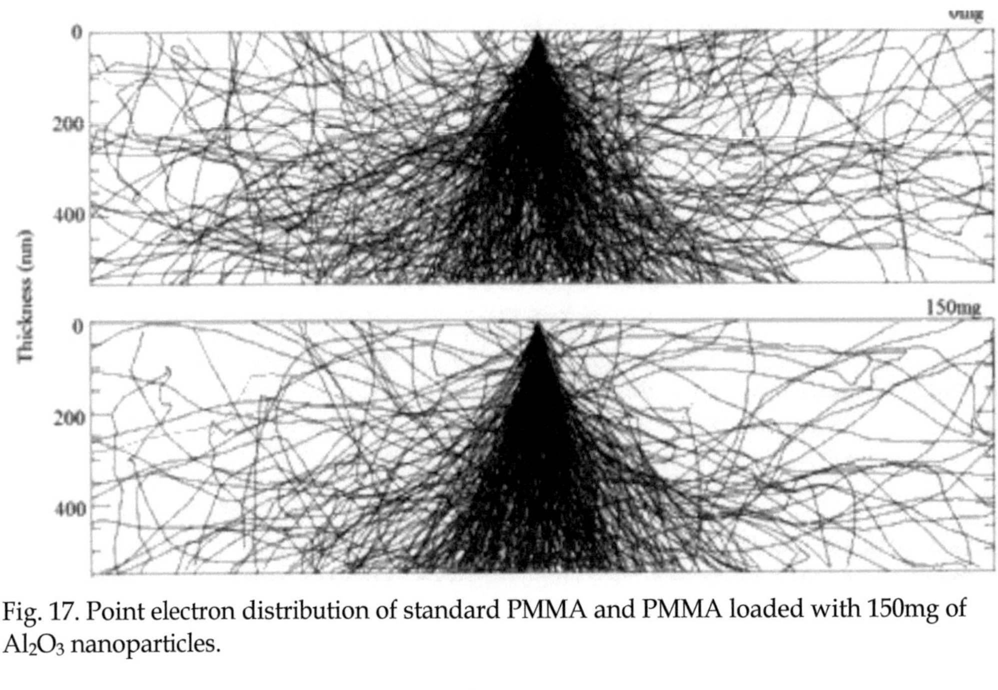

Fig. 17. Point electron distribution of standard PMMA and PMMA loaded with 150mg of Al_2O_3 nanoparticles.

This was achieved when an incident electron collides with the surface of the resist, the Monte Carlo calculations for its trajectory start. As the electron penetrates through the resist, the Monte Carlo procedure was paused at every 50nm in depth (along z axis). At this point the electron's (x, y) coordinates were retrieved. The radial distance of the nth electron at depth z was calculated using,

$$r_n(z) = \sqrt{x^2 + y^2} \qquad \text{for } z = 50,\ 100,\ 150\ldots \text{ nm}, \tag{21}$$

this value is then stored and the procedure continued. This process is repeated for every incident electron, and all of the values of r are stored. Once all of the electrons have entered the resist the values of r_n are recalled for each depth and are averaged. If a total of N electrons have penetrated the resist then the average radial distance at depth z is given by,

$$R(z) = \frac{\sum_{n=1}^{N} r_n(z)}{N}. \tag{22}$$

Figure 18 demonstrates that when the electron spread of the primary electrons had cleared the resist, the average radial distances were approximately 382 and 112nm for the unloaded PMMA and PMMA loaded with 150mg of Al_2O_3 nanoparticle, respectively.

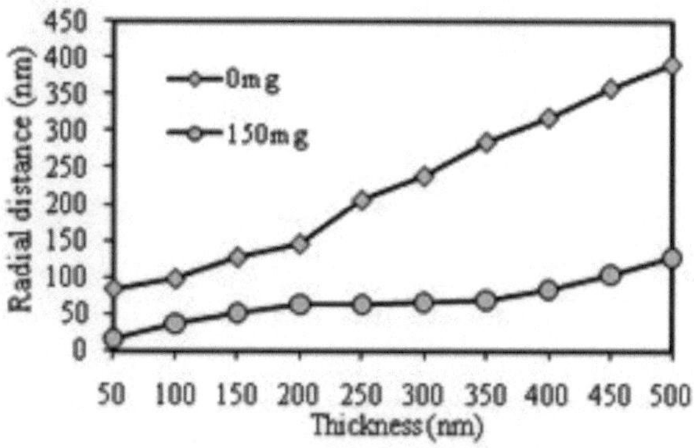

Fig. 18. Average radial distance of standard PMMA and PMMA loaded with 150mg of Al_2O_3 nanoparticles.

From figure 18, if the film thickness was reduced to 400nm of the thickness of the standard PMMA, the electron spread is considerably reduced. This demonstrates that thickness play a major role when exposing the resist to achieve the smallest nano structures.

This result shows that when the nanoparticle are introduced into the PMMA, the electrons incident on the nanocomposite resist experience a greater number of collisions with the atoms in its flight through the nanocomposite resist, hence, the mean free path was decreased. Therefore, the presence of the Al_2O_3 nanoparticles in the PMMA dramatically affects the mean free path and thus controlled the number of collisions that an incident electron encounters in its trajectory. This was due to the fact that the density of the Al_2O_3 nanoparticle was larger than the PMMA molecule.

In order to understand the role of secondary electrons in the 'proximity effect', the number of secondary electrons that are created per primary inserted and the distance that they travel through the resist was simulated by the Mont Carlo model. The number of secondary electrons created is shown in figure 19. From this characteristic, it was found that the number of secondary electrons created were considerably more than the number of the primary electrons incident on the nanocomposite resist surface. It can be seen that number of secondary electrons that were created was 2.0 (per primary electron) when the PMMA was unloaded. Comparing this to the PMMA that was loaded with 150mg of Al_2O_3 nanoparticles, the quantity of secondary electrons that were created was 1.6 (per primary electron). Thus, the presence of the Al_2O_3 nanoparticles in the nanocomposite resist film had the effect of quenching the secondary electrons.

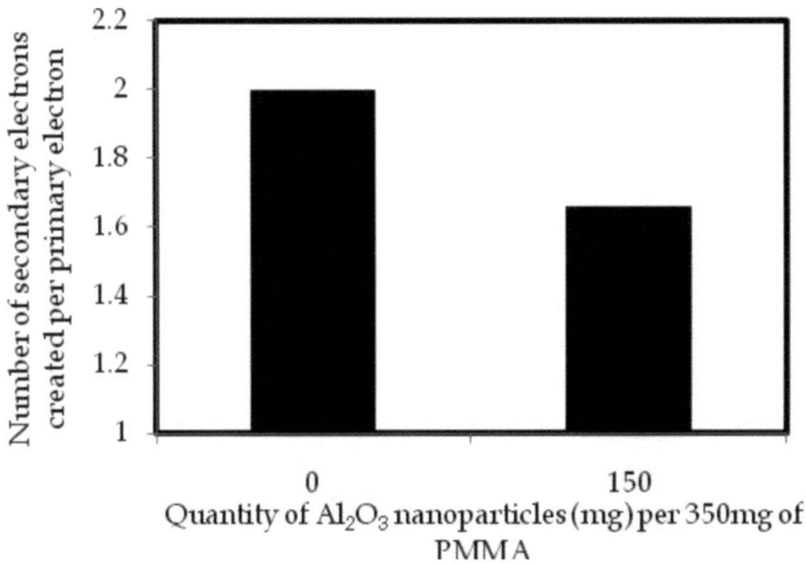

Fig. 19. Number of secondary electrons created per primary electron for a standard PMMA and PMMA based nanocomposite resist.

It was observed that the distance travelled by the secondary electrons decreases as the nanoparticle concentration was introduced and this is shown in figure 20. The secondary electrons travel further in standard PMMA when compared with PMMA based nanocomposite resist. This was due to the stopping power of the Al_2O_3 nanoparticles and is associated with the density of the Al_2O_3 nanoparticles. Note: the atomic numbers associated with the atoms within the nanoparticle are substantially larger than that of the PMMA polymer (Gonsalves et al 2001). The stopping power reduces the lateral distance and energy spread of the electrons travelling through the PMMA polymer.

When the PMMA was loaded with the Al_2O_3 nanoparticles they were able to confine the incident electrons. This was because the Al_2O_3 nanoparticles had a larger density than the PMMA molecule. This resulted in a stronger shielding of the atom which was provided by a larger charge. Therefore, an increased number of collisions took place due to an increasing density of the Al_2O_3 nanoparticles. Therefore, a higher aspect ratio can be achieved by loading the PMMA with a high density material and obtains a better resolution than resists like PMMA. These results agree with the work by A. Jeyakumar on Silica nanoparticles (Jeyakumar 2004).

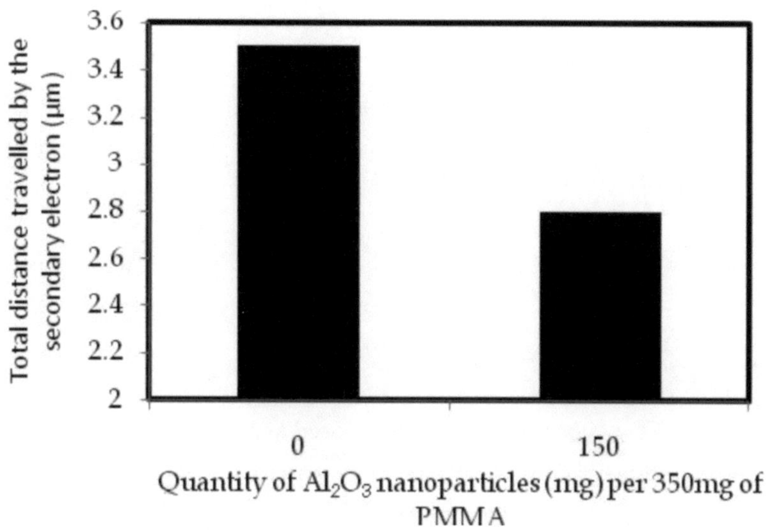

Fig. 20. Total distance travelled by the secondary electrons in a standard PMMA and PMMA loaded with 150mg of Al_2O_3 nanoparticles.

7. Exposure of the nanocomposite resist to the electron beam

To investigate the aspect ratio of developed features a series of 90 nm and 70 nm lines with spacings of 180 nm and 140 nm with a length of 20 µm were exposed. The nanocomposite resist loaded with 150mg of Al_2O_3 nanoparticles and the standard PMMA were exposed with a dose of 265 and 200µC/cm^2 respectively.

Figures 21 and 22 show SEM images that feature nano structures consisting of 90nm and 70nm trenches. Figure 21a shows features that consisted of lines and spaces (trenches) in the PMMA 950K 7% sample loaded with 150mg of nanoparticles. The lines had a width of 185nm and the spacing between each of the lines was 91nm; as the resist thickness was 562nm this gives a high aspect ratio of approximately 6.2:1. The SEM image in Figure 21b shows the unloaded PMMA 950K 7% resist; the aspect ratio is 4.4:1, with line widths of 185nm and spacings of 91nm.

We have been able to produce 90nm trenches in a thicker resist because the Al_2O_3 nanoparticles have contributed to reducing the proximity effect. Also the mechanical properties of the PMMA have been improved by the addition of Al_2O_3 nanoparticles by providing an internal mechanism that bonds the PMMA resist together. Figure 22a shows the lines fabricated in PMMA 950K 7% loaded with 150mg of Al_2O_3 nanoparticles. Each line consisted of feature sizes of 70nm. However, much of the periodic spacing has disappeared, suggesting that the resist has collapsed. Both the primary and secondary electrons have undercut the nanocomposite resist from the silicon surface and consequently the nanocomposite resist has toppled over. The resist was clearly compromised by the proximity effect. From Figure 22, it can be seen clearly that the Al_2O_3 nanoparticles have bound the PMMA polymer together, to make the material more rigid, as the resist has remained intact. In the sample of the standard PMMA 950k 7% (Figure 22b), there are no lines, they have completely collapsed. This is due to the fact that the nanocomposite resist is too thick to control the primary and secondary electrons from transmitting laterally.

Fig. 21. PMMA 950K 7% nanocomposite resist exposed at 265µC/cm^2, a) loaded with 150mg of Al_2O_3 nanoparticles, b) Standard PMMA 950K 7%.

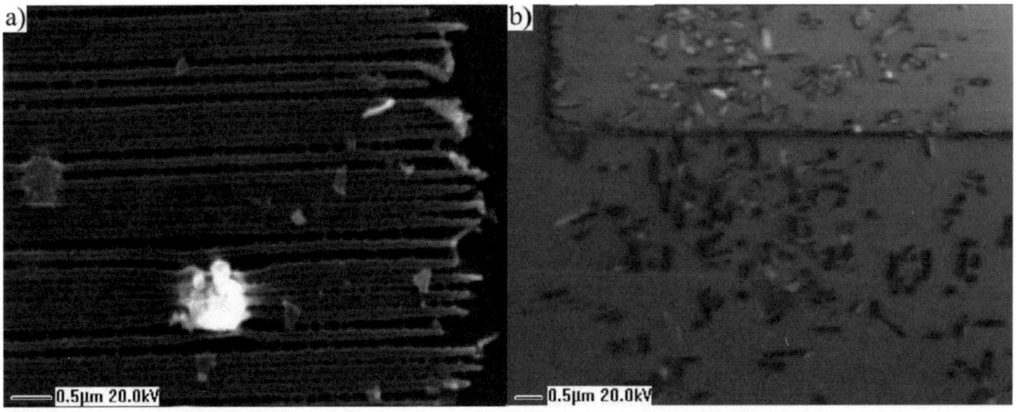

Fig. 22. SEM image of collapsed features in PMMA 950K 7% nanocomposite resist exposed at 265µC/cm^2, a) loaded with 150mg of Al_2O_3 nanoparticles, b) Standard PMMA 950K 7%.

From the SEM images in figures 21a and 22a it can be seen that the Line Edge Roughness (LER) of the resultant features has degraded due to the fact that the Al_2O_3 nanoparticles have aggregated together (Lewis et al 2010).

It is evident that as the quantity of the nanoparticles was increased, the LER is also increased. The LER needs to be as small as possible as to not compromise the pattern. In this case, therefore, reducing the concentration of Al_2O_3 nanoparticles in PMMA 950K 7% will reduce the LER.

8. Summary

A novel electron beam resist has been demonstrated by incorporating Al_2O_3 nanoparticles with a diameter of 15 - 25nm into the PMMA polymer matrix. Oxide nanoparticles were introduced into the polymeric matrix by adding the oxide nanoparticles directly into the PMMA matrix, the PMMA became inhomogeneous. This was indicated by an increase in surface rms roughness measured (via an AFM). This was found to be 402nm produced by Al_2O_3 nanoparticles. This was due to poor particle dispersion as a result of the inter-particle attraction between each nanoparticle caused by the depletion effect. With the surface roughness the resultant pattern developed in the nanocomposite resist would have had a poor edge definition.

Therefore, to improve the homogeneous surface quality, the oxide nanoparticles were suspended in Methanol and Toluene. This was measured by using the zeta potential parameter. It was found that the Methanol suspended the oxide nanoparticles successfully. The zeta potential value was 39.2mV; where the oxide nanoparticle suspension was stable. This method improved the surface roughness of PMMA when the Al_2O_3 was suspended in Methanol. The PMMA with 200mg of Al_2O_3 nanoparticles embedded within it had an rms roughness of 63nm. The improvement here was a factor of 6.34.

The increased number density of Al_2O_3 nanoparticles in the PMMA required an increase in the clearing dose. The presence of the nanoparticles had a quenching effect on the electrons within the resist responsible for 'exposing' the resist. Consequently, this affected the outcome of the developed features, to the extent that in some cases, the features produced were completely deformed. In some instances, nanoparticle clustering affected the quality of the pattern obtained. It was found that as the quantity of the nanoparticles was increased, the line edge roughness (LER) also increased. LER needs to be as small as possible as to not compromise the pattern.

It was found that the Al_2O_3 nanoparticles had prevented the lines from collapsing by blocking the arbitrary scattering secondary electrons from transmitting laterally throughout the nanocomposite resist. Thus, the proximity effect was reduced.

It was found in all samples that, when the line widths were reduced beyond 70nm they had collapsed. However, the lines in the nanocomposite resist had toppled over and thus not completely destroyed. It is suspected that the Al_2O_3 nanoparticles have affected the mechanical properties of the PMMA by causing the molecular chain of the polymer to bond with the Al_2O_3 nanoparticles. All the lines are completely destroyed in the unloaded samples of PMMA 950K 7%.

9. References

R. L. Dean, H. Y. Lem, C. A. Sauer, H. Chang, (1994). PBS resist profile studies for sub - micron mask lithography. *Proc. SPIE*, vol. 2322., 102.

B. Erdem, E. D. Sudol, V. L. Dimonie, M. S. El - Aasser, (2000). Encapsulation of inorganic particles via miniemulsion polymerization. II. Preparation and characterization of sturene minienulsion droplets containing TiO_2 particles. *J. Polym. Sci. Part. A: Polym. Chem* (38) page number 4431 - 40.

K. E. Gonsalves, L. Merhari, H. Wu, Y. Hu, (2001). Organic - inorganic nanocomposite: unique resists for nanolithography. *Advanced Materials*, Vol 13, No 13.

T. Ishii, H. Nozawa, T. Tamamura, (1997). A nano - composite resist system: a new approach to nanometer pattern fabrication. *Microelectronic Engineering*, vol 35., no 1., page numbers 113 - 116.

T. Ishii, H. Nozawa, T. Tamamura, (1997). C60 incorporated nanocomposite resist system. *J. Photopolymer. Sci. Technol*, vol 10., no 4., page numbers 651 - 656.

T. Ishii, T. Tamamura, K. Shigehara, (2000). Fullerene - derivative nanocomposite resist for nanometer pattern fabrication. *Jpn. J. Appl. Phys.* Vol., 39 page numbers L1068 - L1070.

T. Ishii, Y. Murate, K. Shigehara, (2001).'Contrast enhancement of ZEP520 resist by fullerene - derivate incorporation. *Jpn. J. Appl. Phys* Vol., 40 page numbers L478 - L480.

P, Judeinstein, C. Sanchez, (1996). Hybrid organic - inorganic materials: a land of multidisciplinarity. *J. Mater. Chem* (6) page number 511 - 25.

S. Lewis, V. Haynes, R. Wheeler-Jones, J. Sly, R.M. Perks & L. Piccirillo, (2010). Surface characterization of poly(methylmethacrylate) based nanocomposite thin films containing Al_2O_3 and TiO_2 nanoparticle. *Thin Solid Films* Vol., 518, No., 10, page numbers 2683-2687.

S. R. Lu, H. L. Zhang, C. X. Zhao, X. Y. Wang, (2005). Studies on the properties of a new hybrid materials containing chain extended area and SiO_2 - TiO_2 particles. *Polymer* (46) page number 10484 - 92.

L. Merhari, K. E. Gonsalves, Y. Hu, W. He, W. -S. Huang, m. Angelopoulos, W. H. Bruenger, C. Dzionk & M. Torker, (2002). Nanocomposite resist systems for next generation lithography. *Microelectronic Engineering* Vol., 63, page numbers 391 - 403.

R. A. Potyrailo (2006). Polymeric sensor materials: toward an alliance of combinatorial and rational design tools? Angew. Chem. Int. Edit. (45), page numbers 702 - 23.

A. Sidorenko, S. Minko, G. Gafijchuk, S. Voronov, (1999). Radical polymerization initiated from a solid substrate. 3. grafting from the surface of an ultrafine powder. *Macromolecules* (32) page number 4530 - 43.

A. H. Yuwono, B. Liu, J. Xue, J. Wang, H. I. Elim, W. Ji et al, (2005). Controlling the crystallinity and nonlinear optical properties of transparent TiO_2 - PMMA nanohybrids. *J. Mater. Chem* (14) page number 2978 - 87.

22

Polymeric Nanocomposite Materials

Masoud Salavati-Niasari[1,2] and Davood Ghanbari[1]
[1]*Institute of Nano Science and Nano Technology, University of Kashan,*
[2]*Department of Inorganic Chemistry, Faculty of Chemistry, University of Kashan,*
Islamic Republic of Iran

1. Introduction

Nowadays the enormous use of polymer materials is attributed to their extraordinary combination of properties, low weight and ease of processing. However for improvement of some properties such as thermal and mechanical stability, large numbers of additives were added to polymeric matrix and formed polymer matrix composite [1, 2].

A composite is defined as a combination of two or more materials with different physical and chemical properties and distinguishable interface. Composite materials have a wonderful and different range of applications. Important advantages of composites over many metal compounds are high specific stiffness and specific strength, high toughness, corrosion resistance, low density and thermal insulation [1-3].

In most composite materials, one phase is usually continuous and called the matrix, while the other phase called the dispersed phase. On the basis of the nature of the matrices, composites can be classified into four major categories:

1. Polymer matrix composite
2. Metal matrix composite
3. Ceramic matrix composite
4. Carbon matrix composite [4].

Polymer matrix composite can be processed at a much lower temperature, compared to other composite. Depending on the types of polymer matrices, polymer matrix composite are classified as thermosetting composites and thermoplastic composites [4].

2. Polymer matrix nanocomposite

Scientists for improve the properties of composite materials investigate composites with lower and lower fillers size, leading to the development of microcomposites and the recent trend in composite research is nanocomposites. Nanocomposites refer to composites in which one phase has nanoscale morphology such as nanoparticles, nanotubes or lamellar nanostructure [1-5].

The improvement of the properties by the addition of particles can be achieved when:

a. Adequately good interaction between the nanoparticles and the matrix
b. Good dispersion ofparticles within the matrix.

In nanocomposites, covalent bonds, ionic bonds, Vander Waals forces, hydrogen bonding could exist between the matrix and filler components [3,4].

2.1 Classification of nanocomposite

One of the classifications is based on the nanomaterial's dimensional morphology.

1. Zero dimensional nanomaterial such as nanoparticle [6-8].
2. One dimensional nanomaterial such as nanowire and nanotube [9]
3. Two dimensional nanomaterial such as silicate layers
4. Three dimensional nanomaterial such as zeolites [10-12]

2.2 Classification based on kind of synthesis procedure

1. Direct incorporation of nanoscale into a polymer melt or solution, such as addition several type metal oxide and hydroxide to polymeric matrix [5].
2. In situ generation of nanoscale building blocks in a polymer matrix (reduction of metal ions in polymer matrix) [5].

For example for synthesis of polyaniline/cadmium sulfide, polyaniline was dissolved in dimethyl formamide solution. Then dimethyl cadmium was added and simultaneously H_2S was passed through solution (Fig1) [13].

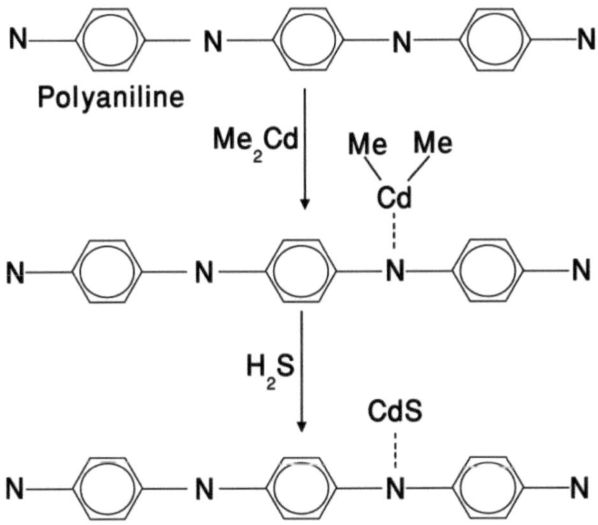

Fig. 1. Poly aniline/cadmium sulfide nanocomposite

3. Polymerization of monomers in the presence of nanoscale building bloks, such as polymerization of methyl metha acrylate or styrene in present nanoparticles [5].
4. A combination of polymerization and formation of nanoscale building blocks (for example intercalation of monomers into layered structures followed by polymerization) [5].

When a material structure has one dimension less than 100 nm, usually appear totally different behavior of its material properties, the properties of nanomaterials can sometimes produce an inverse transition of a property.

3. Some application of polymer matrix nanocomposite:

3.1 Flame retardant nanocomposite

Together with numerous advantages that synthetic polymeric materials provide to society in everyday life, there is one obvious disadvantage related to the high flammability of many synthetic polymers. Consequently, improving the fire retardant behavior of polymers is a major challenge for extending their use to most applications. Nanocomposites are very attractive due to the fact that small amount of nanostructure can lead to great improvement in fire resistant property of nanocomposite.

3.2 Nanocomposite with high mechanical stability, stiffness, strength, toughness

By using minimal addition levels, can be achieved excellent mechanical properties. For example polyamide-6clay nanocomposite, containing 5% clay, shows an increase of 40% in tensile strength, 68% in tensile modulus, 60% in flexural strength and 126% in flexural modulus [14].

3.3 Nanocomposite solar cells

Polymer-based solar cells have the capability of being used to make cheap large flexible panels. The only downside is substantially low efficiency compared to commercial solar cells [15]. Cadmium chalcogenides such as CdS nanoparticles in polymers have been used to make solar cells. Numerous efforts are being taken to improve the efficiency of the cells. In general a higher density of the nanoparticles is advantageous. Control of the band edge of the particles is also very important factor controlling the efficiency of the solar cell. Polymer-semiconductor nanocomposites offer the promise of a new generation of hybrid materials with numerous possibilities of applications such as in optical displays, catalysis, photovoltaics, gas sensors, electrical devices, mechanics, photoconductors and superconductor devices [15].

3.4 Nanocomposite membrane for gas separation and gas permeability

Many inorganic filler have dispersed in polymer matrices to improve pure polymeric membrane properties and constituted mixed-matrix membranes (MMMs). Result of such mixed matrix membrane is to the superior transport properties. With addition of low amount of nanoparticles to polymeric membrane, we achieved great increasing in permeability of membrane. There are many reports of improved penetrate permeability [16].

3.5 Nanocomposite with gas barrier performance

By addition alumina silicate layers to polymer matrix can be obtained gas barrier performance for nanocomposite.

4. Determination of the morphology of nanocomposites

As the nanocomposite properties dramatically depend on the good dispersion of the filler in the matrix, it is necessary to develop technical methods to characterise the nanodispersion. For determination of morphology, polymer matrix nanocomposite were characterized by some techniques such as transmission electron microscopy, TEM, scanning electron microscopy, SEM, atomic force microscopy, AFM and X-ray diffraction, XRD.

4.1 Scanning electron microscopy (SEM) and atomic force microscopy (AFM)

SEM determined morphology and shape of surface of material, in polymer matrix nanocomposite especially when low amount of nanoparticles is added to polymer, polymeric chains cover the nanoparticle, thereupon in surface image of SEM, cannot be determined nanoparticles in matrix. However when high percent of nanoparticles is added to polymer, usually can be seen nanoparticles in matrix. One way for determined of nanoparticles in polymeric matrix is using cross sectional SEM image, for preparation images of cross section of polymeric nanocomposite, a small amount of sample was immersed in liquid nitrogen for nearly some minute to ensure it was completely frozen. It was then removed from the liquid nitrogen and immediately was broken. AFM also determined surface of thin film sample, and in high loading nanoparticle, can be detected fillers in polymeric matrix. For example SEM images of tin and bismuth sulfide nanostructure and polymer matrix nanocomposite of those with acrylonitrile butadiene styrene (ABS) copolymer are shown in Fig 2, that confirm SEM images cannot perfectly determine nanoparticles in polymer matrix[17,18]. Also AFM images of pure copolymer and ABS/tin sulfide are shown in Fig 3, which shows the same results.

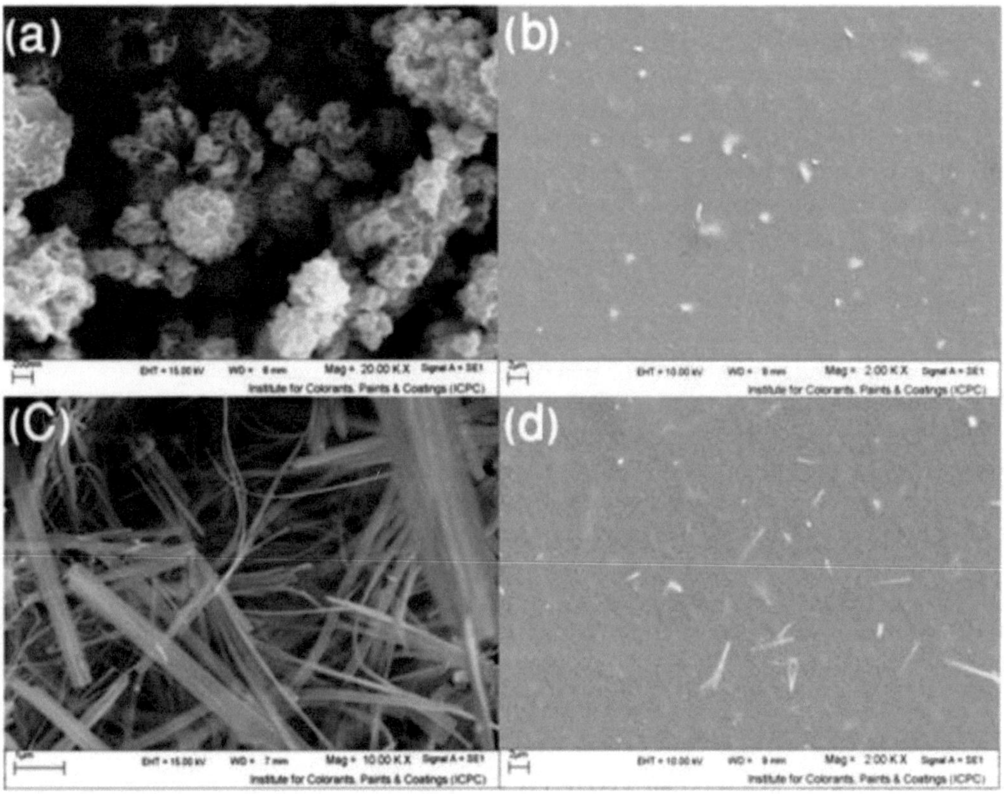

Fig. 2. SEM images of a) SnS nanoflowers b) ABS/SnS c) Bi_2S_3 nanorods d) ABS/Bi_2S_3 ref:17, 18

4.2 X-Ray Diffraction (XRD) technique

XRD enables the determination of the *d*-spacing of the clay. An immiscible system is obtained if the *d*-spacing in the presence of the polymer is unchanged from that of the pristine clay. If the *d*-spacing increases, this indicates that intercalation has occurred. Since the registry between the clay layers is lost in an exfoliated system, no peak is expected. Unfortunately, this same situation will occur if the clay has extensively disordered, so XRD

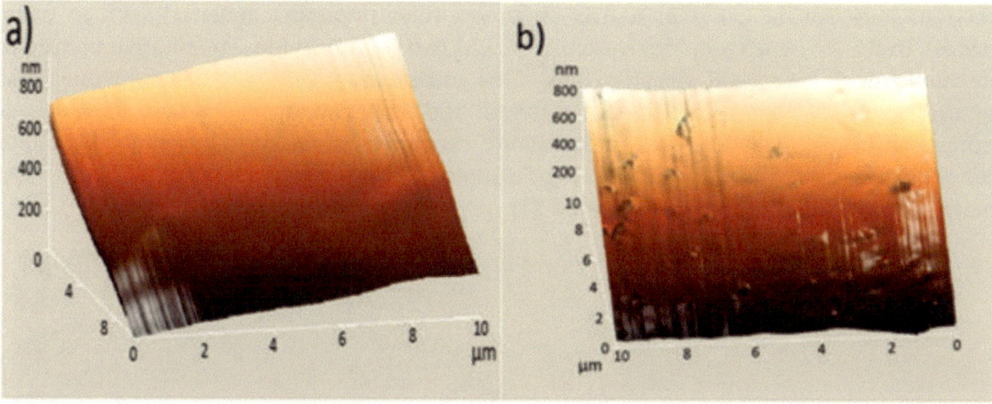

Fig. 3. AFM images of a) pure ABS b) ABS/SnS nanocomposit

information alone is not enough to identify the morphology. For these reasons, in general, XRD analysis is not used alone in studies to deduce the nanocomposite morphology [14].
XRD pattern ABS–clay hybrid nanocomposites are shown in Fig 4. With exchange Na^+ by alkyl ammonium, distance between layers increase and XRD peak shift to lower angle. XRD patterns of ABS-OMMT because of intercalation of single chains of ABS in the galleries of the host shows an increasing interlayer distance and peak shift to lower angle [19].

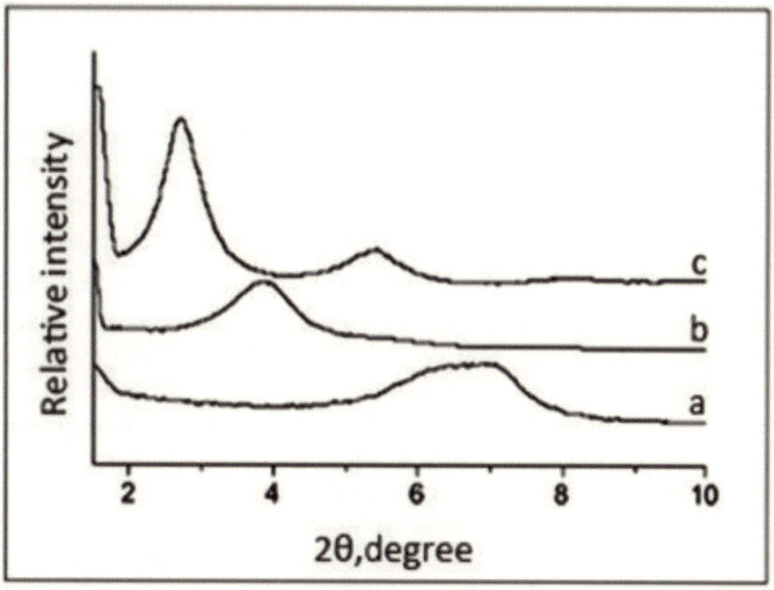

Fig. 4. XRD patterns of a) pristine sodium montmorillonite b) modified montmorillonite (OMT) c) ABS/OMT nanocomposite. ref:19

4.3 Transmission electron microscopy

Transmission electron microscopy (TEM) is widely employed, in its simplest bright-field mode, as a tool for direct visualization of the nanocomposite structure of polymer nanocomposites. This is possible because there exists sufficient contrast for the transmitted electrons between the polymer matrix and most filler (inorganic material such as metal oxides). In the extreme case, high-resolution TEM can even provide a qualitative picture of the inorganic filler crystal structure [20]. One must remember that to obtain one TEM requires only a miniscule piece of material and one cannot be certain that this is representative of the whole. The morphology can only be clearly determined by either sampling enough of the material [14]. Fig 5a shows TEM image of PMMA/TiO2 and TEM image of ABS/clay is shown in Fig 5b [19, 21].

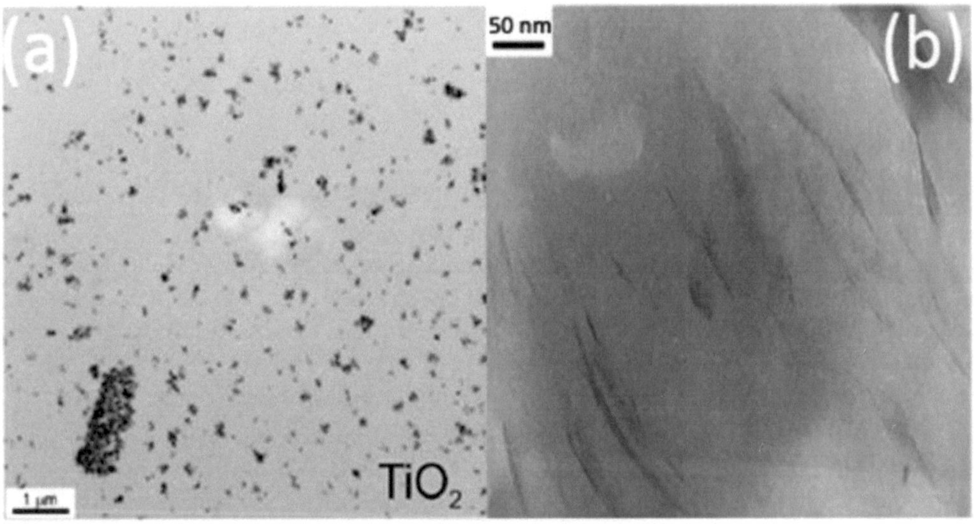

Fig. 5. TEM images of a) PMMA/TiO_2 nanocomposite b) ABS/OMMT nanocomposite ref:19,21

5. Flame retardancy

Flammability of polymers is one of their unpleasant properties that most often accompanied by the production of corrosive or toxic gases and smoke during combustion. Consequently, improving the fire retardant behavior of polymers is a major challenge for extending their use to most applications Flame retardation is a process by which the normal degradation or combustion of polymers is altered by addition of certain chemicals. Various flame retardant additives, such as halogenated additives, are being phased out for their proven or suspected adverse effects on the environment. The combined challenge thus consists in developing effective and environmentally friendly flame retardant systems for polymer materials. Most researches have been done for improving thermal stability by addition of nanoparticles [20, 22]. Any type of inorganic filler, even inert, can influence the reaction of polymers to fire for several reasons:

1. Reduction of content of combustible products
2. Changing the viscosity of the resulting material
3. Modification the thermal conductibility of the resulting material [22].

All these actions have an indirect incidence on the polymer's fire performance. Nevertheless, some minerals are more specifically used as flame retardants owing to their behavior at high temperature [22].

6. Some of laboratory fire testing

6.1 ThermoGravimetric Analysis (TGA)

TGA is an experimental technique in which the weight of a sample is measured as a function of sample temperature or time. The sample is typically heated at a constant heating rate. The results of a TGA measurement are usually displayed as a TGA curve in which percent weight is plotted against temperature. Different effects can cause a sample to lose, or even gain, mass, these include the following:

1. Evaporation of volatile constituents; such as desorption of gases and moisture
2. Thermal decomposition in an inert atmosphere with the formation of gaseous products
3. Oxidation of constituents[23].

TGA are often equipped with DTA (differential thermal analysis). In addition to showing the energetic nature of weight loss events, the DTA signal can also show thermal effects that are not accompanied by a change in mass, e.g. melting, crystallization or a glass transition [23].

6.2 Limited Oxygen Index: LOI

LOI is one of the most important and representative parameter for describing the flame-retardant properties of plastics because of it is relatively simple method. The measurement consists of determining the minimum concentration of oxygen in a nitrogen-oxygen mixture that will sustain combustion. The higher LOI better the flame retardant property. This does not mean that a material with a high oxygen index will not burn; the test measures the ease of extinction of the fire [22].

The value of the LOI is defined as the minimal oxygen concentration [O_2] in the oxygen/nitrogen mixture [O_2/N_2] ,As air contains 21% oxygen, materials with an LOI below 21 are classified as "combustible" whereas those with an LOI above 21 are classified as "self-extinguishing", because their combustion cannot be sustained at ambient temperature without an external energy contribution [22].

6.3 Differential scanning calorimetry: DSC

DSC is the most widely used of the thermal techniques available to the analyst and provides a fast and simple method for obtaining beneficial information about a material. A DSC analyzer measures the energy changes that occur as a sample is heated, cooled or held isothermally. The energy changes enable the user to find and measure the transitions that occur in the sample quantitatively such as measurement of glass transitions [23].

6.4 Cone calorimeter

Cone calorimeter is one of the most effective medium-sized polymer fire behavior tests. The cone calorimeter measures the rate at which heat is released in a fire. This instrument calculate the quantity of heat released per unit of time and surface area: HRR (heat release rate) expressed in kW/m2. The evolution of the HRR over time, in particular the value of its peak (pHRR), is usually taken into account in order to evaluate the fire properties [22].

Currently, numerous fire retardant fillers for the preparation of polymer nanocomposites have been proposed, some famous flame retardant are:

7. Polymer/metal sulfide nanocomposite

With incorporation of a variety of nanoparticles in the polymer matrices, nanocomposites are much different from the conventional composites. Exfoliated metal sulfide filler particles

have significant barrier effect to slow down product volatilization and thermal transport during decomposition of the polymer, which assists composites with high thermal stability. For example influence of the content of cadmium sulfide (CdS) on the thermal degradation of poly styrene/ cadmium sulfide composites were examined [24]. TG curves (Fig 6), shows that thermal stability of PS/CdS composites is higher than in pure PS. Improved thermal stability of composites with respect to the pure PS can be assigned to partially altered molecular mobility of the polymer chains due to their adsorption on the surface of the filler particles. Simultaneously, adsorption of polymer chains onto the surface of CdS filler particles results in restriction of segmental mobility and serves to suppress chain transfer reactions .Polymer adsorption on the surface of filler particles is a crucial step for explanation of changed thermal stability of polymer/metal sulfide composites [25, 26].
Much effort has been made to develop novel methods to prepare nanostructured metal chalcogenide. Hydrothermal method and thermal decomposition provides a more promising way for the synthesis of crystals due to its low cost, high efficiency and potential for large-scale production [27-34].

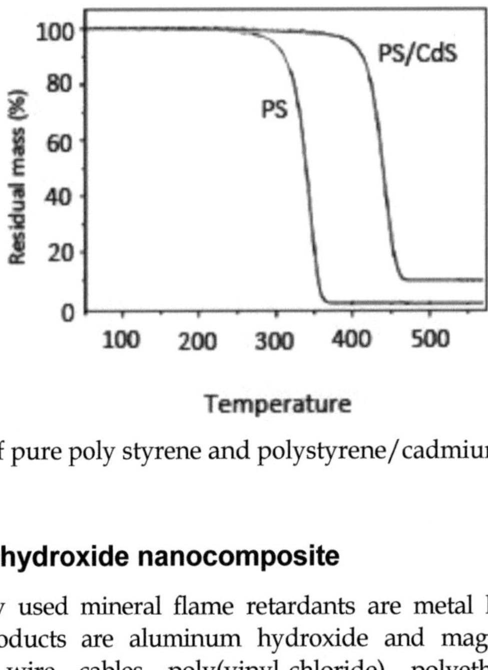

Fig. 6. TGA curves of pure poly styrene and polystyrene/cadmium sulfide nanocomposite from ref:24

8. Polymer/metal hydroxide nanocomposite

The most commonly used mineral flame retardants are metal hydroxides (The two most commonly used products are aluminum hydroxide and magnesium hydroxide) .These material are used in wire , cables , poly(vinyl chloride) , polyethylene, various elastomers , phenolics, epoxies, and unsaturated polyesters. As the temperature rises, aluminum hydroxide decomposes endothermically and therefore absorbs energy. Moreover, they release non-flammable molecules (H_2O, CO_2), which dilute combustible gases upon heating at about 220°C with an endotherm of 1.17 kJ/g, whereas MH starts releasing water at about 330°Cwith an endotherm of 1.356 kJ/g [20]. Since both anhydrous alumina and magnesia are white highly refractory powders, they provide heat insulation by reflecting heat when they accumulate on a surface these flame retardants can also promote the formation of a protective ceramic or vitreous layer. To be used as polymer flame retardants, metal hydroxides need to decompose endothermically and release water at a temperature higher than the polymer processing temperature range, and around the polymer decomposition temperature [20].

The fire properties of ATH-filled polymers are only interesting at high loading levels. To pass flame retardancy tests, 35 to 65 wt% of metal hydroxide is required [20]. Increasing the loading of metal hydroxides will result in a significant loss in physical properties; therefore, the use of nanometric MDH and ATH can enable interesting fire performances to be achieved at lower loading levels. The enhancement of flame retardancy by nanosized $Mg(OH)_2$ was attributed to the good dispersion of the nanoparticles, which leads to the formation of more compact and cohesive char during the combustion test [35].

9. Polymer/metal and metal oxide nanocomposite

Filler particles have significant barrier effect to slow down product volatilization during decomposition of the polymer, which assists composites with high thermal stability. Adsorption of polymer chains onto the surface of filler particles results in restriction of segmental mobility and serves to suppress redistribution and chain transfer reactions [24].
Beside current famous flame retardant additives, many researches have been done about effect of metal oxide and metal nanoparticle to thermal property of polymers. In recent years, synthesis of transition metal nanoparticles is a growing research field in chemical science [36-39]. Metal particles have attracted considerable attention because of application in modern technologies. Metal oxide nanoparticles can be obtained by several methods, via a sol-gel technique a hydrothermal, sonochemistry and co-precipitation method [40-46].
For examples effectiveness of several type of metal oxide such as silica, ferric oxide, titanium oxide and copper oxide for improve thermal property of polymer matrix were surveyed [47-49]. For example influence of the content of ferric oxide and titanium oxide on the thermal degradation of poly methyl methaacrylate was examined. TGA curves, shows that thermal stability of nanocomposites is higher than in pure polymer (Fig7) [21]. Improved thermal stability of composites can be assigned to partially alter molecular mobility of the polymer chains due to their adsorption on the surface of the filler particles [24].

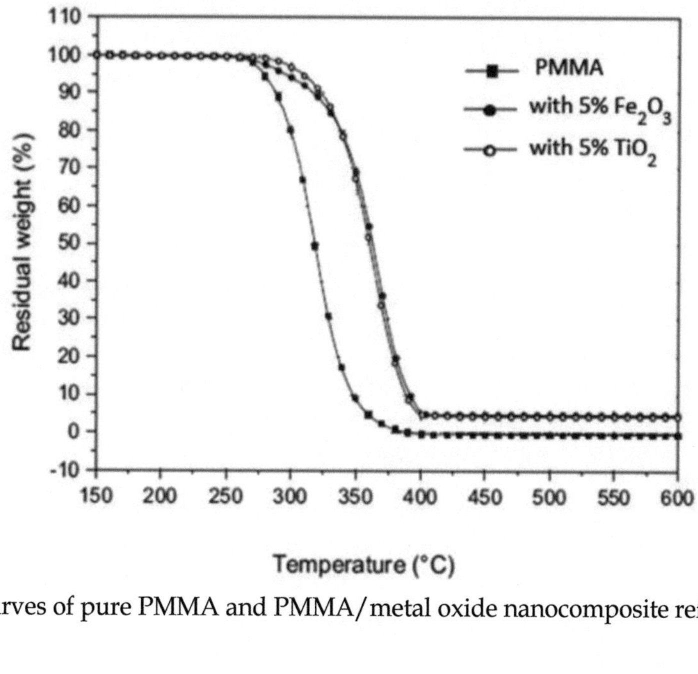

Fig. 7. TGA curves of pure PMMA and PMMA/metal oxide nanocomposite ref: 21

10. Polymer/layered silicate nanocomposite

The formation of well-dispersed polymer nanocomposites leads to modification of the degradation pathway of polymers and this clearly affects the fire behavior of nanocomposites. The currently accepted explanation for enhanced fire properties is that the clay forms a barrier to mass transport and which also insulates the underlying polymer from the heat source (Fig8) [20].

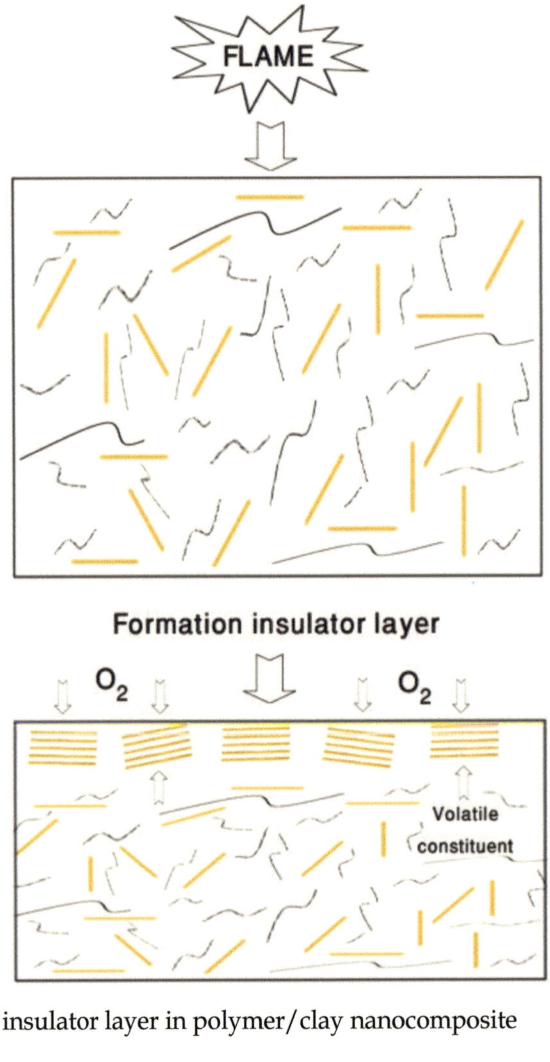

Fig. 8. Formation of insulator layer in polymer/clay nanocomposite

10.1 Crystal structures of some major clay minerals

10.1.1 Kaolinite

Kaolinite is called a 1:1 type of layered mineral that is composed of a tetrahedral silica sheet and an octahedral alumina sheet. The tetrahedral and octahedral sheets are connected with shared oxygen atoms (Fig 9). The charge within the layers is nearly balanced, and the chemical formula is $Al_4[Si_4O_{10}](OH)_8$ or $2Al_2O_3.4SiO_2.4H_2O$ [50].

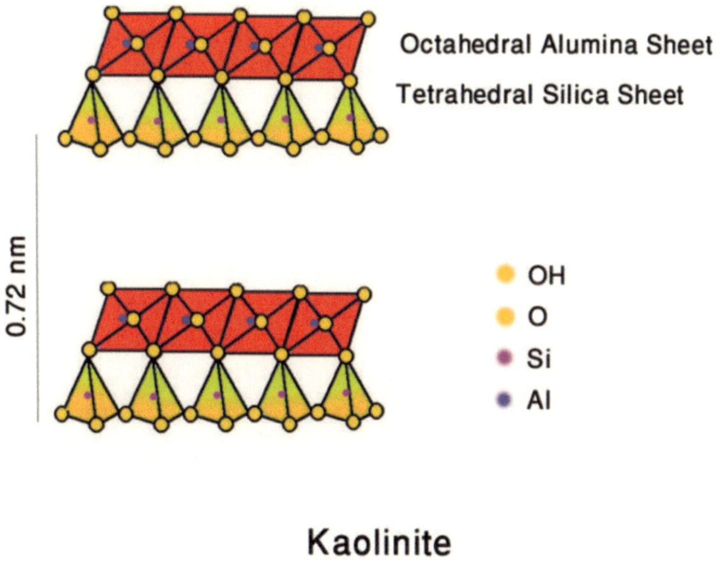

Fig. 9. Structure of kaolinite

10.1.2 Pyrophyllite

The formula for pyrophyllite is $Al_4[Si_4O_{10}](OH)_2$. Pyrophyllite has a 2:1 layer structure consisting of two fused silica tetrahedral sheets sandwiching an edge-shared octahedral sheet of alumina (Fig 10) [50].

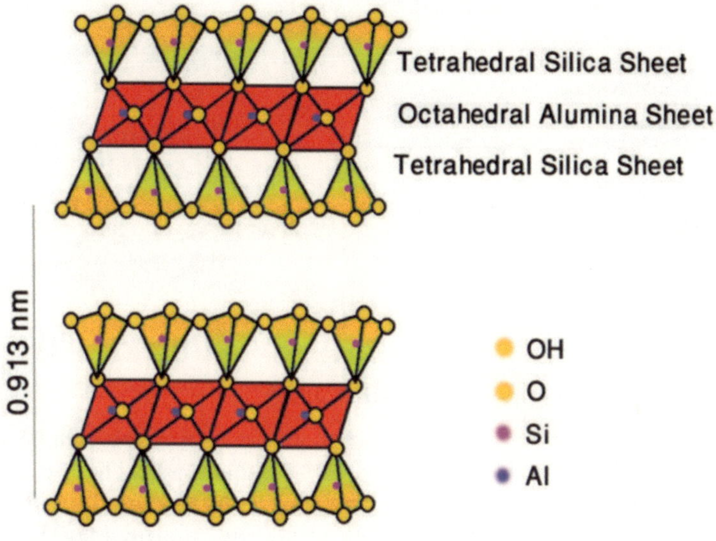

Fig. 10. Structure of pyrophilite

10.1.3 Montmorillonite

The most promising candidate of the layered silicates is montmorillonite (MMT), owing to its natural abundance and high aspect ratio [51, 52]. MMT is a derivative of pyrophyllite.

The difference between the crystal structure of MMT and pyrophyllite is that the latter is neutral, while the former has layer charges due to isomorphic substitution. Substitutions occur in the octahedral sheet with Mg2+and Fe2+for Al3+ (Fig 11) [50].

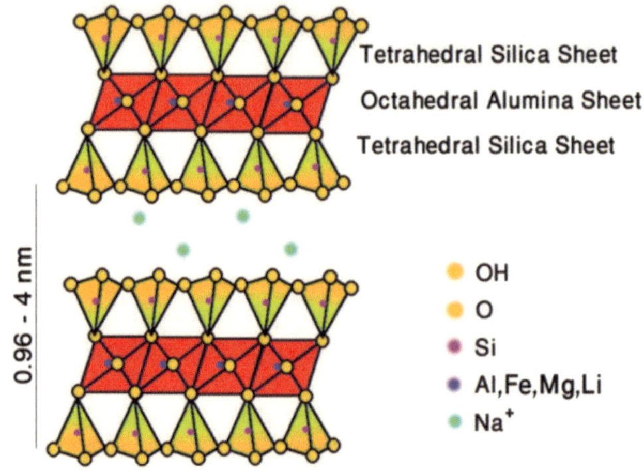

Fig. 11. Structure of sodium montmorillonite.

10.2 Organic clay modification

To make the silicate compatible with the polymer, the clay mineral is modified with an alkyl ammonium salt by cation exchange. Attractive interaction between the surfactant and the polymer greatly enhances the possibility of nano-dispersion of the clay within the polymer. Modification of Na^+ MMT with CTAB is shown in Fig 12.

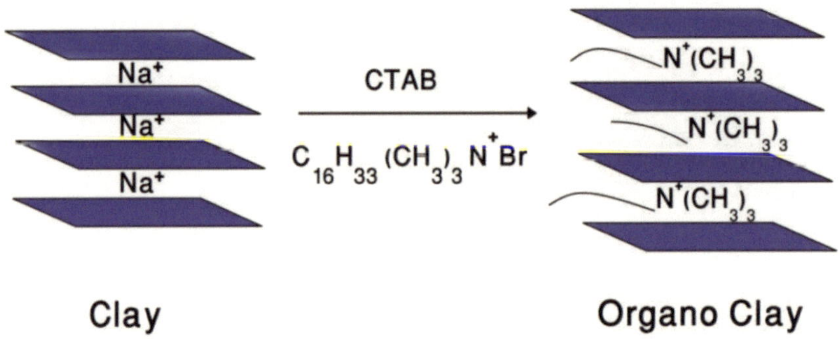

Fig. 12. Modification of Na^+ montmorillonite with CTAB

If the polymer is unable to intercalate between the silicate sheets, the obtained composite can be considered as phase-separated composite. Beyond this classical family of composites when extended polymer chain is inserted in a crystallographically regular fashion between the silicate layers resulting in a well-ordered multilayer morphology built up with alternating polymeric and inorganic layers, the obtained composites can be called as 'intercalated nanocomposites. When the individual silicate layers are completely and uniformly dispersed in a continuous polymer matrix, an exfoliated or delaminated structure is obtained [53]. Different kind of interaction between polymeric chains and clay are shown in Fig 13 .Usually, the clay content of an exfoliated nanocomposite is much lower than that of an intercalated nanocomposite. In the majority of the polymer nanocomposites formed, a combination of delamination and intercalation is obtained [54].

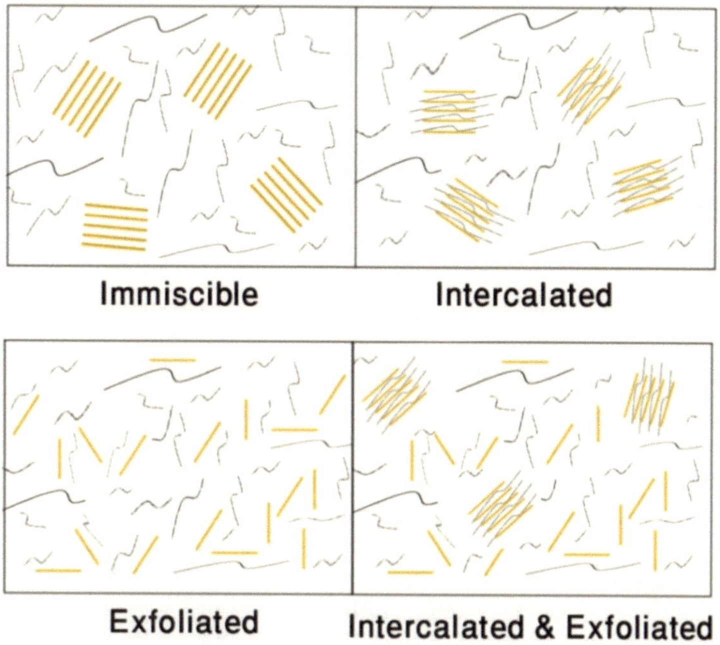

Fig. 13. Different kind of interaction between polymer and clay

11. Polymer/layered double hydroxide nanocomposite

Hydrotalcite compounds, known as layered double hydroxides (LDHs), belong to a class of anionic clay minerals. They have the general molecular formula [M2+1-x M3+x (OH)2]x+A n- x/n.mH2O,where M2+and M3+ are divalent and trivalent metal cations, such as Mg2+, Al3+, respectively, An- is an intercalated anion, such as CO32-, SO42-, and NO3-[55].
The majority of important applications are based on their permanent anion-exchange and adsorption capacity, the mobility of interlayer anions and water molecules, their large surface areas, and the stability and homogeneity of the materials formed by their thermal decomposition. Because of their highly tunable properties, LDHs are considered as a new emerging class of the most favorable layered crystals for preparing the multifunctional polymer/layered crystal nanocomposites [55].
When LDHs were added to a polymer matrix, the LDH releases the CO_2 gas and moisture, diluting the concentration of burning gas. In addition, MgO and Al_2O_3 act as layers of heat insulation. The layer structure of LDH shows in Fig 14.

12. Halogen-containing flame retardant

For explanation of correlation between halogenated flame retardant and polymer matrix nanocomposite firstly we should explain synergistic effect concept:
A synergistic effect occurs when the combined effects of two chemicals are much greater than the sum of the effects of each agent given alone. Halogen-containing flame retardants

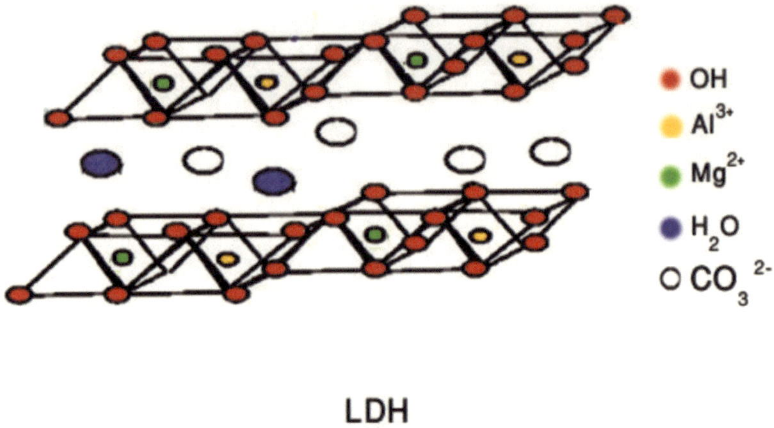

Fig. 14. Structure of layered double hydroxide

release halogen in the form of radical of halogen at temperature of decomposition of the polymer. Chlorine and bromine can be used as halogenated flame retardant. These halogens, because of their low bonding energy with carbon atoms, can readily be released and take part in the combustion process [20].

Very reactive free-radical species such as H and OH are very important for sustaining combustion. The hydrogen radical is responsible for the chain-branching free-radical reactions in the flame whereas the hydroxyl radical is responsible for the oxidation of CO to CO_2, which is a highly exothermic reaction and is responsible for the larger part of the heat generation in the flame. The physical effect of halogen halides is comparable to that of inert gases, CO_2, and water[20].

Sb_2O_3 is synergistic with halogen-containing flame retardants because it facilitates delivery of halogen atoms in the gas phase and prolongs residence of the halogens in the flame zone so that more "hot" radicals can be scavenged. Antimony trioxide reacts with halogen containing compound, forming $SbCl_3$ or $SbBr_3$ [20].

The use of nanometric Sb_2O_3 can enable interesting fire performances to be achieved at lower loading levels. The enhancement of flame retardancy by nanosized Sb_2O_3 was attributed to the good dispersion of the nanoparticles, which leads to the formation of more compact and effective char during the combustion test.

13. Polymer/carbon nano tube nanocomposite

The investigation of carbon nanotubes (CNT) has opened a totally new window for the development of polymer matrix composites with novel properties and applications. The most widely studied nanofibrous materials with respect to polymer flame retardancy are carbon nanotubes (CNTs).The extremely low diameter of CNT and their high aspect ratio make them an ideal material to improve the properties of the polymer matrix. CNTs are an interesting alternative to the use of conventional flame retardants and nanoclays. Their incorporation at low loading rate (<3 wt %) has been reported to improve the flammability of a large range of polymers [5, 23]. For example TGA curve related to PMMA/CNT nanocomposite is shown in Fig 15. Initial degradation onset of pure PMMA and PMMA-SWNT are at 322 and 359 °C, respectively. So, relatively a small amount (0.5 wt %) of SWNT thermally stabilized the PMMA by around 37 °C [56].

13.1 Treatment of carbon nanotubes

Functionalization of carbon nanotubes is an effective way to improve interaction between CNTs and polymer matrix. Treatment has been performed by oxidation using a mixture of sulfuric and nitric acid .The oxidized carbon nanotubes were then functionalized with different types of agents such as amines or $SOCl_2$ (Fig 16).The functionalization has been performed by refluxing a suspension of oxidized CNTs with the amine to ensure a quantitative conversion [57-61].

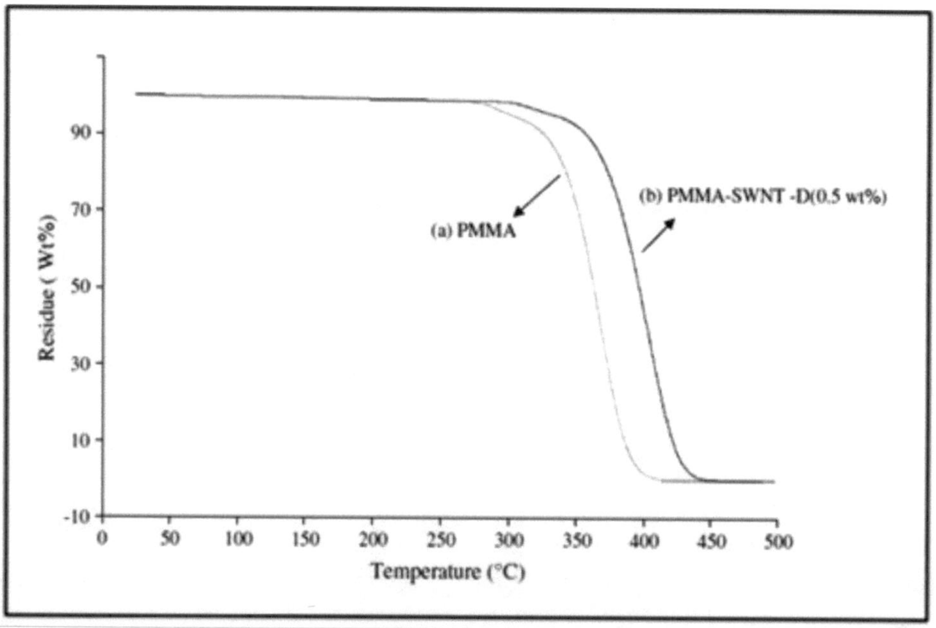

Fig. 15. TGA curves of a)PMMA b)PMMA/CNT nanocomposite ref:56

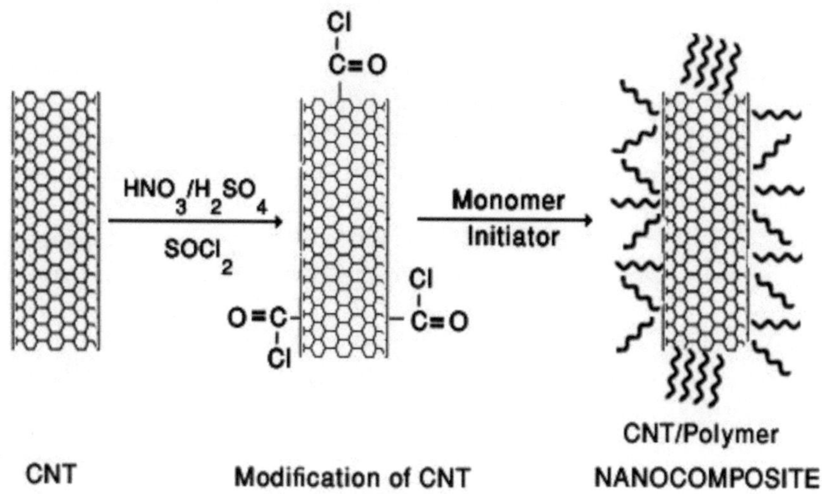

Fig. 16. Modification of carbon nanotube with organig agents

14. References

[1] Mouritz AP, Gibson AG. Fire Properties of Polymer Composite Materials Pub Springer, 2006

[2] Hull TR, Kandola BK. Fire Retardancy of Polymers New Strategies and Mechanisms RSC 2009

[3] Horrocks AR, Price D, Fire retardant materials, Woodhead Publishing ISBN 1 85573 419 2

[4] Ratna D Epoxy Composites: Impact Resistance and Flame Retardancy Rapra Review Reports Volume 16, Number 5, 2005

[5] Friedrich K Fakirov S Zhang Z. Polymer Composite From Nano- to Macro-Scale 2005 Springer

[6] Salavati-Niasari , Davar F, Fereshteh Z. Synthesis and characterization of ZnO nanocrystals from thermolysis of new precursorChemical Engineering Journal 2009;146: 498–502

[7] Salavati-Niasari M, Dadkhah M, Davar F. Pure cubic ZrO_2 nanoparticles by thermolysis of a new precursor. Polyhedron. 2009;28 3005–3009

[8] Salavati-Niasari M, Dadkhah M, Davar F. Synthesis and characterization of pure cubic zirconium oxide nanocrystals by decomposition of bis-aqua, tris-acetylacetonato zirconium(IV) nitrate as new precursor complex. Inorganica Chimica Acta. 2009; 362: 3969–3974

[9] Salavati-Niasari M, Bazarganipour M, Davar F. Hydrothermal synthesis and characterization of bismuth selenide nanorods via a co-reduction route. Inorganica Chimica Acta

[10] Salavati-Niasari M. Template synthesis and characterization of hexaaza macrocycles containing pyridine iron(II) complex nanoparticles dispersed within nanoreactors of zeolite-Y Inorganic Chemistry Communications 2009;12: 359–363

[11] Salavati-Niasari M .Synthesis and characterization of 18- and 20-membered hexaaza macrocycles containing pyridine manganese(II) complex nanoparticles dispersed within nanoreactors of zeolite-Y Polyhedron 2009;28: 2321–2328

[12] Salavati-Niasari M, Shaterian M, Ganjali M R, Norouzi P.Oxidation of cyclohexene with tert-butylhydroperoxide catalysted by host (nanocavity of zeolite-Y)/guest (Mn(II), Co(II), Ni(II) and Cu(II) complexes of N,N_-bis(salicylidene)phenylene-1,3-diamine)nanocomposite materials (HGNM). Journal of Molecular Catalysis A: Chemical. 2007; 261: 147–155

[13] Khanna P.K, Lonkar S.P , Subbarao V.V.V.S , Jun K.W. Polyaniline–CdS nanocomposite from organometallic cadmium precursor Materials Chemistry and Physics 2004;87: 49–52

[14] Bras ML, Wilkie CA, Bourbigot S. Fire Retardancy of Polymers.RSC 2005

[15] Kothurkar N.K. Solid state, transparent, cadmium sulfide-polymer nanocomposite, University of Florida 2004

[16] Kim S, Marand E. High permeability nano-composite membranes based on mesoporous MCM-41 nanoparticles in a polysulfone matrix Microporous and Mesoporous Materials 2008;114: 129–136

[17] Salavati-Niasari M, Ghanbari D, Davar F. Synthesis of different morphologies of bismuth sulfide nanostructures via hydrothermal process in the presence of thioglycolic acid J. Alloy. Compd. 2009; 488: 442–447

[18] Salavati-Niasari M, Ghanbari D, Davar F. Shape selective hydrothermal synthesis of tin sulfide nanoflowers based on nanosheets in the presence of thioglycolic acid. J. Alloy. Compd. 2010; 492:570–575

[19] Wang S, Hua Y, Zonga R, Tanga Y, Chenb Z, Fana W. Preparation and characterization of flame retardant ABS/montmorillonite nanocomposite. Applied Clay Science 2004; 25: 49–55

[20] Morgan A B., Wilkie C A. Flame Retardant Polymer nanocomposite WileyPublishing/ CRC Press;. .2007

[21] Laachachi A, Leroy E, Cochez M, Ferriol M, Lopez Cuesta J.M Use of oxide nanoparticles and organoclays to improve thermal stability and fire retardancy of poly(methyl methacrylate). Polym Degrad Stab. 2005;89: 344-352

[22] Laoutid F , Bonnaud L , Alexandre M , Lopez-Cuesta J.-M , Dubois Ph. New prospects in flame retardant polymer materials:from fundamentals to nanocomposites. Materials Science and Engineering R 63 (2009) 100–125

[23] Gabbott P. Principles and Applications of Thermal Analysis 2008 Blackwell Publishing

[24] Kuljanin J , Marinovic´-Cincovic´M , Stojanovic´Z , Krkljesˇ A, Abazovic´ND , Comor MI, Thermal degradation kinetics of polystyrene/cadmium sulfide composites. Polym Degrad Stab 2009; 94:891–897

[25] Kuljanin J, Vucˇkovic´ M, Comor MI, Bibic´ N, Djokovic´ V, Nedeljkovic´ JM. Influence of CdS-filler on the thermal properties of polystyrene. Eur Polym J. 2002; 38:1659–1662.

[26] Kuljanin J, Comor MI, Djokovic V´, Nedeljkovic JM. Synthesis and characterization of nanocomposite of polyvinyl alcohol and lead sulfide nanoparticles. Materials Chemistry and Physics 2006:95: 67–71

[27] Salavati-Niasari M, Loghman-Estarki M.R, Davar F. Synthesis, thermal stability and photoluminescence of new cadmium sulfide/organic composite hollow sphere nanostructures. Inorg Chim Acta. 2009; 362:3677-3683

[28] Salavati-Niasari M, Loghman-Estarki MR, Davar F. Controllable synthesis of nanocrystalline CdS with different morphologies by hydrothermal process in the presence of thioglycolic acid Chem. Eng. J. 2008;145: 346–350.

[29] Salavati-Niasari M, Davar F, Loghman-Estarki MR. Long chain polymer assisted synthesis of flower-like cadmium sulfide nanorods via hydrothermal process. J. Alloys Compd. 2009; 481:776–780

[30] Salavati-Niasari M, Loghman-Estarki MR, Davar F. Controllable synthesis of wurtzite ZnS nanorods through simple hydrothermal method in the presence of thioglycolic acid. J. Alloys Compd. 2009; 475:782–788.

[31] Salavati-Niasari M, Davar F, Loghman-Estarki MR. Controllable synthesis of thioglycolic acid capped $ZnS(Pn)_{0.5}$ nanotubes via simple aqueous solution route at low temperatures and conversion to wurtzite ZnS nanorods via thermal decompose of precursor. J.Alloy.Compd *2010*; 494:*199-204*

[32] Davar F, Salavati-Niasari M, Mazaheri M. Thermal decomposition of [bis(salicylaldehydato)cadmium(II)] to CdS nanocrystals. Polyhedron. 2009; 28: 3975-3978.

[33] Salavati-Niasari M, Davar F, Mazaheri M. Synthesis, characterization and magnetic properties of NiS_{1+x} nanocrystals from [bis(salicylidene)nickel(II)] as new precursor. Mat ResBulletin.2009; 44:2246-2251

[34] Salavati-Niasari M, Davar F, Mazaheri M. Synthesis and characterization of ZnS nanoclusters via hydrothermal processing from [bis(salicylidene)zinc(II)]. J. Alloys Compd. 2009;470: 502–506.

[35] L Qiu, R Xie, P Ding, B Qu. Preparation and characterization of $Mg(OH)_2$ nanoparticles and flame-retardant property of its nanocomposites with EVA. Composite Structures 2003; 62:391–395.

[36] Salavati-Niasari M, Mir N, Davar F. A novel precursor in preparation and characterization of nickel oxide nanoparticles via thermal decomposition approach. J Alloys Compd 2010; 493:163–168

[37] Salavati-Niasari M , Mir N , Davar F. A novel precursor for synthesis of metallic copper nanocrystals by thermal decomposition approach. Applied Surface Science. 2010; 256:4003–4008.

[38] Salavati-Niasari M, Davar F, Fereshteh Z .Synthesis of nickel and nickel oxide nanoparticles via heat-treatment of simple octanoate precursor. J Alloys Compd. 2010; 494: 410–414

[39] Mohandes F,Davar F, Salavati-Niasari M. Preparation of Co_3O_4 nanoparticles by nonhydrolytic thermolysis of $[Co(Pht)(H_2O)]n$ polymers. J Magnetism Magnetic Materials. 2010;322: 872–877

[40] Salavati-Niasari M , Javidi J , Davar F. Sonochemical synthesis of $Dy_2(CO3)_3$ nanoparticles, $Dy(OH)_3$ nanotubes and their conversion to Dy_2O_3 nanoparticles Ultrasonics Sonochemistry. 2010; 17: 870–877

[41] Salavati-Niasari , N Mir , F Dava.r Synthesis, characterization and optical properties of tin oxide nanoclusters prepared from a novel precursor via thermal decomposition route. Inorganica Chimica Acta 2010; 363: 1719–1726

[42] Davar F, Salavati-Niasari M, Fereshteh Z. Synthesis and characterization of SnO_2 nanoparticles by thermal decomposition of new inorganic precursor. J Alloy Compd. 2010;496: 638–643

[43] Davar F, Salavati-Niasari M , Mir N , Saberyan M, Monemzadeh M , Ahmadi E .Thermal decomposition route for synthesis of Mn_3O_4 nanoparticles in presence of a novel precursor. Polyhedron (2010; 29: 1747–1753

[44] Salavati-Niasari M, Javidi J, Davar F, Amini Fazl A Sonochemical synthesis of $Dy_2(CO_3)_3$ nanoparticles and their conversion to Dy_2O_3 and $Dy(OH)_3$: Effects of synthesis parameters. J Alloy Compd .2010; 503: 500-506

[45] Mohandes F, Davar F, Salavati-Niasari M. Magnesium oxide nanocrystals via thermal decomposition of magnesium oxalate Journal of Physics and Chemistry of Solids

[46] Salavati-Niasari M, Hosseinzadeh G , Davar F Synthesis of lanthanum hydroxide and lanthanum oxide nanoparticles by sonochemical method. J Alloy Compd. 2010; 503: 500-506

[47] Grimes SM, Lateef H , Jafari A J , Mehta L. Studies of the effects of copper, copper(II) oxide and copper(II) chloride on the thermal degradation of poly(vinyl chloride). Polym Degrad Stab 2006;91: 3274-3280

[48] Guo Z, Liang X, Pereira T, Scaffaro R, Hahn H.T. CuO nanoparticle filled vinyl-ester resin nanocomposites: Fabrication, characterization and property analysis. Composites Science and Technology 2007;67: 2036–2044

[49] Tarrıo-Saavedra J, Lo pez-Beceiro J, Naya S, Artiaga R. Effect of silica content on thermal stability of fumed silica/epoxy composites. Polym Degrad Stab. 2008;93: 2133–2137

[50] Ke YC, Stroeve P. Polymer-Layered Silicate and Silica Nanocomposites. 2005 Elsevier

[51] Salavati-Niasari M Synthesis, characterization and catalytic oxidation of cyclohexene with molecular oxygen over host (montmorillonite-K10)/guest (nickel(II) complexes of 12- and 13-membered diaza dioxa Schiff-base macrocyclic ligand) nanocatalyst (HGN) J Molecular Catalysis A: Chemical 2007;263: 247–252

[52] Salavati-Niasari M, Zamani E, Bazarganipour M. Epoxidation of cyclohexene with K10-montmorillonite and Schiff-base macrocyclic copper complexes Applied Clay Science 2007;38: 9–16

[53] Kumara A P, Depana D, Tomerb N S, Singha R P. Nanoscale particles for polymer degradation and stabilization—Trends and future perspectives. Progress in Polymer Science. 2009; 34: 479–515

[54] Wang S, Hu Y, Zong R, Tang Y, Chen Z, Fan W. Preparation and characterization of flame retardant ABS/montmorillonite nanocomposite. Applied Clay Science 2004;25: 49– 55

[55] Du L, Qu B. Effects of Synthesis Conditions on Crystal Morphological Structures and Thermal Degradation Behavior of Hydrotalcites and Flame Retardant and Mechanical Properties of EVA/Hydrotalcite Blends. Polymer Compsites. 2007 131-138

[56] Chowdhury S R, Chen Y, Wang Y, Mitra S .Microwave-induced rapid nanocomposite synthesis using dispersed single-wall carbon nanotubes as the nuclei. J Mater Sci .2009; 44:1245–1250

[57] pitalskya ZS, Tasisb D, Papagelisb K, Galiotis C. Carbon nanotube–polymer composites: Chemistry, processing, mechanical and electrical properties. Progress in Polymer Science. 2010; 35: 357–401

[58] Salavati-Niasari M, Mirsattari SN, Bazarganipour M. Synthesis, characterization and catalytic oxyfunctionalization of cyclohexene with tert-butylhydroperoxide over a manganese(II) complex covalently anchored to multi-wall carbon nanotubes (MWNTs). Polyhedron. 2008; 27: 3653–3661

[59] Salavati-Niasari M, Bazarganipour M. Covalent functionalization of multi-wall carbon nanotubes (MWNTs) by nickel(II) Schiff-base complex: Synthesis, characterization and liquid phase oxidation of phenol with hydrogen peroxide. Applied Surface Science. 2008; 255: 2963–2970

[60] Salavati-Niasari M, Bazarganipour M. Synthesis, characterization and catalytic oxidation properties of multi-wall carbon nanotubes with a covalently attached copper(II) salen complex. Applied Surface Science 2009;255: 7610–7617

[61] Salavati-Niasari M, Bazarganipour M. Effect of single-wall carbon nanotubes on direct epoxidation of cyclohexene catalyzed by new derivatives of *cis*-dioxomolybdenum(VI) complexes with bis-bidentate Schiff-base containing aromatic nitrogen–nitrogen linkers J. Molecular Catalysis A: Chemical 278 2007;278: 173–180

23

Electronic Functionality of Nanocomposites

Pandiyan Murugaraj and David Mainwaring
School of Applied Sciences, College of Science, Engineering and Health, RMIT University
Australia

1. Introduction

Functional polymer nanocomposites have the potential to improve the performance of both active and passive components in advanced electrical devices. Electrically conducting nanocomposites are a growing research focus for such applications in light weight flexible electronics particularly in strategic infrastructure. Such materials also open up a broad range of fabrication techniques which can be easily used in mass manufacturing via imprinting to yield low-cost devices. Here, the mechanisms of electron transport in these polymer nanocomposites are addressed in terms of modifications to existing theories of variable range electron hopping and fluctuation induced tunnelling originally applied to heavily doped semiconductors and granular metal composites. The fabrication and behaviour of two types of nanocomposites that involve dispersed carbon nanoparticles in a polymide medium and carbon nanocluser composites formed by ion irradiation of polyimide are compared through their conduction mechanisms and sensing performance.

2. Electron transport phenomena

2.1 Electron transport mechanisms

Electron transport phenomena in 3-D composite structures remains an area of significant research interest. Earlier work was predominately carried out on granular materials particularly discontinuous metal thin films on dielectric substrates. With increasing interest in electrically functional polymer composites, the existing theoretical approaches are being re-addressed to accommodate transport behaviour these materials. Mott (1969) originally addressed disordered systems where the electronic states are localised near the Fermi level and the electric conduction does not follow classic processes of diffusion but rather a variable range hopping (VRH) mechanism. According to Mott's theory for a 3D system, the temperature dependent conductivity (σ) is governed by the expression:

$$\sigma = \sigma_o \exp\{-(T_o/T)^{1/4}\} \qquad (1)$$

where σ_o and T_o are constants and T the absolute temperature. Here, the density of energy states was considered constant while the interaction between these charge states was not considered (Mott, 1969, Efros & Shklovskii, 1975). Subsequent application of Mott's theory to various systems such as amorphous semiconductors, heavily doped semiconductors, cermets and polymer composites have shown a clear deviation from the $T^{1/4}$ temperature dependent conductivity, even though these systems form spatially distributed localised energy states arising from defects and impurities as envisaged in Mott's original expression (Sheng & Klafter, 1983).

Even though the hopping mechanism is considered to be dominant in electron transport in such disordered materials and conducting composites, the temperature dependence of conductivity is now considered to follow a more generalised Arrhenius relationship $\sigma=\sigma_0 exp\{-(T_0/T)^{1/\gamma}\}$ having the exponent γ ranging from 1 to 4 thus indicating a more complex behaviour involving more than one type of thermally activated conduction mechanism (Sheng & Klafter, 1983, Adkins, 1987). These deviations may arise from the two assumptions above imposed in the Mott's theory as well as from inhomogenous spatial distributions of charge states, particles sizes and morphologies (Devenyi et al., 1972, Neugebauer & Webb, 1962). Additionally, systems in which the electrons are delocalised over large atomic distances such as structured carbon black composites, large metal island films and high aspect ratio structures, have electron transport dominated by large conducting segments (aggregates) rather than by hopping between well localised states (Sheng et al., 1978).

Using a similar approach to Mott, Efros & Shklovskii (Efros, 1976, Efros & Shklovskii, 1975) (ES) introduced a Coulomb gap into the description of the density of states near the Fermi level to account for the electron-electron interactions which diminish the density of states in this region. It was assumed here that the quantum localisation length was much smaller than the distance between the centres of the charged particles in these granular systems resulting in negligible overlap of the wavefunctions. According to Efros' formulation, the Coulomb gap (Δ) plays an important role in the low temperature conductivity where the energy interval responsible for hopping (ε_M) is comparable or greater than the Coulomb gap ($\varepsilon_M < \Delta$) in Mott's expression:

$$\varepsilon_M = T^{3/4} / (\alpha^3 g_o^{1/4}) \qquad (2)$$

here α is the localisation length and g_o is the density of states at the Fermi level. Efros and Shklovski argued that the Coulomb interaction causes the appearance of a gap within the density of states which when incorporated into Mott's theory leads to a $T^{1/2}$ dependence. Thus according to this modified theory, $T^{1/2}$ behaviour should be seen at low temperatures and $T^{1/4}$ behaviour must be observed at higher temperatures, with a transition occurring at T_{ES}. Chui et al. (1981) applied this theory to high resistivity granular samples and found that its temperature dependence was not entirely consistent and concluded that the low density of states in these systems caused such a discrepancy. Entin-Wohlman et al. (1983) subsequently analysed a similar system theoretically accounting for the roles of the metal-insulator transition as well as the diverging dielectric permittivity near this transition and found it qualitatively agreed with the ES theory. Shekhar et al. (2006a) studying the temperature dependent electrical conductivity of an iron carbide/polymer composite system identified an anomalous Mott variable range hopping conduction mechanism having a T_{ES} crossover from one type of ES-VRH to another type with changing composite compositions.

Klafter & Sheng (1984) (KS) also proposed an advance to Mott's theory to account for the $T^{1/2}$ relationship observed with granular materials in terms of a "charging or tunnelling" modification which took into account electron tunnelling from one neutral grain to another thus creating two oppositely charged species. As a consequence of the small but definite capacitance this induces, the overall process involves a charging energy representing the minimum positive energy required for electron transfer between the initial neutral grains which becomes the source of the Coulomb gap for the excitation of charge carriers in each grain. Additionally, this approach used a modified density of states which then demonstrated the $T^{1/2}$ behaviour as an interpolation across the whole temperature range of these Mott type systems. Subsequently McAlister et al. (1984) applied the KS theory to Au/SiO_2 systems over the temperature range 4 - 300K and found it represented the behaviour more exactly than the ES approach.

More recently, Mainwaring et al. (2008) studied the temperature dependent electron transport properties of a carbon nanoparticle polymer composite thin film exhibiting semiconducting behaviour over the temperature range 20 - 450K in terms of Mott's 3-D VRH theory, and found that the behaviour can be explained well by adoption of the KS modification to Mott's approach to electron conduction. This approach took account of the statistical distribution of granular particles which leads to a distribution of gaps and charging energies as well as random potential energies arising from thermally induced voltage fluctuations, as Adkins (1989) predicted for a simple activation type of conduction given by:

$$\sigma \propto \exp(-2\alpha s - W/kT) \tag{3}$$

where α is the exponent of electron wavefunctions in the dielectric medium, s the separation between the grains, and W the charging energy. While the predicted activation energies were of the correct order of magnitude, the temperature exponent (x) varied between $0.3<x<1.0$ for Adkins' system:

$$\sigma \propto \exp(T_o/T)^x \tag{4}$$

Accordingly, Adkins investigated electron - electron correlation and the presence of a Coulomb gap to further modify the theory and found it closely predicted the temperature dependence of metal granular systems, although the optimum hopping distances (R_{opt}) and the barrier heights ($<kT$) obtained from these fits were unrealistically small to accurately consider the mechanism as totally electron hopping.

2.2 Polymer conductors

Conductive polymer composites initially consisted of conducting particles at sufficient concentrations to form percolating transport networks within the insulating medium. Subsequently an increasing interest has grown in intrinsically conductive polymers such as the doped conjugated systems which now provide both metallic and semiconductive behaviour. In organic semiconducting materials such as most conjugated polymers and many amorphous molecular materials conduction is governed and limited by thermally activated hopping mechanisms. Here highly ordered short chain conjugated oligomers allow very high mobilities to be achieved (Garnier, 1998, Hatchett & Josowicz, 2008, Kaiser, 2001). With electron movement along the chains, the electrical conductivity is quasi one dimensional and hence is highly anisotropic depending on the orientation of the polymer chains during polymerization. The conductivity is further influenced by morphological inhomogeneity (Kryszweski & Jeszka, 2003) within the system.

Fluctuation induced tunneling is the dominant mechanism in these polymer structures where extended metallic-like regions separated by narrow imperfect potential barriers prevail as expressed in:

$$\rho(T) = \phi_1 \rho_m \exp\{-(T_m/T)\} + \phi_f \rho_t \exp\{T_t/(T+T_s)\} \quad (5)$$

where ϕ_1 and ϕ_f are weighing factors while T_m, T_t and T_s characteristic temperatures and ρ_m, ρ_t pre-exponential factors of the relative transport processes (Kaiser, 2001). Here, electrical conductivity is modelled in terms of an equivalent circuit consisting of resistors connected in series or parallel depending on the extent of polymer crosslinking. A second mechanism operates when small crystalline regions are dispersed within amorphous regions, electron transport is then represented by the charging energy limited tunnelling VRH as given by Kaiser (2001):

$$\rho(T) = \phi_1 \rho_m \exp\{-(T_m/T)\} + f_f \rho_t \exp\{(T_o/T)^\gamma\} \quad (6)$$

where T_o is a constant dependent on the wavefunction decay and the density of states at the Fermi level and the exponent γ depends on the dimensionality of charge transport. Electrical conductivity is also influenced by chain length as well as defect concentration. The conductivity can then be tuned by chemical manipulation of the polymer backbone, any dopants, as well as by blending with other polymers.

In the metallic-like polymer behaviour, the earlier expressions above predict unrealistic low resistivities below 100K, however in real systems considerable electron scattering exists which results in a residual finite resistance. Kryszewski & Jeszka (2003) modified Kaiser's expression to include this as an additional resistance term (ρ_0).

Electron transport in particulate polymer composite thin films normally involves activated transport between conducting particles separated by potential barriers or dielectric medium (Devenyi et al., 1972, McAlister et al., 1984). At higher particle concentrations, percolating networks form which provide simple metallic-like conduction at ambient temperatures and exhibit anomalous (abrupt) decreases in conductance at elevated temperatures which may arise from structural changes occurring in the polymer medium such as glass transition, crystallization and melting. This unique property of polymer composites has been exploited in current limiting and switching devices as well as self regulating heaters (Shermann et al., 1983, Fournier et al., 1997). Particle concentrations just below the percolation threshold result in isolated particles or particle clusters dominating electron transport. This non-percolating dielectric regime involves thermally activated tunnelling of charge carriers between nearest neighbour particles at sufficiently close proximity to enable overlap of the electron wave functions of each particle similar to the granular films discussed above. At even greater particle separations variable range hopping now provides conductivity through an optimal network of electron hopping sites (Klafter & Sheng, 1984).

2.3 Nanoparticle – polymer composites

Carbon nanoparticle composite films can be produced by either physical blending techniques followed by film casting where the dispersion may be formed either in the molten state or in solution (Hanemann & Szabo, 2010). Film casting techniques are dominated by either spin coating on to substrates or slip casting to produce free standing films. As an alternative, nanoparticle dispersions in polyimide (PI) monomer solutions followed by film formation and thermal curing can be employed to produce highly homogeneous composite thin films which may be either supported or free standing (van Zyl et al., 2002, Mora-Huertas et al., 2004).

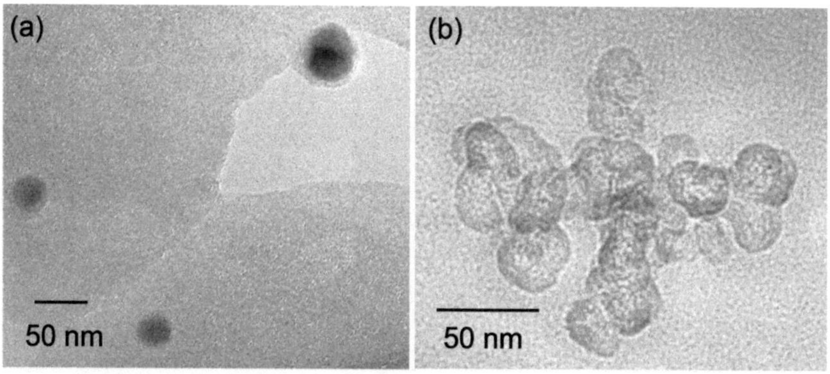

Fig. 1. Electron micrograph (HRTEM) of microtomed sections of C nanoparticle-PI composite film: (a) single particles; (b) aggregate.

Carbon nanoparticle - polyimide (C-PI) thin films can be prepared by the in-situ polymerization technique for a wide range of molecular compositions which yield either single nanoparticle and nanocluser dispersions as shown in Fig. 1 for the polyimide benzophenone tetracarboxylic dianhydride and 4,4'-oxybisbenzenamine (BTDA-ODA). The temperature dependence of resistance of these films over a wide temperature range from 20K to 450K is shown in Fig. 2(a-c). Fig. 2(a) shows the electrical resistance as a function of $1/T^{1/4}$ for a 5 vol.% C-PI film, where the resistance decreased monotonically typical of semiconducting behaviour following Mott's 3-D VRH mechanism according to expression (1). Here, electron transport can be seen to follow the $1/T^{1/4}$ relation well only over the limited temperature range 28 to 63K. Above this, the deviation from $1/T^{1/4}$ increases and forms a transition to the $1/T^{1/2}$ behaviour shown in Fig. 2(b) across the temperature range 53 to 230K. Above this temperature range (Fig. 2(c)), a further deviation produces simple linear thermal activation with a $1/T$ process up to 343K. Above this maximum temperature, a further deviation indicates the on-set of contributions from metallic-like scattering within the nanoparticle clusters.

According to Mott's original treatment of 3-D VRH in amorphous semiconductors, resistance is given by:

$$R = \frac{kT}{e^2 r \omega_0} \exp\left(\frac{W}{kT}\right) \exp\left(\frac{2r}{\xi}\right) \quad (7)$$

here r is the spacing between the localised energy levels, ω_0 the attempt frequency given by the typical phonon frequency, k the Boltzman constant, W the activation energy, and ξ the Anderson's localization length. In equation 7, the overlap function $e^{2r/\xi}$ favours short hops while the activation term $e^{(W/kT)}$ favours long hop distances. Electron transport then results from an optimized competition between short and long hops as observed in conventional doped semiconductors at low temperatures (Mott & Davis, 1979). Fig. 2 indicates that the $1/T^{1/2}$ behaviour observed across the intermediate temperature range arises from a combination of low temperature $1/T^{1/4}$ behaviour and simple Arrhenius activated conductivity at higher temperatures (Mainwaring et al 2008). Overlaps in these temperature intervals provide gradual transitions in behaviour, such that the data over the extended range conforms to the model predicted by Klafter & Sheng (1985).

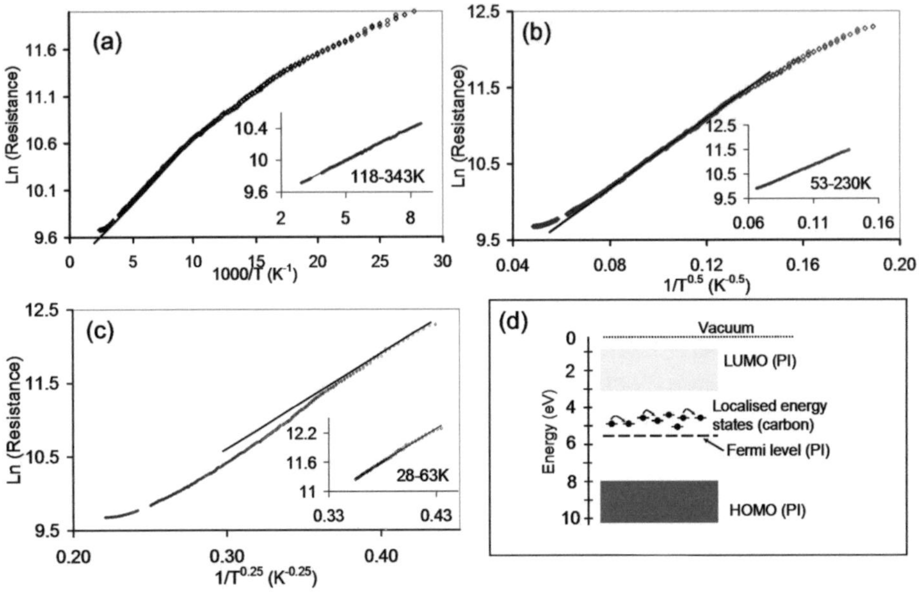

Fig. 2. Temperature dependent electrical resistance of 5 vol% C nanoparticle–PI composite thin film as: (a) $1/T^{0.25}$; (b) $1/T^{0.50}$ and (c) $1/T$. Insets show ranges that conform to T exponents, and (d) schematic representation of the proposed energy level diagram.

These temperature dependencies illustrate that carbon - polyimide nanocomposites have electrical responses very similar to granular metals such as Au - SiO_2 (McAlister et al., 1984) where metal islands form spatially distributed localized electronic states located between the valance band and the conduction band of the oxide dielectric medium. Thermally activated hopping of charge carriers between these localized states give rise to the observed electrical conductivity, which are observed in both polymer systems (Shekar et al., 2006a).

The corresponding energy diagram (Fig. 2(d)) illustrates this dispersion of carbon nanoparticles in the polyimide matrix forming localized electronic states between the highest occupied molecular orbital (HOMO) and lowest unoccupied molecular orbital (LUMO) of the polymer medium. The BTDA-ODA polyimide shown has a band gap of about 7 eV and an ionization potential of 8.97 eV similar to PMDA-ODA (Kafafi et al., 1990), while the work function of the graphitic nanoparticles is about 4.6 eV (Suzuki et al., 2001). Carbon nanoparticles can then be approximated as spatially distributed localized electronic states 2.67 eV below that LUMO of the undoped polyimide, which is ~0.9 eV above the Fermi level.

Differential thermal expansion of polymer nanocomposite thin films significantly influences the total electrical conductivity (σ_t (T)) particularly at higher temperatures. It can be considered to have the following contributions:

$$\sigma_T (T) = \sigma (T) + \Delta_1\sigma (T) + \Delta_2\sigma (T) + \Delta_3\sigma (T) \tag{8}$$

where σ(T) arises from thermal activation of charge carriers which governs both variable range hopping and thermally induced fluctuation induced tunnelling mechanisms and as such is a positive quantity for semiconductor systems. The thermal expansion effect $\Delta_1\sigma$ (T) arises from the differences in the coefficients of thermal expansion between the conducting nanoparticles ($2x10^{-6}$m/°C) and polymer ($150x10^{-6}$ m/°C). Similarly $\Delta_2\sigma$ (T) arises from carrier scattering within the regions of pseudometallic behaviour which is negative and dominates at room temperature and above. Temperature dependent structural properties of the polymer matrix including the degree of crystallinity, glass transition temperature and melting behaviour gives rise to a further contribution $\Delta_3\sigma$ (T). Overall these contributions decrease the effective volume fraction of the conducting regions, decreasing the conductivity across the whole temperature range (Mora-Huertas et al., 2004). Modelling has shown that differential thermal expansion can impose compressive and tensile stresses between well adhered composite particles and the surrounding polymer medium (Klemens et al., 1986). By analogy to other systems (Aneli et al., 1999a), such stresses can alter the local electronic structure of localised energy states through band bending for example and thereby influence electron transport parameters such as barrier height and mobility. When particle adhesion is poor, Sherman et al., (1983) showed that thermal expansion results in delamination causing anomalous changes in conduction with temperature. Particle size within composites also affects conduction such that smaller particles with very small interfacial areas reduce the influence of differential thermal expansion compared to larger particles at similar volume fractions (Sheker et al., 2006b, Zhu et al., 2007) since they effectively provide a more homogeneous system.

2.4 Nanocluster – polymer composites

Various techniques have explored the fabrication of nanocluster polymer composite films. Nanoclusters may be considered as structures that have their interface predominately consisting of surface atoms such that interfacial interactions occur virtually between all atoms, as reviewed by Jena et al. (2001). Such fabrication techniques include:

- physical deposition such as co-deposition of polymer films together with metallic or carbon clusters (Grytsenko & Schrader, 2005, Wei & Eilers, 2008),
- electrochemical deposition forming metallic nanoclusters within conducting polymer matrices combining electropolymerization and controlled electrodeposition (Trung et al., 2005),
- sol gel processing forming metallic or metal oxide nanoclusters within polymers with and without templates (Groehn et al., 2001, Nandi et al., 1990, O'Connor et al. 1997, Di Gianni et al., 2007), and
- ion implantation at low energies (< 100 keV) with metal ions in the fluence range 10^{15} to 10^{17} ions cm^{-2} (Di Girolamo et al., 2010).

High energy ion beam irradiation also produces carbon nanoclusters in polymer films such as polyimide allowing the homogeneous formation of carbonised inclusions within tracks in the polymeric medium (Hioki et al., 1983, Davenas et al., 1988, Murugaraj et al., 2009a). It has also been shown that these clusters always occur when aromatic heterocyclic moieties are present in the original polymeric film (Davenas et al., 1988). High-energy ion penetration of the polymer results in bond cleavage, chain scission and the formation of free radicals along its path as energy dissipates to the surrounding matrix, leaving carbon enriched regions and stable free radicals in the region surrounding the ion track (Murugaraj et al., 2009a, 2010b). Modeling of the ion trajectory and the penetration depth using SRIM numerical simulation (Ziegler et al., 2008) shows that ion tracks critically depend on the irradiation energy parameters as well as the physical and chemical characteristics of the polymer. In general irradiation levels below fluences of 10^{13} ions cm^{-2} produce well separated ion tracks (Fink et al., 1995, Fragala et al., 1998), while fluences above this results in multiple overlapping tracks. Within both the single and overlapping tracks, carbon enriched graphitic clusters become a source of electron transport. Ion induced electrical conductivity can be produced by two differing mechanisms. The first results from the formation of graphitic tracks at sufficient density to yield graphitic-like layers such as those related to the glassy carbon structures (Feurer et al., 1993, De Bonis et al., 1999). The second represents a progression of chemical modification resulting from thermal transformations yielding a gradual modification of the band structure from semiconducting behaviour at low fluences to metallic conduction at high fluences (De Bonis et al., 1999, Costantini et al., 2002). When the ion beam parameters are controlled to prevent bulk carbonisation, it has been shown that irradiation produces randomly distributed graphitic clusters (Fink et al., 1996, Murugaraj et al., 2009a, 2010b).

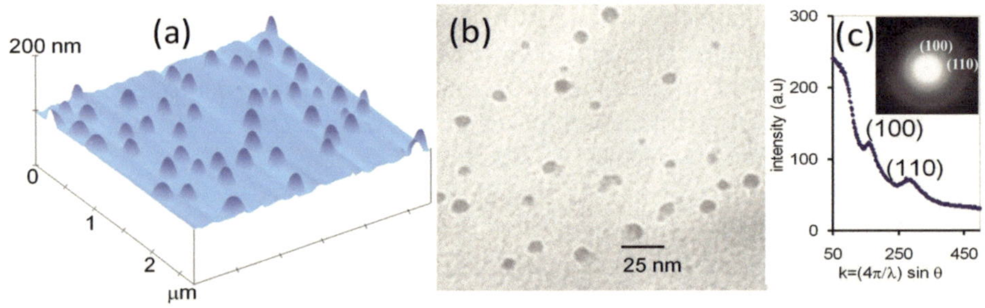

Fig. 3. Microstructure of the C nanocluster-PI composite films (4.5 MeV Cu^{3+} at fluences $<10^{11}$ ions cm^{-2}: (a) AFM tapping mode; (b) HRTEM of planar section and (c) SAED pattern of the ion tracks.

Originally, irradiated polyimide films at low fluences were considered as low-loss insulators (Salvetat et al., 1997) which showed a small increase in dielectric constant (ε) attributable to electrical inhomogeneities originating within the ion tracks themselves rather than structural reorganization in irradiated regions, although recent work has shown that there are significant changes in the dielectric permittivity on ion irradiation (Murugaraj et al. 2009). The 3-D conductivity in the track overlapping region is recognized as a result of the carbon clusters producing spatially distributed charges within the polymer together with specific localized energy states in their electronic structure. Charge transport then results from fluctuation induced tunneling (FIT) at low temperatures (Phillips et al. 1993, Davenas et al., 1988, Salvetat et al., 1997) and hopping via thermal activation process at elevated temperatures (Murugaraj et al., 2009a).

AFM imaging of ion beam irradiated polymer films in both the single track and the multiple overlapping track regime shows the microstructural changes of the surface resulting from track formation. Fig. 3(a) clearly illustrates surface hillocks associated with non-overlapping tracks comprising carbon nanoclusters ejected during escape of volatile gases (O_2, N_2, H_2 and C-H_x) produced within the polyimide medium during the interaction of high energy ions with the molecular chains (Fink et al., 1996, Toulemonde et al., 2004, Murugaraj et al., 2010). High resolution electron microscopy of the planar section of this irradiated film (Fig. 3(b)) reveals 15-25 nm diameter circular cross sections of these carbon channels (Murugaraj et al., 2010). Adla et al. (2003) used chemical staining techniques to delineate such tracks in polymers when they are microtomed and observed by TEM.

Selected area electron diffraction has shown preferential orientation of the carbon nanostructures formed within these channels by the appearance of only (100) and (110) rings of graphitic-like structures when observed in planar section (Fig. 3(c)). The absence of the major diffraction peak (002) due to the graphitic basal planes confirms the orientation of these graphitic nanoclusters parallel to the irradiation direction (ion beam path).

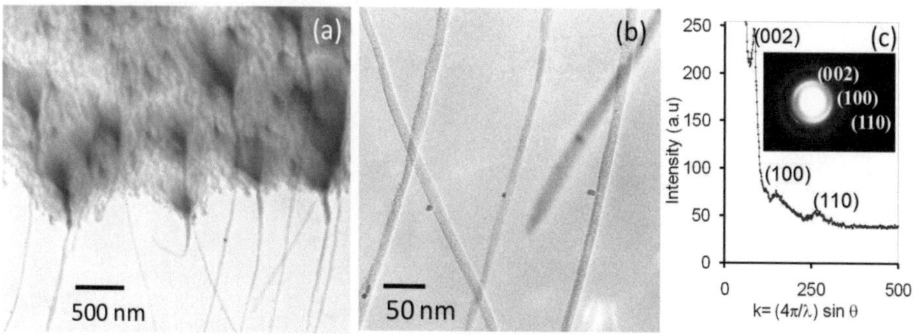

Fig. 4. HRTEM images of (a) transverse cross section of the C nanocluster-PI films (4.5 MeV Cu^{3+}, fluence 4 x 10^{14} ions cm^{-2}); (b) extended single ion-tracks with two tracks separated by a dielectric polymer barrier; and (c) SAED pattern obtained from the ion tracks.

Films irradiated at high fluence clearly provide multiple overlapping tracks that deviate from linear paths as seen in Fig. 4(a) of the HRTEM image of a microtomed cross-section which are consistent with SRIM track simulation (Ziegler et al., 2008). Fig. 4(a) illustrates the undulating pentration depth of the carbon nanocluster channels at the nanoscale due to the initial degree of planarity of the polymer film. Additionally, while the bulk of the ion tracks terminate at a depth between 4 and 4.5 μm, a small number of tracks extend well beyond this, penetrating deeper into the film due to subsequent irradiating ions following previously formed tracks, thereby loosing less energy during transmission. The widths of these single extended channels range between 15 and 25 nm (Fig. 4(b)) where one track can be seen passing in front of another separated by the dielectric polymer medium. Electron diffraction of the carbon clusters in the longitudinal cross-section of these channels (Fig. 4(c)) produces the (002), (100) and (110) rings indicating graphitic nanoclusters with basal spacings. The arc-shaped (002) diffraction ring is indicative of strong preferential orientation of these graphitic cluster layers parallel to the ion beam direction. This orientation could arise from relaxation processes to relieve the local surface stresses within the PI matrix during irradiation as well as forces exerted by the escaping volatile gases aligning the graphitic layers vertically to minimize flow resistance. Srivastava et al., determined the size of such carbon clusters by energy filtered TEM studies of ion irradiated polymer films supported on silicon to be ~5 nm in nanocolumns of ~10 nm diameter (Srivastava et al., 2006), although the planar carbon orientation was not identified.

Carbon bond hybridization is the other important aspect of electron transport by carbon nanocomposites as reflected in the ratio of sp^2 to sp^3 bonding. The low loss and carbon k-edge absorption in parallel electron energy loss spectra (PEELS) of these irradiated polyimide films (here fluence: 5×10^{14} ions cm^{-2}) reveals the presence of sp^2 π bonding in the ~6 eV absorption in the low loss region as well as sp^2 π* anti-bonding in the ~284 eV in the C-k edge region, as shown in Fig. 5(a) (i, ii) which arises from graphitic carbon clusters.

Notably Fig. 5(b) (i, ii) also shows the absence of π bonding in the un-irradiated polyimide films. However, a minor shoulder in the π* absorption region (~284 eV) is clear and arises from the aromatic rings of original PMDA-ODA polyimide molecules. These PEELS spectra of the irradiated films are similar to those obtained when graphitic nanoparticles are embedded in a similar polyimide film (Murugaraj et al., 2010) confirming the graphitic nature of the carbon clusters. Assuming these graphitic carbon nanoparticles possess total sp^2 bonding, the ratio of sp^2 to (sp^2+sp^3) carbon bonding in the un-irradiated and irradiated polyimide can be estimated as 0.074% and 94% respectively.

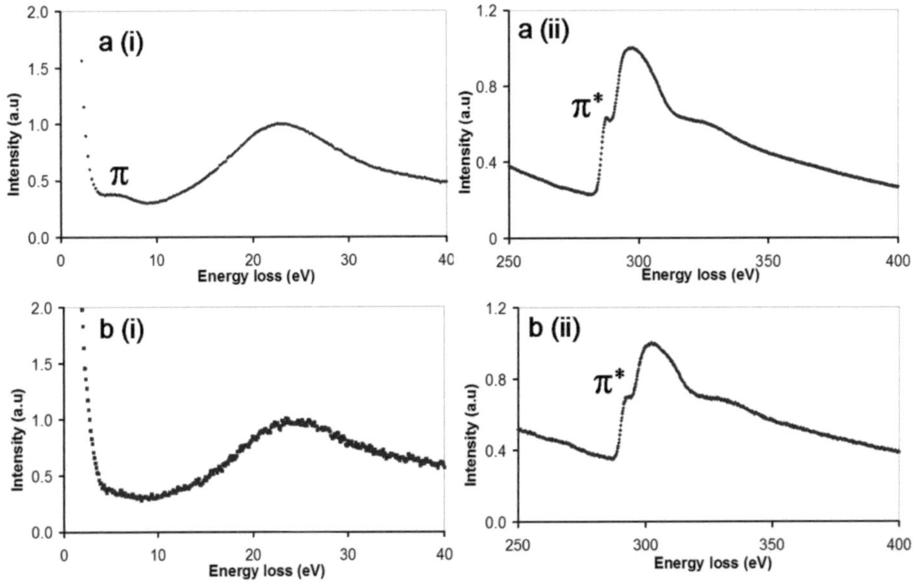

Fig. 5. Low loss (i) and carbon *k*-edge absorption (ii) spectra for: (a) irradiated polyimide film (fluence: 5.0×10^{14} ions cm^{-2}); and (b) pristine polyimide film.

In the single track regime, electron transport is predominantly along the tracks with negligible electronic conduction between the tracks due to the relatively large gaps between neighbouring tracks which would allow thermally activated electron tunneling at ambient temperatures. The electrical resistance of each track is independent of the fluence employed during formation. Along the tracks, the sole conduction mechanism is from electron hopping between carbon clusters within each track. Temperature dependent electron hopping is demonstrated in irradiated 5 μm thick polyimide films supported on gold coated substrates in the isolated ion track regime, where two slopes in the resistance plots during the heating cycle (Fig. 6) clearly indicate the influence of changing dielectric permittivity of the thermally altered polymer surrounding each track (Murugaraj et al., 2009a). The linear log resistance with T^{-1} indicates the dominant semiconducting behaviour of these conducting nanochannels. Hysteresis shown between the slopes of the heating and cooling cycles results from the thermal dissipation of the space charge in the region surrounding each track which accumulated at ambient conditions. Activation energies of these transport processes are provided by their respective slopes of Fig. 6(a-c). Here, the pronounced reduction in activation energies during heating results from dissipation of these space charge effects. The influence of the increased dielectric permittivity seen on irradiation of these polymer films is analogous to bulk heterogeneous systems such as LiAlTi-phosphate with dispersed aluminum oxide (Kumar & Thokchom, 2007). The correlation of this dielectric permittivity enhancement with fluence of the irradiated films signifies a potential to control electron transport in these conducting nanowires by suitably tuning the space charge as well as the dielectric permittivity itself of the surrounding medium.

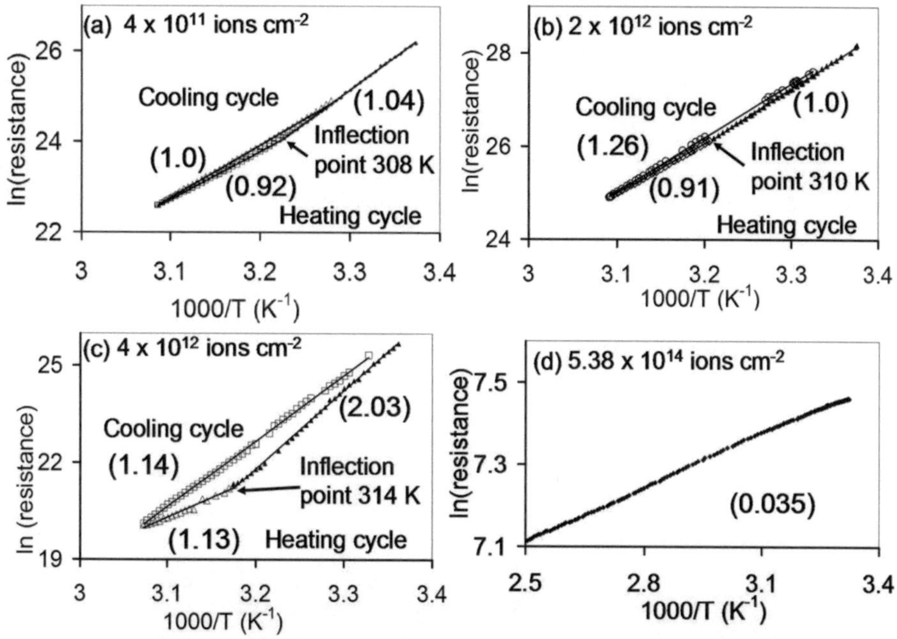

Fig. 6. Temperature dependent resistance of C nanocluster-PI films produced by irradiation: (a-c) irradiated with 55 MeV I^{+11} in the single track regime, and (d) PI film irradiated with 4.5 MeV Cu^{3+} ion in the track overlapping regime. Electron transport activation energies (eV) provided in brackets.

At still higher ion fluences, the multiple overlap regions produce bulk electron conductivity in PI films resulting from both hopping electron transport along the ion channels and electron tunning between neighbouring tracks due to their closer proximity. As irradiation is increased e.g from 9.00×10^{13} to 5.23×10^{14} ion cm^{-2}, the bulk electrical conductance increases from 3.33×10^{-9} to 1×10^{-4} siemens (Ω^{-1}) (Salvetat et al., 1997, Fink et al., 2008, Murugaraj et al., 2010). Their temperature dependence demonstrates Arrhenius thermal activation with conduction activation energy of ~35 meV, a value much smaller than that for the hopping conduction along such ion tracks (~1 eV) indicating the existence of an additional charge transport mechanism in these systems (Fig. 6(d)). That is, tunnelling of electrons in the multiple overlapping track regime provides an additional mode of the transport between neighbouring tracks when in close proximity. Fig. 7 shows schematically multiple overlapping ion tracks together with the conduction mechanisms. Interestingly, carbon nanotubes embedded in polymers provides a similar tunnelling process between neighbouring nanotubes for electron transport, although they are predominantly present as CNT bundles.

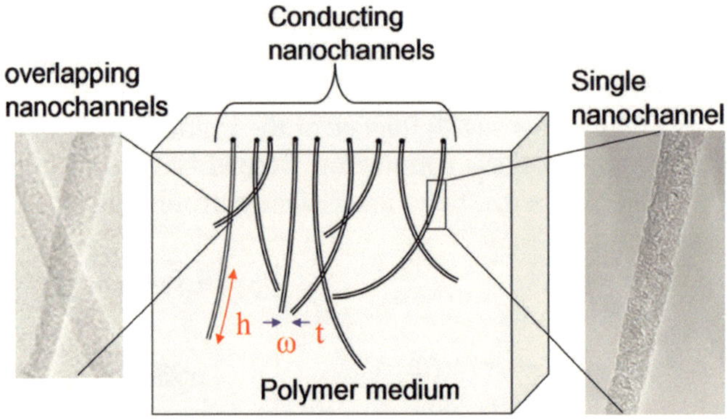

Fig. 7. Schematic of overlapping nanochannels in irradiated PI films showing hopping of electrons ('*h*') along the channels and tunnelling of electrons ('*t*') between neighbouring channels separated by a dielectric of thickness ('ω').

Here, Zhang et al. (2007) confirmed that a fluctuation induced tunneling (FIT) model fitted temperature dependent charge transport well in a functionalised multiwall CNT-polyurethan-urea composite examined. Subsequently, Hu et al. (2009) showed that the electrical resistance varied exponentially with variation of the tunnel gaps using an approach based on equivalent electrical circuits and classical modelling.

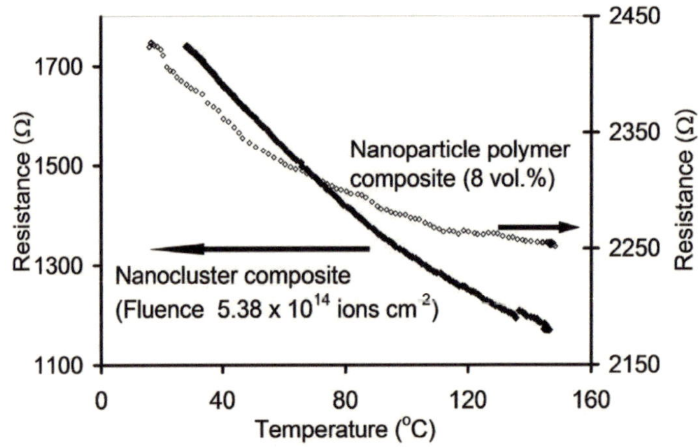

Fig. 8. Comparison of temperature dependent electrical transport of C nanoparticle-PI and C nanocluster-PI composite films.

When comparing the electrical resistance of polymer nanocomposite thin films formed by C nanoparticles and C nanocluster at volume fractions close to their respective percolation thresholds, both show an inverse temperature relationship characteristic of pronounced semiconducting behaviour. Fig. 8 demonstrates that the temperature dependence of the nanoparticle composite decreases with temperature and ultimately becoming independent typical of metallic-like scattering possibly from the nanoparticle aggregates. Whereas the nanocluster film has a significantly higher inverse temperature dependence confirming the absence of such charge scattering, indicating the ability of ion irradiation to produce more homogeneous nanocomposites with significantly lower degrees of aggregation.

3. Influence of imposed stresses

3.1 Thermal stresses

Three types of thermally induced stresses can be identified in carbon polymer nanocomposite thin films that influence the electron transport processes. Initial freestanding nanocomposite thin films prior to thermal annealing exhibit different temperature coefficients of resistance (TCR) which can be defined as the degree of the change of resistance for a given temperature change. The effect of annealing or thermal cycling is shown in Fig. 9 where it can be seen that initial heating to beyond ~90°C results in a coincident path with further heating and subsequent thermal cycling. That is, once annealed to this temperature, no hysteresis results in the electrical behaviour during thermal cycling. Annealing to ~90°C removes residual locked-in internal stresses in the film which results from molecular orientations created during the fabrication process, which in the case of these PI films involved thermal crosslinking at temperatures of 250°C.

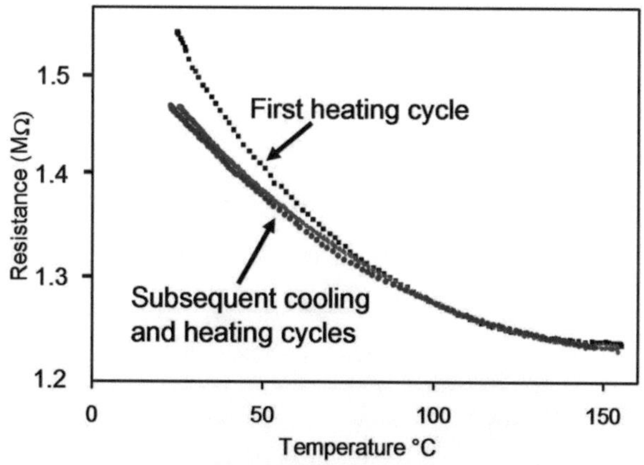

Fig. 9. Temperature dependent electrical resistance of freestanding nanoparticle-PI composite (5 vol% C) obtained during thermal annealing process.

Freestanding nanoparticulate composites also can potentially contain stresses arising from differences in the respective thermal expansion coefficient between the particle and the polymer components as discussed earlier. Thermal gradients can then induce considerable stresses within the interfacial region between particle and polymer medium which is highly dependent on the degree of molecular interaction or adhesion, and which can alter the electronic structure and the resulting electron transport. Murugaraj et al. (2008) examined impact of this effect in terms of the activation energies using Adkins model (1989).

Both nanoparticle and nanocluster composite films, when adhered to substrates, show an additional stress induced behaviour resulting from differential thermal expansion coefficients between the composite film and the substrate, which is reflected in the temperature dependent electrical behaviour of these systems. Fig. 10 shows the impact of nanoparticulate film adhesion to various substrates on the resistance ratios over a wide temperature range, where all these structures have semiconducting behaviour but with very differing negative temperature coefficient of resistance (NTCR) suggesting that the activation parameters for charge are influenced by this differential substrate adhesion. While the TCR for films adhered to rigid substrates (silicon and steel) were smaller than the freestanding film, the relative slopes clearly show that it is higher for the film fabricated on the flexible polyimide substrates. Here, the lower NTCR of the rigidly supported films compared to the freestanding film arises from compressive stresses induced by the substrate on cooling during fabrication resulting in reduced tunnelling gaps through lower interparticle spacings. Nanocomposite films supported on flexible PI substrates experience an extensional deformation with temperature increasing the overall tunnelling gap distribution. Finite element analysis (ANSYS 9.0) of the impact of adhesion and changing temperature showed that stress peaks were very near the constrained (substrate) surface but importantly at a finite distance from it as shown in Fig. 11, which then gradually fall away towards the unconstrained surface (Murugaraj et al., 2008). The lower values of the stress just near the constrained surface suggest the presence of an interface that has material properties different to those of the film and substrate. Ding et al., using these techniques, modelled the micromechanical properties such as the shear and Young's modulus of this interfacial stress region and showed that they conformed to experimental data in terms of nanoparticle, polymer and volume fraction (Ding et al., 2008).

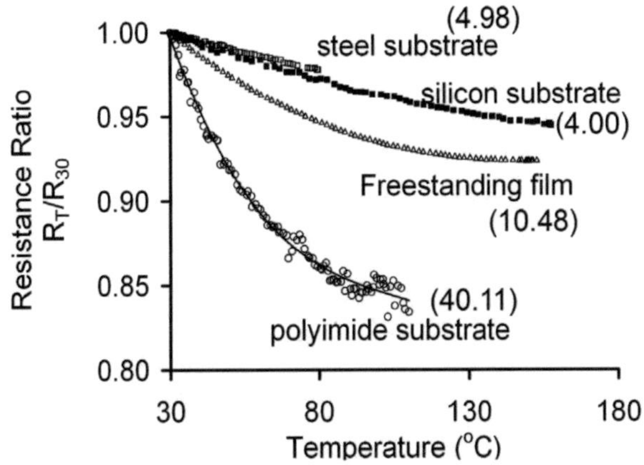

Fig. 10. Temperature dependent electrical resistance of 5 vol% C nanoparticle-PI film supported on various substrates and a freestanding film. Activation energies (meV) for electron transport provided in brackets.

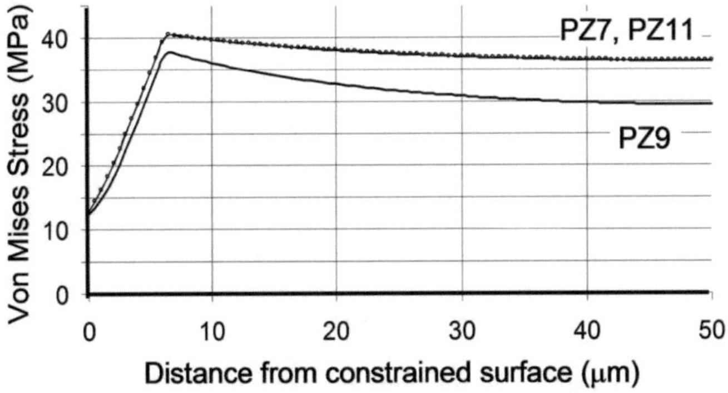

Fig. 11. Finite element analysis of the stress along the thickness of the C nanoparticle-PI film spin coated and cured on a rectangular silicon substrate (PZ9 along the corner, and PZ7 and PZ11 along the edges of the film).

3.2 Mechanical stresses

Conventional polymer composites above the percolation threshold respond non-linearly to the imposition of mechanical stresses due to the destruction and reformation of percolating conducting paths arising from physical contacts among the dispersed particle phase (Aneli et al., 1999b). As a result, their stress dependent resistance is largely confined to extensional deformations rather than compressional deformations. Here, large conducting paths exist where the electrons are delocalised over large atomic distances between, for example, carbon black aggregates, CNTs and dispersed metal islands, where the electron transport is highly dominated by large conducting segments (aggregates) rather than by hopping between well localised states (Sheng et al., 1978, Sheng, 1980). Below the percolation threshold, control of charge transport is critically dependent on electron hop distances or the tunnelling gaps and hence inter-particle spacings (Sheng et al., 1978, Sheng, 1980). When a particle polymer composite is subjected to compressive and extensional stresses, spatial rearrangement of the particles dispersed in the polymer matrix occurs which impacts on the electron transport and the measured electrical conductivity.

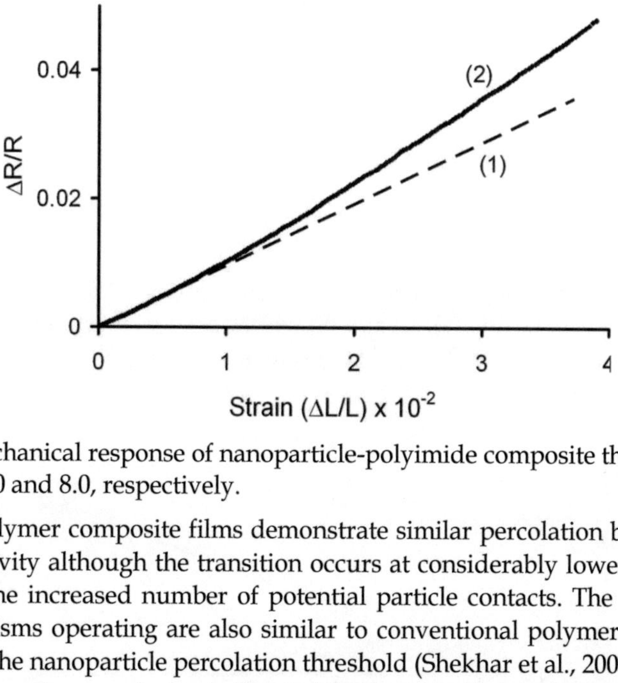

Fig. 12. Electromechanical response of nanoparticle-polyimide composite thin films: (1) and (2) with C vol% 3.0 and 8.0, respectively.

Nanoparticle - polymer composite films demonstrate similar percolation behaviour in their electrical conductivity although the transition occurs at considerably lower particle volume fractions due to the increased number of potential particle contacts. The prevalent charge transport mechanisms operating are also similar to conventional polymer composites both above and below the nanoparticle percolation threshold (Shekhar et al., 2006a). The electrical response to deformation or electromechanical behaviour of freestanding nanocomposite films show a linear response across a wide range of strains when operating well below the percolation threshold as seen in Fig. 12. Approaching the threshold, the response tends to become non-linear (Murugaraj et al., 2009b), while above it, considerable non-linearity is consistently seen. The relative dimensions of these nanoparticle films themselves influence the strain response significantly due to the effect of differential changes in interparticle spacings laterally along the strain axis and transversely across it, as a result of the Poisson's ratio during deformation. Under such an imposed process, lateral extension increases the resistance ratio while transverse contraction tends to reduce it. Fig. 13 shows that as the transverse width of the film decreases, the electromechanical response increases significantly due to these competing effects (Murugaraj et al., 2009b). Fig. 13 also indicates that when the film is adhered to a rigid substrate, the lateral contraction is significantly reduced, which yields an overall increase in electromechanical response. Nanoparticle - polymer films below the percolation threshold clearly demonstrate that the electromechanical response to both extensional and compressive deformation are equivalent, as shown in Fig. 14 since relative particle distances is the primary determinant of electrical response.

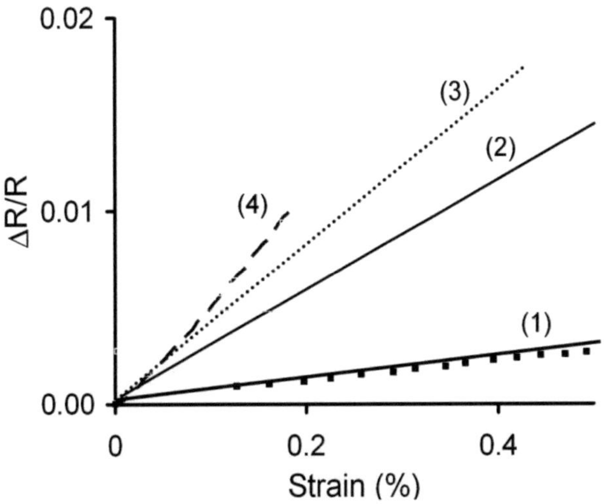

Fig. 13. Electromechanical responses of 8 vol% C nanoparticle-PI freestanding film strips of length 2.2 mm thickness 90 μm: (1), (2) and (3) widths 2.0, 1.0 and 0.7 mm, respectively; (4) Electromechanical response obtained on a similar steel supported film of width 1.5 mm.

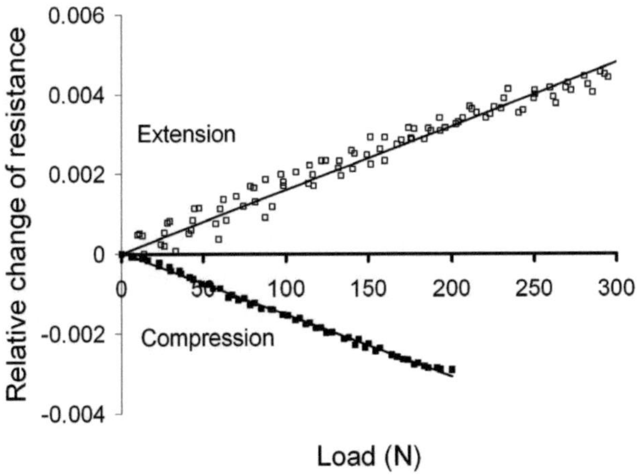

Fig. 14. Electromechanical response of a 5 vol% C nanoparticle-PI film adhered to a carbon fibre composite substrate under extensional and compressive strains.

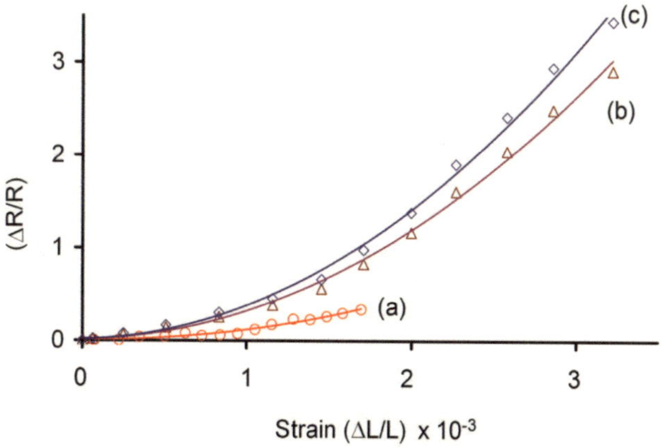

Fig. 15. Electromechanical response of C nanocluster-PI composites fabricated with fluences: (a) 3.28×10^{14}, (b) 5.00×10^{14}, and (c) 5.23×10^{14} ions cm^{-2}.

Irradiation of polymer films inherently produce anisotropic composites with orientated channels consisting of carbon nanoclusters in the ion beam direction. Unlike the dispersed nanoparticle composites, electron transport in these anisotropic nanocluster polymer films occurs through electron hopping across the carbon clusters within the nanochannels together with lateral inter-channel tunnelling as illustrated in Fig. 7. The electrical response to mechanical stress of these films exceeds nanoparticulate films by two orders of magnitude due to the exponential relationship of conductivity to electron tunnelling gap distances shown in Fig. 15. This also indicates that increasing the ion irradiation (fluence) increases the volume fraction of these orientated channels and greatly enhanced electromechanical responses (Murugaraj et al., 2010b). Carbon nanotube - polymer systems, although a similar high aspect ratio nanocomposite, show much reduced electromechanical sensitivities due their inherent random orientations within the three dimensional polymer matrix which results in a wider size dispersion of conducting regions as well as wider variations in the tunnelling gaps between adjacent nanotubes (Zhang et al., 2007, Hu et al., 2009). Whereas, the narrow distribution of tunnelling gaps in the ordered nanocluster composites provides their unique electromechanical enhancement.

3.3 Interfacial adhesion

The interfacial adhesion of polymers is essential for all polymeric devices in electronic applications, particularly those involved in structural sensing devices where the transmission of stress over a wide temperature range is fundamental. The use of conventional polymers such as the polyimides in electronic devices often results in poor adhesion as well as thermal and stiffness mismatches which generate high interfacial stresses resulting in displacements, cracks and delamination (Ho et al., 1989, Ree et al., 1992). Crosslinked polyimides are known to posses weak adhesion characteristics due to stiff molecular chains, closed imide rings and a lack of hydrogen bonding functionality shown by such adhesives as the cyanoacrylates (Ree et al. 1992). In polyimides generally and polyimide based composite films residual stresses due to imperfect adhesion can vary their functional properties such as dielectric constant and electron transport when used as both active and passive device components.

The modification of the polymer surfaces to promote adhesion uses either physical processes including ion beam, photografting, plasma etching, corona discharge and sputtering or chemical treatments such as etching with potassium or ammonium hydroxides and ethylene diamine (Ranucci et al., 2001, Park et al., 2007, Kim et al., 2005, Butoi et al., 2001, Lin et al., 2005). Physical treatments often introduce foreign materials into interfaces which may result in reliability failures whereas chemical based processes, if well controlled, confine reactions to the surface resulting in a more homogenous modification (Lin et al., 2005). Polymer composites, being a biphasic system, add additional complexity to surface treatments needed to enhance interfacial adhesion. Surface treatment, such as etching not only modifies the microstructure but alters the chemical functionality within the interfacial region which may be either the continuous polymer medium or the exposed nanoparticle component. Fig. 16 illustrates the effect of KOH etching on surface microstructure of a carbon nanoparticle - polyimide composite film. Here, the development of surface trenches within the polymer is apparent as etching proceeds, which is followed by increased exposure of the carbon nanoparticles in the final surface (Murugaraj et al., 2010a). The resulting topological structure as well as the chemical modification can be determined by friction force AFM microscopy using a hydrophilic probe in the contact mode to provide imaging and chemical adhesion characteristics simultaneously. Fig. 17(a) shows the topographic image of an etched composite surface where the exposed carbon nanoparticles are visualised together with their heights of about 10-20 nm by line analysis (Murugaraj et al., 2010a). Friction force images of the same area provide information that can be correlated with the wetting and adhesion characteristics of the surface. Fig. 17(b) shows the 3-D image of the lateral force scan, where frictional forces appear to be less over the exposed carbon nanoparticles compared to the surrounding polyimide surface. Since the contact mode AFM tip generates a stronger friction force on more hydrophilic areas, it indicates that the PI polymer surface undergoes significantly greater chemical modification to include polar groups than the exposed carbon nanoparticle areas. Since friction force microscopy provides both microstructural and functional information at the nanoscale, it complements bulk techniques such as FTIR and contact angle measurements particularly when biphasic nanocomposite surfaces are encountered.

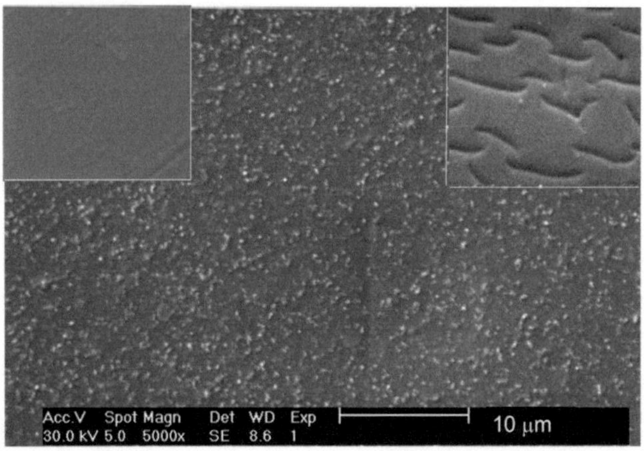

Fig. 16. SEM image of C nanoparticle-PI composite surface treated with 3.0 M KOH at 25 °C for 9 min. Inset on left is untreated composite film surface and on right is composite film treated with 0.1 M KOH at 80 °C for 30 min.

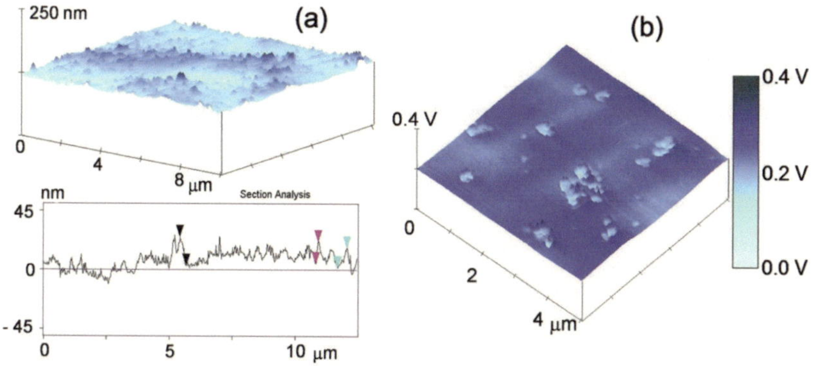

Fig. 17. AFM images of the C nanoparticle-PI composite surface after 3.0 M KOH at 25 °C for 9 min: (a) height scan and (b) friction force.

4. Dielectric behaviour of nanocomposites

Conventional polymer composites employing conductive particles have been shown by Rao & Wong (2002) to behave as a cascade of capacitors connected in series and parallel operating through space charge polarization to achieve increased effective dielectric constants at conductive particle volume fractions close to but not exceeding the percolation threshold. Additionally the dielectric constant of polymer composites has been increased by the incorporation of high dielectric particles such as barium titanate, lead magnesium niobate-lead titanate and titanium dioxide (Rao et al., 2002, Rao & Wong, 2004, Ulrich, 2004, Bai et al., 2000, Liou & Chiou, 1998). Such polymer composites obey Maxwells's mixing rule (Garnett, 1904), where the composite dielectric constant is a linear function of the volume fraction of the two components. Lewis (2005) has recently reviewed the role of nanoparticle interfaces in the dielectric behaviour of nanocomposites in both passive interface systems where the response to the applied electrical field is within the electrical double layer (EDL) and active interface systems where the response is piezomechanical.

Particulate nanocomposites, on the other hand, can exhibit considerably different behaviour in their dielectric permittivities ($\varepsilon_{composite}$) to that predicted by Maxwell's mixing rule for dielectric composites when below their percolation thresholds. Whereas reductions in $\varepsilon_{composite}$ occur when the surface functionality of the nanoparticles such as hydroxyl groups reduce the polymer density and thereby decrease the real part of the dielectric permittivity (Nelson & Fothergill, 2004). Significant enhancement, beyond that predicted by Maxwell's rule, can originate from specific physicochemical interactions between the nanoparticle and the polymer within the interfacial region which provide additional contributions to the overall polarizability (Murugaraj et al., 2005). Such nanocomposites have $\varepsilon_{polymer} < \varepsilon_{particle} < \varepsilon_{composite}$ in contrast to the conventional composites above where $\varepsilon_{polymer} < \varepsilon_{composite} < \varepsilon_{particle}$ which in nanosystems becomes a dominant effect due to greatly increased interfacial surface area above certain volume fractions.

Dielectric nanocomposite thin films consisting of alumina ($\varepsilon = 9.8$) or silica ($\varepsilon = 3.6$) nanoparticles dispersed in polyimide ($\varepsilon = 3.5$) show this monotonically increasing enhancement with increasing nanoparticle content as indicated in Fig. 18. The resultant $\varepsilon_{composite}$ of these systems is independent of frequency up to 10 MHz indicative of its origin as dipolar contributions rather than space charge polarization commonly observed in metal particle composites (Rao et al 2002). Vo and Shi (Vo & Shi, 2002, Todd & Shi, 2003) showed in physical modeling that enhanced dielectric permittivity arise from the interphase region considered as a unique phase with a different dielectric constant depending on particle concentration and size as well as the interaction with the polymer. Fitting the data from the nanoparticle - PI systems to these physical models yields a dielectric permittivity of the interface $\varepsilon_{interface}$ of 280 for the alumina nanocomposites and 19 for their silica counterparts (Murugaraj et al., 2005).

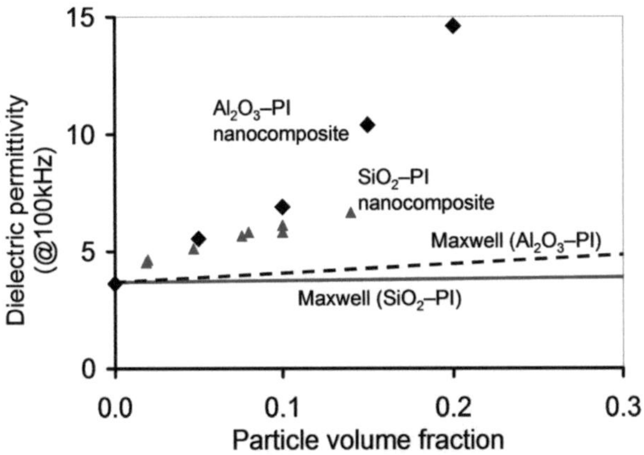

Fig. 18. Composite dielectric permittivity with alumina and silica nanoparticle contents measured at 100 kHz and 25 °C.

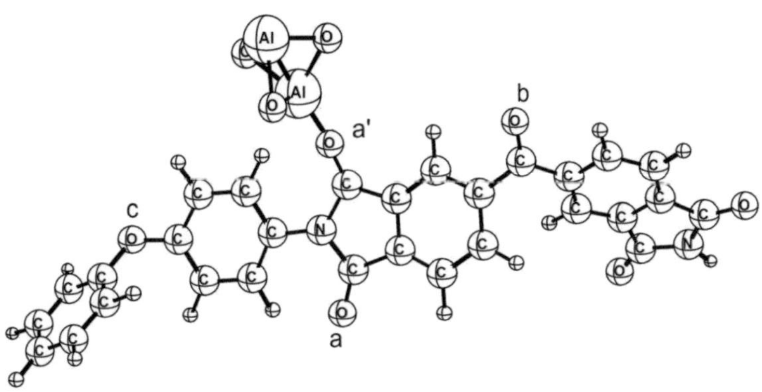

Fig. 19. Optimized molecular structure of the PI unit and interaction with the aluminium: (i) on the carbonyl group of the five membered ring (a and a'), (ii) on the carbonyl group in the chain (*b*) and (iii) the ether oxygen (*c*).

The mechanisms that lead to this enhancement of $\varepsilon_{interface}$ have been examined with *ab initio* molecular orbital calculations of the molecular interactions in the interfacial region. For the alumina - PI system these calculations predict the existence of a strong electrostatic attraction between the positively charged aluminium and the negative oxygens of the polymer functionality leading to polarization of the system and an enhanced dielectric response as illustrated in Fig. 19 (Jacob et al., 2009). The oxygen atoms therefore play a dual role involving covalent bonding with the polymer chain functionality as well as electrostatic interactions with the alumina nanoparticles (Jacob et al. 2009). Shi and Ramprasad (2007, 2008) examined the silica - polymer system using density functional theory (DFT) and provided a spatial profile of the dielectric constant within the interfacial regions, such that calculated optical and static dielectric constants were in close agreement with those determined experimentally.

5. Applications of electron conduction

5.1 Structural health monitoring

Monitoring and thus control of the structural health of critical infrastructure for safety and security as well as economic efficiency is increasingly becoming an important aspect of industrial, defence and national infrastructure. This structural monitoring requires the identification of detrimental changes to the material performance of infrastructure in its current state (diagnostic analysis) or its future condition (prognostic analysis). As such, the current time-based mode of structural and mechanical maintenance is slowly giving way to more cost effective condition based maintenance (Farrar & Worden, 2007). Material changes occur at the local scale within the microstructure which may not be adequately monitored by conventional strain sensors that essentially respond to the overall system. Sensing localised 2-D stress fields requires a high density of direct strain measurements (Loh et al., 2008), which for conventional strain gauges makes the overall systems operationally obtrusive. Increased emphasis is also being given to densely embedded monitoring systems potentially able to utilize wireless interrogation for data transfer (Lynch et al., 2007, Jia & Sun, 2006).

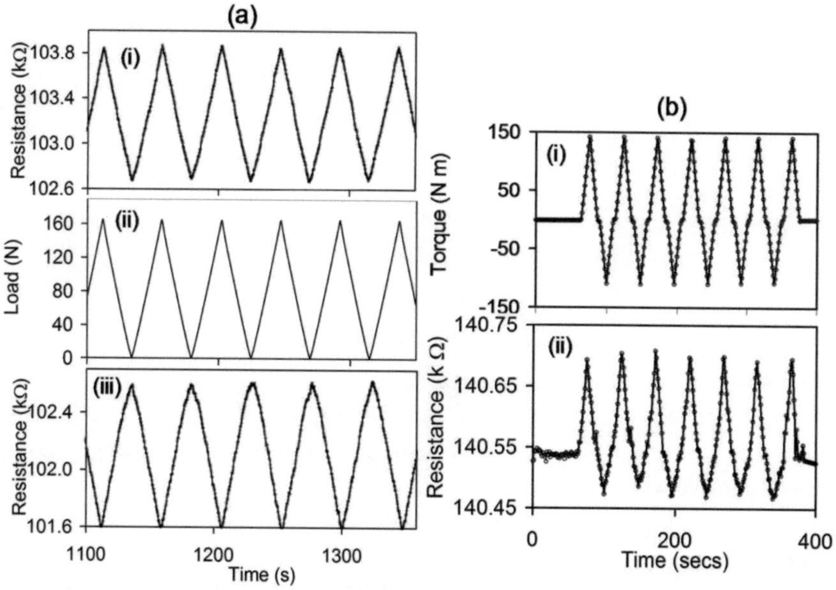

Fig. 20. Electrical resistance of 5 vol% C nanoparticle-PI strain sensing element, (a) cyclic loading: (i) extensional, (iii) compressional deformation, and (ii) is applied cyclic load; (b) cyclic torsional deformation: (i) applied torque and (ii) corresponding resistance.

Polymer based embedded strain monitoring devices such as carbon nanoparticle and CNT nanocomposites are a growing area to meet these demands (Kang et al., 2009, Loh et al., 2007) since they can be fabricated into a variety of structures including 2-D arrays. Carbon nanoparticle films acting as strain sensors can tolerate applied deformations of ±40,000 μstrains (Murugaraj et al., 2009b) where they show a proportional linear response to deformation, in comparison to conventional metal foil and semiconductor based strain sensing elements which are limited to ~10,000 μstrains. Such polymer based sensing films when subjected to cyclic aging exceeding 500 cycles show electromechanical responses that remain in phase with the imposed deformation indicating the absence of material degradation as shown in Fig. 20(a). Also, the electromechanical sensitivity in both extension and compression is equivalent as shown this in Figure, which allows these films to also act as a single element for direct torque measurements (Mainwaring et al., 2008). Fig. 20(b) demonstrates the quantitative relationship between an applied cyclic torque and the measured electrical response. Here it can be seen that this type of nanocomposite is able to electrically track the angular deformation through the whole cycle directly, again without hysteresis, for prolonged periods. Nanoparticle - polymer films having electrical resistances of the order kOhms reduce the power requirements for interrogation by more than two orders of magnitude permitting multiple strain sensors to be used in 2-D strain mapping with minimal electrical interference and joule heating (Mainwaring et al., 2008).
In comparison to nanoparticle composite films, nanocluster films demonstrate electromechanical sensitivities that are two orders of magnitude greater as noted previously, and hence provide proportional increases in strain sensitivities when used in sensing devices. The relative strain sensing response of conventional metal foil gauges, carbon nanoparticle and nanocluster composite films are shown in Fig. 21.

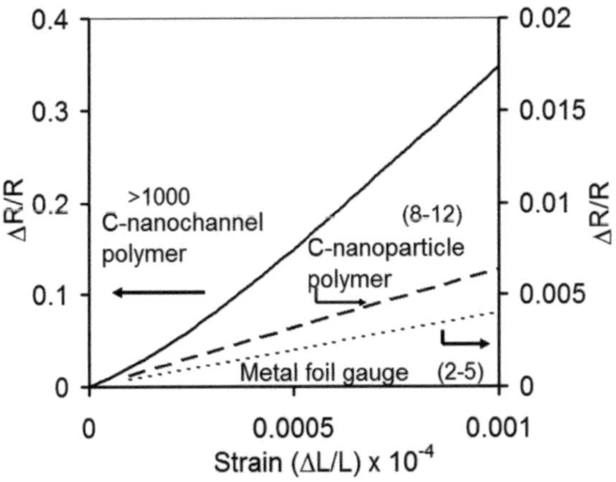

Fig. 21. Electromechanical sensitivities of 3 strain gauge materials, with relative sensing gauge factors in brackets.

5.2 Thermal sensing

Conductive polymer nanocomposite films, whether containing carbon or metal particles exhibit either a positive temperature coefficient of resistance (PTCR) or a negative coefficient (NTCR) depending on the nanoparticle volume fraction. Above the percolation threshold PTCR behaviour has been exploited in such applications as switching and current protection devices and as well as self regulating heater elements (Boiteux et al., 1999). This positive behaviour is also employed in electromagnetic shielding applications (Bigg, 1984) due its metal-like conduction behaviour. In the non-percolating regime nanocomposites have a NTCR due to their semiconducting nature and therefore show good potential for flexible environmentally robust thermal sensing applications (Murugaraj et al., 2006a, 2006b), rather than ceramic based sensors.

Fig. 22(a) shows the behaviour of the C-nanoparticle polyimide films below the percolation threshold (≤ 8 vol %). Here, the two common parameters characteristic of thermistor performance are the thermal sensitivity 'α' and sensitivity index 'β' as given by:

$$\alpha = \frac{1}{R_T}\frac{dR}{dT} \times 100 \tag{9}$$

$$\beta = \frac{\ln\left(\frac{R_1}{R_2}\right)}{\frac{1}{T_1} - \frac{1}{T_2}} \tag{10}$$

where R_T is the resistance at temperature T, and R_1 and R_2 are the resistances at temperature T_1 and T_2 respectively. The thermal sensitivity 'α ' decreases with increasing carbon content due to decreased activation energies at higher volume fractions. Charge carrier or metallic scattering also influences the thermal sensitivity due to increased aggregation at higher carbon fractions while still below the percolation threshold and increased temperatures (<100°C). In contrast, the thermal sensitivity of carbon nanocluster composite films remains constant over a wide temperature range due to the minimum aggregation of such clusters thereby reducing the intraparticle scattering (Murugaraj et al., 2006a and 2006b).

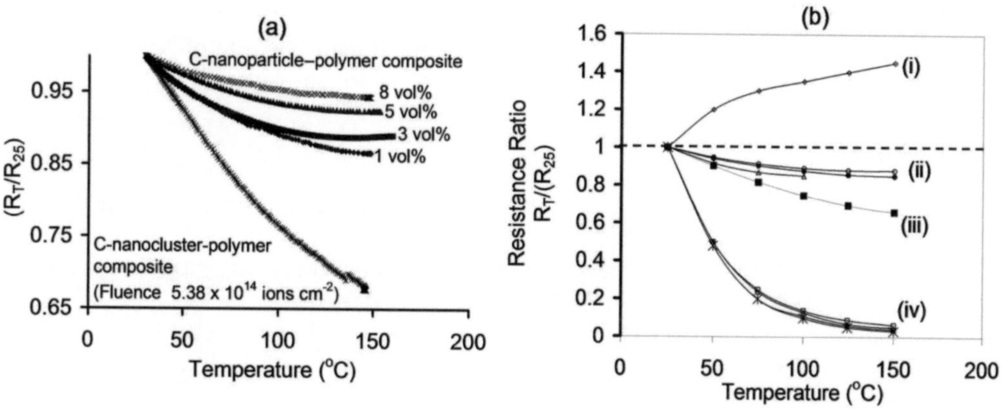

Fig. 22. (a) Resistance ratio of C nanoparticle-PI films and a C nanocluster composite film; and (b) Resistance ratio for different thermistor elements: (i) platinum resistor thermometer (Thermometrics), (ii) C nanoparticle-PI NTC thermistors, (iii) C nanocluster-PI thermistor, and (iv) commercial oxide NTC thermistors (Betatherm).

Fig. 22(b) compares the resistance ratio of polymer nanocomposite thermistor films to conventional thermistor devices. Thermal sensitivities of polymer nanocomposite films fall in the range -0.09 to -0.40 $\%K^{-1}$ with the respective sensitivity indices in the range 95 to 390 K. Whereas ceramic NTC thermistors have α and β values in the range -3 $\%K^{-1}$ to -6 $\%K^{-1}$ and 3000 K to 4000 K respectively (Rousset et al., 1994). Such differences in performance arise from the lower activation energies for electron transport associated with carbon nanocomposite films.

6. Conclusions and future perspectives

Carbon polymer nanocomposites can be produced using ex-situ nanoparticle dispersion within polymer monomers followed by crosslinking to yield highly distributed systems. Alternatively, carbon nanocluster films can be produced in-situ by ion beam irradiation which yields high aspect ratio nanostructures separated by a unaltered dielectric polymer medium. Both systems obey Mott's variable range electron hopping mechanism, but the irradiated nanocluster composites provide an addition mode of charge transfer due to electron tunnelling between isolated nanocluster channels within close proximity. Hence their sensitivity to mechanical deformation is two orders of magnitude higher than the equivalent nanoparticle composites. Electron transport along these nanochannels is modulated by changes in the dielectric behaviour of the immediately surrounding polymer. The dielectric permittivity of nanocomposites can also be locally altered through molecular scale physiochemical interactions between polymer chains and the high surface area of nanoparticles forming the interfacial region. Such carbon nanocomposite films respond to imposed stimuli such as mechanical and thermal stresses allowing their use in various sensing devices. The ability to tune the dielectric characteristics at the nanoscale allows control of electron flow opening new possibilities and novel electronic functionalities not achievable in conventional bulk materials.

7. Acknowledgements

The authors gratefully acknowledge the support of Australian Institute of Nuclear Science and Engineering Inc. in accessing ion beam irradiation facilities and expertise.

8. References

Adkins, C.J. (1987). Conduction in granular metals with potential disorder. *J. Phys. C: Solid State Physics*, 20, 235-244, ISSN 0022-3719

Adkins, C.J. (1989). Conduction in granular metals-variable range hopping in a coulomb gap? *J. Phys. Conden. Mater.* 29, 1253-1259, ISSN 0953-8984

Adla, A.; Fuess, H. & Trautmann, C. (2003). Characterisation of heavy ion tracks in polymers by transmission electron microscopy. *J. Polym. Sci. Part B: Polym. Phys.* 41, 2892-2901, ISSN 0887-6266

Aneli, J. N.; Zaikov, G. E. & Khananashvili, L. M. (1999a). Effects of mechanical deformations on the structurization and electric conductivity of electric conducting polymer composites. *J. Appl. Polym. Sci.*, 74, 601-621, ISSN 0021-8995

Aneli, J. N.; Zaikov, G. E. & Khananashvili, L. M. (1999b). Effects of mechanical deformations on the structurization and electrical conductivity of polymer composites. *Int. J. Polym. Mater.*, 43, 19-61, ISSN 0091-4037

Bai, Y.; Cheng, Z-Y.; Bharti, V.; Xu, H. & Zhang, Q. M. (2000). High-dielectric-constant ceramic-powder polymer composites. *Appl. Phys. Lett.*, 76, 3804-3807, ISSN 0003-6951

Betatherm, www.betatherm.com/news/db/pdf/1102338840.pdf

Bigg, D. M. (1984). An investigation of the effect of carbon black structure, polymer morphology, and processing history on the electrical conductivity of carbon-black-filled thermoplastics. *J. Rheology*, 28, 501-516 ISSN 0148-6055

Boiteux, G.; Fournier, J.; Issotier, D.; Seytre, G.; Marichy, G. (1999). Conductive thermoset composites: PTC effect. *Synthetic Metals*, 102, 1234-1235, ISSN 0379-6779

Butoi, C. I.; Steen, M. L.; Peers, J. R. D. & Fisher, E. R. (2001). Mechanisms and energy transfer for surface generation of NH_2 during NH_3 plasma processing of metal and polymer substrates. *J. Phys. Chem.B*, 105, 5957-5967, ISSN 1089-5647

Chui, T.; Deutscher, G.; Lindenfeld, P. & McLean, W.L. (1981). Conduction in Granular aluminium near metal-insulator transition. *Phy. Rev. B : Condens. Mater.*, 23, 6172-6175, ISSN 0163-1829

Costantini, J.-M.; Couvreur, F.; Salvetat, J.-P. & Bouffard, S. (2002). Micro-Raman study of the carbonization of polyimide induced by swift heavy ion irradiations. *Nucl. Instrum. Methods Phys. Res. Sect. B*, 194, 132-140, ISSN 0168-583x

Davenas, J. Boiteux G.; Xu, X. L. & Adem, E. (1988). Role of the modifications induced by ion beam irradiation in the optical and conducting properties of polyimide. *Nucl. Instrum. Methods Phys. Res. Sect. B*, 32, 136-141 ISSN 0168-583x

De Bonis, A.; Bearzotti A. & Marletta, G. (1999). Structural modifications and electrical properties in ion-irradiated polyimide. *Nucl. Instrum. Methods Phys. Res. Sect. B*, 151, 101-108, ISSN 0168-583x

Devenyi, A.; Manaila-Devenyi, R. & Hill, R.M. (1972). Hopping conduction through localised states in Nb/Al_2O_3 films. *Phy. Rev. Letts.*, 29, 1738-1741, ISSN 0031-9007

Ding, Y.; Tran, K.N.; Gear, J.A.; Mainwaring, D. & Murugaraj, P. (2008). The influence of interphase between nanoparticles and matrix on Young's Modulus of nanocomposites. *Proceedings of the International Conference on Nanoscience and Nanotechnology (ICONN2008)*, pp. 28-31, ISBN 1424415047, Melbourne, Australia, 25-29 February 2008, IEEE Publishing Co., Piscataway, NJ (USA)

Di Gianni, A.; Trabelsi, S.; Rizza, G.; Sangermano, M.; Althues, H.; Kaskel, S.& Voit, B. (2007). Hyperbranched polymer/T_iO_2 hybrid nanoparticles synthesized via an in situ sol-gel process. *Macromol. Chem. Phys.*, 208, 76–86, ISSN 1022-1352

Di Girolamo, G.; Massaro, N.; Piscopiello, M. & Tapfer, L. (2010). Metal ion implantation in inert polymers for strain gauge applications. *Nucl. Instrum. Methods Phys. Res. Sect. B*, 268, 2878–2882, ISSN 0168-583X

Efros, A.L. & Shklovskii, B.I. (1975). Coulomb gap and low temperature conductivity of disordered systems. *J. Phys. C*, 8, L49-L51, ISSN 0022-3719

Efros, A.L. (1976). Coulomb gap in disordered systems. *J. Phys. C*, 9, 2021-2030, ISSN 0022-3179

Entin-Wohlman, O.; Gefen Y. & Shapira, Y. (1983). Variable-range hopping conductivity in granular materials. *J Phys. C*, 16, 1161-1167, ISSN 0022-3719

Farrar, C. R. & Worden, K. (2007). An introduction to structural health monitoring. *Philos. Transact. A: Math. Phys. Eng. Sci.*, 365, 303-315, ISSN 1471-2962

Feurer, T.; Sauerbrey, R.; Smayling M.C. & Story, B.J. (1993). Ultraviolet-laser-induced permanent electrical conductivity in polyimide. *Appl. Phys. A*, 56, 275-278, ISSN 721-7250

Fink, D.; Hu, H.; Klett, R.; Mueller, M.; Zhu, J.; Li, C.; Sun, Y.; Ma, F. & Wang, L. (1995). Conductivity of aged non-overlapping and overlapping tracks in ion irradiated polyimide. *Radiat. Meas.*, 25, 51-54, ISSN 1350-4487

Fink, D.; Klett, R.; Chadderton, L.T.; Cardoso, J.; Montiel, R.; Vazquez, H. & Karanovich, A. A. (1996). Carbonaceous clusters in irradiated polymers as revealed by small angle X-ray scattering and ESR. *Nucl. Instrum. Methods Phys. Res. Sect. B*, 111, 303-314, ISSN 0168-583x

Fink, D.; Chandra, A.; Fahrner, W.R.; Hoppe, K.; Winkelmann, H. Saad, A.; Alegaonkar, P.; Berdinsky, A.; Grasser, D. & Lorenz, R. (2008). Ion track -based electronic elements. *Vacuum*, 82, 900-905, ISSN 0042-207x

Fragala, M. E.; Compagnini, G.; Licciardello, A. & Puglisi, O. (1998). Track overlap regime in ion-irradiated PMMA. *J. Polym. Sci. Part B: Polym. Phys.*, 36, 655-644, ISSN 0887-6266

Fournier, J.; Boiteix, B., Seytre, G. & Marichy, G. (1997). Positive temperature coeffficient in carbon black/epoxy polymer composites. *J. Mater. Sci. Lett.*, 16, 1677-1679, ISSN 0261-8028

Garnett, J. C. M. (1904). Colours in metal glasses and in metallic films. *Phil. Trans. Royal Soc. A:*, 203, 385-420, ISSN 1364-503X

Garnier, F. (1998). Thin film transistors based on organic conjugated semiconductors. *Chem. Phys.* 227, 253-262, ISSN 0301-0104

Groehn, F.; Kim, G.; Bauer, B.J. & Amis, E.J. (2001). Nanoparticle formation within dendrimer-containing polymer networks: Route to new organic-inorganic hybrid materials. *Macromolecules, 34,* 2179-2185, ISSN 0024-9297

Grytsenko, K.P. & Schrader, S. (2005) Nanoclusters in polymer matrices prepared by co-deposition from a gas phase. *Adv. Colloid Interface Sci.*, 116, 263–276, ISSN 0001-8686

Hanemann, T. & Szabó, D.V. (2010). Polymer-nanoparticle composites: From synthesis to modern applications. *Materials*, 3, 3468-3517, ISSN 1996-1944

Hatchett, D. W. & Josowicz, M. (2008). Composites of intrinsically conducting polymers as sensing nanomaterials. *Chem. Rev.*, 108, 746-769, ISSN 0009-2665

Hioki, T.; Noda, S.; Kakeno, M.; Yamada, K. & Kawamoto, J. (1983). Electrical and optical properties of ion - irradiated organic polymer Kapton H. *Appl. Phys. Lett.*, 43, 30-32, ISSN 0003-6951

Ho, P. S. (1989). Chemistry and adhesion of metal-polymer interfaces. *Appl. Surf. Sci.* 41/42, 559-566, ISSN 0169-4332

Hu, N.; Karube, Y.; Yan, C.; Masuda, Z. & Fukunaga, H. (2009). Tunneling effect in a polymer/carbon nanotube nanocomposite strain sensor. *Acta Mater.* 56, 2929-2936, ISSN 1359-6454

Jacob, R.; Jacob, A.P. & Mainwaring, D.E. (2009). Mechanism of the dielectric enhancement in polymer–alumina nano-particle composites. *J. Mol. Struct.*, 933, 77–85, ISSN 0022-2860

Jena, P.; Khanna, S. N. & Rao, B. K. (2001). The role of interface on the properties of cluster assemblies. *J. Cluster Sci.*, 12(3), 443-456, ISSN 1040-7278

Jia, Y. & Sun, K. (2006). Thick film wireless and powerless strain sensor. *Proc. SPIE-Int. Soc. Opt. Eng.*, 6174, 61740Z/1-61740Z/11, ISSN 0277-2668

Kafafi, S.A.; LaFemina, J.P. & Nauss, J.L. (1990). Electronic structure and conformation of polymers from cluster molecular orbital and molecular mechanics calculations: Polyimide. *J. Am. Chem. Soc.*, 112, 8742–8746, ISSN 0002-7863

Kaiser, A.B. (2001). Systematic conductivity behaviour in conducting polymers: Effects of heterogeneous disorder. *Adv. Mater.*, 13, 927-941, ISSN 0935-9648

Kang, J.H.; Park, C.I.; Scholl, J.A.; Brazin, A.H.; Holloway, N.M.; High, J.W.; Lowther, S. E. & Harrison, J. S. (2009). Piezoelectric characteristics of single wall carbon nanotube/polyimide nanocomposites. *J. Polym. Sci. Part B: Polym. Phys.*, 47, 994-1003, ISSN 0887-6266

Kim, S-H.; Cho, S.H.; Lee, N-E.; Kim, H.M.; Nam, Y. W. & Kim, Y-H. (2005). Adhesion properties of Cu/Cr films on polyimide substrate treated by dielectric barrier discharge plasma *Surf. Coat. Technol.*, 193, 101-106, ISSN 0257-8972

Klafter, J. & Sheng, P. (1984). The coulomb gap and the metal-insulator transition in granular films. *J Phys. C*, 17, L93-L96, ISSN 0022-3719

Klemens, P.G. (1986). Thermal expansion of composites. *Int. J. Thermophy.*, 7, 197-206, ISSN 0195-928X

Kryszewski, M. & Jeszka, J. (2003). Charge carrier transport in heterogeneous conducting polymer materials. *Macromol Symp.* 194, 75-86, ISSN 1022-1360

Kumar, B. & Thokchom, J. S. (2007). Space charge signature and its effects on ionic transport in heterogeneous solids. *J. Am. Ceram. Soc.*, 90, 3323-3325, ISSN 0002-7820

Lewis, T. J. (2005). Interfaces: nanometric dielectrics. *J. Phys.D: Appl. Phys.*, 38, 202-212, ISSN 0022-3727 and references therein

Lin, Y-S.; Liu, H-M.& Tsai, C-W. (2005). Nitrogen plasma modification on polyimide films for copper metallization on microelectronic flex substrates *J. Polym Sci. Pt. B: Polym. Phys.*, 43, 2023-2038, ISSN 0887-6266

Liou, J. W. & Chiou, B. S. (1998). Dielectric tunability of barium strontium titanate/silicone-rubber composite. *J. Phys. Condensed Matter.*, 10, 2773-2786, ISSN 0953-8984

Loh, K.J.; Kim, J.; Lynch, J.P.; Kam, N.W.S. & Kotov, N.A. (2007). Multifunctional layer by layer carbon nanotube-polyelectrolyte thin films for strain and corrosion sensing. *Smart Mater. Struct.*, 16, 429-438, ISSN 0964-1726

Loh, K. J.; Lynch, J. P.; Shim, B. S. & Kotov, N. A. (2008). Tailoring piezoresistive sensitivity of multilayer carbon nanotube composite strain sensors. *J. Intell. Mater. Struct.*, 19, 747-764, ISSN 1045-389X

Lynch, J. P. (2007). An overview of wireless structural health monitoring for civil structures. *Philos. Transact. A: Math. Phys Eng. Sci.*, 365, 345-372, ISSN 1364-503X

Mainwaring, D.; Murugaraj, P.; Mora-Huertas, N. & Sethupathi, K. (2008). Enhanced electromechanical response of nonpercolating polymer-nanoparticle composite films. *Appl. Phys. Lett.* 92 (25), 253303/1-253303/3, ISSN 0003-6951

McAlister, S.P.; Inglis, A.D. & Kroeker, D.R. (1984). Crossover between hopping and tunnelling conduction in Au-SiO_2 films. *J. Phys C* 17 (1984). L751-L755, ISSN 0022-3719

Mora-Huertas, N.E.; Murugaraj, P. & Mainwaring, D.E. (2004). Temperature-dependent transport properties in the semiconducting regime of nanoparticle carbon-polyimide composite films. *Physica E (Amsterdam, Neth.)*, 24, 119-123, ISSN 1386-9477

Mott, N.F. (1969). Conduction in noncrystalline materials. *Phi. Mag.*, 19, 160, 835-852, ISSN 0031-8086

Mott, N.F. & Davis, E.A. (1978). *International Series of Monographs on Physics. Electronic Processes in Non-Crystalline Materials.* Oxford University Press, ISBN 0198512880, New York

Murugaraj, P.; Mainwaring, D. & Mora-Huertas, N. (2005). Dielectric enhancement in polymer-nanoparticle composites through interphase polarizability. *J. Appl. Phys.*, 98, 054304/1-054304/6, ISSN 0021-8979

Murugaraj, P.; Mainwaring, D. & Mora-Huertas, N. (2006a). Thermistor behaviour in a semiconducting polymer-nanoparticle composite film. *J. Phys. D: Appl. Phys.*, 39, 2072-2078, ISSN 0022-3727

Murugaraj, P.; Mainwaring, D.E.; Jakubov, T.; Mora-Huertas, N.; Khelil, N.A. & Siegele, R. (2006b). Electron transport in semiconducting nanoparticle and nanocluster carbon-polymer composites. *Solid State Commun.*, 137, 422-426, ISSN 0038-1098

Murugaraj, P.; Mora-Huertas, N.; Mainwaring, D.E.; Ding, Y.; Agrawal, S. (2008). Influence of thermal stresses on electron transport in carbon-polymer nanocomposite films. *Composites, Part A*, 39A, 308-313, ISSN 1359-835X

Murugaraj, P.; Mainwaring, D. & Siegele, R. (2009a). Electron transport properties of irradiated polyimide thin films in single track regime. *Appl. Phys. Lett.*, 94, 122101/1-122101/3, ISSN 0003-6951

Murugaraj, P.; Mainwaring, D.E. & Mora-Huertas, N. (2009b). Electromechanical response of semiconducting carbon-polyimide nanocomposite thin films. *Compos. Sci. Technol.*, 69, 2454-2459 ISSN 0266-3538

Murugaraj, P.; Mainwaring, D.E.; Chen, L.G.; Sawant, P.; Al Kobaisi, M. & Yek, W.M. (2010a). Interfacial aspects of adhesion in polymer nanocomposite thin-film devices. *J. Appl. Polym. Sci.*, 115, 1054-1061, ISSN 0021-8995

Murugaraj, P.; Mainwaring, D.; Khelil, N.A.; Peng, J.L; Siegele, R. & Sawant, P. (2010b). The improved electromechanical sensitivity of polymer thin films containing carbon clusters produced in situ by irradiation with metal ions. *Carbon*, 48, 4230-4237, ISSN 0008-6223

Nandi, M.; Conklin, J.A.; Salvati Jr. L. & Sen, A. (1990). Molecular level ceramic/polymer composites. 1. Synthesis of polymer-trapped oxide nanoclusters of chromium and iron. *Chem. Mater.*, 2, 772-776, ISSN 0897-4756

Nelson, J. K. & Fothergill, J. C. (2004). Internal charge behaviour of nanocomposites. *Nanotechnology*, 15, 586-595, ISSN 0957-4484

Neugebauer, C.A. & Webb, M.B. (1962). Electrical conduction mechanism in ultrathin, evaporated metal films. *J. Appl. Phys.*, 33, 74-82, ISSN 0021-8979

O'Connor, C. J.; Buisson, Y. S. L.; Banerjee, S. Li S.; Premchandran, R.; Baumgartner, T.; John, V. T.; McPherson, G. L.; Akkara, J. A. & Kaplan, D. L. (1997). Ferrite synthesis in microstructured media: Template effects and magnetic properties. *J. Appl. Phys.* 81, 4741-4743, ISSN 0021-8979

Park, S-J.; Lee, E-J. & Kwon S-H. (2007). Influence of surface treatment of polyimide film on adhesion enhancement between polyimide and metal films. *Bull Korean Chem. Soc.*, 28, 188-192, ISSN 0253-2964

Phillips, H.M.; Wahi S. & Sauerbrey, R. (1993). Submicron electrically conducting wires produced in polyimide by ultraviolet laser irradiation. *Appl. Phys. Lett.* 62, 2572-2524, ISSN 0003-6951

Ranucci, E.; Sandgren, A.; Andronova, N. & Albertsson, A-C. (2001). Improved polyimide/metal adhesion by chemical modification approaches. *J. Appl, Polym, Sci,.* 82, 1971-1985, ISSN 0021-8995

Rao, Y. & Wong, C.P. (2002). Novel ultra-high dielectric constant polymer based composite development for embedded capacitor application. *Proc. IEEE Electronic Packag. Technol. Conf.* 2002, 65-69, ISBN 0780374355, 10-12 December 2002, Singapore, IEEE Inc., Piscataway, USA

Rao, Y.; Ogitani, S.; Kohl, P. & Wong, C. P. (2002). Novel polymer–ceramic nanocomposite based on high dielectric constant epoxy formula for embedded capacitor application. *J. Appl. Polym. Sci.*, 83, 1084-1090, ISSN 0021-8995

Rao, Y. & Wong, C. P. (2004). Material characterization of a high-dielectric-constant polymer–ceramic composite for embedded capacitor for RF applications. *J. Appl. Polym. Sci.*, 92, 2228-2231, ISSN 0021-8995

Ree, M.; Chen, K-J.; Kirby, D. P.; Katzenellenbogen, N. & Grischikowsky, D. (1992). Anisotropic properties of high - temperature polyimide thin films: Dielectric and thermal - expansion behaviors. *J. Appl. Phys.*, 72, 2014-2021, ISSN 0021-8979

Rousset, A.; Legros, R. & Lagrange, A. (1994). Recent progress in the fabrication of ceramic negative temperature coefficient thermistors. J. Eur. Ceram. Soc., 13, 185-195, ISSN 0955-2219

Salvetat, J-P; Costantini, J-M.; Brisard, F.& Zuppiroli, L. (1997). Onset and growth of conduction in polyimide kapton induced by swift heavy-ion irradiation. *Phys. Rev. B: Condens. Matter*, 55, 6238-6248, ISSN 0163-1829

Shekhar, S.; Prasad, V. & Subramanyam, S.V. (2006a). Anomalous Efros-Shklovskii variable range hopping conduction in composites of polymer and iron carbide nanoparticles embedded in carbon. *Phys. Lett. A.* 360, 390-393, ISSN 0375-9601

Shekhar, S.; Prasad, V. and Subramanyam, S.V. (2006b). Structural and electrical properties of composites of polymer–iron carbide nanoparticles embedded in carbon. *Mater. Sci. Eng. B*, 133, 108–112, ISSN 0921-5107

Sheng, P.; Sichel, E.K. & Gittleman, J.L. (1978). Fluctuation induced tunnelling conduction in carbon-polyvinylchloride composites. *Phys. Rev. Letts.* 40, 1197-1200, ISSN 0031-9007

Sheng, P. (1980). Fluctuation-induced tunneling conduction in disordered materials. *Phy. Rev. B: Condens. Mater.*, 21, 2180-2195, ISSN 0163-1829

Sheng, P. & Klafter, J. (1983). Hopping conductivity in granular disordered systems. *Phy. Rev. B: Condens. Mater.*, 27, 2583-2586, ISSN 1094-1622

Sherman, R.D.; Middleman, L.M. & Jacobs, S.M. (1983). Electron transport processes in conductor-filled polymers. *Polym. Eng. Sci.*, 23, 37-47, ISSN 0032-3888

Shi, N & Ramprasad, R. (2007). Dielectric properties of nanoscale multi-component systems: A first principles computional study. *J. Comput.-Aided Mater. Des.*, 14, 133-139, ISSN 0928-1045

Shi, N & Ramprasad, R. (2008). Local properties at interfaces in nanodielectrics: An *ab initio* computational study. *IEEE Trans. Dielectr. Electr. Insul.*, 15, 170-177, ISSN 1090-9878

Srivastava, S.K.; Avasthi, D.K. & Pippel, E. (2006). Swift heavy ion induced formation of nanocolumns of C clusters in a Si based polymer. *Nanotechnology,* 17, 2518-2522, ISSN 0957-4484

Suzuki. S.; Bower, C.; Kiyokura, T.; Nath, K.G.; Watanabe,Y. & Zhou, O. (2001). Photoemission spectroscopy of single walled carbon nanotube bundles. *J. Electron. Spectrosc. Relat. Phenom.,* 114–116, 225–228, ISSN 0368-2048

Thermometrics, www.thermometrics.com/assets/imges/ntcnotes.pdf

Todd M. G. & Shi, F. G. (2003). Characterizing the interphase dielectric constant of polymer composite materials: Effect of chemical coupling agents. *J. Appl. Phys.,* 94, 4551-4557, ISSN 0021-8979

Toulemonde, M.; Trautmann, C.; Balanzat, E.; Hjort, K. & Weidinger, A. (2004). Track formation and fabrication of nanostructures with MeV-ion beams. *Nucl. Instrum. Methods Phys. Res. Sect. B,* 216, 1-8, ISSN 0168-583X

Trung, T.; Trung, T.H. & Hab, C-S. (2005). Preparation and cyclic voltammetry studies on nickel-nanoclusters containing polyaniline composites having layer-by-layer structures. *Electrochim. Acta,* 51, 984–990, ISSN 0013-4686

Ulrich, R. (2004). Matching embedded capacitor dielectrics to applications. *Circuit World,* 30, 20-24, ISSN 0305-6120 and references therein

Van Zyl, W. E.; Garcia, M.; Schruwen, B. A. G.; Kooi, B. J.; Hosson, J. T. M. & Verweij, H. (2002). Hybrid polyamide/silica nanocomposites: synthesis and mechanical testing. *Macromol. Mater. Eng.,* 287, 106-110, ISSN 1438-7492

Vo, H.T. & Shi, F. G. (2002). Towards model-based engineering of optoelectronic packaging materials: dielectric constant modeling. *Microelectron. J.,* 33, 409-415, ISSN 0026-2692 and references therein

Wei, H. & Eilers, H. (2008). Electrical conductivity of thin-film composites containing silver nanoparticles embedded in a dielectric fluoropolymer matrix. *Thin Solid Films,* 517, 575–581, ISSN 0040-6090

Zhang, R.; Baxendale, M. & Peijs, T. (2007). Universal resistivity-strain dependence of carbon nanotube/polymer composites. *Phys. Rev. B: Condens. Matter Mater. Phys.,* 76, 195433/1-195433/5, ISSN 1098-0121

Ziegler, J.F.; Biersack, J.P. & Ziegler, M.D. (2008) http://www.srim.org/ SRIM - SRIM Co.. ISBN 0-9654207-1-X

Zhu, D.; Bin, Y. & Matsuo, M. (2007) Electrical conducting behaviors in polymeric composites with carbonaceous fllers. *Journal of Polym. Sci. Part B*: Polym. Phys., 45, 1037–1044, ISSN 0887-6266